SOLUTIONS MANUAL FOR

INTRODUCTION TO ELECTRICAL ENGINEERING

Mulukutla S. Sarma

New York Oxford
OXFORD UNIVERSITY PRESS
2001

Oxford University Press

Oxford New York
Athens Auckland Bangkok Bogotá Buenos Aires Calcutta
Cape Town Chennai Dar es Salaam Delhi Florence Hong Kong Istanbul
Karachi Kuala Lumpur Madrid Melbourne Mexico City Mumbai
Nairobi Paris São Paulo Shanghai Singapore Taipei Tokyo Toronto Warsaw

and associated companies in
Berlin Ibadan

Published by Oxford University Press, Inc.
198 Madison Avenue, New York, New York 10016
http://www.oup-usa.org

ISBN 0-19-514260-8 (paper)

Printing (last digit): 9 8 7 6 5 4 3 2 1

Printed in the United States of America
on acid-free paper.

Contents

Chapter 1	Circuit Concepts	1
Chapter 2	Circuit Analysis Techniques	29
Chapter 3	Time-Dep. Circuit Analysis	56
Chapter 4	3-Phase Circuits & Res. Wiring	161
Chapter 5	Analog Building Blocks & OP-Amps	172
Chapter 6	Digital Building Blocks & Comp. Sys.	201
Chapter 7	Semiconductor Devices	255
Chapter 8	Transistor Amplifiers	292
Chapter 9	Digital Circuits	320
Chapter 10	AC Power Systems	338
Chapter 11	Mag. Circuits & Trans.	352
Chapter 12	Electromechanics	381
Chapter 13	Rotating Machines	402
Chapter 14	Signal Processing	448
Chapter 15	Communication Systems	473
Chapter 16	Control Systems	506

CHAPTER 1

1.1.1.

$$F = \frac{Q_1 Q_2}{4\pi \epsilon_0 R^2} = \frac{1 \times 1}{4\pi (10^{-9}/36\pi) 1^2} = 9 \times 10^9 \text{ N}$$

SINCE 1 TON-FORCE = 2000 POUND-FORCE (lbf) = 2000 × 4.5 N

$$F = \frac{9 \times 10^9}{2000 \times 4.5} = 1.01 \times 10^6 \text{ TON-FORCE}$$

1.1.2.

$Q_1 = Q_2 = Q_3 = \sqrt{4\pi\epsilon_0}$ C

LET US FIND FORCE ON Q_3 DUE TO Q_1 AND Q_2.

DUE TO Q_1: $F_{13} = \frac{4\pi\epsilon_0}{(4\pi\epsilon_0) a^2} = \frac{1}{a^2}$ N

DUE TO Q_2: $F_{23} = \frac{4\pi\epsilon_0}{(4\pi\epsilon_0) a^2} = \frac{1}{a^2}$ N

Q_1, Q_2, Q_3, a, $\bar{F}_{23}$, $30°$, $30°$, $\bar{F}_{13}$, $\bar{F}_3$

RESULTANT FORCE $F_3 = (\sqrt{3}/a^2)$ N

(DIRECTED AWAY FROM THE CENTER OF THE TRIANGLE AS SHOWN IN FIGURE)

SIMILARLY, $\bar{F}_1$ AND $\bar{F}_2$ COULD BE FOUND.

1.1.3.

$Q_1 = 5 \times 10^{-6}$ C $q = 2 \times 10^{-6}$ C $Q_2 = -Q_1 = -5 \times 10^{-6}$ C x

5 m 5 m

$$\bar{F}_{1q} = \bar{a}_x \frac{(5 \times 10^{-6})(2 \times 10^{-6})}{4\pi (10^{-9}/36\pi) 5^2} = 3.6 \times 10^{-3} \, \bar{a}_x \text{ N}$$

$$\bar{F}_{2q} = \bar{a}_x \, 3.6 \times 10^{-3} \text{ N}$$

$$\bar{F}_q = \bar{F}_{1q} + \bar{F}_{2q} = 7.2 \times 10^{-3} \, \bar{a}_x \text{ N}$$

(WHERE $\bar{a}_x$ IS A UNIT VECTOR IN x-DIRECTION)

1.1.4. GIVEN $[(-\bar{a}_x - \bar{a}_y + \bar{a}_z)/\sqrt{12}]\ \frac{V}{m}$ AT $(0,0,1)$; $[6\bar{a}_z]\ \frac{V}{m}$ AT $(2,2,0)$

LET Q BE LOCATED AT (x_1, y_1, z_1)

NOTING THAT $\bar{E} = \frac{Q}{4\pi\epsilon_0 R^2}\bar{a}_R$ AND $\epsilon_0 = 10^{-9}/(36\pi)$, WE HAVE

AT $(0,0,1)$:
$$\bar{E} = \frac{Q}{4\pi\epsilon_0[x_1^2 + y_1^2 + (1-z_1)^2]^{3/2}}\{(0-x_1)\bar{a}_x + (0-y_1)\bar{a}_y + (1-z_1)\bar{a}_z\}$$

AT $(2,2,0)$:
$$\bar{E} = \frac{Q}{4\pi\epsilon_0[(2-x_1)^2 + (2-y_1)^2 + (0-z_1)^2]^{3/2}}\{(2-x_1)\bar{a}_x + (2-y_1)\bar{a}_y + (0-z_1)\bar{a}_z\}$$

COMPARING WITH THE GIVEN $\bar{E}$ AT $(2,2,0)$, SINCE IT IS ONLY Z-DIRECTED, ONE GETS

$2 - x_1 = 0$ OR $x_1 = 2$; $2 - y_1 = 0$ OR $y_1 = 2$

NOW AT $(0,0,1)$:
$$\frac{Q}{4\pi\epsilon_0[x_1^2 + y_1^2 + (1-z_1)^2]^{3/2}}[-x_1\bar{a}_x - y_1\bar{a}_y + (1-z_1)\bar{a}_z] = \frac{1}{\sqrt{12}}(-\bar{a}_x - \bar{a}_y + \bar{a}_z)$$

AT $(2,2,0)$:
$$\frac{Q}{4\pi\epsilon_0[(2-x_1)^2 + (2-y_1)^2 + z_1^2]^{3/2}}[(2-x_1)\bar{a}_x + (2-y_1)\bar{a}_y - z_1\bar{a}_z] = 6\bar{a}_z$$

THE ABOVE EQUATIONS YIELD $Q = 24\pi\epsilon_0$; $x_1 = 2$; $y_1 = 2$; $z_1 = -1$

$\therefore$ LOCATION IS GIVEN BY $(2,2,-1)$; VALUE $Q = (24\pi\epsilon_0)$ C ←

1.1.5

$$i = \frac{dq}{dt} = \frac{dq}{dl}\cdot\frac{dl}{dt} = \frac{dq}{dt}\cdot v$$

$dl = v\cdot dt$ IS THE DISTANCE TRAVELED BY A CARRIER IN TIME dt, AND dq EQUALS THE TOTAL AMOUNT OF CHARGE IN LENGTH dl.

IF THE CARRIER DENSITY IS n PARTICLES PER UNIT VOLUME, THEN THE VOLUME $(A\cdot dl)$ CONTAINS $(nA\,dl)$ CARRIERS AT ANY INSTANT OF TIME;

SO $dq = q_0\, n\, A\, dl$ OR $\frac{dq}{dl} = q_0\, n\, A$

WHERE q_0 IS THE CHARGE CARRIED BY EACH CHARGED PARTICLE.

THEN, THE CURRENT EXPRESSED IN TERMS OF THE CARRIER'S CHARGE, VOLUME DENSITY, AND VELOCITY IS GIVEN BY $i = q_0\, n\, A\, v$

SINCE i IS GIVEN TO BE 50 mA, q IN ONE SECOND $= (50\times 10^{-3})$ C

1.1.5. CONTINUED

$\therefore$ NO. OF ELECTRONS THAT PASS A GIVEN POINT IN ONE SECOND IS GIVEN BY

$$\frac{50 \times 10^{-3}}{1.6 \times 10^{-19}} = 312.5 \times 10^{15} \,/\, \text{SEC.} \leftarrow$$

SUBSTITUTING NUMBERS IN EQ. $i = q_0 n A v$, ONE GETS

$$50 \times 10^{-3} = \left|-1.6 \times 10^{-19}\right| (10)^{30} \, (1 \times 10^{-6}) \, v$$

OR AVERAGE VALUE OF $v = \dfrac{50 \times 10^{-3}}{1.6 \times 10^{-19} \times 10^{30} \times 10^{-6}} = 0.3125 \times 10^{-6}$ m/s $\leftarrow$

1.1.6.

I: $3 \times 1.6 \times 10^{-19} \times 5 \times 10^{15} = 24 \times 10^{-4}$ DUE TO POSITIVELY CHARGED PARTICLES

II: $2 \times 1.6 \times 10^{-19} \times 10 \times 10^{15} = 32 \times 10^{-4}$ DUE TO NEGATIVELY CHARGED PARTICLES

DIFFERENCE $= 8 \times 10^{-4}$

$\therefore$ CURRENT FROM B TO A $= 8 \times 10^{-3}$ A OR 8 mA $\leftarrow$

1.1.7.

$$q(t) = (50 + t)\,c \;; \quad i = \frac{dq}{dt} = 1 \text{ A} \leftarrow$$

1.1.8.

$$i = \frac{dq}{dt}$$

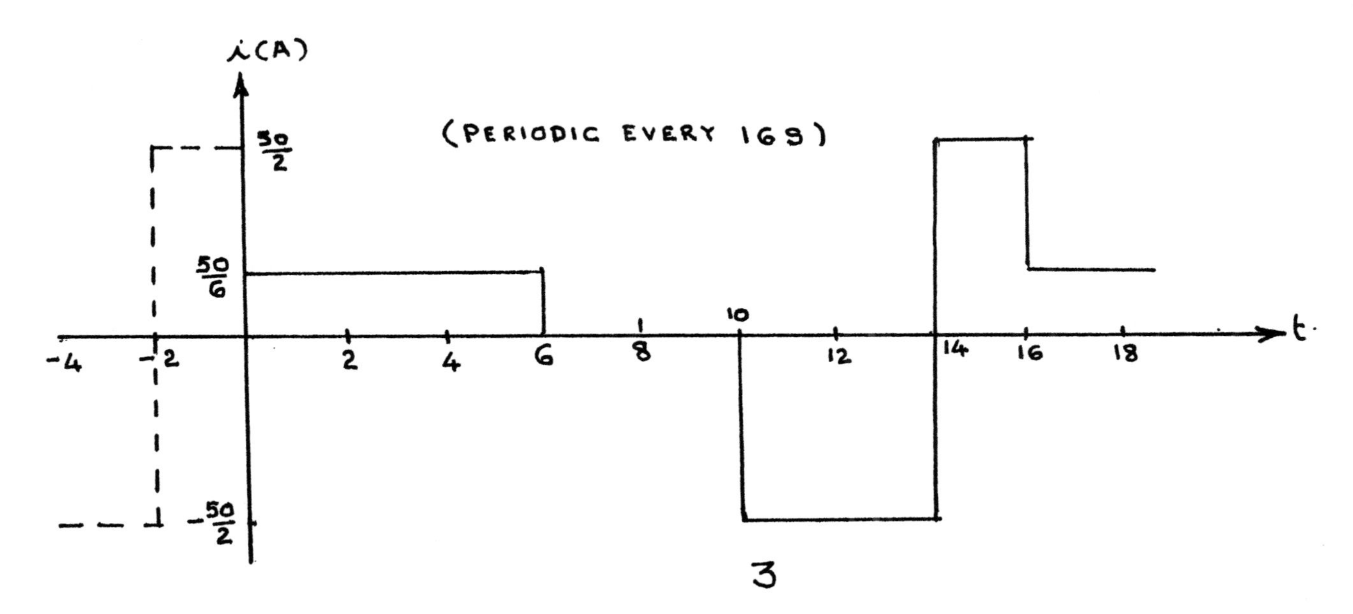

1.1.9. $i(t) = 20 \cos 120\pi t$ A

$$Q(t) = \int_0^t (20 \cos 120\pi\tau)\, d\tau = \frac{20}{120\pi} \sin 120\pi\tau \Big]_0^t = \frac{20}{120\pi} \sin 120\pi t \text{ C} \leftarrow$$

No. OF ELECTRONS PER SEC. $= \dfrac{20}{1.6 \times 10^{-19}} |\cos 120\pi t| = 1.25 \times 10^{20} |\cos 120\pi t| \leftarrow$

1.1.10.

$$d\bar{F}_{21} = I_1 d\bar{l}_1 \times \bar{B}_2 = I_1 d\bar{l}_1 \times \left(\frac{\mu_0}{4\pi} I_2 d\bar{l}_2 \times \bar{a}_{21} \right)$$

WHERE $\bar{a}_{21}$ IS A UNIT VECTOR FROM POINT 2 (0,1,0) TO 1 (0,0,1).

$$= (10^4 dz\, \bar{a}_z) \times \left[\frac{4\pi \times 10^{-7}}{4\pi} 5 \times 10^3 dx\, \bar{a}_x \times \frac{\{(0-0)\bar{a}_x + (0-1)\bar{a}_y + (1-0)\bar{a}_z\}}{\{0 + (-1)^2 + (1)^2\}^{3/2}} \right]$$

$$= (10^4 dz\, \bar{a}_z) \times \left[\frac{5 \times 10^{-4}}{2\sqrt{2}} dx(-\bar{a}_z - \bar{a}_y) \right] = \frac{5}{2\sqrt{2}} dx\, dz\, \bar{a}_x \text{ N} \leftarrow$$

$$d\bar{F}_{12} = I_2 d\bar{l}_2 \times \bar{B}_1 = I_2 d\bar{l}_2 \times \left(\frac{\mu_0}{4\pi} I_1 d\bar{l}_1 \times \bar{a}_{12} \right)$$

WHERE $\bar{a}_{12}$ IS A UNIT VECTOR FROM POINT 1 (0,0,1) TO POINT 2 (0,1,0).

$$= (5 \times 10^3 dx\, \bar{a}_x) \times \left[\frac{4\pi \times 10^{-7}}{4\pi} 10^4 dz\, \bar{a}_z \times \frac{\{(0-0)\bar{a}_x + (1-0)\bar{a}_y + (0-1)\bar{a}_z\}}{\{0^2 + 1^2 + (-1)^2\}^{3/2}} \right]$$

$$= (5 \times 10^3 dx\, \bar{a}_x) \times \left[\frac{10^{-3}}{2\sqrt{2}} dz(-a_x) \right] = 0 \leftarrow$$

1.1.11.

$$d\bar{F} = I d\bar{l} \times \bar{B}$$

$$= (5 \times 0.001\, \bar{a}_z) \times [(y\bar{a}_x - x\bar{a}_y)/(x^2+y^2)]$$

$$= 0.005 [y\bar{a}_y + x\bar{a}_x]/(x^2+y^2)$$

AT (3,4,2) : $-\dfrac{0.005}{25} [4\bar{a}_y + 3\bar{a}_x] = -0.0002[3\bar{a}_x + 4\bar{a}_y]$ N $\leftarrow$

$$(0.005\, a_z) \times \left(\frac{4a_x}{25} - \frac{3a_y}{25} \right)$$

$$\frac{1}{25} \begin{vmatrix} a_x & a_y & a_z \\ 4 & -3 & 0 \\ 0 & 0 & 0.005 \end{vmatrix}$$

4

$$\frac{1}{25}\left[a_x(-3 \times 0.005) - a_y(4 \times 0.005)\right]$$

$$= \frac{0.005}{25}[-3a_x - 4a_y]$$

1.1.12.

NET FORCE = 0

$$\bar{F} = q\,(\bar{E} + \bar{v} \times \bar{B})$$

$$0 = q\,[\bar{E} + v_0\,(\bar{a}_x + \bar{a}_y - \bar{a}_z) \times B_0\,(\bar{a}_x - 2\bar{a}_y + 2\bar{a}_z)]$$

$$\therefore \bar{E} = -v_0 B_0\,(-2\bar{a}_z - 2\bar{a}_y - \bar{a}_z + 2\bar{a}_x - \bar{a}_y - 2\bar{a}_x)$$

$$= -v_0 B_0\,(-3\bar{a}_z - 3\bar{a}_y)$$

$$= 3 v_0 B_0\,(\bar{a}_y + \bar{a}_z) \quad \longleftarrow$$

1.1.13.

Let us consider a point on the xy-plane specified by the cylindrical coordinates $(r, \phi, 0)$, as shown in Fig.S1.1.13(a). Then the solution for the magnetic flux density at $(r, \phi, 0)$ can be obtained by considering a differential length dz of the wire at the point $(0, 0, z)$ and using superposition. Applying Biot–Savart law [Eq.(1.1.9)] to the geometry in Fig.S1.1.13(a), we obtain the magnetic flux density at $(r, \phi, 0)$ due to the current element $I\,dz\,\mathbf{i}_z$ at $(0, 0, z)$ to be

$$[d\mathbf{B}]_{(r,\phi,0)} = \frac{\mu_0}{4\pi}\frac{I\,dz\,\mathbf{i}_z \times \mathbf{i}_R}{R^2}$$

$$= \frac{\mu_0 I\,dz}{4\pi}\frac{\sin\alpha}{R^2}\mathbf{i}_\phi$$

$$= \frac{\mu_0 I\,dz}{4\pi}\frac{r}{R^3}\mathbf{i}_\phi$$

$$= \frac{\mu_0 Ir\,dz}{4\pi(z^2 + r^2)^{3/2}}\mathbf{i}_\phi$$

1.1.13 CONTINUED

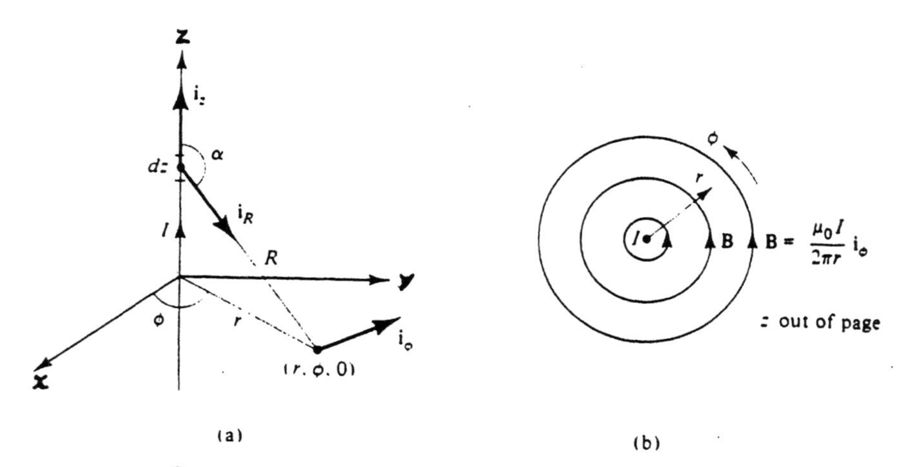

Figure S1.1.13 (a) Determination of magnetic field due to an infinitely long, straight wire of current I A. (b) Magnetic field due to the wire of (a).

The magnetic flux density due to the entire wire is then given by

$$
\begin{aligned}
[\mathbf{B}]_{(r,\phi,0)} &= \int_{z=-\infty}^{\infty} d\mathbf{B} \\
&= \int_{z=-\infty}^{\infty} \frac{\mu_0 I r}{4\pi (z^2 + r^2)^{3/2}}\, dz\, \mathbf{i}_\phi \\
&= \frac{\mu_0 I r}{4\pi} \left[\frac{z}{r^2\sqrt{z^2 + r^2}} \right]_{z=-\infty}^{\infty} \mathbf{i}_\phi \\
&= \frac{\mu_0 I}{2\pi r} \mathbf{i}_\phi
\end{aligned}
$$

Now, since the origin can be chosen to be anywhere on the wire without changing the geometry, this result is valid everywhere. Thus the required magnetic flux density is

$$\mathbf{B} = \frac{\mu_0 I}{2\pi r} \mathbf{i}_\phi \quad \longleftarrow$$

which has the magnitude $\mu_0 I/2\pi r$ and surrounds the wire, as shown by the cross-sectional view in Fig. S1.1.13(b).

1.1.14.

$\overline{dF} = I\,\overline{d\ell} \times \overline{B}$; $\overline{dF}_{21} = I_2\,dz\,\bar{a}_z \times \frac{\mu_0}{4\pi}\frac{I_1}{r}\,\bar{a}_\phi$

THE FORCE PER UNIT LENGTH ALONG THE WIRE IS THEN GIVEN BY

$$\overline{F}_{21} = -\frac{\mu_0}{2\pi}\frac{I_1 I_2}{r}\,\bar{a}_r \ \text{N/m} \quad \longleftarrow$$

WHERE $\bar{a}_r$ IS A UNIT VECTOR FROM $\overline{d\ell}_1$ TO $\overline{d\ell}_2$.

IF I_2 AND I_1 ARE IN THE SAME DIRECTION, THE FORCE PULLS THE WIRES TOGETHER; WHEN THE WIRES CARRY CURRENT IN OPPOSITE DIRECTIONS, THE WIRES ARE REPELLED BY THE MAGNETIC FORCE.

1.1.15.

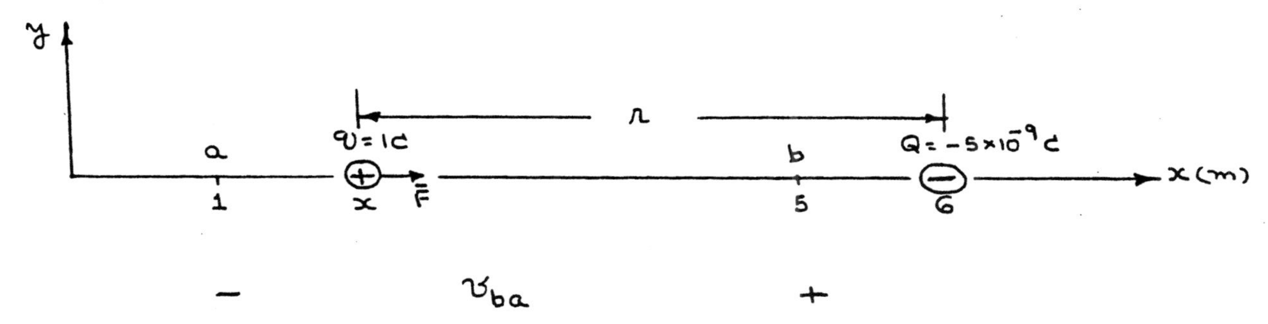

$$\overline{F} = \frac{Qq}{4\pi\epsilon_0 r^2}\,\bar{a}_x = \frac{36\pi\times10^9}{4\pi}\,\frac{5\times10^{-9}}{(6-x)^2}\,\bar{a}_x = \frac{45}{(6-x)^2}\,\bar{a}_x \ \text{N}$$

ENERGY EXPENDED IN MOVING q FROM POINT a TO POINT b

$$w_{ba} = -\int_1^5 F\,dx = -\int_1^5 \frac{45}{(6-x)^2}\,dx = -36\ \text{J}$$

$$\therefore\ v_{ba} = \frac{w_{ba}}{q} = -36\ \text{V} \quad \text{OR} \quad v_{ab} = 36\ \text{V} \quad \longleftarrow$$

POINT a IS AT A HIGHER POTENTIAL WITH RESPECT TO b. ⟵

1.1.16.

$$dw = v\,dq = 6\,(0.1) = 0.6\ \text{J} \quad \longleftarrow$$

SINCE A POSITIVE CHARGE IS MOVING FROM NEGATIVE TO POSITIVE TERMINALS, THE CHARGE HAS GAINED ENERGY. ⟵

NEGATIVE SIGN, IMPLYING THAT BATTERY IS GENERATING POWER. ⟵

1.1.17.

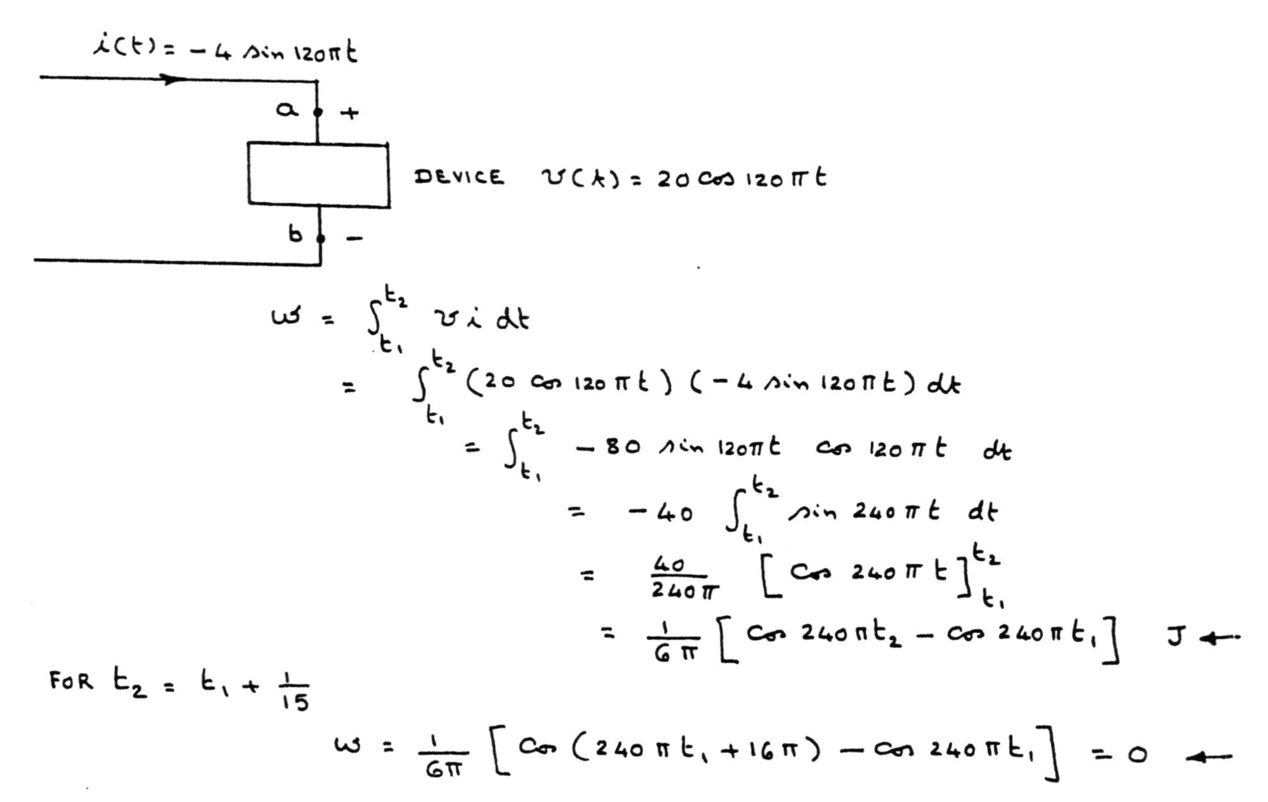

$$w = \int_{t_1}^{t_2} v\,i\,dt$$
$$= \int_{t_1}^{t_2} (20\cos 120\pi t)(-4\sin 120\pi t)\,dt$$
$$= \int_{t_1}^{t_2} -80 \sin 120\pi t \,\cos 120\pi t \,dt$$
$$= -40 \int_{t_1}^{t_2} \sin 240\pi t \,dt$$
$$= \frac{40}{240\pi}\left[\cos 240\pi t\right]_{t_1}^{t_2}$$
$$= \frac{1}{6\pi}\left[\cos 240\pi t_2 - \cos 240\pi t_1\right] \text{ J} \leftarrow$$

FOR $t_2 = t_1 + \frac{1}{15}$

$$w = \frac{1}{6\pi}\left[\cos(240\pi t_1 + 16\pi) - \cos 240\pi t_1\right] = 0 \leftarrow$$

1.1.18.

$$p = vi = -80 \sin 120\pi t \,\cos 120\pi t = -40 \sin 240\pi t \text{ W} \leftarrow$$

NEGATIVE (−) SIGN : DEVICE IS PROVIDING POWER TO THE EXTERNAL CIRCUIT; IT DOES SO HALF THE TIME AND ABSORBS DURING THE OTHER HALF OF THE TIME. ←

1.1.19.

$$P_{av} = \frac{1}{t_1 + \frac{m}{60} - t_1}\int_{t_1}^{t_2 = t_1 + \frac{m}{60}} v\,i\,dt = \frac{60}{m}\int_{t_1}^{t_1 + \frac{m}{60}} 2200\cos^2(120\pi t)\,dt$$
$$= \frac{60 \times 2200}{m}\int_{t_1}^{t_1 + \frac{m}{60}} \left(\tfrac{1}{2} + \tfrac{1}{2}\cos 240\pi t\right) dt = 1100 \text{ W} \leftarrow$$

1.1.20.

$P = VI$; $I = P/V = 6/12 = 0.5$; $\frac{115}{0.5} = 230$ hours ←

SINCE $1\text{ Ah} = 3600\text{ C}$, ENERGY = $[115 \times 3600]\,12 \simeq 5 \times 10^6 \text{ J} = 5 \text{ MJ}$ ←

1.2.1.

$$R_{dc,\,20^\circ C} = \frac{\rho_{20^\circ C}\, l}{A}$$

$\rho_{20^\circ C} = 28 \times 10^{-9}\ \Omega \cdot m$ FROM TABLE 1.2.1

$l = 1000\text{ ft} = 1000 \times 12 \times \frac{2.54}{100} = 304.8\text{ m}$

$$A = 1113\text{ kcmil} = 1113 \times 10^3\text{ cmil} \times \frac{(\pi/4)\text{ sq.mil}}{1\text{ cmil}}\ \frac{(1\text{ in})^2}{(1000\text{ mil})^2} \left(\frac{0.0254\text{ m}}{1\text{ in}}\right)^2$$

$$= 5.64 \times 10^{-4}\text{ m}^2$$

(a) $$R_{dc,\,20^\circ C} = \frac{28 \times 10^{-9} \times 304.8 \times 1.03}{5.64 \times 10^{-4}} = 0.01558\ \Omega\text{ (PER 1000 ft.)} \leftarrow$$

(b) $$R_{dc,\,50^\circ C} = R_{dc,\,20^\circ C}\left(\frac{50+T}{20+T}\right), \quad \text{WHERE } T = 228.1$$

$$= 0.01558\left(\frac{278.1}{248.1}\right) = 0.01746\ \Omega\text{ (PER 1000 ft.)} \leftarrow$$

(c) $$\frac{R_{60Hz,\,50^\circ C}}{R_{dc,\,50^\circ C}} = \frac{0.0956\ \Omega/\text{mi}}{\left(0.01746\ \frac{\Omega}{1000'}\right)\left(5.28\ \frac{1000'}{1\text{ mi}}\right)} = \frac{0.0956}{0.0922} = 1.037$$

WHICH IMPLIES 3.7% INCREASE DUE TO SKIN EFFECT. ←

1.2.2.

$$795\text{ MCM} = (795 \times 10^3\text{ cmil})\ \frac{(\pi/4)\text{ sq mil}}{1\text{ cmil}} \left(\frac{1\text{ in}}{1000\text{ mil}}\right)^2 \left(\frac{0.0254\text{ m}}{1\text{ in}}\right)^2$$

$$= 4.0283 \times 10^{-4}\text{ m}^2 \leftarrow$$

$$R_{60Hz,\,50^\circ C} = R_{60Hz,\,75^\circ C}\left(\frac{50+T}{75+T}\right), \quad \text{WHERE } T = 228.1$$

$$= (0.0880)\left(\frac{278.1}{303.1}\right) = 0.0807\ \Omega/\text{km} \leftarrow$$

1.2.3.

$d = 0.1328\text{ in} \times \frac{1000\text{ mil}}{1\text{ in}} = 132.8\text{ mil}$; STRAND AREA $= d^2$ cmil

$\therefore\ A = 12\,d^2 = 12\,(132.8)^2 = 211{,}600$ cmil, FOR 12-STRAND CONDUCTOR ←

$R_{dc,\,20^\circ C} = \frac{\rho_{20^\circ C}\, l}{A}$; $\rho_{20^\circ C} = 17 \times 10^{-9}\ \Omega \cdot m$ FROM TABLE 1.2.1

$$= \frac{17 \times 10^{-9}\ \Omega \cdot m \times 1000\text{ m}}{(1.072 \times 10^{-4})\text{ m}^2\ (1\text{ km})} \times 1.02 = 0.16175\ \Omega/\text{km} \leftarrow$$

$$\left(\because\ A = 211{,}600\text{ cmil}\ \frac{(\pi/4)\text{ sq mil}}{1\text{ cmil}} \left(\frac{1\text{ in}}{1000\text{ mil}}\right)^2 \left(\frac{0.0254\text{ m}}{1\text{ in}}\right)^2 = 1.072 \times 10^{-4}\text{ m}^2\right)$$

1.2.4. $R_{60Hz, 50°C} = 0.1185 \frac{\Omega}{mi} \times \frac{1\,mi}{1.609\,km} = 0.07365 \frac{\Omega}{km}$; FOR A LINE OF FOUR CONDUCTORS IN PARALLEL, $R_{60Hz, 50°C} = \frac{0.07365}{4} = 0.01841 \frac{\Omega}{km}$ ←

1.2.5.

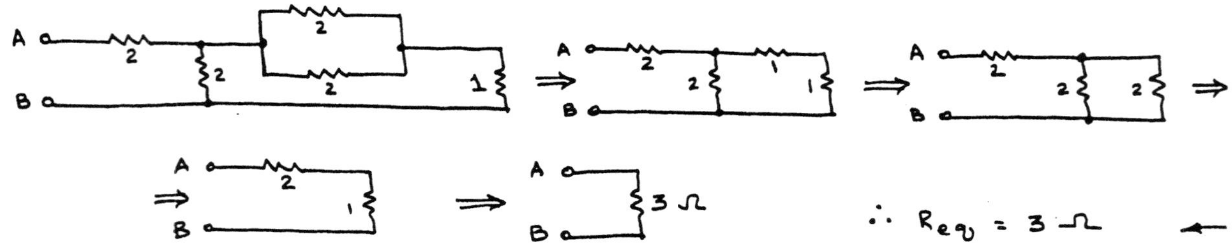

A, B, 2, 2, 2, 2, 1 ⇒ A, B, 2, 2, 1, 1 ⇒ A, B, 2, 2, 2 ⇒ A, B, 2, 1 ⇒ A, B, 3 Ω

$\therefore R_{eq} = 3\,\Omega$ ←

1.2.6

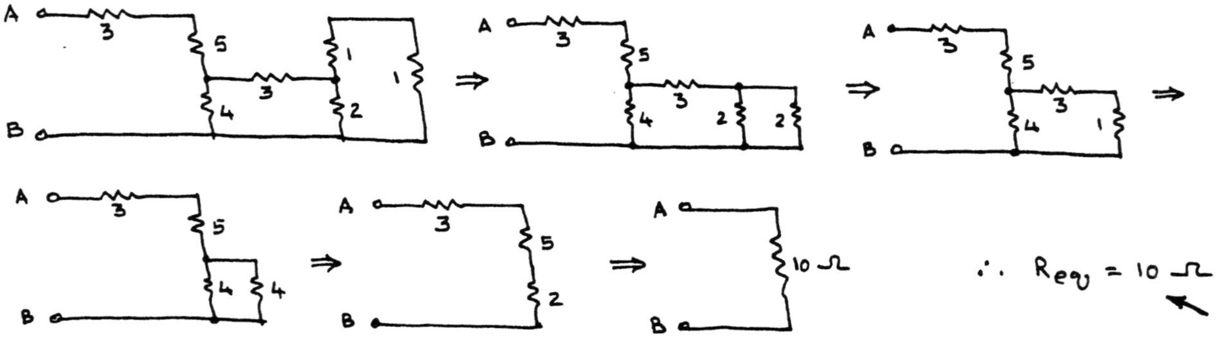

A, B, 3, 5, 3, 1, 1, 4, 2 ⇒ A, B, 3, 5, 3, 4, 2, 2 ⇒ A, B, 3, 5, 3, 4, 1 ⇒ A, B, 3, 5, 4, 4 ⇒ A, B, 3, 5, 2 ⇒ A, B, 10 Ω

$\therefore R_{eq} = 10\,\Omega$ ←

1.2.7

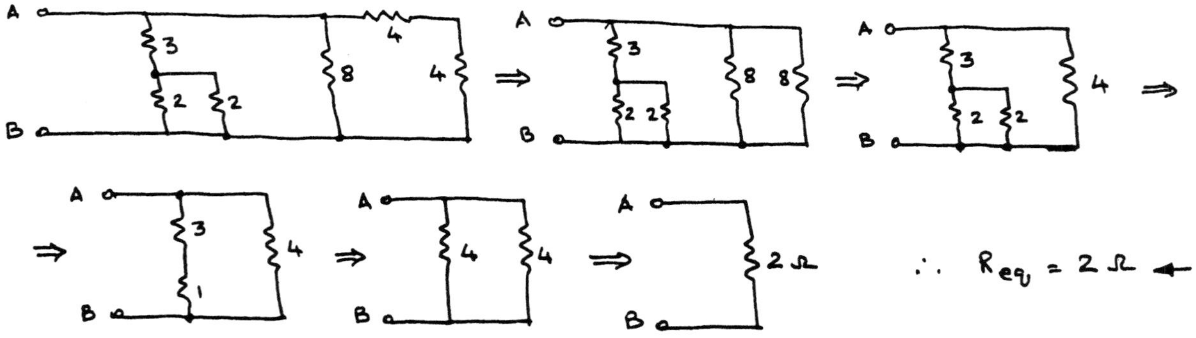

A, B, 3, 2, 2, 8, 4, 4 ⇒ A, B, 3, 2, 2, 8, 8 ⇒ A, B, 3, 2, 2, 4 ⇒ A, B, 3, 1, 4 ⇒ A, B, 4, 4 ⇒ A, B, 2 Ω

$\therefore R_{eq} = 2\,\Omega$ ←

1.2.8

2, 4, 2, 2, 4 ⇒ 2, 4, 4, 4 ⇒ 2, 2, 4 ⇒ 8 Ω

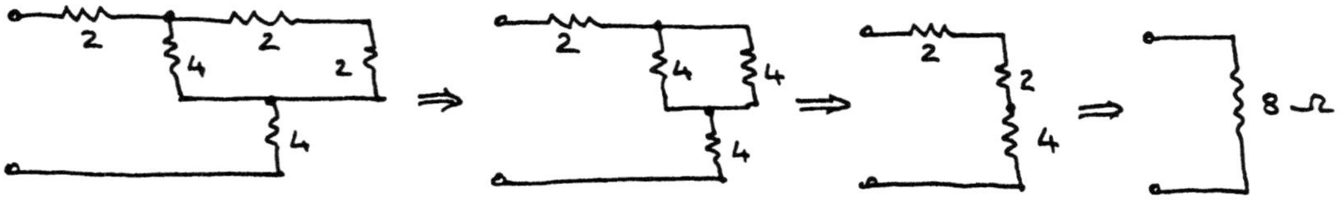

$\therefore R_{eq} = 8\,\Omega$ ←

1.2.9.

WITH THE 16-Ω SPEAKER, THE POWER DELIVERED BY THE AMPLIFIER IS CALCULATED AS

$$V_L = \frac{R_L}{R_L + R_S} V_S = \frac{16}{16+8} V_S = \frac{2}{3} V_S$$

$$P_L = \frac{V_L^2}{R_L} = \frac{4}{9R_L} V_S^2 = \frac{4}{(9)(16)} V_S^2 = 0.0278\, V_S^2$$

IF THE LOAD RESISTANCE IS MATCHED TO THAT OF THE SOURCE, i.e., $R_L = R_S = 8\,\Omega$,

$$V_L' = \frac{V_S}{2}$$

AND THE POWER DELIVERED TO THE SPEAKER, P_L', UNDER MATCHED LOAD CONDITIONS IS GIVEN BY

$$P_L' = \frac{(V_L')^2}{R_L} = \frac{V_S^2}{4R_L} = \frac{V_S^2}{(4)(8)} = 0.03125\, V_S^2$$

THEREFORE, THE INCREASE IN POWER UNDER MATCHED CONDITIONS IS GIVEN BY

$$\frac{0.03125 - 0.0278}{0.0278} \times 100 = 12.4\% \longleftarrow$$

1.2.10.
(a) LOAD CURRENT $I_L = \dfrac{10}{500 + R_L}$ A

POWER ABSORBED BY THE LOAD $= P_L = I_L^2 R_L = \left(\dfrac{10}{500 + R_L}\right)^2 R_L$

$$= \frac{100\, R_L}{R_L^2 + 1000\, R_L + 250{,}000} \text{ W} \longleftarrow$$

(b) THE PLOT OF P_L VS R_L IS SHOWN BELOW;

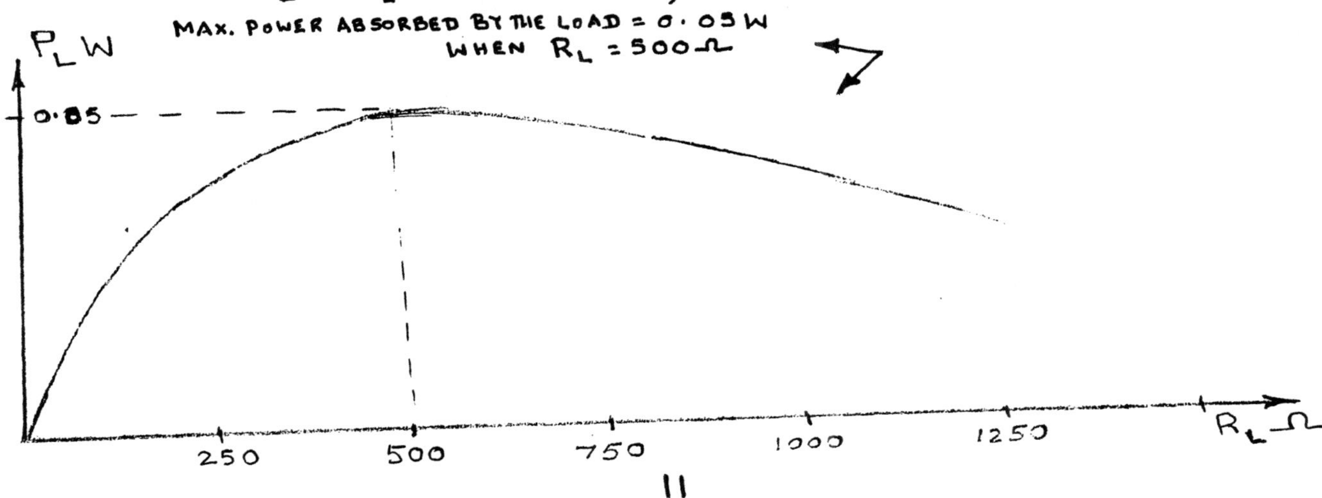

1.2.11.

REFERRING TO FIGURE 1.2.4 (a)

$$v_{int} = i_L R_S = R_S \frac{v}{R_S + R_L} = \frac{1.2 \times 30}{1.2 + R_L} = \frac{36}{R_L + 1.2}$$

THIS SHOULD BE NO MORE THAN 2 PERCENT OF 30V, i.e., $30 \times 0.02 = 0.6\text{V}$

$$\therefore \quad \frac{36}{R_L + 1.2} \leq 0.6$$

THE MINIMUM R_L RESULTS WHEN $\frac{36}{R_L + 1.2} = 0.6$

Then $R_{L,\,min} = 58.8\,\Omega$ ←

1.2.12.

REFERRING TO FIGURE 1.2.4 (b)

$$i_L = i - i_{int} = i - \frac{v_L}{R_S} = (200 \times 10^{-3}) - \frac{V_L}{12 \times 10^3}$$

BUT $V_L = I_L R_L = \frac{I R_S}{R_S + R_L} R_L = \frac{(200 \times 10^{-3})\, 12 \times 10^3}{(12{,}000 + 200)} (200) = 39.34\text{V}$

THEN $i_L = (200 \times 10^{-3}) - \frac{37.34}{12 \times 10^3} = (200 - 3.28) = 196.72\ \text{mA}$

$\therefore$ PERCENTAGE DROP IN LOAD CURRENT WITH RESPECT TO THE SOURCE SHORT-CIRCUIT CURRENT IS GIVEN BY

$$\frac{200 - 196.72}{200} \times 100 = 1.64\,\%$$ ←

1.2.13.

(a)

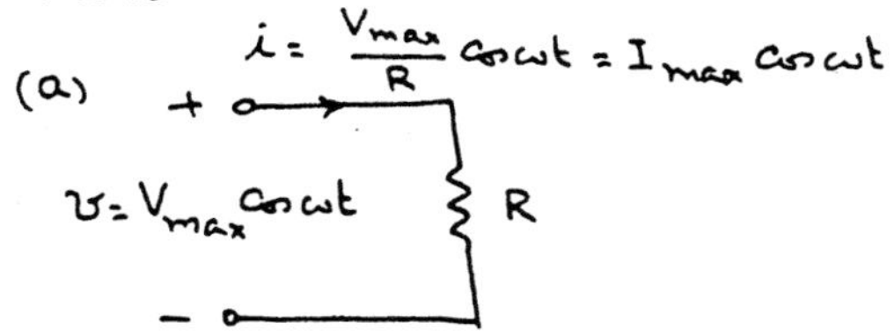

$$p(t) = vi = V_{max} I_{max} \cos^2 \omega t$$
$$= \frac{V_{max} I_{max}}{2} \left[1 + \cos 2\omega t \right]$$

$$P_{av} = \frac{V_{max} I_{max}}{2} = \frac{V_{max}}{\sqrt{2}} \frac{I_{max}}{\sqrt{2}} = V_{RMS} I_{RMS}$$
$$= I_{RMS}^2 R = V_{RMS}^2 / R = V_{RMS}^2 G \quad \leftarrow$$

(b)

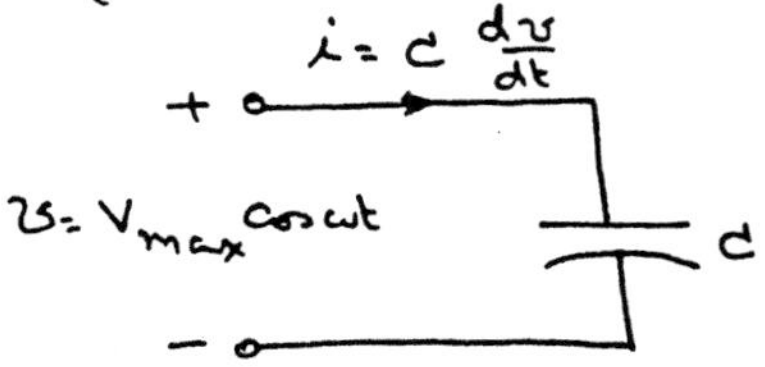

$$i = C \frac{dv}{dt} = -C V_{max} \omega \sin \omega t$$
$$p(t) = vi = -\omega C V_{max}^2 \sin \omega t \cos \omega t$$
$$= -\frac{\omega C V_{max}^2}{2} \sin 2\omega t$$
$$P_{av} = 0 \quad \leftarrow$$

(c)

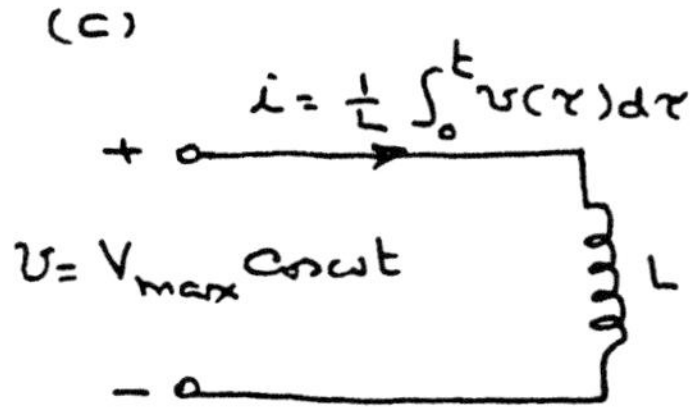

$$i = \frac{1}{L}\int_0^t v(\tau)\,d\tau = \frac{1}{L}\int_0^t V_{max}\cos\omega\tau\,d\tau$$
$$= \frac{1}{\omega L} V_{max} \sin \omega t$$
$$p(t) = vi = \frac{1}{\omega L} V_{max}^2 \sin \omega t \cos \omega t = \frac{V_{max}^2}{2\omega L} \sin 2\omega t$$
$$P_{av} = 0 \quad \leftarrow$$

1.2.14.

(a) FROM PROB. 1.2.5, $R_{eq} = 3\,\Omega$

$V_{RMS} = 120\,V$; $P_{av} = V_{RMS}^2 / R = (120)^2/3 = 4800\,W = 4.8\,kW \leftarrow$

(b) FROM PROB. 1.2.6, $R_{eq} = 10\,\Omega$

$V_{RMS} = 120\,V$; $P_{av} = V_{RMS}^2 / R = (120)^2/10 = 1440\,W = 1.44\,kW \leftarrow$

(c) FROM PROB. 1.2.7, $R_{eq} = 2\,\Omega$

$V_{RMS} = 120\,V$; $P_{av} = V_{RMS}^2 / R = (120)^2/2 = 7200\,W = 7.2\,kW \leftarrow$

(d) FROM PROB. 1.2.8, $R_{eq} = 8\,\Omega$

$V_{RMS} = 120\,V$; $P_{av} = V_{RMS}^2 / R = (120)^2/8 = 1800\,W = 1.8\,kW \leftarrow$

1.2.15.

(a) $I_{dc}^2 R = 4800\,W$ or $I_{dc} = \sqrt{4800/3} = 40\,A$; $I_{eff} = I_{RMS} = \frac{V_{RMS}}{R} = \frac{120}{3} = 40\,A$ ←

(b) $I_{dc}^2 R = 1440\,W$ or $I_{dc} = \sqrt{1440/3} = 12\,A$; $I_{eff} = I_{RMS} = \frac{V_{RMS}}{R} = \frac{120}{10} = 12\,A$ ←

(c) $I_{dc}^2 R = 7200\,W$ or $I_{dc} = \sqrt{7200/2} = 60\,A$; $I_{eff} = I_{RMS} = \frac{V_{RMS}}{R} = \frac{120}{2} = 60\,A$ ←

(d) $I_{dc}^2 R = 1800\,W$ or $I_{dc} = \sqrt{1800/8} = 15\,A$; $I_{eff} = I_{RMS} = \frac{V_{RMS}}{R} = \frac{120}{8} = 15\,A$ ←

1.2.16.

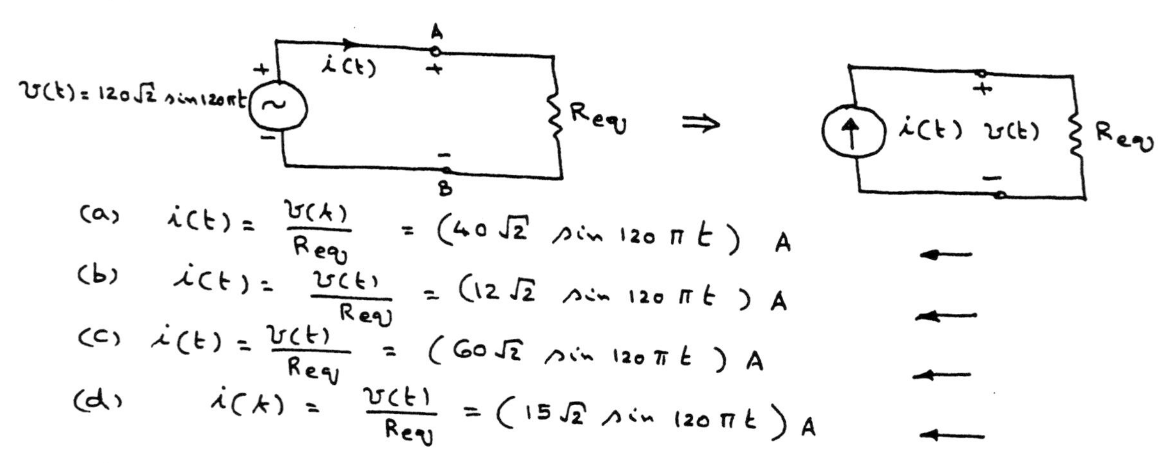

(a) $i(t) = \frac{v(t)}{R_{eq}} = (40\sqrt{2}\sin 120\pi t)\ A$ ←

(b) $i(t) = \frac{v(t)}{R_{eq}} = (12\sqrt{2}\sin 120\pi t)\ A$ ←

(c) $i(t) = \frac{v(t)}{R_{eq}} = (60\sqrt{2}\sin 120\pi t)\ A$ ←

(d) $i(t) = \frac{v(t)}{R_{eq}} = (15\sqrt{2}\sin 120\pi t)\ A$ ←

1.2.17.

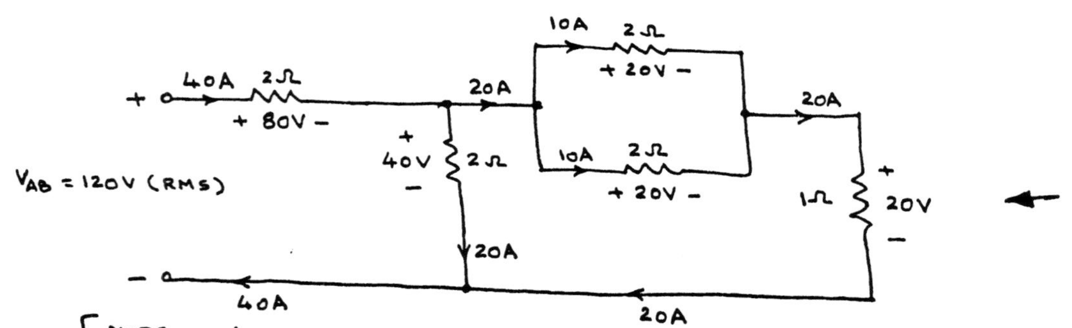

[NOTE: ALL VOLTAGE AND CURRENT VALUES ARE RMS QUANTITIES.

SEE SOLUTION OF PROB. 1.2.5 AND RETRACE STEPS BACKWARDS IN ORDER TO ARRIVE AT THE NUMBERS SHOWN ABOVE.]

1.2.18.

(a) $V_x = \frac{2}{3+1+2} \times 12 = \frac{2}{6} \times 12 = 4\,V$ ←

(b) $V_x = -\left(\frac{2\|2}{4+(2\|2)}\right) \times 15 = -\frac{1}{5} \times 15 = -3\,V$ ←

(c) $V_x = \frac{3}{3+2+(2\|2)} \times 6 = \frac{3}{6} \times 6 = 3\,V$ ←

1.2.18. CONTINUED

(d)

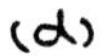

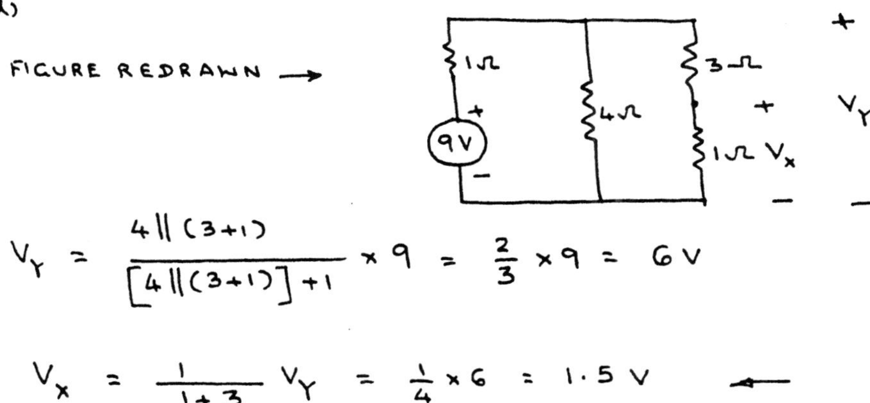

$$V_Y = \frac{4 \| (3+1)}{[4\|(3+1)]+1} \times 9 = \frac{2}{3} \times 9 = 6\text{ V}$$

$$V_X = \frac{1}{1+3} V_Y = \frac{1}{4} \times 6 = 1.5\text{ V} \leftarrow$$

1.2.19.

(a) $I_X = -\frac{3}{3+6} \times 6 = -\frac{3}{9} \times 6 = -2\text{ A} \leftarrow$

(b)

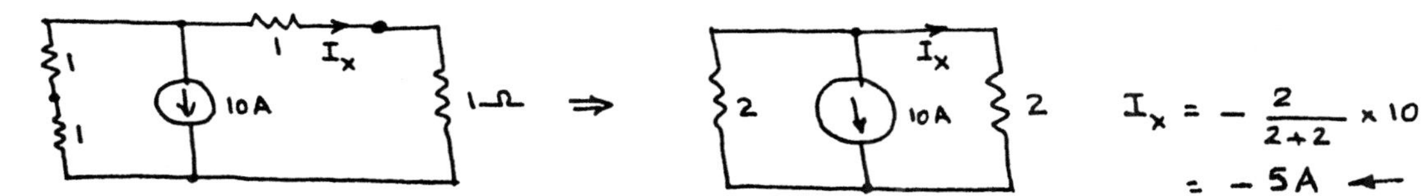

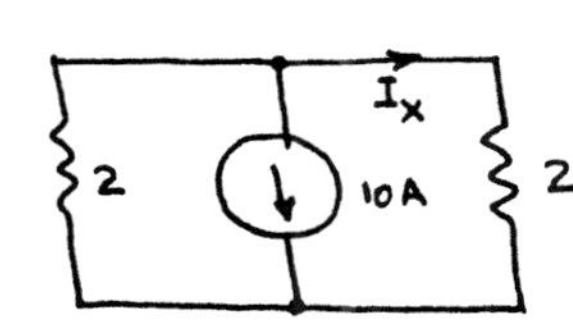

$$I_X = -\frac{2}{2+2} \times 10 = -5\text{A} \leftarrow$$

(c)

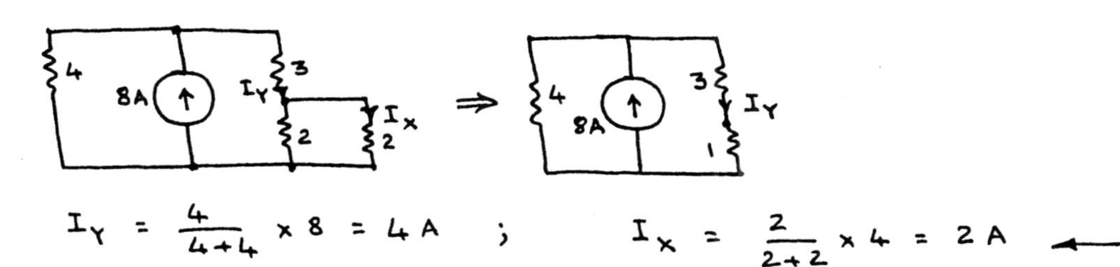

$$I_Y = \frac{4}{4+4} \times 8 = 4\text{ A} \quad ; \quad I_X = \frac{2}{2+2} \times 4 = 2\text{ A} \leftarrow$$

(d)

FIGURE REDRAWN BELOW:

$$I_Y = \frac{5}{5+[3+(4\|4)]} \times 6 = \frac{5}{10} \times 6 = 3\text{ A}$$

$$I_X = \frac{4}{4+4} I_Y = \frac{1}{2} \times 3 = 1.5\text{ A} \leftarrow$$

1.2.20.

$$v(t) = \frac{1}{C}\int_{-\infty}^{t} i(t)\,dt = \frac{1}{100 \times 10^{-12}} \int_{-\infty}^{t} i(t)\,dt$$

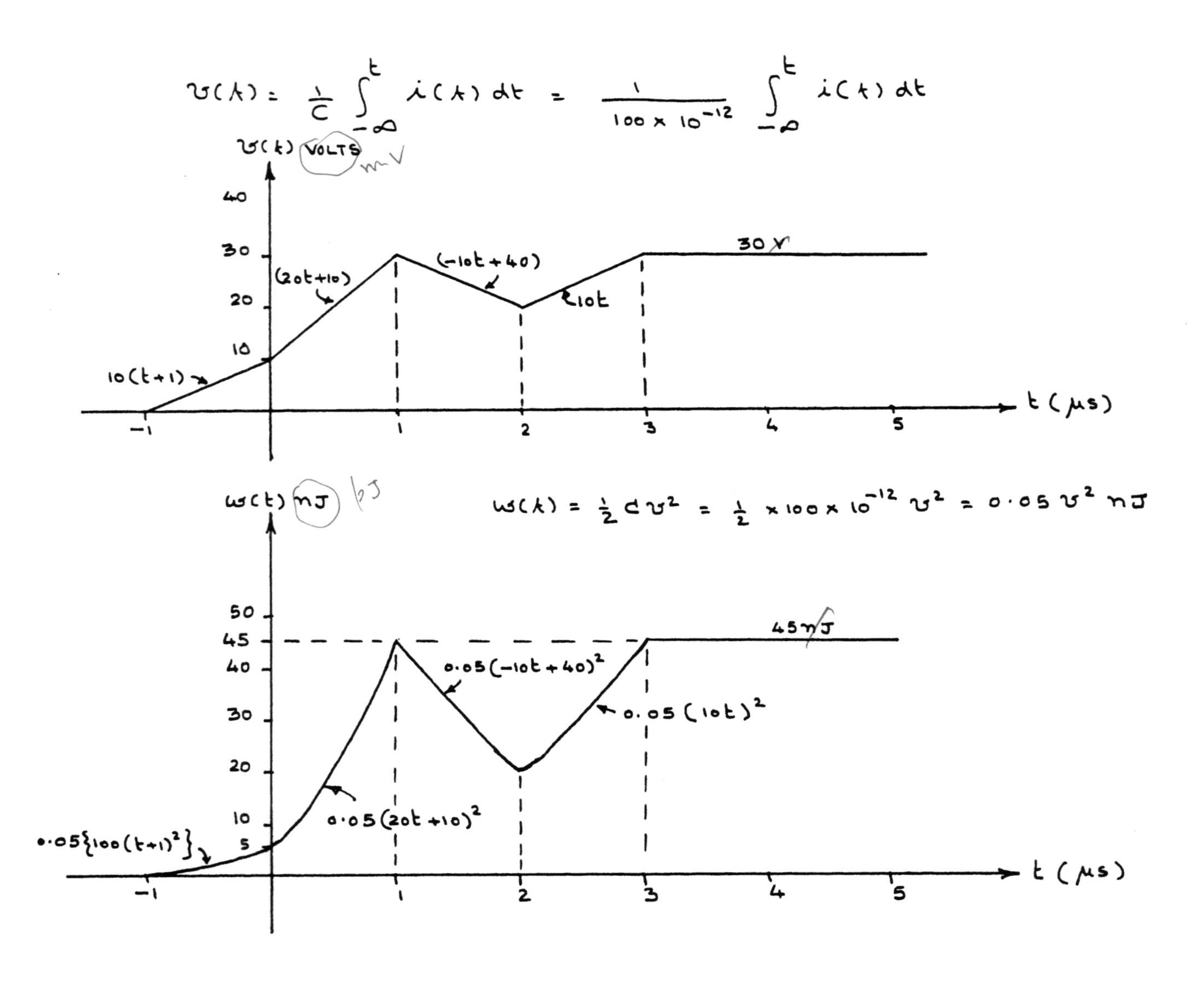

1.2.21.

$$i(t) = C\frac{dv}{dt} = 3\times10^{-6}\frac{dv}{dt}$$

i(t) mA

3

−3

−6

−1 1 2 3 4 5 t μs

$$w(t) = \frac{1}{2}Cv^2 = \frac{1}{2}\times3\times10^{-6}v^2 = \frac{3}{2}\times10^{-6}v^2$$

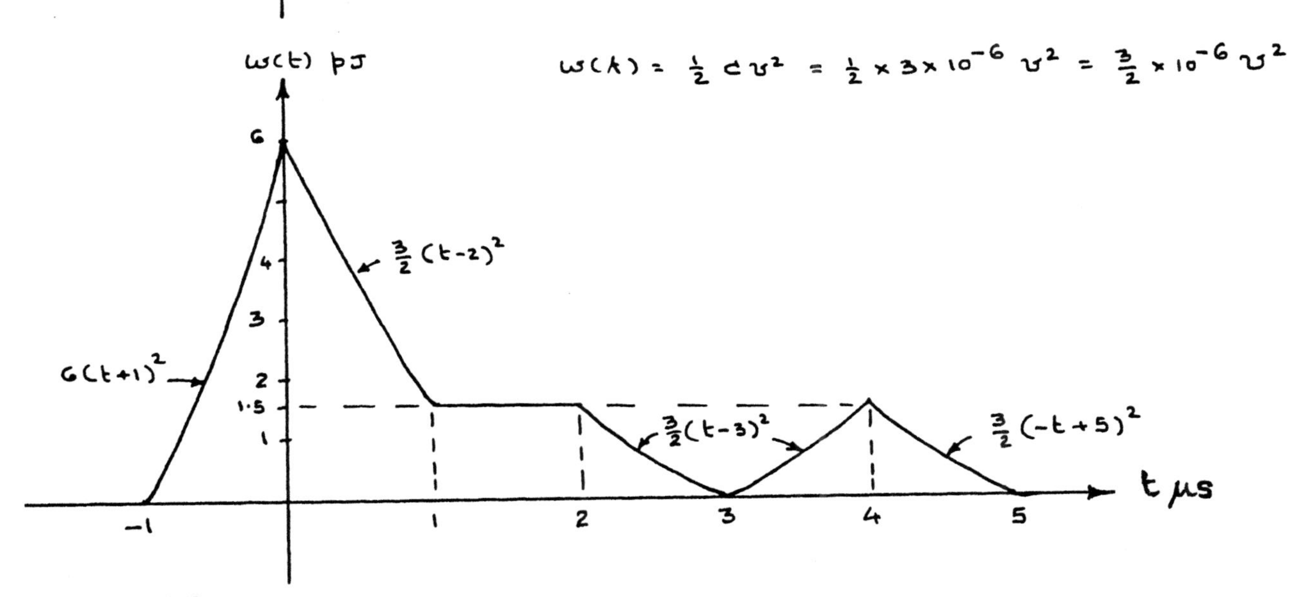

1.2.22.

$$w(t) = \frac{1}{2}C\,v_c^2(t) = \frac{1}{2}\times2\,v_c^2(t) = 9e^{-2t} \quad \text{OR} \quad v_c(t) = \pm 3e^{-t}\ \text{V}$$

AT $t = 1\,\text{s}$, $\quad v_c(t) = \pm 3e^{-1} = \pm 3(2.718) = \pm 8.154\ \text{V}$ ←

$$i_c(t) = C\frac{dv(t)}{dt} = 2\left[\pm(-3)e^{-t}\right] = \mp 6e^{-t}\,\mu\text{A}$$

AT $t = 1\,\text{s}$, $\quad i_c(t) = \mp 6e^{-1} = \mp 6(2.718) = \mp 16.308\,\mu\text{A}$ ←

1.2.23.

$$C = 8.854\times10^{-12}\times\frac{A}{1} = 1\times10^{-12}\ \text{F}$$

$$\therefore A = \frac{1\times10^{-12}}{8.854\times10^{-12}} = 0.113\ \text{m}^2 \quad \text{OR} \quad 1130\ \text{cm}^2$$ ←

1.2.24.

(a) GIVEN CIRCUIT ⇒

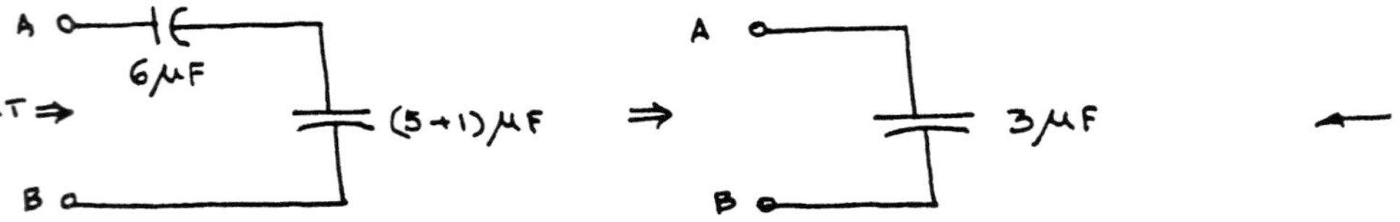

(b) GIVEN CIRCUIT ⇒

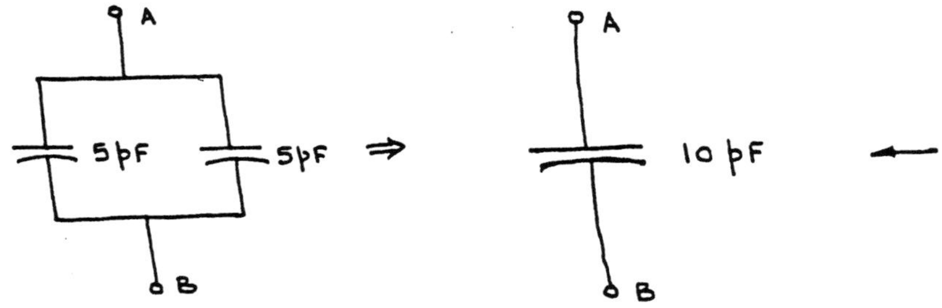

(c) GIVEN CIRCUIT ⇒

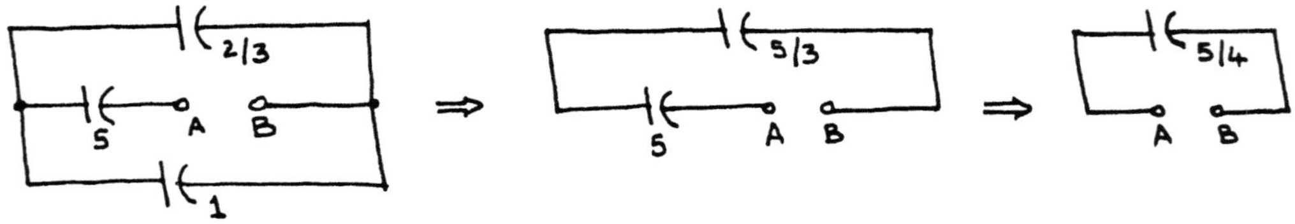

$C_{eq} = \frac{5}{4} = 1.25\ \mu F$ ←

1.2.25.

$$i(t) = \frac{1}{L}\int_{-\infty}^{t} v(\tau)\,d\tau = \frac{1}{3\times10^{-6}}\int_{-\infty}^{t} v(\tau)\,d\tau$$

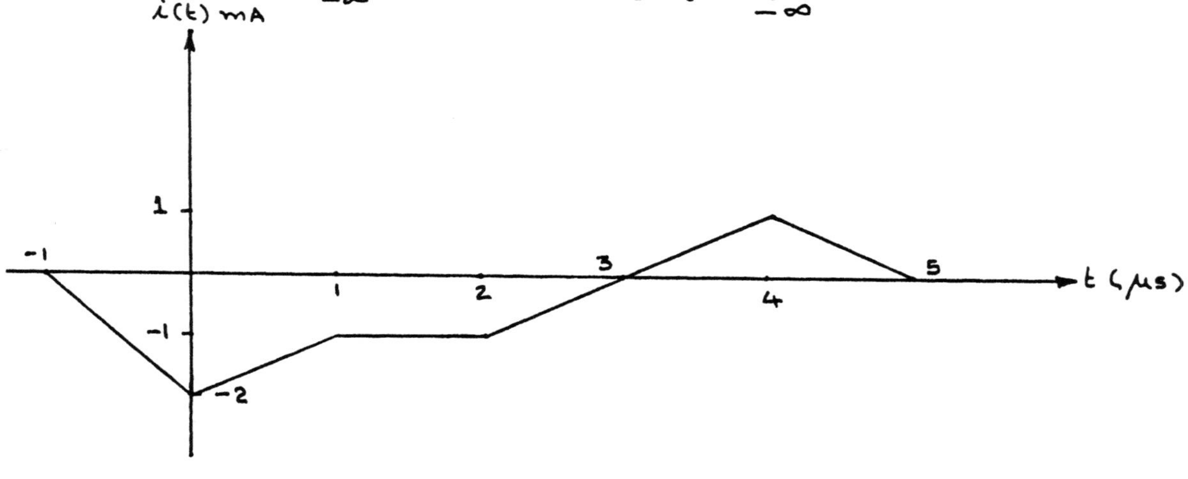

$$w(t) = \frac{1}{2} L i^2 = \frac{1}{2}\times 3\times10^{-6}\ i^2(t)$$

w(t) IS SAME AS THAT OF PR. 1.2.17.

DUALITY PRINCIPLE IS SATISFIED (SEE PR. 1.2.17)

1.2.26.

$$v(t) = L\,\frac{di(t)}{dt} = 100 \times 10^{-12}\,\frac{di(t)}{dt}$$

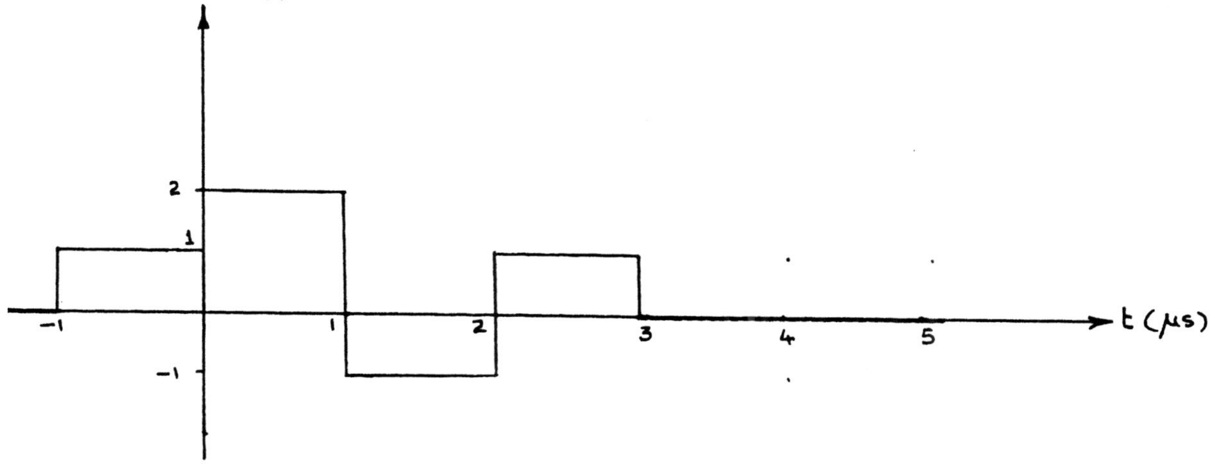

$v(t)$ mV

t (μs)

$$w(t) = \frac{1}{2} L i^2 = \frac{1}{2} \times 100 \times 10^{-12}\, i^2(t) = 50 \times 10^{-12}\, i^2(t)$$

WHICH IS SAME AS THAT OF PR. 1.2.16.

DUALITY PRINCIPLE IS SATISFIED. (SEE PR. 1.2.16)

1.2.27.

$$w(t) = \frac{1}{2} L\, i_L^2(t) = \frac{1}{2} \times 2 \times i_L^2(t) = 9 e^{-2t} \quad \text{OR} \quad i_L(t) = \pm 3 e^{-t}$$

AT $t = 1\,s$, $i_L(t) = \pm 3 e^{-1} = \pm 3\,(2.718) = \pm 8.154$ A ←

$$v_L(t) = L\,\frac{di_L(t)}{dt} = 2\left[\pm(-3)\,e^{-t}\right] = \mp 6 e^{-t}\ \text{V}$$

AT $t = 1\,s$, $v_L(t) = \mp 6 e^{-1} = \mp 6\,(2.718) = \mp 16.308$ V ←

DUALITY PRINCIPLE SATISFIED (SEE PR. 1.2.18)

1.2.28.

$$L = \mu_0 \frac{d}{w} \;; \quad L = 4\pi \times 10^{-7} \times \frac{1}{0.113} = 11.12 \times 10^{-6}\ \text{H/m} = 11.12\ \mu\text{H/m}$$ ←

FROM PROB. 1.2.**23**, C PER UNIT LENGTH = $\epsilon_0\, w/d$

L PER UNIT LENGTH = $\mu_0\, d/w$

∴ THE PRODUCT IS $(\mu_0 \epsilon_0)$ ←

1.2.29.

(a)

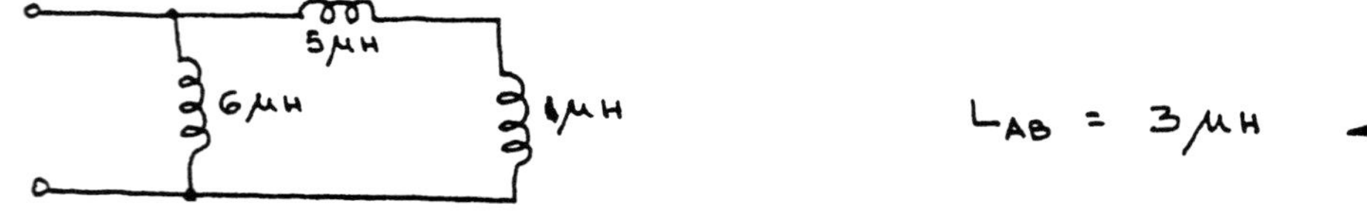

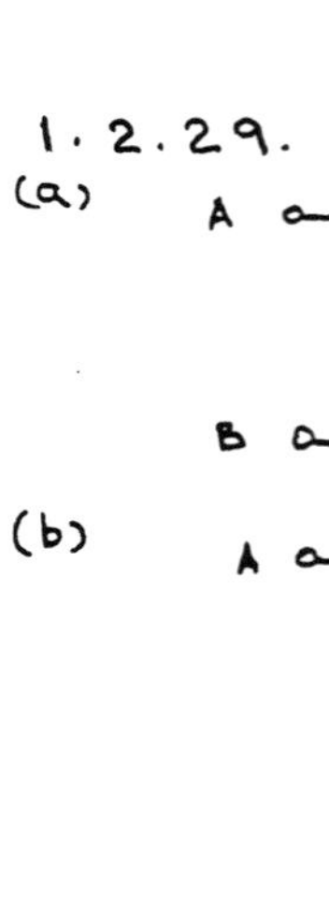

$L_{AB} = 3\,\mu H$ ←

(b)

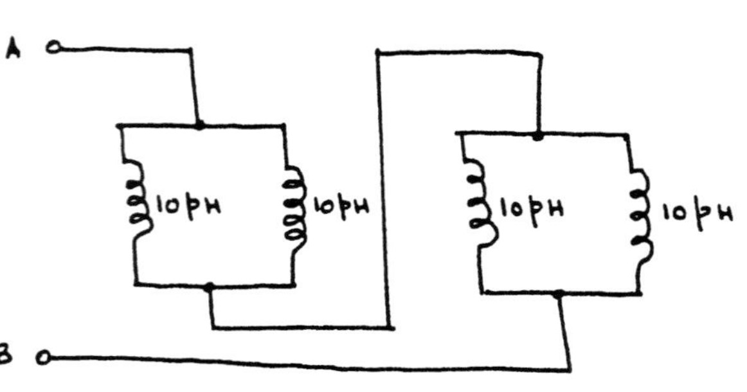

$L_{AB} = 10\,pH$ ←

(c)

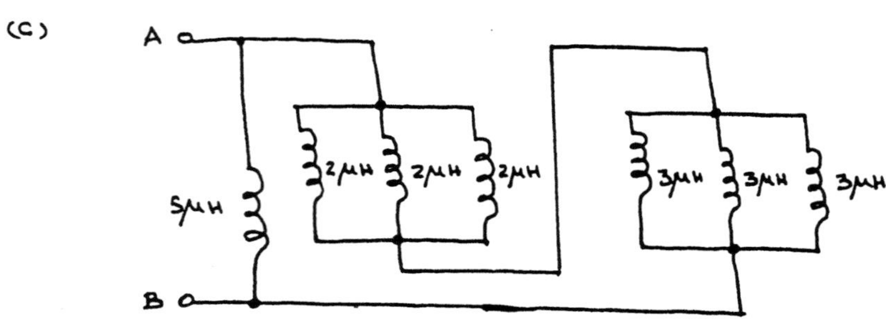

$L_{AB} = 1.25\,\mu H$ ←

1.2.30.

(i) $p_1 = v_1 i_1 = \left(L_{11}\frac{di_1}{dt} + L_{12}\frac{di_2}{dt}\right) i_1$; $p_2 = v_2 i_2 = \left(L_{21}\frac{di_1}{dt} + L_{22}\frac{di_2}{dt}\right) i_2$

At $t = t_0$, no energy is stored in the magnetic field, since the magnetic flux is zero; i.e., $W(t_0) = 0$

From t_0 to t_1, $i_2 = 0$ and $\frac{di_2}{dt} = 0$

$$W_1 = \int_{t_0}^{t_1} (p_1 + p_2)\,dt = \int_{t_0}^{t_1} L_{11} i_1 \frac{di_1}{dt}\,dt = \int_0^{\text{at } t=t_1,\ i_1 = I_1} L_{11} i_1\,di_1 = \frac{1}{2} L_{11} I_1^2$$

From t_1 to t_2, $i_1 = I_1$; $\frac{di_1}{dt} = 0$

$$W_2 = \int_{t_1}^{t_2} \left(L_{12} I_1 \frac{di_2}{dt} + L_{22} i_2 \frac{di_2}{dt}\right) dt = \int_0^{\text{at } t=t_2,\ i_2 = I_2} (L_{12} I_1 + L_{22} i_2)\,di_2 = L_{12} I_1 I_2 + \frac{1}{2} L_{22} I_2^2$$

Thus the energy stored at time t_2 is $W(t_2) = W(t_0) + W_1 + W_2$

or $W(t_2) = \frac{1}{2} L_{11} I_1^2 + L_{12} I_1 I_2 + \frac{1}{2} L_{22} I_2^2$ ←

(ii) Using the same steps as before, $W(t_2) = \frac{1}{2} L_{11} I_1^2 + L_{21} I_1 I_2 + \frac{1}{2} L_{22} I_2^2$

Comparing, $L_{12} = L_{21} = M$ ←

1.2.31.

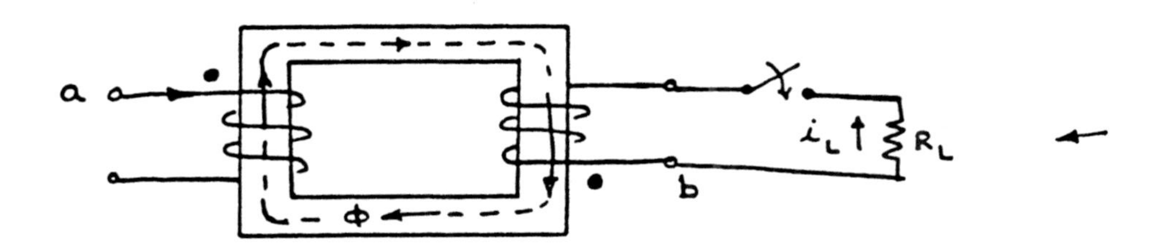

1.2.32.

(a) $v_1 = L_{11}\frac{di_1}{dt} - M\frac{di_2}{dt}$; $v_2 = -M\frac{di_1}{dt} + L_{22}\frac{di_2}{dt}$ ←

(b) $v_1 = -L_{11}\frac{di_1}{dt} - M\frac{di_2}{dt}$; $v_2 = M\frac{di_1}{dt} + L_{22}\frac{di_2}{dt}$ ←

1.2.33.

Series case:

$$e = (L_{11}+M)\frac{di}{dt} + (L_{22}+M)\frac{di}{dt} = (L_{11}+L_{22}+2M)\frac{di}{dt}$$

$L_{series} = L_{11}+L_{22}+2M$; M can be either +ve or −ve. If the fluxes due to the two coils aid each other, it would be positive; otherwise negative.

Hence $L_{series} = L_{11}+L_{22} \pm 2M$, where M is a positive quantity. ←

Parallel Case:

$e = L_{11}\frac{di_1}{dt} + M\frac{di_2}{dt}$ —① ; $e = L_{22}\frac{di_2}{dt} + M\frac{di_1}{dt}$ —②

or $(L_{11}-M)\frac{di_1}{dt} = (L_{22}-M)\frac{di_2}{dt}$ —③

$e = L_{par}\left(\frac{di_1}{dt} + \frac{di_2}{dt}\right)$ —④ where L_{par} is the equiv. ind.

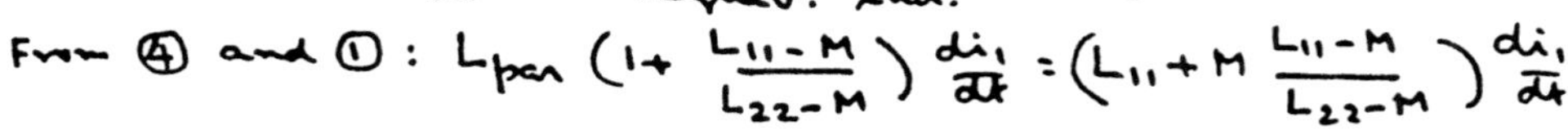

From ④ and ①: $L_{par}\left(1 + \frac{L_{11}-M}{L_{22}-M}\right)\frac{di_1}{dt} = \left(L_{11} + M\frac{L_{11}-M}{L_{22}-M}\right)\frac{di_1}{dt}$

or $$L_{par} = \frac{L_{11}(L_{22}-M)+M(L_{11}-M)}{(L_{22}-M)+(L_{11}-M)} = \frac{L_{11}L_{22}-M^2}{L_{11}+L_{22}-2M}$$

M can be positive or negative. If the mmfs of the windings oppose, M will be negative; otherwise positive.

Thus, $L_{par} = \frac{L_{11}L_{22}-M^2}{L_{11}+L_{22}\mp 2M}$, where M is a positive quantity. ←

1.2.34.

$$\left.\begin{aligned} v_1 &= -L_{11}\frac{di_1}{dt} + L_{12}\frac{di_2}{dt} + L_{13}\frac{di_3}{dt} \\ v_2 &= -L_{12}\frac{di_1}{dt} + L_{22}\frac{di_2}{dt} + L_{23}\frac{di_3}{dt} \\ v_3 &= L_{13}\frac{di_1}{dt} - L_{23}\frac{di_2}{dt} - L_{33}\frac{di_3}{dt} \end{aligned}\right\} \leftarrow$$

1.2.35.

$$\left(\frac{N_1}{N_2}\right)^2 8 = 72 \quad \text{OR} \quad \frac{N_1}{N_2} = \sqrt{9} = 3 \leftarrow$$

1.2.36.

3:1

1:2

18 Ω

$v_{in} = 18 \sin 10t$

v_1

36 Ω

v_2

2 Ω

v_3

8 Ω

$v_{out}(t)$

18 Ω

36 Ω

2 Ω

$$v_1 = \frac{18 \sin 10t}{2} = 9 \sin 10t \; ; \; v_2 = \frac{1}{3} v_1 = 3 \sin 10t \; ; \; v_3 = \frac{1}{2} v_2 = 1.5 \sin 10t$$

$$v_{out} = -2 v_3 = (-3 \sin 10t)\ \text{V} \leftarrow$$

1.2.37.

(a) $a = N_1/N_2 = 2400/240 = 10 \leftarrow$

(b) $V_2 I_2 = 100 \times 10^3$ VA OR $I_2 = \frac{100 \times 10^3}{240} = 416.67$ A (RMS) $\leftarrow$

$I_1 = I_2/a = 416.67/10 = 41.67$ A (RMS) $\leftarrow$

$R_L = V_2/I_2 = 240/416.67 = 0.576\ \Omega \leftarrow$

(c) $R_L' = (N_1/N_2)^2 R_L = a^2 R_L = 10^2 \times 0.576 = 57.6\ \Omega \leftarrow$

1.2.38.

(a) $N_1/N_2 = a = 220/110 = 2$; $I_1 = \frac{10 \times 10^3}{220} = 45.45$ A (RMS) $\leftarrow$

$I_2 = \frac{10 \times 10^3}{110} = 90.91$ A (RMS) $\leftarrow$

(b) $I_2 = \frac{110}{2} = 55$ A (RMS) ON LV SIDE $\leftarrow$; $I_1 = \frac{55}{2} = 27.5$ A (RMS) ON HV SIDE $\leftarrow$

(c) $V_1/I_1 = a^2 R_2 = (2)^2 2 = 8\ \Omega \leftarrow$

1.3.1.

KVL: LOOP L_1: $v_C - 10 + 8 - 6 = 0$ OR $v_C = 8\text{ V}$

LOOP L_2: $-8 + v_G - 4 - v_B = 0$ OR $v_G - v_B = 12\text{ V}$

LOOP L_3: $4 - v_G + 10 = 0$ OR $v_G = 14\text{ V}$

THUS $v_B = v_G - 12 = 14 - 12 = 2\text{ V}$; $v_C = 8\text{ V}$; $v_G = 14\text{ V}$ ←

KCL: NODE a: $i_A = -1\text{ A}$

NODE b: $i_D + i_A + 2 = 0$ OR $i_D = -i_A - 2 = 1 - 2 = -1\text{ A}$

NODE c: $i_E = 2\text{ A}$

NODE d: $i_F = 1 + i_H$

NODE e: $i_F = i_D + 3 = -1 + 3 = 2\text{ A}$

NODE f: $i_E + i_H = 3$ OR $2 + i_H = 3$ OR $i_H = 3 - 2 = 1\text{ A}$

THUS $i_A = -1\text{A}$; $i_D = -1\text{A}$; $i_E = 2\text{A}$; $i_F = 2\text{A}$; $i_H = 1\text{A}$ ←

POWER DELIVERED TO EACH ELEMENT IS GIVEN BELOW:

A:	$6(-i_A) = 6(1) = 6\text{ W}$;	SINK
B:	$v_B(2) = 2(2) = 4\text{ W}$;	SINK
C:	$v_C(-1) = 8(-1) = -8\text{ W}$;	SOURCE
D:	$8(-i_D) = 8(1) = 8\text{ W}$;	SINK
E:	$4(i_E) = 4(2) = 8\text{ W}$;	SINK
F:	$10(i_F) = 10(2) = 20\text{ W}$;	SINK
G:	$v_G(-3) = 14(-3) = -42\text{ W}$;	SOURCE
H:	$4(i_H) = 4(1) = 4\text{ W}$;	SINK

←

THE ALGEBRAIC SUM OF POWERS DELIVERED IS ZERO. ←

CONSERVATION OF POWER IS SATISFIED FOR THE NETWORK. ←

1.3.2.

FOLLOWING THE CLOSED PATH (LOOP) ABCDEFIHA, KVL YIELDS

$$-4 + 6 - (-4) + v - (-12) - (12) + (-2) - (6) = 0$$

OR $v + 22 - 24 = 0$ OR $v = 2\text{ V}$ ←

1.3.3.

KCL AT NODE C : $i_x = 3+2 = 5\text{ A}$

KCL AT NODE B : $i_x + i = 6$ OR $5+i=6$ OR $i = 1\text{ A}$ ←

KVL AROUND LOOP BGHEDCB : $v + (-1) + (4) - (-2) = 0$ OR $v = -5\text{ V}$ ←

KCL AT NODE E : $i_y = 2 + i = 2+1 = 3\text{ A}$

KCL AT NODE F : $i_z = 3 + i_y = 3+3 = 6\text{ A}$

KVL AROUND CDEFC : $-4 - (-1) + 1 - v_{CF} = 0$ OR $v_{CF} = -2\text{ V}$

KVL AROUND ABCFA : $v_{AB} - 2 + v_{CF} - 3 = 0$ OR $v_{AB} = 7\text{ V}$

ELEMENT	POWER DELIVERED	
S (G–H)	$vi = (-5)1$	$= -5\text{ W}$
A – B	$v_{AB}(6) = 7 \times 6$	$= 42\text{ W}$
B – C	$(-2)i_x = -2 \times 5$	$= -10\text{ W}$
C – D	$4(-2)$	$= -8\text{ W}$
D – E	$(-1)(-2)$	$= 2\text{ W}$
E – F	$(1)i_y = 1 \times 3$	$= 3\text{ W}$
F – A	$3 \times (-6)$	$= -18\text{ W}$
C – F	$v_{CF}(3) = (-2)(3)$	$= -6\text{ W}$
	ALGEBRAIC SUM	$= 0$

∴ CONSERVATION OF POWER IS SATISFIED. ←

1.3.4.

$v_1 = i_1(10) = 4 \times 10 = 40\text{ V}$ ←

$$v_3 = 2\frac{di_3}{dt} = 2\frac{d}{dt}(5e^{-t}) = (-10e^{-t})\text{ V}$$ ←

$$v_4 = 1\frac{di_4}{dt} = \frac{d}{dt}(10\cos 2t) = (-20\sin 2t)\text{ V}$$ ←

KVL AROUND ABCDA : $v_1 - v_2 - v_3 + v_4 = 0$

OR $v_2 = v_1 - v_3 + v_4 = 40 - (-10e^{-t}) + (-20\sin 2t) = (40 + 10e^{-t} - 20\sin 2t)\text{ V}$ ←

$$i_2 = 0.1\frac{dv_2}{dt} = 0.1\frac{d}{dt}\left[40 + 10e^{-t} - 20\sin 2t\right] = (-e^{-t} - 4\cos 2t)\text{ A}$$ ←

KCL AT NODE B : $i_5 = i_1 + i_2 = (4 - e^{-t} - 4\cos 2t)\text{ A}$ ←

1.3.5.

$V_{AC} = V_{AB} + V_{BC}$; $V_{BD} = V_{BC} + V_{CD}$

$\therefore\ V_{AC} - V_{BD} = 10 - 20 = V_{AB} - V_{CD} = -10\ V$

$V_{AB} = 2I$; $V_{CD} = 4I$

$\therefore\ V_{AB} - V_{CD} = 2I - 4I = -2I = -10$ or $I = 5A$

KVL AROUND ABCDA: $V_{AB} + V_{BC} + V_{CD} + V_{DA} = 0$

or $V_{AC} + 4I - V_1 + I = 0$

or $V_1 = 10 + 4(5) + 5 = 35\ V$ ←

$V_{BD} = 3I + V_2 + 4I = 20$ or $V_2 = 20 - 7(5) = -15\ V$ ←

POWER ABSORBED BY RESISTANCES: $(2+3+4+1)\ I^2 = 10 \times 5^2 = 250\ W$

POWER DELIVERED BY SOURCE V_2 : $-V_2 I = 15 \times 5 = 75\ W$

POWER DELIVERED BY SOURCE V_1 : $V_1 I = 35 \times 5 = 175\ W$

$\therefore$ POWER ABSORBED BY RESISTANCES = POWER DELIVERED BY BOTH SOURCES

HENCE CONSERVATION OF POWER IS SATISFIED. ←

1.3.6.

(a)

KCL: $I_A + I_B = I_1 + I_2 + I_3$

$30 + 50 = V\left(\frac{1}{R_1} + \frac{1}{R_2} + \frac{1}{R_3}\right) = V\left(\frac{1}{20} + \frac{1}{40} + \frac{1}{80}\right) = V\left(\frac{7}{80}\right)$

or $V = \frac{80 \times 80}{7} = 914.3\ V$ ←

(b)

$I_1 = V/R_1 = \frac{6400}{7} \times \frac{1}{20} = \frac{320}{7} = 45.71\ V$ ←

$I_2 = V/R_2 = \frac{6400}{7} \times \frac{1}{40} = \frac{160}{7} = 22.86\ V$ ←

$I_3 = V/R_3 = \frac{6400}{7} \times \frac{1}{80} = \frac{80}{7} = 11.43\ V$ ←

(c)

POWER SUPPLIED BY SOURCE A: $I_A V = 30\left(\frac{6400}{7}\right) = 27{,}428.6\ W$ ←

POWER SUPPLIED BY SOURCE B: $I_B V = 50\left(\frac{6400}{7}\right) = 45{,}714.3\ W$ ←

POWER ABSORBED BY RESISTORS $= I_1^2 R_1 + I_2^2 R_2 + I_3^2 R_3$

$= \left(\frac{320}{7}\right)^2 20 + \left(\frac{160}{7}\right)^2 40 + \left(\frac{80}{7}\right)^2 80 = 73{,}142.9\ W$

YES, CONSERVATION OF POWER IS SATISFIED. ←

1.3.7.

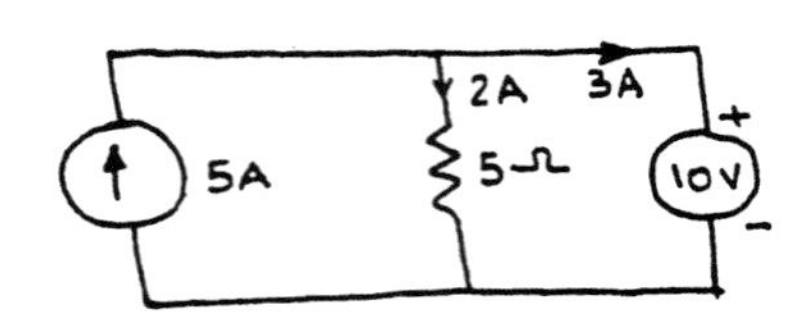

POWER ABSORBED BY 5-Ω RESISTOR = $(2)^2\,5$ = 20 W

POWER DELIVERED BY 5-A SOURCE = 5×10 = 50 W

POWER ABSORBED BY 10-V SOURCE = 3×10 = 30 W

TOTAL POWER ABSORBED = TOTAL POWER DELIVERED ←

1.3.8.

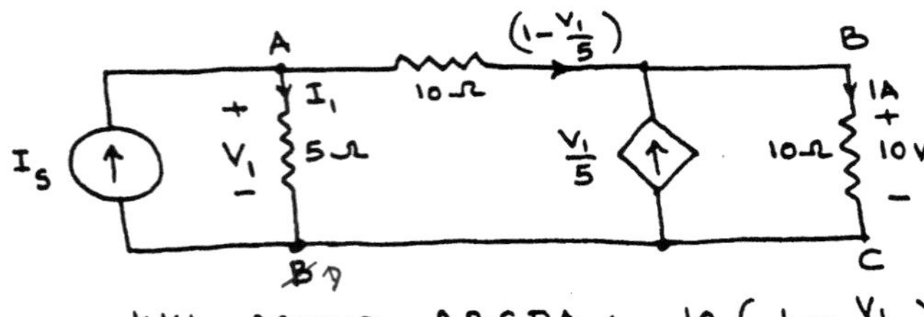

KVL AROUND ABCDA : $10\left(1-\frac{V_1}{5}\right)+10 = V_1$

OR $10-2V_1+10 = V_1$ OR $3V_1 = 20$ OR $V_1 = (20/3)$ V

$$1-\frac{V_1}{5} = 1-\frac{20}{3}\cdot\frac{1}{5} = 1-\frac{4}{3} = -\frac{1}{3}\text{ A}$$

$$I_1 = \frac{V_1}{5} = \frac{20}{3}\cdot\frac{1}{5} = \frac{4}{3}\text{ A}$$

$$I_S = \frac{4}{3}+\left(-\frac{1}{3}\right) = 1\text{ A} \quad \leftarrow$$

1.3.9.

(a)

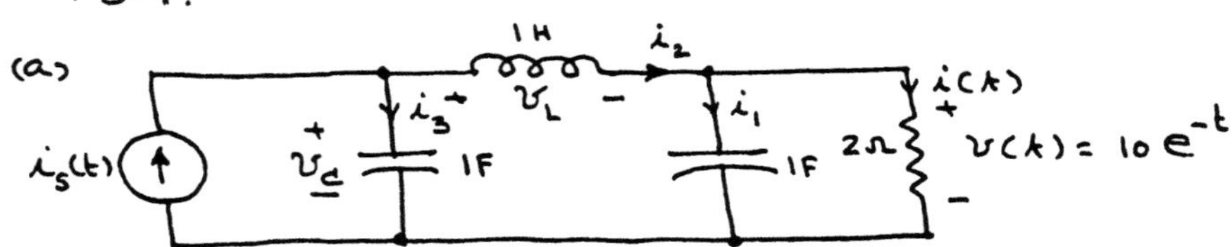

$$i(t) = \frac{v(t)}{2} = 5e^{-t}\ ;\quad i_1(t) = C\frac{dv}{dt} = 1\frac{d}{dt}\left(10e^{-t}\right) = -10e^{-t}$$

$$i_2 = i_1 + i = 5e^{-t} - 10e^{-t} = -5e^{-t}$$

$$v_L = L\frac{di_2}{dt} = 1\frac{d}{dt}\left(-5e^{-t}\right) = +5e^{-t}$$

$$v_c = v_L + v = 5e^{-t} + 10e^{-t} = 15e^{-t}$$

$$i_3 = C\frac{dv_c}{dt} = 1\frac{d}{dt}\left(15e^{-t}\right) = -15e^{-t}$$

$$i_s = i_3 + i_2 = -15e^{-t} + \left(-5e^{-t}\right) = -20e^{-t}\ \text{V} \quad \leftarrow$$

1.3.9. CONTINUED

(b)

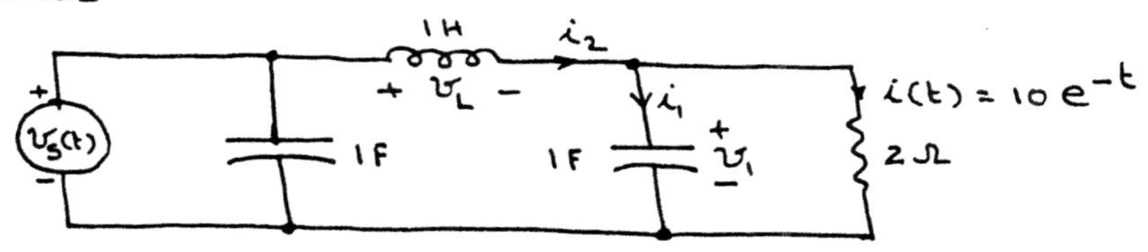

$$v_1 = Ri = 2(10e^{-t}) = 20e^{-t}$$

$$i_1 = C\frac{dv_1}{dt} = 1\frac{d}{dt}(20e^{-t}) = -20e^{-t}$$

$$i_2 = i_1 + i = -20e^{-t} + 10e^{-t} = -10e^{-t}$$

$$v_L = L\frac{di_2}{dt} = 1\frac{d}{dt}(-10e^{-t}) = 10e^{-t}$$

$$v_s(t) = v_L + v_1 = 10e^{-t} + 20e^{-t} = 30e^{-t}\ \text{V} \quad \longleftarrow$$

1.3.10.

$$V_{out} = -10^5 V_1$$

KVL: $5{,}000I + 10{,}000I - 10^5 V_1 - 2.5 = 0$ OR $10^5 V_1 = -2.5 + 15{,}000I$ —— ①

KVL: $5{,}000I + V_1 - 2.5 = 0$ OR $V_1 = 2.5 - 5{,}000I$ —— ②

SOLVING ① AND ②, $V_1 = \dfrac{5}{3+10^5}$

$$V_{out} = -10^5 V_1 = -10^5\left(\frac{5}{3+10^5}\right) \simeq -5\text{V} \quad \longleftarrow$$

$$I = \frac{2.5 - V_1}{5{,}000} = \frac{2.5 - \{5/(3+10^5)\}}{5{,}000} = 0.5\times10^{-3}\ \text{A} = 0.5\ \text{mA}$$

$$P = VI = 2.5\times0.5\times10^{-3} = 1.25\times10^{-3}\ \text{W} = 1.25\ \text{mW} \quad \longleftarrow$$

1.4.1.

0.1% OF 1,000 V = 1 V

$$\pm\frac{1}{100}\times100 = \pm1\% \quad \longleftarrow$$

1.4.2.

MAX. ERROR = ± 0.5% × 100 = ± 0.5 A

FOR A READING OF 65A, $\pm\dfrac{0.5}{65}\times100 = \pm0.77\%$

MAX. PROBABLE % ERROR = 2 × 0.77 = 1.54% ⟵

1.4.3.

DIGITAL ERROR: $(0.07\%\times5) + (0.05\%\times10) + (0.005\%\times5\times20) = 0.0135\ \text{A}$

$\%\ \text{ERROR} = \dfrac{0.0135}{5}\times100 = 0.27\%$ FOR THE DIGITAL METER ⟵

ANALOG ERROR: $(0.5\%\times10) + (0.001\%\times5\times20) = 0.051\text{A}$

$\%\ \text{ERROR} = \dfrac{0.051}{5}\times100 = 1.02\%$ FOR THE ANALOG METER ⟵

1.4.4.

(a) $\dfrac{5}{\sqrt{2}} = (0.707)\,5 = 3.536\text{ A}$ ←

(b) 5A ←

(c) $\dfrac{5}{\sqrt{3}} = 2.887\text{ A}$ ←

(d) $\dfrac{5}{\sqrt{3}} = 2.887\text{ A}$ ←

1.4.5.

$$R_1 R_4 = R_2 R_3 \quad \text{OR} \quad R_4 = \frac{R_2 R_3}{R_1} = \frac{48 \times 10}{24} = 20\,\text{k}\Omega \leftarrow$$

1.4.6.

(a) $R_1 R_4 = R_2 R_3 \quad \text{OR} \quad R_4 = \dfrac{R_2 R_3}{R_1} = \dfrac{8 \times 40}{16} = 20\,\Omega$ ←

(b)

$$I_1 = \frac{24}{12 + (56 \| 28)} = \frac{24}{12 + \frac{56}{3}} = \frac{24 \times 3}{92} = 0.7826\text{A}$$

$$I_2 = 0.7826 \times \frac{56}{84} = 0.5217\text{ A} \leftarrow$$

1.4.7.

(a)

(i) $\pm 3 \times 5 = \pm 15\text{ mV}$ OR MAX. VALUE IS 15 mV ←

(ii) TIME PERIOD $T = 4 \times 50 \times 10^{-3}\text{ s} = 0.2\text{ s}$

FREQUENCY $f = \dfrac{1}{T} = \dfrac{1}{0.2} = 5\text{ Hz}$ ←

(b)

(i) $\pm 4 \times 5 = \pm 20\text{ mV}$; MAX. VALUE IS 20 mV ←

(ii) $T = 4 \times 50 \times 10^{-3}\text{ s} = 0.2\text{ s}$

$f = \dfrac{1}{T} = \dfrac{1}{0.2} = 5\text{ Hz}$ ←

(c)

(i) $2.5 \times 5 = 12.5\text{ mV}$ ←

(ii) $T = 3 \times 50 \times 10^{-3}\text{ s} = 0.15\text{ s}$

$f = \dfrac{1}{T} = \dfrac{1}{0.15} = 6.67\text{ Hz}$ ←

CHAPTER 2

2.1.1.

(a)

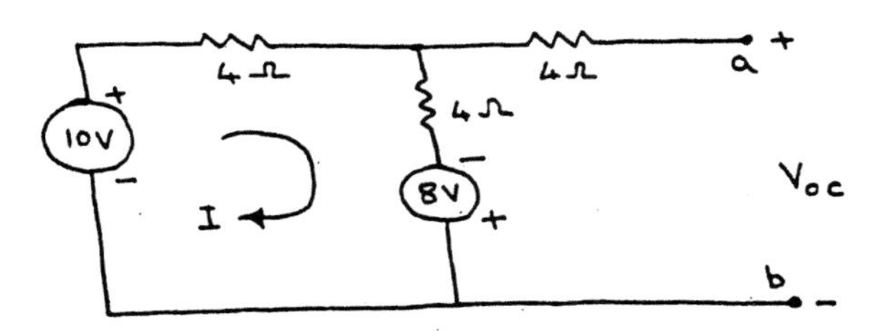

KVL: $10 - 4I - 4I + 8 = 0$ OR $8I = 18$ OR $I = 2.25$ A

$V_{oc} = 4I - 8 = (4 \times 2.25) - 8 = 1$ V

TO FIND R_{Th} :

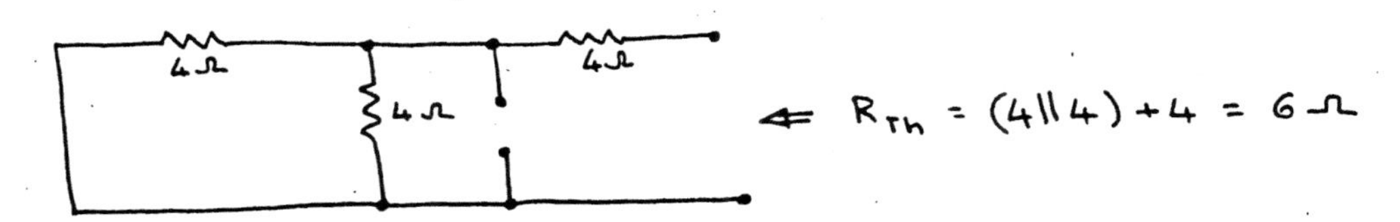

$\Leftarrow R_{Th} = (4 \| 4) + 4 = 6\ \Omega$

THÉVENIN EQUIVALENT CIRCUIT :

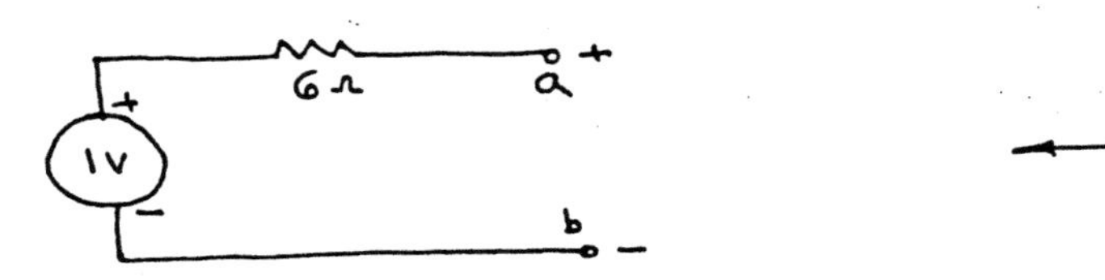

←

NORTON EQUIVALENT CIRCUIT :

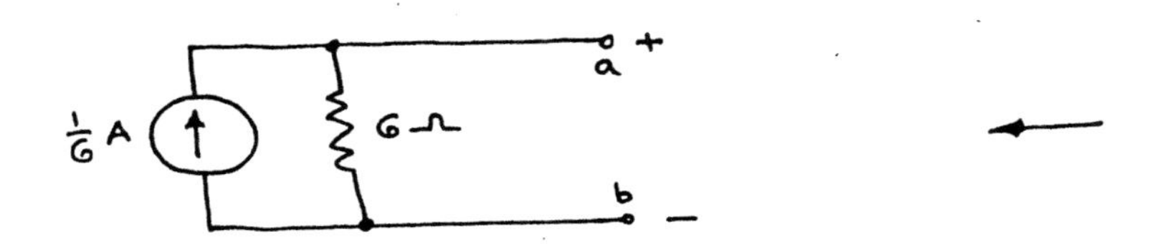

←

(b)

R = 6 Ω FOR RESISTANCE MATCHING ←

(C)

$P = I^2 R = \left(\frac{1}{6+6}\right)^2 6 = \frac{1}{144} \times 6 = \frac{1}{24}$ W ←

2.1.2.

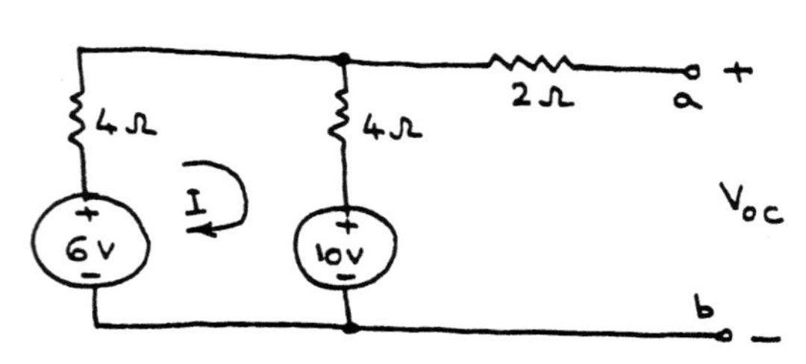

KVL : $6 - 4I - 4I - 10 = 0$ OR $8I = -4$ OR $I = -\frac{1}{2}$ A

$V_{oc} = 4(-\frac{1}{2}) + 10 = 8$ V

TO FIND R_{Th} :

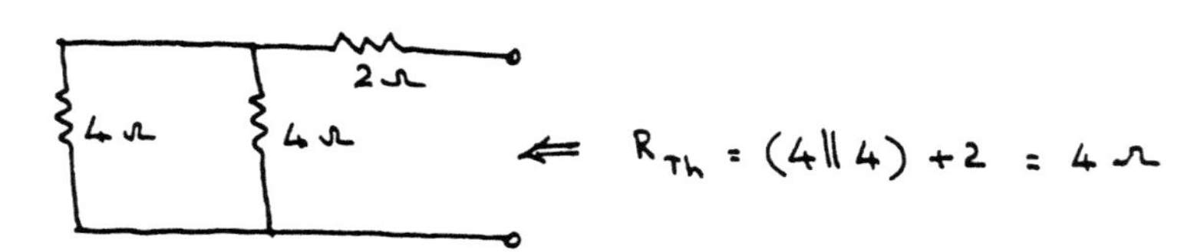

⇐ $R_{Th} = (4 \| 4) + 2 = 4\,\Omega$

THÉVENIN EQUIVALENT CIRCUIT :

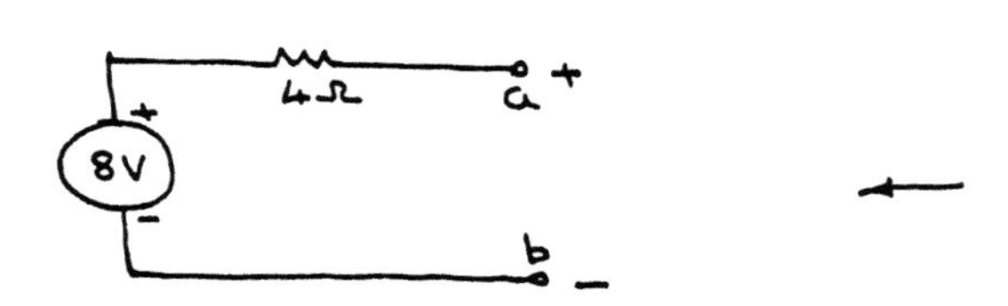

NORTON EQUIVALENT CIRCUIT:

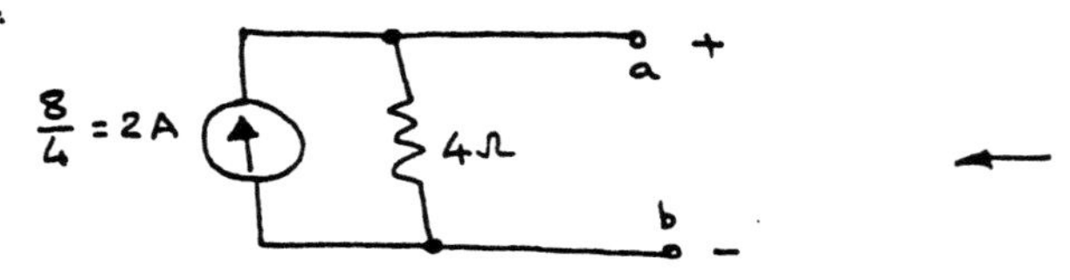

2.1.3.

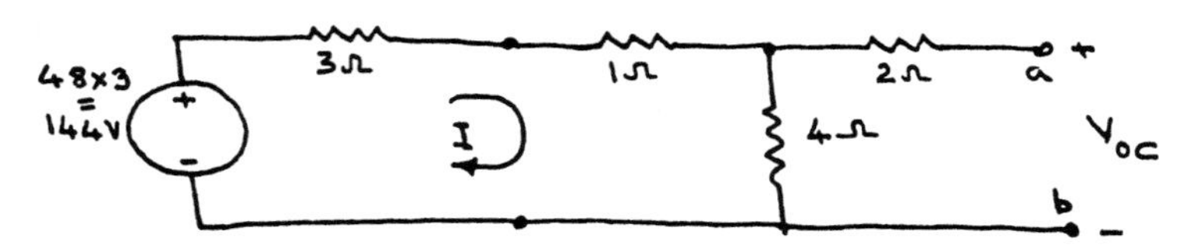

KVL : $I(3+1+4) = 144$ OR $I = 144/8 = 18$ A

$V_{oc} = 18 \times 4 = 72$ V

TO FIND R_{Th}:

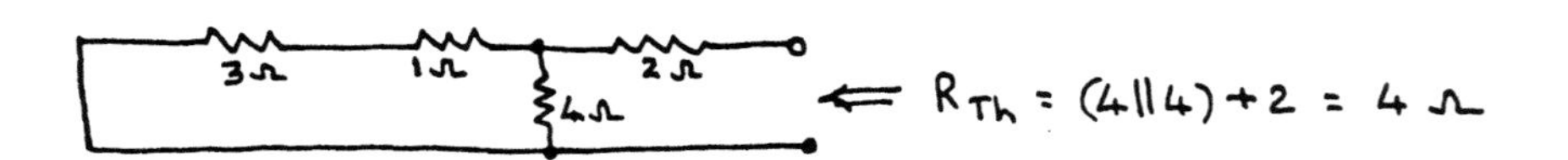

⇐ $R_{Th} = (4\|4) + 2 = 4\,\Omega$

THÉVENIN EQUIVALENT CIRCUIT:

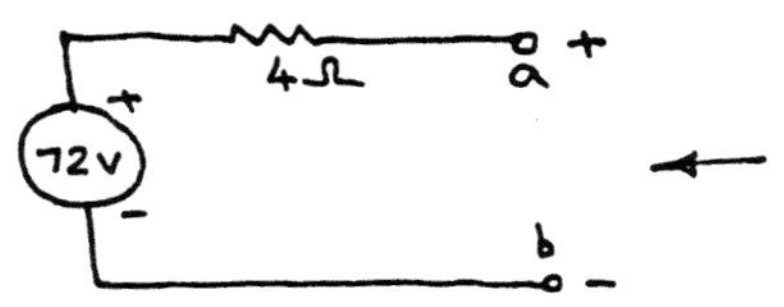

NORTON EQUIVALENT CIRCUIT :

$\frac{72}{4} = 18$A

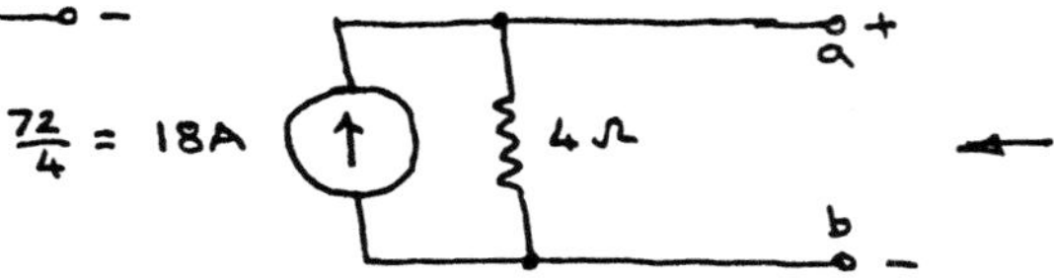

2.1.4.

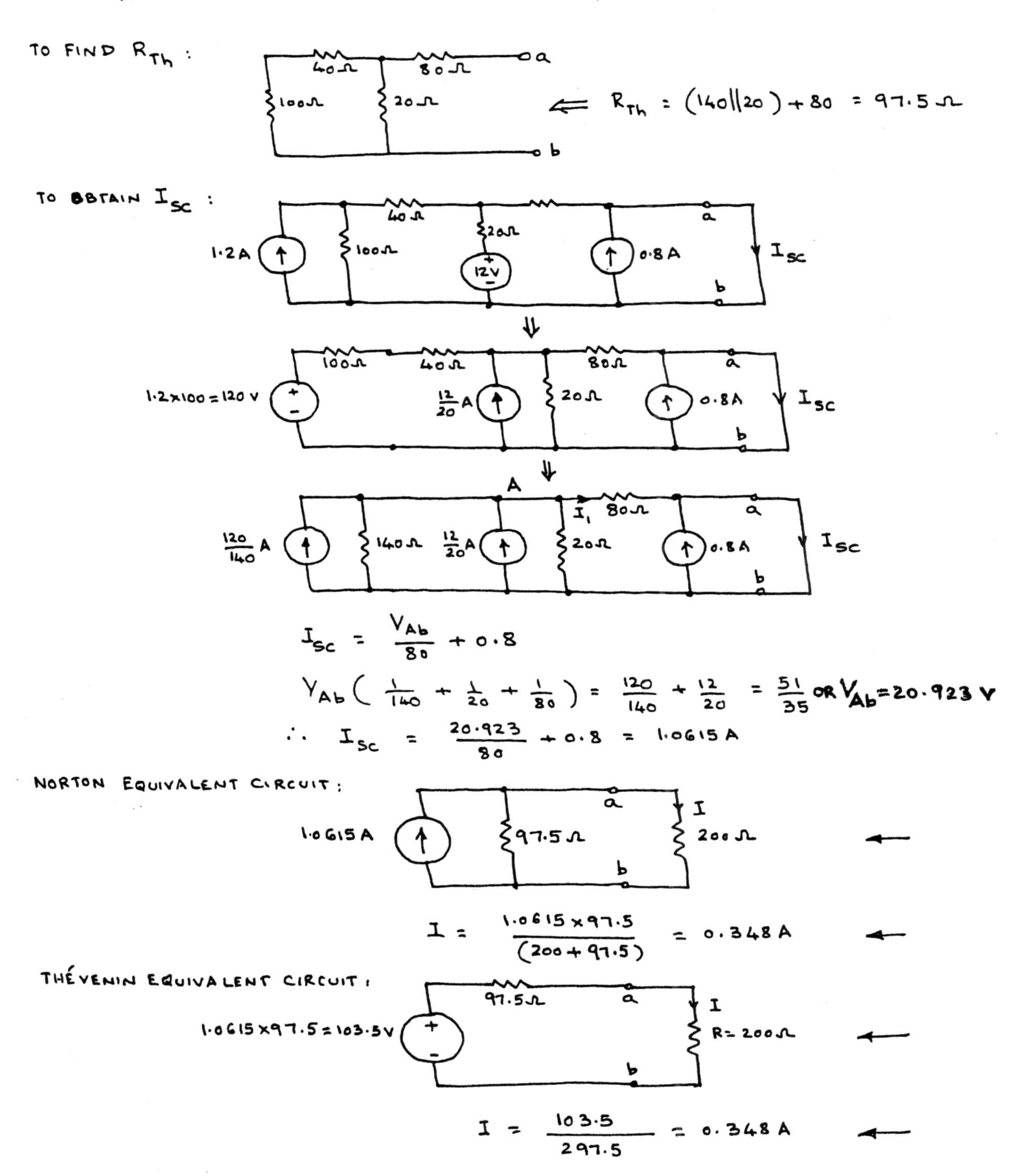
TO FIND R_Th :
40 Ω
80 Ω
100 Ω
20 Ω
a
b
R_Th = (140||20) + 80 = 97.5 Ω
TO OBTAIN I_SC :
1.2 A
100 Ω
40 Ω
2 Ω
12 V
0.8 A
I_SC
1.2×100 = 120 V
12/20 A
A
I_1
120/140 A
140 Ω
I_SC = V_Ab/80 + 0.8
V_Ab (1/140 + 1/20 + 1/80) = 120/140 + 12/20 = 51/35 OR V_Ab = 20.923 V
∴ I_SC = 20.923/80 + 0.8 = 1.0615 A
NORTON EQUIVALENT CIRCUIT :
1.0615 A
97.5 Ω
I
200 Ω
I = 1.0615×97.5/(200 + 97.5) = 0.348 A
THÉVENIN EQUIVALENT CIRCUIT :
1.0615×97.5 = 103.5 V
R = 200 Ω
I = 103.5/297.5 = 0.348 A

2.1.5.

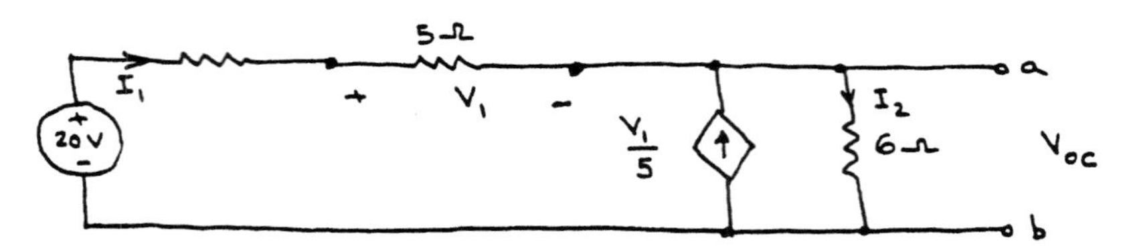

$I_2 = I_1 + \frac{V_1}{5}$; $V_1 = 5 I_1$; $I_2 = I_1 + I_1 = 2 I_1$

KVL FOR OUTER LOOP : $-20 + I_1(4+5) + 6 I_2 = 0$ OR $9I_1 + 6(2I_1) = 20$

OR $I_1 = \frac{20}{21}$ A

$V_{oc} = 6 I_2 = 12 I_1 = 12 \times \frac{20}{21} = \frac{80}{7} = 11.43$ V

TO FIND I_{sc} :

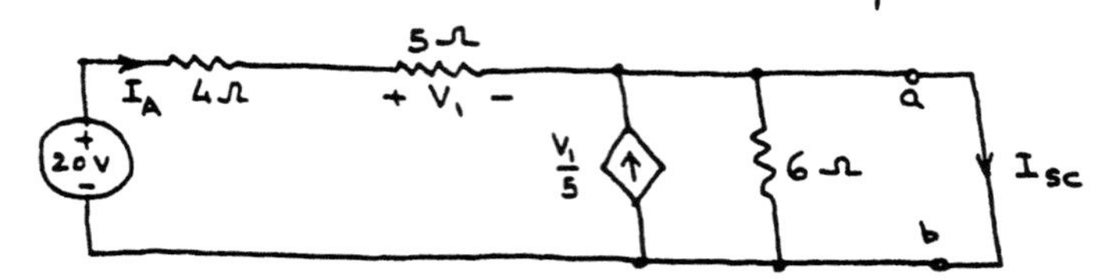

$I_{sc} = I_A + \frac{V_1}{5}$; $V_1 = 5 I_A$; $I_{sc} = 2 I_A$

KVL : $-20 + I_A(4+5) = 0$ OR $I_A = \frac{20}{9}$ A

$\therefore\ I_{sc} = 2\left(\frac{20}{9}\right) = \frac{40}{9}$ A

$R_{Th} = \frac{V_{oc}}{I_{sc}} = \frac{80}{7}\left(\frac{9}{40}\right) = \frac{18}{7}\ \Omega$

THÉVENIN EQUIVALENT CIRCUIT :

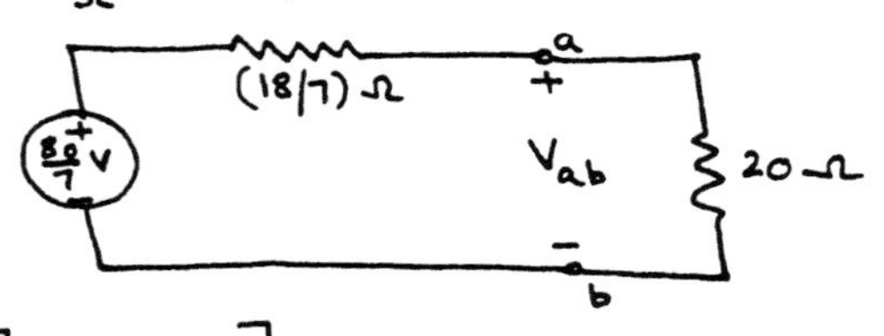

$$V_{ab} = 20\left[\frac{80/7}{(18/7) + 20}\right] = 20 \times \frac{80}{7} \times \frac{7}{158} = 10.13\ \text{V} \leftarrow$$

2.1.6.

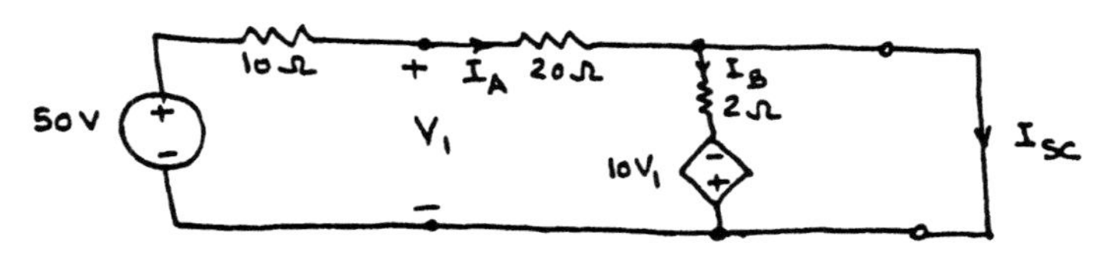

$I_{sc} = I_A - I_B$; $I_A = \frac{50}{30}$ A ; $I_B = \frac{10 V_1}{2} = 5V_1$

$V_1 = 50 - 10 I_A = 50 - 10\left(\frac{50}{30}\right) = \frac{100}{3}$ V

$I_{sc} = \frac{5}{3} - 5\left(\frac{100}{3}\right) = -165$ A

2.1.6. CONTINUED

TO FIND V_{oc} :

50V, 10Ω, 20Ω, V_1, I_1, 2Ω, $10V_1$, a, b, $V_{oc} = V_{ab}$

$$V_{ab} = V_{oc} = 2I_1 - 10V_1 \; ; \quad V_1 = -10I_1 + 50$$

$$\therefore \; V_{ab} = 2I_1 - 10(50 - 10I_1) = 102I_1 - 500$$

KVL: $\quad -50 + I_1(10 + 20 + 2) - 10V_1 = 0$

OR $\quad 32I_1 - 10(50 - 10I_1) = 50 \quad$ OR $\quad I_1 = \frac{550}{132}$ A

THUS $\quad V_{oc} = V_{ab} = 102\left(\frac{550}{132}\right) - 500 = -75$ V

$$\therefore \; R_{Th} = \frac{V_{oc}}{I_{sc}} = \frac{75}{165} = 0.4545\,\Omega$$

NORTON EQUIVALENT CIRCUIT:

165A, 0.4545Ω, a, b, I, 5Ω

$$I = -\frac{165 \times 0.4545}{5.4545} = -13.75\text{A}$$

2.1.7.

I_1, 1kΩ, 1mv, 1kΩ, $10^5 I_1$, 20kΩ, $I_2 = 0$, a, b, V_{oc}

$$V_{oc} = -10^5 I_1 + I_1(10^3)$$

$$I_1 = \frac{10^{-3}}{(1+1)\,10^3} = 0.5 \times 10^{-6}\text{ A}$$

$$V_{oc} = 0.5 \times 10^{-6}(10^3 - 10^5) = -49.5\text{ mV}$$

TO FIND I_{sc} :

1mv, 1kΩ, I_1, 1kΩ, $10^5 I_1$, 20kΩ, I_{sc}, a, b

MESH (LOOP) EQUATIONS: $\quad I_1(10^2 + 10^3) - I_{sc}(10^3) = 10^{-3}$;

$$I_1(10^5 - 10^3) + I_{sc}(20+1)10^3 = 0$$

SOLUTION OF SIMULTANEOUS EQUATIONS YIELDS: $\quad I_{sc} = -0.702\,\mu$A

$$\therefore \; R_{Th} = \frac{V_{oc}}{I_{sc}} = \frac{-49.5\text{ mV}}{-0.702\,\mu\text{A}} = 70.5\text{ k}\Omega$$

2.1.7. CONTINUED

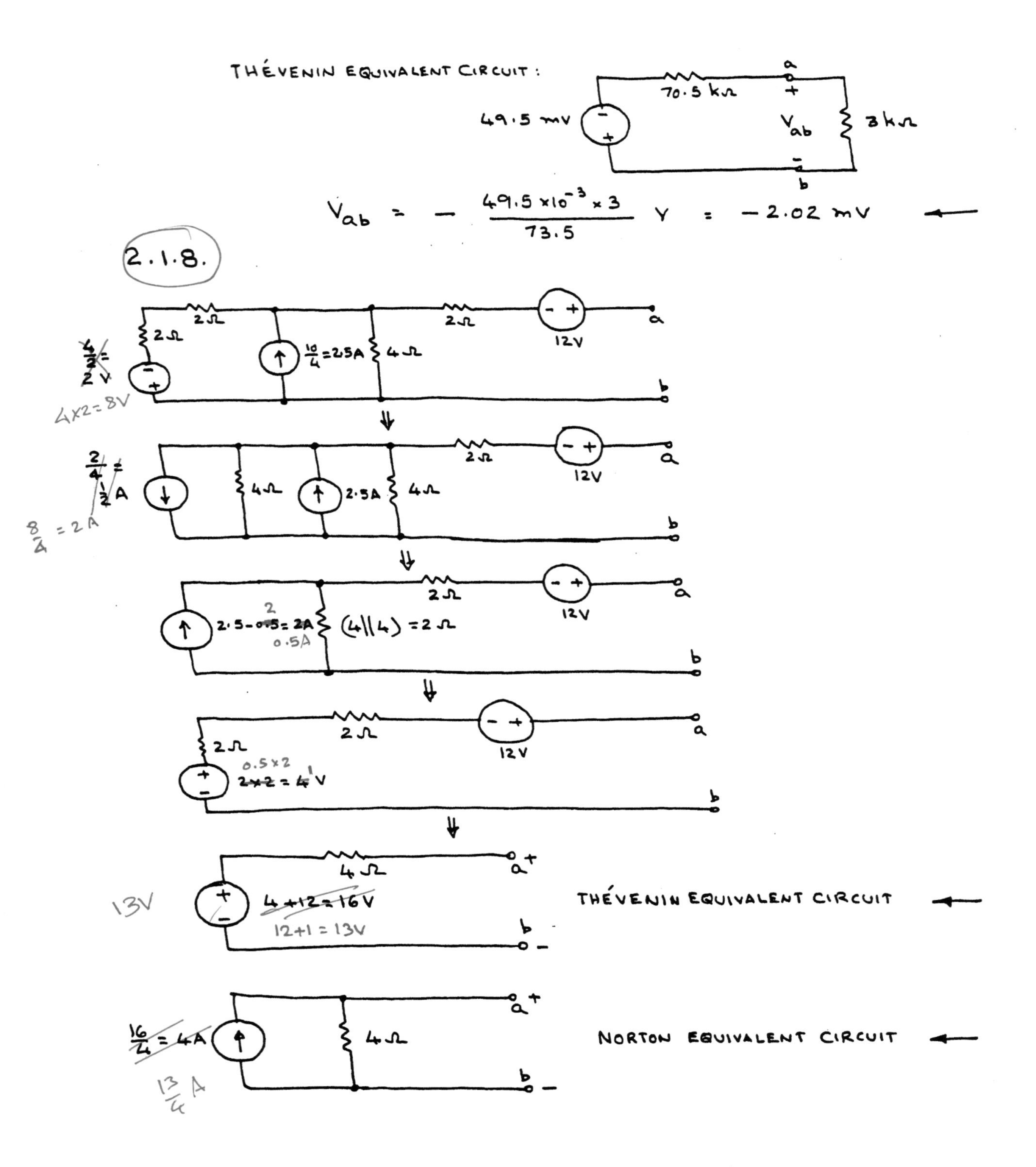

2.1.9.

(a)

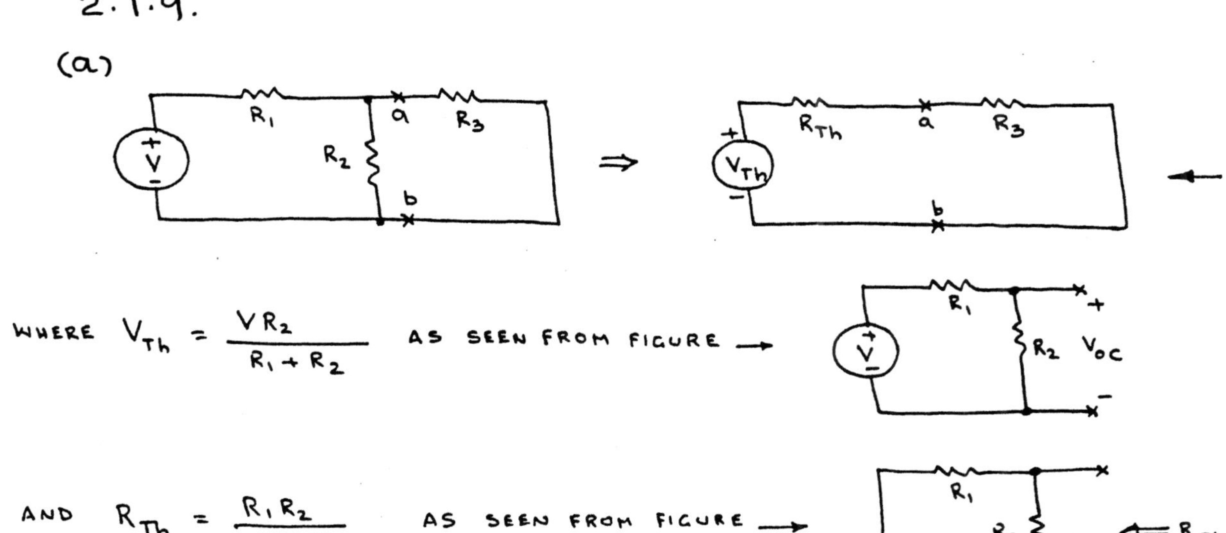

WHERE $V_{Th} = \dfrac{V R_2}{R_1 + R_2}$ AS SEEN FROM FIGURE →

AND $R_{Th} = \dfrac{R_1 R_2}{R_1 + R_2}$ AS SEEN FROM FIGURE →

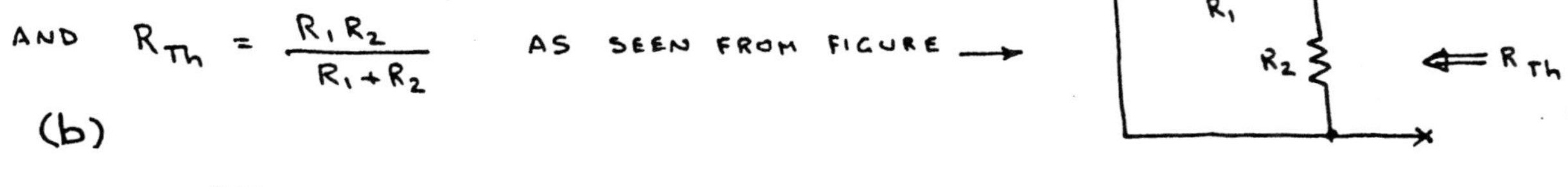

(b)

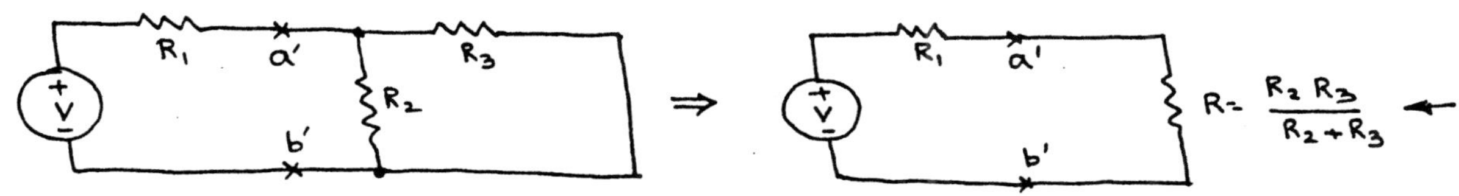

$R = \dfrac{R_2 R_3}{R_2 + R_3}$ ←

2.1.10.

(a)

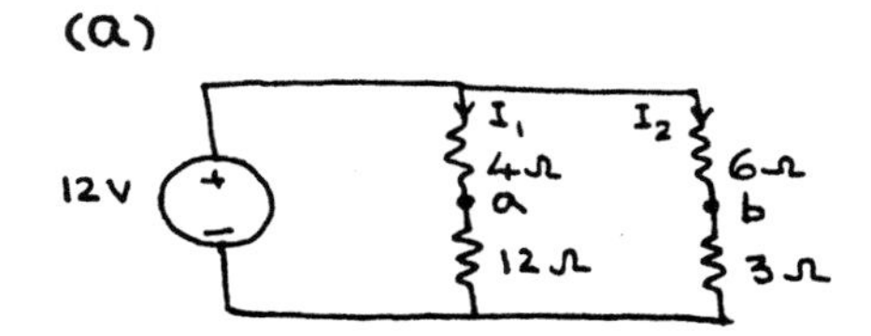

$I_1 = \frac{12}{16} = \frac{3}{4}\,A$; $I_2 = \frac{12}{9} = \frac{4}{3}\,A$

$V_{oc} = 12\,I_1 - 3\,I_2 = 12\left(\frac{3}{4}\right) - 3\left(\frac{4}{3}\right) = 5\,V$

TO FIND R_{Th}:

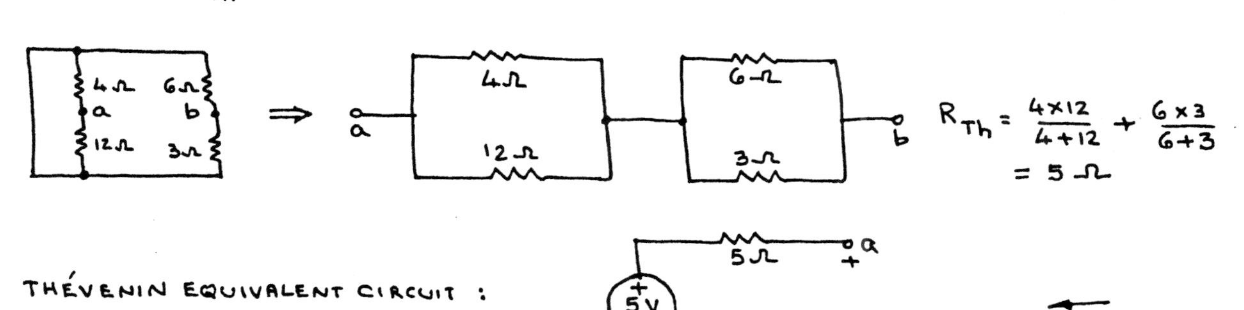

$$R_{Th} = \frac{4 \times 12}{4+12} + \frac{6 \times 3}{6+3} = 5\,\Omega$$

THÉVENIN EQUIVALENT CIRCUIT : ←

(b)

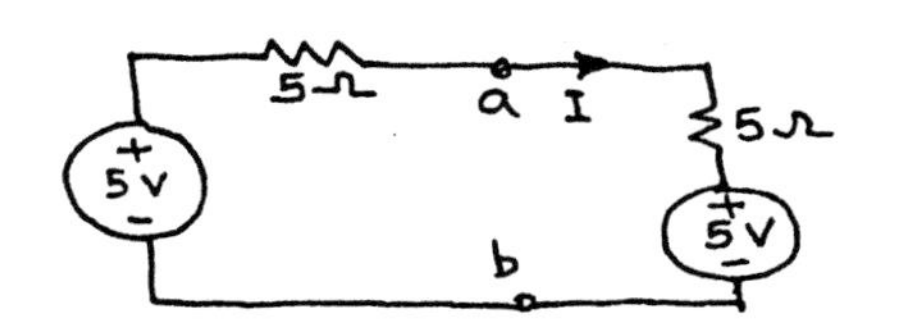

$$I = \frac{5-5}{5+5} = 0$$ ←

2.2.1.

(a)

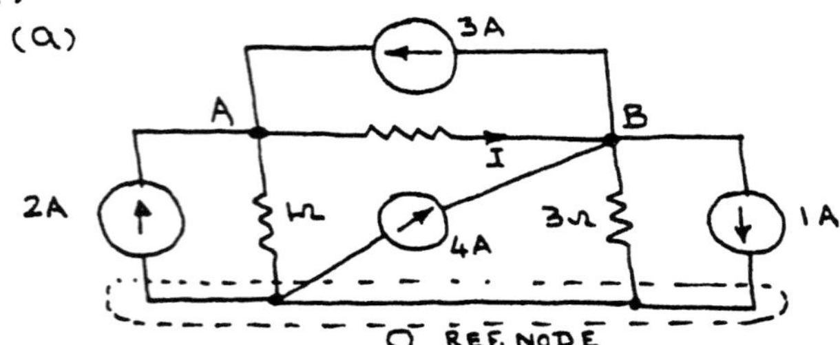

KCL AT NODE A : $\left(\frac{1}{1}+\frac{1}{2}\right)V_A - \frac{1}{2}V_B = 2+3 = 5$ OR $3V_A - V_B = 10$

KCL AT NODE B : $-\frac{1}{2}V_A + \left(\frac{1}{2}+\frac{1}{3}\right)V_B = 4-3-1=0$ OR $-3V_A + 5V_B = 0$

APPLYING CRAMER'S RULE, ONE GETS

$$V_A = \frac{\begin{vmatrix} 10 & -1 \\ 0 & 5 \end{vmatrix}}{\begin{vmatrix} 3 & -1 \\ -3 & 5 \end{vmatrix}} = \frac{50}{12} = \frac{25}{6}\text{ V} \; ; \qquad V_B = \frac{\begin{vmatrix} 3 & 10 \\ -3 & 0 \end{vmatrix}}{\begin{vmatrix} 3 & -1 \\ -3 & 5 \end{vmatrix}} = \frac{30}{12} = \frac{5}{2}\text{ V}$$

$$\therefore\ I = \frac{V_A - V_B}{2} = \frac{1}{2}\left(\frac{25}{6} - \frac{5}{2}\right) = \frac{5}{6}\text{ A} \longleftarrow$$

(b)

CONVERT ALL CURRENT SOURCES WITH PARALLEL RESISTORS INTO THEIR THÉVENIN EQUIVALENT.

$$(1+2+3)I_L = 2-6-9 \quad \text{OR} \quad I_L = -\frac{13}{6}\text{ A}$$

$$V_A - V_B = \left(-\frac{13}{6}\right)2 + 6 = \frac{5}{3}\text{ V}$$

SINCE NODE VOLTAGES REMAIN TO BE THE SAME AS IN THE ORIGINAL CIRCUIT,

$$I = \frac{V_A - V_B}{2} = \frac{5}{6}\text{ A} \quad \text{WHICH IS SAME AS IN PART (a)} \longleftarrow$$

2.2.2.

$$I = I_L = \frac{15-4-4}{(2+3+2)} = \frac{7}{7} = 1\text{ A}$$

2.2.3.

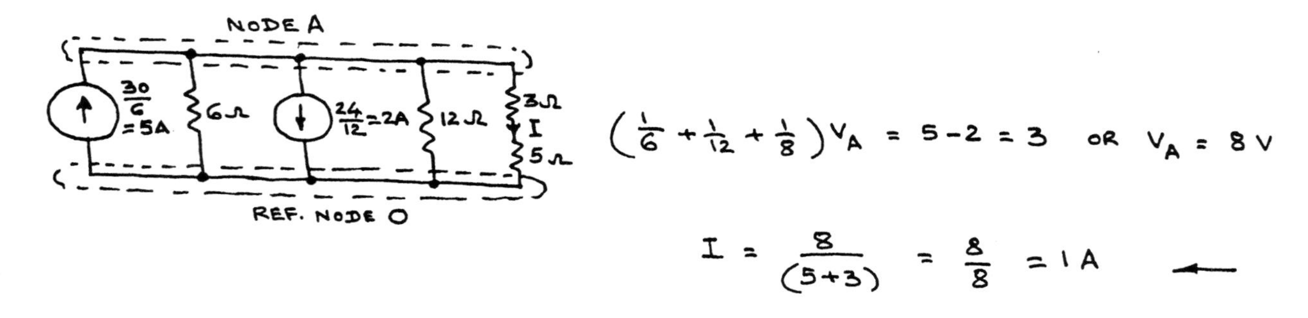

$$\left(\frac{1}{6}+\frac{1}{12}+\frac{1}{8}\right)V_A = 5-2 = 3 \quad \text{OR} \quad V_A = 8\text{ V}$$

$$I = \frac{8}{(5+3)} = \frac{8}{8} = 1\text{ A} \longleftarrow$$

2.2.4.

NODE-VOLTAGE METHOD:

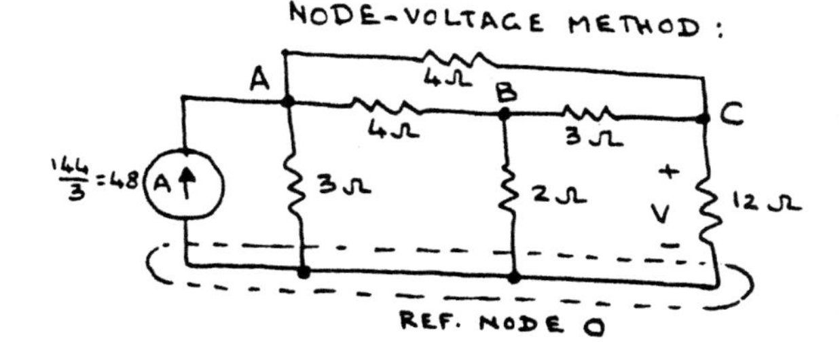

MESH ANALYSIS:

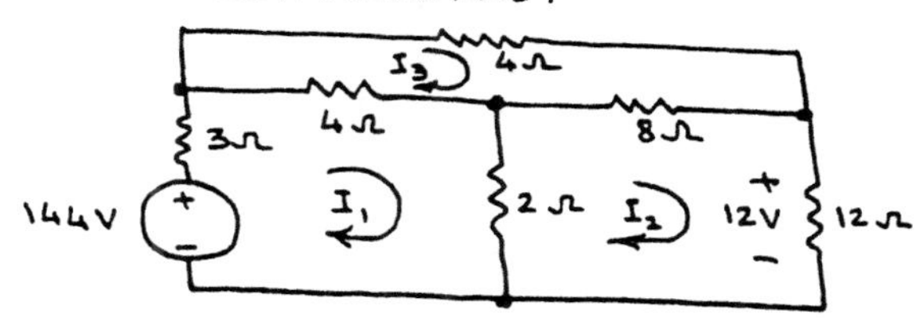

$$V_A\left(\frac{1}{3}+\frac{1}{4}+\frac{1}{4}\right)-\frac{1}{4}V_B-\frac{1}{4}V_C=48$$

$$-\frac{1}{4}V_A+V_B\left(\frac{1}{4}+\frac{1}{8}+\frac{1}{2}\right)-\frac{1}{8}V_C=0$$

$$-\frac{1}{4}V_A-\frac{1}{8}V_B+\left(\frac{1}{4}+\frac{1}{8}+\frac{1}{12}\right)V_C=0$$

SIMULTANEOUS SOLUTION YIELDS:

$$V_C = 54\text{ V} = V \leftarrow$$

$$I_1(3+4+2)-2I_2-4I_3=144$$

$$-2I_1+(2+8+12)I_2-8I_3=0$$

$$-4I_1-8I_2+(4+4+8)I_3=0$$

SIMULTANEOUS SOLUTION YIELDS:
$I_2 = 4.5\text{A}$

$\therefore V = 12I_2 = 12\times 4.5 = 54\text{ V} \leftarrow$

SAME AS GIVEN BY NODE-VOLTAGE METHOD

2.2.5.

NODE-VOLTAGE METHOD:

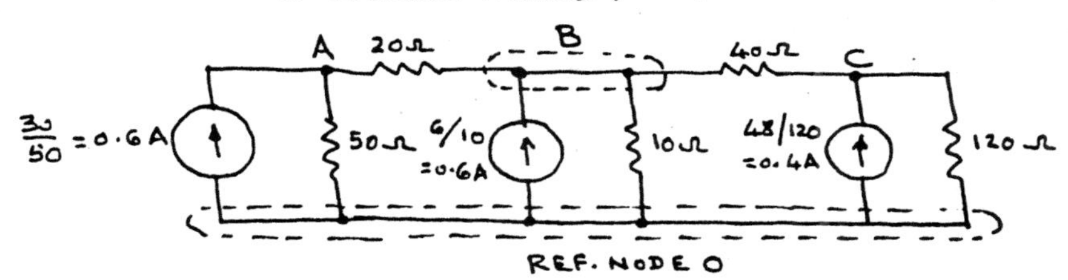

$$\left(\frac{1}{50}+\frac{1}{20}\right)V_A-\frac{1}{20}V_B = 0.6$$

$$-\frac{1}{20}V_A+\left(\frac{1}{20}+\frac{1}{10}+\frac{1}{40}\right)V_B-\frac{1}{40}V_C=0.6$$

$$-\frac{1}{40}V_B+\left(\frac{1}{40}+\frac{1}{120}\right)V_C=0.4$$

SIMULTANEOUS SOLUTION YIELDS $V_B = 11\text{ V}$

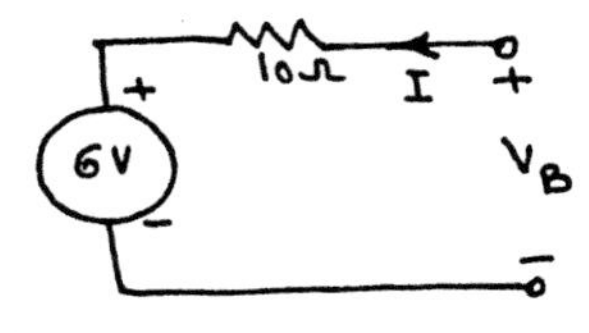

$$V_B = 10I+6 \quad\text{OR}\quad I=\frac{V_B-6}{10}=\frac{11-6}{10}=0.5\text{A} \leftarrow$$

MESH ANALYSIS:

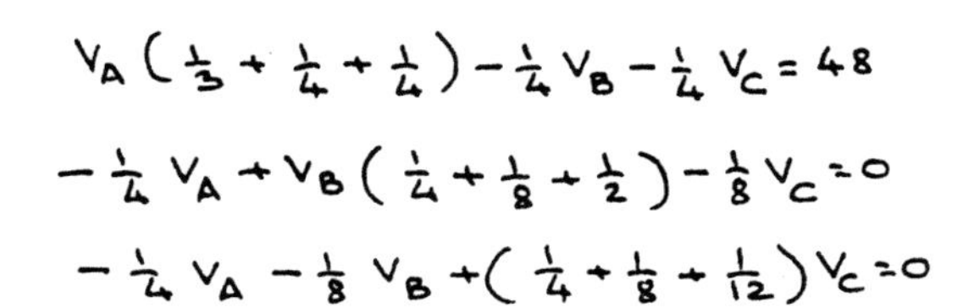

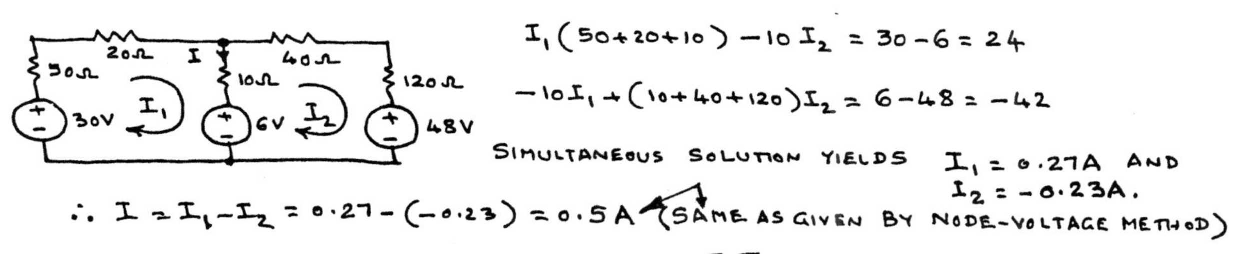

$$I_1(50+20+10)-10I_2=30-6=24$$

$$-10I_1+(10+40+120)I_2=6-48=-42$$

SIMULTANEOUS SOLUTION YIELDS $I_1 = 0.27\text{A}$ AND $I_2=-0.23\text{A}$.

$\therefore I = I_1 - I_2 = 0.27-(-0.23) = 0.5\text{A} \leftarrow$ (SAME AS GIVEN BY NODE-VOLTAGE METHOD)

2.2.6. (a)

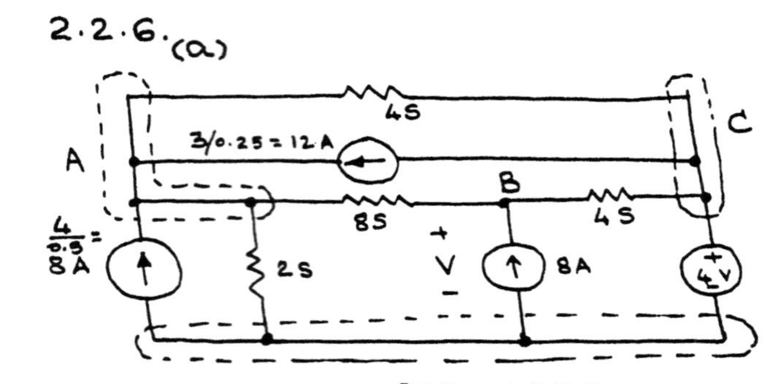

$$(2+8+4)V_A - 8V_B - 4V_C = 8+12 = 20$$

$$-8V_A + (8+4)V_B - 4V_C = 8$$

$$V_C = 4$$

SOLUTION YIELDS $V_B = 6\text{ V}$ ←

(b)

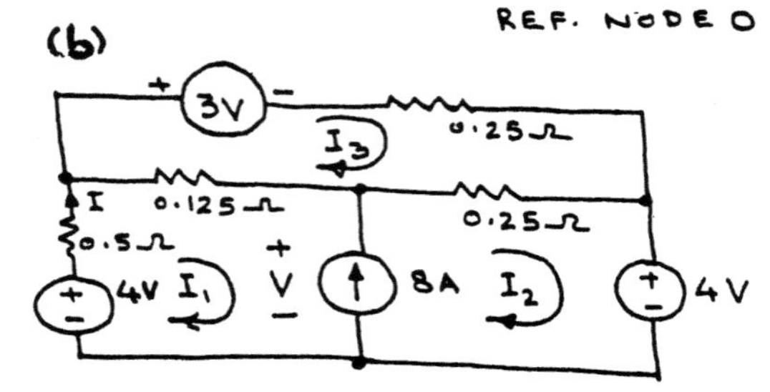

$$I_1 - I_2 = -8 \quad \text{OR} \quad I_2 = 8 + I_1$$

$$I_1(0.5+0.125) + 0\,I_2 - 0.125\,I_3 + V = 4 \quad — ①$$

$$0.1\,I_1 + 0.25\,I_2 - 0.25\,I_3 - V = -4 \quad — ②$$

$$-0.125\,I_1 - 0.25\,I_2 + (0.25+0.25+0.125)I_3 = -3 \quad — ③$$

ADDING ① AND ②, AND SUBSTITUTING FOR I_2 GIVES

$$0.875\,I_1 - 0.375\,I_3 = -2 \; ; \; -0.375\,I_1 + 0.625\,I_3 = -1$$

WHOSE SOLUTION YIELDS $I_1 = -4\text{A} = I$ ←

2.2.7.

MESH-CURRENT METHOD:

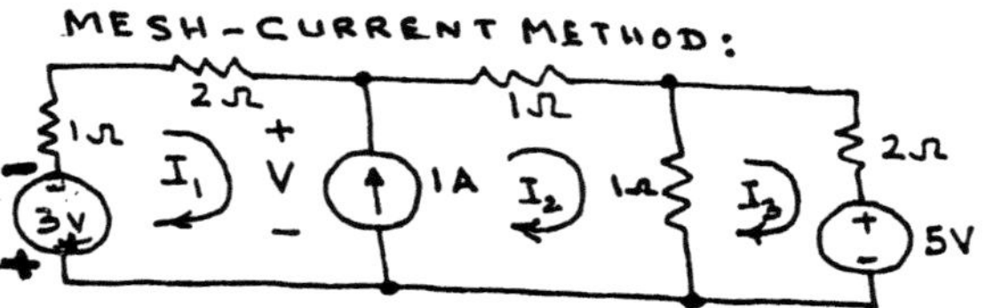

$$I_1 - I_2 = -1 \quad \text{OR} \quad I_1 = I_2 - 1$$

$$(1+2)\,I_1 + V = -3$$

$$-V + (1+1)\,I_2 - I_3 = 0 \;\Rightarrow\; V = 2I_2 - I_3$$

$$-I_2 + (2+1)\,I_3 = -5$$

SOLUTION YIELDS $I_2 = -\frac{5}{14}\text{ A}$; $I_3 = -\frac{25}{14}\text{ A}$

THEN $V = 2I_2 - I_3 = -\frac{10}{14} + \frac{25}{14} = \frac{15}{14}\text{ V}$ ←

NODAL ANALYSIS:

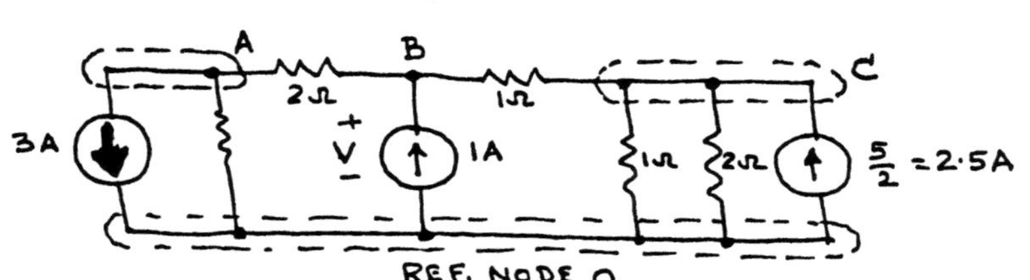

KCL AT NODE A: $\left(1+\frac{1}{2}\right)V_A - \frac{1}{2}V_B = -3$ OR $3V_A - V_B = -6$

KCL AT NODE B: $-\frac{1}{2}V_A + \left(\frac{1}{2}+1\right)V_B - V_C = 1$ OR $-V_A + 3V_B - 2V_C = 2$

KCL AT NODE C: $-V_B + \left(1+1+\frac{1}{2}\right)V_C = 2.5$ OR $-2V_B + 5V_C = 5$

SOLUTION YIELDS $V_B = \frac{15}{14}\text{ V} = V$ ← (SAME AS BY MESH-CURRENT ANALYSIS)

2.2.8.

MESH ANALYSIS:

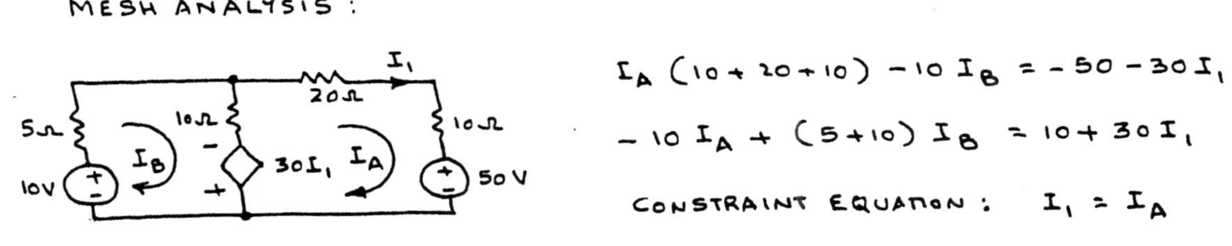

$$I_A(10+20+10) - 10I_B = -50 - 30I_1$$

$$-10I_A + (5+10)I_B = 10 + 30I_1$$

CONSTRAINT EQUATION: $I_1 = I_A$

THUS WE HAVE $70I_A - 10I_B = -50$; $-40I_A + 15I_B = 10$

SOLUTION YIELDS $I_B = -2\text{ A}$; $I_A = -1\text{ A}$

$$\therefore \quad I_A = I_1 = -1\text{ A} \quad \longleftarrow$$

NODAL ANALYSIS:

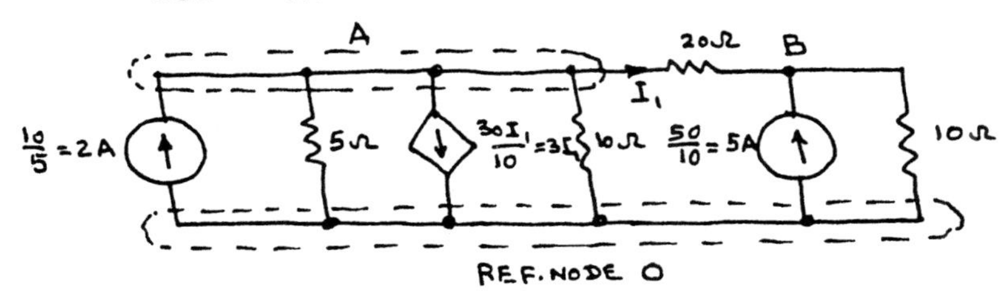

KCL AT NODE A: $\left(\frac{1}{5}+\frac{1}{10}+\frac{1}{20}\right)V_A - \frac{1}{20}V_B = 2 - 3I_1$

KCL AT NODE B: $-\frac{1}{20}V_A + \left(\frac{1}{20}+\frac{1}{10}\right)V_B = 5$

CONSTRAINT EQUATION: $I_1 = (V_A - V_B)/20$

$\frac{7}{20}V_A - \frac{1}{20}V_B = 2 - \frac{3(V_A - V_B)}{20}$ OR $5V_A - 2V_B = 20$

$-\frac{1}{20}V_A + \frac{3}{20}V_B = 5$ OR $-V_A + 3V_B = 100$

SOLUTION YIELDS $V_A = 20\text{ V}$; $V_B = 40\text{ V}$

$$\therefore \quad I_1 = \frac{V_A - V_B}{20} = \frac{20-40}{20} = -1\text{A} \quad \longleftarrow$$

(SAME AS GIVEN BY MESH ANALYSIS)

2.2.9.

NODAL ANALYSIS:

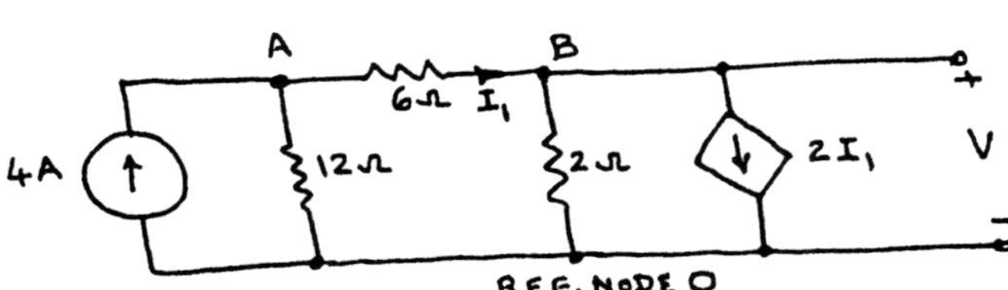

KCL AT NODE A: $\left(\frac{1}{12}+\frac{1}{6}\right)V_A - \frac{1}{6}V_B = 4$ OR $3V_A - 2V_B = 48$

KCL AT NODE B: $-\frac{1}{6}V_A + \left(\frac{1}{6}+\frac{1}{2}\right)V_B = -2I_1$ OR $-V_A + 4V_B = -12I_1$

CONSTRAINT EQUATION: $\frac{V_A - V_B}{6} = I_1$

SOLUTION YIELDS $V_A = 12\text{V}$; $V_B = -6\text{V}$; THEN $V_B = V = -6\text{V} \quad \longleftarrow$

2.2.9. CONTINUED

MESH ANALYSIS:

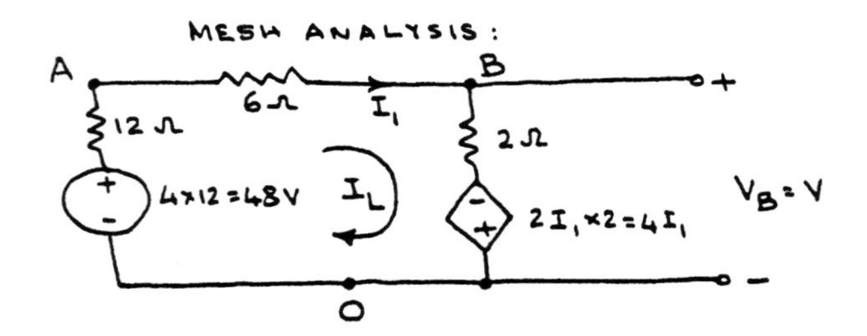

$I_L(12+6+2) = 48 + 4I_1$; CONSTRAINT EQ.: $I_1 = I_L$

$\therefore\ 16 I_L = 48$ OR $I_L = 3\text{A}$

$V_B = 2I_L - 4I_1 = 2I_L - 4I_L = -2I_L = -6\text{ V}$ ←

WHICH IS SAME AS GIVEN BY NODAL ANALYSIS

2.2.10.

MESH ANALYSIS:

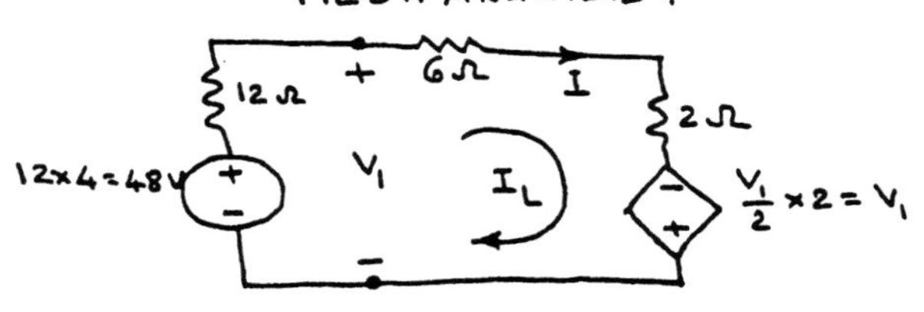

$(12+6+2)I_L = 48 + V_1$; CONSTRAINT EQ.: $V_1 = 48 - 12I_L$

THUS $20I_L = 48 + 48 - 12I_L$ OR $I_L = 3\text{A}$

$\therefore\ I = I_L = 3\text{A}$ ←

NODAL ANALYSIS:

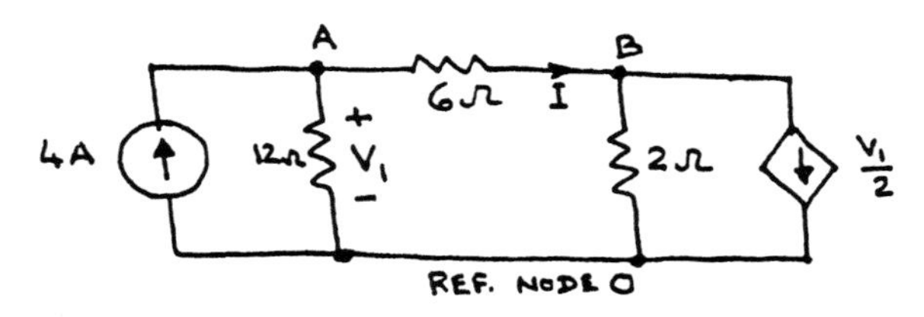

$\left(\frac{1}{12} + \frac{1}{6}\right)V_A - \frac{1}{6}V_B = 4 \Rightarrow 3V_A - 2V_B = 48$

$-\frac{1}{6}V_A + \left(\frac{1}{6} + \frac{1}{2}\right)V_B = -\frac{V_1}{2} \Rightarrow -V_A + 4V_B = -3V_1$

CONSTRAINT EQUATION: $V_1 = V_A$

SOLUTION YIELDS $V_A = 12\text{ V}$ AND $V_B = -6\text{ V}$

$\therefore\ I = \frac{V_A - V_B}{6} = \frac{12-(-6)}{6} = 3\text{A}$ ←

(SAME AS THAT GIVEN BY MESH ANALYSIS)

2.2.11. CHOOSE THE REFERENCE NODE AS THE GROUND, SO THAT $V_3 = -15\text{ V}$.

FORM A SUPERNODE WHICH INCLUDES THE 10-V SOURCE, NODES 1 AND 2.

THE CIRCUIT IS REDRAWN BELOW:

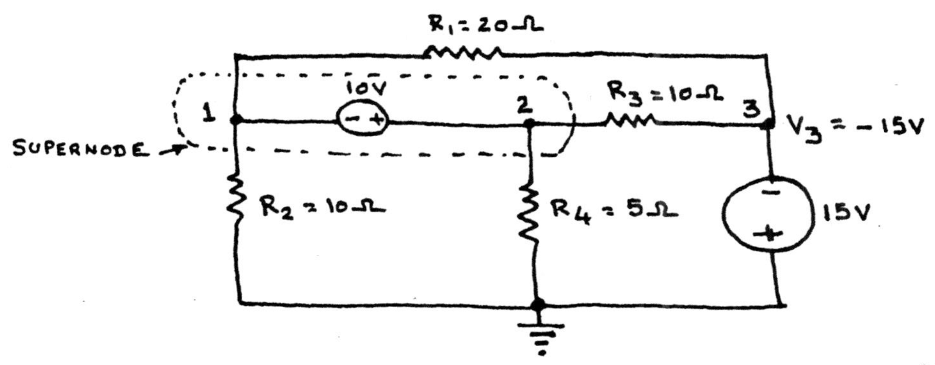

THE CONSTRAINT EQUATION RELATING V_1 AND V_2 IS GIVEN BY $V_2 - V_1 = 10$

KCL EQUATION FOR THE SUPERNODE YIELDS THE FOLLOWING:

$$\frac{V_1}{R_2} + \frac{V_1 - V_3}{R_1} + \frac{V_2}{R_4} + \frac{V_2 - V_3}{R_3} = 0$$

OR $$\frac{V_1}{10} + \frac{V_1 + 15}{20} + \frac{V_2}{5} + \frac{V_2 + 15}{10} = 0$$

SUBSTITUTING $V_2 = V_1 + 10$, WE GET

$$\frac{V_1}{10} + \frac{V_1 + 15}{20} + \frac{V_1 + 10}{5} + \frac{V_1 + 25}{10} = 0 \quad \text{OR} \quad V_1 = -\frac{35}{3}\text{ V} \leftarrow$$

AND $V_2 = V_1 + 10 = -\frac{35}{3} + 10 = -\frac{5}{3}\text{ V} \leftarrow$

WE ALREADY KNOW THAT $V_3 = -15\text{ V} \leftarrow$

MESH-CURRENT ANALYSIS:

$$10 I_1 - 10 + 5(I_1 - I_2) = 0$$

$$5(I_2 - I_1) + 10(I_2 - I_3) - 15 = 0$$

$$20 I_3 + 10(I_3 - I_2) + 10 = 0$$

SOLUTION YIELDS $I_1 = \frac{7}{6}\text{ A}$; $I_2 = \frac{3}{2}\text{ A}$; $I_3 = \frac{1}{6}\text{ A}$

$$\therefore V_1 = -10 I_1 = -\frac{35}{3}\text{ V} \leftarrow$$

$$V_2 = 5(I_1 - I_2) = -\frac{5}{3}\text{ V} \leftarrow$$

$$V_3 = -15\text{ V} \leftarrow$$

SAME AS BEFORE.

2.2.12.

FORM A SUPERNODE INCLUDING 12-V SOURCE, NODES 1 AND 2.

KCL FOR THE SUPERNODE:

$$\frac{V_1}{1} - 2 + \frac{V_2}{1} + \frac{V_2}{2} = 0$$

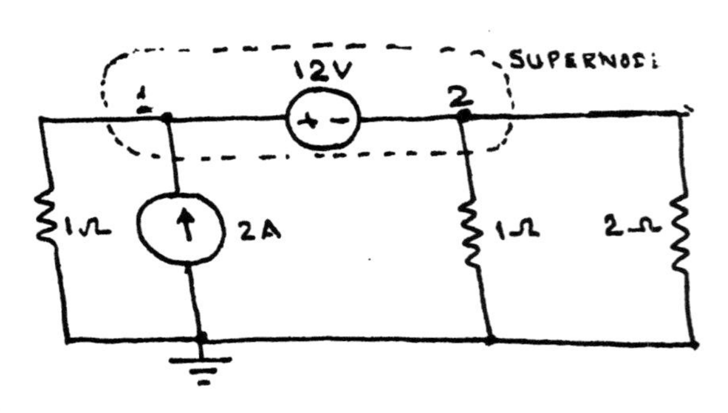

CONSTRAINT EQUATION: $V_1 - V_2 = 12$

SOLVING FOR V_2, WE GET

$$V_2 = -4\text{ V} \leftarrow$$

MESH ANALYSIS BY SOURCE TRANSFORMATION:

REPLACE THE 2-A CURRENT SOURCE WITH SHUNT RESISTANCE OF 1Ω BY A VOLTAGE SOURCE OF 2V WITH SERIES RESISTANCE OF 1Ω; THE CIRCUIT IS REDRAWN:

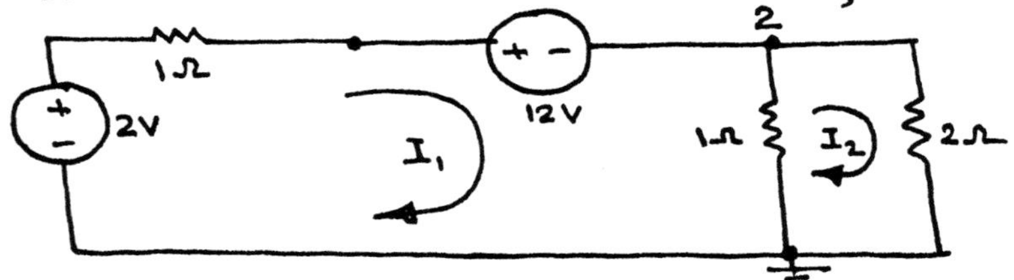

KVL YIELDS $\quad -2 + I_1 + 12 + (I_1 - I_2) = 0 \quad$ OR $\quad 2I_1 - I_2 = -10$

$(I_2 - I_1) + 2I_2 = 0 \quad$ OR $\quad 3I_2 - I_1 = 0$

SOLVING $\quad I_1 = -6\text{A}; \quad I_2 = -2\text{A}$

$\therefore \quad V_2 = (I_1 - I_2)1 = -6 + 2 = -4\text{ V} \leftarrow$ SAME AS BEFORE.

MESH ANALYSIS BY SUPERMESH CONCEPT:

CONSTRAINT EQUATION:

$$I_2 - I_1 = 2$$

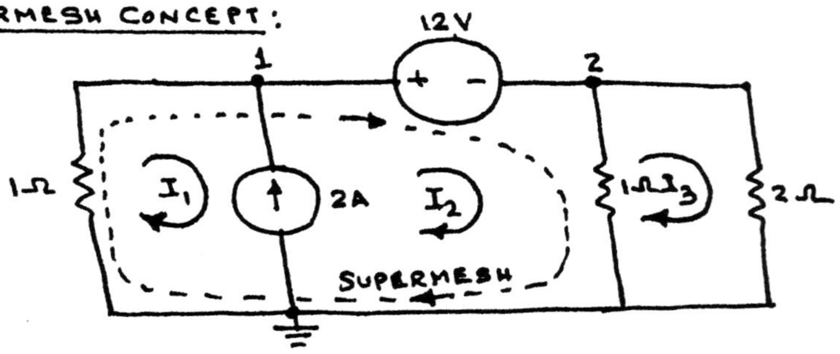

SUPERMESH EQUATION:

$$I_1 + 12 + (I_2 - I_3) = 0$$

OR $\quad I_1 + I_2 - I_3 = -12$

MESH 3 EQUATION: $\quad (I_3 - I_2) + 2I_3 = 0 \quad$ OR $\quad 3I_3 - I_2 = 0$

SOLVING THE EQUATIONS, WE GET

$$I_1 = -8\text{A}; \quad I_2 = -6\text{A}; \quad I_3 = -2\text{A}$$

$\therefore \quad V_2 = (I_2 - I_3)1 = -4\text{V} \leftarrow$ SAME AS BEFORE.

2.2.13.

MESH-CURRENT ANALYSIS USING SUPERMESH CONCEPT

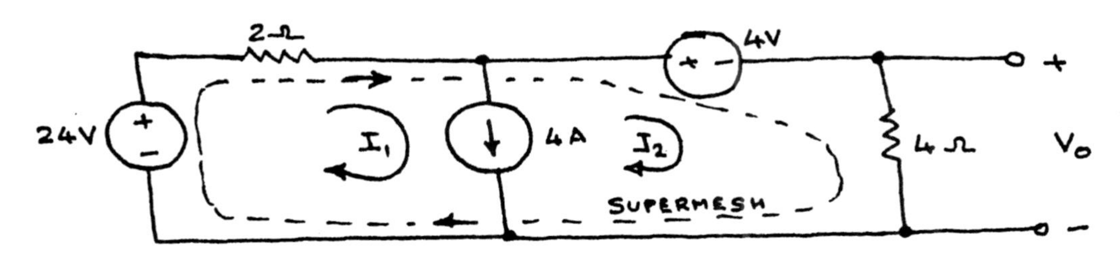

CONSTRAINT EQUATION : $I_1 - I_2 = 4$

SUPERMESH KVL EQUATION : $-24 + 2I_1 + 4 + 4I_2 = 0$

OR $2I_1 + 4I_2 = 20$

SOLUTION OF THE EQUATIONS YIELDS $I_1 = 6A$; $I_2 = 2A$

$\therefore \quad V_o = 4I_2 = 8V$ ⟵

NODAL ANALYSIS

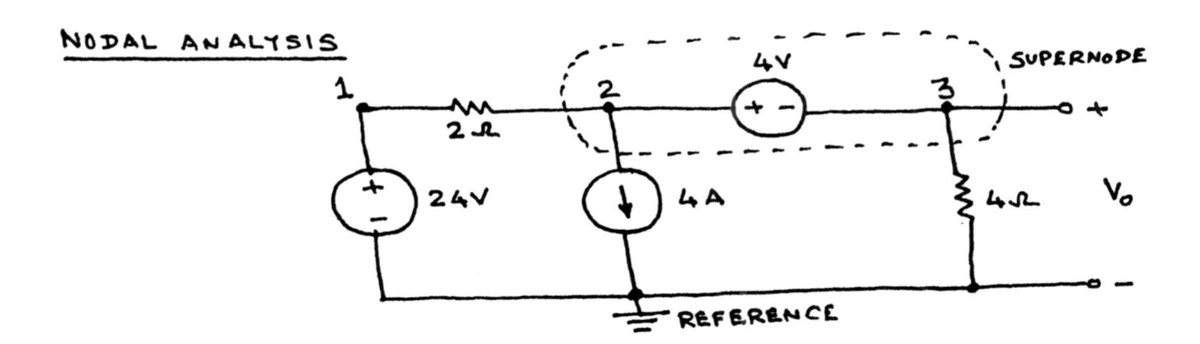

BY CHOOSING THE REFERENCE AS SHOWN, $V_1 = 24V$

CONSTRAINT EQUATION : $V_2 - V_3 = 4$

KCL FOR THE SUPERNODE : $\frac{V_2 - V_1}{2} + 4 + \frac{V_3}{4} = 0$

OR $\frac{V_2 - 24}{2} + 4 + \frac{V_3}{4} = 0$

OR $2V_2 - 48 + 16 + V_3 = 0$

OR $2V_2 + V_3 = 32$

SOLVING THE EQUATIONS, WE HAVE

$V_2 = 12V$; $V_3 = 8V$

$\therefore \quad V_o = V_3 = 8V$ ⟵ SAME AS BEFORE

2.2.14.

MESH-CURRENT ANALYSIS

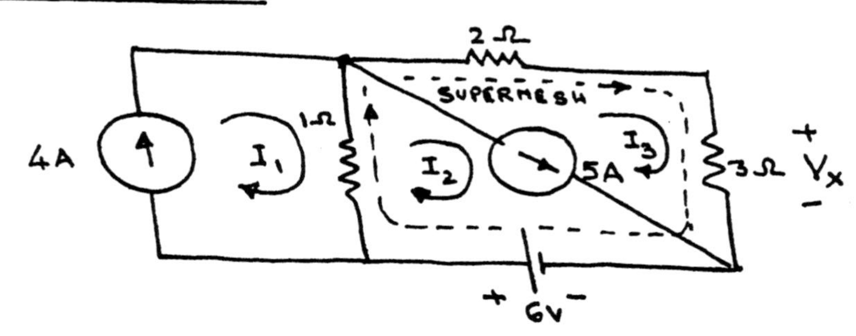

$I_1 = 4A$

CONSTRAINT EQUATION: $I_2 - I_3 = 5$

SUPERMESH KVL EQUATION: $2I_3 + 3I_3 - 6 + (1)(I_2 - I_1) = 0$

OR $5I_3 + I_2 - I_1 = 6$

OR $5I_3 + I_2 = 6 + I_1 = 10$

SOLVING THE EQUATIONS, WE GET $I_3 = \frac{5}{6}$ A

$\therefore V_x = 3I_3 = 2.5$ V ←

NODAL ANALYSIS

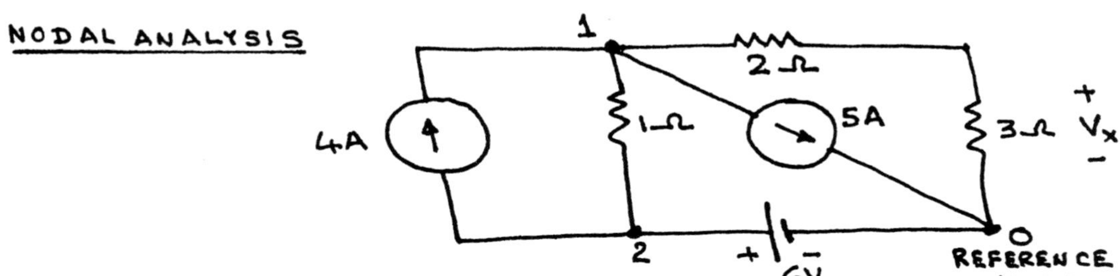

BY CHOOSING THE REFERENCE NODE AS SHOWN IN FIGURE, WE HAVE $V_2 = 6V$

KCL AT NODE 1: $-4 + \frac{V_1 - V_2}{1} + 5 + \frac{V_1}{5} = 0$ OR $6V_1 - 5V_2 = -5$

NOTE THAT BY COMBINING 2 Ω AND 3 Ω IN SERIES, AN INTERMEDIATE NODE IS AVOIDED, THEREBY SIMPLIFYING THE ANALYSIS.

THE UNKNOWN VOLTAGE V_x CAN, HOWEVER, BE DETERMINED BY A VOLTAGE-DIVIDER.

SOLUTION OF THE EQUATIONS YIELDS $V_1 = \frac{25}{6}$ V

HENCE, FROM THE VOLTAGE DIVIDER, $V_x = \frac{V_1 \times 3}{(2+3)} = \frac{25}{6} \times \frac{3}{5}$

(SAME AS BEFORE) $= \frac{5}{2} = 2.5V$ ←

2.3.1.

(i)

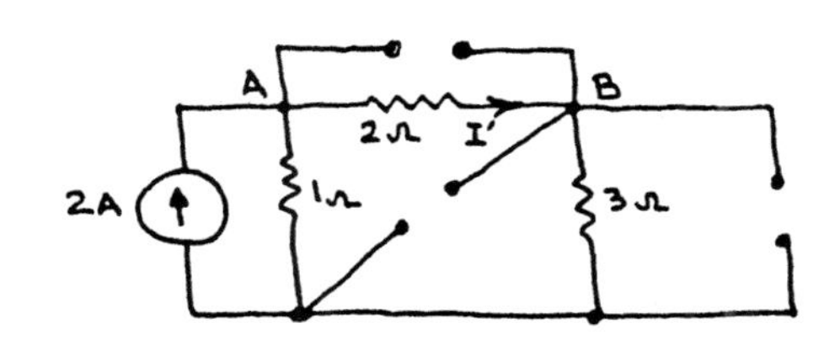

NODE A: $\left(\frac{1}{1}+\frac{1}{2}\right)V_A' - \frac{1}{2}V_B' = 2$ OR $3V_A' - V_B' = 4$

NODE B: $-\frac{1}{2}V_A' + \left(\frac{1}{2}+\frac{1}{3}\right)V_B' = 0$ OR $-3V_A' + 5V_B' = 0$

SOLUTION YIELDS $V_B' = 1\text{V}$; $V_A' = \frac{5}{3}\text{V}$

$$I' = \frac{V_A' - V_B'}{2} = \frac{1}{3}\text{A}$$

(ii)

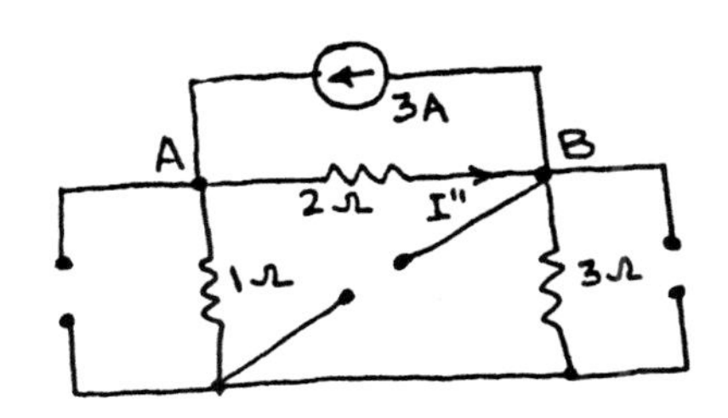

NODE A: $\left(\frac{1}{1}+\frac{1}{2}\right)V_A'' - \frac{1}{2}V_B'' = 3$ OR $3V_A'' - V_B'' = 6$

NODE B: $-\frac{1}{2}V_A'' + \left(\frac{1}{2}+\frac{1}{3}\right)V_B'' = -3$ OR $-3V_A'' + 5V_B'' = -18$

SOLVING, ONE GETS $V_B'' = -3\text{V}$; $V_A'' = 1\text{V}$

$$I'' = \frac{1-(-3)}{2} = 2\text{A}$$

(iii)

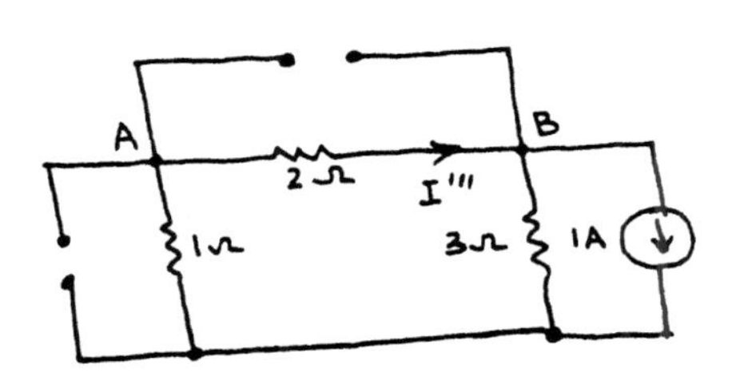

NODE A: $\left(\frac{1}{1}+\frac{1}{2}\right)V_A''' - \frac{1}{2}V_B''' = 0$ OR $3V_A''' - V_B''' = 0$

NODE B: $-\frac{1}{2}V_A''' + \left(\frac{1}{2}+\frac{1}{3}\right)V_B''' = -1$ OR $-3V_A''' + 5V_B''' = -6$

SOLUTION YIELDS $V_B''' = -1.5\text{V}$; $V_A''' = -0.5\text{V}$

$$I''' = \frac{-0.5-(-1.5)}{2} = \frac{1}{2}\text{A}$$

2.3.1. CONTINUED

(iv)

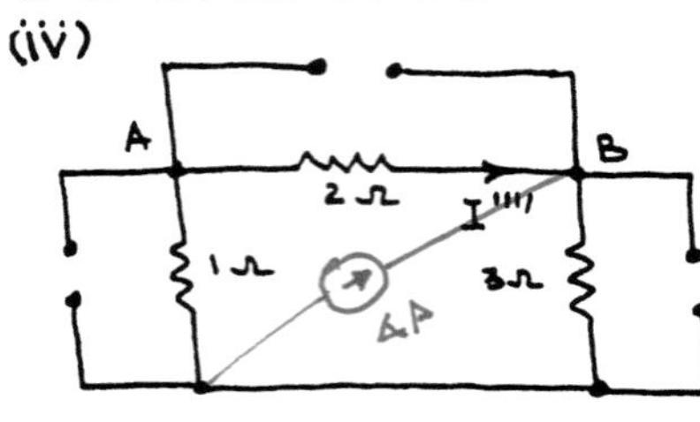

NODE A: $\left(\frac{1}{1}+\frac{1}{2}\right) V_A'''' - \frac{1}{2} V_B'''' = 0$ OR $3V_A'''' - V_B'''' = 0$

NODE B: $-\frac{1}{2} V_A'''' + \left(\frac{1}{2}+\frac{1}{3}\right) V_B'''' = 4$ OR $-3V_A'''' + 5V_B'''' = 24$

ON SOLVING, ONE GETS $V_B'''' = 6\text{ V}$; $V_A'''' = 2\text{V}$

$$I'''' = \frac{2-6}{2} = -2\text{ A}$$

BY SUPERPOSITION, $I = I' + I'' + I''' + I'''' = \frac{1}{3} + 2 + \frac{1}{2} - 2 = \frac{5}{6}\text{ A}$ ←

2.3.2.

(i)

$$I_A = \frac{30}{6+(12\|8)} = \frac{30}{6+\frac{24}{5}} = \frac{25}{9}$$

$$I' = \frac{25}{9} \times \frac{12}{20} = \frac{5}{3}\text{ A}$$

(ii)

$$I_B = -\frac{24}{12+(6\|8)} = -\frac{24}{12+\frac{24}{7}} = -\frac{14}{9}$$

$$I'' = -\frac{14}{9} \times \frac{6}{14} = -\frac{2}{3}\text{ A}$$

BY SUPERPOSITION, $I = I' + I'' = \frac{5}{3} - \frac{2}{3} = 1\text{ A}$ ←

2.3.3.

(i)

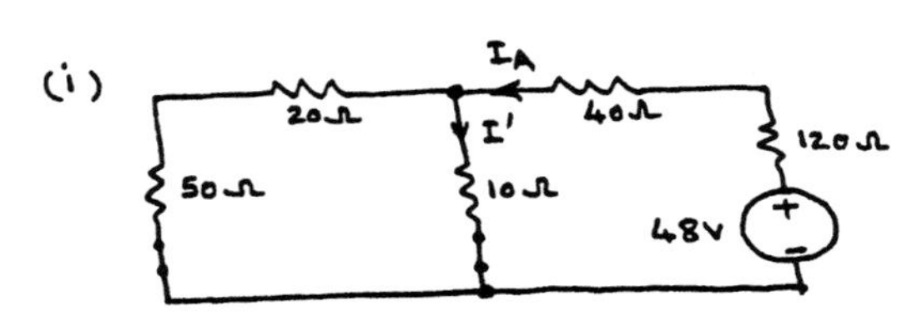

$$I_A = \frac{48}{160+(10\|70)} = \frac{48}{160+\frac{35}{4}} = \frac{192}{675}$$

$$I' = \frac{192}{675} \times \frac{70}{80} = \frac{168}{675} = \frac{56}{225}\text{ A}$$

(ii)

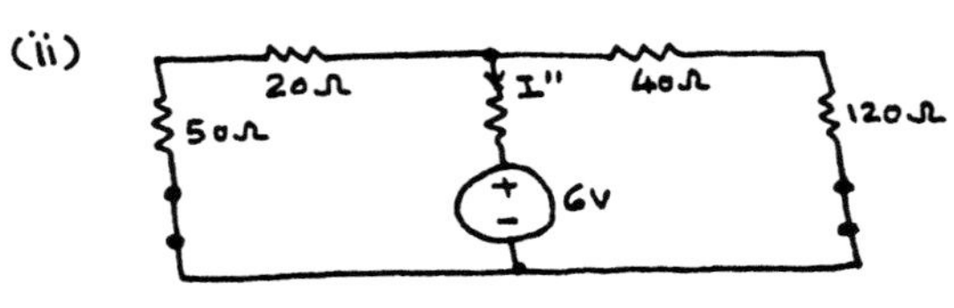

$$I'' = -\frac{6}{10+(160\|70)} = -\frac{6}{10+\frac{1120}{23}} = -\frac{23}{225}\text{ A}$$

(iii)

$$I_B = \frac{30}{70+(10\|160)} = \frac{30}{70+\frac{160}{17}} = \frac{17}{45}$$

$$I''' = \frac{17}{45} \times \frac{160}{170} = \frac{16}{45}\text{ A}$$

BY SUPERPOSITION, $I = I' + I'' + I''' = \frac{56}{225} - \frac{23}{225} + \frac{16}{45} = \frac{339}{675} \simeq 0.5\text{ A}$ ←

2.3.4.

(i)

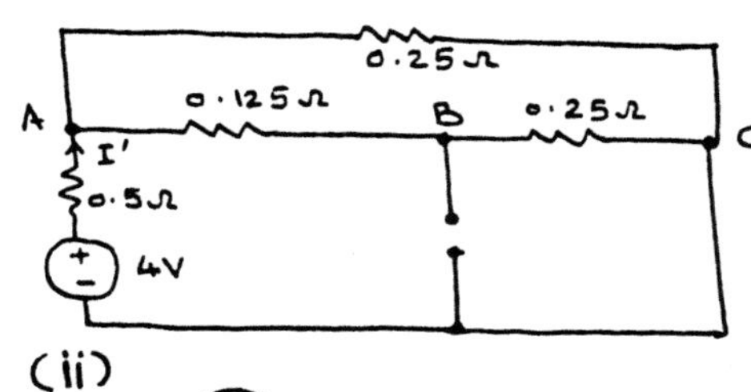

$$I' = \frac{4}{0.5 + (0.375 \| 0.25)} = \frac{4}{0.5 + 0.15} = \frac{80}{13} = 6.1538 \text{ A}$$

$$V_B' = V_{BC} = 0.25 \times \frac{80}{13} \times \frac{0.25}{0.625} = \frac{8}{13} = 0.6154 \text{ V}$$

(ii)

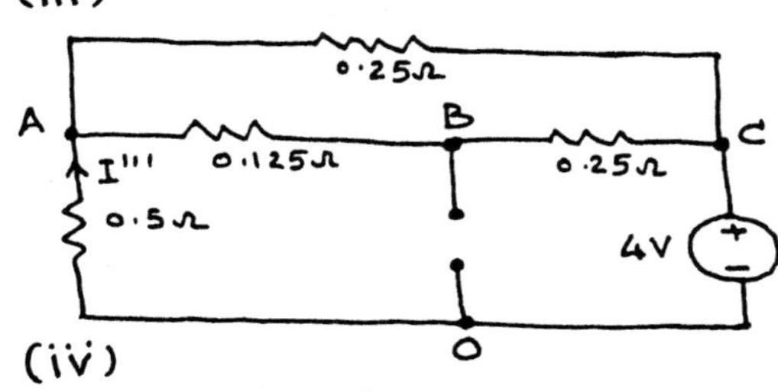

$$I_A = \frac{3}{0.25 + (0.5 \| 0.375)} = \frac{3}{0.25 + 0.2143} = 6.4613 \text{ A}$$

$$I'' = -\frac{6.4613 \times 0.375}{0.875} = -2.7691 \text{ A}$$

$$V_B'' = V_{BC} = 0.25(6.4613 - 2.7691) = 0.9231 \text{ V}$$

(iii)

0.25Ω
A
B
C
I'''
0.125Ω
0.25Ω
0.5Ω
4V
O

$$I''' = -\frac{4}{0.5 + (0.25 \| 0.375)} = -\frac{4}{0.5 + 0.15} = -6.1538 \text{ A}$$

$$V_B''' = V_{BO} = V_{BC} + V_{CO}$$

$$= 0.25\left(-6.1538 \times \frac{0.25}{0.625}\right) + 4 = 3.3846 \text{ V}$$

(iv)

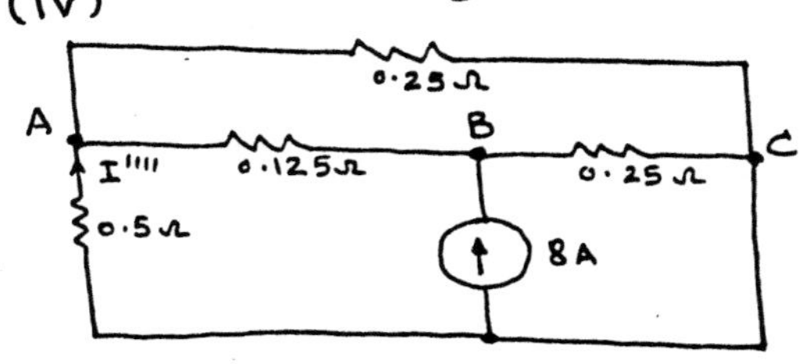

⇒

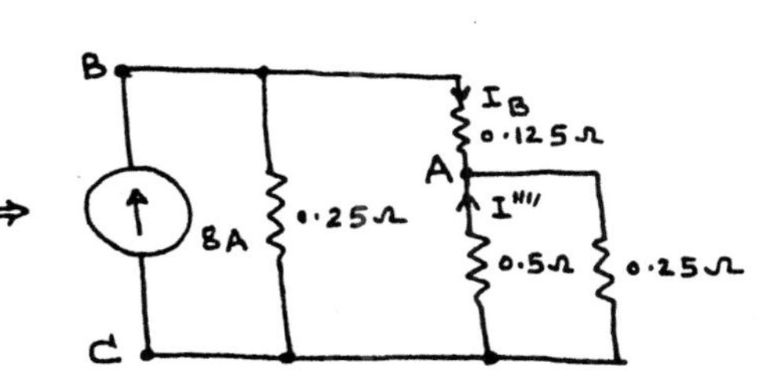

NOTING THAT $0.5 \| 0.25 = 0.1667$ AND $0.125 + 0.1667 = 0.2917$

$$I_B = \frac{8 \times 0.25}{(0.25 + 0.2917)} = 3.692 \text{ A} \; ; \quad I'''' = -\frac{3.692 \times 0.25}{0.75} = -1.231 \text{ A}$$

$$V_B'''' = V_{BC} = 0.25\left(\frac{8 \times 0.2917}{0.5417}\right) = 1.077 \text{ V}$$

BY SUPERPOSITION,

$$I = I' + I'' + I''' + I'''' = 6.1538 - 2.7691 - 6.1538 - 1.231 = -4 \text{ A} \quad \leftarrow$$

$$V_B = V_B' + V_B'' + V_B''' + V_B'''' = 0.6154 + 0.9231 + 3.3846 + 1.077 = 6 \text{ V} \quad \leftarrow$$

2.3.5. (i)

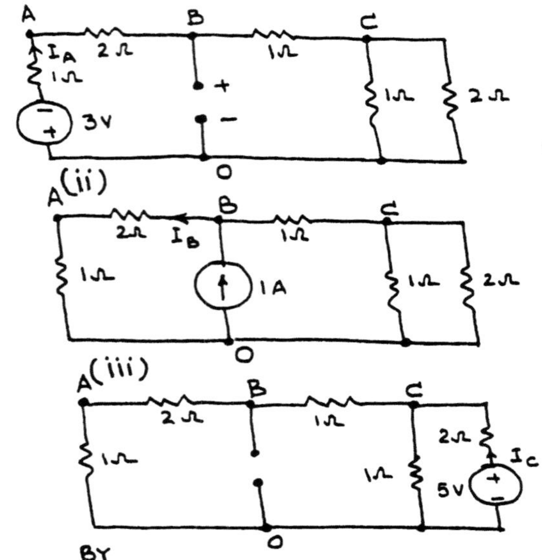

$$I_A = -\frac{3}{4+(1\|2)} = -\frac{3}{4+\frac{2}{3}} = -\frac{9}{14} = -0.6429\,\text{A}$$

KVL: $3 I_A + V_{BO} + 3 = 0$

$$V_B' = V_{BO} = -3 - 3 I_A = -3 + 1.9287 = -1.0713\,\text{V}$$

NOTING THAT $1\|2 = \frac{2}{3}$ AND $1 + \frac{2}{3} = \frac{5}{3}$

$$I_B = 1\,\frac{(5/3)}{(\frac{5}{3}+3)} = \frac{5}{14} = 0.3571\,\text{A}$$

$$V_B'' = V_{BO} = 3 I_B = 3(0.3571) = 1.0713\,\text{V}$$

$$I_C = \frac{5}{2+(1\|4)} = \frac{5}{2+\frac{4}{5}} = \frac{25}{14}$$

$$V_B''' = V_{BO} = 3\left(\frac{25}{14}\times\frac{1}{5}\right) = \frac{15}{14} = 1.0714\,\text{V}$$

BY SUPERPOSITION, $V_{BO} = V_B' + V_B'' + V_B''' = -1.0713 + 1.0713 + 1.0714 = 1.0714\,\text{V}$ ←

2.3.6.

(i)

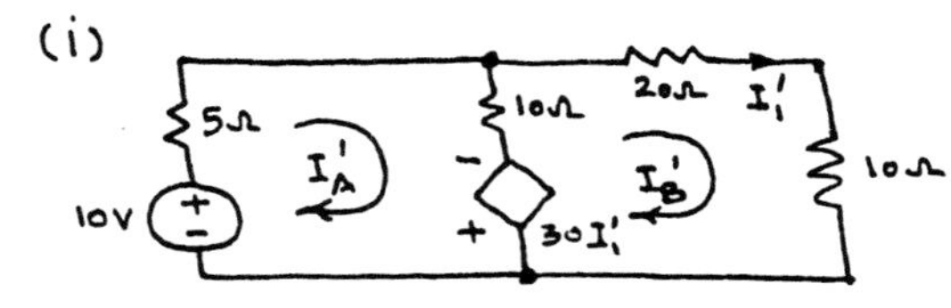

$$I_A'(15) - 10 I_B' = 10 + 30 I_1'$$

$$-10 I_A' + (10+20+10) I_B' = -30 I_1'$$

$$I_1' = I_B'$$

SOLUTION YIELDS $I_1' = I_B' = \frac{2}{13}\,\text{A}$; $I_A' = 7 I_B' = \frac{14}{13}\,\text{A}$

(ii)

$$15 I_A'' - 10 I_B'' = 30 I_1''$$

$$-10 I_A'' + 40 I_B'' = -50 - 30 I_1''$$

$$I_1'' = I_B''$$

SOLUTION YIELDS $I_1'' = I_B'' = -\frac{15}{13}\,\text{A}$; $I_A'' = \frac{8}{3} I_B'' = -\frac{40}{13}\,\text{A}$

BY SUPERPOSITION, $I_1 = I_1' + I_1'' = \frac{2}{3} - \frac{15}{13} = -1\,\text{A}$ ←

2.4.1.

VIEWED FROM TERMINALS A−B:

FROM FIG. 2.4.1 (a): $R_{eq\,A-B} = (R_A + R_B)$

FROM FIG. 2.4.1 (b): $R_{eq\,A-B} = R_{AB} \| (R_{CA} + R_{BC}) = \dfrac{R_{AB}(R_{BC}+R_{CA})}{R_{AB}+R_{BC}+R_{CA}}$

FOR EQUIVALENCE, $R_A + R_B = \dfrac{R_{AB}(R_{BC}+R_{CA})}{R_{AB}+R_{BC}+R_{CA}}$

TWO SIMILAR EQUATIONS CAN BE WRITTEN FOR THE OTHER TWO TERMINAL PAIRS.

SIMULTANEOUS SOLUTION YIELDS EQS. (2.4.1) AND (2.4.2) ←

2.4.2.

CONVERTING THE Y INTO Δ BY USING EQ. (2.4.4), ONE GETS

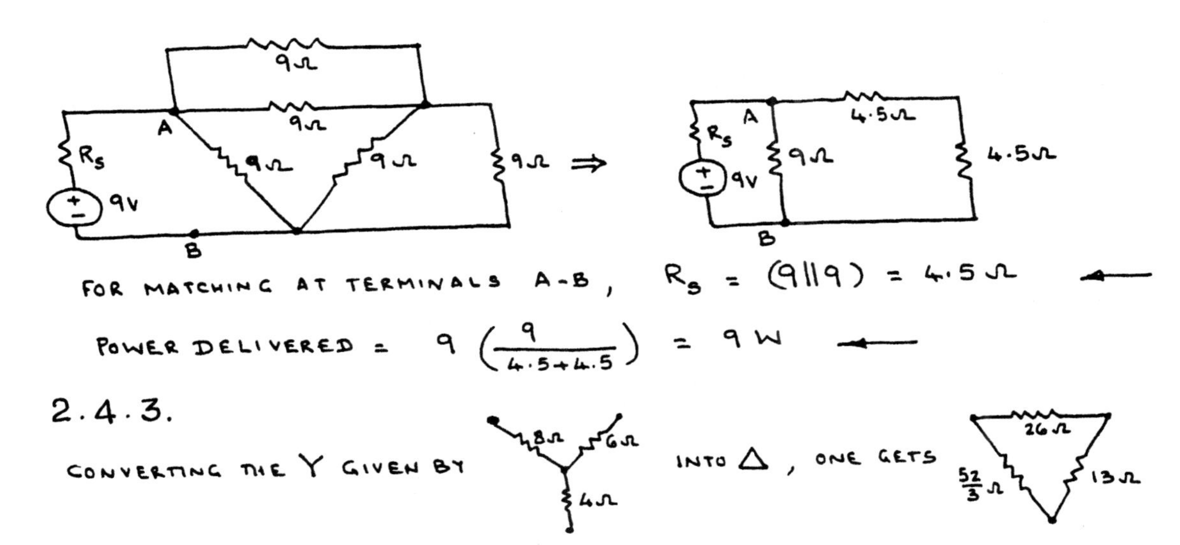

FOR MATCHING AT TERMINALS A-B, $R_s = (9 \| 9) = 4.5\,\Omega$ ←

POWER DELIVERED $= 9\left(\dfrac{9}{4.5+4.5}\right) = 9\,W$ ←

2.4.3.

CONVERTING THE Y GIVEN BY [Y: 8Ω, 6Ω, 4Ω] INTO Δ, ONE GETS [Δ: 26Ω, $\frac{52}{3}$Ω, 13Ω]

USING THE ABOVE, ONE CAN REDRAW FIG. P2.4.3:

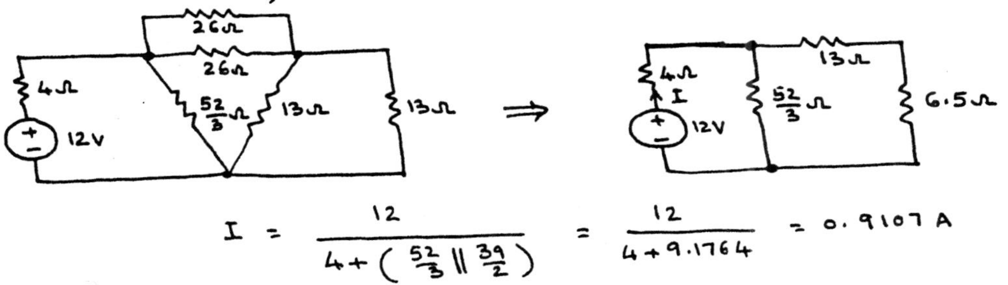

$$I = \frac{12}{4 + \left(\frac{52}{3} \| \frac{39}{2}\right)} = \frac{12}{4 + 9.1764} = 0.9107\,A$$

POWER DELIVERED BY SOURCE $= 12 \times 0.9107 = 10.93\,W$ ←

2.5.1. The program listing is given below:

```
PROBLEM 2.5.1

*THE CIRCUIT DIAGRAM IS SHOWN IN FIGURE P2.5.1
*CIRCUIT DESCRIPTION:
VS   1  0  15V
R1   1  2  5
R2   2  0  10
IS   0  2  1
*ANALYSIS REQUEST:
.DC VS 15 15 1
*OUTPUT REQUEST:
.PRINT DC V(1) V(2) I(VS) I(R1) I(R2)
*END STATEMENT:
.END
```

We must create a file containing this program and then execute PSpice. The details of this process vary depending on the type of computer and the version of PSpice being used. In any case, an output file is produced by PSpice that contains the program listing, error messages, information about the program execution, and finally, a tabulation of the results requested by the .PRINT command. The results for this program are as follows:

```
VS         V(1)       V(2)       I(VS)      I(R1)      I(R2)
1.500E+01 1.500E+01 1.333E+01 -3.333E-01 3.333E-01 1.333E+00
```

2.5.2.

The program listing is given below:

```
PROBLEM 2.5.2

*THE CIRCUIT DIAGRAM IS SHOWN IN FIGURE P2.5.2
*CIRCUIT DESCRIPTION:
IS  0  1  1A
R1  1  0  10
R2  1  2  20
*
EVCVS  2  0  1  0  2.5
*ANALYSIS REQUEST:
.DC  IS  1  1  1
*OUTPUT REQUEST:
.PRINT DC V(1) V(2) I(R1) I(R2) I(EVCVS)
*END STATEMENT:
.END
```

ECVS IS THE VOLTAGE-CONTROLLED VOLTAGE SOURCE

When this program is executed, an output file is created that contains the following results:

```
IS         V(1)       V(2)       I(R1)      I(R2)      I(EVCVS)
1.000E+00 4.000E+01 1.000E+02 4.000E+00 -3.000E+00 -3.000E+00
```

2.5.3.

THE PROGRAM LISTING IS GIVEN BELOW:

```
PROBLEM 2.5.3

*THE CIRCUIT DIAGRAM IS SHOWN IN FIGURE          P2.5.3
*CIRCUIT DESCRIPTION:
VS       1  0  10V
HCCVS    3  2  VSENSE  20
VSENSE   2  4  0V
R1       1  2  10
R2       3  0  10
R3       4  0  15
*ANALYSIS REQUEST:
.DC  VS  10  10  1
*OUTPUT REQUEST:
.PRINT  DC  V(2)  V(3)  I(R1)  I(R2)  I(R3)
*END STATEMENT:
.END
```

After running the program, the following results are found in the output file generated by PSpice:

```
VS          V(2)        V(3)        I(R1)       I(R2)       I(R3)
1.000E+01 2.500E+00 5.833E+00 7.500E-01 5.833E-01 1.667E-01
```

These results can be verified by manual circuit analysis.

2.5.4.

The program listing is given below:

```
PROBLEM 2.5.4

*THE CIRCUIT DIAGRAM IS IN FIGURE          P2.5.4
VS  2  1  10
IS  0  3  2
R1  1  0  7
R2  1  3  3
R3  2  3  5
R4  2  0  9
.DC  VS  10  10  1
.PRINT  DC  V(1)  V(2)  V(3)
.END
```

After running this program, we find the results V(1) = 3.5, V(2) = 13.5, and V(3) = 11 V in the output file.

2.5.5.

PROGRAM LISTING FOR PROBLEM 2.5.5

```
* THE CIRCUIT DIAGRAM IS SHOWN IN FIG. P 2.5.5
V1 A O 15
R1 A B 60
R2 B O 90
R3 B C 50
R4 A C 90
R5 C O 60
.DC V1 15 15 1
.PRINT DC V(A) V(B) V(C) I(V1)
.END
```

FROM THE OUTPUT FILE, $V_A = 15V$; $V_B = 8.1148V$; $V_C = 6.8852V$;

AND VOLTAGE SOURCE CURRENT = − 0.2049 A ←

NOTE THAT THE MINUS SIGN IN THE VOLTAGE-SOURCE CURRENT MEANS THAT THE ACTUAL DIRECTION OF THE CURRENT IS DIRECTED IN AT THE MINUS TERMINAL AND OUT AT THE PLUS TERMINAL, AS ONE WOULD EXPECT.

THE TOTAL POWER SUPPLIED BY THE SOURCE IS THEN GIVEN BY

$$15 \times 0.2049 = 3.07\ W \leftarrow$$

THE VOLTAGE ACROSS THE 50-Ω RESISTOR IS FOUND AS

$$V_x = V_B - V_C = 8.1148 - 6.8852 = 1.2296V$$

2.6.1.

```
function problem261
clc

syms t L

% Given Inductance
L = 2.5e-3

% Given Inductor Current
i = 10*exp(-500*t)*sin(2000*t);

subplot(221)
ezplot(i,[0,0.004])
title('Current')
xlabel('t')
ylabel('i(t)')

% Inductor Voltage
v=L*diff(i,t)
subplot(222)
ezplot(v,[0,0.004])
title('Voltage')
xlabel('t')
ylabel('v(t)')

% Inductor Power
p=v*i
subplot(223)
ezplot(p,[0,0.004])
title('Power')
xlabel('t')
ylabel('p(t)')

% Stored Energy
w=int(p)
subplot(224)
ezplot(w,[0,0.004])
title('Energy')
xlabel('t')
ylabel('w(t)')
```

2.6.1 (CONTD.)

```
L =

    0.0025

v =

-25/2*exp(-500*t)*sin(2000*t)+50*exp(-500*t)*cos(2000*t)

p =

10*(-25/2*exp(-500*t)*sin(2000*t)+50*exp(-500*t)*cos(2000*t))*exp(-500*t)*sin(20

w =

1/16*exp(-500*t)^2-1/16*exp(-1000*t)*cos(4000*t)
```

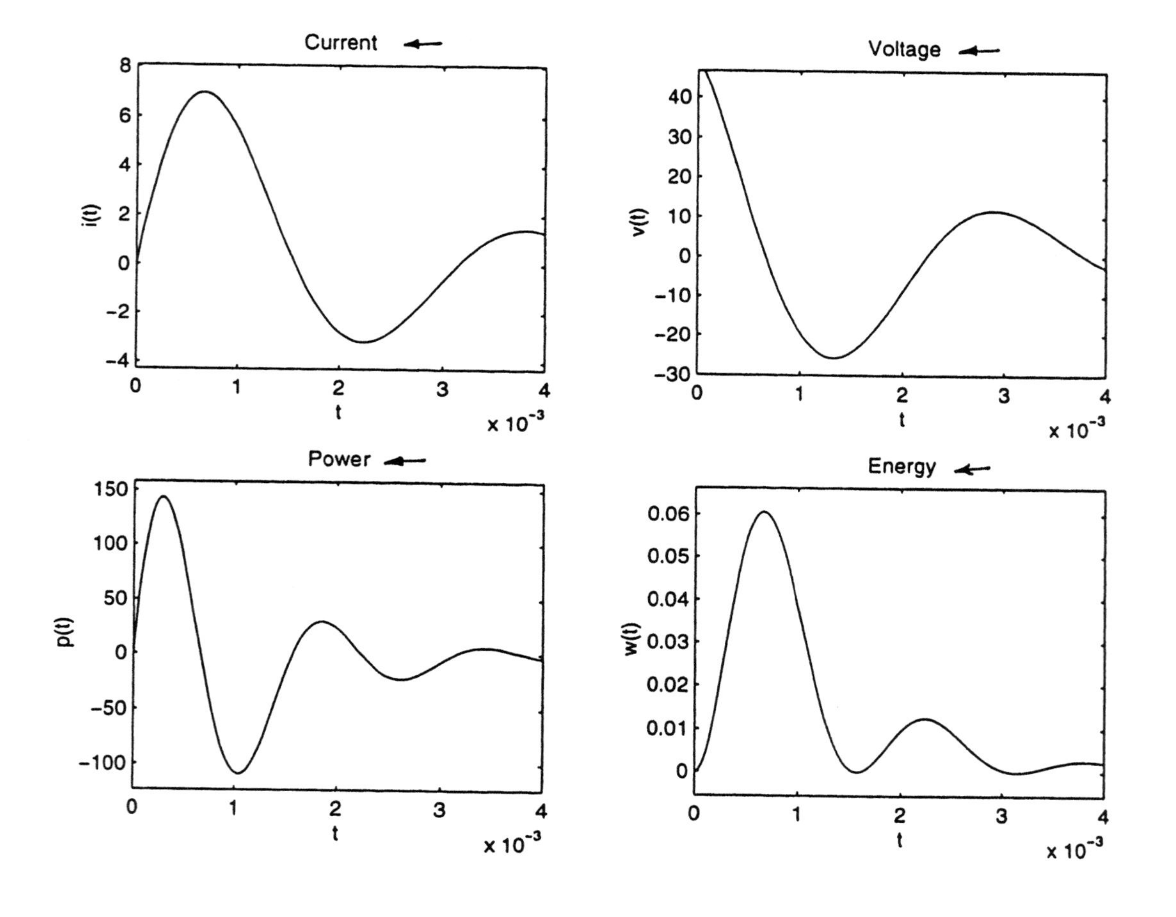

2.6.2.

```
function problem262
clc

% Design equation for terminals C-D
eqn01 = '(R1+300)*R2/(R1+300+R2)=50';

% Design equation for terminals A-B
eqn02 = 'R1+50*R2/(R2+50)=300';

% Solve equations
sol = solve(eqn01,eqn02,'R1,R2');

% Solutions
R1 = eval(sol.R1)
R2 = eval(sol.R2)
```

```
R1 =

  273.8613   <--
 -273.8613

R2 =

   54.7723   <--
  -54.7723
```

CHAPTER 3

3.1.1.

(a)

		FORCE–CURRENT ANALOG	FORCE–VOLTAGE ANALOG
(i.) NEWTON'S LAW	$F = M\frac{du}{dt}$	C	L
(F: FORCE; M: MASS; u = VELOCITY)		$i = C\frac{dv}{dt}$	$v = L\frac{di}{dt}$
(ii.) HOOKE'S LAW	$F = \frac{1}{C_m}\int u\,dt$	L	C
(C_m: COMPLIANCE = 1/STIFFNESS)		$i = \frac{1}{L}\int v\,dt$	$v = \frac{1}{C}\int i\,dt$
(iii.) VISCOUS FRICTION LAW	$F = Du$	G	R
(D: VISCOUS FRICTION)		$i = Gv$	$v = Ri$

(b)

$$\text{VOLUME} = Ah$$

$$\frac{d}{dt}(Ah) = F_i - F_o = F_i - \frac{h}{R}$$

OR

$$A\frac{dh}{dt} + \frac{h}{R} = F_i \quad \longleftarrow$$

WHERE F_i is the forcing function.

FOR AN RC CIRCUIT EXCITED BY $i(t) = I$

$$C\frac{dv_c}{dt} + \frac{v_c}{R} = I \quad \longleftarrow$$

WHICH IS ANALOGOUS TO THE ABOVE.

3.1.2.

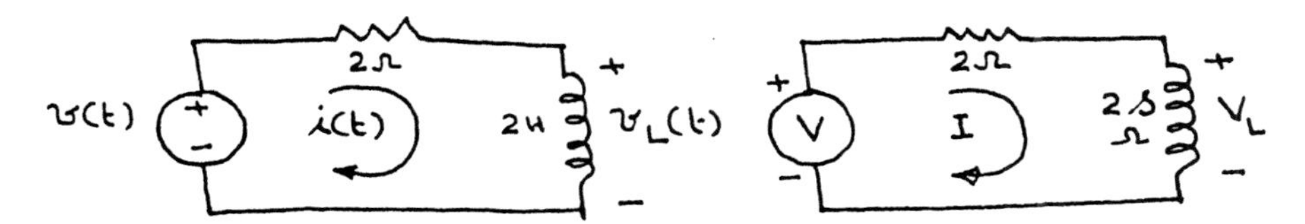

TIME-DOMAIN NETWORK FREQUENCY-DOMAIN TRANSFORMED NETWORK

(a)

KVL: $V = (2+2s)I$ OR $I = V/(2+2s)$

$V_L = 2s(I) = \dfrac{s}{1+s}V$; WITH $s=-2$ AND $V=20$, $V_L = \dfrac{-2}{1-2}(20) = 40\,V$

FROM WHICH $v_L(t) = 40e^{-2t}$ V ←

(b)

CONSTANT EXCITATION IS REPRESENTED BY $20e^{0t}$.

SO WITH $s=0$, $V_L = 0$; THUS $v_L(t) = 0$ ←

SIGNIFYING THE SHORT-CIRCUIT BEHAVIOR OF THE INDUCTANCE IN THE DC STEADY STATE.

(NOTE THAT THE IMPEDANCE OF THE INDUCTOR IS ZERO, FOR $s=0$.)

3.1.3.

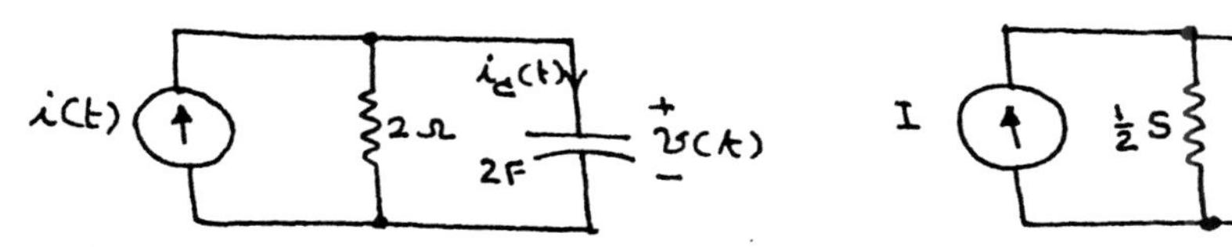

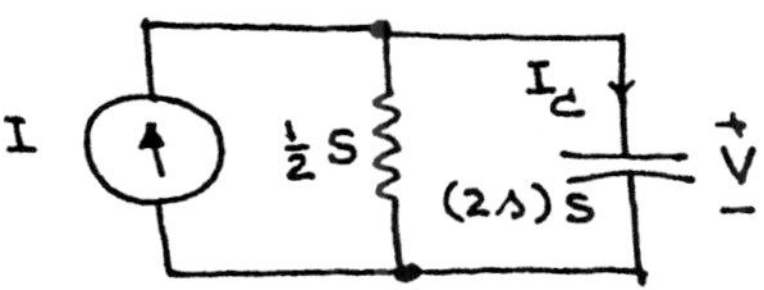

TIME-DOMAIN NETWORK FREQUENCY-DOMAIN TRANSFORMED NETWORK

(a)

KCL: $I = V(\tfrac{1}{2}) + 2s(V)$ OR $V = \dfrac{2}{1+4s}I$

$I_c = (2s)V = \dfrac{4s}{1+4s}I$; WITH $s=-2$ AND $I=20$, $I_c = \dfrac{-8}{1-8}(20) = \dfrac{160}{7}$ A

FROM WHICH $i_c(t) = \dfrac{160}{7}e^{-2t}$ A ←

3.1.3. CONTINUED

(b) CONSTANT EXCITATION IS REPRESENTED BY $20\, e^{0t}$.

SO WITH $s = 0$, $I_c = 0$; THUS $i_c(t) = 0$ ←

SIGNIFYING THE OPEN-CIRCUIT BEHAVIOR OF THE CAPACITOR IN THE DC STEADY STATE.

(NOTE THAT THE ADMITTANCE OF THE CAPACITOR IS ZERO, FOR $s = 0$.)

3.1.4.

(i) FORCE-CURRENT ANALOG:

$$i(t) = 20 \frac{dv}{dt} + 4v + \frac{1}{8}\int v\, dt \;;$$ CORRESPONDING CIRCUIT IS SHOWN BELOW:

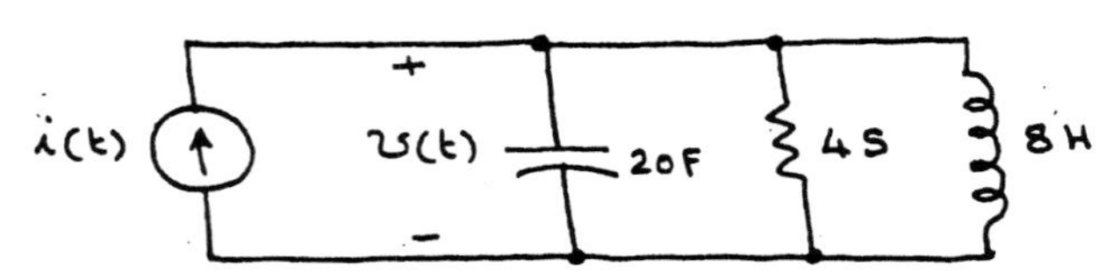

TRANSFORMED NETWORK, FOR $s = -\frac{1}{4}$, IS SHOWN BELOW:

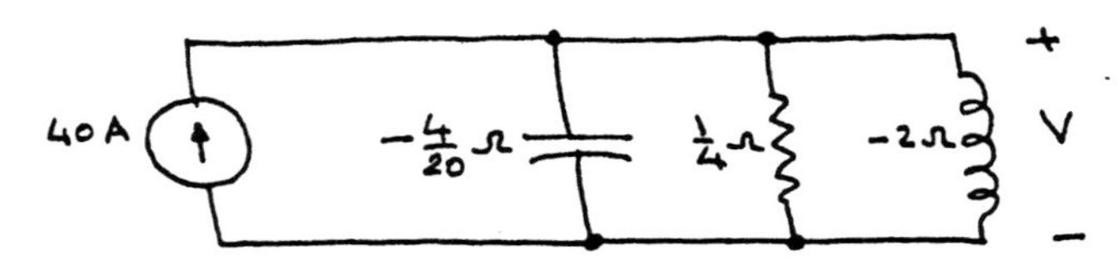

KCL: $$-40 + \frac{V}{-4/20} + \frac{V}{1/4} + \frac{V}{-2} = 0 \quad \text{OR} \quad V = -80/3$$

$$v(t) = -\frac{80}{3} e^{-t/4} \text{ V}$$

SINCE $v(t)$ IS THE ANALOG OF $u(t)$

$$u(t) = -\frac{80}{3} e^{-t/4} \text{ m/s} \quad \leftarrow$$

(ii) FORCE-VOLTAGE ANALOG:

L R v(t) i(t) C

$v \to F$; $i \to u$; $L \to M$; $R \to D$; $C \to C_m$

FOR $v(t) = 40 e^{-t/4}$, $V = 40$; $s = -\frac{1}{4}$

$$I = \frac{V}{Z} = \frac{40}{[(20 \times -0.25) + 4 + (-4/8)]} = -\frac{80}{3}$$

$$i(t) = -\frac{80}{3} e^{-t/4} \quad \text{OR} \quad u(t) = -\frac{80}{3} e^{-t/4} \text{ m/s} \quad \leftarrow$$

3.1.5.

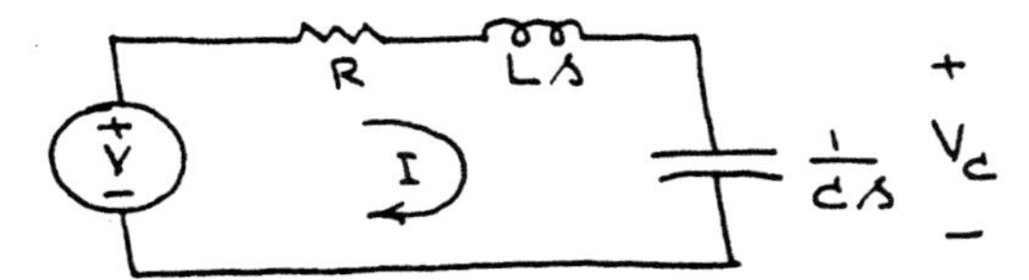

FOR THE GIVEN VALUES, $I = \dfrac{V_c}{(1/sC)} = sC V_c = (-10)(0.1)\,5 = -5\text{ A}$

$$V = \left[10 + 1(-10) + \frac{1}{(-10)(0.1)}\right](-5) = 5$$

$$\therefore\ v(t) = 5\,e^{-10t}\ \text{V} \quad \longleftarrow$$

3.1.6.

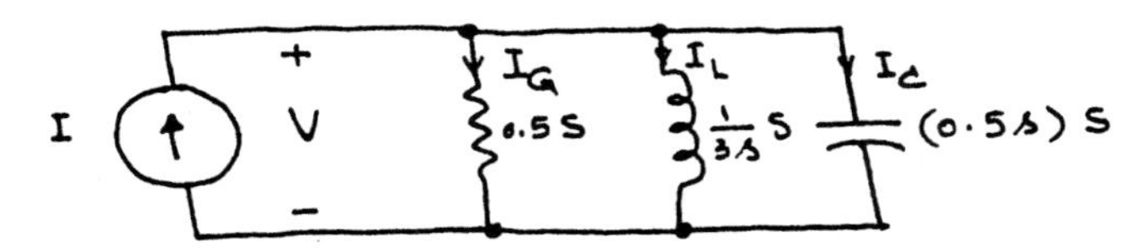

$$I = GV + \frac{1}{Ls}V + CsV\ ; \quad I_L = \frac{1}{Ls}V = \frac{1}{3(-0.5)}\,12 = -8\text{ A}$$

$$I_G = GV = 0.5 \times 12 = 6\text{A}\ ; \quad I_c = CsV = 0.5(-0.5)\,12 = -3\text{ A}$$

$$I = 6 - 8 - 3 = -5\text{ A}$$

$$\therefore\ i(t) = -5\,e^{-0.5t} \quad \longleftarrow$$

3.1.7.

$\bar{I}$ + $\bar{V}$ −   0.5S $\bar{I}_L = 2\angle 0^\circ$ $\frac{1}{j\omega L} = -jS$ $j\omega C = j\left(\frac{1}{3} \times 0.5\right) = \left(\frac{j}{6}\right)$ S

NOTING THAT $\omega = \frac{1}{3}$, $\bar{V} = \bar{I}_L(j\omega L) = 2j\left(\frac{1}{3}\right)3 = j2 = 2\angle 90^\circ$ V

$$v(t) = 2\cos\left(\frac{t}{3} + 90^\circ\right)\ \text{V}$$

$$\bar{I} = \left(0.5 - j + \frac{j}{6}\right)(j2) = j + 2 - \frac{1}{3} = \frac{5}{3} + j = 1.94\angle 30.96^\circ\ \text{V}$$

$$\therefore\ i(t) = 1.94\cos\left(\frac{t}{3} + 30.96^\circ\right)\ \text{A} \quad \longleftarrow$$

3.1.8.

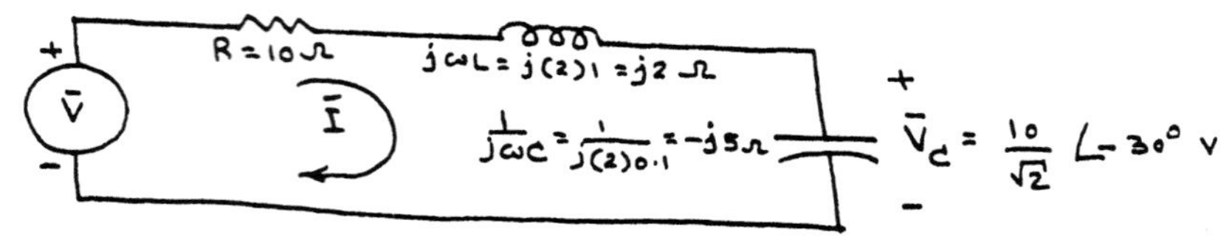

NOTING THAT $\omega = 2$, $\bar{I} = \dfrac{(10/\sqrt{2})\angle -30^\circ}{-j5} = \dfrac{2}{\sqrt{2}}\angle 60^\circ$ V

$\bar{V} = (10 + j2 - j5)\,\bar{I} = (10 - j3)\,\dfrac{2}{\sqrt{2}}\angle 60^\circ = (10.44\angle -16.7^\circ)\left(\dfrac{2}{\sqrt{2}}\angle 60^\circ\right)$

$= \dfrac{20.88}{\sqrt{2}}\angle 43.3^\circ$ V

$v(t) = 20.88\cos(2t + 43.3^\circ)$ V ⟵

3.1.9.

$i(t) = 0.02\cos 5000t$; $\omega = 5000$; $\bar{I} = \dfrac{0.02}{\sqrt{2}}\angle 0^\circ$

$\bar{Z} = R + j\omega L = (100 + j100)\,\Omega = 100\sqrt{2}\angle 45^\circ\,\Omega$ ⟵

$\bar{V} = \bar{I}\bar{Z} = \left(\dfrac{0.02}{\sqrt{2}}\angle 0^\circ\right)\left(100\sqrt{2}\angle 45^\circ\right) = 2\angle 45^\circ$ V

$v(t) = 2\sqrt{2}\cos(5000t + 45^\circ)$ V ⟵

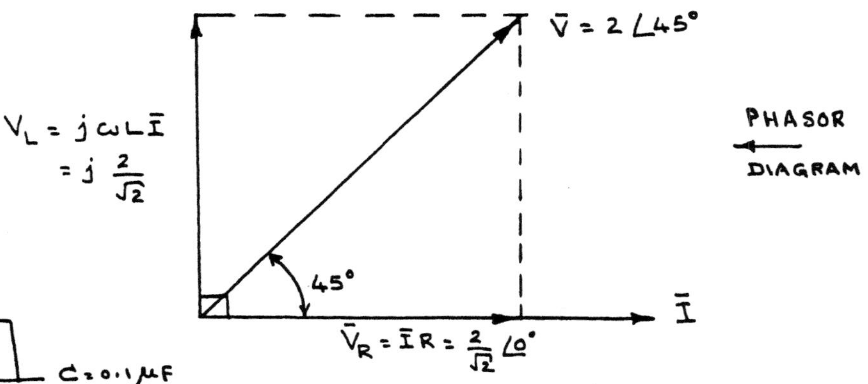

3.1.10.

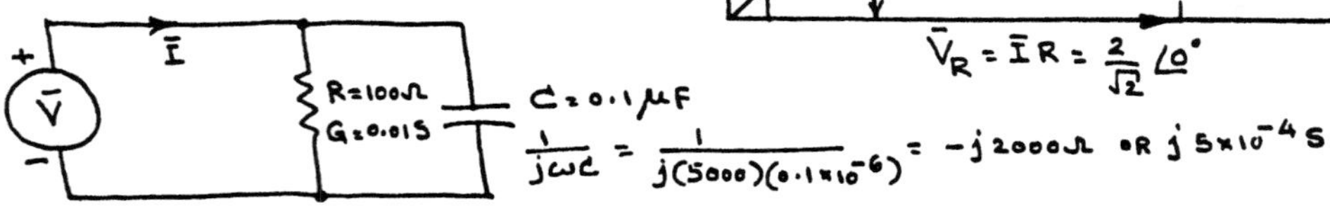

$v(t) = 10\cos(5000t + 30^\circ)$; $\omega = 5000$ rad/s; $\bar{V} = \dfrac{10}{\sqrt{2}}\angle 30^\circ$ V

$\bar{Y} = G + j\omega C = (0.01 + j5\times 10^{-4})$ S $= 0.01\angle 2.86^\circ$ S ⟵

$\bar{I} = \bar{V}\bar{Y} = \left(\dfrac{10}{\sqrt{2}}\angle 30^\circ\right)(0.01\angle 2.86^\circ) = \dfrac{0.1}{\sqrt{2}}\angle 32.86^\circ$ A

$\therefore\ i(t) = 0.1\cos(5000t + 32.86^\circ)$ A ⟵

$\bar{I}_R = \dfrac{\bar{V}}{R} = \dfrac{(10/\sqrt{2})\angle 30^\circ}{100} = \dfrac{0.1}{\sqrt{2}}\angle 30^\circ$ A; $\bar{I}_C = \dfrac{10\angle 30^\circ}{\sqrt{2}}(5\times 10^{-4}\angle 90^\circ) = \dfrac{5\times 10^{-3}}{\sqrt{2}}\angle 120^\circ$ A

3.1.10. CONTINUED

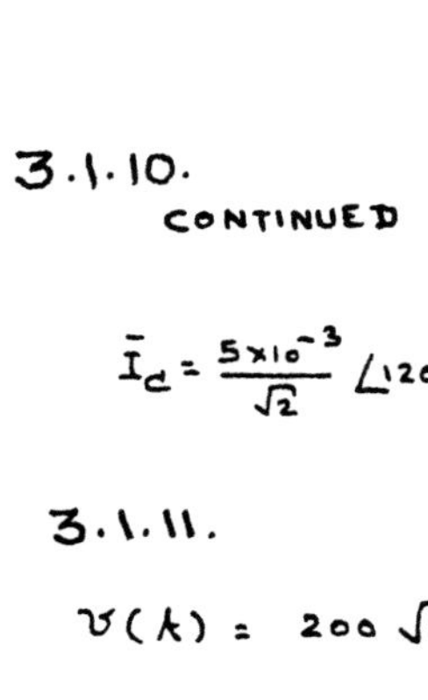
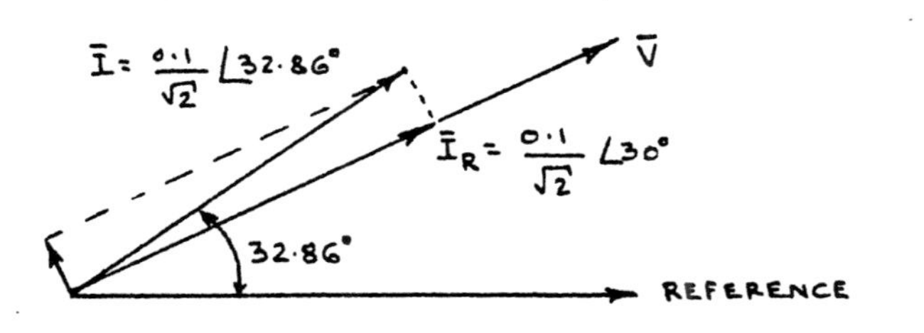

PHASOR DIAGRAM ←

3.1.11.

$v(t) = 200\sqrt{2}\cos(377t + 60°) \Rightarrow \bar{V} = 200\angle 60°$ V

$i(t) = 10\sqrt{2}\cos(377t + 30°) \Rightarrow \bar{I} = 10\angle 30°$ A

$\bar{S} = \bar{V}\bar{I}^* = (200\angle 60°)(10\angle -30°) = 2000\angle 30° = P + jQ$

(a)

$S = 2000$ VA $= 2$ kVA ←

$P = 2000\cos 30° = 1732$ W $= 1.732$ kW ←

$Q = 2000\sin 30° = 1000$ VAR $= 1$ kVAR ←

(b)

$\bar{Z} = \frac{\bar{V}}{\bar{I}} = \frac{200\angle 60°}{10\angle 30°} = 20\angle 30° = (17.32 + j10)\,\Omega$

$\omega L = 10$ OR $L = \frac{10}{377} = 0.0265$ H $= 26.5$ mH

A — $R = 17.32\,\Omega$ — $L = 26.5$ mH — B ←

(c) $\bar{Z}_{Th} = 20\angle 30° = (17.32 + j10)\,\Omega$ ←

3.1.12.

(a)

$\bar{S} = \bar{V}\bar{I}^* = P + jQ$

$P = VI\cos\phi = 100 \times 10 \times 0.8 = 800$ W $= 0.8$ kW, ABSORBED BY NETWORK ←

$Q = VI\sin\phi = 100 \times 10 \times 0.6 = 600$ VAR $= 0.6$ kVAR, DELIVERED BY NETWORK ←
(OR -0.6 kVAR, ABSORBED BY NETWORK)

$S = VI = 100 \times 10 = 1000$ VA $= 1$ kVA ←

(b)

$\bar{Z} = \frac{\bar{V}}{\bar{I}} = \frac{100}{10\angle\cos^{-1}0.8} = \frac{100}{10\angle 36.87°} = 10\angle -36.87° = (8 - j6)\,\Omega$

$\frac{1}{j\omega C} = -j6$ OR $\frac{1}{\omega C} = 6$ OR $C = \frac{1}{(2\pi \times 60)\,6} = 0.000442$ F $= 0.442$ mF (corrected by hand from 0.00053 F $= 0.53$ mF)

A — $R = 8\,\Omega$ — $C = 0.442$ mF (corrected by hand from 0.53 mF) — B ←

(c) $\bar{Z}_{Th} = 10\angle -36.87° = (8 - j6)\,\Omega$ ←

3.1.13.

(a) MESH ANALYSIS:

$$(2\angle -30^\circ)(8 - j4) = (2\angle -30^\circ)(8.94\angle -26.6^\circ) = 17.88\angle -56.6^\circ$$

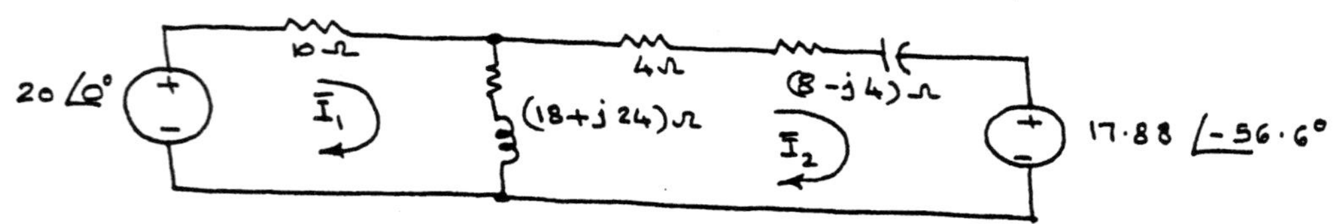

KVL:

$$\bar{I}_1(10 + 18 + j24) - \bar{I}_2(18 + j24) = 20\angle 0^\circ$$

$$-\bar{I}_1(18 + j24) + \bar{I}_2(18 + j24 + 4 + 8 - j4) = -17.88\angle -56.6^\circ$$

SIMULTANEOUS SOLUTION YIELDS $\bar{I}_2 = 0.96\angle 78.5^\circ$ A ⟵

(b) NODAL ANALYSIS:

$$20\angle 0^\circ \div 10 = 2\angle 0^\circ$$

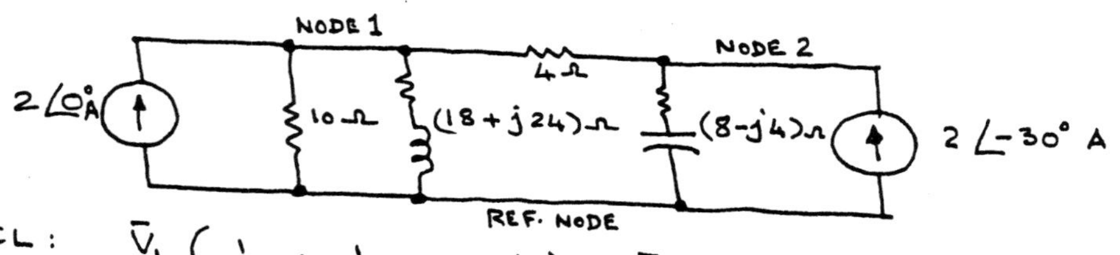

KCL:

$$\bar{V}_1\left(\frac{1}{10} + \frac{1}{18 + j24} + \frac{1}{4}\right) - \bar{V}_2\left(\frac{1}{4}\right) = 2\angle 0^\circ$$

$$-\bar{V}_1\left(\frac{1}{4}\right) + \bar{V}_2\left(\frac{1}{8 - j4} + \frac{1}{4}\right) = 2\angle -30^\circ$$

SIMULTANEOUS SOLUTION YIELDS $\bar{V}_1$ AND $\bar{V}_2$, FROM WHICH $\bar{V}_1 - \bar{V}_2 = 3.84\angle 78.5^\circ$ V

AND THE CURRENT THROUGH THE 4-Ω RESISTOR $= \dfrac{\bar{V}_1 - \bar{V}_2}{4} = 0.96\angle 78.5^\circ$ A ⟵

3.1.14.

(a) MESH ANALYSIS

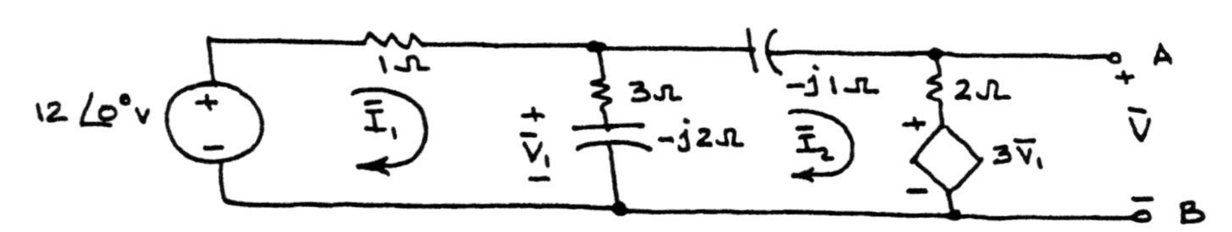

KVL:

$$\bar{I}_1(1 + 3 - j2) - \bar{I}_2(3 - j2) = 12$$

$$-\bar{I}_1(3 - j2) + \bar{I}_2(3 - j2 - j1 + 2) = -3\bar{V}_1$$

CONSTRAINT EQUATION: $\bar{V}_1 = -j2(\bar{I}_1 - \bar{I}_2)$

AFTER SUBSTITUTING THE CONSTRAINT EQUATION, SIMULTANEOUS SOLUTION YIELDS

$$\bar{I}_1 = 7.1\angle 54.9^\circ \text{ A} \quad ; \quad \bar{I}_2 = 6.09\angle 77.1^\circ \text{ A}$$

THEN $\bar{V} = 2\bar{I}_2 + 3\bar{V}_1 = 2\bar{I}_2 - j6(\bar{I}_1 - \bar{I}_2) = 4.87\angle 66^\circ$ V ⟵

3.1.14. CONTINUED

(b) NODAL ANALYSIS:

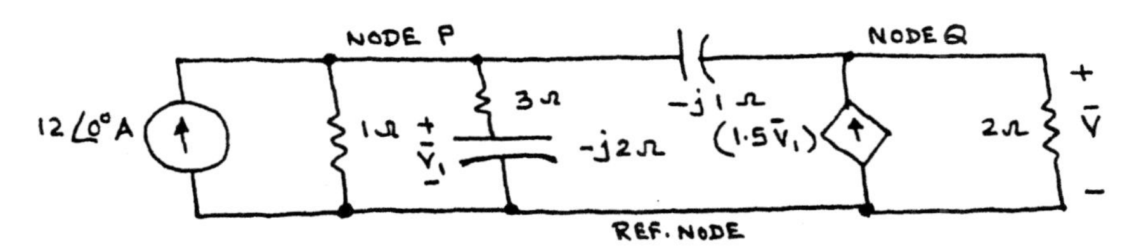

KCL: $\bar{V}_P\left(\frac{1}{1}+\frac{1}{3-j2}+\frac{1}{-j1}\right)-\bar{V}_Q\left(\frac{1}{-j1}\right)=12\angle 0^\circ$

$$-\bar{V}_P\left(\frac{1}{-j1}\right)+\bar{V}_Q\left(\frac{1}{-j1}+\frac{1}{2}\right)=1.5\,\bar{V}_1$$

CONSTRAINT EQUATION: $\bar{V}_1=\dfrac{-j2}{3-j2}\,\bar{V}_P$

AFTER SUBSTITUTING THE CONSTRAINT EQUATION, SIMULTANEOUS SOLUTION YIELDS

$$\bar{V}_Q=\bar{V}=4.87\angle 66^\circ \text{ V} \quad \leftarrow$$

3.1.15.

(a) OPEN-CIRCUIT VOLTAGE ACROSS TERMINALS A-B IS FOUND IN PROB. 3.1.14 AS $\bar{V}=4.87\angle 66^\circ$

THEN LET US FIND THE SHORT-CIRCUIT CURRENT.

KVL: $\bar{I}_1(1+3-j2)-\bar{I}_2(3-j2)=12\angle 0^\circ$

$$-\bar{I}_1(3-j2)+\bar{I}_2(3-j2-j1+2)-\bar{I}_3(2)=-3\bar{V}_1$$

$$-\bar{I}_2(2)+\bar{I}_3(2)=3\bar{V}_1$$

CONSTRAINT EQUATION: $\bar{V}_1=-j2(\bar{I}_1-\bar{I}_2)$

UPON SUBSTITUTION, SIMULTANEOUS SOLUTION YIELDS

$$\bar{I}_3=\bar{I}_{SC}=17.7\angle 67.66^\circ \text{ A}$$

$$\therefore\ \bar{Z}_{Th}=\frac{4.87\angle 66^\circ}{17.7\angle 67.66^\circ}=0.275\angle -1.66^\circ\ \Omega \quad \leftarrow$$

THÉVENIN EQUIVALENT CIRCUIT IS THEN GIVEN BY THE FOLLOWING:

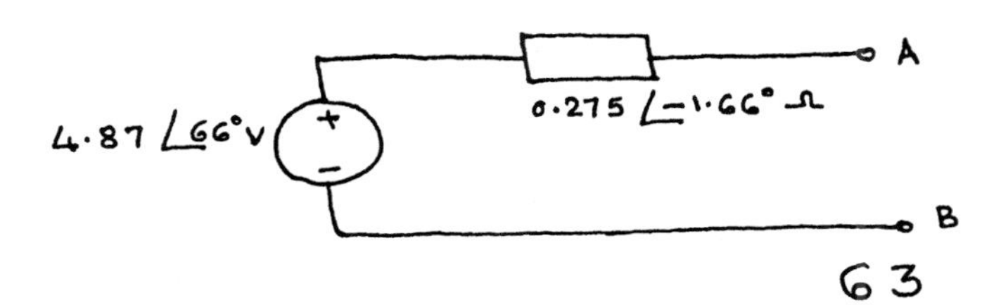

3.1.15. CONTINUED

(b) $\bar{Z}_L = \bar{Z}^*_{Th} = 0.275 \angle 1.66^\circ\ \Omega = (0.275 + j\,0.008)\ \Omega$ ←

(c) $P_{MAX} = \left(\frac{|\bar{V}_{oc}|}{\bar{Z}_L + \bar{Z}_{Th}}\right)^2 (\text{Re}\ \bar{Z}_L)$

$= \left(\frac{4.87}{2 \times 0.275}\right)^2 (0.275) = \frac{4.87^2}{4 \times 0.275} = 21.56\ \text{W}$ ←

3.1.16.

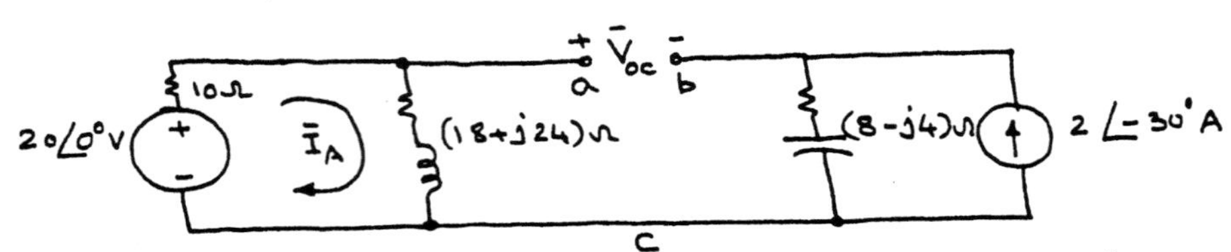

$\bar{V}_{ac} = \bar{I}_A (18 + j24)$; $\bar{I}_A = \frac{20 \angle 0^\circ}{10 + 18 + j24} = \frac{20 \angle 0^\circ}{36.88 \angle 40.6^\circ} = 0.542 \angle -40.6^\circ\ \text{A}$

$\bar{V}_{ac} = (0.542 \angle -40.6^\circ)(18 + j24) = 16.26 \angle 12.53^\circ\ \text{V}$

$\bar{V}_{bc} = (2 \angle -30^\circ)(8 - j4) = 17.9 \angle -56.57^\circ\ \text{V}$

$\bar{V}_{oc} = \bar{V}_{ac} - \bar{V}_{bc} = 16.26 \angle 12.53^\circ - 17.9 \angle -56.57^\circ = 19.42 \angle 72^\circ\ \text{V}$

TO OBTAIN $\bar{Z}_{Th}$, SUPPRESS THE SOURCES; THEN WITH RESPECT TO TERMINALS a–b,

$\bar{Z}_{Th} = [10 \,\|\, (18 + j24)] + (8 - j4) = (15.94 - j\,2.24)\ \Omega$

THE CURRENT THROUGH THE 4-Ω RESISTOR $= \frac{\bar{V}_{oc}}{\bar{Z}_{Th} + 4} = \frac{19.42 \angle 72^\circ}{15.94 - j2.24 + 4}$
$= 0.967 \angle 78.3^\circ\ \text{A}$ ←

3.1.17.

(a) LOAD A: $P_A = 100\ \text{kW}$; $Q_A = P_A \tan(\cos^{-1} 0.6) = 133.33\ \text{kVAR}$

LOAD B: $P_B = 100 \times 0.8 = 80\ \text{kW}$; $Q_B = 100 \sin(\cos^{-1} 0.8) = 60\ \text{kVAR}$

COMBINED LOAD: $P_{TOTAL} = 100 + 80 = 180\ \text{kW}$; $Q_{TOTAL} = 133.33 + 60 = 193.33\ \text{kVAR}$

$(VA)_{TOTAL} = \sqrt{P_T^2 + Q_T^2} = \sqrt{(180)^2 + (193.33)^2} = 264.15\ \text{kVA}$ ←

$I_L = \frac{(VA)_{TOTAL}}{V_L} = \frac{264.15}{6.6} = 40\ \text{A}$ ←

(b) FOR UNITY POWER FACTOR ON SUPPLY END, THE CAPACITANCE NEEDS TO DELIVER 193.33 kVAR

$\because\ Q_C = \frac{V_L^2}{X_C}$ OR $X_C = \frac{V_L^2}{Q_C} = \frac{(6.6 \times 1000)^2}{193.33 \times 10^3} = 225.3\ \Omega$

$C = \frac{1}{\omega X_C} = \frac{1}{2\pi \times 60 \times 225.3} = 11.8 \times 10^{-6}\ \text{F} = 11.8\ \mu\text{F}$ ←

3.1.18.

LOAD A: $P_A = 10 \times 0.8 = 8\,kW$; $Q_A = -10[\sin(\cos^{-1} 0.8)] = -6\,kVAR$

LOAD B: $P_B = 15\,kW$; $Q_B = P_B \tan(\cos^{-1} 0.6) = 20\,kVAR$

LOAD C: $P_C = 5\,kW$; $Q_C = 0$

COMBINED LOAD: $P_T = 8 + 15 + 5 = 28\,kW$ ←

$$Q_T = -6 + 20 + 0 = 14\,kVAR \leftarrow$$

$$(VA)_T = \sqrt{P_T^2 + Q_T^2} = \sqrt{28^2 + 14^2} = 31.3\,kVA \leftarrow$$

$$I_L = \frac{(VA)_T}{V_L} = \frac{31.3 \times 10^3}{400} = 78.25\,A \leftarrow$$

FOR THE SUPPLY LINE TO OPERATE AT 0.9 LEADING POWER FACTOR,

$$Q_L / P_L = -\tan(\cos^{-1} 0.9)\ ; \quad Q_L = -28 \tan 25.84^\circ = -13.56\,kVAR$$

$$\therefore Q_x = Q_L - (Q_A + Q_B) = -13.56 - (-6 + 20) = -27.56\,kVAR$$

$$\therefore X_C = \frac{(400)^2}{27.56 \times 10^3} = 5.806\,\Omega$$

$$\text{OR} \quad C = \frac{1}{2\pi \times 60 (5.806)} = 456.9 \times 10^{-6}\,F \simeq 457\,\mu F \leftarrow$$

3.1.19.

(a) BECAUSE $v(t)$ IS AN EVEN FUNCTION, $b_n = 0$ FOR ALL n. ←

$$a_n = \frac{2}{T}\int_{-T/2}^{T/2} v(t)\cos(n\omega t)\,dt = \frac{2}{T}\int_{-\tau}^{0}\left(A + \frac{At}{\tau}\right)\cos(n\omega t)\,dt + \frac{2}{T}\int_{0}^{\tau}\left(A - \frac{At}{\tau}\right)\cos(n\omega t)\,dt$$

[NOTE: $\omega = 2\pi f = 2\pi/T$]

$$\therefore a_n = \frac{2A}{T}\int_{-\tau}^{\tau}\cos(n\omega t)\,dt + \frac{2A}{\tau}\int_{-\tau}^{0}\frac{t}{T}\cos(n\omega t)\,dt - \frac{2A}{\tau}\int_{0}^{\tau}\frac{t}{T}\cos(n\omega t)\,dt$$

LET $x = n\omega t$; $dx = n\omega\,dt$

THEN
$$a_n = \frac{2A}{T}\left[\frac{\sin(n\omega\tau) - \sin(-n\omega\tau)}{n\omega}\right] + \frac{2A}{\tau}\frac{T}{(n2\pi)^2}\int_{-n2\pi\frac{\tau}{T}}^{0} x\cos x\,dx - \frac{2A}{\tau}\frac{T}{(n2\pi)^2}\int_{0}^{n2\pi\tau/T} x\cos x\,dx$$

or
$$a_n = \frac{2A\tau}{T}\,\frac{\sin^2(n\omega\tau/2)}{(n\omega\tau/2)^2}, \quad \text{FOR } n \neq 0 \quad ←$$

FOR $n = 0$,
$$a_0 = \frac{1}{T}\int_0^T f(t)\,dt \quad \text{or} \quad \frac{1}{T}\int_{-T/2}^{T/2} f(t)\,dt$$

WHICH IS ALSO THE AVERAGE ORDINATE OR THE DC COMPONENT OF THE WAVE.

$$a_0 = \frac{\text{AREA}}{\text{BASE}} = \frac{A\tau}{T} \quad ←$$

(b) SINCE $i(t)$ IS ODD, $a_n = 0$ FOR ALL n. ←

$$b_n = \frac{2}{T}\int_{-\tau}^{\tau}\frac{A}{\tau}t\sin(n\omega t)\,dt = \frac{2A}{\tau}\int_{-\tau}^{\tau}\frac{n2\pi t}{T}\sin(n\omega t)(n\omega\,dt)\frac{1}{n\omega}\cdot\frac{1}{n2\pi}$$

[NOTE: $\omega = 2\pi/T$]

$$b_n = \frac{2A}{\tau}\,\frac{1}{(n\omega)^2 T}\int_{-n\omega\tau}^{n\omega\tau} x\sin x\,dx \quad \text{by LETTING } x = n\omega t;\ dx = n\omega\,dt$$

$$\therefore b_n = \frac{4A\tau}{T}\,\frac{1}{(n\omega\tau)^2}\left[\sin(n\omega\tau) - n\omega\tau\cos(n\omega\tau)\right], \quad ←$$

$n = 1, 2, \ldots$

3.1.20.

$$\bar{C}_n = \frac{1}{T}\int_0^T f(t)\, e^{-jn\omega t}\, dt \qquad (3.1.49)\ ; \ \text{NOTE: } \omega = 2\pi/T$$

$$\bar{C}_n = \frac{1}{T}\int_{-\tau}^{\tau} A\cos\left(\frac{\pi}{2}\frac{t}{\tau}\right) e^{-jn\omega t}\, dt = \frac{A}{2T}\int_{-\tau}^{\tau}\left(e^{j\frac{\pi t}{2\tau}} + e^{-j\frac{\pi t}{2\tau}}\right) e^{-jn\omega t}\, dt$$

$$= \frac{A\tau}{2T}\left[\frac{e^{-j(n\omega - \frac{\pi}{2\tau})\tau} - e^{j(n\omega - \frac{\pi}{2\tau})\tau}}{-j\left(n\omega - \frac{\pi}{2\tau}\right)\tau} + \frac{e^{-j(n\omega + \frac{\pi}{2\tau})\tau} - e^{j(n\omega + \frac{\pi}{2\tau})\tau}}{-j\left(n\omega + \frac{\pi}{2\tau}\right)\tau}\right]$$

$$= \frac{A\tau}{T}\left[\frac{\sin\left(n\omega\tau - \frac{\pi}{2}\right)}{n\omega\tau - \frac{\pi}{2}} + \frac{\sin\left(n\omega\tau + \frac{\pi}{2}\right)}{n\omega\tau + \frac{\pi}{2}}\right]$$

$$\bar{C}_n = \frac{A\tau\pi\cos(n\omega\tau)}{T\left[\left(\frac{\pi}{2}\right)^2 - (n\omega\tau)^2\right]}, \quad n = 0, \pm 1, \pm 2, \ldots$$

THEN $$i(t) = \sum_{-\infty}^{\infty} \bar{C}_n\, e^{jn\omega t}$$ ←

FOR $T = 2\tau$,

$$\bar{C}_n = \frac{A\pi}{2}\,\frac{\cos(n\pi)}{\left[\left(\frac{\pi}{2}\right)^2 - (n\pi)^2\right]} = \frac{2A\cos(n\pi)}{\pi(1-4n^2)}, \quad n = 0, \pm 1, \pm 2, \ldots$$ ←

3.1.21.

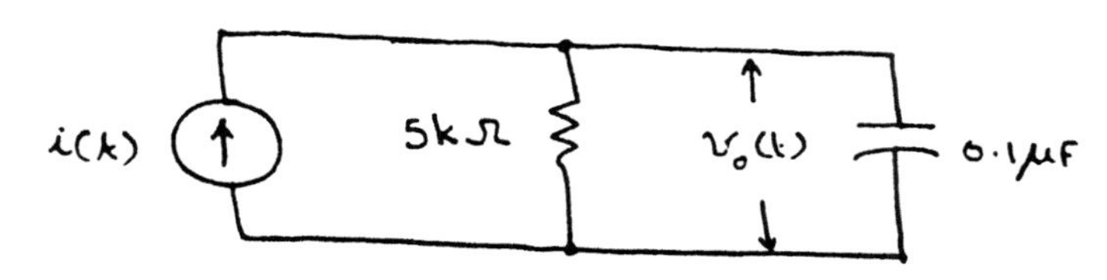

TIME DOMAIN CIRCUIT

$\bar{I}$; $5k\Omega = \frac{1}{5\times 10^3}\,S$; $\bar{V}_o$; $(-j10^7/\omega)\,\Omega = (j\omega 10^{-7})\,S$

FREQUENCY-DOMAIN CIRCUIT

$$\bar{V}_o\left(\frac{1}{5\times 10^3} + j\omega 10^{-7}\right) = \bar{I} \quad \text{or} \quad \bar{V}_o = \frac{\bar{I}\ 5\times 10^3}{(1 + j\omega\, 5\times 10^{-4})}$$

3.1.21. CONTINUED

THE FOUR COMPONENTS OF $\bar{V}_o$ ARE GIVEN BY

1. FOR $\omega = 2\pi/T = 10^3(2\pi)$ AND $\bar{I}_1 = \frac{2 I_m}{\pi\sqrt{2}} = \frac{2\times 15\times 10^{-3}}{\pi\sqrt{2}} = \frac{9.55\times 10^{-3}}{\sqrt{2}}$

$$\bar{V}_{01} = \frac{9.55\times 10^{-3}}{\sqrt{2}}\left[\frac{5\times 10^3}{1+j\pi}\right] = \frac{9.55\times 10^{-3}\times 5\times 10^3}{\sqrt{2}}\left[0.092 - j0.289\right]$$

$$= \frac{1}{\sqrt{2}}\left[4.393 - j\,13.8\right] = \frac{14.4}{\sqrt{2}}\angle -72.34^\circ \text{ V}$$

2. FOR $\omega = 4\pi/T = 10^3(4\pi)$ AND $\bar{I}_2 = -\frac{I_m}{\pi\sqrt{2}} = -\frac{15\times 10^{-3}}{\pi\sqrt{2}}$

$$\bar{V}_{02} = -\frac{15\times 10^{-3}}{\pi\sqrt{2}}\left[\frac{5\times 10^3}{1+j2\pi}\right] = -\frac{4.775\times 10^{-3}}{\sqrt{2}}\times 5\times 10^3\left[0.025 - j\,0.155\right]$$

$$= -\frac{1}{\sqrt{2}}\left[0.597 - j3.7\right] = -\frac{3.75}{\sqrt{2}}\angle -80.8^\circ \text{ V}$$

3. FOR $\omega = 6\pi/T = 10^3(6\pi)$ AND $\bar{I}_3 = \frac{2 I_m}{3\pi\sqrt{2}} = \frac{10\times 10^{-3}}{\pi\sqrt{2}}$

$$\bar{V}_{03} = \frac{3.183\times 10^{-3}}{\sqrt{2}}\left[\frac{5\times 10^3}{1+j\,3\pi}\right] = \frac{15.915}{\sqrt{2}}\left[0.011 - j\,0.105\right]$$

$$= \frac{1}{\sqrt{2}}(0.175 - j\,1.671) = \frac{1.68}{\sqrt{2}}\angle -84.02^\circ$$

4. FOR $\omega = 8\pi/T = 10^3(8\pi)$ AND $\bar{I}_4 = -\frac{I_m}{2\pi\sqrt{2}} = -\frac{15\times 10^{-3}}{2\pi\sqrt{2}}$

$$\bar{V}_{04} = -\frac{2.387\times 10^{-3}}{\sqrt{2}}\left[\frac{5\times 10^3}{1+j\,4\pi}\right] = -\frac{11.935}{\sqrt{2}}\left[0.006 - j0.079\right]$$

$$= -\frac{1}{\sqrt{2}}\left[0.072 - j0.943\right] = -\frac{0.946}{\sqrt{2}}\angle -85.6^\circ$$

THEN

$$v_o(t) = 14.4\sin(2\pi\times 10^3 t - 72.34^\circ) - 3.75\sin(4\pi\times 10^3 t - 80.8^\circ)$$
$$+ 1.68\sin(6\pi\times 10^3 t - 84.02^\circ) - 0.946\sin(8\pi\times 10^3 t - 85.6^\circ) \text{ V} \leftarrow$$

3.1.22.

THE TRANSFORMED CIRCUIT IS SHOWN BELOW:

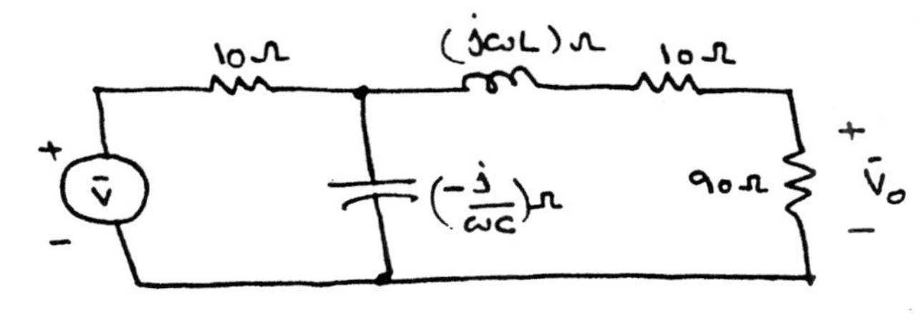

FOR $\omega = 0$ (dc), THE INDUCTOR IS A SHORT, AND THE CAPACITOR IS OPEN.

$$\therefore \quad V_o \Big|_{\omega=0} = \frac{90}{10+10+90} \frac{2V_m}{\pi} = \frac{90}{110} \frac{2(100)}{\pi} = 52.1 \text{ V}$$

FOR $\omega = 2\pi \times 120$ AND $\omega = 2\pi \times 240$, ONE CAN CALCULATE THE VOLTAGE COMPONENTS EITHER BY MESH-EQUATION METHOD OR NODAL-EQUATION METHOD. THE CIRCUIT IS REDRAWN FOR NODAL METHOD OF ANALYSIS.

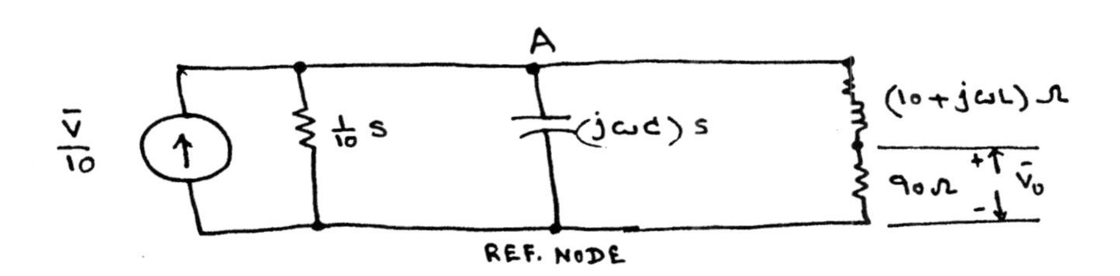

$$\bar{V}_A \left(\frac{1}{10} + j\omega C + \frac{1}{100 + j\omega L} \right) = \frac{\bar{V}}{10} \quad ; \quad \bar{V}_o = \frac{90\,\bar{V}_A}{100 + j\omega L}$$

$$\bar{V}_A = \frac{\bar{V}(100 + j\omega L)}{110 - 10\omega^2 LC + j\omega(L + 1000C)} \quad ; \quad \bar{V}_o = \frac{90\,\bar{V}}{110 - 10\omega^2 LC + j\omega(L + 1000C)}$$

USING $C = 500\,\mu F = 5 \times 10^{-4}$ F AND $L = 5$ mH $= 5 \times 10^{-3}$ H,

FOR $\omega = 2\pi \times 120$, WITH $\bar{V}_1 = -\frac{4V_m}{3\pi} \frac{1}{\sqrt{2}} = -\frac{42.44}{\sqrt{2}} \angle 0°$

$$\bar{V}_{o1} = -\frac{90(42.44)/\sqrt{2}}{110 - 10(240\pi)^2(5 \times 10^{-3})(5 \times 10^{-4}) + j(240\pi)\{5 \times 10^{-3} + (1000 \times 5 \times 10^{-4})\}}$$

3.1.22. CONTINUED

$$\bar{V}_{01} = -\frac{3819.6/\sqrt{2}}{95.79 + j\,380.76} = -\frac{3819.6/\sqrt{2}}{392.63\,\angle 75.9^\circ} = -\frac{9.73}{\sqrt{2}}\angle -75.9^\circ$$

FOR $\omega = 2\pi \times 240$, WITH $\bar{V}_2 = \frac{4V_m}{15\pi}\,\frac{1}{\sqrt{2}} = \frac{8.49}{\sqrt{2}}\angle 0^\circ$

$$\bar{V}_{02} = \frac{90\,(8.49)/\sqrt{2}}{110 - 10\,(480\pi)^2(5\times10^{-3})(5\times10^{-4}) + j\,(480\pi)\{5\times10^{-3} + (1000\times5\times10^{-4})\}}$$

$$= \frac{764.1/\sqrt{2}}{53.15 + j\,761.52} = \frac{764.1/\sqrt{2}}{763.37\,\angle 86^\circ} = \frac{1}{\sqrt{2}}\angle -86^\circ$$

THUS THE TRUNCATED FOURIER SERIES FOR THE OUTPUT VOLTAGE IS GIVEN BY

$$v_o(t) = \left[52.1 - 9.73\cos(2\pi\times120t - 75.9^\circ) + 1\cos(2\pi\times240t - 86^\circ)\right]\,\text{V}$$

3.2.1.

$i(0^-) = i(0^+) = 0$; $i_{ss}(t) = \frac{20}{8} = 2.5$; $\frac{R}{L} = \frac{8}{6} = \frac{4}{3} = a$

$$i(t) = \left[i(0^+) - i_{ss}(0^+)\right]e^{-at} + i_{ss}(t)$$

$$= -2.5\,e^{-\frac{4}{3}t} + 2.5 = 2.5\left(1 - e^{-\frac{4}{3}t}\right)\text{A}, \quad \text{FOR } t \geq 0$$

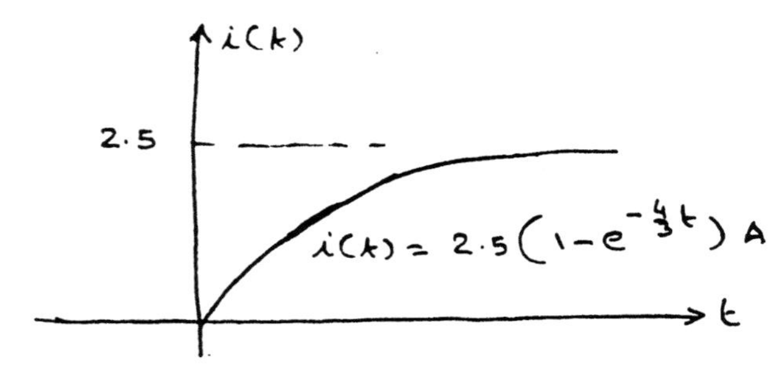

3.2.2.

$i(0^-) = i(0^+) = 0$

TO FIND $i_{ss}(t)$:

4Ω
24V 6Ω 10 H ⇒ $R_{TH} = 6 \| 4 = \frac{12}{5}\,\Omega$
i
$V_{TH} = \frac{6}{6+4} \times 24 = \frac{72}{5}\,V$ 10H

$$i_{ss} = \frac{72}{5} \div \frac{12}{5} = 6 \quad ; \quad \frac{R}{L} = \frac{12}{5} \times \frac{1}{10} = \frac{6}{25}$$

$$\therefore \; i(t) = -6\, e^{-\frac{6}{25}t} + 6 = 6\left(1 - e^{-\frac{6}{25}t}\right) A, \quad \text{FOR } t > 0 \quad \leftarrow$$

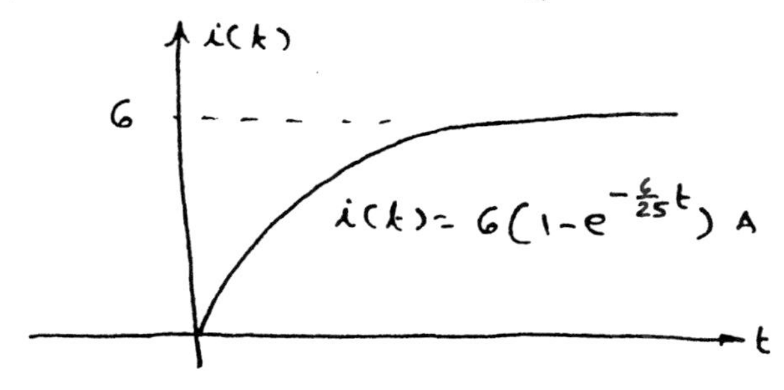

3.2.3.

$v_c(0^-) = v_c(0^+) = 0 \quad ; \quad v_{ss}(t) = 10 \times 4 = 40\,V \quad ; \quad \frac{1}{RC} = \frac{1}{4 \times \frac{2}{3}} = 3/8 = a$

$$v(t) = \left[v(0^+) - v_{ss}(0^+)\right] e^{-at} + v_{ss}(t)$$

$$= -40\, e^{-\frac{3}{8}t} + 40 = 40\left(1 - e^{-\frac{3}{8}t}\right) V, \quad \text{FOR } t > 0 \quad \leftarrow$$

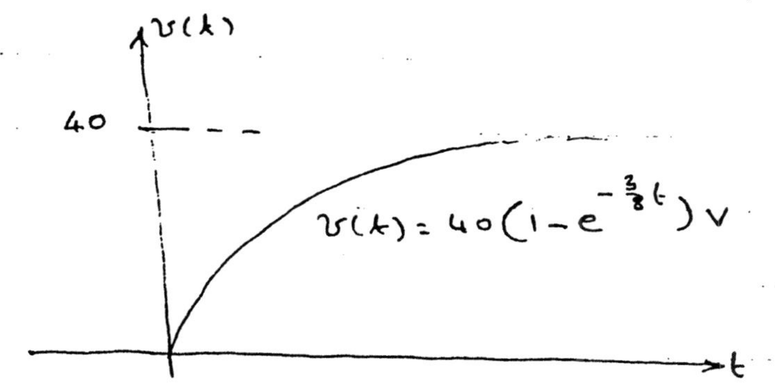

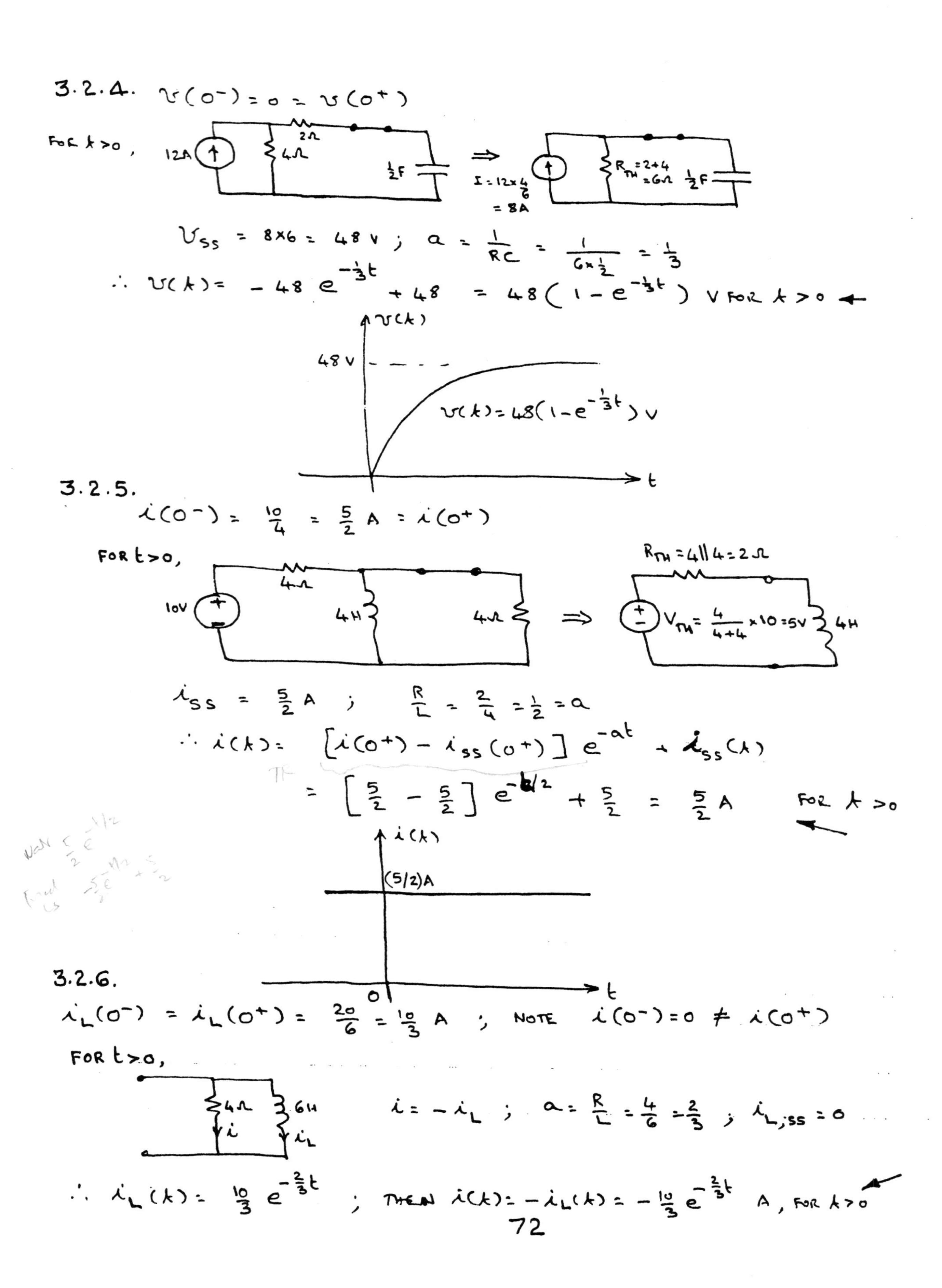

3.2.4. $v(0^-) = 0 = v(0^+)$

For $t > 0$,

$v_{ss} = 8 \times 6 = 48\text{ V}$; $a = \frac{1}{RC} = \frac{1}{6 \times \frac{1}{2}} = \frac{1}{3}$

$\therefore v(t) = -48 e^{-\frac{1}{3}t} + 48 = 48(1 - e^{-\frac{1}{3}t})$ V for $t > 0$

3.2.5.

$i(0^-) = \frac{10}{4} = \frac{5}{2}\text{ A} = i(0^+)$

For $t > 0$,

$i_{ss} = \frac{5}{2}\text{ A}$; $\frac{R}{L} = \frac{2}{4} = \frac{1}{2} = a$

$\therefore i(t) = [i(0^+) - i_{ss}(0^+)]e^{-at} + i_{ss}(t)$

$= \left[\frac{5}{2} - \frac{5}{2}\right]e^{-t/2} + \frac{5}{2} = \frac{5}{2}\text{ A}$ for $t > 0$

3.2.6.

$i_L(0^-) = i_L(0^+) = \frac{20}{6} = \frac{10}{3}\text{ A}$; NOTE $i(0^-) = 0 \neq i(0^+)$

For $t > 0$,

$i = -i_L$; $a = \frac{R}{L} = \frac{4}{6} = \frac{2}{3}$; $i_{L,ss} = 0$

$\therefore i_L(t) = \frac{10}{3} e^{-\frac{2}{3}t}$; then $i(t) = -i_L(t) = -\frac{10}{3} e^{-\frac{2}{3}t}$ A, for $t > 0$

3.2.6. CONTINUED

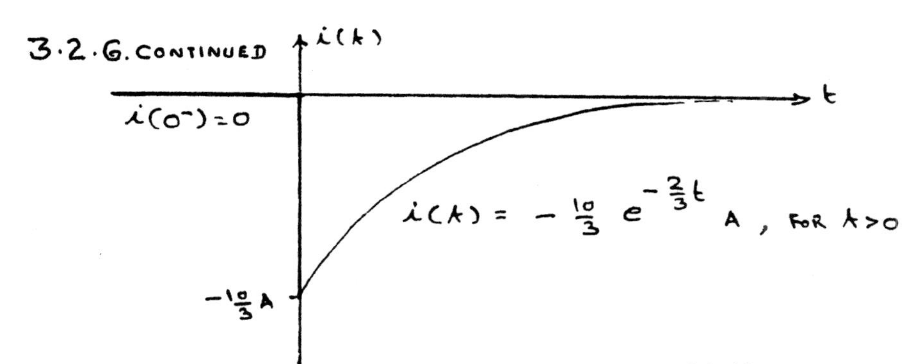

3.2.7. FOR $t = 0^-$,

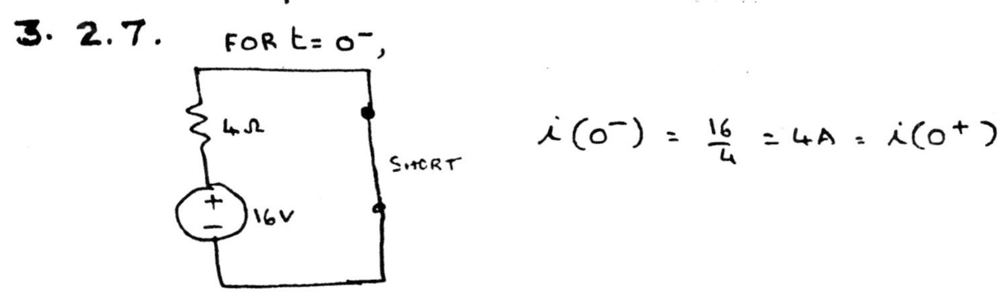

$$i(0^-) = \frac{16}{4} = 4A = i(0^+)$$

FOR $t > 0$,

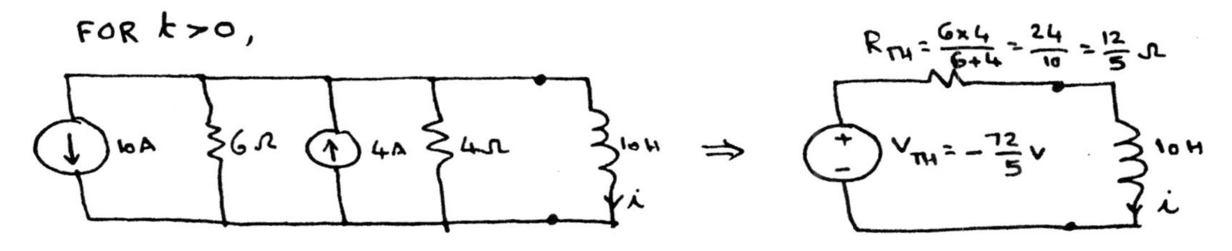

$$R_{TH} = \frac{6 \times 4}{6+4} = \frac{24}{10} = \frac{12}{5}\ \Omega$$

$$V_{TH} = (6 \| 4)(4-10) = \frac{24}{10}(-6) = -\frac{72}{5}$$

$$i_{ss} = \left(-\frac{72}{5}\right) \div \frac{12}{5} = -6\ A\ ; \quad a = \frac{R}{L} = \frac{12}{5} \times \frac{1}{10} = \frac{12}{50} = \frac{6}{25}$$

$$\therefore i(t) = (4+6)\, e^{-\frac{6}{25}t} - 6 = \left(10\, e^{-\frac{6}{25}t} - 6\right) A, \quad \text{FOR } t > 0. \leftarrow$$

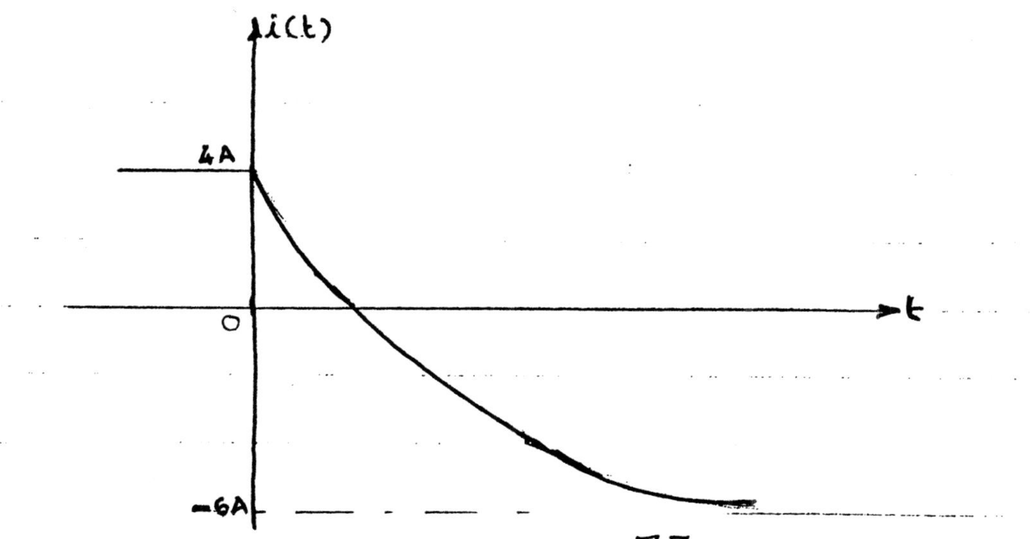

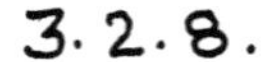

3.2.8.

For $t \geq 0$,

DIFFERENTIAL EQUATION

DE: $\frac{v}{4} + \int v\,dt = 24e^{-t}$; AFTER DIFFERENTIATION, $\frac{1}{4}\frac{dv}{dt} + v = -24e^{-t}$

THE NATURAL RESPONSE IS $v_n(t) = Ae^{-4t}$ (NOTE $\frac{R}{L} = \frac{4}{1} = 4$)

THE FORCED RESPONSE IS $v_f(t) = V_f e^{-t}$ SO THAT

$-\frac{V_f}{4}e^{-t} + V_f e^{-t} = -24e^{-t}$ or $-\frac{V_f}{4} + V_f = -24$ or $V_f = -32$; $v_f(t) = -32e^{-t}$

THE TOTAL RESPONSE IS THEN $v(t) = -32e^{-t} + Ae^{-4t}$

FOR $t < 0$, $i(t) = 5A$; IN THE DC STEADY STATE $i_L(0^-) = 5A = i_L(0^+)$ ←

AT $t = 0^+$, THE CIRCUIT IS GIVEN BELOW:

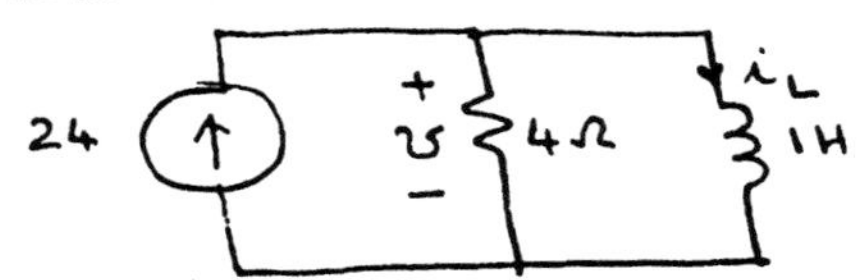

KCL REQUIRES $\frac{v(0^+)}{4} + 5 = 24$ or $v(0^+) = 76\text{ V}$;

$v_L(0^+) = v(0^+) = 76 = \frac{di_L}{dt}(0^+)$

or $\frac{di_L}{dt}(0^+) = 76\text{ A/s}$ ←

AT $t = 0^+$, THE DIFFERENTIAL EQUATION YIELDS $\left\{\frac{1}{4}\frac{dv}{dt}(0^+) + v(0^+)\right\} = -24$

or $\frac{1}{4}\frac{dv}{dt}(0^+) + 76 = -24$ or $\frac{dv}{dt}(0^+) = -400\text{ V/s}$ ←

3.2.9.

AT $t = 0^-$,

8Ω, 24V, 24Ω, v_C (+, −)

$v_C(0^-) = \frac{24}{8+24} \times 24 = 18\text{V} = v_C(0^+)$

AT $t = 0^+$,

8Ω, i_1, 12V, i_2, 24Ω, C, i_C, 18V (+, −)

$i_1 = \frac{12-18}{8} = -\frac{3}{4}\text{A}$; $i_2 = 18/24 = 3/4\text{A}$

$i_C(0^+) = i_1 - i_2 = -\frac{3}{4} - \frac{3}{4} = -\frac{3}{2}\text{A}$ ←

$\frac{dv_C}{dt}(0^+) = \frac{1}{C}i_C(0^+) = -\frac{3}{2}\text{ V/s}$ ←

3.2.9. CONTINUED

FOR $t \geq 0$,

$$i_1 = \frac{v - v_c}{8} \; ; \quad i_2 = \frac{v_c}{24} \; ; \quad i_c = i_1 - i_2$$

$$\therefore \quad \frac{di_c}{dt} = \frac{di_1}{dt} - \frac{di_2}{dt}$$

AT $t = 0^+$, $\frac{di_1}{dt} = \frac{1}{8}\left[\frac{dv}{dt} - \frac{dv_c}{dt}\right] = \frac{1}{8}\left(0 + \frac{3}{2}\right) = \frac{3}{16}$ A/s

AT $t = 0^+$, $\frac{di_2}{dt} = \frac{1}{24}\frac{dv_c}{dt}(0^+) = \frac{1}{24}\left(-\frac{3}{2}\right) = -\frac{3}{48} = -\frac{1}{16}$ A/s

$$\therefore \quad \frac{di_c}{dt}(0^+) = \frac{3}{16} + \frac{1}{16} = \frac{4}{16} = \frac{1}{4} \text{ A/s} \quad \leftarrow$$

3.2.10.

FOR $t < 0$,

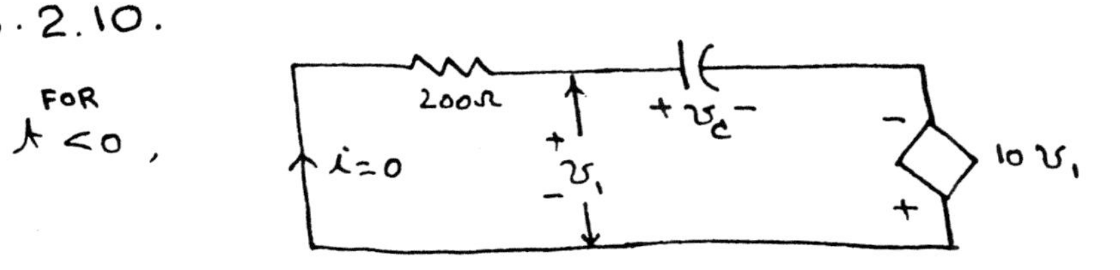

$v_1 = 0$ AND $v_c(0^-) = 0$

FOR $t = 0^+$,

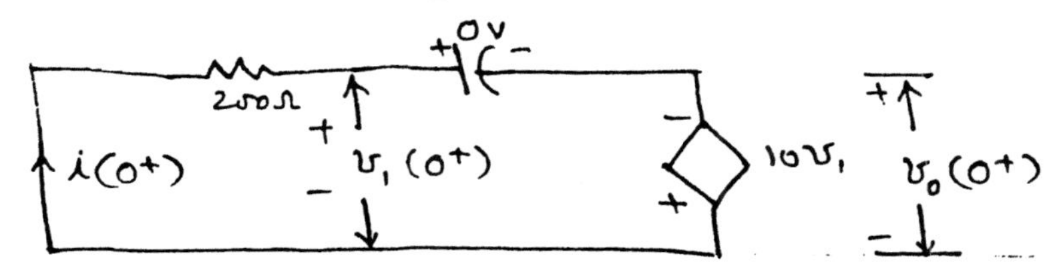

$$v_1(0^+) = 0 - 10\,v_1(0^+) \quad \text{or} \quad v_1(0^+) = 0$$

$$v_o(0^+) = -10\,v_1(0^+) = 0 \quad \leftarrow$$

$$i(0^+) = C\,\frac{dv_c}{dt}(0^+)$$

$$v_1(0^+) = -200\,i(0^+) = 0 \; ; \quad \frac{dv_c}{dt}(0^+) = 0 \; ; \quad \frac{dv_o}{dt} = -10\,\frac{dv_1}{dt}(0^+)$$

FROM $v_1 = v_c - 10v_1$ or $v_1 = \frac{v_c}{11}$, $\frac{dv_1}{dt}(0^+) = \frac{1}{11}\frac{dv_c}{dt}(0^+) = 0$

$$\therefore \quad \frac{dv_o}{dt}(0^+) = 0 \quad \leftarrow$$

3.2.11. FOR NATURAL RESPONSE:

WITH $v(t)$ SUPPRESSED, DE : $200\, i(t) + \frac{1}{2\times10^{-8}}\int i\,dt - 10 v_1 = 0$

CONSTRAINT EQ. IS $v_1 = -200\, i$

SUBSTITUTION GIVES: $2200\, i(t) + \frac{10^8}{2}\int i\,dt = 0 \Rightarrow 10^8 i(t) + 4400\frac{di}{dt} = 0$

CHARACTERISTIC EQ. IS $10^8 + 4400\,S = 0$ or $S = -\frac{10^8}{4400} = -\frac{10^6}{44}$

$\therefore\; i(t) = A_1 e^{-\frac{10^6}{44}t}$ AND $v_n(t) = A\, e^{-\frac{10^6}{44}t}$

DE CAN BE EXPRESSED AS $\frac{v - v_1}{200} = \frac{10^{-8}}{2}\frac{d}{dt}\left[v_1 - (-10 v_1)\right]$

or $v(t) = v_1 + 11\times10^{-6}\frac{dv_1}{dt}$ AND $v_0 = -10\, v_1$

FOR $v(t) = 0.4t$, THE FORCED RESPONSE IS OF THE FORM $v_{1,f}(t) = k_1 t + k_2$

SUBSTITUTION YIELDS: $0.4t = (k_1 t + k_2) + 11\times10^{-6}k_1$

FROM WHICH $k_1 = 0.4$ AND $k_2 = -(4.4)10^{-6}$

THE TOTAL RESPONSE IS THEN $v_0(t) = -10\left[A e^{-\frac{10^6}{44}t} + 0.4t - (4.4)10^{-6}\right]$

FROM PROB. 3.2.10, WE HAVE $v_0(0^+) = 0 \Rightarrow A = 4.4\times10^{-6}$

THUS $v_0(t) = -4.4\times10^{-5} e^{-\frac{10^6}{44}t} - 4t + 4.4\times10^{-5}$

or $v_0(t) = 4.4\times10^{-5}\left(1 - e^{-\frac{10^6}{44}t}\right) - 4t$, FOR $t \geq 0$ ←

3.2.12.
(a) $s^2 + \frac{R}{L}s + \frac{1}{LC} = 0$; $\frac{R}{L} = \frac{8}{2} = 4$; $\frac{1}{LC} = \frac{1}{2\times\frac{1}{6}} = 3$

$\therefore\; s^2 + 4s + 3 = 0$ OR $(s+1)(s+3) = 0 \Rightarrow s_1 = -1$ AND $s_2 = -3$

$i_L(t) = A_1 e^{-t} + A_2 e^{-3t}$ ($\because i_{L,ss} = 0$)

$v_C(t) = B_1 e^{-t} + B_2 e^{-3t} + 10$

AT $t = 0^-$, SWITCH OPEN: $i_L(0^-) = 0 = i_L(0^+)$; $v_C(0^-) = 0 = v_C(0^+)$

AT $t = 0^+$,

$A_1 + A_2 = 0$; $B_1 + B_2 + 10 = 0$

$v_L(0^+) = L\frac{di_L}{dt}(0^+) = L(-A_1 - 3A_2) = 10 - 8\, i_L(0^+) - v_C(0^+)$

OR $A_1 + 3A_2 = -5$

$i_C(0^+) = i_L(0^+) = C\frac{dv_C}{dt}(0^+) = C(-B_1 - 3B_2) = 0$ OR $B_1 + 3B_2 = 0$

SIMULTANEOUS SOLUTION YIELDS: $A_1 = \frac{5}{2}$; $A_2 = -\frac{5}{2}$; $B_1 = -15$; $B_2 = 5$

$\therefore\; i_L(t) = \frac{5}{2}(e^{-t} - e^{-3t})$, FOR $t > 0$; $v_C(t) = -15e^{-t} + 5e^{-3t} + 10$, FOR $t > 0$. ←

3.2.12. CONTINUED

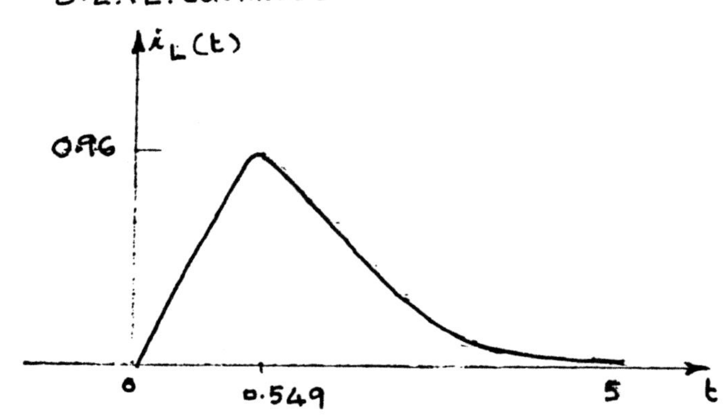

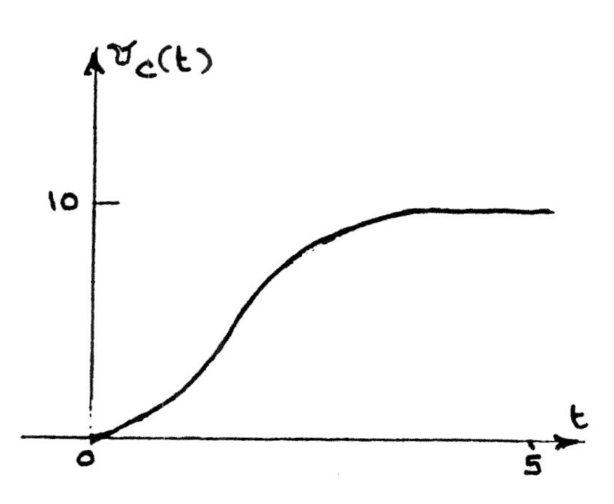

3.2.12.(b) $R = 8\,\Omega$; $L = 2\,H$; $C = \frac{1}{8}\,F$; $V_S = 10\,V$

$$s^2 + \frac{R}{L}s + \frac{1}{LC} = s^2 + 4s + 4 = 0 \quad \text{or} \quad (s+2)^2 = 0 \quad \text{or} \quad s_1 = s_2 = -2$$

$$i_{L,ss} = 0 \quad ; \quad v_{C,ss} = 10\,V$$

$$\therefore \quad i_L(t) = A_1 e^{-2t} + A_2 t\, e^{-2t} \quad ; \quad v_C(t) = B_1 e^{-2t} + B_2 t e^{-2t} + 10$$

AT $t = 0^-$,

$$i_L(0^-) = 0 = i_L(0^+) \quad ; \quad v_C(0^-) = 0 = v_C(0^+)$$

AT $t = 0^+$,

SWITCH CLOSED 8Ω 2H
10V
1/8 F

$$A_1 = 0 \quad ; \quad B_1 + 10 = 0$$

$$V_L(0^+) = 10 - 8\,i_L(0^+) - V_C(0^+) = 10$$

$$= L\frac{di_L}{dt}(0^+) = 2(-2A_1 + A_2)$$

$$\text{or} \quad -2A_1 + A_2 = 5$$

$i_C(0^+) =$

$$= i_L(0^+) = C\frac{dv_C}{dt}(0^+) = C(-2B_1 + B_2) = 0 \Rightarrow -2B_1 + B_2 = 0$$

SIMULTANEOUS SOLUTION YIELDS: $A_1 = 0$; $A_2 = 5$; $B_1 = -10$; $B_2 = -20$

THEREFORE

$$i_L(t) = 5te^{-2t}, \text{ FOR } t > 0 \quad ; \quad v_C(t) = -10e^{-2t} - 20te^{-2t} + 10, \text{ FOR } t > 0.$$

iL(t)
0.919
0.5
t

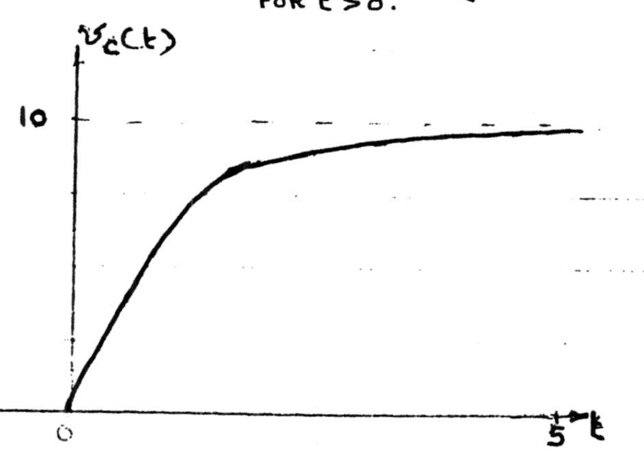

3.2.12. (c)

$R = 8\,\Omega$; $L = 2H$; $C = \frac{1}{26}F$; $V_S = 10V$

$s^2 + 4s + 13 = 0 \Rightarrow s_1 = -2 + j3$; $s_2 = -2 - j3$

$i_{L,ss} = 0$; $v_{C,ss} = 10$

$\therefore\ i_L(t) = e^{-2t}(A_1 \cos 3t + A_2 \sin 3t)$; $v_C(t) = e^{-2t}(B_1 \cos 3t + B_2 \sin 3t) + 10$

At $t = 0^-$, $i_L(0^-) = 0 = i_L(0^+)$; $v_C(0^-) = 0 = v_C(0^+)$

$v_L(0^+) = 10 - 8\, i_L(0^+) - v_C(0^+) = 10 = L \frac{di_L}{dt}(0^+) \Rightarrow \frac{di_L}{dt}(0^+) = 5$

$\therefore\ A_1 = 0$; $-2A_1 + 3A_2 = 5 \Rightarrow A_1 = 0$; $A_2 = \frac{5}{3}$

$i_C(0^+) = i_L(0^+) = 0 = C \frac{dv_C}{dt}(0^+) \Rightarrow \frac{dv_C}{dt}(0^+) = 0$

$\therefore\ B_1 + 10 = 0$; $-2B_1 + 3B_2 = 0 \Rightarrow B_1 = -10$; $B_2 = -\frac{20}{3}$

$\therefore\ i_L(t) = \frac{5}{3} e^{-2t} \sin 3t$, for $t > 0$; $v_C(t) = e^{-2t}\left(-10 \cos 3t - \frac{20}{3} \sin 3t\right) + 10$, for $t > 0$. ←

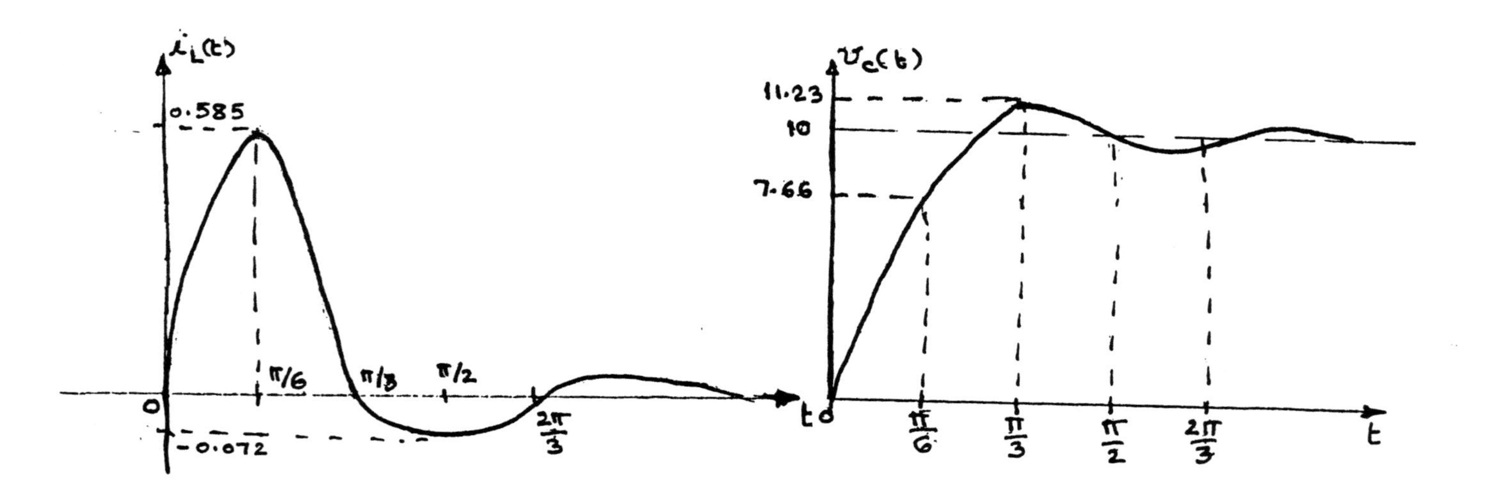

3.2.13. (a) $i_S = 5\,A$; $R = 1\,\Omega$; $L = \frac{3}{4}\,H$; $C = \frac{1}{6}\,F$

$$s^2 + \frac{1}{RC}s + \frac{1}{LC} = 0 \quad \text{or} \quad s^2 + 6s + 8 = 0 \Rightarrow s_1 = -2 \text{ AND } s_2 = -4$$

$$i_{L,ss} = i_s = 5 \; ; \quad V_{C,ss} = 0$$

$$\therefore i_L(t) = A_1 e^{-2t} + A_2 e^{-4t} + 5 \; ; \quad v_C(t) = B_1 e^{-2t} + B_2 e^{-4t}$$

AT $t = 0^-$,

$$i_L(0^-) = 0 = i_L(0^+) \; ; \quad v_C(0^-) = 0 = v_C(0^+)$$

AT $t = 0^+$,

5A, R, L, i_L, i_C, C, $+$ v_C $-$

$$i_C(0^+) = 5 - \frac{v_C(0^+)}{R} - i_L(0^+) = 5$$

$$i_C(0^+) = C\frac{dv_C}{dt}(0^+) = \frac{1}{6}\frac{dv_C}{dt}(0^+) \Rightarrow$$

$$\Rightarrow \frac{dv_C}{dt}(0^+) = 30$$

$$0 = V_L(0^+) = V_C(0^+) = L\frac{di_L}{dt}(0^+) \Rightarrow$$

$$\Rightarrow \frac{di_L}{dt}(0^+) = 0 = -2A_1 - 4A_2$$

THUS

$$A_1 + A_2 + 5 = 0 \text{ AND } -2A_1 - 4A_2 = 0 \Rightarrow A_1 = -10 \; ; \quad A_2 = 5$$

$$B_1 + B_2 = 0 \text{ AND } -2B_1 - 4B_2 = 30 \Rightarrow B_1 = 15 \; ; \quad B_2 = -15$$

THEREFORE

$$i_L(t) = -10e^{-2t} + 5e^{-4t} + 5, \text{ FOR } t > 0 \; ; \quad v_C(t) = 15e^{-2t} - 15e^{-4t}, \text{ FOR } t > 0.$$

$i_L(t)$: 5, 0, 3, t

$v_C(t)$: $\frac{15}{4}$, 0, 0.347, 3, t

3.2.13. (b) $i_S = 5\,A$; $R = 1\,\Omega$; $L = \frac{2}{3}\,H$; $C = \frac{1}{6}\,F$

$$s^2 + \frac{1}{RC}s + \frac{1}{LC} = 0 \quad \text{or} \quad s^2 + 6s + 9 = 0 \quad \text{or} \quad (s+3)^2 = 0$$

$$\Rightarrow s_1 = s_2 = -3$$

$$i_{L,ss} = 5 \; ; \quad V_{C,ss} = 0$$

AT $t = 0^-$, $i_L(0^-) = i_L(0^+) = 0$; $V_C(0^-) = V_C(0^+) = 0$

$$\dot{i}_L(0^+) = \frac{1}{L}v_C(0^+) = 0 \; ; \quad \dot{v}_C(0^+) = \frac{1}{C}i_S = \frac{5}{1/6} = 30$$

THUS $i_L(t) = A_1 e^{-5t} + A_2 t e^{-5t} + 5$; $v_C(t) = B_1 e^{-5t} + B_2 t e^{-5t}$

THEN

$$A_1 + 5 = 0 \; ; \quad -5A_1 + A_2 = 0 \Rightarrow A_1 = -5 \; ; \quad A_2 = -25$$

$$B_1 = 0 \; ; \quad -5B_1 + B_2 = 30 \Rightarrow B_1 = 0 \; ; \quad B_2 = 30$$

$$\therefore i_L(t) = -5e^{-5t} - 25te^{-5t} + 5, \text{ FOR } t > 0 \; ; \quad v_C(t) = 30te^{-5t}, \text{ FOR } t > 0.$$

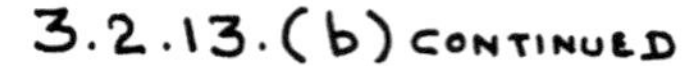

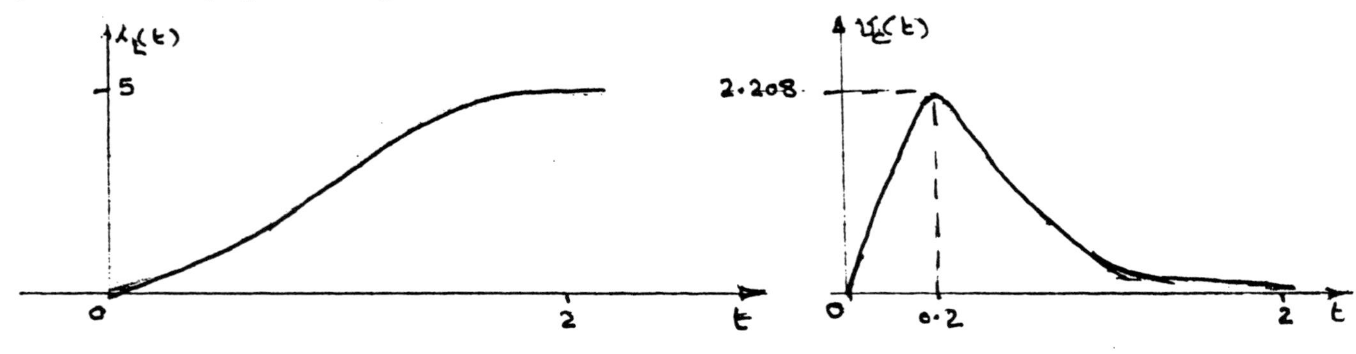

3.2.13.(c)

$i_s = 5A$; $R = 1\Omega$; $L = \frac{3}{17}H$; $C = \frac{1}{6}F$

$s^2 + \frac{1}{RC}s + \frac{1}{LC} = 0$ or $s^2 + 6s + 34 = 0 \Rightarrow s_1 = -3 + j5$; $s_2 = -3 - j5$

$i_{L,ss} = 5$; $V_{C,ss} = 0$

$\therefore i_L(t) = e^{-3t}(A_1 \cos 5t + A_2 \sin 5t) + 5$; $v_C(t) = e^{-3t}(B_1 \cos 5t + B_2 \sin 5t)$

$i_L(0^+) = i_L(0^-) = 0$; $v_C(0^+) = v_C(0^-) = 0$

$0 = v_C(0^+) = v_L(0^+) = L\frac{di}{dt}(0^+)$; $C\frac{dv_C}{dt}(0^+) = i_C(0^+) = i_S$; $\dot{v}_C(0^+) = 30$

Thus $A_1 + 5 = 0$; $-3A_1 + 5A_2 = 0 \Rightarrow A_1 = -5$; $A_2 = -3$

$B_1 = 0$; $-3B_1 + 5B_2 = 30 \Rightarrow B_1 = 0$; $B_2 = 6$

$\therefore i_L(t) = e^{-3t}\left[-5\cos 5t - 3\sin 5t\right] + 5$, for $t > 0$; $v_C(t) = 6\, e^{-3t} \sin 5t$, for $t > 0$.

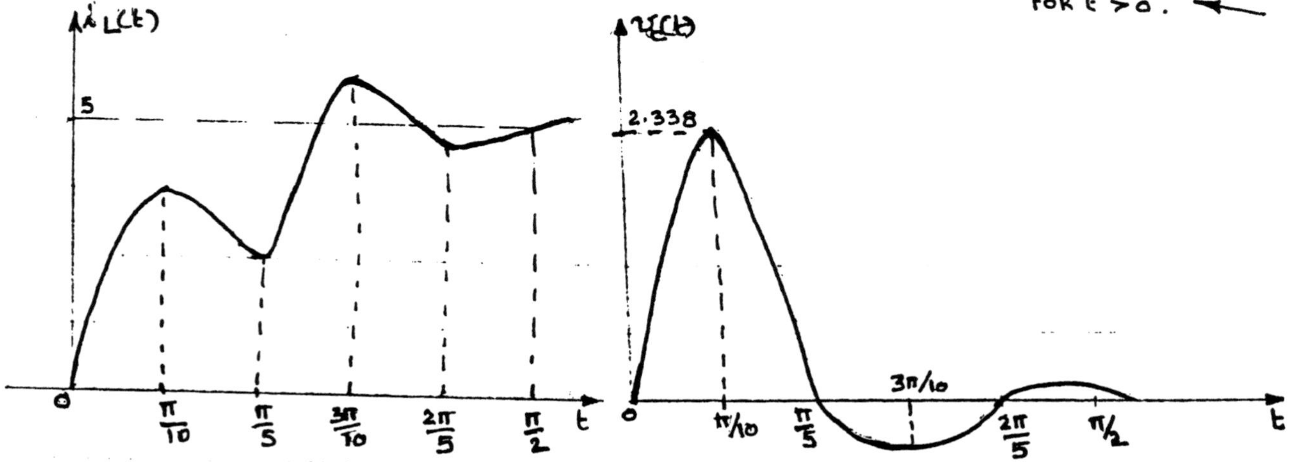

3.2.14.

$$i_L(0^-) = v_c(0^-) = 0$$

FOR $t \geq 0$, KVL EQ. FOR OUTER LOOP: $10^{-3}\frac{di}{dt} + 100i + 10^7\int(i + 99i)\,dt = 10\cos 10^6 t$

ALSO $10^{-3}\frac{di}{dt} + 100i + v_c = 10\cos 10^6 t$; $100i = 10^{-7}\frac{dv_c}{dt} \Rightarrow \frac{di}{dt} = 10^{-9}\frac{d^2 v_c}{dt^2}$

SUBSTITUTION YIELDS: $10^{-12}\frac{d^2 v_c}{dt^2} + 10^{-7}\frac{dv_c}{dt} + v_c = 10\cos 10^6 t$

CHARACTERISTIC EQ: $10^{-12}s^2 + 10^{-7}s + 1 = 0$ or $s^2 + 10^5 s + 10^{12} = 0$

ROOTS ARE $s_1 = (-0.05 + j0.999)10^6$; $s_2 = (-0.05 - j0.999)10^6$

$$v_{c,n}(t) = A_1 e^{-5\times10^4 t}\cos(0.999\times10^6 t + \theta)$$

THE FORCED RESPONSE IS OBTAINED BY USING PHASOR QUANTITIES.

TRANSFORMED NETWORK WITH PEAK VALUES IS SHOWN BELOW:

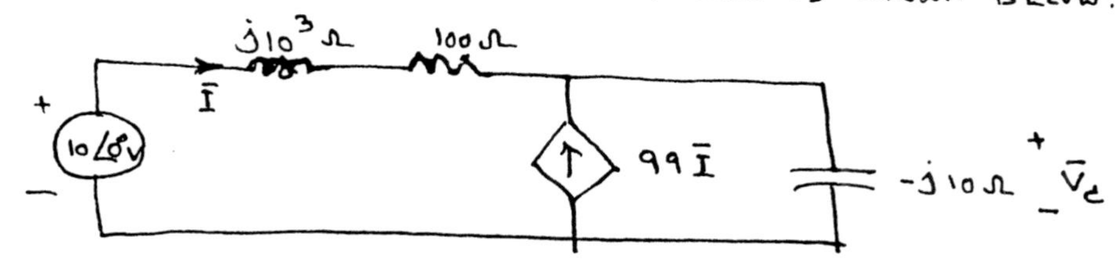

$\bar{I}(100 + j1000) + (1+99)\bar{I}(-j10) = 10 \Rightarrow \bar{I} = 0.1\angle 0°$;

$V_c = 100\bar{I}(-j10) = 100\angle -90°$

$v_{c,f}(t) = 100\cos(10^6 t - 90°) = 100\sin 10^6 t$

THE COMPLETE SOLUTION IS THEN $v_c(t) = A_1 e^{-5\times10^4 t}\cos(0.999\times10^6 t + \theta) + 100\sin 10^6 t$

AT $t = 0^+$,

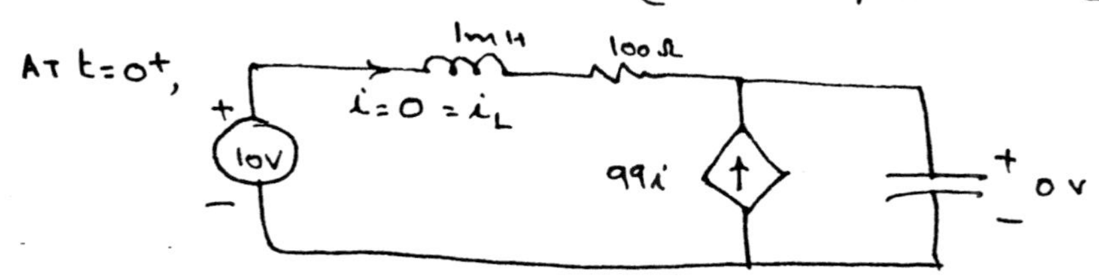

USING CONTINUITY PRINCIPLE, $v_c(0^+) = 0$; $i_L(0^+) = 0$; $i_c(0^+) = 0 \Rightarrow \frac{dv_c}{dt}(0^+) = 0$

THEN
$A_1\cos\theta = 0$; $-5\times10^4 A_1\sin\theta - A_1\cos\theta + 10^6(100) = 0$

FROM WHICH $\theta = \pi/2$ AND $A_1 = 2\times10^3$

$\therefore\ v_c(t) = 2\times10^3 e^{-5\times10^4 t}\cos(0.999\times10^6 t + \frac{\pi}{2}) + 100\sin 10^6 t$, FOR $t \geq 0$

3.2.15.

AT $t = 0^-$,

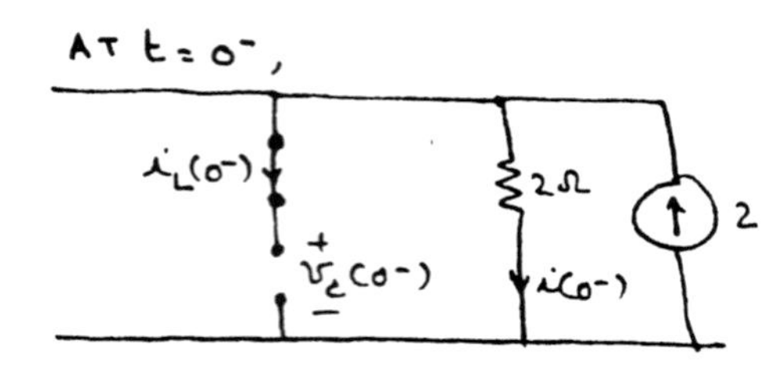

$i_L(0^-) = 0 = i_L(0^+)$

$v_c(0^-) = v_c(0^+) = 2 \times 2 = 4\,V$

$i(0^-) = 2 - i_L(0^-) = 2\,A$

AT $t = 0^+$,

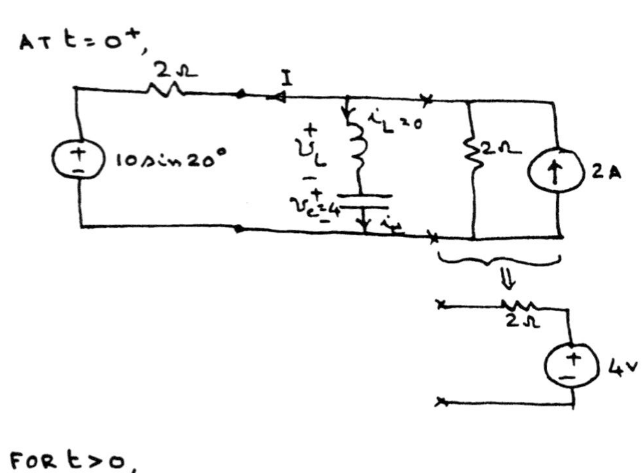

$v_L(0^+) = L\frac{di_L}{dt}(0^+) = -2I + 4 - 4 = -2I$

$2I + 2I + 10 \sin 20° = 4 \quad \text{(KVL)}$

$\therefore I = 1 - \frac{10}{4}\sin 20° = 0.145$

$v_L(0^+) = -2I = -0.29; \quad \frac{di_L}{dt}(0^+) = \frac{-0.29}{L} = -0.87$

$i_c(0^+) = C\frac{dv_c}{dt}(0^+) = i_L(0^+) = 0$

$\therefore \frac{dv_c}{dt}(0^+) = 0$

FOR $t > 0$,

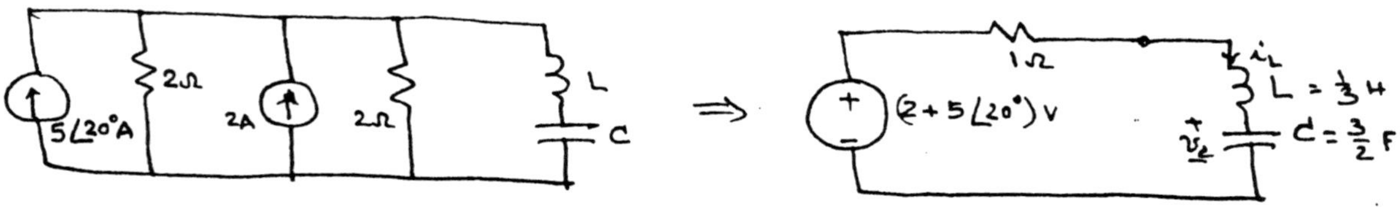

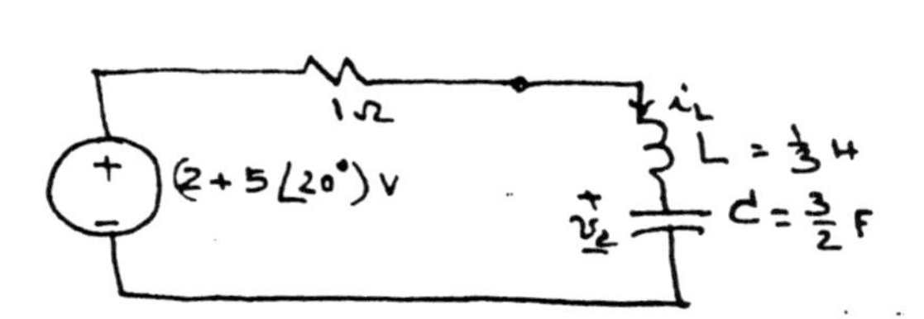

$L = \frac{1}{3}H; \quad C = \frac{3}{2}F; \quad R = 1\,\Omega; \quad \frac{R}{L} = 3; \quad \frac{1}{LC} = 2$

$s^2 + 3s + 2 = 0$ OR $(s+2)(s+1) = 0 \Rightarrow s_1 = -2$ AND $s_2 = -1$

$\therefore i_L(t) = A_1 e^{-t} + A_2 e^{-2t} + i_{L,ss}$

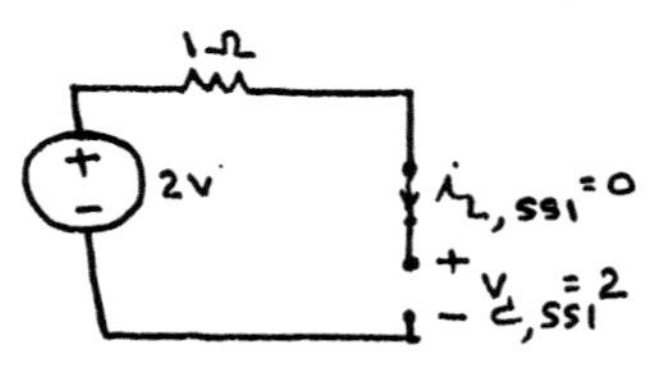

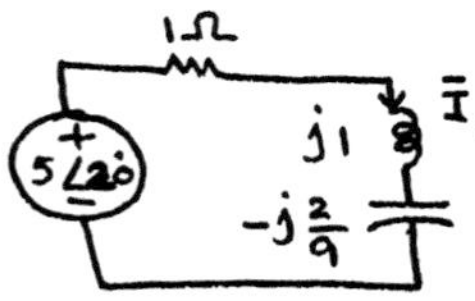

$i_{L,ss1} = 0$, $v_{c,ss1} = 2$

$\bar{I}_{L,ss2} = \frac{5\angle 20°}{1 + (1 - \frac{2}{9}j)} = 3.947\angle -17.87°$

$\bar{V}_{c,ss2} = \frac{-j\,2/9}{1 + j\frac{7}{9}}\, 5\angle 20° = 0.877\angle -107.87°$

SUPERPOSITION: $i_{L,ss} = 3.947 \sin(3t - 17.87°); \quad v_{c,ss} = 2 + 0.877 \sin(3t - 107.87°)$

THUS

$i_L(t) = A_1 e^{-t} + A_2 e^{-2t} + 3.947 \sin(3t - 17.87°); \quad v_c(t) = B_1 e^{-t} + B_2 e^{-2t} + 2 + 0.877 \sin(3t - 107.87°)$

$$\left.\begin{aligned} i_L(0^+) &= 0 = A_1 + A_2 + 3.947 \sin(-17.87°) \\ \frac{di_L}{dt}(0^+) &= -0.87 = -A_1 - 2A_2 + 11.841 \cos(-17.87°) \end{aligned}\right\} \Rightarrow A_1 = -9.656; \; A_2 = 10.867$$

$$\left.\begin{aligned} v_c(0^+) &= 4 = B_1 + B_2 + 2 + 0.877 \sin(-107.87°) \\ \frac{dv_c}{dt}(0^+) &= 0 = -B_1 - 2B_2 + 2.631 \cos(-107.87°) \end{aligned}\right\} \Rightarrow B_1 = 6.477; \; B_2 = -3.642$$

$$\left.\begin{aligned} \therefore i_L(t) &= -9.656 e^{-t} + 10.867 e^{-2t} + 3.947 \sin(3t - 17.87°) \\ v_c(t) &= 6.477 e^{-t} - 3.642 e^{-2t} + 2 + 0.877 \sin(3t - 107.87°) \end{aligned}\right\} \text{FOR } t > 0.$$

3.2.16.

AT $t=0^-$:

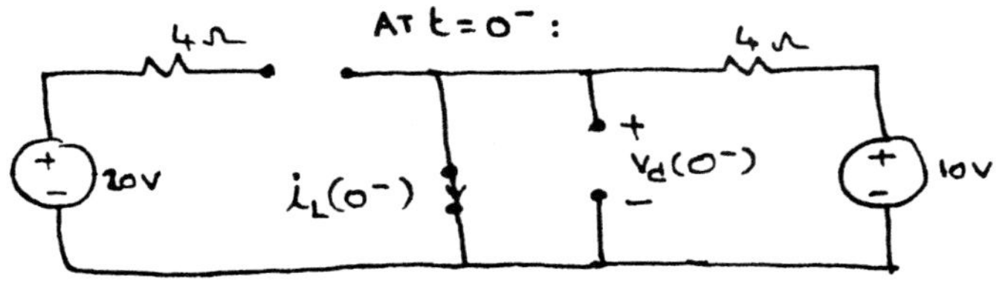

$v_c(0^-) = v_c(0^+) = 0$

$i_L(0^-) = i_L(0^+) = \frac{10}{4}$

FOR $t>0$:

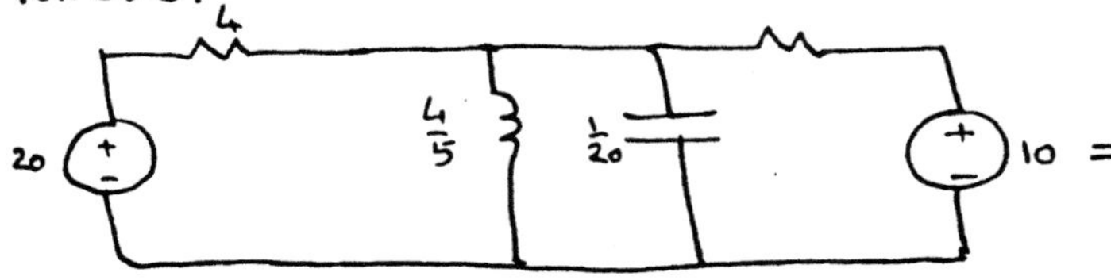

$\Rightarrow$

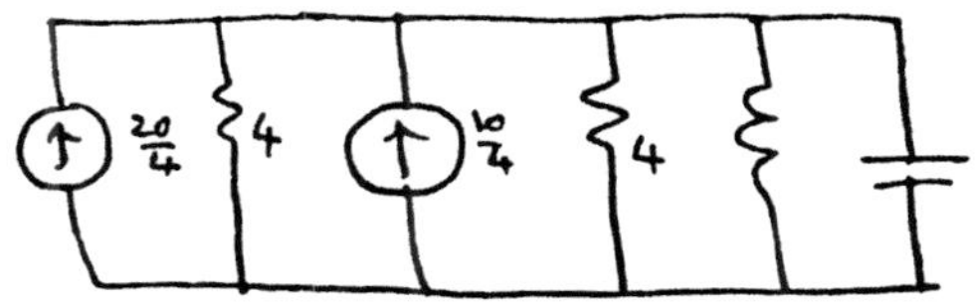

$\Rightarrow$

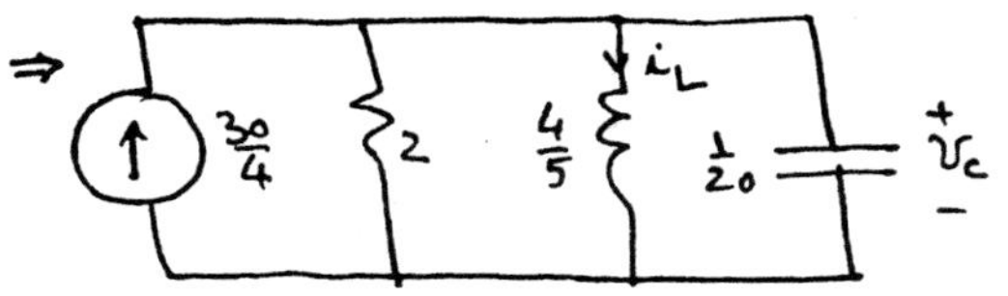

$V_{c,ss} = 0 \; ; \; i_{L,ss} = \frac{30}{4}$

$s^2 + \frac{1}{RC}s + \frac{1}{LC} = 0 \Rightarrow s^2 + 10s + 25 = 0$

$\Rightarrow s_1 = s_2 = -5$

$i_L(t) = A_1 e^{-5t} + A_2 t e^{-5t} + \frac{30}{4} \quad ; \quad v_c(t) = B_1 e^{-5t} + B_2 t e^{-5t}$

$v_L(0^+) = v_c(0^+) = 0 = L \frac{di_L}{dt}(0^+) = L(-5A_1 + A_2) = 0$

$i_c(0^+) = c\frac{dv_c}{dt}(0^+) = c(-5B_1 + B_2) = \frac{30}{4} - 0 - \frac{10}{4} = \frac{20}{4} = 5 \; ; \; \frac{dv_c}{dt}(0^+) = 100$

THUS $\quad A_1 + \frac{30}{4} = \frac{10}{4}$ AND $-5A_1 + A_2 = 0 \Rightarrow A_1 = -5$ & $A_2 = -25$

$B_1 = 0$ AND $-5B_1 + B_2 = 100 \Rightarrow B_1 = 0$ & $B_2 = 100$

$\therefore i_L(t) = -5e^{-5t} - 25te^{-5t} + \frac{15}{2}$, FOR $t>0$; $v_c(t) = 100t\,e^{-5t}$, FOR $t>0$ ←

3.2.17.

$f(t) = 5u(t) + 5u(t-1) + 5u(t-2) + 5u(t-3) + 5u(t-4)$
$\quad - 5u(t-5) - 5u(t-6) - 5u(t-7) - 5u(t-8) - 5u(t-9)$

$f(t) = 5\sum_{n=0}^{4} u(t-n) - 5\sum_{n=5}^{9} u(t-n)$ ←

3.2.18.

WAVEFORM: $v(t) = 10u(t) - 20u(t-1) + 10u(t-2)$

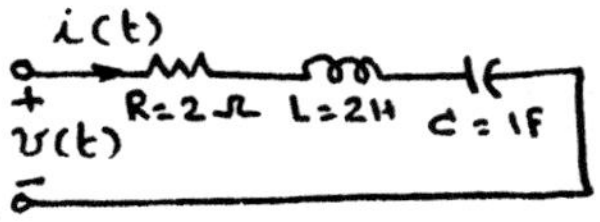

INTEGRODIFFERENTIAL EQUATION: $\quad 2i + 2\frac{di}{dt} + \int i\,dt = v(t)$

FIRST, RESPONSE TO $10u(t)$ IS DETERMINED. LATER USING SUPERPOSITION AND LINEARITY, OTHER RESPONSES CAN BE OBTAINED.

3.2.18. CONTINUED — WITH DC APPLIED, FORCED RESPONSE = 0.

TO GET NATURAL RESPONSE: $2 + 2s + \frac{1}{s} = 0$ or $2s^2 + 2s + 1 = 0 \Rightarrow s_1 = -\frac{1}{2} + j\frac{1}{2}$; $s_2 = -\frac{1}{2} - j\frac{1}{2}$

TOTAL RESPONSE: $i(t) = A\, e^{-t/2} \cos\left(\frac{t}{2} + \theta\right)$

AT $t = 0^-$, $i_L(0^-) = 0 = i_L(0^+)$; $v_c(0^-) = 0 = v_c(0^+)$; $v_L(0^+) = 10$

$L \frac{di}{dt}(0^+) = v_L(0^+)$ or $\frac{di}{dt}(0^+) = \frac{10}{2} = 5$ A/s.

SUBSTITUTION & SIMULTANEOUS SOLUTION YIELD $A = 10$ AND $\theta = -\pi/2$

$\therefore i(t) = \left[10\, e^{-t/2} \sin \frac{t}{2}\right] u(t)$ WHICH IS THE RESPONSE FOR EXCITATION $10\, u(t)$.

FOR $-20\,u(t-1)$ EXCITATION, THE RESPONSE IS
$i(t) = \left[-20\, e^{-(t-1)/2} \sin \frac{t-1}{2}\right] u(t-1)$
NOTE THE CONSTANT MULTIPLICATIVE FACTOR (−2) AND TIME DELAY BY 1 s.

FOR $10\,u(t-2)$ EXCITATION, THE RESPONSE IS
$i(t) = \left[10\, e^{-(t-2)/2} \sin \frac{t-2}{2}\right] u(t-2)$

TOTAL RESPONSE IS THEN GIVEN BY

$$i(t) = \left[10\, e^{-t/2} \sin \frac{t}{2}\right] u(t) + \left[10\, e^{-(t-2)/2} \sin \frac{t-2}{2}\right] u(t-2) - \left[20\, e^{-(t-1)/2} \sin \frac{t-1}{2}\right] u(t-1) \leftarrow$$

3.2.19.

(a)

$i(t)$; 3 Ω ; $\frac{1}{2}$ F ; $+\ v_c(t)\ -$

DE: $\frac{v_c}{3} + \frac{1}{2}\frac{dv_c}{dt} = i(t)$

FOR $i(t) = \delta(t)$, $v_c(t) = A\, e^{-2t/3}$ FOR $t \geq 0$; $v_c(t) = 0$ FOR $t < 0$.

INITIAL VOLTAGE $v_c(0^+)$ IS THE IMPULSE OF CURRENT DIVIDED BY C : $v_c(0^+) = \frac{1}{1/2} = 2$ V

THEN A COMES OUT AS 2, AND $v_c(t) = 2e^{-2t/3}\, u(t) \leftarrow$

(b) FOR $i(t) = \delta(t-3)$, THE RESPONSE IS SAME AS THAT OF (a), BUT DELAYED BY 3 s.

$\therefore v_c(t) = 2\, e^{-2(t-3)/3} \cdot u(t-3) \leftarrow$

3.2.20.

(a)

$v(t)$; $i(t)$; 20 Ω ; 10 mH

NATURAL RESPONSE: $20 + 0.01s = 0$ or $s = -2000$

$i(t) = A\, e^{-2000t}$. THE INITIAL CURRENT IS $\frac{1}{10^{-2}} = 100$

$\therefore i(t) = 100\, e^{-2000t}\, u(t) \leftarrow$

(b) $i(t) = 100\, e^{-2000t}\, u(t) + 100\, e^{-2000(t-3)}\, u(t-3) \leftarrow$

3.3.1. (a) $F_1(s) = \mathcal{L}[u(t)] = \int_0^\infty e^{-st}\,dt = -\frac{1}{s} e^{-st}\Big|_0^\infty = \frac{1}{s}$ ←

(b)
$$F_2(s) = \mathcal{L}\left[e^{-at}\right] = \int_0^\infty e^{-at} e^{-st}\,dt = -\frac{1}{s+a} e^{-(s+a)t}\Bigg|_0^\infty = \frac{1}{(s+a)} \leftarrow$$

(c)
$$F_3(s) = \mathcal{L}\left[\frac{df(t)}{dt}\right] = \int_0^\infty \frac{df(t)}{dt} e^{-st}\,dt$$

BY USING INTEGRATION BY PARTS $\int u\,dv = uv - \int v\,du$, WHERE $u = e^{-st}$ AND $dv = df(t)$, AND $dv = df(t)$

ONE OBTAINS $du = -s e^{-st}\,dt$ AND $v = f(t)$

AND $\int_0^\infty u\,dv = f(t)e^{-st}\Big|_0^\infty - \int_0^\infty f(t)\left[-s e^{-st}\,dt\right] = -f(0+) + s\int_0^\infty f(t)e^{-st}\,dt$

RECOGNIZING $\int_0^\infty f(t)\,e^{-st}\,dt$ IS $\mathcal{L}\left[f(t)\right] = F(s)$, THE DESIRED RESULT IS

$$F_3(s) = \mathcal{L}\left[\frac{df(t)}{dt}\right] = sF(s) - f(0+). \leftarrow$$

(d)
$$F_4(s) = \mathcal{L}[t] = \int_0^\infty t\,e^{-st}\,dt$$

INTEGRATING BY PARTS, ONE GETS

$$F_4(s) = \left[\frac{t\,e^{-st}}{s} - \frac{1}{s^2} e^{-st}\right]_0^\infty = \frac{1}{s^2} \leftarrow$$

(e)
$$F_5(s) = \mathcal{L}\left[\sin\omega t\right] = \int_0^\infty e^{-st}\sin\omega t\,dt$$

RECALLING $\sin\omega t = \dfrac{e^{j\omega t} - e^{-j\omega t}}{2j}$ AND INTEGRATING, ONE OBTAINS

$$F_5(s) = \frac{e^{-st}\left(s\sin\omega t - \omega\cos\omega t\right)}{(s^2+\omega^2)}\Bigg|_0^\infty = \frac{\omega}{(s^2+\omega^2)} \leftarrow$$

(f)
$$\cos(\omega t+\theta) = \cos\omega t\cos\theta - \sin\omega t\sin\theta$$

$$\mathcal{L}\left[\cos\omega t\right] = \frac{s}{(s^2+\omega^2)} \;;\; \mathcal{L}\left[\sin\omega t\right] = \frac{\omega}{s^2+\omega^2}$$

THUS $\mathcal{L}\left[\cos(\omega t+\theta)\right] = F_6(s) = \dfrac{s\cos\theta - \omega\sin\theta}{(s^2+\omega^2)}$ ←

(g)
$$F_7(s) = \mathcal{L}\left[t e^{-at}\right] = \int_0^\infty t e^{-at} e^{-st}\,dt = \int_0^\infty t\,e^{-t(s+a)}\,dt$$

THE RESULT IS THAT OF PART (d) WITH s REPLACED BY $(s+a)$.

$$\therefore F_7(s) = \mathcal{L}\left[t e^{-at}\right] = \frac{1}{(s+a)^2} \leftarrow$$

3.3.1. CONTINUED (h)

$$F_8(s) = \mathcal{L}[\sinh t] = \int_0^\infty \sinh t \; e^{-st} dt$$

RECALLING THAT $\sinh t = \frac{e^t - e^{-t}}{2}$, AND PERFORMING INTEGRATION, ONE GETS

$$F_8(s) = \frac{1}{(s^2-1)} \longleftarrow$$

(i)

$$F_9(s) = \mathcal{L}[\cosh t] = \int_0^\infty \cosh t \; e^{-st} dt$$

RECALLING THAT $\cosh t = \frac{e^t + e^{-t}}{2}$, AND PERFORMING INTEGRATION, ONE OBTAINS

$$F_9(s) = \frac{s}{(s^2-1)} \longleftarrow$$

(j)

$$F_{10}(s) = \mathcal{L}\left[5 f(t) + 2 \frac{d f(t)}{dt}\right] = \int_0^\infty \left[5 f(t) + \frac{2 d f(t)}{dt}\right] e^{-st} dt$$

$$= 5 \int_0^\infty f(t) e^{-st} dt + 2 \int_0^\infty \frac{d f(t)}{dt} e^{-st} dt$$

FROM THE RESULT OF PART (c) AND EQ. (3.3.1), ONE CAN WRITE

$$\mathcal{L}\left[5 f(t) + 2 \frac{d f(t)}{dt}\right] = 5 F(s) + 2\left[s F(s) - f(0^+)\right] \longleftarrow$$

3.3.2.

(a)

$$\mathcal{L}[t e^{-t}] = \mathcal{L}[t f(t)] \quad \text{WHERE } f(t) = e^{-t}$$

FROM THE PROPERTY OF FREQUENCY DIFFERENTIATION (MULTIPLICATION BY t),

$$\mathcal{L}[t f(t)] = -\frac{dF(s)}{ds} \quad \text{AND} \quad F(s) = \mathcal{L}[e^{-t}] = \frac{1}{s+1}$$

$$\therefore \mathcal{L}[t e^{-t}] = -\frac{dF(s)}{ds} = \frac{1}{(s+1)^2} \longleftarrow$$

ALTERNATIVELY THE FREQUENCY-SHIFTING (EXPONENTIAL TRANSLATION) PROPERTY CAN BE USED:

$$\mathcal{L}[t e^{-t}] = \mathcal{L}[e^{-t} f(t)] \quad \text{WHERE } f(t) = t$$

$$\mathcal{L}[t] = \frac{1}{s^2} = F(s)$$

$$\therefore \mathcal{L}[t e^{-t}] = F(s+1) = \frac{1}{(s+1)^2} \longleftarrow$$

3.3.2. CONTINUED

(b) LET $f(t) = t e^{-t}$

USING THE FREQUENCY DIFFERENTIATION (MULTIPLICATION BY t),

$$\mathcal{L}[t f(t)] = \mathcal{L}[t^2 e^{-t}] = -\frac{dF}{ds} = -\frac{d}{ds}\left[\frac{1}{(s+1)^2}\right] = \frac{2}{(s+1)^3}$$

(c) LET $f(t) = e^{-2t}\sin 2t$; $F(s) = \dfrac{2}{(s+2)^2+(2)^2} = \dfrac{2}{s^2+4s+8}$

USING TABLE 3.3.1)

$$\mathcal{L}\{t f(t)\} = \mathcal{L}[t e^{-2t}\sin 2t] = -\frac{dF(s)}{ds} = \frac{4(s+2)}{(s^2+4s+8)^2}$$

WHICH CAN ALSO BE VERIFIED FROM TABLE 3.3.1

(d) LET $f(t) = 1-\cos t$; $F(s) = \dfrac{1}{s} - \dfrac{s}{s^2+1}$

USING THE FREQUENCY INTEGRATION PROPERTY (DIVISION BY t),

$$\mathcal{L}\left[\frac{f(t)}{t}\right] = \mathcal{L}\left[\frac{1-\cos t}{t}\right] = \int_s^\infty \left(\frac{1}{s} - \frac{s}{s^2+1}\right) ds = \int_s^\infty \frac{ds}{s(s^2+1)}$$

$$= \frac{1}{2}\ln \frac{s^2}{s^2+1}\Bigg|_s^\infty = -\frac{1}{2}\ln\frac{s^2}{s^2+1}$$

(e) LET $f(t) = e^{-2t}\sin 2t$; $F(s) = \dfrac{2}{s^2+4s+8}$ [FROM TABLE 3.3.1]

USING THE FREQUENCY INTEGRATION PROPERTY (DIVISION BY t),

$$\mathcal{L}\left[\frac{f(t)}{t}\right] = \mathcal{L}\left[\frac{e^{-2t}\sin 2t}{t}\right] = \int_s^\infty \frac{2\,ds}{(s+2)^2+(2)^2}$$

$$= \left[\frac{2}{2}\tan^{-1}\frac{s+2}{2}\right]_s^\infty = \frac{\pi}{2} - \tan^{-1}\frac{s+2}{2}$$

3.3.3.

$$i(0^+) = \lim_{s\to\infty} s I(s) = \lim_{s\to\infty}\frac{10s}{2s+5} = 5\text{ A}$$

$$i(\infty) = \lim_{s\to 0} s I(s) = \lim_{s\to 0}\frac{10s}{2s+5} = 0$$

(WHICH IS THE STEADY-STATE VALUE)

3.3.4.

(a)

$$F(s) = \frac{6(s+3)}{s^2+2s+10} = \frac{6(s+1)+12}{(s+1)^2+(3)^2} = \frac{6(s+1)}{(s+1)^2+(3)^2} + \frac{12}{(s+1)^2+(3)^2}$$
$$= F_1(s) + F_2(s)$$

$$\mathcal{L}^{-1}[F(s)] = \mathcal{L}^{-1}[F_1(s)+F_2(s)] = 6e^{-t}\cos 3t + 4e^{-t}\sin 3t$$
$$= 2e^{-t}[3\cos 3t + 2\sin 3t]$$

RECALLING THAT $A\cos x + B\sin x = \sqrt{A^2+B^2}\cos\left(x - \tan^{-1}\frac{B}{A}\right)$
$= \sqrt{A^2+B^2}\sin\left(x + \tan^{-1}\frac{A}{B}\right)$

ONE GETS

$$f(t) = 2e^{-t}\sqrt{13}\cos\left(3t - \tan^{-1}\frac{2}{3}\right) = 2e^{-t}\sqrt{13}\sin\left(3t + \tan^{-1}\frac{3}{2}\right)$$

or $f(t) = 2\sqrt{13}\,e^{-t}\cos(3t - 33.7°) = 2\sqrt{13}\,e^{-t}\sin(3t + 56.3°)$ ←

(b)

$$F(s) = \frac{s+2}{(s+1)(s+3)(s+4)} = \frac{k_1}{s+1} + \frac{k_2}{s+3} + \frac{k_3}{s+4}$$

WHERE $k_1 = (s+1)F(s)\Big|_{s=-1} = \frac{(s+2)}{(s+3)(s+4)}\Big|_{s=-1} = \frac{-1+2}{(-1+3)(-1+4)} = \frac{1}{6}$

$k_2 = (s+3)F(s)\Big|_{s=-3} = \frac{(s+2)}{(s+1)(s+4)}\Big|_{s=-3} = \frac{(-3+2)}{(-3+1)(-3+4)} = \frac{-1}{-2} = \frac{1}{2}$

$k_3 = (s+4)F(s)\Big|_{s=-4} = \frac{s+2}{(s+1)(s+3)}\Big|_{s=-4} = \frac{(-4+2)}{(-4+1)(-4+3)} = \frac{-2}{3}$

$$\therefore f(t) = \frac{1}{6}e^{-t} + \frac{1}{2}e^{-3t} - \frac{2}{3}e^{-4t}$$ ←

3.3.4. (c)

$$F(s) = \frac{8(s+1)}{s(s^2+2s+2)} = \frac{k_1}{s} + \frac{k_2}{s+1-j1} + \frac{k_3}{s+1+j1}$$

(NOTE THAT k_3 WILL BE EQUAL TO k_2^*)

WHERE

$$k_1 = sF(s)\Big|_{s=0} = \frac{8(s+1)}{(s^2+2s+2)}\Bigg|_{s=0} = \frac{8}{2} = 4$$

$$k_2 = (s+1-j1)\,F(s)\Big|_{s=-1+j1} = \frac{8(s+1)}{s(s+1+j1)}\Bigg|_{s=-1+j1} = \frac{8j}{(-1+j1)(j2)} = \frac{4}{-1+j1} = 2\sqrt{2}\,e^{-j135°}$$

$$k_3 = k_2^* = 2\sqrt{2}\,e^{j135°}$$

$$\therefore\ f(t) = 4 + 2\sqrt{2}\,e^{-j135°}\,e^{(-1+j)t} + 2\sqrt{2}\,e^{j135°}\,e^{(-1-j)t}$$

$$= 4 + 2\sqrt{2}\,e^{-t}\left[e^{j(t-135°)} + e^{-j(t-135°)}\right]$$

$$= 4 + 4\sqrt{2}\,e^{-t}\cos(t-135°) \quad \text{BY USE OF EULER'S IDENTITY.}$$

OR

$$f(t) = 4\left[1 + \sqrt{2}\,e^{-t}\cos(t-135°)\right]$$

(d)

$$F(s) = \frac{s^2+4s+5}{s^2+3s+2} = 1 + \frac{s+3}{s^2+3s+2} = 1 + F_1(s)$$

WHERE

$$F_1(s) = \frac{s+3}{s^2+3s+2} = \frac{k_1}{s+1} + \frac{k_2}{s+2}$$

WHERE

$$k_1 = \frac{s+3}{s+2}\Bigg|_{s=-1} = \frac{-1+3}{-1+2} = \frac{2}{1} = 2$$

$$k_2 = \frac{s+3}{s+1}\Bigg|_{s=-2} = \frac{-2+3}{-2+1} = \frac{1}{-1} = -1$$

THUS

$$F(s) = 1 + \frac{2}{s+1} - \frac{1}{s+2}$$

$$\therefore\ f(t) = \delta(t) + 2e^{-t} - e^{-2t}$$

WHERE $\delta(t)$ IS THE UNIT IMPULSE FUNCTION.

3.3.4.(e)

$$F(s) = \frac{2(s+2)}{s(s+1)^3} = \frac{k_0}{s} + \frac{k_1}{(s+1)^3} + \frac{k_2}{(s+1)^2} + \frac{k_3}{(s+1)}$$

WHERE $k_0 = sF(s)\Big|_{s=0} = \frac{2(s+2)}{(s+1)^3}\Big|_{s=0} = \frac{2\times 2}{1} = 4$

$$k_1 = (s+1)^3 F(s)\Big|_{s=-1} = \frac{2(s+2)}{s}\Big|_{s=-1} = \frac{2(-1+2)}{-1} = -2$$

$$k_2 = \frac{d}{ds}\left[(s+1)^3 F(s)\right]\Big|_{s=-1} = \frac{d}{ds}\left[\frac{2(s+2)}{s}\right]\Big|_{s=-1} = \frac{-4}{s^2}\Big|_{s=-1} = -4$$

$$k_3 = \frac{1}{2}\frac{d^2}{ds^2}\left[(s+1)^3 F(s)\right]\Big|_{s=-1} = \frac{1}{2}\frac{d^2}{ds^2}\left[\frac{2(s+2)}{s}\right]\Big|_{s=-1} = \frac{1}{2}\left[\frac{8}{s^3}\right]\Big|_{s=-1} = -4$$

THUS

$$F(s) = \frac{4}{s} - \frac{2}{(s+1)^3} - \frac{4}{(s+1)^2} - \frac{4}{(s+1)}$$

$$\therefore\ f(t) = 4 - t^2 e^{-t} - 4te^{-t} - 4e^{-t} = 4 - e^{-t}(t^2 + 4t + 4) \leftarrow$$

(f)

$$F(s) = \frac{s-1}{s^3 + 4s^2 + 3s} = \frac{s-1}{s(s^2+4s+3)} = \frac{s-1}{s(s+1)(s+3)}$$

$$= \frac{k_1}{s} + \frac{k_2}{s+1} + \frac{k_3}{s+3}$$

WHERE

$$k_1 = \frac{s-1}{(s+1)(s+3)}\Big|_{s=0} = \frac{-1}{(1)(3)} = -\frac{1}{3}$$

$$k_2 = \frac{s-1}{s(s+3)}\Big|_{s=-1} = \frac{(-1-1)}{(-1)(-1+3)} = \frac{-2}{-2} = 1$$

$$k_3 = \frac{s-1}{s(s+1)}\Big|_{s=-3} = \frac{(-3-1)}{(-3)(-3+1)} = \frac{-4}{6} = -\frac{2}{3}$$

$$\therefore\ f(t) = -\frac{1}{3} + e^{-t} - \frac{2}{3}e^{-3t} \leftarrow$$

3.3.4. (g) $F(s) = \dfrac{10(s^2+3s+2)}{3(s^2+2s+2)} = \dfrac{10}{3}\left[1 + \dfrac{s}{(s^2+2s+2)}\right]$

$$= \frac{10}{3} + \frac{10}{3} \frac{s}{(s+1-j)(s+1+j)}$$

$$= \frac{10}{3} + \frac{10}{3} F_1(s)$$

WHERE $F_1(s) = \dfrac{s}{(s+1-j)(s+1+j)} = \dfrac{k_1}{(s+1-j)} + \dfrac{k_1^*}{(s+1+j)}$

WHERE $k_1 = \left.\dfrac{s}{s+1+j}\right|_{s=-1+j} = \dfrac{-1+j}{-1+j+1+j} = \dfrac{-1+j}{2j} = \dfrac{1}{2} + j\dfrac{1}{2}$

$$= \frac{\sqrt{2}}{2} e^{j45^\circ}$$

$$k_1^* = \frac{\sqrt{2}}{2} e^{-j45^\circ}$$

THUS $F(s) = \dfrac{10}{3} + \dfrac{10}{3} \dfrac{\sqrt{2}}{2} e^{j45^\circ} \dfrac{1}{(s+1-j)} +$

$$+ \frac{10}{3} \frac{\sqrt{2}}{2} e^{-j45^\circ} \frac{1}{(s+1+j)}$$

$$\therefore f(t) = \frac{10}{3}\delta(t) + \frac{5\sqrt{2}}{3} e^{j45^\circ} e^{(-1+j)t} + \frac{5\sqrt{2}}{3} e^{-j45^\circ} e^{(-1-j)t}$$

$$= \frac{10}{3}\delta(t) + \frac{5\sqrt{2}}{3} e^{-t}\left[e^{j(t+45^\circ)} + e^{-j(t+45^\circ)}\right]$$

$$= \frac{10}{3}\delta(t) + \frac{5\sqrt{2}}{3} e^{-t}\left[2\cos(t+45^\circ)\right]$$

$$= \frac{10}{3}\left[\delta(t) + \sqrt{2}\, e^{-t} \cos(t+45^\circ)\right] \longleftarrow$$

(h)

$F(s) = \dfrac{(s+1)(s+2)}{s^2(s^2+2s+2)} = \dfrac{k_1}{s^2} + \dfrac{k_2}{s} + \dfrac{k_3}{s+1-j} + \dfrac{k_3^*}{s+1+j}$

WHERE $k_1 = \left.\dfrac{(s+1)(s+2)}{s^2+2s+2}\right|_{s=0} = \dfrac{1\times 2}{2} = 1$

$$k_2 = \frac{d}{ds}\left[\frac{(s+1)(s+2)}{s^2+2s+2}\right]\Bigg|_{s=0} = \frac{1}{2}$$

$$k_3 = \left.\frac{(s+1)(s+2)}{s^2(s+1+j)}\right|_{s=-1+j} = \frac{-1+j}{4} \; ; \; k_3^* = \frac{-1-j}{4}$$

3.3.4.(h) CONTINUED

$$\therefore f(t) = t + \frac{1}{2}u(t) + \frac{-1+j}{4} e^{(-1+j)t} + \frac{-1-j}{4} e^{(-1-j)t} \longleftarrow$$

OR NOTING THAT $-1+j = \sqrt{2}\, e^{j135°}$ AND $-1-j = \sqrt{2}\, e^{-j135°}$

$$f(t) = t + \frac{1}{2}u(t) + \frac{\sqrt{2}e^{-t}}{4}\left[e^{j(t+135°)} + e^{-j(t+135°)} \right]$$

$$= t + \frac{1}{2}u(t) + \frac{\sqrt{2}}{4} e^{-t} \left[2\cos(t+135°) \right]$$

$$= t + \frac{1}{2}u(t) + \frac{\sqrt{2}}{2} e^{-t} \cos(t+135°) \longleftarrow$$

3.3.4.(i)

$$F(s) = \frac{4s}{(s^2+4)(s+2)} = \frac{4s}{(s+2j)(s-2j)(s+2)} = \frac{K_1}{s+2j} + \frac{K_1^*}{s-2j} + \frac{K_2}{s+2}$$

WHERE $K_1 = \left.\frac{4s}{(s-2j)(s+2)}\right|_{s=-2j} = \frac{-8j}{(-4j)(-2j+2)} = \frac{2}{-2j+2} = \frac{1}{1-j} = \frac{\sqrt{2}}{2} e^{j\pi/4}$

$$K_1^* = \frac{\sqrt{2}}{2} e^{-j\pi/4}$$

$$K_2 = \left.\frac{4s}{s^2+4}\right|_{s=-2} = \frac{-8}{4+4} = -1$$

THUS $F(s) = \frac{\sqrt{2}}{2} e^{j\pi/4} \frac{1}{s+2j} + \frac{\sqrt{2}}{2} e^{-j\pi/4} \frac{1}{s-2j} - \frac{1}{s+2}$

$$= \frac{s+2}{s^2+4} - \frac{1}{s+2}$$

OR $F(s) = \frac{s}{s^2+4} + \frac{2}{s^2+4} - \frac{1}{s+2}$

$$\therefore f(t) = \cos 2t + \sin 2t - e^{-2t} \longleftarrow$$

3.3.4.(j)

$$F(s) = \frac{4s}{[(s+2)^2+4]^2} = \frac{4s}{\{(s+2)^2+(2)^2\}^2} = \frac{K_1}{s+2-j2} + \frac{K_1^*}{s+2+j2} + \frac{K_2}{s+2-j2} + \frac{K_2^*}{s+2+j2}$$

WHERE $K_1 = \left.(s+2-j2)^2 F(s)\right|_{s=-2+j2} = \frac{\sqrt{2}}{2} e^{-j45°}$; $K_1^* = \frac{\sqrt{2}}{2} e^{j45°}$

$K_2 = \left.\frac{d}{ds}\left[\frac{4s}{(s+2+j2)^2}\right]\right|_{s=-2+j2} = -4j$; $K_2^* = 4j$

$$\therefore f(t) = 12\sqrt{2}\, t\, e^{-2t} \cos(2t-45°) + 8e^{-2t}\cos(2t-90°) \longleftarrow$$

3.3.5.

(a)

TAKING THE LAPLACE TRANSFORM OF THE DE

$$s^2 V(s) - s\,v(0^+) - \frac{dv}{dt}(0^+) + 5sV(s) - 5v(0^+) + 4V(s) = \frac{10}{s}$$

OR WITH THE GIVEN CONDITIONS, $s^2 V(s) + 5sV(s) + 4V(s) = \frac{10}{s}$

$$\text{OR} \quad V(s) = \frac{10}{s(s^2+5s+4)} = \frac{10}{s(s+1)(s+4)} = \frac{K_1}{s} + \frac{K_2}{s+1} + \frac{K_3}{s+4}$$

WHERE

$$K_1 = \left.\frac{10}{(s+1)(s+4)}\right|_{s=0} = \frac{10}{1\times 4} = \frac{5}{2}$$

$$K_2 = \left.\frac{10}{s(s+4)}\right|_{s=-1} = \frac{10}{(-1)(-1+4)} = \frac{10}{-3} = -\frac{10}{3}$$

$$K_3 = \left.\frac{10}{s(s+1)}\right|_{s=-4} = \frac{10}{(-4)(-4+1)} = \frac{10}{12} = \frac{5}{6}$$

$$\therefore\ v(t) = \underbrace{\frac{5}{2}}_{\text{FORCED RESPONSE}} \underbrace{- \frac{10}{3}e^{-t} + \frac{5}{6}e^{-3t}}_{\text{NATURAL RESPONSE}} \quad \leftarrow$$

(b)

APPLYING THE LAPLACE TRANSFORM TO THE DE

$$3s^2 I(s) - 3s\,i(0^+) - 3\frac{di}{dt}(0^+) + 7sI(s) - 7i(0^+) + 2I(s) = 10s/(s^2+4)$$

WITH $i(0^+) = 4$ AND $\frac{di}{dt}(0^+) = -4$

$$3s^2 I(s) - 12s + 12 + 7s\,I(s) - 28 + 2I(s) = \frac{10s}{s^2+4}$$

$$I(s)\left[3s^2 + 7s + 2\right] = \frac{10s}{s^2+4} + 12s + 16$$

$$I(s) = \frac{12s^3 + 16s^2 + 58s + 64}{(s^2+4)(3s^2+7s+2)} = \frac{12s^3 + 16s^2 + 58s + 64}{(s-j2)(s+j2)(3s+1)(s+2)}$$

$$= \frac{K_1}{s-j2} + \frac{K_1^*}{s+j2} + \frac{K_2}{s+\frac{1}{3}} + \frac{K_3}{s+2}$$

3.3.5.(b) CONTINUED

WHERE $K_1 = \left.\dfrac{12s^3+16s^2+58s+64}{(s+j2)(3s+1)(s+2)}\right|_{s=2j} = \dfrac{-96j-64+116j+64}{(4j)(6j+1)(2j+2)}$

$= 20j/[8j(6j+1)(j+1)]$

$= \dfrac{5}{2}\;\dfrac{1}{7j-5}$

$= -\dfrac{5\sqrt{74}}{2\times 74}\angle 54.5°$

$K_1^* = \dfrac{-5\sqrt{74}}{2\times 74}\angle -54.5°$

$K_2 = \left.\dfrac{12s^3+16s^2+58s+64}{3(s^2+4)(s+2)}\right|_{s=-\frac{1}{3}} = \dfrac{-\frac{12}{27}+\frac{16}{9}-\frac{58}{3}+64}{3\left(\frac{1}{9}+4\right)\left(-\frac{1}{3}+2\right)} = \dfrac{1306}{27}\times\dfrac{27}{555}$

$= \dfrac{1306}{555} = 2.35$

$K_3 = \left.\dfrac{12s^3+16s^2+58s+64}{(s^2+4)(3s+1)}\right|_{s=-2} = \dfrac{-96+64-116+64}{(8)(-6+1)} = \dfrac{-84}{-40} = \dfrac{21}{10}$

$= 2.1$

$\therefore\; i(t) = -\dfrac{5\sqrt{74}}{148}e^{j54.5°}e^{j2t} - \dfrac{5\sqrt{74}}{148}e^{-j54.5°}e^{-j2t} + 2.35e^{-\frac{1}{3}t} + 2.1e^{-2t}$

$= -\dfrac{5\sqrt{74}}{148}\left[e^{j(2t+54.5°)} - e^{-j(2t+54.5°)}\right] + 2.35e^{-\frac{t}{3}} + 2.1e^{-2t}$

$= -\dfrac{10\sqrt{74}}{148}\cos(2t+54.5°) + 2.35e^{-t/3} + 2.1e^{-2t}$

$= \underbrace{-0.58\cos(2t+54.5°)}_{\text{FORCED RESPONSE}} + \underbrace{2.35e^{-t/3}+2.1e^{-2t}}_{\text{NATURAL RESPONSE}}$

3.3.5.(c)

APPLYING THE LAPLACE TRANSFORM TO THE DE

$s^2 I(s) - s\,i(0^+) - \dfrac{di}{dt}(0^+) + 2sI(s) - 2i(0^+) + 2I(s) = \dfrac{1}{s^2+1} - \dfrac{1}{s+2}$

WITH $i(0^+) = 0$ AND $\dfrac{di}{dt}(0^+) = 4$

$s^2 I(s) - 4 + 2sI(s) + 2I(s) = \dfrac{1}{s^2+1} - \dfrac{1}{s+2}$

$I(s)\left[s^2+2s+2\right] = \dfrac{1}{s^2+1} - \dfrac{1}{s+2} + 4 = \dfrac{4s^3+7s^2+5s+9}{(s^2+1)(s+2)}$

$I(s) = \dfrac{4s^3+7s^2+5s+9}{(s^2+1)(s+2)(s^2+2s+2)}$

$= (4s^3+7s^2+5s+9)/[(s+2)(s-j)(s+j)(s+1-j)(s+1+j)]$

$= \dfrac{K_1}{s+2} + \dfrac{K_2}{s-j} + \dfrac{K_2^*}{s+j} + \dfrac{K_3}{s+1-j} + \dfrac{K_3^*}{s+1+j}$

3.3.5.(c) CONTINUED

$$\text{WHERE } K_1 = \left.\frac{4s^3+7s^2+5s+9}{(s^2+1)(s^2+2s+2)}\right|_{s=-2} = \frac{-32+28-10+9}{(+4+1)(4-4+2)} = \frac{-5}{10} = -\frac{1}{2}$$

$$K_2 = \left.\frac{4s^3+7s^2+5s+9}{(s+2)(s+j)(s^2+2s+2)}\right|_{s=j} = \frac{-4j-7+5j+9}{(2+j)\,2j\,(-1+2j+2)} = \frac{2+j}{-10}$$

$$= -\frac{1}{10}(2+j) = -\frac{\sqrt{5}}{10}\angle 26.6^\circ$$

$$K_2^* = -\frac{\sqrt{5}}{10}\angle -26.6^\circ$$

$$K_3 = \left.\frac{4s^3+7s^2+5s+9}{(s+2)(s^2+1)(s+1+j)}\right|_{s=-1+j} = \frac{8(1+j)+7(-2j)-5+5j+9}{(1+j)(1-2j)(2j)} = \frac{(12-j)}{(2+6j)}$$

$$= \frac{1}{20}(9-37j) = \frac{\sqrt{1450}}{20}\angle -76.3^\circ$$

$$= 1.9\angle -76.3^\circ$$

$$K_3^* = 1.9\angle +76.3^\circ$$

$$\therefore\ i(t) = -\frac{1}{2}e^{-2t} - \frac{\sqrt{5}}{5}\cos(t+26.6^\circ) + 3.8\,e^{-t}\cos(t-76.3^\circ)$$

$$= \underbrace{-0.5\,e^{-2t} - 0.45\cos(t+26.6^\circ)}_{\text{FORCED RESPONSE}} + \underbrace{3.8\,e^{-t}\cos(t-76.3^\circ)}_{\text{NATURAL RESPONSE}}$$

3.3.6.

$$f(t) = 10\,\delta(t) + 10u(t) - 20u(t-1) + 20u(t-2)$$

$$\therefore F(s) = 10 + \frac{10}{s} - \frac{20}{s}e^{-s} + \frac{20}{s}e^{-2s}$$

$$= 10\left(1+\frac{1}{s}\right) - \frac{10}{s}e^{-s}\left(2 - e^{-s}\right)$$

3.3.7. WITH THE SWITCH OPEN, THE CIRCUIT IS IN THE STEADY STATE AT $t=0^-$.

$$i_1(0^-) = \frac{100}{2+8} = 10 \text{ A}$$

THE TRANSFORMED NETWORK FOR THE CIRCUIT IS SHOWN BELOW:

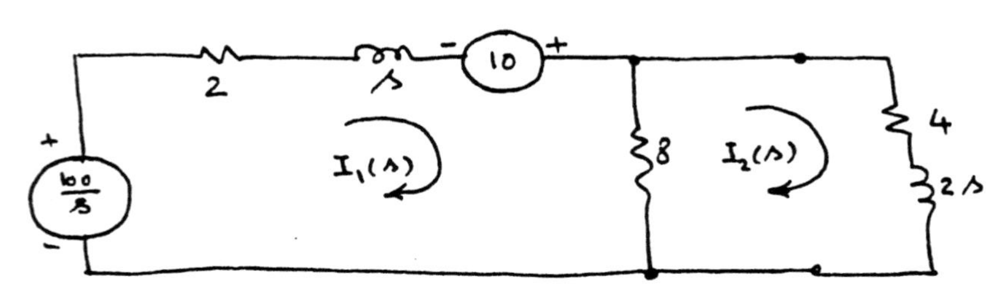

KVL EQUATIONS ARE GIVEN BY $I_1(s)[10+s] - 8I_2(s) = \frac{100}{s} + 10$;

$-8I_1(s) + I_2(s)[12+2s] = 0$

SIMULTANEOUS SOLUTION GIVES $I_1(s) = \dfrac{10(s+10)(s+6)}{s(s+2)(s+14)}$;

$I_2(s) = [40(s+10)]/[s(s+2)(s+14)]$

NOW $$I_1(s) = \frac{10(s+10)(s+6)}{s(s+2)(s+14)} = \frac{k_1}{s} + \frac{k_2}{s+2} + \frac{k_3}{s+14}$$

WHERE $$k_1 = \left.\frac{10(s+10)(s+6)}{(s+2)(s+14)}\right|_{s=0} = \frac{10\times10\times6}{2\times14} = \frac{150}{7}$$

$$k_2 = \left.\frac{10(s+10)(s+6)}{s(s+14)}\right|_{s=-2} = \frac{10\times8\times4}{(-2)(12)} = -\frac{40}{3}$$

$$k_3 = \left.\frac{10(s+10)(s+6)}{s(s+2)}\right|_{s=-14} = \frac{10(-4)(-8)}{(-14)(-12)} = \frac{40}{21}$$

$$\therefore\ i_1(t) = \frac{150}{7} - \frac{40}{3}e^{-2t} + \frac{40}{21}e^{-14t}, \text{ FOR } t \geq 0 \leftarrow$$

$$I_2(s) = \frac{40(s+10)}{s(s+2)(s+14)} = \frac{k_1}{s} + \frac{k_2}{s+2} + \frac{k_3}{s+14}$$

WHERE $$k_1 = \left.\frac{40(s+10)}{(s+2)(s+14)}\right|_{s=0} = \frac{40(10)}{2\times14} = \frac{100}{7}$$

$$k_2 = \left.\frac{40(s+10)}{s(s+14)}\right|_{s=-2} = \frac{40(8)}{(-2)(12)} = -\frac{40}{3}$$

$$k_3 = \left.\frac{40(s+10)}{s(s+2)}\right|_{s=-14} = \frac{40(-4)}{(-14)(-12)} = -\frac{20}{21}$$

$$\therefore\ i_2(t) = \frac{100}{7} - \frac{40}{3}e^{-2t} - \frac{20}{21}e^{-14t}, \text{ FOR } t \geq 0 \leftarrow$$

3.3.8. $i(t) = 10\,t\,e^{-t}\,u(t)$; $I(s) = \dfrac{10}{(s+1)^2}$

THE TRANSFORMED NETWORK IS SHOWN BELOW:

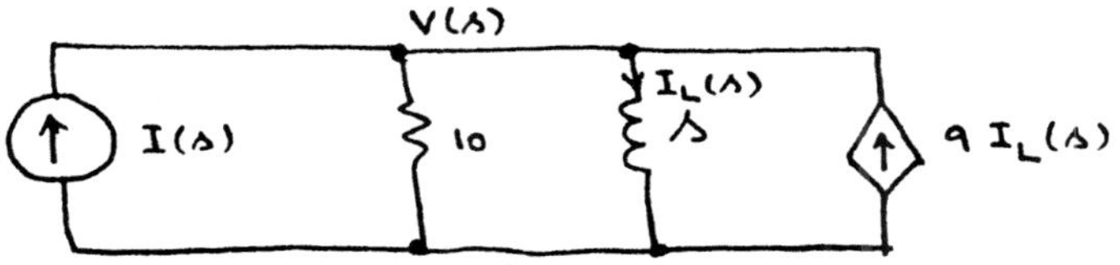

V(s)

I(s) 10 $I_L(s)$ s $9\,I_L(s)$

REF. NODE

NODAL EQUATION: $V(s)\left[\frac{1}{10} + \frac{1}{s}\right] = I(s) + 9\,I_L(s) = \dfrac{10}{(s+1)^2} + 9\,I_L(s)$

$$I_L(s) = \frac{V(s)}{s}$$

$$\therefore\ V(s)\,\frac{s+10}{10s} = \frac{10}{(s+1)^2} + \frac{9\,V(s)}{s}\ ;\ V(s)\left[\frac{s+10}{10s} - \frac{9}{s}\right] = \frac{10}{(s+1)^2}$$

$$\text{OR}\quad V(s) = \frac{100\,s}{(s+1)^2\,(s-80)} = \frac{k_1}{(s+1)^2} + \frac{k_2}{(s+1)} + \frac{k_3}{(s-80)}$$

WHERE $k_1 = \left.\dfrac{100\,s}{(s-80)}\right|_{s=-1} = \dfrac{-100}{-81} = \dfrac{100}{81} = 1.23$

$$k_2 = \frac{d}{ds}\left[(s+1)^2\,\frac{100\,s}{(s+1)^2(s-80)}\right]\Bigg|_{s=-1} = \frac{-8000}{(81)^2} = -\,1.22$$

$$k_3 = \left.\frac{100\,s}{(s+1)^2}\right|_{s=80} = \frac{100\times 80}{(81)^2} = 1.22$$

$$\therefore\ v(t) = 1.23\,t\,e^{-t} - 1.22\,e^{-t} + 1.22\,e^{80t} \leftarrow$$

$$I_L(s) = \frac{100}{(s+1)^2(s-80)} = \frac{k_1}{(s+1)^2} + \frac{k_2}{(s+1)} + \frac{k_3}{(s-80)}$$

WHERE $k_1 = \left.\dfrac{100}{(s-80)}\right|_{s=-1} = \dfrac{100}{-81} = -1.23$

$$k_2 = \frac{d}{ds}\left[\frac{100}{s-80}\right]\Bigg|_{s=-1} = \left.\frac{-100}{(s-80)^2}\right|_{s=-1} = \frac{-100}{(-81)^2} = -\,0.015$$

$$k_3 = \left.\frac{100}{(s+1)^2}\right|_{s=80} = \frac{100}{(81)^2} = 0.015$$

$$\therefore\ i_L(t) = -1.23\,t\,e^{-t} - 0.015\,e^{-t} + 0.015\,e^{80t} \leftarrow$$

3.3.9. AT $t = 0^-$, $v_c(0^-) = 10$ V

FOR $t \geq 0$, THE TRANSFORMED NETWORK IS SHOWN BELOW:

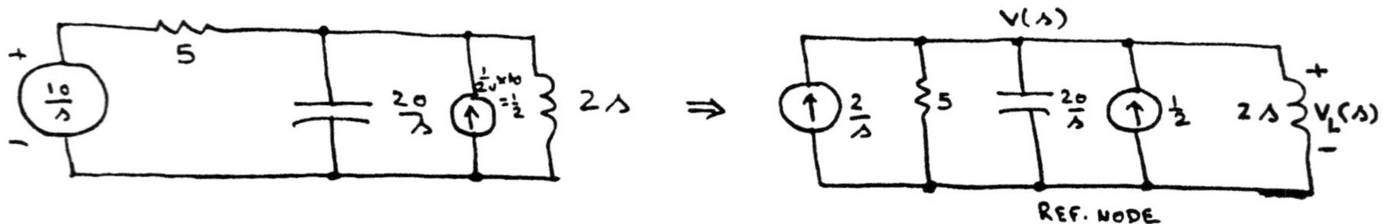

NOTE THAT $V_L(s) = V(s)$

NODAL EQUATION: $V(s)\left[\frac{1}{5} + \frac{s}{20} + \frac{1}{2s}\right] = \frac{2}{s} + \frac{1}{2} = \frac{4+s}{2s}$

OR $V(s) = \frac{10(s+4)}{s^2+4s+10} = \frac{K_1}{s+2-j\sqrt{6}} + \frac{K_1^*}{s+2+j\sqrt{6}}$

WHERE $K_1 = \left.\frac{10(s+4)}{s+2+j\sqrt{6}}\right|_{s=-2+j\sqrt{6}} = \frac{10(2+j\sqrt{6})}{2\sqrt{6}\,j} = 10\left(0.5 - j\frac{1}{\sqrt{6}}\right)$

$= 5 - j\,4.08$

$= 6.45\angle -39.2^\circ$

$K_1^* = 6.45\angle +39.2^\circ$

$\therefore\ v_L(t) = v(t) = 6.45\,e^{-j39.2^\circ}e^{(-2+j\sqrt{6})t} + 6.45\,e^{j39.2^\circ}e^{(-2-j\sqrt{6})t}$

$= 6.45\,e^{-2t}\left[e^{j(\sqrt{6}t-39.2^\circ)} + e^{-j(\sqrt{6}t-39.2^\circ)}\right]$

$= 6.45\,e^{-2t}\left[2\cos(\sqrt{6}t - 39.2^\circ)\right]$

$= 12.9\,e^{-2t}\cos(\sqrt{6}t - 39.2^\circ)$ V ←

3.3.10.

$i(t) = 100\times10^{-6}\left[u(t) - u(t-10^{-5})\right]$

$I(s) = \frac{100\times10^{-6}}{s}\left[1 - e^{-10^{-5}s}\right]$

THE TRANSFORMED NETWORK IS SHOWN BELOW:

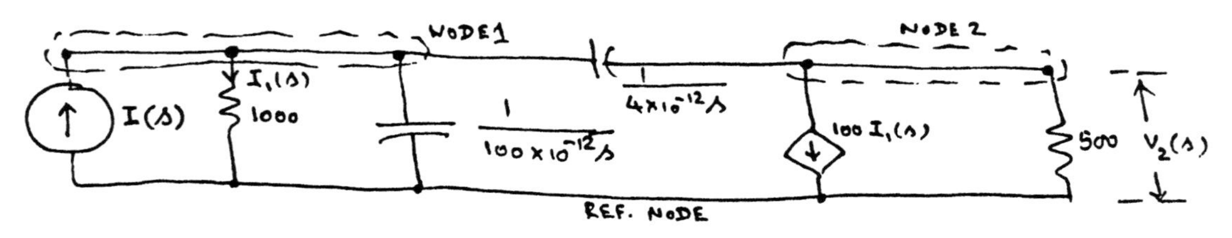

$I_1(s) = \frac{V_1(s)}{1000}$

3.3.10. CONTINUED

NODAL EQUATIONS:

$$V_1(s)\left[\frac{1}{1000} + (104\times10^{-12})s\right] - \left[(4\times10^{-12})s\right]V_2(s) = I(s)$$

$$V_1(s)\left[-4\times10^{-12}s\right] + V_2(s)\left[\frac{1}{500} + (4\times10^{-12})s\right] = -100\,I_1(s) = -V_1(s)/10$$

OR

$$V_1(s)\left[0.1 - 4\times10^{-12}s\right] + V_2(s)\left[2\times10^{-3} + 4\times10^{-12}s\right] = 0$$

$$V_1(s)\left[10^{-3} + (1.04)10^{-10}s\right] - V_2(s)\left[4\times10^{-12}s\right] = I(s)$$

SIMULTANEOUS SOLUTION YIELDS

$$V_2(s) = \frac{10^6\left(1 - e^{-10^{-5}s}\right)\left(s - 2.5\times10^{10}\right)}{s\left(s^2 + 15.3\times10^8 s + 0.5\times10^{16}\right)} = \underbrace{F_1(s) - F_1(s)\,e^{-10^{-5}s}}_{\text{(NOTE THE FORMAT)}}$$

WHERE

$$F_1(s) = \frac{K_1}{s} + \frac{K_2}{s + 3.275\times10^6} + \frac{K_3}{s + 15.27\times10^8}$$

WHERE

$$K_1 = \frac{-2.5\times10^{16}}{0.5\times10^{16}} = -5$$

$$K_2 \simeq \frac{10^6\left(-3.275\times10^6 - 2.5\times10^{10}\right)}{\left(-3.275\times10^6\right)\left(-3.275\times10^6 + 15.27\times10^8\right)} \simeq 5.01$$

$$K_3 \simeq \frac{10^6\left(-15.27\times10^8 - 2.5\times10^{10}\right)}{\left(-15.27\times10^8\right)\left(-15.27\times10^8 + 3.275\times10^6\right)} \simeq \frac{-10^{14}\times 234.73}{10^{14}(23267)} \simeq -0.01$$

THEREFORE

$$v(t) = v_2(t) = \left[-5 + 5.01\,e^{-3.275\times10^6 t} - 0.01\,e^{-15.27\times10^8 t}\right]u(t)$$
$$+\left[-2 + 5.01\,e^{-3.275\times10^6(t-10^{-5})} - 0.01\,e^{-15.27\times10^8(t-10^{-5})}\right]\cdot u(t - 10^{-5})$$

3.3.11.

(a)

TRANSFORMED NETWORK:

KVL EQUATIONS:

$$V_i(s) = R_1 I_1(s) + \frac{1}{C_1 s}(I_1(s)) - \frac{1}{C_1 s} I_2(s)$$

$$0 = -\frac{1}{C_1 s} I_1(s) + R_2 I_2(s) + \frac{1}{C_1 s} I_2(s) + \frac{1}{C_2 s} I_2(s)$$

$$I_2(s) = \frac{\frac{1}{C_1 s} I_1}{R_2 + \frac{1}{C_1 s} + \frac{1}{C_2 s}} \quad \text{OR } I_1(s) = I_2(s)\left[C_1 s\left(R_2 + \frac{1}{C_1 s} + \frac{1}{C_2 s}\right)\right]$$

$$V_i(s) = I_2(s)\left[C_1 s\left(R_2 + \frac{1}{C_1 s} + \frac{1}{C_2 s}\right)\left(R_1 + \frac{1}{C_1 s}\right) - \frac{1}{C_1 s}\right]$$

$$V_o(s) = \frac{1}{C_2 s} I_2(s)$$

$$\therefore \quad G(s) = \frac{V_o(s)}{V_i(s)} = \frac{1}{(R_1 R_2 C_1 C_2)s^2 + (R_1 C_1 + R_2 C_2 + R_1 C_2)s + 1}$$

(b)

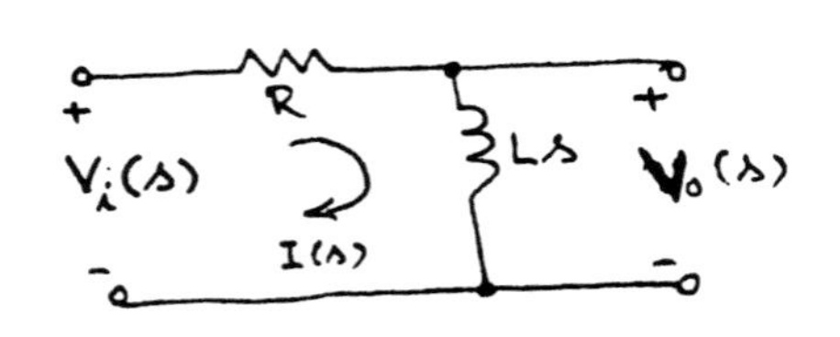

$$V_i(s) = (R + Ls)\, I(s) \;;\quad V_o(s) = (Ls)\, I(s)$$

$$\therefore G(s) = \frac{V_o(s)}{V_i(s)} = \frac{Ls}{R + Ls} = \frac{s\tau}{1 + s\tau}$$

WHERE $\tau = L/R$.

(c)

$$V_o(s) = \left(R_2 + \frac{1}{Cs}\right) I(s) \;;$$

$$V_i(s) = \left(R_1 + R_2 + \frac{1}{Cs}\right) I(s) \;;$$

$$\therefore G(s) = \frac{V_o(s)}{V_i(s)} = \frac{R_2 + \frac{1}{Cs}}{(R_1 + R_2) + \frac{1}{Cs}} = \frac{1 + R_2 C s}{1 + (R_1 + R_2)Cs}$$

$$= \frac{1 + \tau_1 s}{1 + \tau_2 s}, \quad \text{WHERE } \tau_1 = R_2 C \text{ AND } \tau_2 = (R_1 + R_2)C$$

(d)

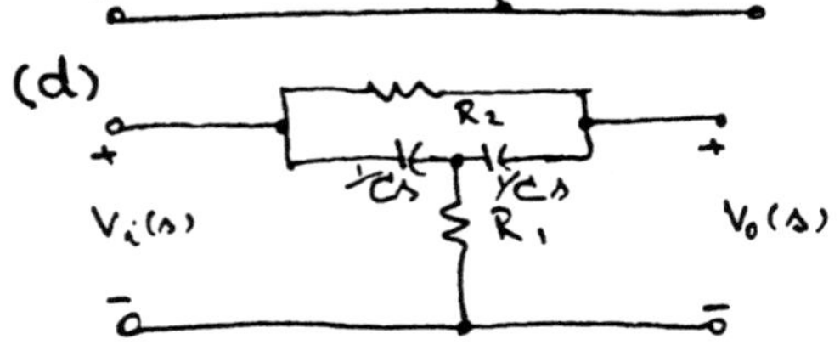

3.3.11. (d) CONTINUED

R_2 I_2 + V_i I_1 $\frac{1}{Cs}$ R_1 $\frac{1}{Cs}$ V_o + − −

$$V_i = I_1\left(\frac{1}{Cs}+R_1\right) - I_2\left(\frac{1}{Cs}\right)$$

$$0 = -I_1\left(\frac{1}{Cs}\right) + I_2\left(R_2+\frac{2}{Cs}\right)$$

$$V_o = I_2\left(\frac{1}{Cs}\right) + I_1 R_1$$

$$I_1 = I_2\left(R_2+\frac{2}{Cs}\right)Cs$$

$$= I_2\left(R_2 Cs + 2\right)$$

$$V_i = I_2\left[(R_2Cs+2)\left(\frac{1}{Cs}+R_1\right) - \frac{1}{Cs}\right] = I_2\left[R_2 + R_1R_2Cs + \frac{2}{Cs} + 2R_1 - \frac{1}{Cs}\right]$$

$$\therefore I_2 = \frac{V_i\,(Cs)}{R_1R_2C^2s^2 + (2R_1+R_2)Cs + 1}$$

$$V_o = \frac{V_i}{R_1R_2C^2s^2+(2R_1+R_2)Cs+1} + \frac{V_i\,Cs\,R_1\,(R_2Cs+2)}{R_1R_2C^2s^2+(2R_1+R_2)Cs+1}$$

$$\frac{V_o}{V_i} = \frac{1+2R_1Cs+R_1R_2C^2s^2}{R_1R_2C^2s^2+(2R_1+R_2)Cs+1}$$

$$= \frac{R_1R_2C^2s^2+2R_1Cs+1}{R_1R_2C^2s^2+2R_1Cs+R_2Cs+1} = 1 - \frac{R_2Cs}{R_1R_2C^2s^2+2R_1Cs+R_2Cs+1}$$

LETTING $R_1C = \tau_1$ & $R_2C = \tau_2$

$$\frac{V_o}{V_i} = \frac{1+2\tau_1 s+\tau_1\tau_2 s^2}{1+2\tau_1 s+\tau_2 s+\tau_1\tau_2 s^2} \longleftarrow$$

3.3.11. (e)

$$V_i(s) = I(s)\left[R_1 + R_2 + \frac{1}{Cs} + Ls\right]$$

$$V_o(s) = I(s)\left[R_2 + \frac{1}{Cs} + Ls\right]$$

$$\therefore G = \frac{V_o}{V_i} = \frac{R_2 + \frac{1}{Cs} + Ls}{R_1 + R_2 + \frac{1}{Cs} + Ls} = \frac{R_2Cs + 1 + LCs^2}{(R_1+R_2)Cs + 1 + LCs^2}$$

$$= \frac{1 + s\tau_1 + s^2LC}{1 + s\tau_2 + s^2LC} \quad \text{WHERE } \tau_1 = RC \text{ AND } \tau_2 = (R_1 + R_2)C$$

3.3.11. (f)

KVL EQ. AROUND OUTER BRANCHES: $-V_i + I_1X + V_o + I_1X = 0$ OR $I_1 = \frac{V_i - V_o}{2X}$

KVL EQ. ALONG INNER ZIG-ZAG BRANCHES: $-V_i + I_2Y - V_o + I_2Y = 0$ OR $I_2 = \frac{V_i + V_o}{2Y}$

NOTE:

RESULTANT IMPEDANCE $= \frac{\frac{1}{Cs}\cdot Ls}{\frac{1}{Cs} + Ls} = \frac{Ls}{1 + s^2LC} = X$ (say)

RESULTANT IMPEDANCE $= \frac{1}{Cs} + Ls = \frac{1 + LCs^2}{Cs} = Y$ (say)

ALSO $V_o = (I_1 - I_2)R = R\left[\frac{V_i - V_o}{2X} - \frac{V_i + V_o}{2Y}\right] = R\left[V_i\left(\frac{1}{2X} - \frac{1}{2Y}\right) - V_o\left(\frac{1}{2X} + \frac{1}{2Y}\right)\right]$

$$V_o\left[1 + \frac{R}{2X} + \frac{R}{2Y}\right] = V_i\left[\frac{R}{2X} - \frac{R}{2Y}\right]$$

THUS $\frac{V_o}{V_i} = \frac{R(Y-X)}{2XY} \frac{2XY}{2XY + RY + RX} = \frac{R(Y-X)}{2XY + R(X+Y)}$

$$Y - X = \frac{1 + LCs^2}{Cs} - \frac{Ls}{1 + LCs^2} = \frac{(1+LCs^2)^2 - LCs^2}{Cs(1 + LCs^2)} = \frac{1 + LCs^2 + L^2C^2s^4}{Cs(1 + LCs^2)}$$

$$XY = \frac{Ls}{1 + LCs^2} \cdot \frac{1 + LCs^2}{Cs} = \frac{L}{C}$$

$$X + Y = \frac{Ls}{1 + LCs^2} + \frac{1 + LCs^2}{Cs} = \frac{LCs^2 + 1 + LCs^2 + LCs^2 + L^2C^2s^4}{Cs(1 + LCs^2)}$$

$$= (1 + 3LCs^2 + L^2C^2s^4)/[Cs(1 + LCs^2)]$$

THEN: $\frac{V_o}{V_i} = \frac{R(1 + LCs^2 + L^2C^2s^4)}{Cs(1 + LCs^2)} \div \left[2\frac{L}{C} + R\left(\frac{1 + 3LCs^2 + L^2C^2s^4}{Cs(1 + LCs^2)}\right)\right]$

3.3.11. (f) CONTINUED

$$\frac{V_o}{V_i} = \frac{R(1 + LCs^2 + L^2C^2s^4)}{2\frac{L}{C}Cs(1+LCs^2) + R(1 + 3LCs^2 + L^2C^2s^4)}$$

$$= \frac{RL^2C^2\left(s^4 + \frac{1}{LC}s^2 + \frac{1}{L^2C^2}\right)}{2L\,LC\left(s^3 + \frac{s}{LC}\right) + RL^2C^2\left(s^4 + \frac{3}{LC}s^2 + \frac{1}{L^2C^2}\right)}$$

WITH $\omega_o^2 = \frac{1}{LC}$, $\frac{V_o}{V_i} = \frac{R(s^4 + \omega_o^2 s^2 + \omega_o^4)}{2L\omega_o^2(s^3 + s\omega_o^2) + R(s^4 + 3s^2\omega_o^2 + \omega_o^4)}$

OR $\frac{V_o}{V_i} = \frac{s^4 + \omega_o^2 s^2 + \omega_o^4}{s^4 + 2\frac{L}{R}\omega_o^2 s^3 + 3\omega_o^2 s^2 + 2\frac{L}{R}\omega_o^4 s + \omega_o^4}$

WITH $\frac{L}{R} = \tau$ AND $\omega_o^2 = \frac{1}{LC}$

$$\frac{V_o}{V_i} = \frac{s^4 + \omega_o^2 s^2 + \omega_o^4}{s^4 + 2\tau\omega_o^2 s^3 + 3\omega_o^2 s^2 + 2\tau\omega_o^4 s + \omega_o^4}$$

3.3.12.

$$v(t) = (5 - 3e^{-t} + 2e^{-2t})u(t)$$

$$V(s) = \frac{5}{s} - \frac{3}{s+1} + \frac{2}{s+2} = \frac{5(s+1)(s+2) - 3(s+2)s + 2s(s+1)}{s(s+1)(s+2)} = \frac{4s^2+11s+10}{s(s+1)(s+2)}$$

SINCE $V(s) = H(s)\, I(s)$ AND $I(s) = \frac{1}{s}$

$$H(s) = \frac{4s^2+11s+10}{(s+1)(s+2)} = 4 - \frac{(s-2)}{(s+1)(s+2)} \quad \longleftarrow$$

3.3.13. (a)

$$y(t) = (4 - 10e^{-t} + 8e^{-2t})u(t)$$

$$Y(s) = \frac{4}{s} - \frac{10}{s+1} + \frac{8}{s+2} = \frac{2s^2+8}{s(s+1)(s+2)}$$

SINCE $Y(s) = H(s) \cdot \frac{1}{s}$, $H(s) = \frac{2(s^2+4)}{(s+1)(s+2)}$ $\longleftarrow$

(b) $H(s)$ HAS ZEROS AT $s = \pm j2$

$\therefore$ THE RESPONSE IS ZERO FOR EXCITATION FREQUENCY $\omega = 2$ rad/s. $\longleftarrow$

3.3.14.

$$h(t) = 5e^{-t}\cos(2t - 30^\circ) = 5e^{-t}\left[\cos 2t \cos 30^\circ + \sin 2t \sin 30^\circ\right]$$

$$= 4.33\, e^{-t}\cos 2t + 2.5\, e^{-t}\sin 2t$$

$$H(s) = \frac{4.33(s+1)}{(s+1)^2+4} + \frac{2.5(2)}{(s+1)^2+4} = \frac{4.33s+9.33}{s^2+2s+5}$$

$$= \frac{4.33(s+2.15)}{s^2+2s+5} \quad \longleftarrow$$

3.3.15.

$$y(t) = (t+2)e^{-t}\, u(t)$$

$$Y(s) = \frac{1}{(s+1)^2} + \frac{2}{s+1} = \frac{2s+3}{(s+1)^2}$$

$\because$ $Y(s) = H(s)\, X(s)$ AND $X(s) = \frac{1}{s+2}$

$$H(s) = \frac{(2s+3)(s+2)}{(s+1)^2} = 2 + \frac{(3s+4)}{(s+1)^2} \quad \longleftarrow$$

3.3.16. (a)

THE TRANSFORMED NETWORK IS SHOWN BELOW:

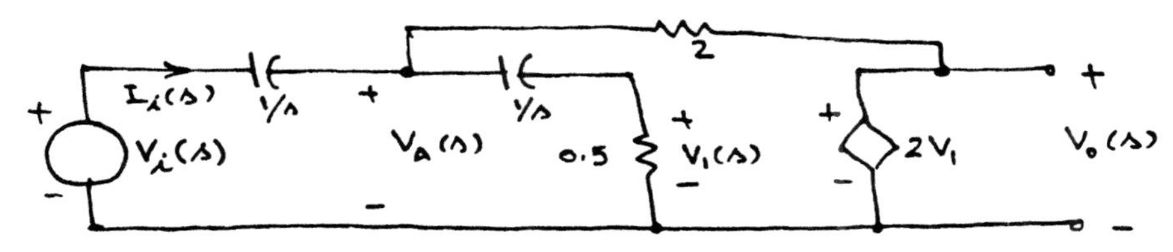

$$V_1(s) = \frac{0.5}{0.5 + \frac{1}{s}} V_A(s) = \frac{s}{s+2} V_A(s)$$

$$V_o(s) = 2V_1(s) = \frac{2s}{s+2} V_A(s)$$

CONVERSION OF BOTH VOLTAGE SOURCES TO THEIR CURRENT SOURCE EQUIVALENTS GIVES:

$V_A(s)$

$sV_i(s)$ 1/s 2 V_1 1/s 0.5

$$V_A(s)\left[s + \frac{1}{2} + \frac{1}{\frac{1}{s} + \frac{1}{2}}\right] = V_1(s) + s\,V_i(s)$$

$$= \frac{s}{s+2} V_A(s) + s\,V_i(s)$$

$$V_A(s)\left[s + \frac{1}{2} + \frac{2s}{s+2} - \frac{s}{s+2}\right] = s\,V_i(s)$$

$$V_A(s)\left[\frac{2s^2+4s+s+2+2s}{2(s+2)}\right] = s\,V_i(s)$$

$$V_A(s) = \frac{2s(s+2)}{2s^2+7s+2} V_i(s) = \frac{s(s+2)}{s^2+3.5s+1} V_i(s)$$

$$V_o(s) = \frac{2s}{s+2}\,\frac{s(s+2)}{s^2+3.5s+1} V_i(s) = \frac{2s^2}{s^2+3.5s+1} V_i(s)$$

$$\therefore \frac{V_o(s)}{V_i(s)} = \frac{2s^2}{s^2+3.5s+1} \longleftarrow$$

3.3.16.(b)

FOR $V_i(s) = \frac{1}{s}$, $V_o(s) = \frac{2s}{s^2+3.5s+1} = \frac{K_1}{s+0.3138} + \frac{K_2}{s+3.186}$

WHERE $K_1 = \frac{2(-0.3138)}{(3.186-0.3138)} = -0.219$

& $K_2 = \frac{2(-3.186)}{(0.3138-3.186)} = 2.219$

$$v_o(t) = 2.219\, e^{-3.186t} - 0.219\, e^{-0.3138t} \leftarrow$$

3.3.16.(c)

$$I_i(s) = \frac{V_i(s) - V_A(s)}{1/s} = s\left[V_i(s) - \frac{s(s+2)}{s^2+3.5s+1} V_i(s)\right]$$

$$= V_i(s)\quad s\left[1 - \frac{s^2+2s}{s^2+3.5s+1}\right]$$

$$= V_i\quad s\left[\frac{1.5s+1}{s^2+3.5s+1}\right]$$

DRIVING-POINT IMPEDANCE $= \frac{V_i(s)}{I_i(s)} = \frac{s^2+3.5s+1}{1.5s^2+s} = \frac{2s^2+7s+2}{3s^2+2}$

$$= \frac{2}{3}\,\frac{s^2+3.5s+1}{s^2+\frac{2}{3}} = \frac{2}{3}\left[1 + \frac{3.5s+\frac{1}{3}}{s^2+\frac{2}{3}}\right] \leftarrow$$

3.4.1. (a)

$$\bar{H}(j\omega) = \frac{\bar{V}_{out}}{\bar{V}_{in}} = \frac{j\omega}{10 + j\omega} = \frac{j\omega}{10\left(1 + j\frac{\omega}{10}\right)} = \frac{\omega(\omega + 10j)}{100 + \omega^2}$$

$$H(\omega) = \frac{\omega}{\sqrt{100 + \omega^2}} \quad ; \quad \theta(\omega) = \tan^{-1}\frac{10}{\omega}$$

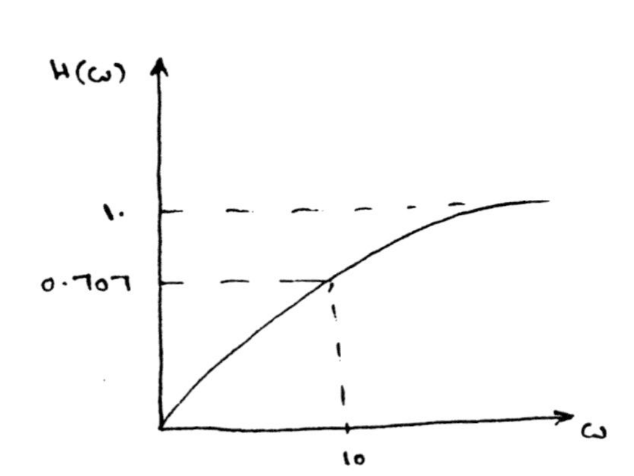

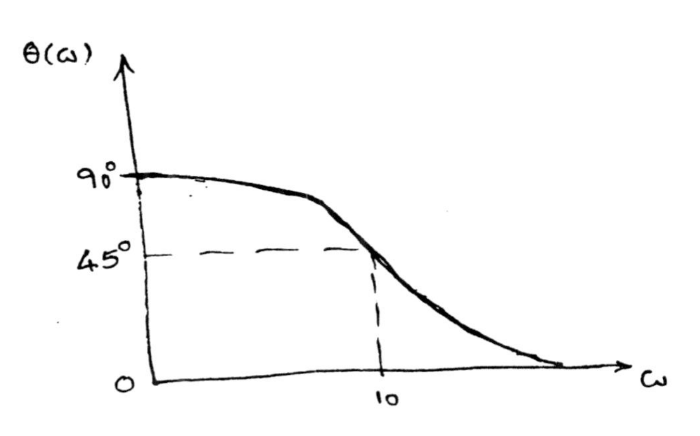

3.4.1. (b)

$$\bar{H}(j\omega) = \frac{30/j\omega}{10 + \frac{30}{j\omega}} = \frac{30}{j\omega 10 + 30} = \frac{3}{j\omega + 3}$$

$$H(\omega) = \frac{3}{\sqrt{9 + \omega^2}} \quad ; \quad \theta(\omega) = -\tan^{-1}\frac{\omega}{3}$$

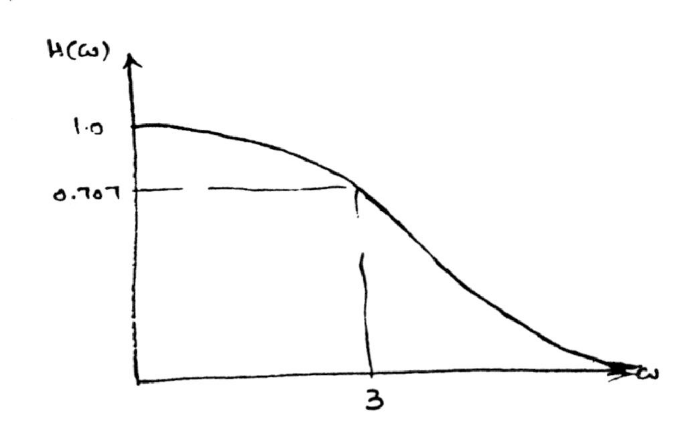

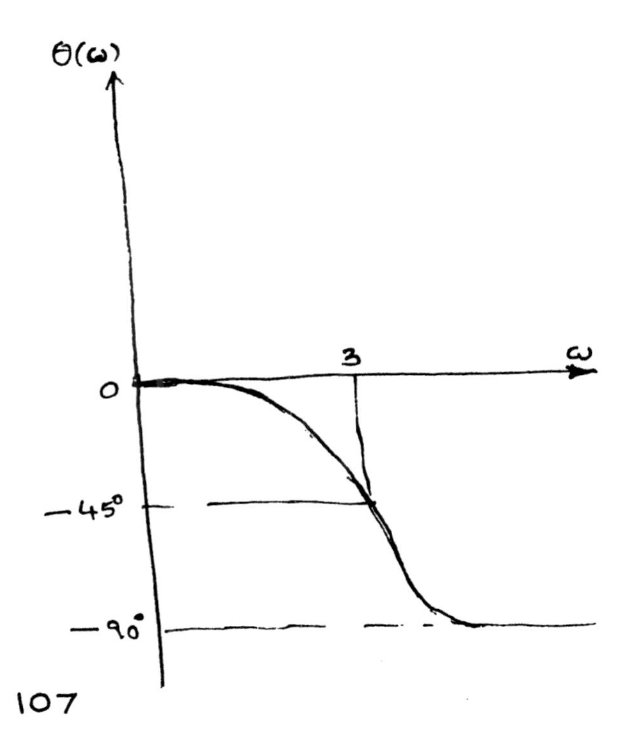

3.4.1.(c) COMBINATION OF L & C :

$$Z = \frac{(j\omega 4)\left(\frac{1000}{j\omega}\right)}{j\omega 4 + \frac{1000}{j\omega}} = \frac{1000\, j\omega}{250 - \omega^2}$$

$$\frac{\bar{V}_{out}}{\bar{V}_{in}} = \frac{Z}{Z + 10} = \frac{100\, j\omega}{250 + 100 j\omega - \omega^2} = \bar{H}(j\omega)$$

$$H(\omega) = \frac{100\,\omega}{\sqrt{(250-\omega^2)^2 + (100\omega)^2}} \; ; \quad \theta(\omega) = \tan^{-1} \frac{250-\omega^2}{100\,\omega}$$

3.4.2.

$$\bar{H}(j\omega) = \frac{R_L}{R_S + R_L + j\omega L} = \frac{30}{30 + 30 + j\omega L} = \frac{30}{60 + j\omega L}$$

$$H(\omega) = \frac{1}{60} \frac{30}{\sqrt{1 + \left(\frac{L\omega}{60}\right)^2}}$$

TO HAVE A HALF-POWER FREQUENCY OF 10 kHz, $L\omega = 60$

OR $2\pi\, 10^4\, L = 60$ or $L = \frac{60}{2\pi \times 10^4} \times 10^3$ mH

$= 0.955$ mH

3.4.3.

$$\bar{H}(j\omega) = \frac{R_L}{R_S + R_L + \frac{1}{j\omega C}} = \frac{j\omega C R_L}{1 + j\omega C(R_L + R_S)}$$

TO HAVE A HALF-POWER FREQUENCY OF 1 MHz, $\omega C(R_L + R_S) = 1$

or $C = \frac{1}{2\pi \times 10^6 (50+50)} = \frac{1}{2\pi \times 10^8}$ or $\frac{10^{12}}{2\pi \times 10^8}$ pF $= 1592$ pF

3.4.4.

$$\bar{H}(j\omega) = \frac{R_L}{R_S + R_L + j\omega L + \frac{1}{j\omega C}} = \frac{j\omega C R_L}{1 + (R_S + R_L) j\omega C - LC\omega^2}$$

$$\omega_0 = \frac{1}{\sqrt{LC}} = 2\pi \times 10^6 \;; \quad \text{BANDWIDTH} = 2\pi \times 10^4 = \frac{R_S + R_L}{L} = \frac{100}{L}$$

$$\therefore \quad L = \frac{100}{2\pi \times 10^4} \text{ or } \frac{100 \times 10^3}{2\pi \times 10^4} \text{ mH} = 1.6 \text{ mH} \leftarrow$$

$$\frac{1}{C} = (2\pi \times 10^6)^2 \, (1.6 \times 10^{-3})$$

$$\text{or } C = \frac{10^{12}}{(2\pi)^2 \times 10^{12} \times 1.6 \times 10^{-3}} \text{ pF} = 15.8 \text{ pF} \leftarrow$$

$$Q = \frac{\omega_0}{\text{BANDWIDTH}} = \frac{2\pi \times 10^6}{2\pi \times 10^4} = 100 \leftarrow$$

3.4.5.

$$\omega_0 = 2\pi \times 100 \times 10^3 = \frac{1}{\sqrt{LC}} \;; \quad \text{BANDWIDTH} = \frac{1}{C(R_s + R_L)} = \frac{1}{200C} = 2\pi \times 5 \times 10^3$$

$$\therefore \quad C = \frac{10^6}{200 \times 2\pi \times 5 \times 10^3} \, \mu\text{F} = 0.16 \, \mu\text{F} \leftarrow$$

$$L = \frac{1}{(0.16 \times 10^{-6})\,(2\pi \times 10^5)^2}$$

$$\text{or } \frac{10^3 \times 10^6}{0.16\,(2\pi)^2 \times 10^{10}} \text{ mH} = 0.016 \text{ mH or } 16 \, \mu\text{H} \leftarrow$$

$$Q = \frac{\omega_0}{\text{BANDWIDTH}} = \frac{2\pi \times 10^5}{2\pi \times 5 \times 10^3} = 20 \leftarrow$$

3.4.6.

$$\frac{\bar{V}_L}{\bar{V}_S} = \bar{H}(j\omega) = \frac{R_L\left(\frac{1}{j\omega C}\right) / \left(R_L + \frac{1}{j\omega C}\right)}{\frac{R_L\left(\frac{1}{j\omega C}\right)}{R_L + \frac{1}{j\omega C}} + R_L + R_S} = \frac{R_L}{R_L + R_S + j\omega(R_S R_L C + L) - LCR_L\omega^2}$$

$$= \frac{R_L}{R_S + R_L} \; \frac{1}{1 + j\omega\left[\frac{L + CR_L R_S}{R_S + R_L}\right] - \frac{CLR_L}{R_S + R_L}\omega^2} \leftarrow$$

$$\omega_0 = \sqrt{(R_S + R_L)/(CLR_L)} \leftarrow$$

3.4.7.

COMBINATION OF L & R_L : $Z = \dfrac{j\omega L R_L}{j\omega L + R_L}$

$$\bar{H}(j\omega) = \frac{Z}{Z + R_S + \frac{1}{j\omega C}} = \frac{j\omega C Z}{j\omega C Z + j\omega C R_S + 1}$$

$$= \frac{-CLR_L\,\omega^2}{R_L + j\omega L + j\omega C R_S R_L - \omega^2 (CLR_S + CLR_L)}$$

$$= \frac{R_L}{R_L + R_S} \; \frac{-\omega^2}{\frac{1}{LC} + j\omega\left(\frac{L + CR_SR_L}{CL(R_L+R_S)}\right) - \omega^2} \leftarrow$$

$\omega_0 = 1/\sqrt{LC}$ ←

3.4.8.

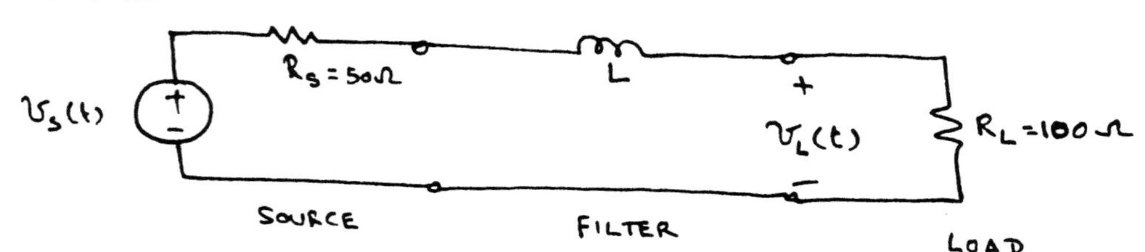

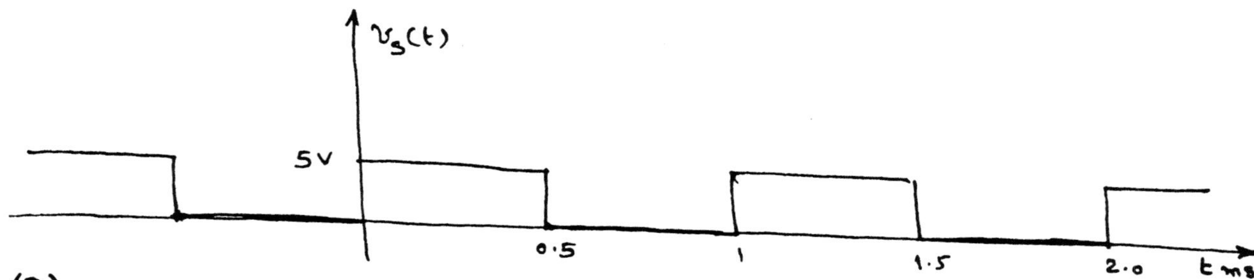

(a)

THE FOURIER SERIES OF THE SQUARE WAVE IS FOUND AS

$$v_S(t) = 2.5 + \frac{10}{\pi}\sin\omega_0 t + \frac{10}{3\pi}\sin 3\omega_0 t + \frac{10}{5\pi}\sin 5\omega_0 t + \ldots$$

WHERE $\omega_0 = 2\pi f_0$ AND $f_0 = 1/T = 1\text{kHz}$

SELECTING THE HALF-POWER POINT OF THE FILTER TO BE 5 kHz,

$$2\pi \times 5\times10^3 = \frac{R_S + R_L}{L} = \frac{150}{L}$$

or $L = \dfrac{1}{2\pi}\,\dfrac{150}{5\times10^3} \times 10^3 \text{ mH} = 4.77 \text{ mH}$ ←

(b) $\bar{V}_L = \dfrac{R_L}{R_S + R_L + j\omega L}\,\bar{V}_S$

3.4.8.(b) CONTINUED

DC COMPONENT $(f=0)$ OF 2.5V IS REDUCED TO

$$\frac{R_L}{R_g+R_L} \times 2.5 = 1.67\text{ V}$$

FUNDAMENTAL FREQUENCY COMPONENT $(f = 1\text{kHz})$ OF $\frac{10}{\pi} = 3.18\text{V}$ IS REDUCED TO

$$\bar{V}_{L1} = \frac{100}{150 + j\,2\pi\times10^3\times4.77\times10^{-3}} \times 3.18\angle 0^\circ = 2.08\angle -11.3^\circ$$

SIMILARLY, THE THIRD HARMONIC $(f = 3\text{kHz})$ OF $\frac{10}{3\pi} = 1.06\text{V}$ IS REDUCED TO

$$\bar{V}_{L3} = 0.607\angle -31^\circ$$

AND THE FIFTH HARMONIC $(f = 5\text{kHz})$ OF $\frac{10}{5\pi} = 0.636\text{V}$ IS REDUCED TO

$$\bar{V}_{L5} = 0.3003\angle -45^\circ$$

AND THE SEVENTH HARMONIC $(f = 7\text{kHz})$ OF $\frac{10}{7\pi} = 0.454\text{V}$ IS REDUCED TO

$$\bar{V}_{L7} = 0.176\angle -54.5^\circ.$$

THE AMPLITUDE SPECTRA OF $v_s(t)$ AND $v_L(t)$ ARE SKETCHED BELOW:

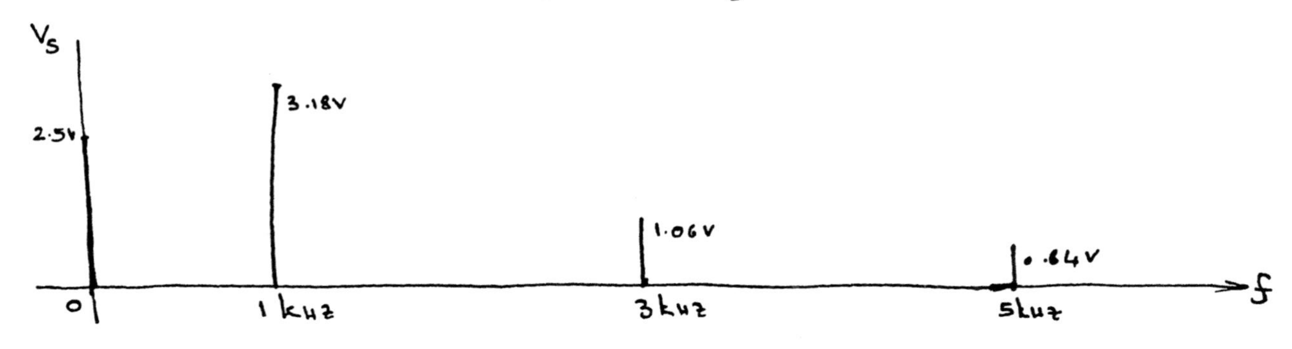

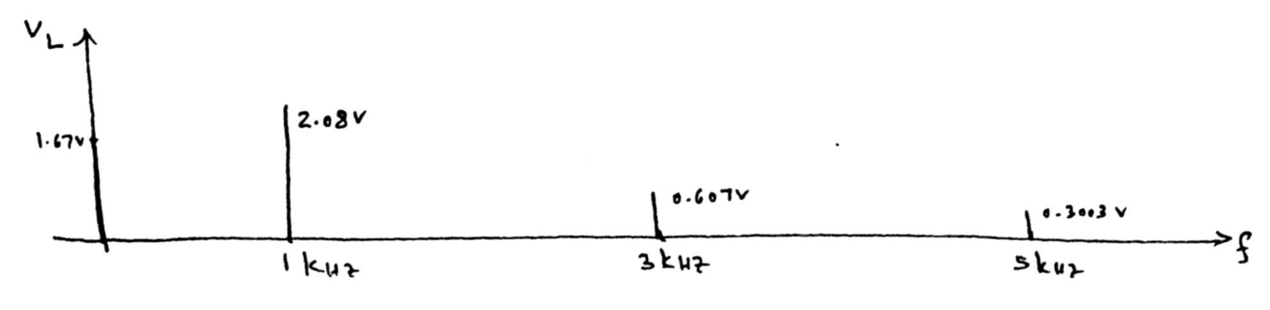

(c)

$$v_L(t) = 1.67 + 2.08\sin(2\pi\times10^3 t - 11.3^\circ) + 0.607\sin(6\pi\times10^3 t - 31^\circ)$$
$$+ 0.3003\sin(10\pi\times10^3 t - 45^\circ) + 0.176\sin(14\pi\times10^3 t - 54.5^\circ) + \cdots$$

3.4.9.

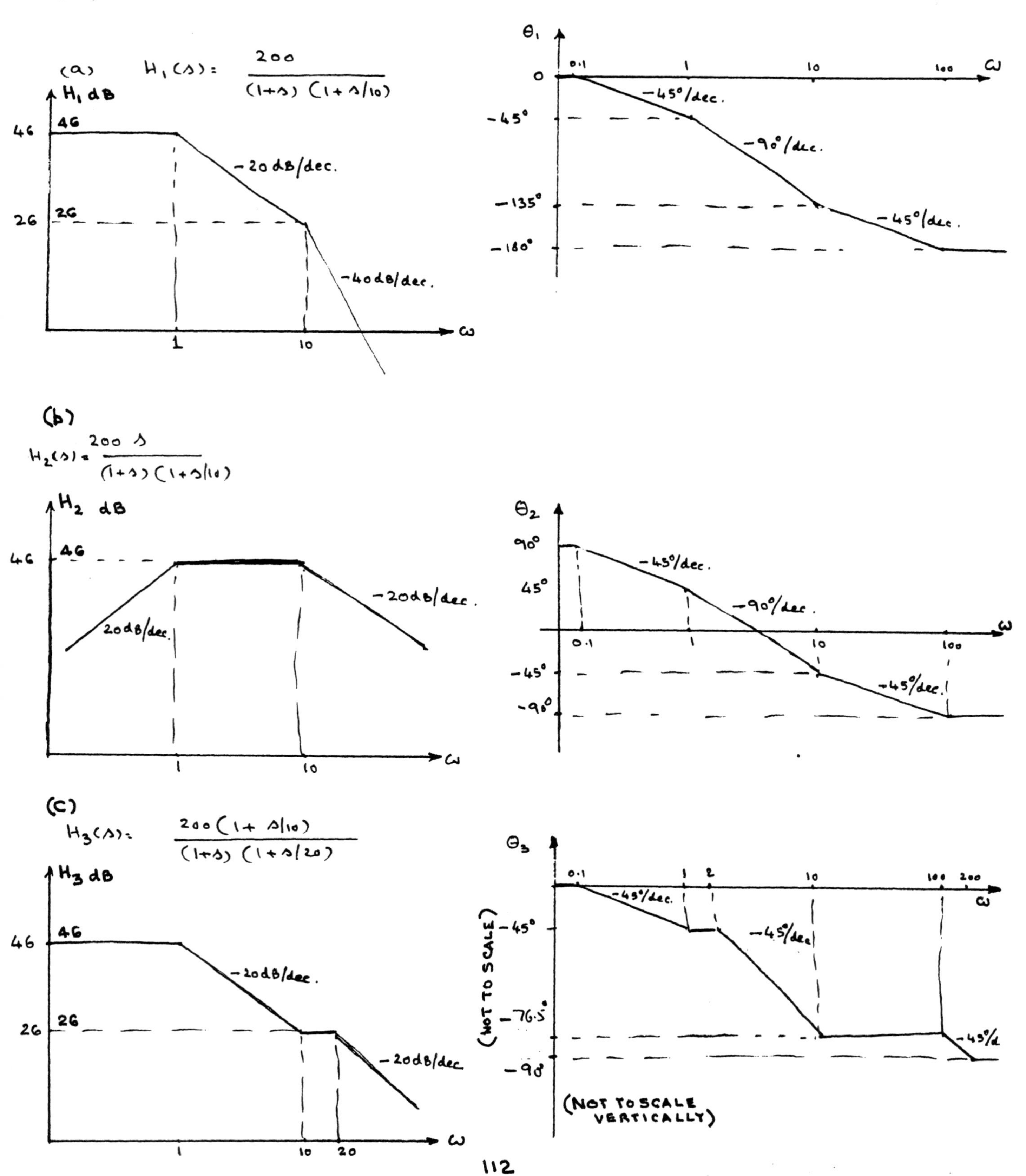

(a) $H_1(s) = \frac{200}{(1+s)(1+s/10)}$
H_1 dB
46
46
26
26
−20 dB/dec.
−40 dB/dec.
1
10
ω
θ_1
0
0.1
1
10
100
ω
−45°/dec.
−45°
−90°/dec.
−135°
−45°/dec.
−180°
(b)
$H_2(s) = \frac{200\,s}{(1+s)(1+s/10)}$
H_2 dB
46
46
20 dB/dec.
−20 dB/dec.
1
10
ω
θ_2
90°
−45°/dec.
45°
−90°/dec.
ω
0.1
1
10
100
−45°
−45°/dec.
−90°
(c)
$H_3(s) = \frac{200(1+s/10)}{(1+s)(1+s/20)}$
H_3 dB
46
46
−20 dB/dec.
26
26
−20 dB/dec
1
10
20
ω
θ_3
0.1
1
2
10
100
200
ω
−45°/dec.
−45°
−45°/dec
−76.5°
−45°/d
−90°
(NOT TO SCALE)
(NOT TO SCALE VERTICALLY)

3.4.9. CONTINUED

(d)

$$H_4(s) = \frac{0.5(1+s/10)}{(1+s)^2(1+s/40)}$$

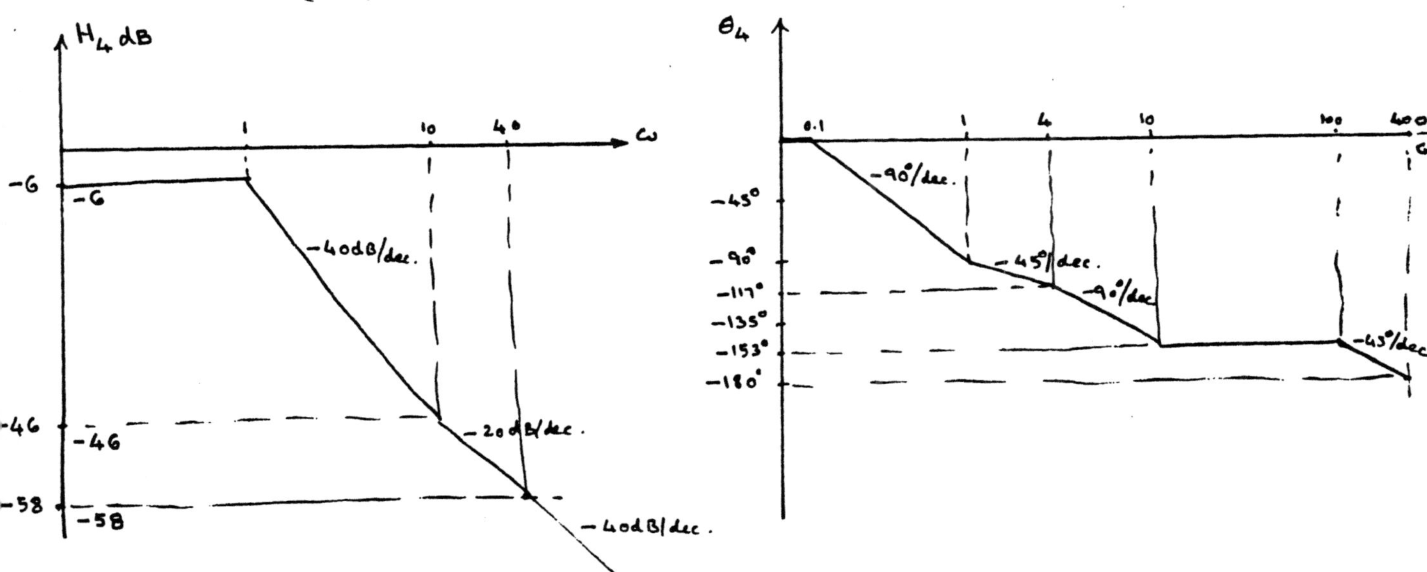

(e)

$$H_5(s) = \frac{0.5(1+s)^2}{s(1+s/10)(1+s/50)}$$

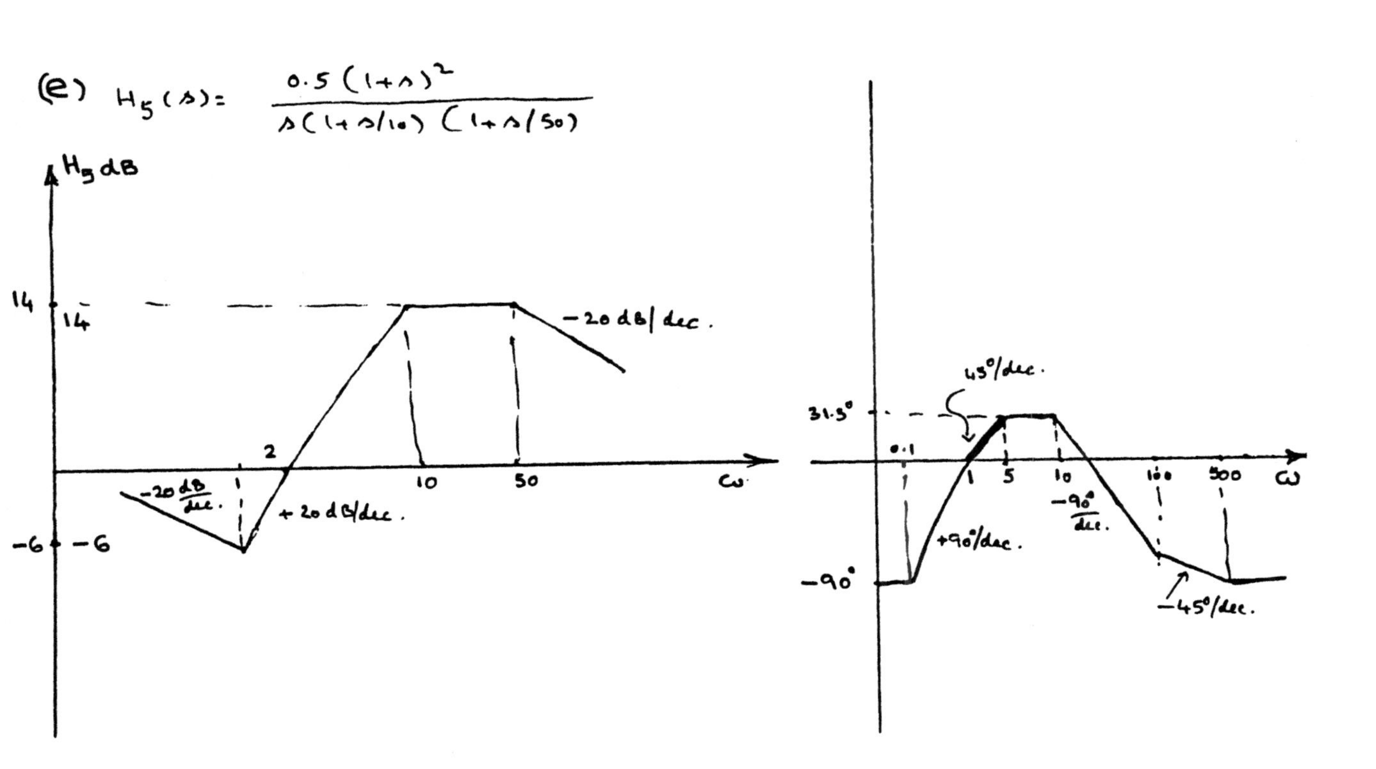

3.4.9. CONTINUED

(f)

$$H_6(s) = \frac{20(1+s/8)}{(1+s)(1+s/10)^2(1+s/40)}$$

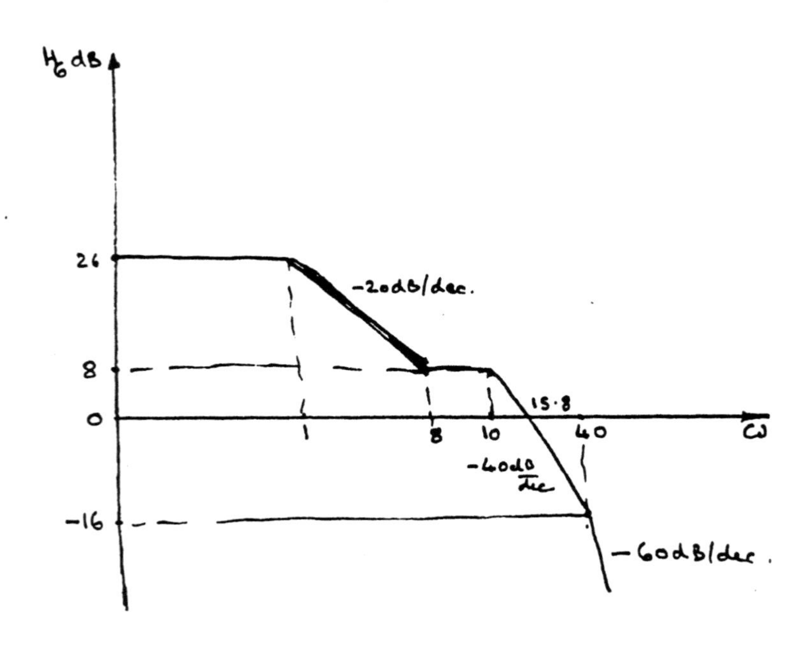

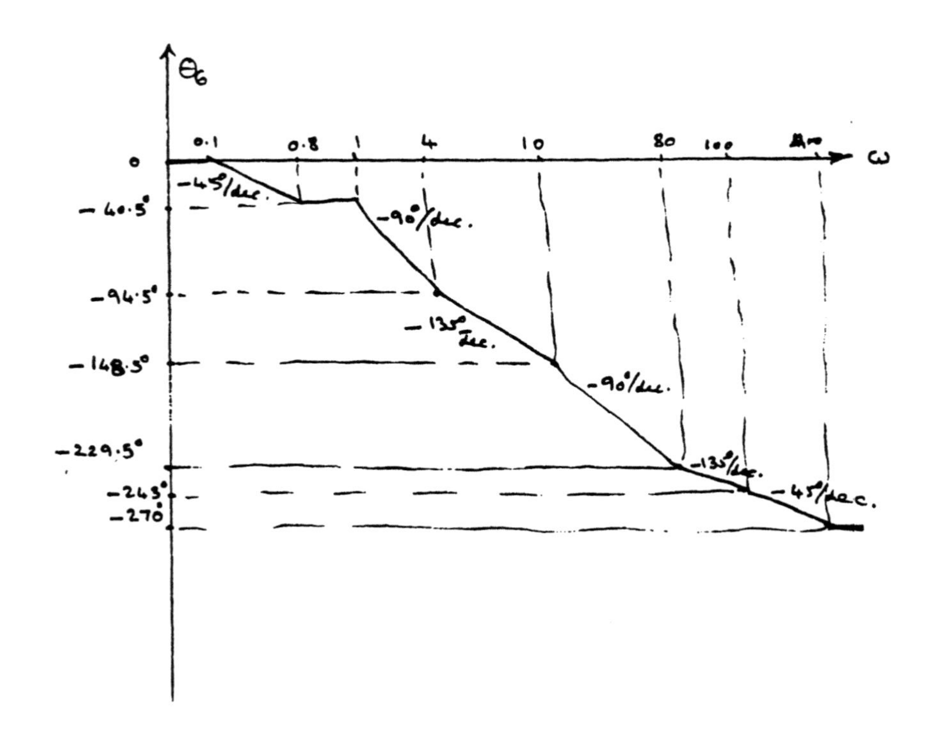

3.4.10. (a)

SEE SOLUTION OF PROBLEM 3.4.9.

(i)

AT $\omega = 1$, $H_1 = 46$ dB ; AT $\omega = 10$, $H_1 = 26$ dB

LINEAR WITH RESPECT TO LOG-SCALE (i.e. log 1 to log 10, or 0 to 1)

CORRESPONDING TO $\omega = 5$, $\log 5 \simeq 0.7$; $H_1 = 46 - 20(0.7) = 46 - 14 = 32$ dB ←

SIMILARLY, AT $\omega = 1$, $\theta_1 = -45°$; AT $\omega = 10$, $\theta_1 = -135°$

LINEAR W.R.T. LOG-SCALE (i.e. log 1 to log 10 or 0 to 1)

CORRESPONDING TO $\omega = 5$, $\log 5 = 0.7$; $\theta_1 = -45 - 90(0.7) = -45 - 63 = -108°$ ←

(ii)

AT $\omega = 5$, $H_2 = 46$ dB ; $\theta_2 = +45 - 90(0.7) = 45 - 63 = -18°$ ←

(iii)

AT $\omega = 5$, $H_3 = 46 - 20(0.7) = 32$ dB ;

$$\theta_3 = -45 - 45(\log 5 - \log 2) = -45 - 45(0.7 - 0.3)$$

$$= -45 - 45(0.4) = -63° \leftarrow$$

3.4.10. (b)

AT $\omega = 5$, $H_4 = -6 - 40(0.7) = -6 - 28 = -34$ dB

ONE-HALF THE MAGNITUDE CORRESPONDS TO A DECREASE OF 6 dB.

∴ CORRESPONDING TO $H_4 = -40$ dB, $-6 - 40(x) = -40$

or $x = \frac{34}{40} = \frac{17}{20} = 0.85$

ω = inv. log 0.85 = 7 rad/sec. ←

3.4.10. (c)

$H_6(\omega) = 0$ AT 0.2 DECADES AWAY FROM $\omega = 10$ rad/s.

or ω = inv. log 1.2 = 15.8 rad/s. ←

$\theta_6(\omega) = -180°$ AT $\frac{180 - 148.5}{90} = 0.35$ decades away from $\omega = 10$

or ω = inv. log 1.35 = 22.38 rad/sec. ←

3.4.10. (d)

AT $\omega = 20$, $H_5(\omega) = 14$ dB AND $\theta_5(\omega) = 31.5° - (0.3)90 = 4.5°$

SINCE inv. log$\left(\frac{14}{20}\right) = 5$, $v_2(t) = 0.5 \cos(20t + 4.5°)$ V ←

3.4.11.

$$\overline{H}(j\omega) = \frac{10(1+j2\omega)}{(1+j10\omega)(1+j0.25\omega)} = \frac{10(1+j\omega/0.5)}{(1+j\omega/0.1)(1+j\omega/4)}$$

H(ω) dB
10
-20dB/decade
ω
0.01 0.1 0.5 1 4 10 100
-4
-20dB/decade
-10
-12
-20

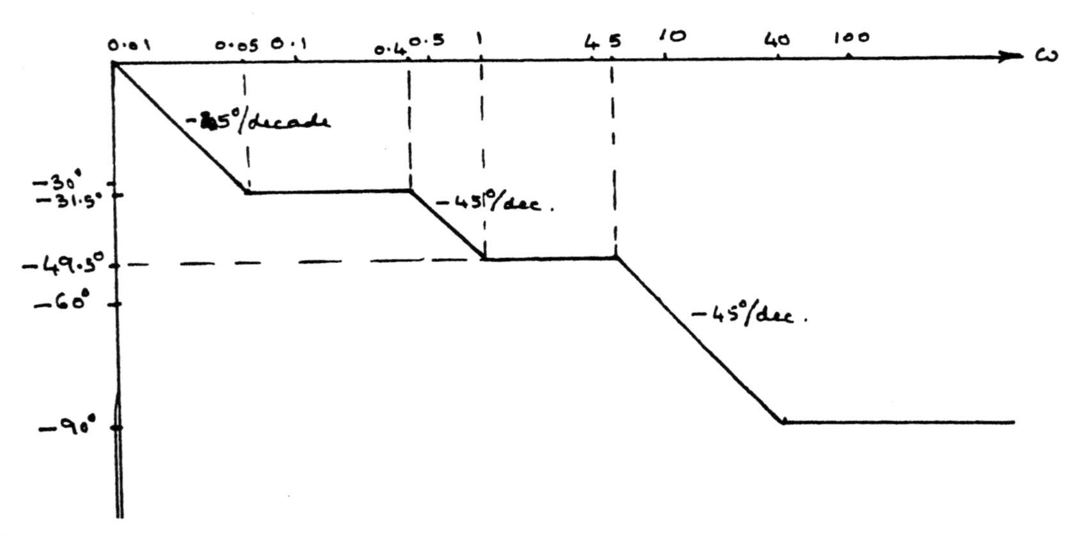

H(ω) = 0 AT 0.5 DECADES AWAY FROM ω = 0.1

OR ω = INV. LOG (−1 + 0.5) = INV. LOG (−0.5) = 0.32 RAD/S ←

θ(ω) = −60°, $\quad -60° = -49.5° - 45(x)$ OR $x = 0.23$

ω = INV. LOG (LOG 5 + 0.23) = INV. LOG (0.7 + 0.23) = 8.5 RAD/S ←

3.4.12.

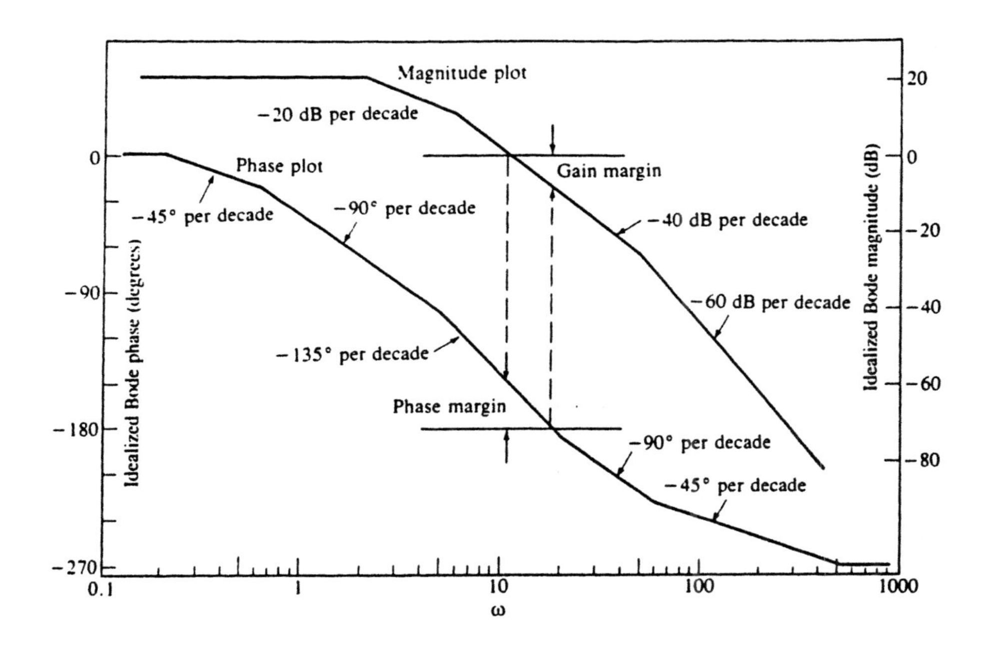

LOOP-GAIN MAGNITUDE IS UNITY (0 dB) AT $\omega_u = 10.8$ rad/s, WHERE THE PHASE MARGIN IS FOUND TO BE 30°. ←

THE PHASE OF THE LOOP GAIN IS $-\pi$ or -180° AT $\omega_\pi = 17.7$ rad/s. WHERE THE GAIN MARGIN IS 8.5 dB. ←

3.4.13. (a)

$$G_1(s)H_1(s) = \frac{10^4(1+s/0.7)}{(1+s/0.003)(1+s/0.04)(1+s/7)}$$

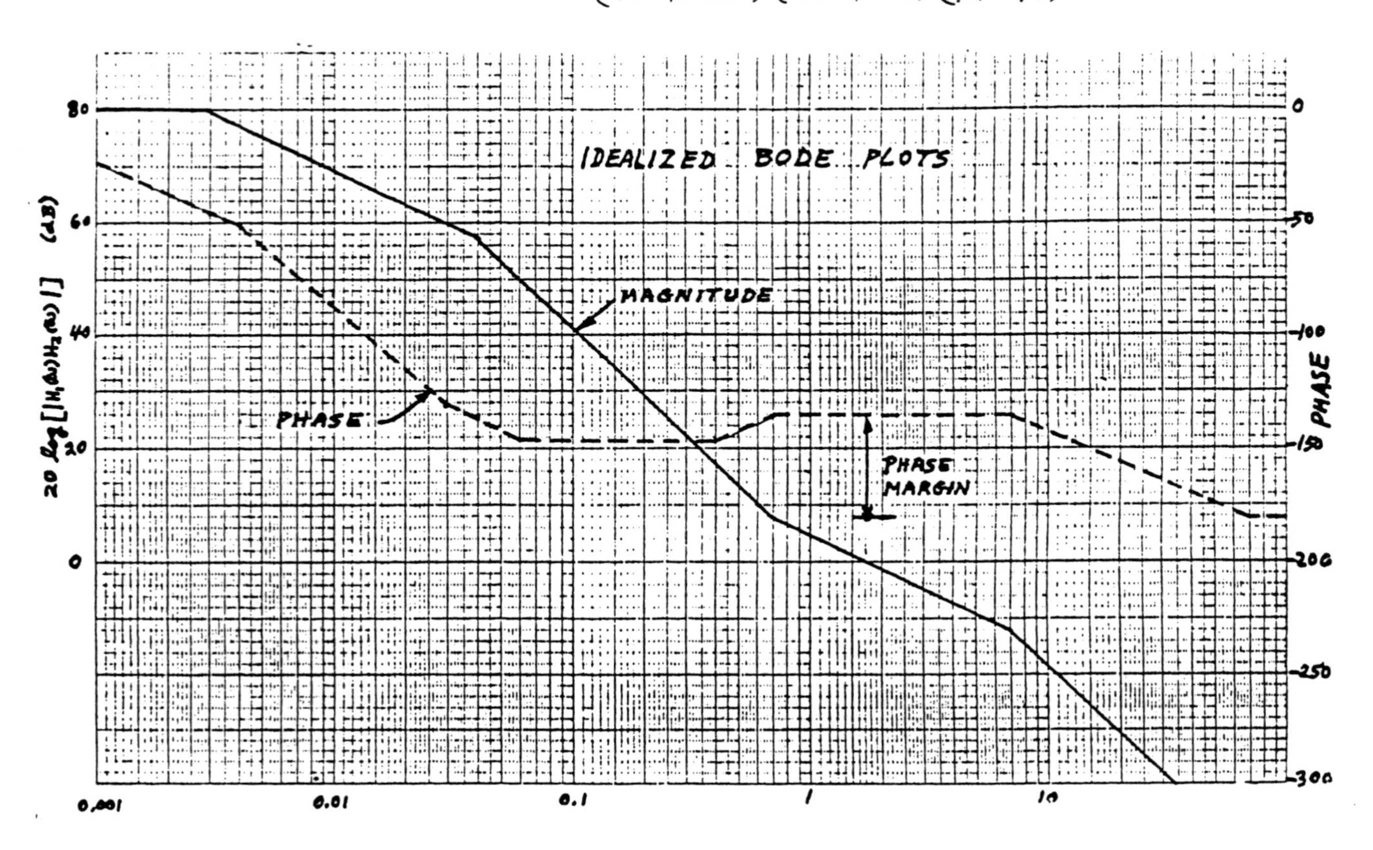

PHASE MARGIN ≃ 45°

GAIN MARGIN ≃ 52.3 dB

(NOTE THE PHASE OF THE LOOP GAIN IS −180° AT ω_π = 70 rad/s;)
CORRESPONDING GAIN IS −52.26 dB.

3.4.13.(b)

$$G_2(s)\,H_2(s) = \frac{100\,(1+s/3.9607)}{s(1+2s)\,(1+s/39.607)}$$

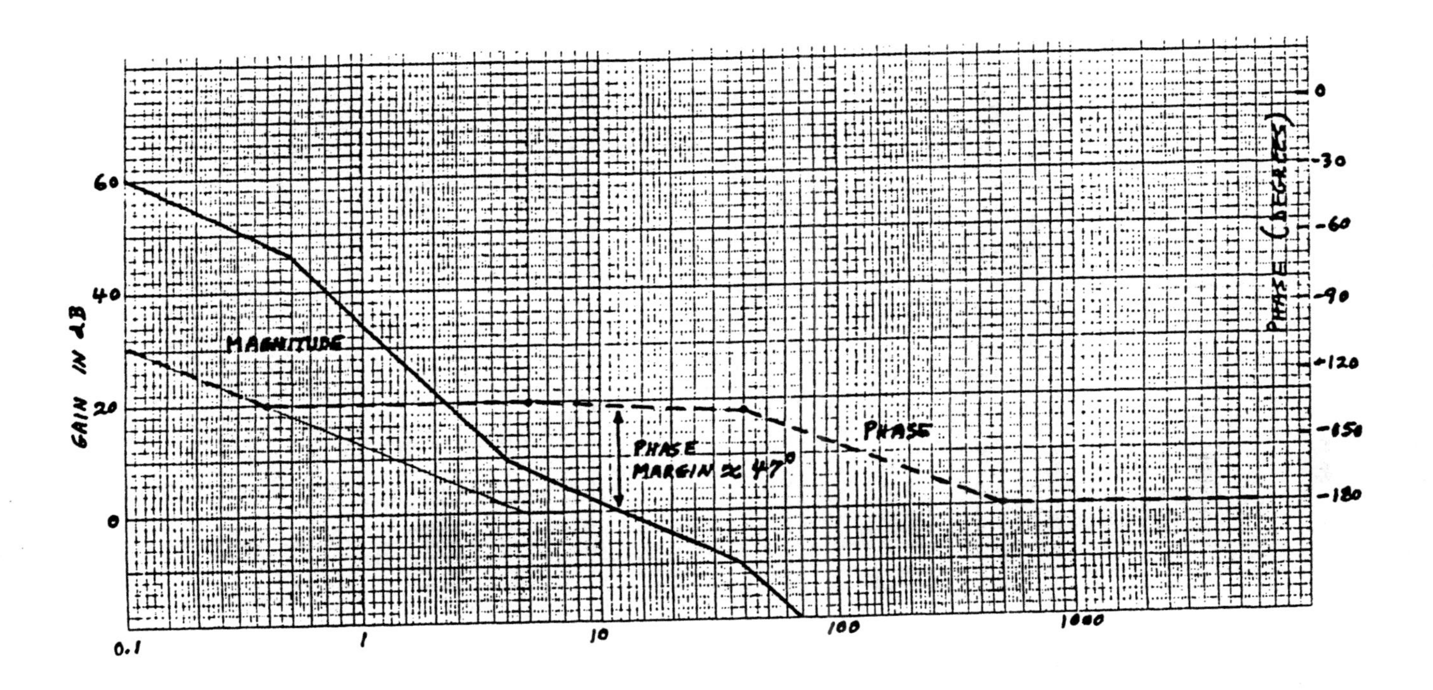

PHASE MARGIN $\simeq 47°$; GAIN MARGIN $\simeq \infty$

3.4.14

(a)

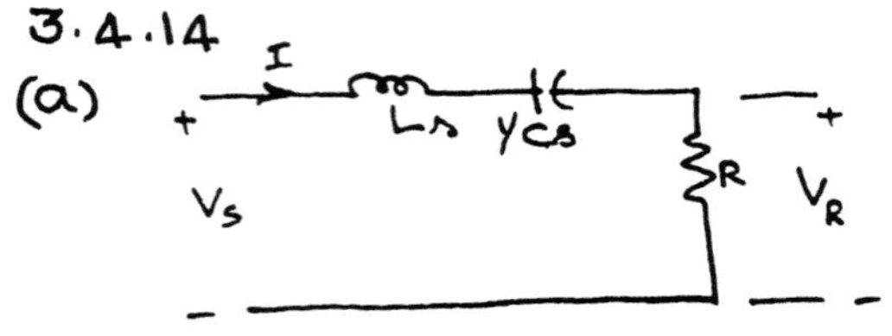

$$I = \frac{V_s}{R + Ls + \frac{1}{Cs}}$$

$$V_R = \frac{R\,V_s}{R + Ls + \frac{1}{Cs}}$$

$$\frac{V_R}{V_s} = \frac{R}{R + Ls + \frac{1}{Cs}} = \frac{(R/L)}{(R/L) + s + \frac{1}{LCs}} = \frac{(R/L)\,s}{s^2 + (R/L)\,s + \frac{1}{LC}} \leftarrow$$

SERIES RESONANT FREQUENCY $\omega_0 = \frac{1}{\sqrt{LC}}$ ←

BANDWIDTH : $\omega_0/Q = \omega_0 \Big/ \left(\frac{\omega_0 L}{R}\right) = R/L$ ←

(b)

BY COMPARING $\frac{10^3}{s^2 + 10^3 s + 10^{10}}$ WITH THE RESULT OF PART (a), ONE GETS

$$\omega_0^2 = \frac{1}{LC} = 10^{10} \quad \text{or} \quad \omega_0 = 10^5 \text{ rad/s} \leftarrow$$

AND $BW = \frac{R}{L} = 10^3$ rad/s ←

3.4.15.(a)

WITH A RESONANT FREQUENCY OF 1000 kHz, SINCE $\omega_0 = \frac{1}{\sqrt{LC}}$,

$$2\pi\,(1000)\,10^3 = \frac{1}{\sqrt{50\times10^{-6}\,C}}$$

$$\text{or} \quad C = 507\times10^{-12}\ \text{F} = 507\ \text{pF}. \leftarrow$$

THE RELATIVE RESPONSE IS TO BE 0.25 AT A FREQUENCY OF 1020 kHz FOR AN ATTENUATION OF 4 TIMES.

SINCE $\frac{\bar{Y}(j\omega)}{\bar{Y}_0} = \frac{1}{1 + j2\delta Q_s}$ NEAR RESONANCE (3.4.10)

THE MAGNITUDE OF THE RELATIVE RESPONSE IS OBTAINED BELOW:

$$\frac{1}{4} = \frac{1}{\sqrt{1 + (2\delta Q)^2}} \quad \text{or} \quad \delta Q = 1.94$$

THE VALUE OF δ, FROM $\delta = \frac{\omega - \omega_0}{\omega_0}$, IS

$$\delta = \frac{2\pi\,(1020)\,10^3 - 2\pi\,(1000)\,10^3}{2\pi\,(1000)\,10^3} = 0.02$$

3.4.15. CONTINUED

(a) $\therefore Q = \frac{1.94}{0.02} = 97$

FROM $Q = \omega_0 C / G$, $97 = \frac{2\pi(1000)10^3 \times 507\times10^{-12}}{G}$

or $G = 32.9 \times 10^{-6}$ S $= 32.9\ \mu$S. ←

(b)

$BW = \omega_2 - \omega_1 = \frac{\omega_0}{Q} = \frac{2\pi(1000)10^3}{97} = 64.8\times10^3$ rad/s

& 10.3 KHZ ←

3.4.16.

(a) THE NETWORK IS SHOWN BELOW WITH SOME MODIFICATIONS:

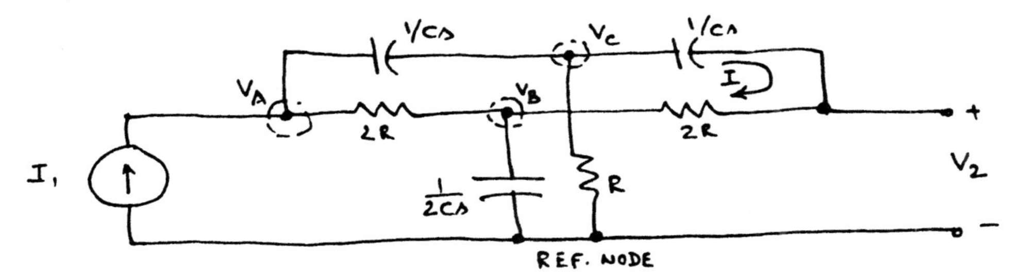

$V_2 = 2RI + V_B$; $I = (V_C - V_B)/(2R + 1/Cs)$

NODAL EQUATIONS ARE

$$V_A\left(\frac{1}{2R} + Cs\right) - V_B\left(\frac{1}{2R}\right) - V_C(Cs) = I_1$$

$$-V_A\left(\frac{1}{2R}\right) + V_B\left(2Cs + \frac{1}{2R} + \frac{1}{2R + 1/Cs}\right) - V_C\left(\frac{1}{2R + 1/Cs}\right) = 0$$

$$-CsV_A - V_B\frac{1}{2R + 1/Cs} + V_C\left(\frac{1}{2R + 1/Cs} + \frac{1}{R} + \frac{1}{Cs}\right) = 0$$

SIMULTANEOUS SOLUTION FOR V_B & V_C FOLLOWED BY SUBSTITUTION INTO THE EXPRESSION FOR I AND V_2 YIELDS

$$\frac{V_2}{I_1} = \frac{4R^2C^2s^2 + 1}{4Cs(2RCs + 1)} \quad \leftarrow$$

(b)

ZEROS OCCUR WHEN $4R^2C^2s^2 + 1 = 0$ or when $s = \pm j/2RC$.

IN THE VICINITY OF $\omega_0 = \frac{1}{2RC}$, THERE IS HIGH ATTENUATION. ←

AT FREQUENCIES REMOVED FROM ω_0, SIGNALS ARE UNATTENUATED.

3.4.17.

(a)

FOR THE PORTION OF THE CIRCUIT TO THE RIGHT OF TERMINALS 1-1', THE INPUT IMPEDANCE IS

$$Z_{in} = \frac{V_A}{-g V_B} \quad \text{WHERE} \quad V_B = -g V_A \frac{R}{1 + R C_2 s}$$

$$\therefore Z_{in} = \frac{1 + R C_2 s}{g^2 R} = \frac{1}{g^2 R} + \frac{C_2}{g^2} s$$

WHICH IS OF THE FORM $R_L + Ls$, IDENTIFYING $R_L = \frac{1}{g^2 R}$ AND $L = \frac{C_2}{g^2}$ ←

THE EQUIVALENT CIRCUIT SHOWN BELOW APPLIES:

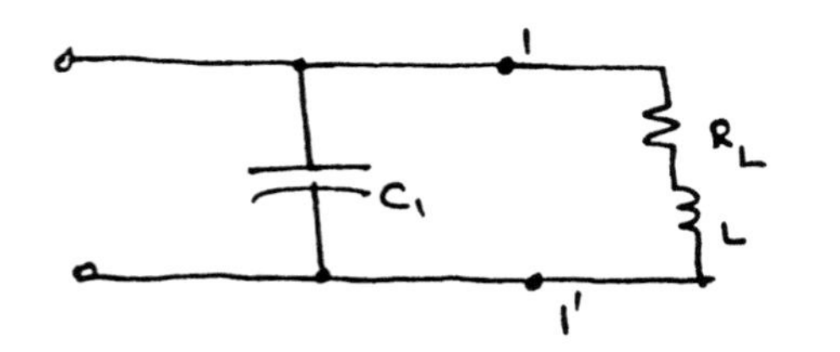

←

(b)

$$R_L = \frac{1}{g^2 R} = \frac{1}{10^{-6} \times 10^5} = 10\ \Omega \ ; \quad L = \frac{C_2}{g^2} = \frac{100 \times 10^{-12}}{10^{-6}} = 10^{-4}\ \text{H} = 100\ \mu\text{H}.$$

AS IN EXAMPLE 3.4.2,

$$\omega_o = \frac{1}{\sqrt{LC_1}} = \frac{1}{\sqrt{10^{-4} \times 10^{-10}}} = 10^7 \text{ rad/s.} \quad \leftarrow$$

$$BW = \frac{\omega_o}{Q_o} \quad \text{WHERE} \quad Q_o = \omega_o L / R_L = 10^7 \times 10^{-4} / 10 = 100$$

$$\therefore BW = 10^7 / 100 = 10^5 \text{ rad/s} \quad \leftarrow$$

3.4.18.

(a)

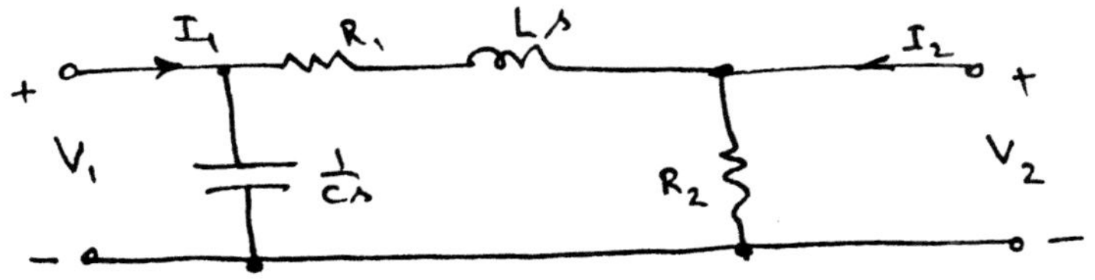

$$Z_{11} = \left.\frac{V_1}{I_1}\right|_{I_2=0} \quad ; \quad Z_{21} = \left.\frac{V_2}{I_1}\right|_{I_2=0}$$

$$Z_{11} = \frac{\frac{1}{Cs}(R_1+R_2+Ls)}{\frac{1}{Cs}+R_1+R_2+Ls} = \frac{R_1+R_2+Ls}{LCs^2+C(R_1+R_2)s+1} \leftarrow$$

$$V_1 = I_1 Z_{11} \quad \text{AND} \quad V_2 = \frac{R_2}{R_1+R_2+Ls} V_1$$

THEN,
$$V_2 = \frac{R_2}{R_1+R_2+Ls} \times \frac{R_1+R_2+Ls}{LCs^2+C(R_1+R_2)s+1} I_1$$

FROM WHICH

$$Z_{21} = \frac{R_2}{LCs^2+C(R_1+R_2)s+1} \leftarrow$$

$$Z_{22} = \left.\frac{V_2}{I_2}\right|_{I_1=0} \quad \text{AND} \quad Z_{12} = \left.\frac{V_1}{I_2}\right|_{I_1=0}$$

$$Z_{22} = \frac{R_2(R_1+Ls+1/Cs)}{R_1+R_2+Ls+1/Cs} = \frac{R_2(LCs^2+R_1Cs+1)}{LCs^2+C(R_1+R_2)s+1} \leftarrow$$

$$V_2 = I_2 Z_{22} \quad \text{AND} \quad V_1 = \frac{1/Cs}{R_1+Ls+1/Cs} V_2$$

SUBSTITUTION AND DIVISION BY I_2 YIELDS

$$Z_{12} = \frac{R_2}{LCs^2+sC(R_1+R_2)+1} \leftarrow$$

(b)

WHEN $I_2 = 0$, $V_1 = Z_{11} I_1$ AND $V_2 = Z_{21} I_1$

$$\therefore \frac{V_2}{V_1} = \frac{Z_{21}}{Z_{11}} = \frac{R_2}{R_1+R_2+Ls} \leftarrow$$

3.4.19. (a) To calculate y_{11} and y_{21}, excite port 1 by a voltage source V_1 and short port 2. Let $\mu = g_m / g_d$

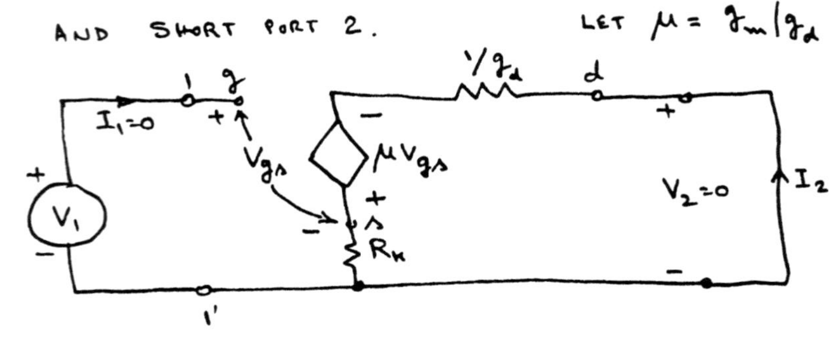

$I_1 = 0$; KVL equation for the output loop is

$$I_2 \frac{1}{g_d} - \mu V_{gs} + I_2 R_K = 0$$

Constraint equation for the controlled source is

$$V_{gs} = V_1 - I_2 R_K$$

Combining: $$I_2 \left[(1+\mu) R_K + \frac{1}{g_d} \right] = \mu V_1$$

$$y_{11} = \left. \frac{I_1}{V_1} \right|_{V_2 = 0} = 0 \quad \leftarrow$$

$$y_{21} = \left. \frac{I_2}{V_1} \right|_{V_2 = 0} = \frac{\mu}{R_K(1+\mu) + 1/g_d} \quad \leftarrow$$

To calculate y_{22} and y_{12}, short port 1 and excite port 2 by a voltage source V_2.

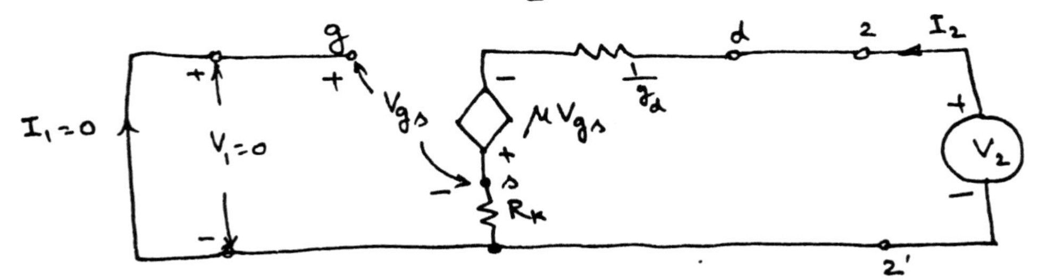

Because of the opencircuit between terminals g and s, $I_1 = 0$

$$y_{12} = \left. \frac{I_1}{V_2} \right|_{V_1 = 0} ; \quad y_{22} = \left. \frac{I_2}{V_2} \right|_{V_1 = 0}$$

$$\therefore y_{22} = \frac{1}{(1+\mu) R_K + 1/g_d} ; \quad y_{12} = 0 \quad \leftarrow$$

3.4.19. (b)

THE y - PARAMETER EQUIVALENT CIRCUIT IS SHOWN BELOW:

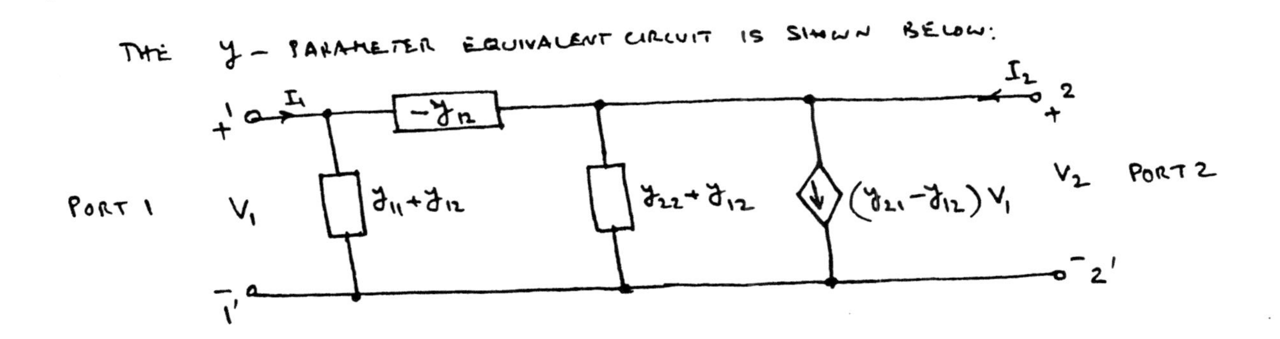

$$y_{11} + y_{12} = 0 \quad ; \quad y_{12} = 0$$

$$y_{21} - y_{12} = y_{21} = \frac{\mu}{R_k(1+\mu) + 1/g_d}$$

$$y_{22} + y_{12} = y_{22} = \frac{1}{(1+\mu)R_k + 1/g_d}$$

FOR $\mu = g_m/g_d >> 1$,

$$y_{21} = \frac{g_m}{1 + g_m R_k} \quad ; \quad y_{22} = \frac{g_d}{1 + g_m R_k}$$

THE RESULTANT y-PARAMETER EQUIVALENT CIRCUIT IS SHOWN BELOW:

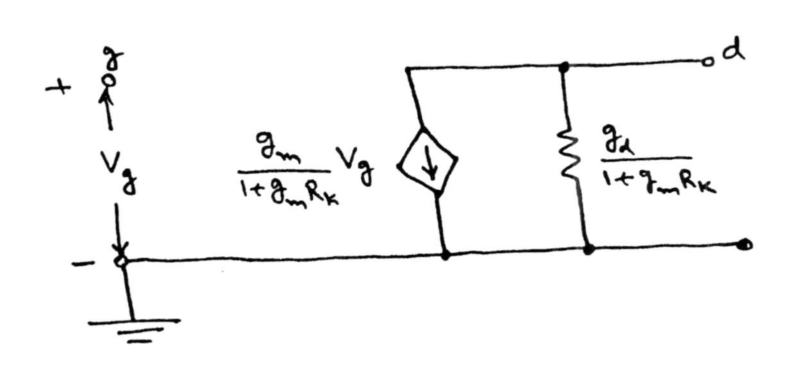

3.4.20.(a)

$$Z_1 = \frac{R_1}{1 + R_1 C_1 s}$$

$$Z_{11} = \left.\frac{V_1}{I_1}\right|_{I_2=0} \quad ; \quad Z_{21} = \left.\frac{V_2}{I_1}\right|_{I_2=0}$$

CONVERTING gV AND R_2 TO ITS VOLTAGE SOURCE EQUIVALENT, THE KVL EQ. IS

$$V_1 = I_1 Z_1 + I_1 R_2 + g R_2 V \quad ; \quad V = I_1 Z_1 \quad \text{SO THAT}$$

$$Z_{11} = \frac{V_1}{I_1} = Z_1(1 + gR_2) + R_2 = R_2(1 + gZ_1) + Z_1 \quad \longleftarrow$$

$$V_2 = (I_1 + gV) R_2 = R_2 I_1 (1 + gZ_1) \quad \text{AND}$$

$$Z_{21} = R_2 (1 + gZ_1) \quad \longleftarrow$$

TO FIND Z_{22} AND Z_{12}, THE FOLLOWING CIRCUIT IS USEFUL.

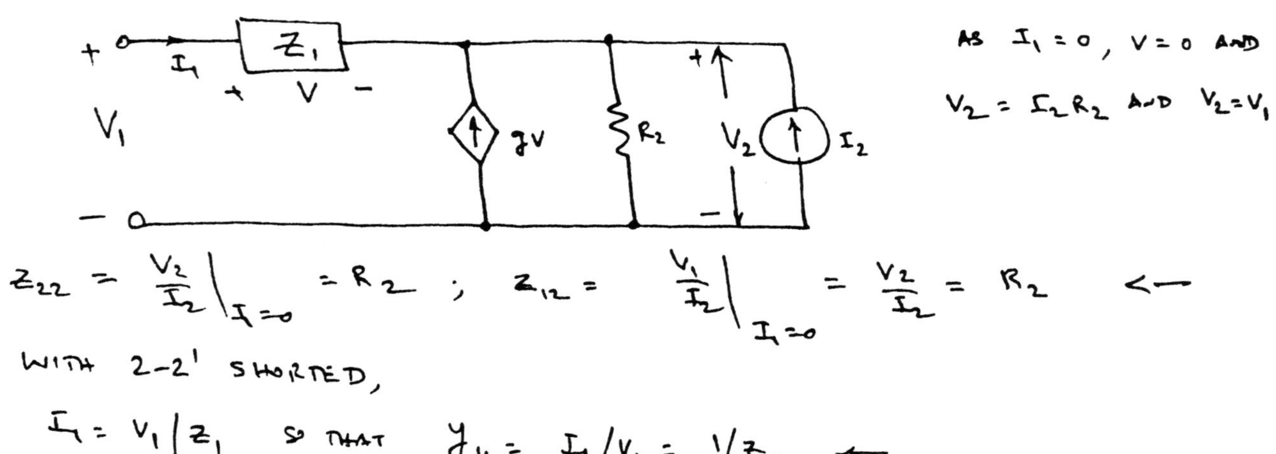

AS $I_1 = 0$, $V = 0$ AND $V_2 = I_2 R_2$ AND $V_2 = V_1$

$$Z_{22} = \left.\frac{V_2}{I_2}\right|_{I_1=0} = R_2 \quad ; \quad Z_{12} = \left.\frac{V_1}{I_2}\right|_{I_1=0} = \frac{V_2}{I_2} = R_2 \quad \longleftarrow$$

WITH 2-2' SHORTED,

$$I_1 = V_1 / Z_1 \quad \text{SO THAT} \quad y_{11} = I_1 / V_1 = 1/Z_1 \quad \longleftarrow$$

$$V = V_1 = I_1 Z_1 \quad \text{AND} \quad I_2 = I_1 + gV = I_1(1 + gZ_1)$$

$$y_{21} = I_2 / V_1 = I_1(1 + gZ_1) / I_1 Z_1 = (1 + gZ_1)/Z_1 \quad \longleftarrow$$

WITH 1-1' SHORTED, $V = -V_2$ AND

$$I_2 = \frac{1}{R_2} V_2 - gV + \frac{V_2}{Z_1} \quad \text{AND} \quad y_{22} = \frac{I_2}{V_2} = \frac{1}{R_2} + g + \frac{1}{Z_1} \quad \longleftarrow$$

$$I_1 = V_2 / Z_1 \quad \text{SO THAT} \quad y_{12} = I_1 / V_2 = 1/Z_1 \quad \longleftarrow$$

3.4.21.
(b)

SETTING $V_2 = 0$, SOLVING FOR I_2 / I_1 YIELDS (SEE EQ. 3.4.14)

$$\left.\frac{I_2}{I_1}\right|_{V_2=0} = -\frac{z_{21}}{z_{22}} = -\frac{R_2(1+gZ_1)}{R_2} = -(1+gZ_1) \longleftarrow$$

3.4.21.

TO OBTAIN h-PARAMETERS:

h_{11} & h_{21}: APPLY I_1 AND MEASURE V_1 & I_2 WITH PORT 2 SHORTED.

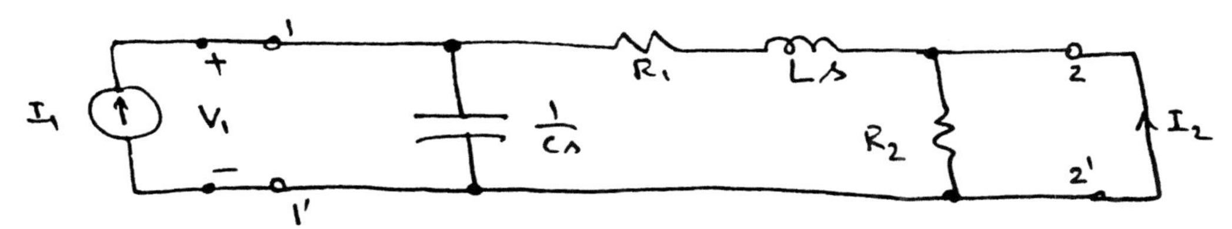

NO CURRENT EXISTS IN R_2. BY CURRENT-DIVIDER TECHNIQUES,

$$I_2 = -I_1 \frac{1/Cs}{1/Cs + Ls + R_1}$$

AND $$h_{21} = \left.\frac{I_2}{I_1}\right|_{V_2=0} = -\frac{1}{LCs^2 + R_1Cs + 1} \longleftarrow$$

$$V_1 = I_1 \frac{R_1 + Ls}{1/Cs + Ls + R_1} \times \frac{1}{Cs} \quad \text{AND}$$

$$h_{11} = \left.\frac{V_1}{I_1}\right|_{V_2=0} = \frac{R_1 + Ls}{LCs^2 + R_1Cs + 1} \longleftarrow$$

TO OBTAIN h_{22} & h_{12}: APPLY V_2 AND MEASURE I_2 & V_1 WHEN PORT IS OPEN-CIRCUITED.

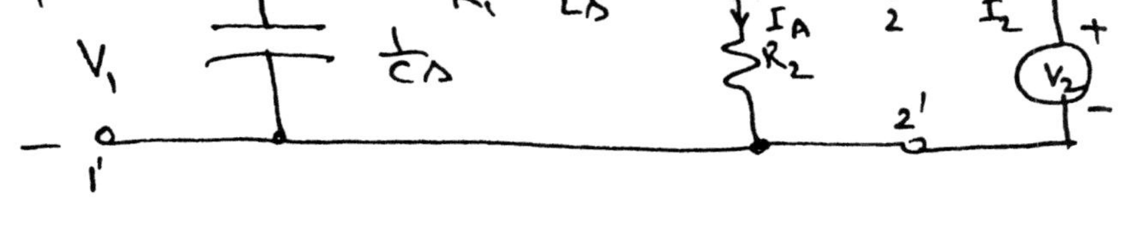

3.4.21. CONTINUED

$$I_2 = I_A + I_B = \frac{V_2}{R_2} + \frac{V_2}{R_1 + Ls + 1/Cs}$$

$$h_{22} = \left.\frac{I_2}{V_2}\right|_{I_1=0} = \frac{1}{R^2} + \frac{Cs}{LCs^2 + R_1Cs + 1} \quad \longleftarrow$$

$$V_1 = I_B \cdot \frac{1}{Cs} = \frac{V_2}{R_1 + Ls + 1/Cs} \times \frac{1}{Cs}$$

$$h_{12} = \left.\frac{V_1}{V_2}\right|_{I_1=0} = \frac{1}{LCs^2 + R_1Cs + 1} \quad \longleftarrow$$

3.4.22.

FOR h_{11} & h_{21}: APPLY I_1 AND MEASURE V_1 & I_2 WITH PORT 2 SHORTED.

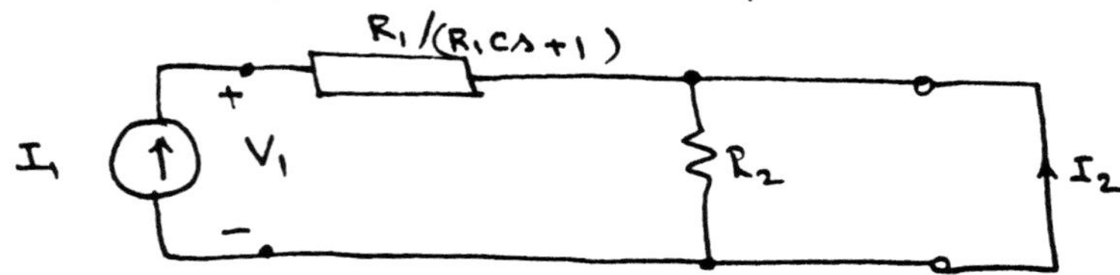

$$V_1 = I_1 \frac{R_1}{R_1Cs+1} \quad \text{AND} \quad \left.\frac{V_1}{I_1}\right|_{V_2=0} = h_{11} = \frac{R_1}{R_1Cs+1} \quad \longleftarrow$$

$$I_2 = -I_1 \quad \text{AND} \quad h_{21} = \left.\frac{I_2}{I_1}\right|_{V_2=0} = -1 \quad \longleftarrow$$

FOR h_{22} AND h_{12}, APPLY V_2 AND MEASURE I_2 & V_1 WHEN PORT 1 IS OPEN-CIRCUITED.

+ V_1 − R1/(R1Cs+1) I_2 R_2 V_2 + −

AS $I_1 = 0$, $V_2 = R_2 I_2$ AND $h_{22} = \left.\frac{I_2}{V_2}\right|_{I_1=0} = \frac{1}{R_2}$ ⟵

$V_1 = V_2$ AND $h_{12} = \left.\frac{V_1}{V_2}\right|_{I_1=0} = 1$ ⟵

3.4.22. CONTINUED

FROM h- PARAMETER EQNS

$$I_2 = h_{21} I_1 + h_{22} V_2 \quad \text{AND} \quad V_1 = h_{11} I_1 + h_{12} V_2$$

WITH $I_2 = 0$, $\quad I_1 = -V_2 \left(h_{22} / h_{21} \right)$

$$V_1 = -\frac{h_{11} h_{22}}{h_{21}} V_2 + h_{12} V_2$$

OR

$$\frac{V_2}{V_1} = -\frac{h_{21}}{h_{11} h_{22} - h_{21} h_{12}}$$

$$= -\frac{-1}{\dfrac{R_1}{R_1 Cs + 1} \cdot \dfrac{1}{R_2} - (-1)(1)} = \frac{1}{\dfrac{R_1}{R_2 (R_1 Cs + 1)} + 1}$$

$$= \frac{R_2 (R_1 Cs + 1)}{R_1 + R_2 (1 + R_1 Cs)} = \frac{R_2 (R_1 Cs + 1)}{R_1 + R_2 + R_1 R_2 Cs}$$

3.4.23.

(a) FOR h_{11} & h_{12}:

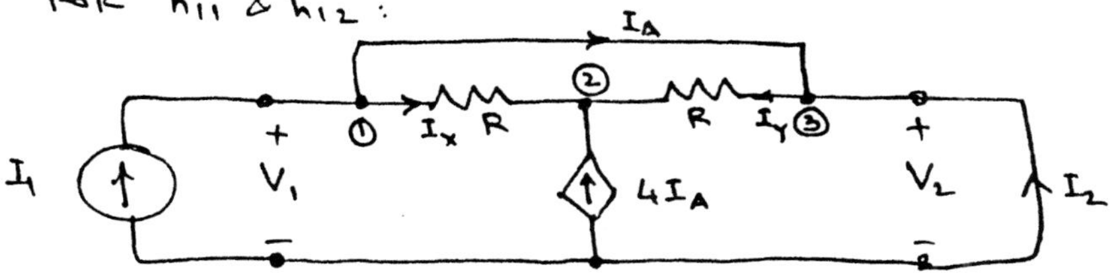

KCL: $\quad I_1 = I_X + I_A \quad$ AT NODE ①

$I_X + I_Y = -4 I_A \quad$ " " ②

$I_2 = I_Y - I_A \quad$ " " ③

KVL AROUND OUTER LOOP: $\quad V_1 = 0$

AROUND LOOPS CONTAINING THE TWO RESISTORS: $\quad I_X R - I_Y R = 0 \quad$ OR $I_X = I_Y$

THUS FROM ②, $\quad 2 I_X = -4 I_A \quad$ or $\quad I_X = -2 I_A = I_Y$

SUBSTITUTION INTO ① AND ③ GIVES

$$I_1 = -2 I_A + I_A = -I_A \quad \text{AND} \quad I_2 = -2 I_A - I_A = -3 I_A$$

$$\therefore h_{11} = \left.\frac{V_1}{I_1}\right|_{V_2=0} = 0 \; ; \quad h_{21} = \left.\frac{I_2}{I_1}\right|_{V_2=0} = \frac{-3 I_A}{-I_A} = 3$$

3.4.23. CONTINUED (a)

FOR h_{12} & h_{22}:

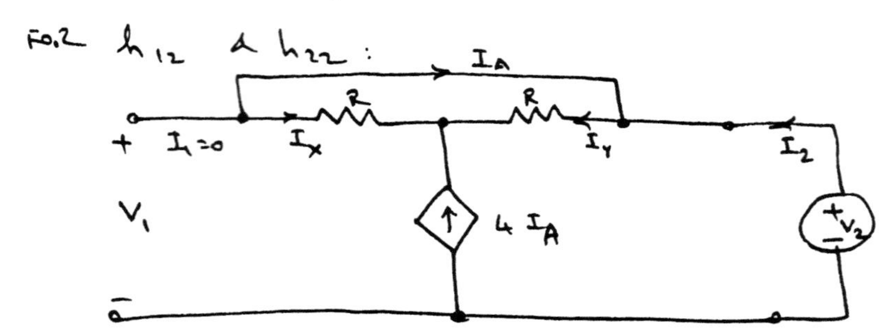

AROUND OUTSIDE LOOP: $V_1 = V_2$

KCL: $I_2 = I_Y - I_A$; $I_A = -I_X$; $I_X + I_Y = -4I_A$

KVL FOR THE TOPMOST LOOP: $I_X R - I_Y R = 0$ & $I_X = I_Y$

$\because I_A = -I_X$, $2I_X = -4I_A$ & $I_X = -2I_A$

THE ONLY VALUE THAT SATISFIES BOTH CONDITIONS IS

$$I_X = I_Y = I_A = 0$$

$$\therefore I_2 = 0$$

$$h_{22} = \left.\frac{I_2}{V_2}\right|_{I_1=0} = 0 \quad \text{AND} \quad h_{12} = \left.\frac{V_1}{V_2}\right|_{I_1=0} = 1$$

3.4.23. (b)

h-PARAMETERS DEFINE THE FOLLOWING EQUATIONS,

$V_1 = h_{12} V_2$ AND $I_2 = h_{21} I_1$

CONNECTING R ACROSS PORT 2 CAUSES $V_2 = -R_2 I_2$

SUBSTITUTION YIELDS

$$V_1 = h_{12}(-R I_2) = h_{12}(-R\, h_{21} I_1)$$

$$Z_{in} = \frac{V_1}{I_1} = -h_{12} h_{21} R = -(1)(3)R = -3R$$

3.4.24.

1. MOVE 3rd SUMMING JUNCTION TO RIGHT

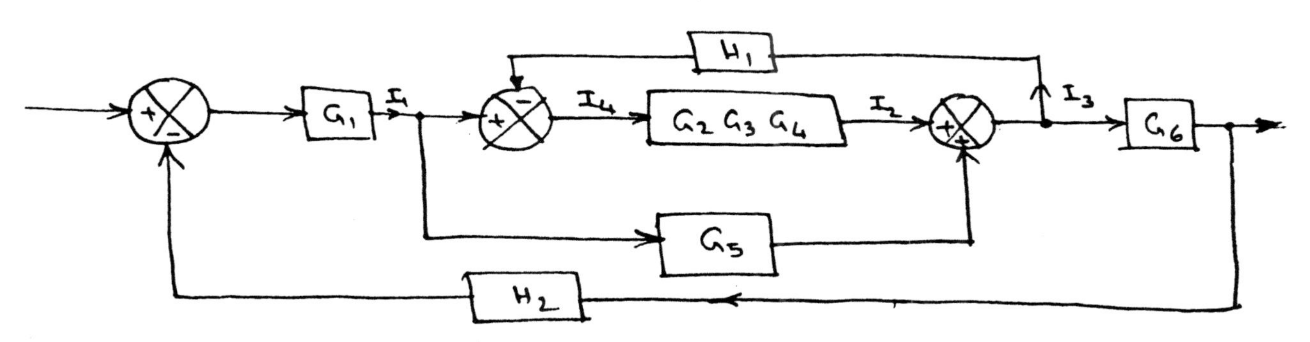

2. LET $G_2G_3G_4 = G_x$

3. SET UP EQNS AT JUNCTIONS:

$$I_4 = I_1 - H_1 I_3$$

$$I_2 = G_x I_4$$

$$I_3 = I_2 + G_5 I_1$$

4. ELIMINATE I_4 & I_2:

$$\frac{I_3}{I_1} = \frac{G_x + G_5}{1 + G_x H_1} = G_y$$

5. REDRAW SIMPLIFIED BLOCK DIAGRAM

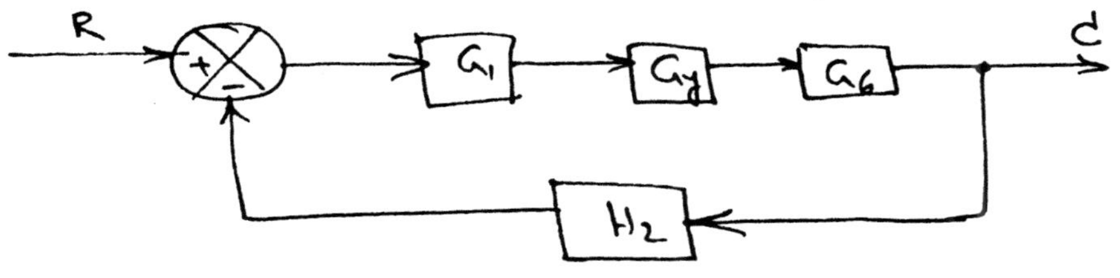

6.

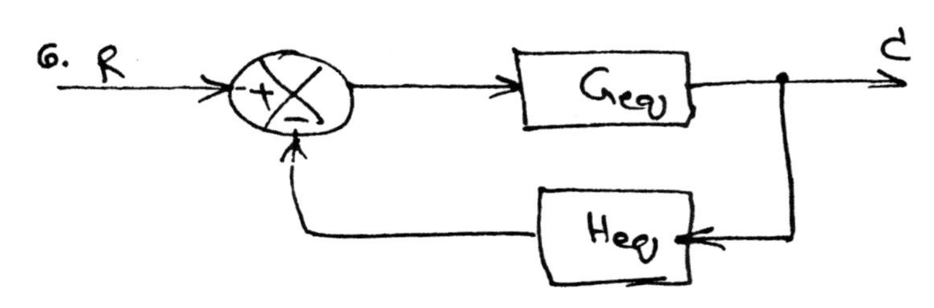

$$G_{eq} = G_1 G_6 G_y; \quad H_{eq} = H_2$$

3.4.25.

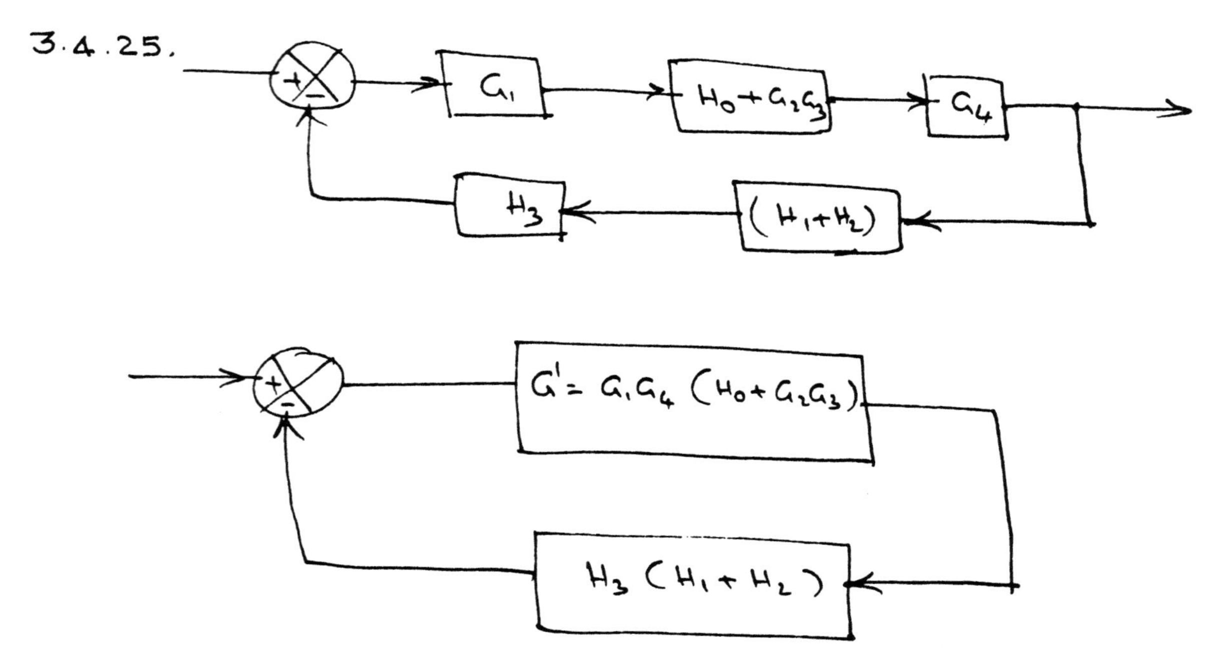

3.4.26.

(a)

1. IDENTIFY X_1 AND X_2 AS INPUTS TO G_1 & G_2, RESPECTIVELY.
2. BY REDUCTION, THE $G_2 - G_3 - H_2$ COMBINATION CAN BE REPLACED BY A SINGLE BLOCK WHOSE TRANSFER FN IS

$$G_2G_3/(1 - G_2G_3H_2)$$

3. THIS BLOCK IS IN CASCADE WITH G_1 AND H_1 IS A FEEDBACK TRANSMISSION AROUND THE CASCADE.

THUS
$$\frac{C(s)}{R(s)} = \frac{G_1G_2G_3/(1 - G_2G_3H_2)}{1 + G_1G_2G_3H_1/(1 - G_2G_3H_2)}$$

$$= \frac{G_1G_2G_3}{1 - G_2G_3H_2 + G_1G_2G_3H_1} \longleftarrow$$

(b) FOR $G_2G_3H_2 = 1$

$$\frac{C(s)}{R(s)} = \frac{1}{H_1} \longleftarrow$$

3.4.27. G_3 AND H_3 CAN BE COMBINED TO GIVE

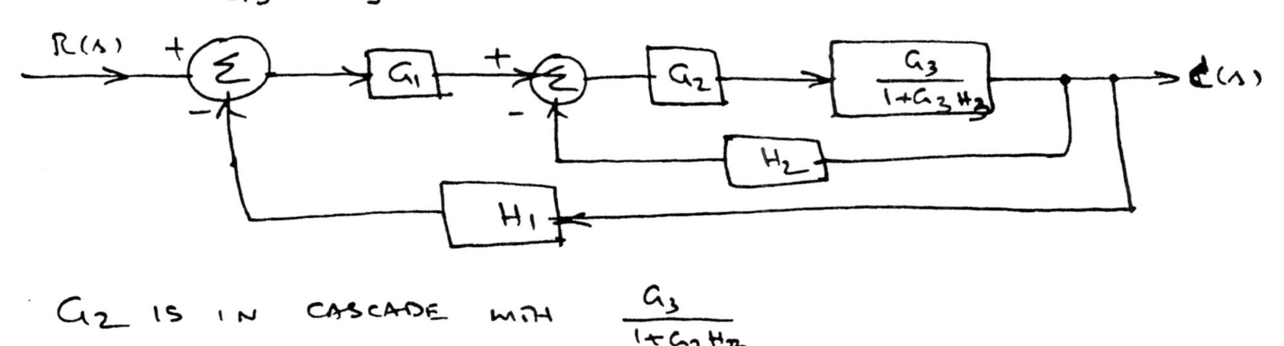

G_2 IS IN CASCADE WITH $\dfrac{G_3}{1+G_3H_3}$

THE CASCADE IS IN A FEEDBACK ARRANGEMENT WITH H_2

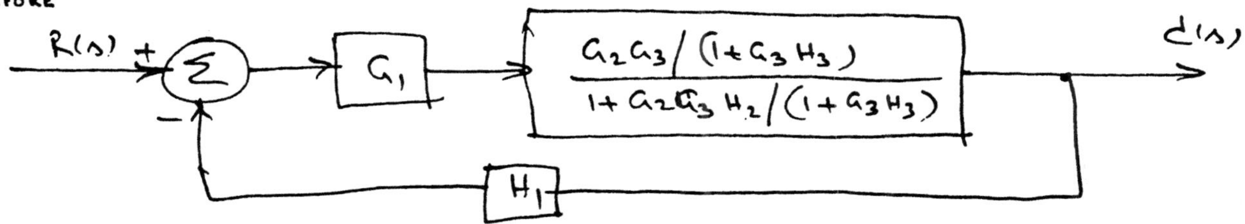

THE EQUIV. BLOCK IS OBTAINED FROM THE CASCADE OF THE TWO FORWARD BLOCKS IN A FEEDBACK CONNECTION WITH H_1. THE RESULTING TR. FN. IS
(TRANSFER FUNCTION)

$$\frac{C(s)}{R(s)} = \frac{G_1G_2G_3}{1+G_3H_3+G_2G_3H_2+G_1G_2G_3H_1} \longleftarrow$$

3.4.28. (a)

FROM FIRST EQ.

$$I_1 = \frac{1}{Z_{11}}V_1 - \frac{Z_{12}}{Z_{11}}I_2$$

FROM 2nd EQ.

$$I_2 = -\frac{Z_{21}}{Z_{22}+Z_L}I_1$$

IDENTIFY VARIABLES V_1, I_1, I_2, AND V_2.

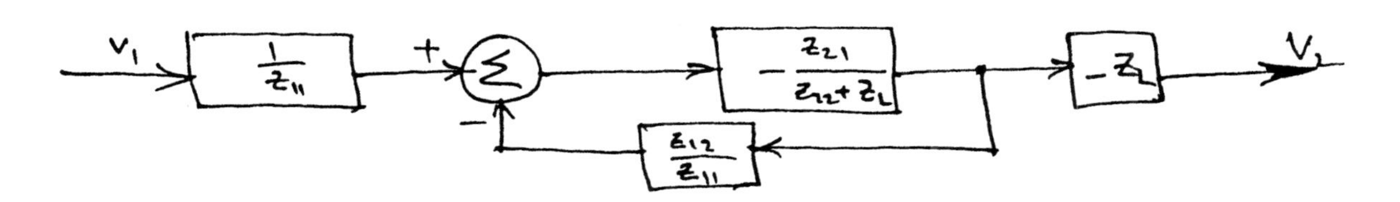

3.4.28. (b)

The blocks $\frac{-z_{21}}{z_{22}+Z_L}$ & $\frac{z_{12}}{z_{11}}$ FORM A FEEDBACK LOOP, THE EQUIV. OF WHICH IS CASCADED WITH $\frac{1}{z_{11}}$ AND $-Z_L$.

$$\frac{V_2}{V_1} = \frac{1}{z_{11}} \times \frac{-z_{21}/(z_{22}+Z_L)}{1+\left[-z_{21}/(z_{22}+z_L)\right](z_{12}/z_{11})} \times (-z_L)$$

$$= \frac{z_{21} Z_L}{z_{11}(z_{22}+Z_L) - z_{12} z_{21}} \quad \longleftarrow$$

3.5.1.

THE CIRCUIT DIAGRAM IS GIVEN BELOW:

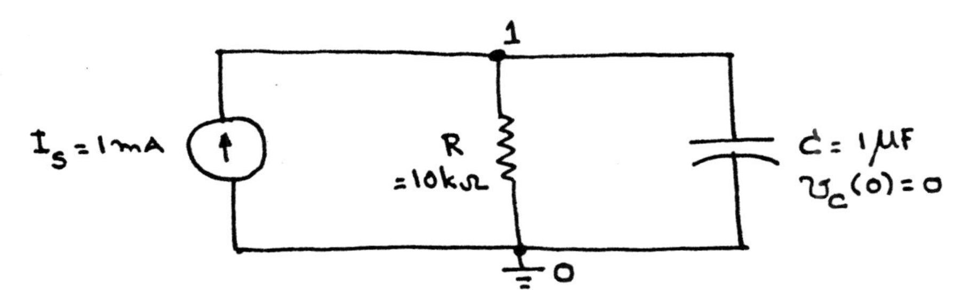

The program listing is:

```
PROBLEM 3.5.1
IS    0  1  1MA
R     1  0  10K
C     1  0  1UF   IC=0V
*TIME CONSTANT = RC = 10MS
*CHOOSE TSTEP = 0.01*(TIME CONSTANT)
*CHOOSE TMAX = 5*(TIME CONSTANT)
.TRAN 0.1MS 50MS 0 0.1MS UIC
.PROBE
.END
```

After running the program, Probe was used to obtain a plot of the voltage across the capacitor [i.e., the voltage at node 1 denoted as V(1)].

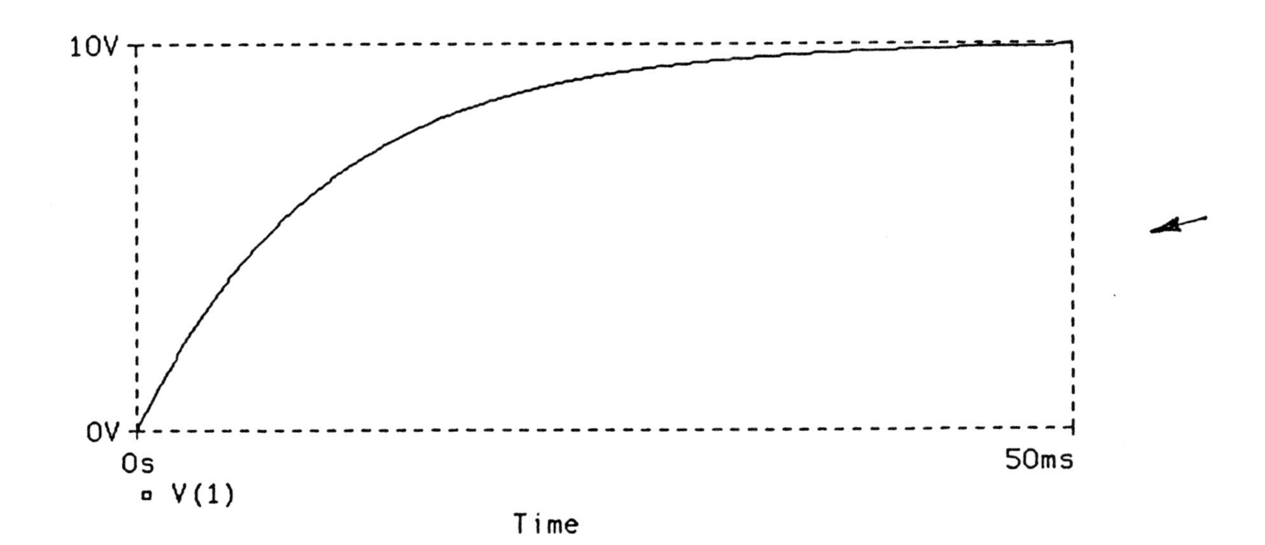

3.5.2.

HERE WE HAVE A SINUSOIDAL SOURCE GIVEN BY $v_s(t) = 2\sin(200t)$

THE PARAMETERS ARE THEN SELECTED AS FOLLOWS:

VDC $= 0$

VPEAK $= 2$

TD $= 0$

2π FREQ $= 200$ (thus FREQ $= 200/2\pi = 31.83$)

DF $= 0$

PHASE $= 0$

THE PROGRAM LISTING IS GIVEN BELOW:

```
PROBLEM 3.5.2
*THE CIRCUIT IS SHOWN IN FIGURE F3.5.2
VS  1  0  SIN(0 2 31.83 0 0 0)
R  1  2  5K
C  2  0  1UF  IC=1
.TRAN  0.2MS  80MS  0  0.2MS  UIC
.PROBE
.END
```

AFTER RUNNING THIS PROGRAM, WE START PROBE AND USE THE MENU COMMANDS TO ADD A TRACE OF I(R), WHICH IS THE CURRENT FLOWING THROUGH THE RESISTANCE R REFERENCED FROM THE FIRST NODE TO THE SECOND IN THE COMPONENT STATEMENT FOR R. THE RESULTING PLOT IS SHOWN BELOW:

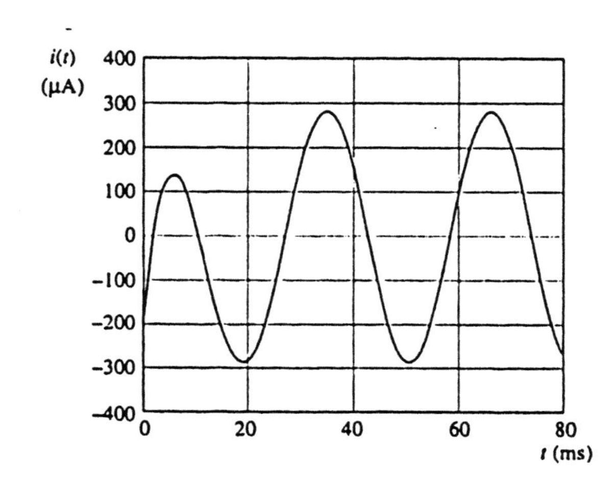

3.5.3.

THE PSPICE PROGRAM IS GIVEN BELOW:

```
PROBLEM 3.5.3

*THE CIRCUIT IS SHOWN IN FIGURE P3.5.3
C  1  0  1UF  IC=10V
R  1  0  2MEG
.TRAN  0.02  10  0  0.02  UIC
.PROBE
.END
```

AFTER RUNNING THE PSPICE PROGRAM, PROBE IS STARTED. THEN THE MENU COMMANDS ARE USED TO ADD A TRACE OF V(1), WHICH IS THE VOLTAGE OF NODE 1 OF THE CIRCUIT. THE RESULTING SCREEN DISPLAY IS SHOWN BELOW. AS EXPECTED, THE VOLTAGE DECAYS EXPONENTIALLY TO ZERO WITH A 2-SEC. TIME CONSTANT.

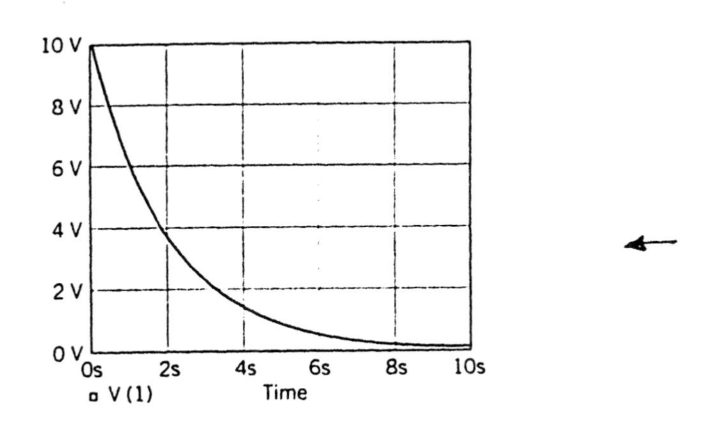

3.5.4.

THE CIRCUIT DIAGRAM IS SHOWN BELOW:

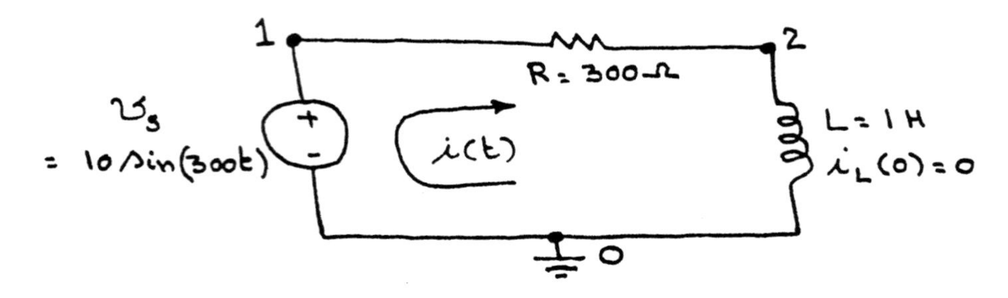

The program listing is:

```
PROBLEM 3.5.4
VS    1   0   SIN(0 10 47.57HZ)
R     1   2   300
L     2   0   1H IC=0
.TRAN 0.1MS 100MS 0 0.1MS UIC
.PROBE
.END
```

We have selected the time step to be small compared to the period of the sinewave (which is 20.9 ms) and small compared to the circuit time constant (which is 3.33 ms). Also we picked TMAX to include about 5 cycles of the source.

After running the program, Probe was used to obtain a plot of the current, denoted by Probe as I(R).

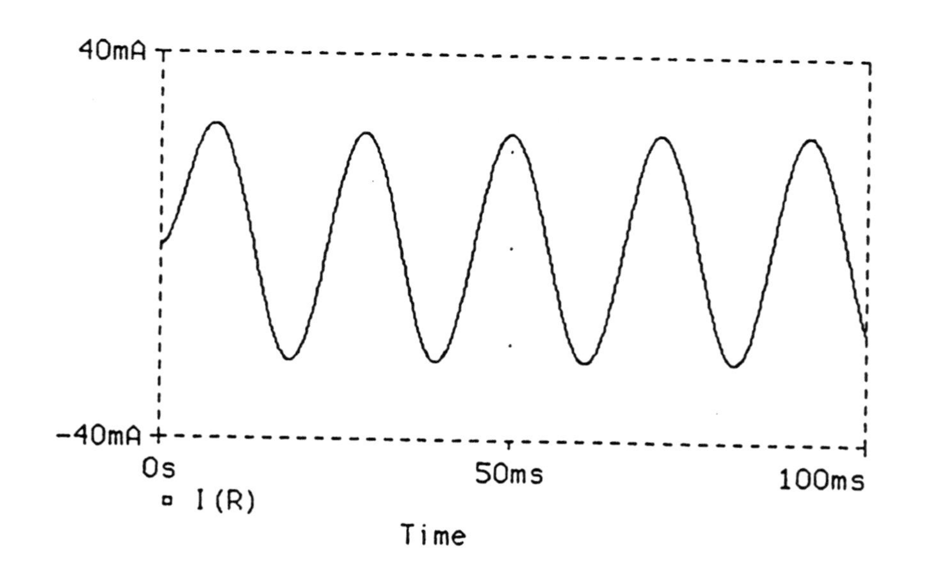

3.5.5.

THE CIRCUIT DIAGRAM IS SHOWN BELOW:

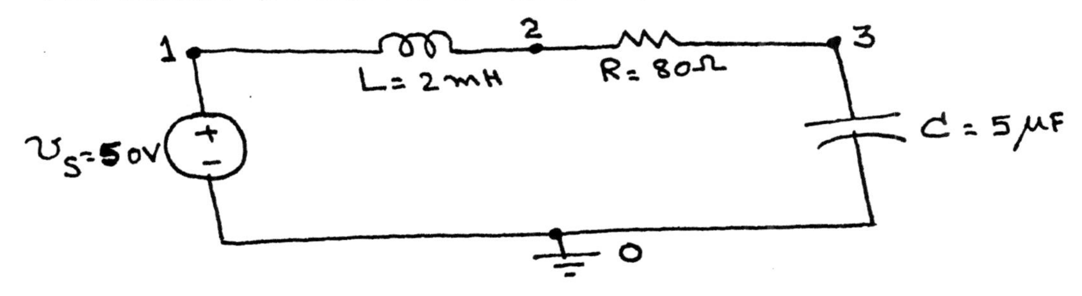

The program listing is:

```
PROBLEM P3.5.5
VS    1   0   50V
L     1   2   2MH
R     2   3   80
C     3   0   5UF
.TRAN 1US 2MS 0 1US UIC
.PROBE
.END
```

After running the program, Probe was used to obtain a plot of the voltage across the capacitance, denoted by Probe as V(3). The program can be easily modified to obtain results for R = 40 Ω and R = 20 Ω.

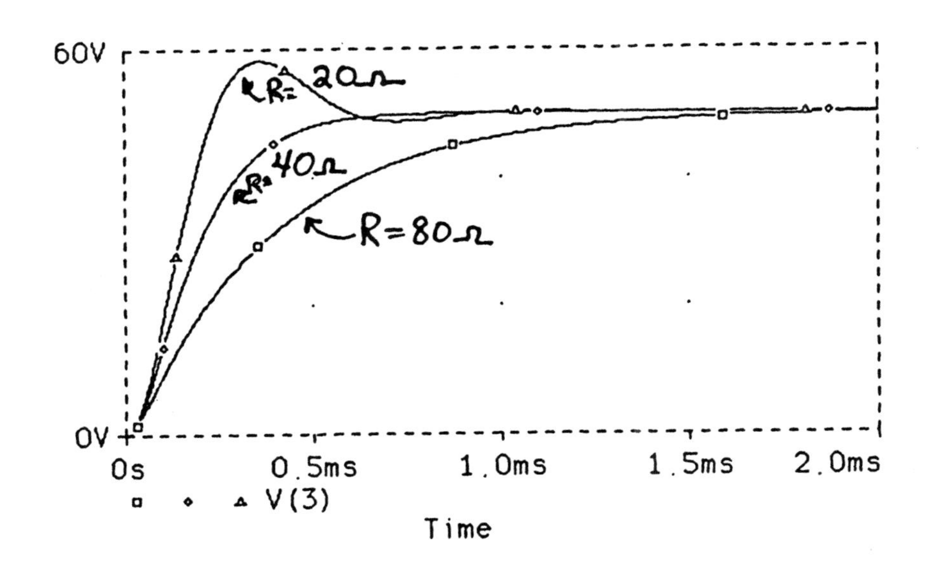

3.5.6.

THE PROGRAM LISTING IS GIVEN BELOW

```
PROBLEM 3.5.6

I1  0  1  AC  2     -90
I2  0  2  AC  1.5  0
R    1  0  10
C    1  2  2000UF
L    2  0  0.1H
.AC  LIN  1  15.92  15.92
.PRINT  AC  VM(1)  VP(1)  VM(2)  VP(2)
.END
```

The output file contains the results

FREQ	V(1)	VP(1)	V(2)	VP(2)
1.592E+01	1.613E+01	2.971E+01	2.802E+01	2.028E+00

THUS THE PHASOR VOLTAGE AT NODE 1 IS $\bar{V}_1 = 16.13 \angle 29.71^\circ$

AND $\bar{V}_2 = 28.02 \angle 2.03^\circ$

3.5.7.

THE CIRCUIT DIAGRAM IS SHOWN BELOW WITH NODE NUMBERS:

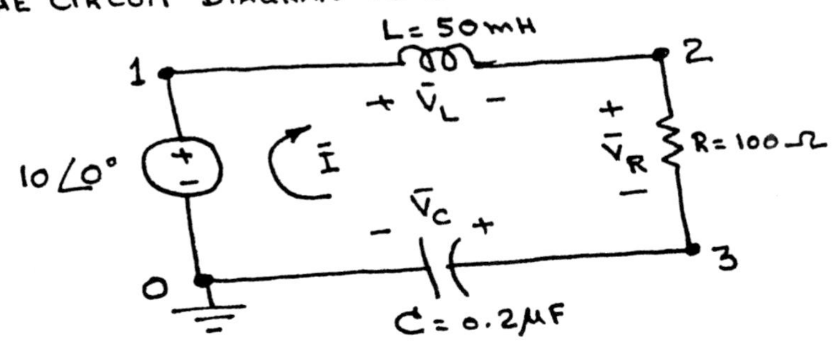

The program listing is

```
PROBLEM P3.5.7
VS   1   0   AC   10V   0
L    1   2   50MH
R    2   3   100OHMS
C    3   0   0.2UF
.AC   LIN   1   1591.5 1591.5
.PRINT AC   IM(R) IP(R) VM(1,2) VP(1,2)
.PRINT AC   VM(2,3) VP(2,3) VM(3) VP(3)
.END
```

In the output file we find the following results:

FREQ	IM(R)	IP(R)	VM(1,2)	VP(1,2)
1.592E+03	1.000E-01	1.780E-02	5.000E+01	9.002E+01

FREQ	VM(2,3)	VP(2,3)	VM(3)	VP(3)
1.592E+03	1.000E+01	1.780E-02	5.000E+01	-8.998E+01

Thus $\bar{I} = 0.1\angle 0.018^\circ$ A

$\bar{V}_L = 50\angle 90.02^\circ$ V

$\bar{V}_R = 10\angle 0.018^\circ$ V

$\bar{V}_C = 50\angle -89.98^\circ$ V

3.5.8. THE CIRCUIT DIAGRAM INCLUDING NODE NUMBERS IS SHOWN BELOW:

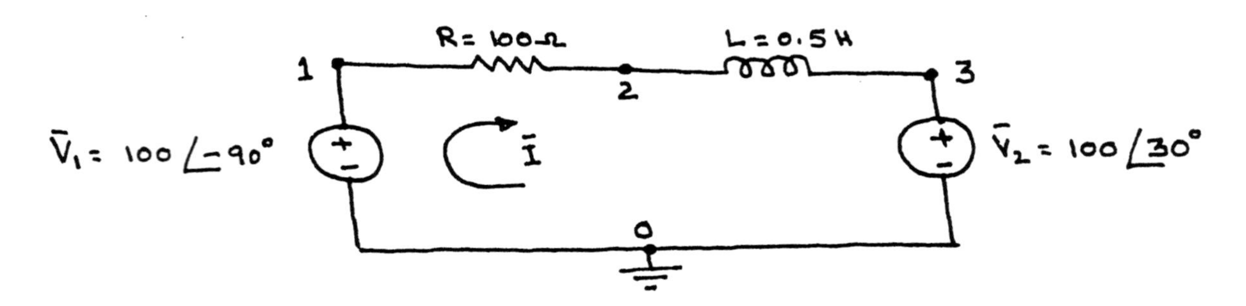

The program listing is

```
PROBLEM 3.5-8
V1   1  0  AC   100V   -90DEGREES
R    1  2  100
L    2  3  0.5H
V2   3  0  AC   100V   30DEGREES
.AC   LIN  1  15.915 15.915
.PRINT AC   IM(R) IP(R)
.END
```

In the output file we find the following results:

FREQ	IM(R)	IP(R)
1.592E+01	1.549E+00	-1.466E+02

Thus $\bar{I} = 1.549\angle -146.6^\circ$ ←

3.5.9. The program listing is:

```
PROBLEM 3.5.9
VIN    1    0    AC   1
R1     1    2    200
R2     2    3    200
C1     2    0    1UF
C2     3    0    1UF
.AC   DEC 100    10HZ    100KHZ
.PROBE
.END
```

Requesting a plot of VDB(3) produces the Bode magnitude plot:

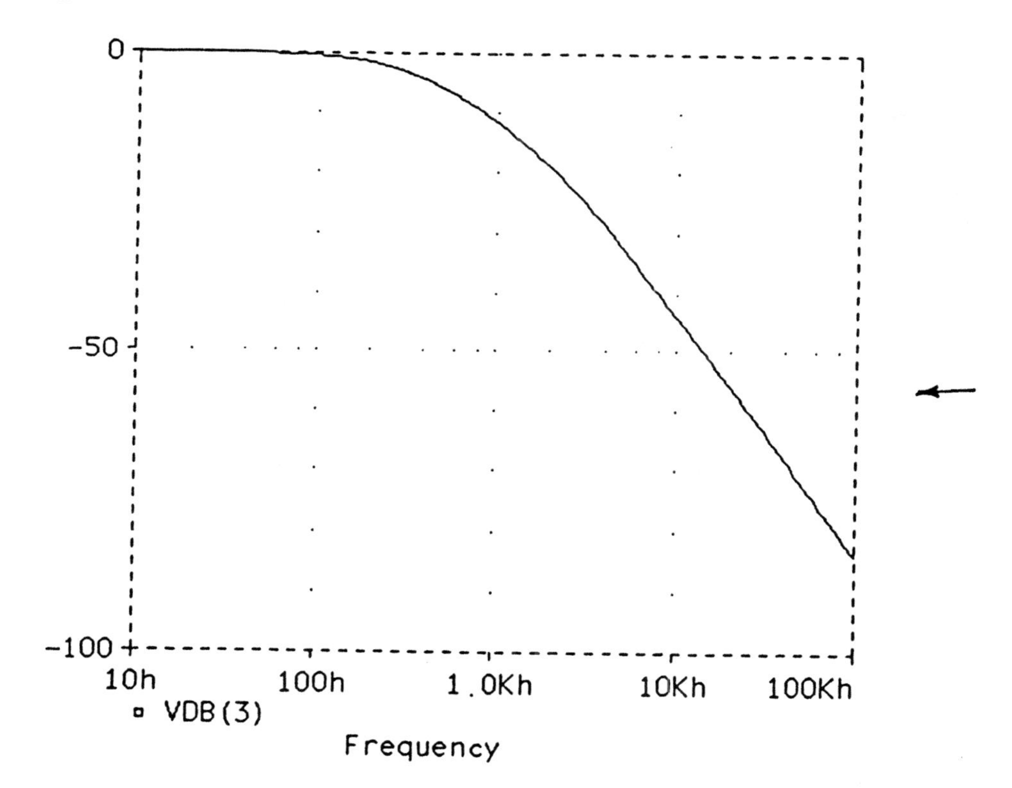

Using the cursor feature of Probe, we find that the 3-dB frequency is approximately 296 Hz. The rate of the roll off is 40 dB per decade.

3.5.10. The program listing is:

```
PROBLEM 3.5.10
VIN    1    0    AC  1
C1     1    2    0.1UF
C2     2    3    0.1UF
R1     2    0    2K
R2     3    0    2K
.AC   DEC 100   10HZ   100KHZ
.PROBE
.END
```

Requesting a plot of VDB(3) produces the Bode magnitude plot:

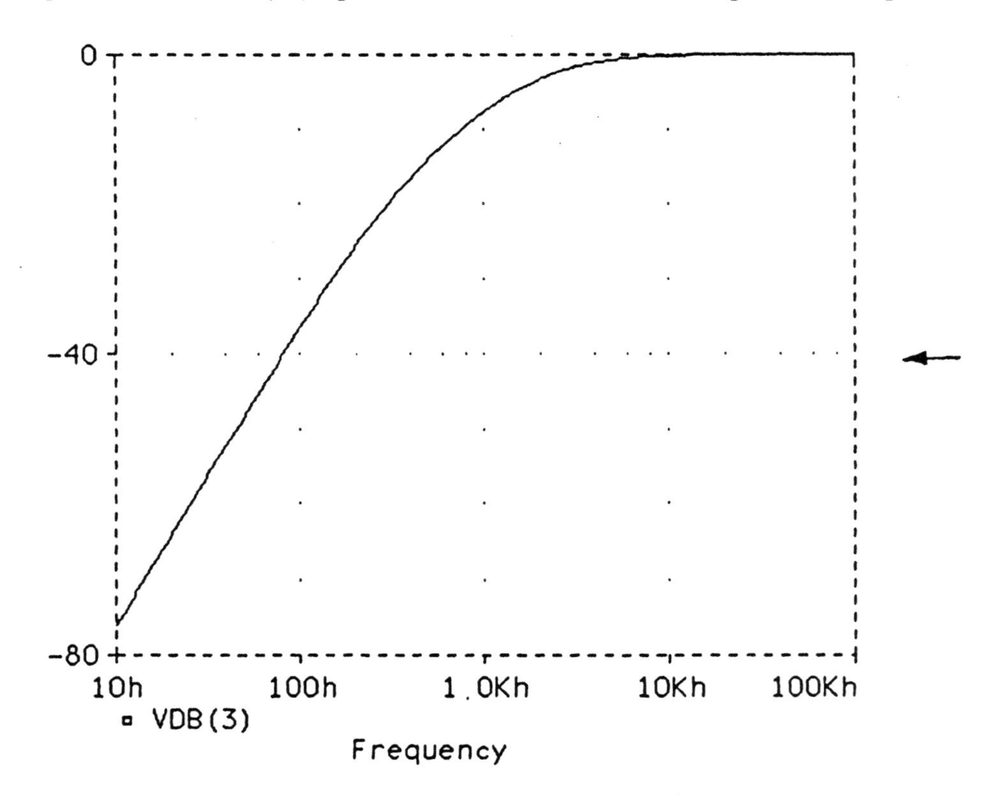

Using the cursor feature of Probe, we find that the 3-dB frequency is approximately 2.13 kHz. The rate of the roll off is 40 dB per decade.

3.5.11.

The program listing is:

```
PROBLEM 3.5.11
VIN    1    0    AC   1
R1     1    2    200
R2     3    0    2K
C1     2    0    10UF
C2     2    3    0.1UF
.AC   DEC 100   1HZ   1MEGHZ
.PROBE
.END
```

Requesting a plot of VDB(3) produces the Bode magnitude plot:

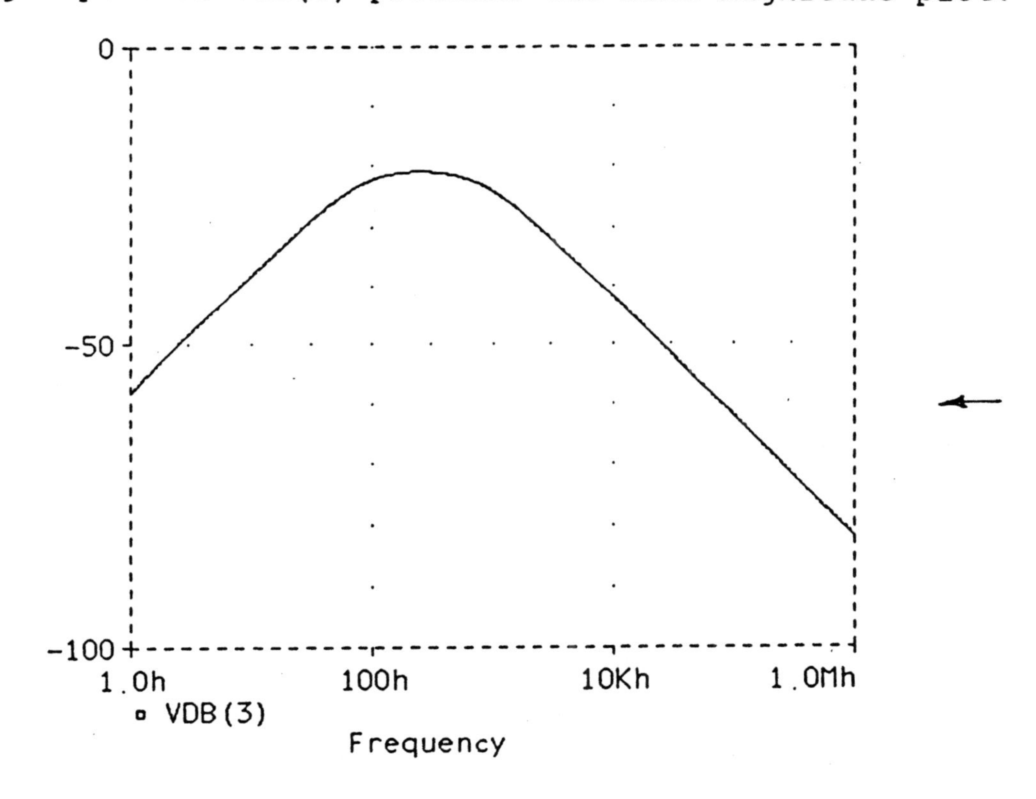

This is a bandpass filter. Using the cursor feature of Probe, we find that the lower 3-dB frequency is approximately 66.5 Hz and the upper 3-dB frequency is approximately 950 Hz. The rate of the roll off on both ends is 20 dB per decade.

3.6.1.

```
function problem361
clc

syms t Tc Va To

% Waveform parameters
Tc = 0.002
Va = 10
To = 5*Tc

% Waveform for 0<t<To
v = Va*exp(-t/Tc);

% RMS calculation and value
Vrms = eval(sqrt(int(v*v,t,0,To)/To))
```

```
Tc =

    0.0020


Va =

    10


To =

    0.0100


Vrms =

    3.1622        ←
```

3.6.2.

```
function problem362
clc

syms t s
% Given waveform f(t) for t>0

ft = 200*t*exp(-25*t)+10*exp(-50*t)*sin(40*t)

% F(s)
Fs = factor(laplace(ft))

[num,den] = numden(Fs);

% Pole Locations
P = eval(solve(den,s))

% Zero Locations
Z = eval(solve(num,s))

% Pole-Zero Diagram
plot(P,'x');hold on
plot(Z,'o');hold off
axis([-100 10 -50 50])
grid on
xlabel('Re(p_i), Re(z_k)')
ylabel('Im(p_i), Im(z_k)')
```

```
ft =

200*t*exp(-25*t)+10*exp(-50*t)*sin(40*t)


Fs =

200*(3*s^2+200*s+5350)/(s+25)^2/(s^2+100*s+4100)  ←


P =

 -25.0000
 -25.0000
 -50.0000 +40.0000i
 -50.0000 -40.0000i


Z =

 -33.3333 +25.9272i
 -33.3333 -25.9272i
```

3.6.2. (CONTD.)

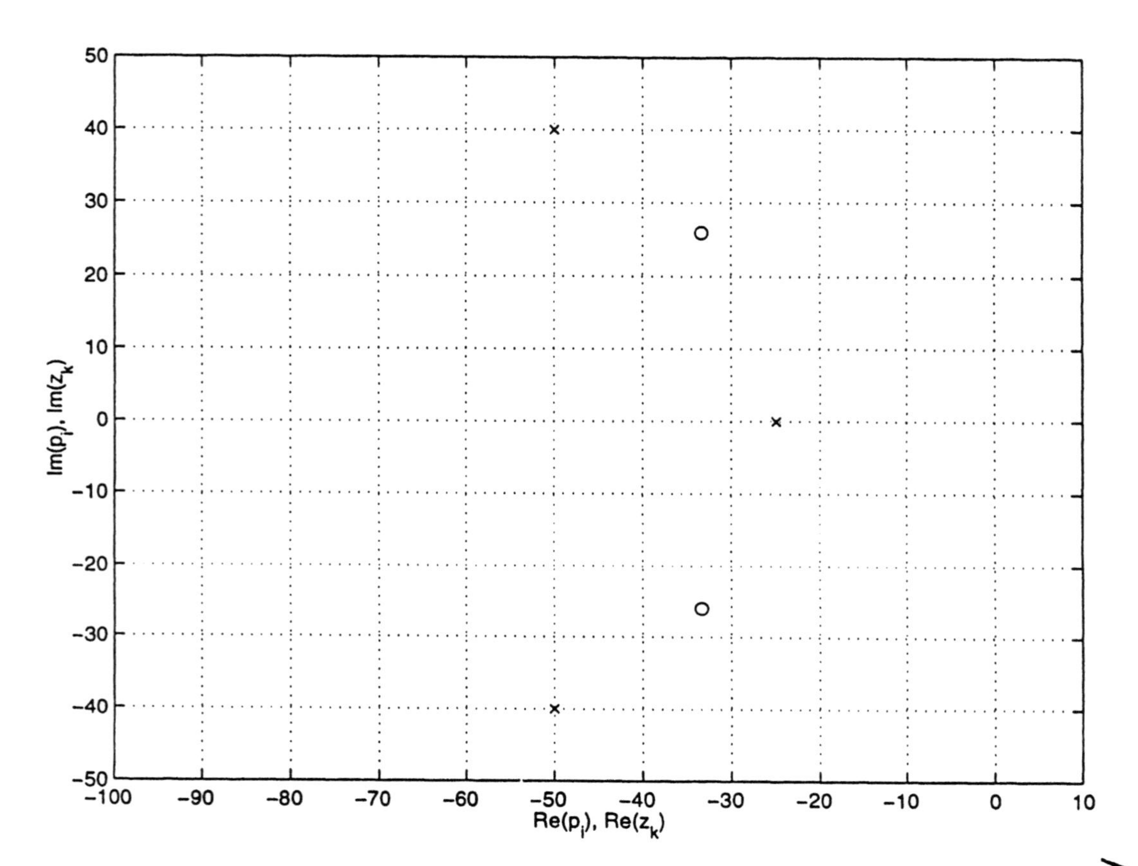

POLE-ZERO PLOT OF THE TRANSFORM $F(s)$

3.6.3.

```
function problem363
clc
digits(3)

syms t s K

% Given F(s) with unknown K
Fs = K*(s+400)/(s^2+400^2)^2/(s+1000)

% Evaluating K from F(0)
equ01 = 'K*400/(400^2)^2/1000=2e-4'
K = solve(equ01,'K')

% Given Transform F(s)
Fs = K*(s+400)/(s^2+400^2)^2/(s+1000)

% Generated Inverse Laplace Transform ; f(t)
ft = ilaplace(Fs)

% Plot f(t)
ezplot(ft,[0,0.06])
xlabel('t')
ylabel('f(t)')
```

```
Fs =

K*(s+400)/(s^2+160000)^2/(s+1000)


equ01 =

K*400/(400^2)^2/1000=2e-4


K =

12800000.


Fs =

(12800000.*s+.512e10)/(s^2+160000)^2/(s+1000)


ft =

-.571e-2*exp(-1000.*t)+.571e-2*cos(400.*t)+.340e-1*sin(400.*t)-19.3*t*cos(400.*t)
                                                  +8.28*t*sin(400.*t)
```

3.6.3. CONTINUED

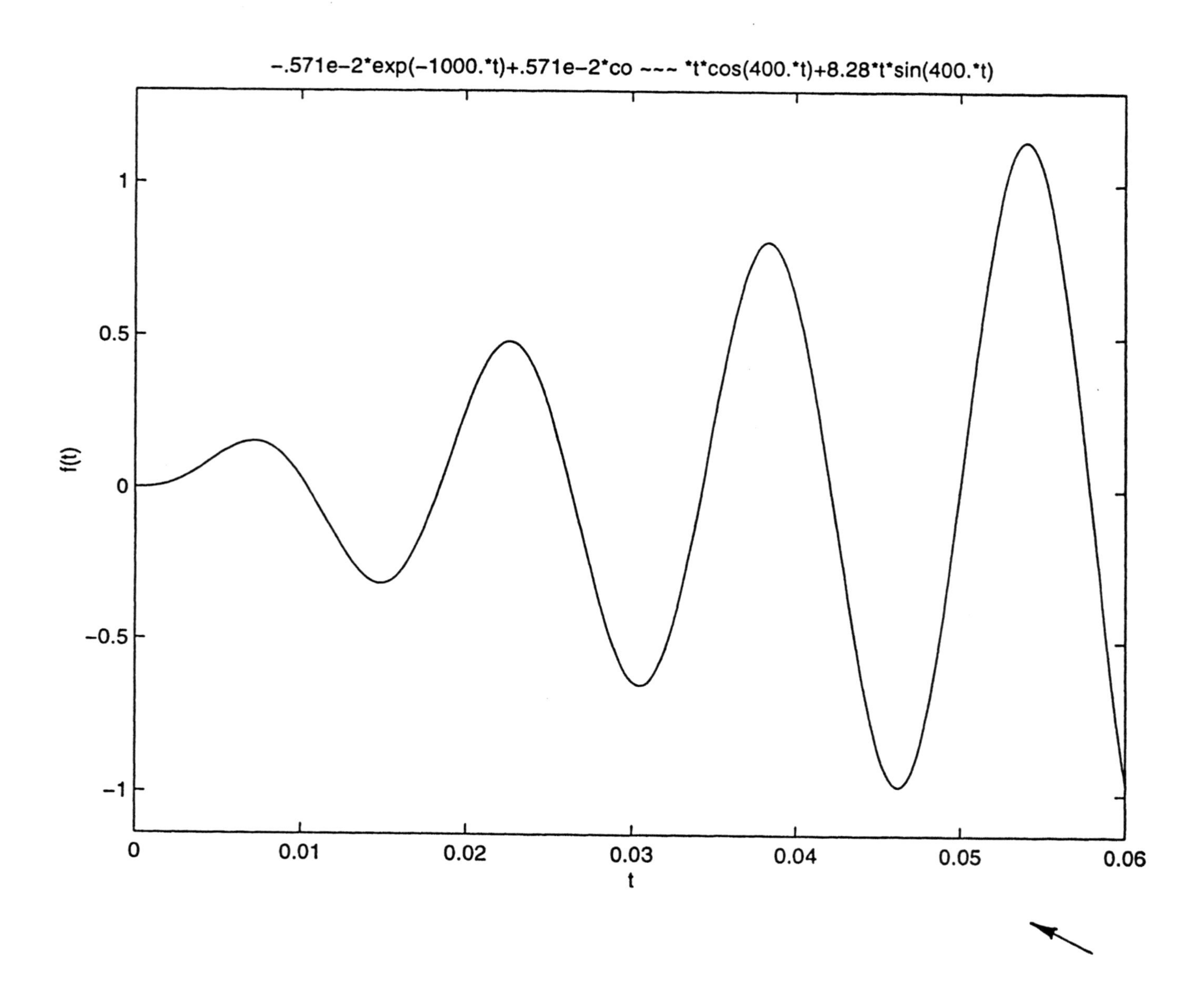

3.6.4.

```
function problem364
clc

% Circuit Parameters
R = 1.59e3
C = 10e-9
L = 12e-6
wo = 1/sqrt(L*C)

% Frequency Range
w = [0.1*wo:0.1*wo:10*wo];

% Circuit Transfer Function |T(jw)|
Ts = abs(L/R*j*w./(L*C*(j*w).^2+L/R*j*w+1));

% Straight Line Gain
TSL = L/R*w.*(w<=wo)+1/R/C./w.*(wo<w);

% Plot
loglog(w,Ts,'-.',w,TSL)
xlabel('w')
ylabel('|Ts(jw)|, TSL(w)')
```

```
R =

        1590


C =

   1.0000e-08


L =

   1.2000e-05


wo =

   2.8868e+06
```

3.6.4. CONTINUED

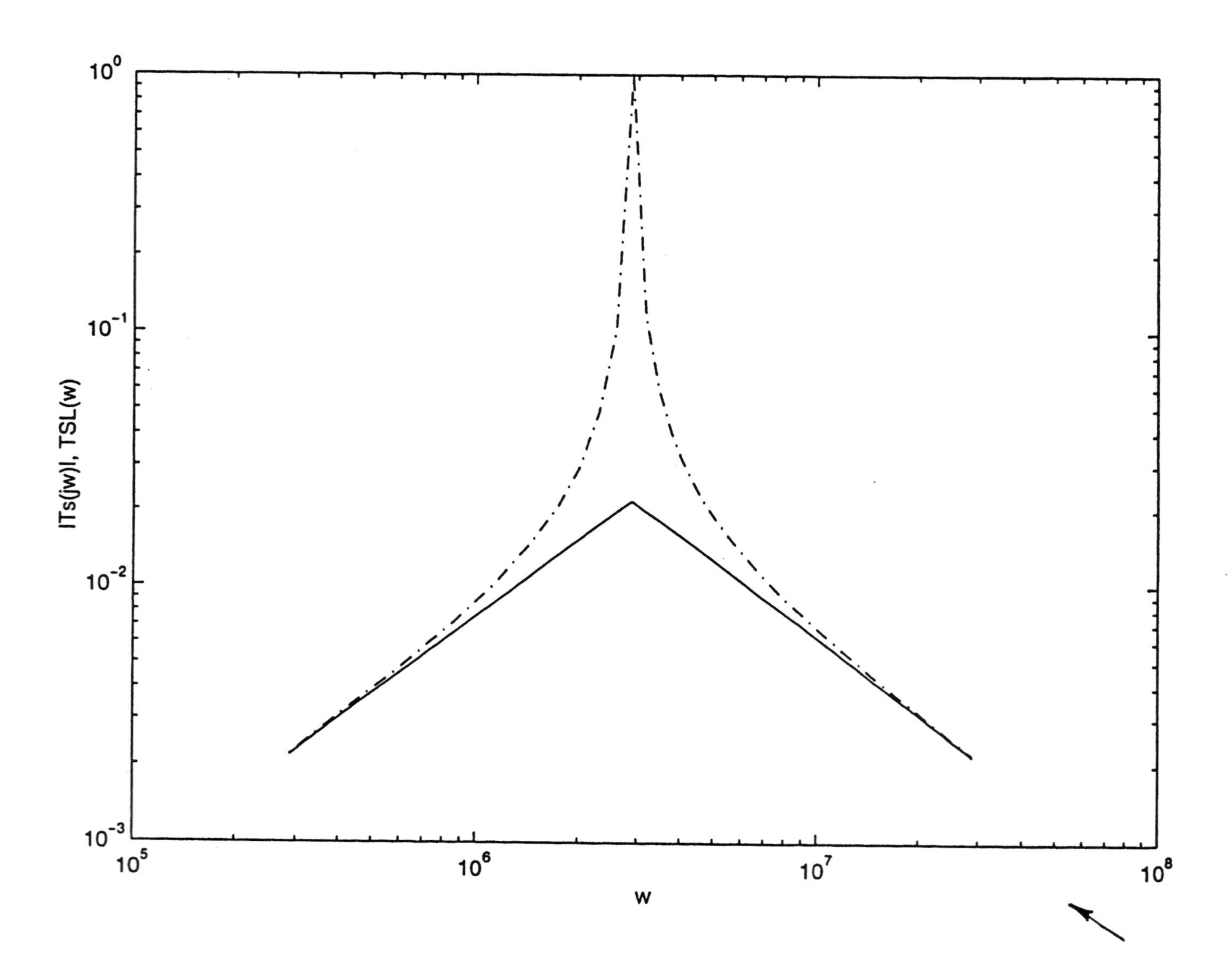

3.6.5.

```
function problem365
clc

% Transfer Function
Ts = tf(5000*[1 100],[1 400 500^2])

% Gain and Phase response
bode(Ts)
```

```
Transfer function:
  5000 s + 500000
--------------------
s^2 + 400 s + 250000
```

3.6.5. CONTINUED

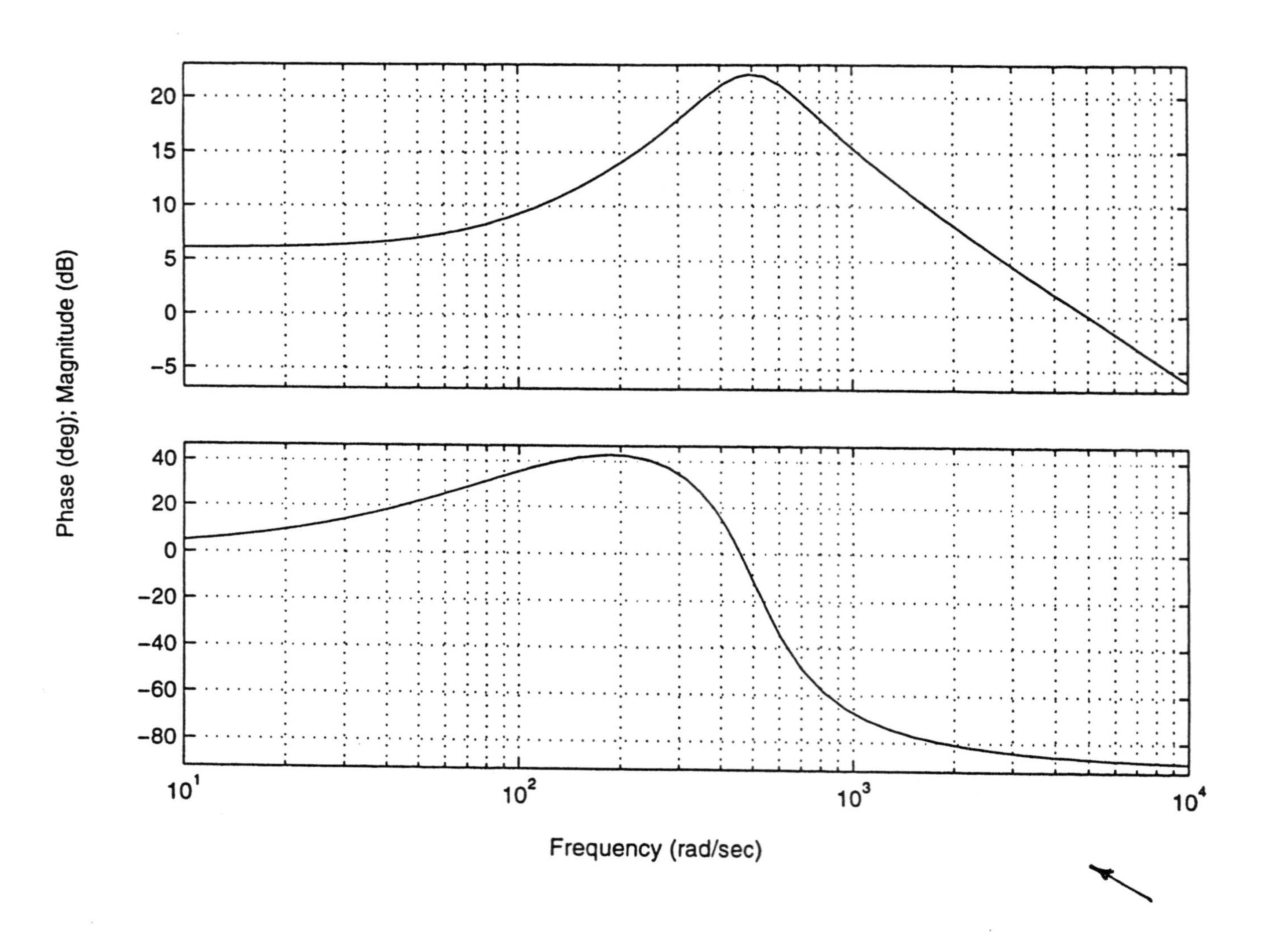

3.6.6.

```
function problem366
clc

% 2nd order Butterworth low pass wc=10
TLPs = tf(1,[1/10^2 sqrt(2)/10 1])

% 2nd order Butterworth high pass wc=50
THPs=tf([1/50^2 0 0],[1/50^2 sqrt(2)/50 1])

% 4th order Butterworth bandpass wc1=10 wc2=50
TBPs = TLPs+THPs

bode(TBPs,{1,1e3});
```

```
Transfer function:
           1
-----------------------
0.01 s^2 + 0.1414 s + 1


Transfer function:
        0.0004 s^2
-------------------------
0.0004 s^2 + 0.02828 s + 1


Transfer function:
4e-06 s^4 + 5.657e-05 s^3 + 0.0008 s^2 + 0.02828 s + 1
------------------------------------------------------
4e-06 s^4 + 0.0003394 s^3 + 0.0144 s^2 + 0.1697 s + 1
```

3.6.6. CONTINUED

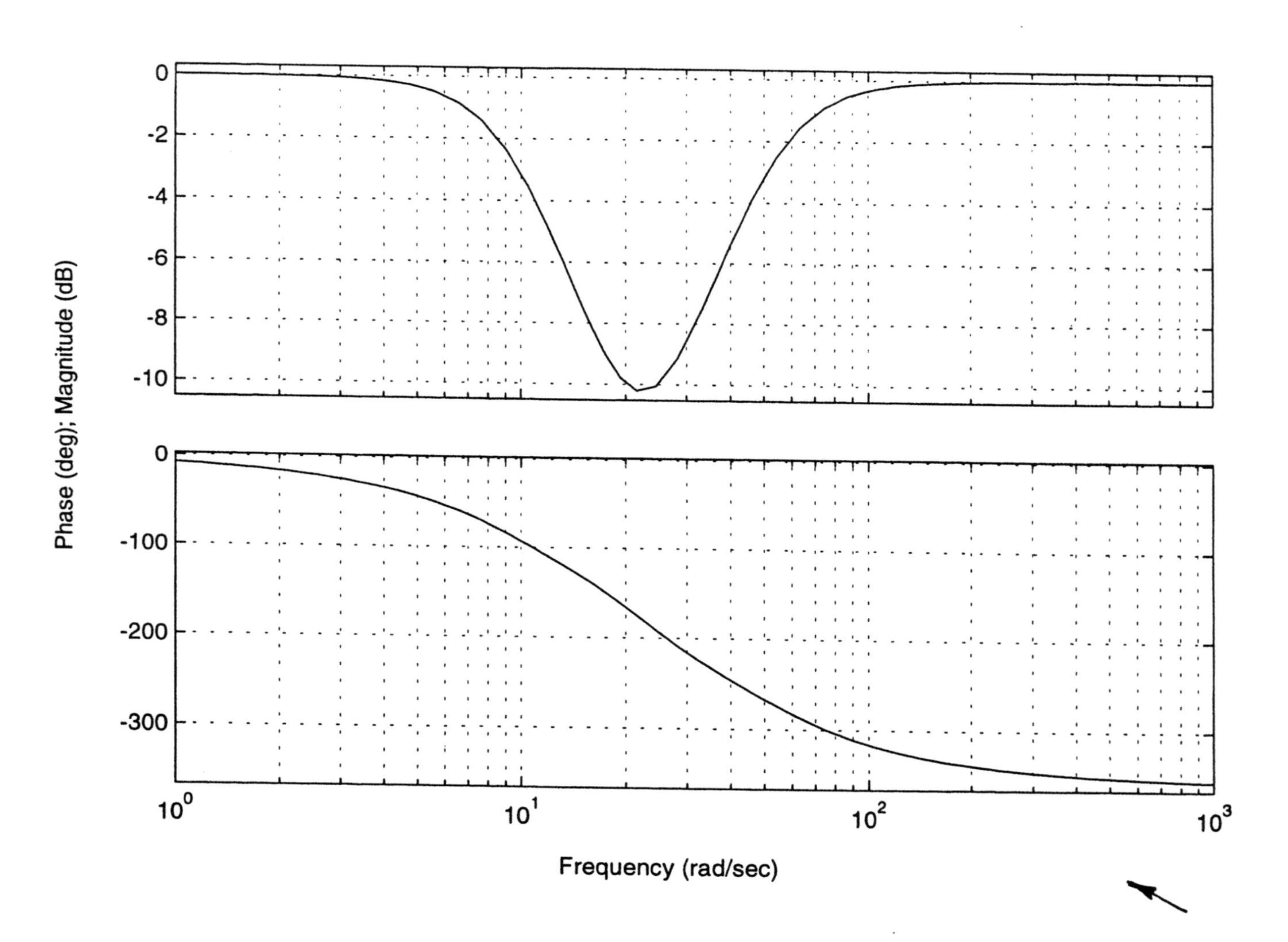

3.6.7.

```
function problem367
clc

% Waveform period
To = 2e-3

% time step and interval
t = 0:To/400:2*To;

subplot(121)
plot(t,fun(5,t,To))
title('dc+first 5 harmonics')
xlabel('t')
ylabel('f(5,t)')

subplot(122)
plot(t,fun(10,t,To))
title('dc+first 10 harmonics')
xlabel('t')
ylabel('f(10,t)')

function f=fun(k,t,To)
%  Waveform Amplitude
A=10;

% Fourier Coefficients
n  = 1:20;
ao = A/2;
a  = n*0;
b  = -A./n/pi;

% Fourier Series
f = ao;
for m = 1:k
  f = f+a(m)*cos(2*pi*m*t/To)+b(m)*sin(2*pi*m*t/To);
end
```

3.6.7. CONTINUED

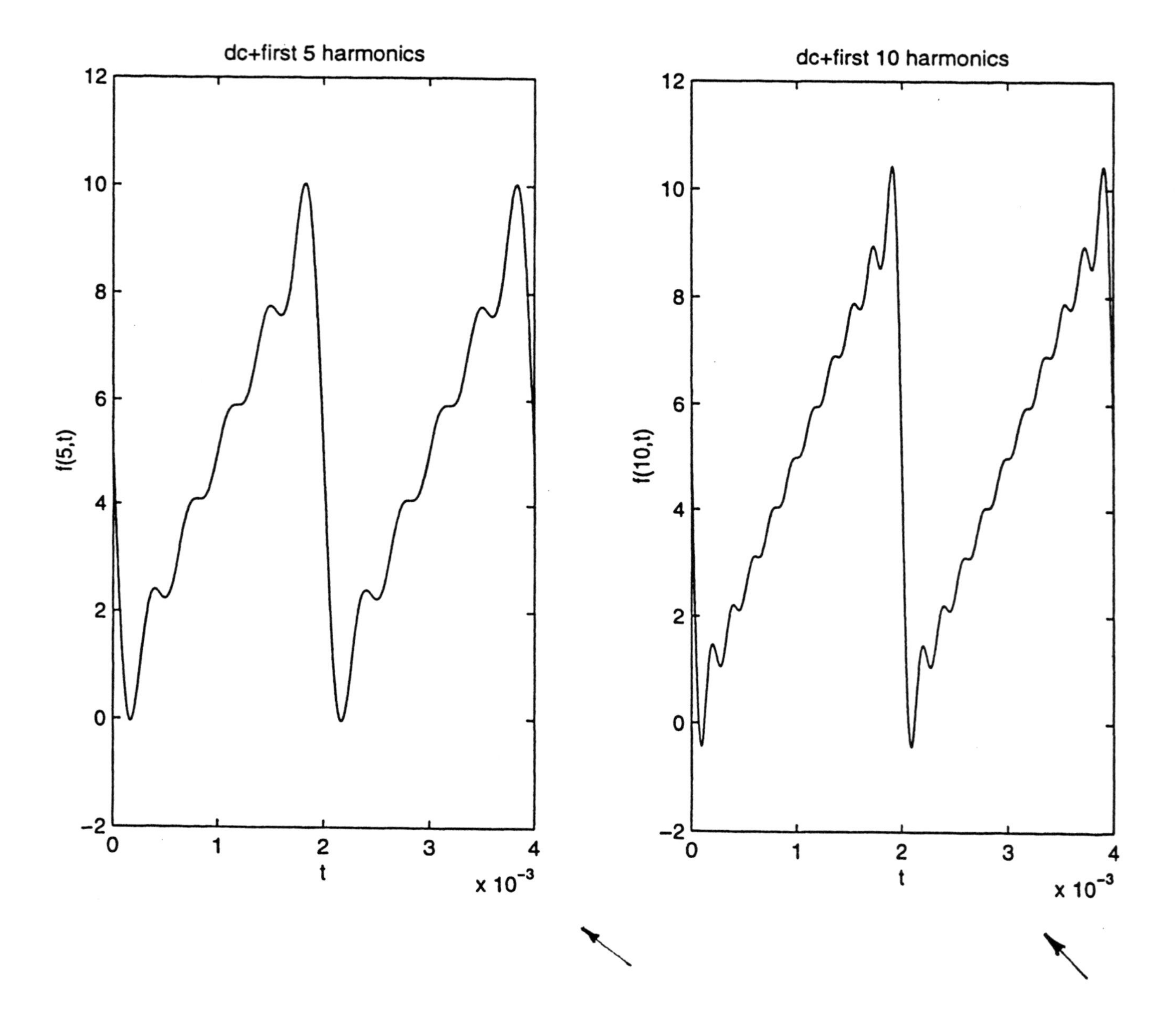

3.6.8.

```
function problem368
clc

% Waveform period
To = 5e-6

% time step and interval
t = 0:To/400:2*To;

subplot(121)
plot(t,fun(5,t,To))
axis([0 2*To 0 0.5])
title('dc+first 5 harmonics')
xlabel('t')
ylabel('i(5,t)')

subplot(122)
plot(t,fun(10,t,To))
axis([0 2*To 0 0.5])
title('dc+first 10 harmonics')
xlabel('t')
ylabel('i(10,t)')

function f=fun(k,t,To)
%  Waveform Amplitude
VA = 25;
wo = 2*pi/To;

% Circuit Parameters
L = 40e-6;
R = 50;

% Fourier Coefficients
n  = 1:20;
Io = VA/2/R;
I = VA/R./(pi*n.*(1+(n*wo*L/R).^2).^(1/2));
theta  = atan(n*wo*L/R);

% Fourier Series
f = Io;
for m = 1:k
  f = f+I(m)*cos(m*wo*t+0.5*pi-theta(m));
end
```

3.6.8. CONTINUED

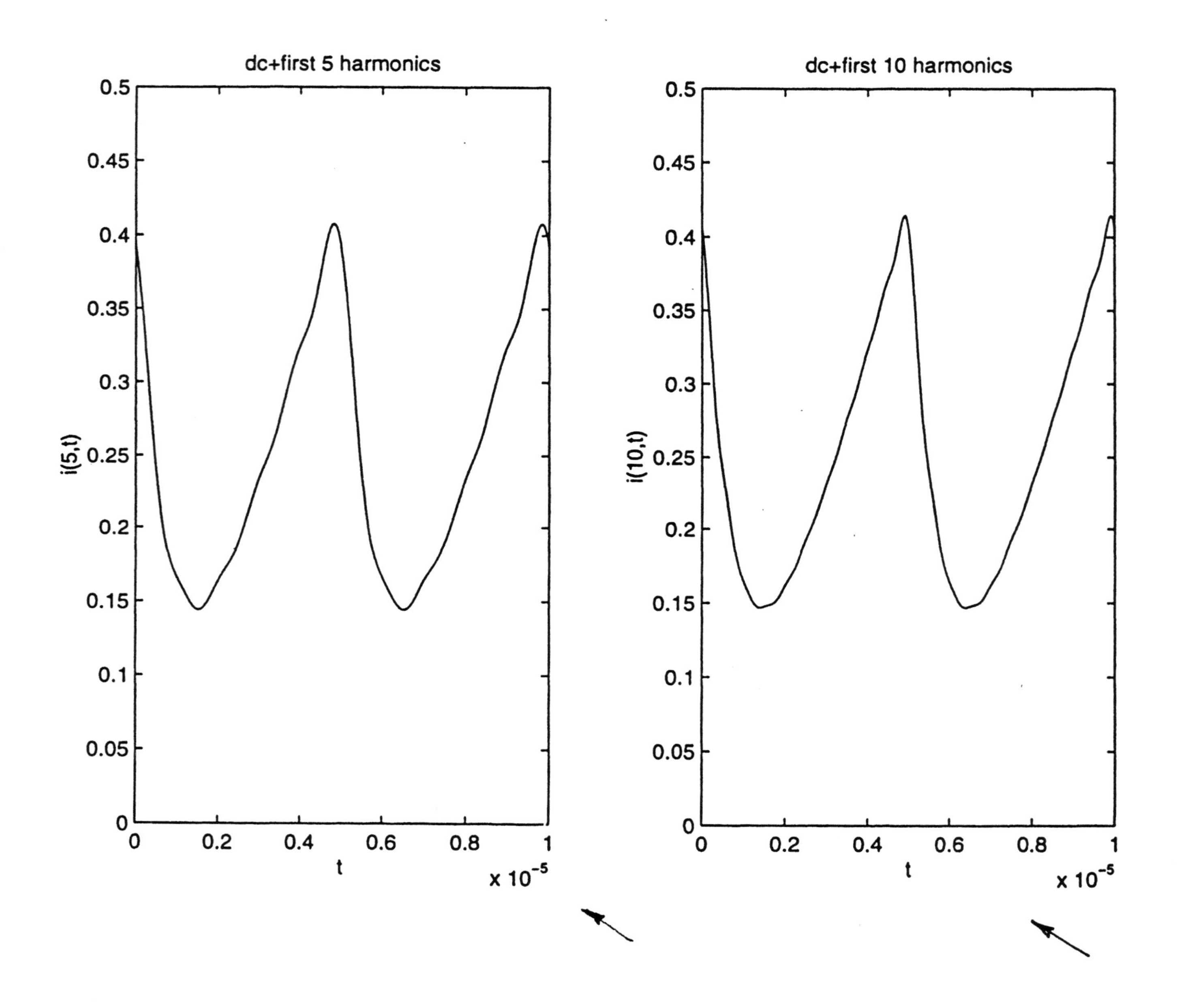

CHAPTER 4

4.1.1.

(a)

$$\bar{V}_{AN} = \frac{100}{\sqrt{3}} \angle -30^\circ$$

$$\bar{V}_{BN} = \frac{100}{\sqrt{3}} \angle -150^\circ$$

$$\bar{V}_{CN} = \frac{100}{\sqrt{3}} \angle 90^\circ$$

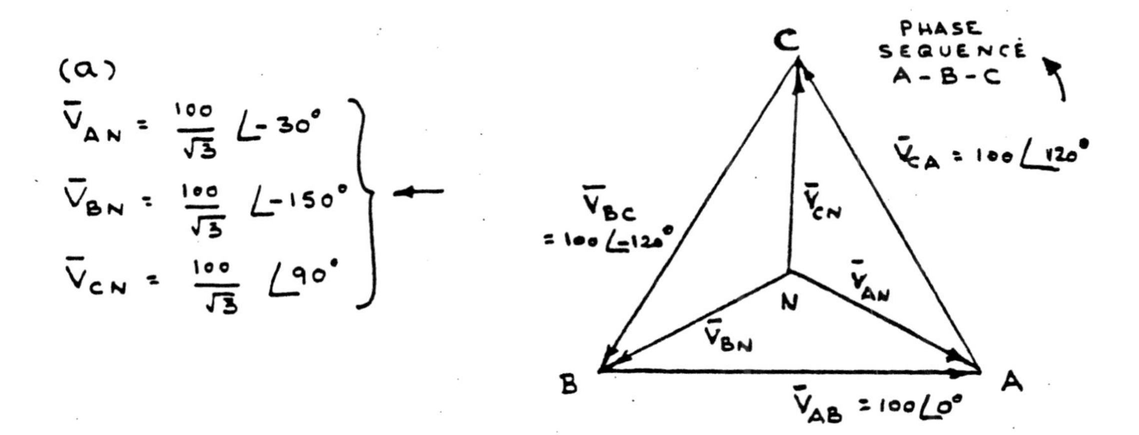

(b)

B A C N

$\bar{V}_{AB} = 100\angle 0^\circ$

$\bar{V}_{BN}$ $\bar{V}_{AN}$ $\bar{V}_{CN}$

$\bar{V}_{BC} = 100\angle 120^\circ$

PHASE SEQUENCE A-C-B OR C-B-A

$\bar{V}_{CA} = 100\angle -120^\circ$

$$\bar{V}_{AN} = \frac{100}{\sqrt{3}} \angle 30^\circ$$

$$\bar{V}_{BN} = \frac{100}{\sqrt{3}} \angle 150^\circ$$

$$\bar{V}_{CN} = \frac{100}{\sqrt{3}} \angle -90^\circ$$

4.2.1.

Let $\bar{V}_{BC} = 208\angle 0^\circ$; $\bar{V}_{CA} = 208\angle -120^\circ$; $\bar{V}_{AB} = 208\angle +120^\circ$

(FOR THE PHASE SEQUENCE A-B-C)

Then $\bar{I}_A = 10\angle 0^\circ$; $\bar{I}_B = 10\angle -120^\circ$; $\bar{I}_C = 10\angle +120^\circ$

(a) For the wye-connected load (Refer to Fig. 2.2.2)

$$\bar{V}_{AN} = \frac{208}{\sqrt{3}}\angle 90^\circ;\ \bar{I}_A = \frac{\bar{V}_{AN}}{Z} = \frac{120\angle 90^\circ}{Z} = 10\angle 0^\circ;\ Z = 12\angle 90^\circ = j12\ \Omega$$

(b) For the Delta-connected load (See Fig. 2.2.3)

$$\bar{I}_A = \bar{I}_{AB} + \bar{I}_{AC} = \frac{\bar{V}_{AB}}{Z} - \frac{\bar{V}_{CA}}{Z} = \frac{1}{Z}(\bar{V}_{AB} - \bar{V}_{CA}) = \frac{1}{Z}(208\angle 120^\circ - 208\angle -120^\circ)$$

$$\text{or } Z = \frac{j\,208\sqrt{3}}{\bar{I}_A} = \frac{j\,208\sqrt{3}}{10} = j\,36\ \Omega \leftarrow \qquad \text{CHECK: } Z_Y = \frac{1}{3} Z_\Delta$$

4.2.2.

(a) Wye-connected case:

$\sqrt{3}\, V_L I_L = 25 \times 10^3$; $V_L = 440\,V$

$\therefore I_L = \frac{25 \times 10^3}{\sqrt{3} \times 440} = 32.8\,A = I_{ph}$ ←

(b) Delta-connected case:

$I_L = 32.8\,A$ as in part (a) ←

$I_{ph} = \frac{I_L}{\sqrt{3}} = 18.94\,A$ ←

4.2.3.

(a) $kVA = \frac{\sqrt{3}\, V_L I_L}{1000} = \frac{\sqrt{3} \times 440 \times 20}{1000} = 15.24$ ←

(b) $kW = 15.24 \times 0.8 = 12.19$ ←

(c) $kVAR = 15.24 \times 0.6 = 9.14$ ←

4.2.4.

(a) $V_L = 133\,V$; $Z_{ph} = (4 + j3)\,\Omega$; $|Z_{ph}| = 5$

$V_{ph} = \frac{173}{\sqrt{3}} = 100\,V$; $I_{ph} = \frac{V_{ph}}{|Z_{ph}|} = \frac{100}{5} = 20\,A = I_L$ ←

power factor $= \cos\phi = 4/5 = 0.8$ (lagging) ←

$VA = \sqrt{3}\, V_L I_L = \sqrt{3} \times 173 \times 20 = 6{,}000$ or $6\,kVA$ ←

$kW = 6 \times 0.8 = 4.8$; $kVAR = 6 \times 0.6 = 3.6$ ←

(b) Fig. 2.2.2 (b) of the Text applies except that the diagram as a whole must be rotated such that $\bar{V}_{AB}$ becomes horizontal (i.e. a reference) with 0° angle.

(c) Zero; because the load and supply are balanced.

4.2.5.

(a) $V_L = 173\,V$; $Z_{ph} = (12 + j9)\,\Omega$; $|Z_{ph}| = \sqrt{12^2 + 9^2} = 15$

$V_{ph} = V_L = 173\,V$; $I_{ph} = \frac{V_{ph}}{|Z_{ph}|} = \frac{173}{15}\,A$; $I_L = \sqrt{3}\, I_{ph} = 20\,A$ ←

power factor $= \cos\phi = 12/15 = 0.8$ (lagging) ←

$VA = \sqrt{3}\, V_L I_L = 6\,kVA$; $kW = 4.8$; $kVAR = 3.6$ ←

(b) Since $Z_\Delta = 3 Z_Y$ for a balanced case, the loads are equivalent.

(c) Fig. 2.2.3 (b) of the Text applies except that the diagram as a whole must be rotated such that $\bar{V}_{AB}$ becomes horizontal (i.e. a ref.) with 0° angle.

4.2.6.

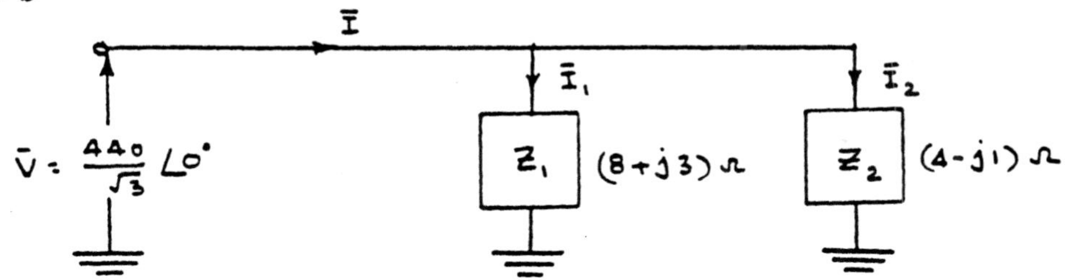

$$\bar{I}_1 = \frac{\bar{V}}{\bar{Z}_1} = \frac{440}{\sqrt{3}} \times \frac{1}{(8+j3)} = \underset{\angle -\tan^{-1} 3/8}{29.73\,\text{A}} \;;\quad \bar{I}_2 = \frac{\bar{V}}{\bar{Z}_2} = \frac{440}{\sqrt{3}} \times \frac{1}{4-j1} = \underset{\angle +\tan^{-1} 1/4}{61.62\,\text{A}}$$

(a) Real power delivered to inductive load $\Big\} = 3 V_{ph} I_{ph} \cos\phi = 3 \times \frac{440}{\sqrt{3}} \times 29.73 \times \frac{8}{\sqrt{64+9}} = 21.2\,\text{kW}$ ←

(b) Real power delivered to capacitive load $\Big\} = 3 \times \frac{440}{\sqrt{3}} \times 61.62 \times \frac{4}{\sqrt{16+1}} = 45.56\,\text{kW}$ ←

4.2.7.

$\bar{I}$, $\bar{I}_1$, $\bar{I}_2$, $\frac{300}{\sqrt{3}}$ V, L_1, L_2

$\because P = \sqrt{3}\, V_L I_L \cos\phi$, $\quad P_1 = \sqrt{3} \times 300 \times I_1 \times 0.8 = 10 \times 10^3$ or $I_1 = 24.06\,\text{A}$

$$\bar{I}_1 = 24.06 \angle -\cos^{-1} 0.8 = 24.06 \angle -36.9^\circ$$

$P_2 = \sqrt{3} \times 300 \times I_2 \times 0.9 = 15 \times 10^3$ or $I_2 = 32.08\,\text{A}$

$$\bar{I}_2 = 32.08 \angle +\cos^{-1} 0.9 = 32.08 \angle +25.84^\circ$$

$\bar{I} = \bar{I}_1 + \bar{I}_2 = 48.12 \angle -0.54^\circ \;;\quad I = 48.12\,\text{A}$ ←

4.2.8.

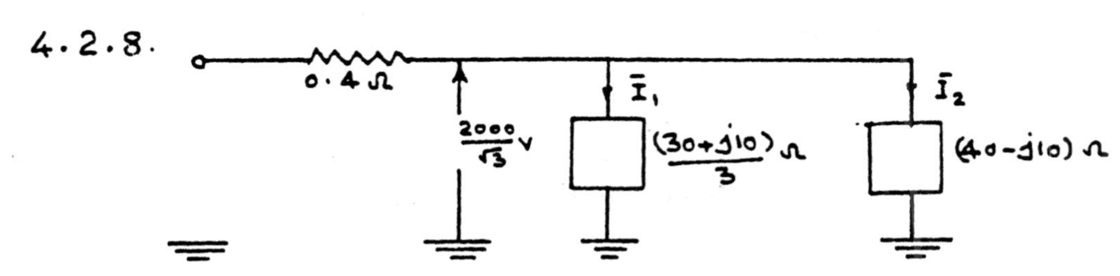

4.2.8. CONTD.

$$\bar{I}_1 = \frac{(2000/\sqrt{3})\,3}{(30+j10)} = 109.56\angle -18.43^\circ\,;\quad \bar{I}_2 = \frac{(2000/\sqrt{3})}{(40-j10)} = 28\angle 14.04^\circ$$

(a) $P_1 = \sqrt{3} \times 2000 \times 109.56 \times \cos 18.43^\circ = 360.54$ kW ←

(b) $P_2 = \sqrt{3} \times 2000 \times 28 \times \cos 14.04^\circ = 94.08$ kW ←

(c) $\bar{I} = \bar{I}_1 + \bar{I}_2 = 134.16\angle -11.97^\circ$

Power loss in line conductors $= 3 \times (134.16)^2 \times 0.4 = 21.6$ kW ←

4.2.9.

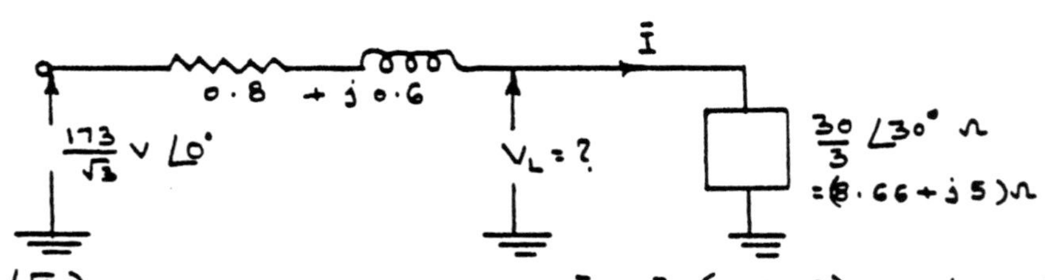

$$\bar{I} = \frac{(173/\sqrt{3})}{(0.8+j0.6)+(8.66+j5)} = 9.1\angle -30.6^\circ\,;\quad \bar{V}_L = \bar{I}\,(10\angle 30^\circ) = 91\angle -0.6^\circ$$

The line-to-line voltage at the terminals of the load $= \sqrt{3} \times 91 = 157.6$ V ←

4.2.10.

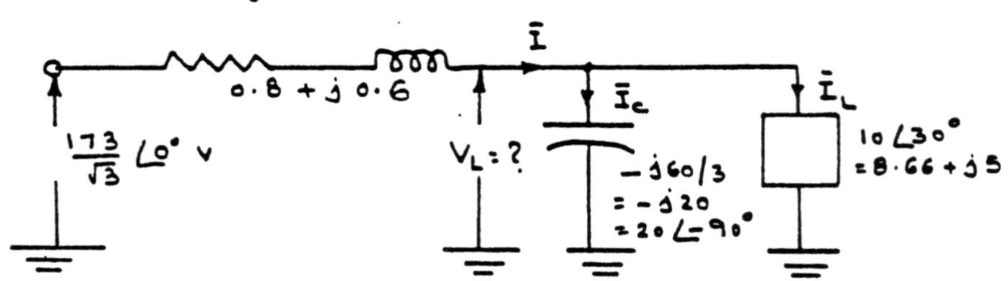

Z_p of the parallel combination $= \dfrac{(10\angle 30^\circ)(20\angle -90^\circ)}{(8.66+j5) - j20} = 11.55\angle 0^\circ$

$$\bar{I} = \frac{(173/\sqrt{3})}{(0.8+j0.6)+11.55} = 8.09\angle -2.78^\circ\,;\quad \bar{V}_L = \bar{I}\,(11.55\angle 0^\circ) = 93.44\angle -2.78^\circ$$

The corresponding line-to-line voltage $= 93.44\sqrt{3} \approx 162$ V ←

4.2.11. (a) $\bar{I}_1 = 20\,($~~0.9 + j0.4359~~ (0.8 + j0.6)$)$ = ~~18 + j8.718~~ (16 + j12) A

$\bar{I}_2 = 30\,(0.8 - j0.6) = 24 - j18$ A

$\bar{I}_{TOTAL} =$ ~~42 − j9.282~~ (40 − j6) A ; $I_{TOTAL} =$ ~~43~~ (40.45) A ←

SUPPLY PF = ~~42/43 = 0.976~~ (0.988) LAGGING ←

(b) $P_{3\phi} = \sqrt{3}\,(400)$ ~~43~~ (40.45) × ~~0.976~~ (0.988) = ~~29.1~~ (27.6) kW ←

4.2.12.

$\bar{S}_1 = 15(0.8 + j0.6) = 12 + j9$

$\bar{S}_2 = 20(0.6 - j0.8) = 12 - j16$

$\bar{S}_{TOTAL} = 24 - j7$

(a) $P_{TOTAL} = 24\ kW$ ←

(b) $kVA_{TOTAL} = \sqrt{24^2 + 7^2} = 25$ ←

(c) OVERALL PF = 24/25 = 0.96 LEADING ←

4.2.13. (a)

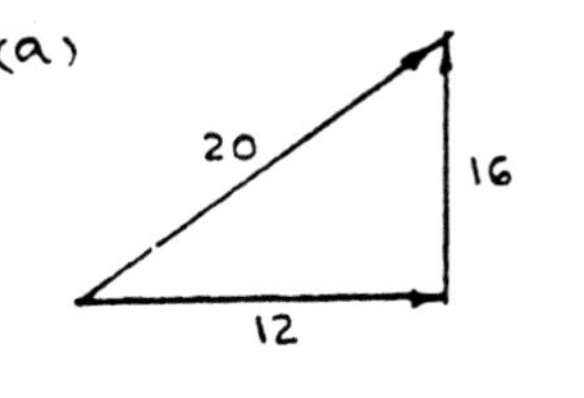

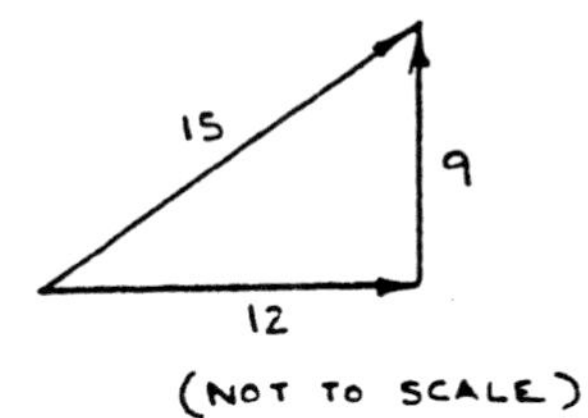

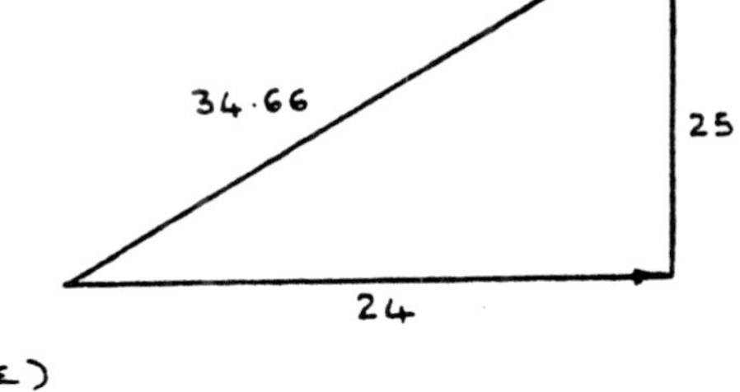

(NOT TO SCALE)

(b) 34.66 kVA ←

(c) SUPPLY PF = 24/34.66 = 0.692 LAGGING ←

(d) $I_1 = \dfrac{12000}{\sqrt{3} \times 208 \times 0.6} = 55.5\ A$ ←

$I_2 = \dfrac{15000}{\sqrt{3} \times 208} = 41.6\ A$ ←

$I_{SUPPLY} = \dfrac{34.66 \times 10^3}{\sqrt{3} \times 208} = 96.2\ A$ ←

(e) $V_L^2 / X_\phi = 25000$ OR $X_\phi = 208^2 / 25000 = 1.73 = 1/\omega C$

WITH $\omega = 2\pi \times 60$, $C = \dfrac{1000}{2\pi(60)\,1.73} = 1.533\ mF$ ←

4.2.14.

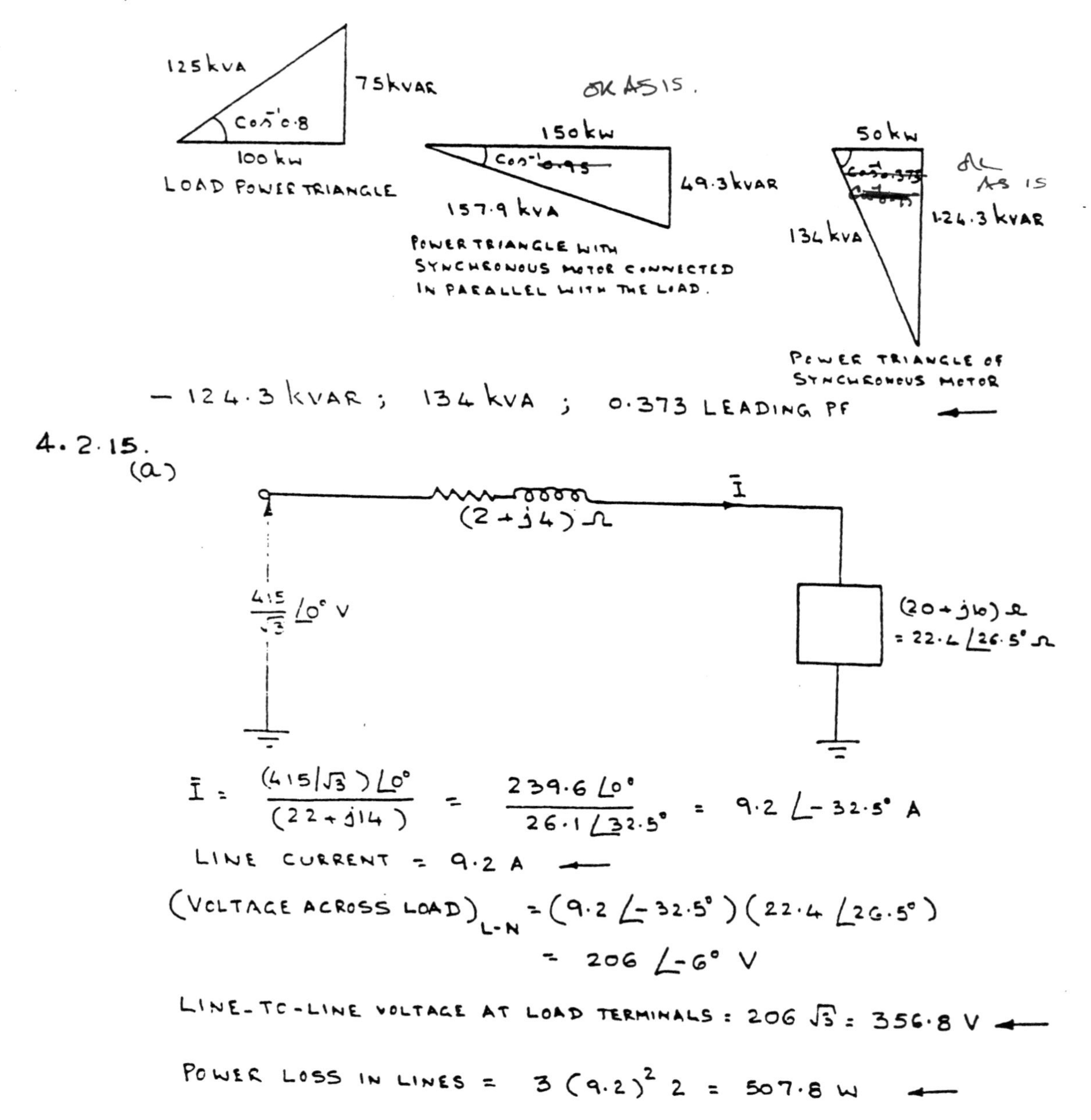

-124.3 kVAR ; 134 kVA ; 0.373 LEADING PF ⟵

4.2.15.

(a)

$$\bar{I} = \frac{(415/\sqrt{3})\angle 0°}{(22 + j14)} = \frac{239.6 \angle 0°}{26.1 \angle 32.5°} = 9.2 \angle -32.5°\ \text{A}$$

LINE CURRENT = 9.2 A ⟵

$$(\text{VOLTAGE ACROSS LOAD})_{L-N} = (9.2 \angle -32.5°)(22.4 \angle 26.5°)$$

$$= 206 \angle -6°\ \text{V}$$

LINE-TO-LINE VOLTAGE AT LOAD TERMINALS : $206\sqrt{3} = 356.8$ V ⟵

POWER LOSS IN LINES = $3(9.2)^2\, 2 = 507.8$ W ⟵

4.2.15. CONTD.

(b)

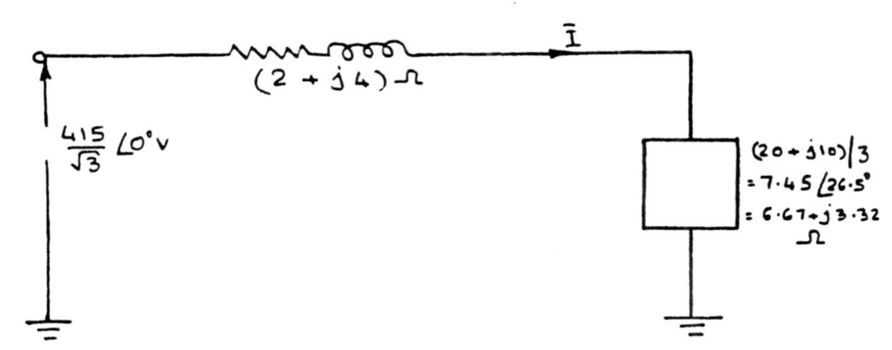

$$\bar{I} = \frac{(415/\sqrt{3})\angle 0°}{(8.67 + j7.32)} = \frac{239.6 \angle 0°}{11.35 \angle 40.2°} = 21.1 \angle -40.2° \text{ A}$$

LINE CURRENT = 21.1 A ←

$(\text{VOLTAGE ACROSS LOAD})_{L-N}$ FROM THE MODEL = $(21.1 \angle -40.2°)(7.45 \angle 26.5°)$
$= 157.2 \angle -13.7°$ V

LINE-TO-LINE VOLTAGE AT LOAD TERMINALS = $157.2\sqrt{3}$ = 272.3 V ←

POWER LOSS IN LINES : $3(21.1)^2 \, 2$ = 2671.3 W ←

4.2.16. (a) APPARENT POWER = 9.8/0.8 = 12.25 MVA ←

REACTIVE POWER = 12.25 (0.6) = 7.35 MVAR ←

(b) WITH ADDITIONAL LOAD INSTALLED

TOTAL REAL POWER = 9.8 + 1.5 = 11.3 MW

TOTAL REACTIVE POWER = 7.35 + 0.7 = 8.05 MVAR

TOTAL APPARENT POWER = $\sqrt{11.3^2 + 8.05^2}$ = 13.874 MVA

MAXIMUM LOAD THAT THE LINE CAN CARRY IS

$$\sqrt{3} \times 11 \times 0.66 = 12.575 \text{ MVA}$$

TO BRING THE TOTAL APPARENT POWER WITHIN THE LOAD CARRYING CAPACITY OF THE LINE, A CAPACITOR MUST BE INSTALLED DRAWING LEADING VARS EQUIVALENT TO $8.05 - \sqrt{12.575^2 - 11.3^2}$ = 2.53 MVAR

∴ RATING OF CAPACITOR = 2.53 MVAR ←

SYSTEM POWER FACTOR = 11.3/12.575 = 0.9 LAGGING ←

(c) EACH CAPACITOR SECTION WILL ACCOUNT FOR 2.53/3 = 0.843 MVAR

$$V_L^2 / X_C = \frac{(11 \times 10^3)^2}{X_C} = 0.843 \times 10^6 \text{ OR } X_C = 143.5\,\Omega$$

$$\therefore C = \frac{1}{2\pi(60)143.5} = 18.5\,\mu\text{F} \leftarrow$$

4.2.17.

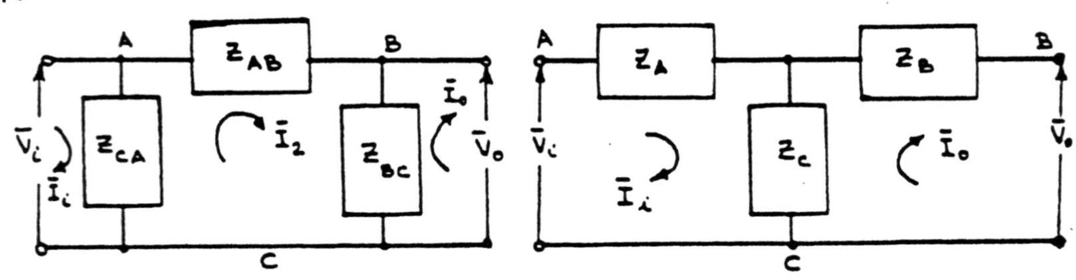

The two circuits are equivalent if their respective input, output, and transfer impedances are equal.
ALTERNATIVELY, as suggested in the Text, the impedance between terminals A and B must be the same for either network:
for the wye-connected load : $Z_A + Z_B$
for the Delta-connected load : $[Z_{AB} \cdot (Z_{BC} + Z_{CA})]/(Z_{AB} + Z_{BC} + Z_{CA})$

$$\therefore Z_A + Z_B = [Z_{AB}(Z_{BC} + Z_{CA})]/(Z_{AB} + Z_{BC} + Z_{CA})$$

Similarly $Z_B + Z_C = [Z_{BC}(Z_{CA} + Z_{AB})]/(Z_{BC} + Z_{CA} + Z_{AB})$

$Z_C + Z_A = [Z_{CA}(Z_{AB} + Z_{BC})]/(Z_{CA} + Z_{AB} + Z_{BC})$

Solving the above, one gets

$$Z_A = \frac{Z_{AB} Z_{CA}}{Z_{AB} + Z_{BC} + Z_{CA}} \; ; \; Z_B = \frac{Z_{BC} Z_{AB}}{Z_{AB} + Z_{BC} + Z_{CA}} \; ; \; Z_C = \frac{Z_{CA} Z_{BC}}{Z_{AB} + Z_{BC} + Z_{CA}}$$

AND

$$Z_{AB} = \frac{Z_A Z_B + Z_B Z_C + Z_C Z_A}{Z_C} \; ; \; Z_{BC} = \frac{Z_A Z_B + Z_B Z_C + Z_C Z_A}{Z_A} \; ; \; Z_{CA} = \frac{(Z_A Z_B + Z_B Z_C + Z_C Z_A)}{Z_B}$$

The results for the balanced case simply follow.

4.2.18.

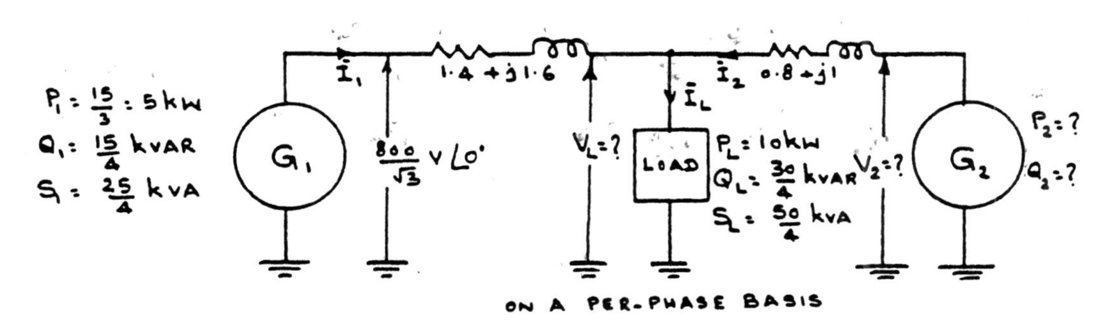

ON A PER-PHASE BASIS

For G_1: $\sqrt{3} \times 800 \times I_1 \times 0.8 = 15 \times 10^3$ or $I_1 = 13.53$; $\bar{I}_1 = 13.53 \angle -\cos^{-1} 0.8$

$\bar{V}_L = \frac{800}{\sqrt{3}} \angle 0° - 13.53 \angle -36.9° (1.4 + j1.6) = 433.8 \angle -0.78°$

Corresponding line-to-line voltage at the load terminals $= 433.8\sqrt{3} = 751.34$ V

For load: $\sqrt{3} \times 751.34 \times I_L \times 0.8 = 30 \times 10^3$ or $I_L = 28.8$; $\bar{I}_L = 28.8 \angle -(0.78 + 36.9)°$

$\bar{I}_2 = \bar{I}_L - \bar{I}_1 = 15.22 \angle -38.38°$; $\bar{V}_2 = \bar{V}_L + 15.22 \angle -38.38° (0.8 + j1) = 452.75 \angle -0.22°$

Corresponding line-to-line voltage at G_2-terminals $= 452.75\sqrt{3} = 784.2$ V ←

P_L + Loss in line resistances $= P_1 + P_2$ or $(10 \times 10^3) + (13.53)^2 1.4 + (15.22)^2 0.8 = (5 \times 10^3) + P_2$

or $P_2 = 5441.6$ W; on a 3-ph. basis, $3 \times 5441.6 = 16.325$ kW ←

$Q_L + (I^2 x)$ in line inductances $= Q_1 + Q_2$; $\frac{30}{4} \times 10^3 + (13.53)^2 1.6 + (15.22)^2 1 = (\frac{15}{4} \times 10^3) + Q_2$

or $Q_2 = 4274.5$ VAR; on a 3-ph. basis, $3 \times 4274.5 = 12.824$ kVAR ←

4.3.1.

$W_1 = V_L I_L \cos(30 + \phi) = 173 \times 20 \times \cos(30° + 36.9°) = 1357.36$ W

$W_2 = V_L I_L \cos(30 - \phi) = 173 \times 20 \times \cos(30° - 36.9°) = 3434.94$ W

$W_1 + W_2 = 4792$ W or 4.8 kW, same as in Prob. 2-5.

4.3.2.

$W_1 = 1200$ W; $W_2 = 400$ W; $V_L = 440$ V

For balanced 3-ph. loads, $\tan\phi = \pm\sqrt{3}\,\frac{1200 - 400}{1200 + 400} = \pm 0.866$ or $\phi = \pm 40.89°$

Note that ± is written since the location of the meters is not known.

Total real power $= 1600$ W $= \sqrt{3} V_L I_L \cos\phi = \sqrt{3} \times 440 \times I_L \times 0.756$ or $I_L = 2.78$ A

$Z_Y = \frac{(440/\sqrt{3}) \angle 0°}{2.78 \angle \pm 40.89} = 91.4 \angle \pm 40.89°$ or $Z_\Delta = 3 Z_Y = 274.2 \angle \pm 40.89°$ ← ohms per phase

NO ←

4.3.3. $W_C = 836$ W ; $W_A = 224$ W ; $V_L = 100$ V

$$\tan\phi = \sqrt{3}\,\frac{836-224}{836+224} \simeq 1.0 \text{ or } \phi = 45°$$

Note that the sign of the angle is determined since the meter locations are known.

$P = 1060$ W $= \sqrt{3} \times 100 \times I_L \times \cos 45°$ or $I_L = 8.66$ A; $\bar{I}_L = 8.66\angle -45°$

$$Z_Y = \frac{(100/\sqrt{3})\angle 0°}{8.66\angle -45°} = 6.67\angle +45° \text{ ohms per phase}$$ ← YES, it is an inductive load impedance ←

4.3.4. (a) $P = W_A + W_C = -500 + 1300 = 800$ W or 0.8 kW

$$\tan\phi = \sqrt{3}\,\frac{1300-(-500)}{1300+(-500)} = 3.897 \text{ or } \phi = 75.6° \text{ or } pf = \cos\phi = 0.25 \text{ (lagging)}$$

$P = 800 = \sqrt{3} \times 120 \times I_L \times 0.25$ or $I_L = 15.4$ A ; $\bar{I}_L = 15.4\angle -75.6°$

$$Z_Y = \frac{(120/\sqrt{3})\angle 0°}{15.4\angle -75.6°} = 4.5\angle 75.6° \text{ or } Z_\Delta = 3Z_Y = 13.5\angle 75.6°\ \Omega/\text{ph.}$$ ←

(b) $P = W_A + W_C = 800$ W ; $\tan\phi = \sqrt{3}\,\dfrac{(-500)-1300}{-500+1300} = -3.897$ or $\phi = -75.6°$

$pf = \cos\phi = 0.25$ (leading)

$P = 800 = \sqrt{3} \times 120 \times I_L \times 0.25$ or $I_L = 15.4$ A ; $\bar{I}_L = 15.4\angle +75.6°$

$$Z_Y = \frac{(120/\sqrt{3})\angle 0°}{15.4\angle +75.6°} = 4.5\angle -75.6° \text{ or } Z_\Delta = 13.5\angle -75.6°\ \Omega/\text{ph.}$$ ←

4.3.5.

$W_A = V_{AB} \cdot I_A \cdot \cos\theta_A$

$= V_L\ I_L \cos(30° + \phi)$ FOR BALANCED LOADS

$= (272.3)(21.1)\cos 56.5° = 3171$ W ←

$W_C = V_{CB} \cdot I_C \cdot \cos\theta_C$

$= V_L\ I_L \cos(30° - \phi)$ FOR BALANCED LOADS

$= (272.3)(21.1)\cos 3.5° = 5735$ W ←

$W_A + W_C = 8906$ W ←

TOTAL REAL POWER $= 3\left(\dfrac{21.1}{\sqrt{3}}\right)^2 20 = 8905$ W ←

WHICH AGREES WITH THE POWER MEASURED BY THE TWO WATTMETERS.

4.4.1.

$\omega = 2\pi f = 120\pi$ rad/s.

$v_{BN}(t) = 120\sqrt{2}\cos\omega t$ ←

$v_{RN}(t) = -120\sqrt{2}\cos\omega t$ ←

$v_{BR}(t) = v_{BN}(t) + v_{NR}(t) = v_{BN}(t) - v_{RN}(t)$

$= 240\sqrt{2}\cos\omega t$ ←

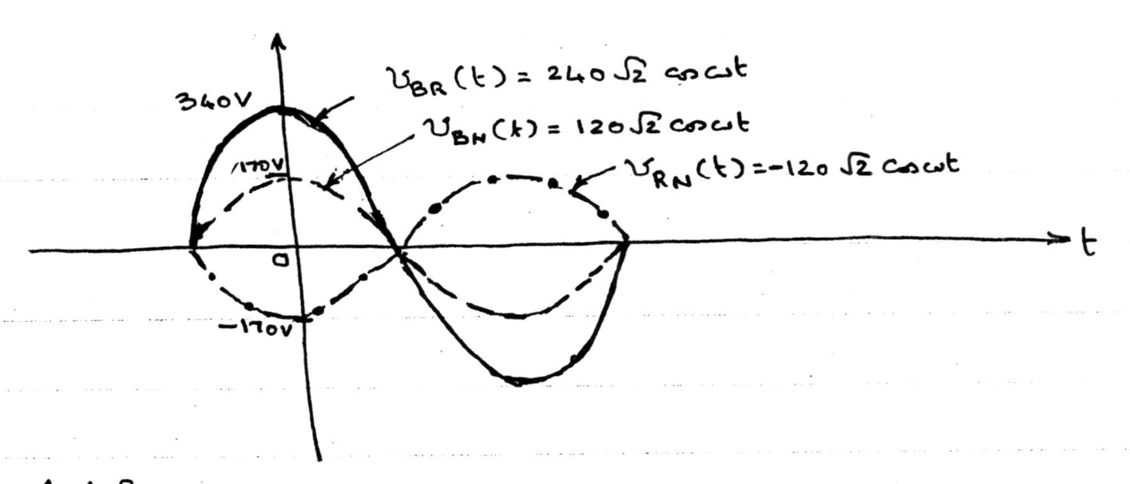

4.4.2.

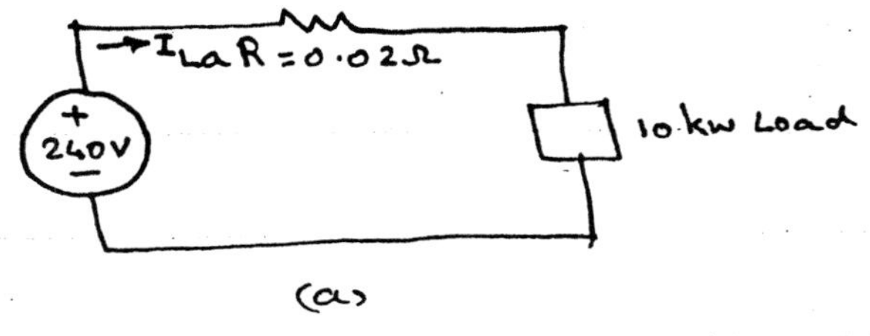

I_{Lb} $\frac{R}{2}$ = 0.01Ω

120V

10 kW Load

(b)

$I_{La} \simeq \frac{10 \times 10^3}{240} = 41.67$ A

$I_{Lb} \simeq \frac{10 \times 10^3}{120} = 83.33$A

$I_{La}^2 R = (41.67)^2\, 0.02 = 34.7$ W ←

$(I_{Lb})^2 \left(\frac{R}{2}\right) = (83.33)^2\, 0.01 = 69.4$W ←

I^2R losses are higher with 120-V supply, in spite of the reduced line resistance. ←

4.4.3.

Current through the human body $= \frac{2400}{10 \times 10^3} = 240$ mA ←

which is very dangerous indeed, and may cause possible death. ←

It would have been safer, if the individual remained in the car. ←

CHAPTER 5

5.1.1. $v_A = A i_1 R_2$

$i_1 = -V_i / R_1$

$\therefore v_A = -A V_i R_2 / R_1 \leftarrow$

(b) $i_2 = A i_1 = -A V_i / R_1 = -100 \times 10 / (10 \times 10^3) = -\frac{1}{10}$ A

$= -100$ mA $\leftarrow$

5.1.2. (a)

$v_1 = V_i = 1$ V

$A v_1 = 3$ V

$i_2 = \dfrac{A v_1}{R_2} = \dfrac{3}{2000} = 1.5 \times 10^{-3}$ A $= 1.5$ mA $\leftarrow$

(b) i_2 IS SAME AS IN PART (a) i.e. $i_2 = 1.5$ mA $\leftarrow$

5.1.3.

(a) $v_{OUT} = A i_1 R_3$

NODE EQ AT NODE A: $\dfrac{V_i - v_A}{R_1} - \dfrac{v_A}{R_2} - A i_1 = 0$

BUT $v_A = -i_1 R_2$

$\therefore v_{OUT} = \dfrac{A R_3 V_i}{R_1 [A - 1 - (R_2/R_1)]} \leftarrow$

(b)

$v_{OUT} = \dfrac{100 (5 \times 10^3) 2}{(2.5 \times 10^3) [100 - 1 - 1]} = \dfrac{400}{98} = 4.$ V $\leftarrow$

5.1.4.

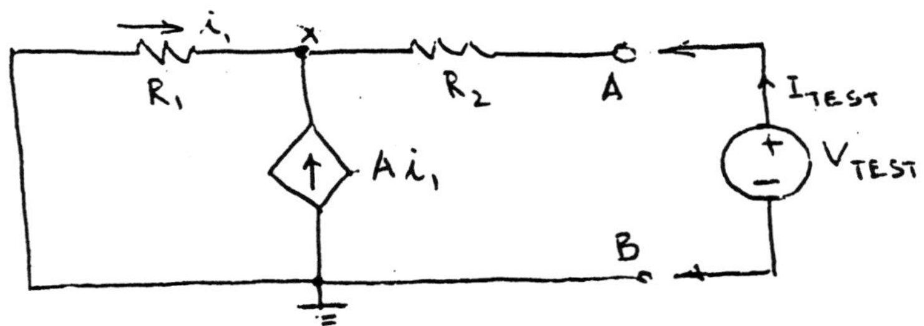

SET INDEPENDENT SOURCE $v_1(t) = 0$; KEEP THE DEPENDENT SOURCE.

$R_{Th} = V_{TEST} / I_{TEST}$

NODE EQ AT X: $i_1 + A i_1 + I_{TEST} = 0$ & $I_{TEST} = -(A+1) i_1$

NOTING $v_x = -i_1 R_1$ AND $V_{TEST} = v_x + I_{TEST} R_2$

5.1.4. CONTINUED

$$i_1 = \frac{I_{TEST} R_2 - V_{TEST}}{R_1}$$

$$I_{TEST} = -(A+1)\, i_1 = \frac{A+1}{R_1}\left(V_{TEST} - I_{TEST} R_2\right)$$

SOLVING FOR I_{TEST}, WE HAVE

$$R_{Th} = \frac{V_{TEST}}{I_{TEST}} = \frac{R_1 + (A+1) R_2}{A+1} \quad \longleftarrow$$

5.1.5.

CONVERT THE OUTPUT PART OF THE HYBRID MODEL TO ITS THÉVENIN'S EQUIVALENT

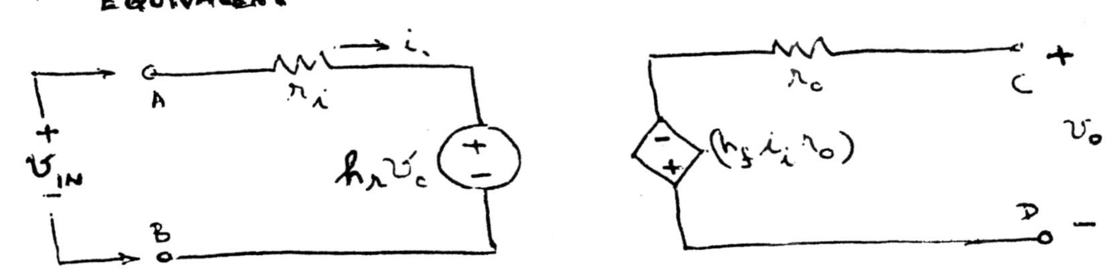

COMPARING THIS TO FIG. 5.1.1 OF THE TEXT, WE HAVE:

$$r_i = R_i \quad ; \quad r_o = R_o \quad \longleftarrow$$

$$h_r = 0 \quad ; \quad \longleftarrow$$

$$h_f\, i_i\, r_o = -A v_{IN}$$

$$\text{or} \quad h_f \frac{v_{IN}}{r_i} r_o = -A v_{IN}$$

$$\text{or} \quad h_f = -\frac{A r_i}{r_o} = -\frac{A R_i}{R_o} \quad \longleftarrow$$

5.1.6.

$$G_p = \frac{4 A^2 R_L R_i^2 R_{Th}}{(R_o + R_L)^2 (R_i + R_{Th})^2} \qquad \text{FROM EXAMPLE 5.1.2}$$

WITH $A = 1$ AND $R_i = R_o$

$$G_p = \frac{4 R_L R_o^2 R_{Th}}{(R_o + R_L)^2 (R_o + R_{Th})^2}$$

FOR ANY VALUES OF R_{Th} AND R_L, IT CAN BE SEEN THAT $G_p < 1$. ←

WITH $R_i = R_o = R_L = R_{Th}$,

THE MAXIMUM VALUE OF $G_p = \dfrac{4 R_L^4}{16 R_L^4} = \dfrac{1}{4}$ ←

5.1.7. (a)

SUBSTITUTE THE MODEL OF FIG. 5.1.1 OF THE TEXT. WE THEN HAVE.

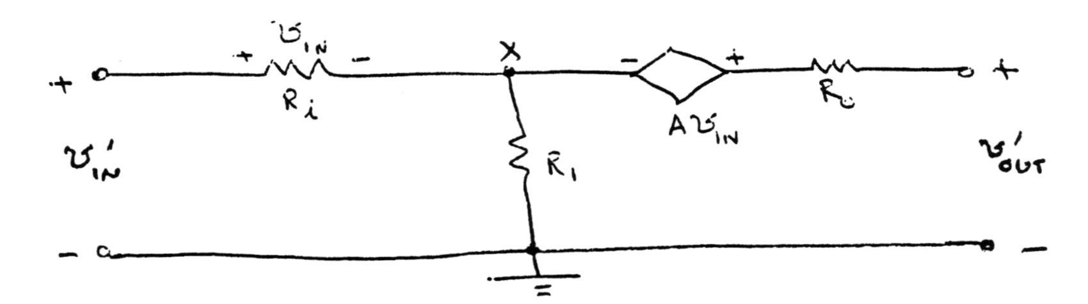

$$v_x = \frac{R_1}{R_1 + R_i} v'_{IN}$$

$$v'_{OUT} = v_x + A v_{IN} = v_x + A(v'_{IN} - v_x)$$

SOLVING, ONE GETS $\quad A' = \dfrac{v'_{OUT}}{v'_{IN}} = \dfrac{R_1 + A R_i}{R_1 + R_i}$ ←

5.1.7. (b)

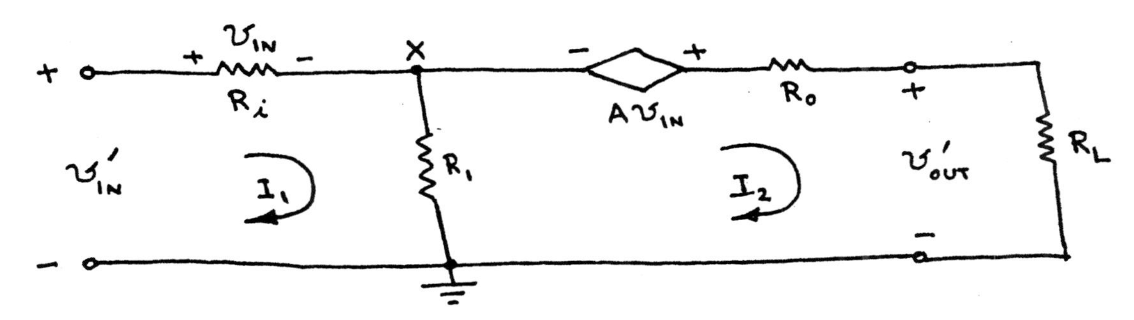

$$v_{IN} = v'_{IN} - v_x$$

LOOP EQUATIONS

$$v'_{IN} = (R_i + R_1) I_1 - R_1 I_2$$

$$0 = -R_1 I_1 - A v_{IN} + (R_o + R_L + R_1) I_2$$

ALSO

$$v_x = R_1 (I_1 - I_2)$$

$$v_{IN} = v'_{IN} - v_x = v'_{IN} - R_1 (I_1 - I_2)$$

THUS

$$v'_{IN} = (R_i + R_1) I_1 - R_1 I_2 \quad \text{—} \; (1)$$

$$0 = -R_1 I_1 - A [v'_{IN} - R_1 (I_1 - I_2)] + (R_o + R_L + R_1) I_2$$

OR

$$A v'_{IN} = - I_1 (R_1 - A R_1) + I_2 (R_o + R_L + R_1 - A R_1) \quad \text{—} \; (2)$$

ELIMINATING I_2, ONE GETS

$$v'_{IN} [R_o + R_L + R_1] = I_1 [R_i (R_o + R_L + R_1 - A R_1) + R_1 (R_o + R_L)]$$

$$R'_i = \frac{v'_{IN}}{I_1} = \frac{(R_o + R_L)(R_1 + R_i) - (A - 1) R_1 R_i}{(R_o + R_L + R_1)} \quad \longleftarrow$$

5.1.8 (a)

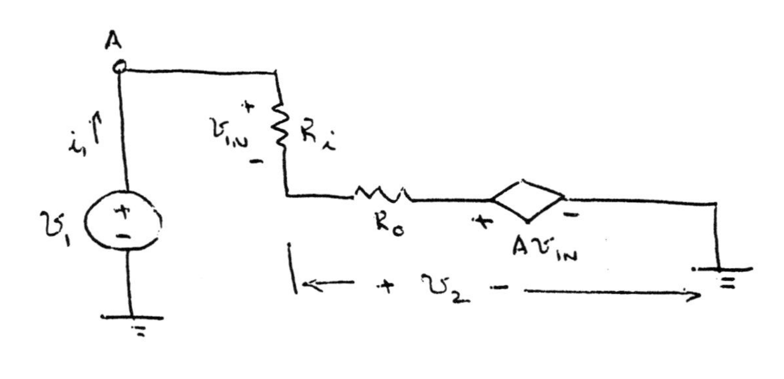

$$v_1 - Av_{in} = (R_i + R_o)\, i_1$$

$$v_{IN} = R_i\, i_1$$

$$v_1 = AR_i\, i_1 + (R_i + R_o)\, i_1 = i_1\left[R_i + R_o + AR_i\right]$$

$$v_2 = R_o i_1 + Av_{in} = R_o i_1 + AR_i i_1 = i_1\left[R_o + AR_i\right]$$

$$\therefore \quad \frac{v_2}{v_1} = \frac{R_o + AR_i}{R_o + (A+1)R_i} \quad \longleftarrow$$

5.1.8.(b)

OUTPUT RESISTANCE $= \dfrac{V_{TEST}}{I_{TEST}}$ IN THE FOLLOWING CIRCUIT

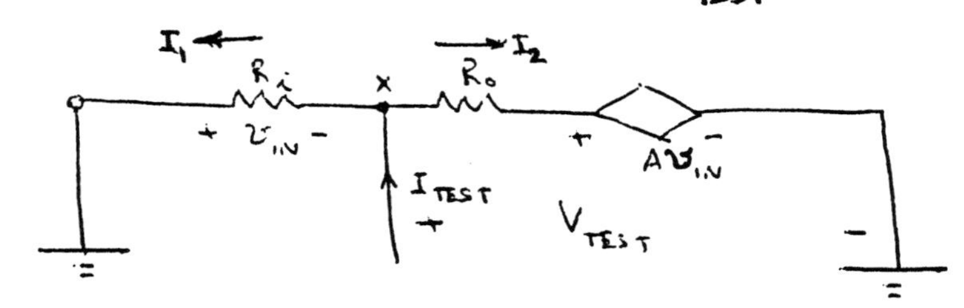

$$I_{TEST} = I_1 + I_2$$

$$v_{IN} = -R_i I_1$$

$$V_{TEST} = R_o I_2 + Av_{IN} = R_o I_2 + A(-R_i I_1)$$

$$= R_o(I_{TEST} - I_1) - AR_i I_1$$

$$= R_o I_{TEST} - I_1(R_o + AR_i)$$

5.1.8.(b) CONTINUED

$$v_X = R_i I_1 = R_o I_2 + A v_{IN} = R_o I_2 + A(-R_i I_1)$$

$$I_1 [R_i + A R_i] = R_o I_2$$

$$I_1 = \frac{R_o}{(1+A) R_i} I_2 = \frac{R_o}{(1+A) R_i} (I_{TEST} - I_1)$$

$$I_1 \left[1 + \frac{R_o}{(1+A) R_i}\right] = \frac{R_o}{(1+A) R_i} I_{TEST}$$

$$I_1 = \frac{R_o}{(1+A) R_i + R_o} I_{TEST}$$

$$V_{TEST} = R_o I_{TEST} - (R_o + A R_i) \frac{R_o}{(1+A) R_i + R_o} I_{TEST}$$

$$= I_{TEST} \left[R_o - \frac{R_o (R_o + A R_i)}{R_o + (1+A) R_i} \right]$$

$$= I_{TEST} \left[\frac{R_i R_o}{R_o + (1+A) R_i} \right]$$

$$\therefore \text{ OUTPUT RESISTANCE } = \frac{V_{TEST}}{I_{TEST}} = \frac{R_i R_o}{R_o + (1+A) R_i}$$

5.1.9. (a)

$$\frac{v_x}{v_2} = \frac{A_2 R_L}{(R_{o2} + R_L)}$$ See Eq 5.1.1 of Text.

but $$\frac{v_2}{v_1} = \frac{A_1 R_{i2}}{R_{o1} + R_{i2}}$$

$$\therefore \quad \frac{v_x}{v_1} = A_1 A_2 \frac{R_L}{(R_L + R_{o2})} \cdot \frac{(R_{i2}/R_{o1})}{1 + (R_{i2}/R_{o1})} \leftarrow$$

(b)

For $\frac{R_{i2}}{R_{o1}} \to 0$, $\frac{v_x}{v_1} \to 0$ ←

For $\frac{R_{i2}}{R_{o1}} \to \infty$, $\frac{v_x}{v_1} = A_1 A_2 \frac{R_L}{(R_L + R_{o2})}$ ←

5.1.10.

$$G_{p2} = \frac{A_2^2 R_L R_{i2}}{(R_{o2} + R_L)^2} = \frac{A_2^2 R_L R_i}{(R_o + R_L)^2}$$

$$G_{p1} = \frac{A_1^2 R_{i2} R_{i1}}{(R_{o1} + R_{i2})^2} = \frac{A_1^2 R_i^2}{(R_o + R_i)^2}$$

$$G_p = G_{p1} \cdot G_{p2} = A_1^2 A_2^2 R_i R_L \left(\frac{1}{R_L + R_o}\right)^2 \left(\frac{R_i}{R_i + R_o}\right)^2 \leftarrow$$

5.1.11.

FROM EXAMPLE 5.1.4 OF THE TEXT

$$P = (V_{ps} - V_{OUT}) I_{ps} = (V_{ps} - V_{OUT}) I_L$$

THE MAXIMUM OCCURS WHEN $V_{OUT} = V_{ps}/2$ ←

5.1.12. $p(t) = (V_{ps} - v_{OUT})\, i_L$

$i_L = v_{OUT}/R_L$

$v_{OUT} = A v_{IN} = 10\,(1 - 0.5 \cos \omega t) = 10 - 5 \cos \omega t$

$\therefore i_L = \dfrac{10 - 5 \cos \omega t}{100}$

$p(t) = (20 - 10 + 5 \cos \omega t)\, \dfrac{10 - 5 \cos \omega t}{100}$

$= \dfrac{(10 + 5 \cos \omega t)(10 - 5 \cos \omega t)}{100} = \dfrac{100 - 25 \cos^2 \omega t}{100}$

$= (1 - 0.25 \cos^2 \omega t)$

$P_{av} = 1 - \dfrac{0.25}{2} = 0.375\ \text{W}$ ←

5.1.13.

FOR MAXIMUM POWER $v_{OUT} = V_{ps}/2$ (See Pr. 5.1.11)

$I_L = v_{OUT}/R_L = V_{ps}/(2R_L)$

$P = (V_{ps} - v_{OUT})\, I_L = (V_{ps}/2)\, \dfrac{V_{ps}}{2R_L}$

OR $\dfrac{V_{ps}^2}{4R_L} < (50 \times 10^{-3})$

FOR $V_{ps} = 20\ \text{V}$

$R_L > \dfrac{20^2}{4 \times 50 \times 10^{-3}}$

$> \dfrac{400 \times 10^3}{200}$

> 2000

i.e. SMALLEST VALUE OF $R_L = 2000\ \Omega$ ←

5.1.14.(a)

USING PHASOR ANALYSIS AND GENERALIZED VOLTAGE-DIVIDER FORMULA, NOTING THAT R_i IS IN PARALLEL WITH C, ONE GETS

$$\frac{\bar{V}_{IN}}{\bar{V}_S} = \frac{R_i \parallel (j\omega C)^{-1}}{R_i \parallel (j\omega C)^{-1} + R_S}$$

$$= \frac{R_i}{R_i + R_S + j\omega C R_i R_S} = \frac{R_i}{R_i + R_S} \; \frac{1}{1 + j\omega C R_{\parallel}}$$

WHERE $R_{\parallel} = R_i \parallel R_S = \dfrac{R_i R_S}{R_i + R_S}$

$$\left| \frac{\bar{V}_{IN}}{\bar{V}_S} \right| = \frac{R_i}{R_i + R_S} \; \frac{1}{\sqrt{1 + (\omega R_{\parallel} C)^2}} \qquad (1)$$

THE FUNCTION THAT NEEDS TO BE PLOTTED IS (SINCE $R_o = 0$)

$$G_V(f) = \left| \frac{\bar{V}_{IN}}{\bar{V}_S} \right| \cdot \left| \bar{A}(f) \right|$$

WHERE $|\bar{A}(f)|$ IS GIVEN IN FIG. 5.1.3(a).
$|\bar{A}(f)|$ AS WELL AS $G_V(f)$ ARE SKETCHED BELOW:

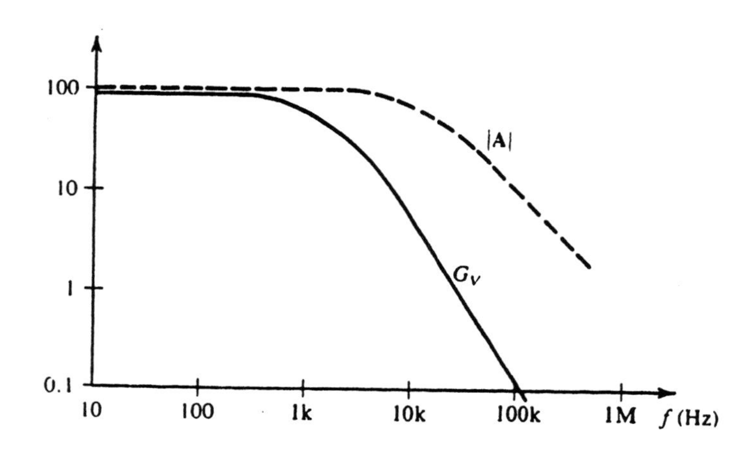

5.1.14.(a) CONTINUED

DUE TO THE BYPASSING EFFECT OF C, THE AMPLIFIER'S GAIN IS CLEARLY DECREASED CONSIDERABLY. THE RESPONSE IS FLAT OR CONSTANT UP TO FREQUENCIES OF THE ORDER OF $(2\pi R_{11} C)^{-1} = 875$ HZ. THEN IT GOES DOWN AT APPROXIMATELY -20 dB PER DECADE, UNTIL THE VICINITY OF 10 kHZ, DUE TO THE ROLLOFF OF $|\bar{V}_{IN}|/|\bar{V}_S|$. IN THIS FREQUENCY RANGE, NOTE THAT $|\bar{A}|$ HAS NOT YET BEGUN TO DECREASE. ABOVE 10kHZ, G_V DECREASES FASTER BECAUSE $|\bar{V}_{IN}/\bar{V}_S|$ AND $|\bar{A}|$ ARE BOTH DECREASING. SINCE EACH OF THE TWO FACTORS DECREASES AT -20dB PER DECADE, G_V DECREASES AT -40dB PER DECADE IN THE HIGHEST-FREQUENCY REGION.

5.1.14.(b)

IN THE RANGE OF $875 \text{ HZ} < f < 10 \text{ kHZ}$, $|\bar{A}|$ IS NEARLY CONSTANT AND $|\bar{V}_{IN}|/|\bar{V}_S|$ IS DECREASING. THEREFORE, THE 6-dB POINT IS DETERMINED PRIMARILY BY THE LATTER. SINCE A CHANGE OF 6 dB CORRESPONDS TO A DECREASE OF 50%, THE 6-dB ROLLOFF FREQUENCY IS THE FREQUENCY ω_{6dB} AT WHICH

[SEE EQ.(1) ABOVE IN PART(a)]

$$\frac{1}{\sqrt{1 + (\omega_{6dB} R_{11} C)^2}} = \frac{1}{2}$$

THUS

$$2\pi f_{6dB} R_{11} C = \sqrt{3}$$

or

$$f_{6dB} = \frac{\sqrt{3}}{2\pi R_{11} C} = 1516 \text{ HZ} \quad \leftarrow$$

5.2.1.

USING THE IDEAL OP-AMP CHARACTERISTICS, NO CURRENT FLOWS THROUGH EITHER RESISTOR R_1 OR R_2.

$$\therefore \quad v_i = v_+ \quad \text{AND} \quad v_o = v_-$$

ALSO $\quad v_+ = v_- \quad (\because v_d = 0$ FOR AN IDEAL OP-AMP$)$

THUS $\quad v_o = v_i$

AND $\quad A' = 1 \quad \leftarrow$

5.2.2.

$$v_o = -I R_F \quad \leftarrow$$

5.2.3.

(a) $\quad v_o = v_i \quad \leftarrow$

(b) $\quad v_A = (v_i - 2)\,V \quad \leftarrow$

5.2.4.

AT NODE B: $\quad \dfrac{v_a}{R_2} = -\dfrac{v_o}{R_{F2}}$ WHERE v_a IS THE VOLTAGE AT NODE A

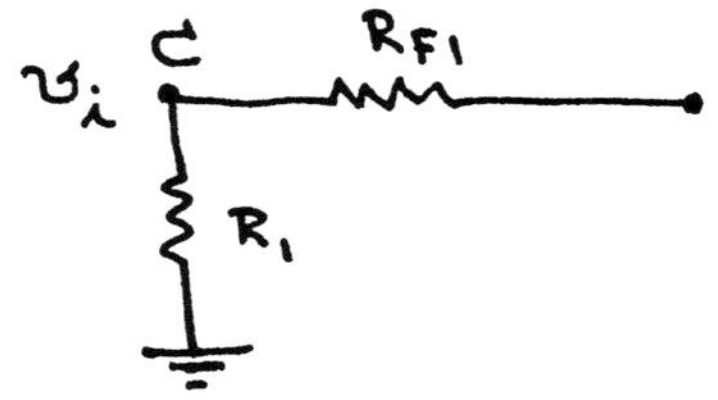

$$v_a = -\frac{R_2}{R_{F2}} v_o$$

AT NODE C: $\quad \dfrac{v_i}{R_1} = \dfrac{v_a - v_i}{R_{F1}}$

$$v_a R_1 - v_i R_1 = v_i R_{F1}$$

$$v_a = \frac{v_i (R_{F1} + R_1)}{R_1}$$

$$v_o = -v_a \frac{R_{F2}}{R_2} = -\frac{R_{F2}(R_1 + R_{F1})}{R_1 R_2} v_i \quad \leftarrow$$

5.2.5.

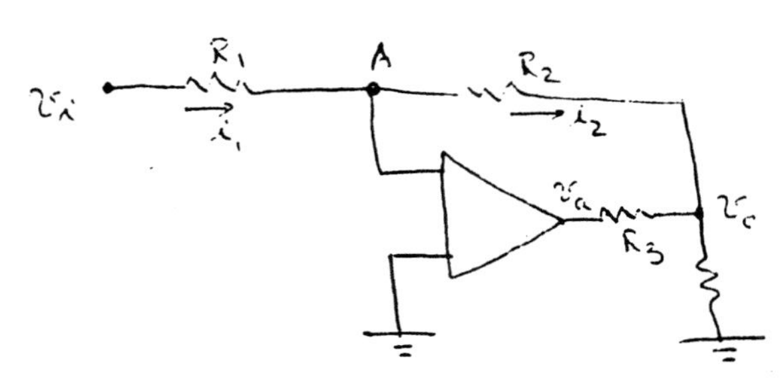

POINT A IS A VIRTUAL GROUND.

$$i_1 = \frac{v_i}{R_1} \quad ; \quad i_2 = -\frac{v_c}{R_2} = i_1 = \frac{v_i}{R_1}$$

$$\therefore \frac{v_c}{v_i} = -\frac{R_2}{R_1}$$ WHICH IS INDEPENDENT OF R_3 AND R_4. ←

5.2.6.

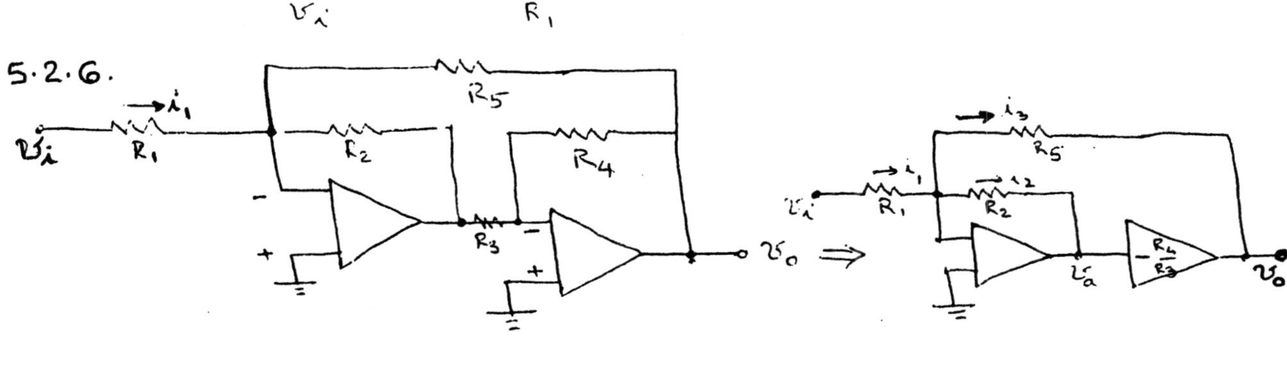

$$i_1 = \frac{v_i}{R_1} = i_2 + i_3 = -\frac{v_a}{R_2} - \frac{v_o}{R_5} = -\frac{1}{R_2}\left(-\frac{R_3}{R_4}\right)v_o - \frac{v_o}{R_5}$$

$$\therefore \frac{v_o}{v_i} = \frac{1}{R_1} \frac{1}{\frac{R_3}{R_2R_4} - \frac{1}{R_5}} = \frac{R_5}{R_1\left(\frac{R_3R_5}{R_2R_4} - 1\right)}$$

SUBSTITUTING VALUES $\quad \frac{v_o}{v_i} = \frac{2}{0.5\left(\frac{0.5\times 2}{1(2)} - 1\right)} = \frac{4}{-\frac{1}{2}}$

$= -8$ ←

5.2.7.

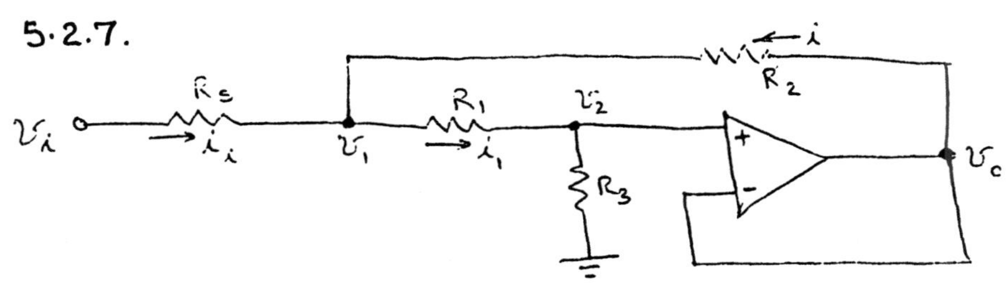

$$i_1 = i_i + i = \frac{v_i - v_1}{R_S} + \frac{v_o - v_1}{R_2} \quad ; \quad v_1 = i_1(R_1 + R_3)$$

$$v_1 = (R_1 + R_3)\left[\frac{v_o}{R_2} - v_1\left(\frac{1}{R_S} + \frac{1}{R_2}\right) + \frac{v_i}{R_S}\right] = \frac{(R_1+R_3)\left[\frac{v_o}{R_2} - \frac{v_i}{R_S}\right]}{1 + (R_1+R_3)\left(\frac{R_S + R_2}{R_S R_2}\right)}$$

5.2.7. CONTINUED

$$v_o = v_2 = v_1 R_3/(R_1+R_3) = \frac{R_3\left(\frac{v_o}{R_2}+\frac{v_i}{R_s}\right)}{1+(R_1+R_3)\left(\frac{R_s+R_2}{R_sR_2}\right)}$$

$$v_o\left[1+(R_1+R_3)\left(\frac{R_s+R_2}{R_sR_2}\right)-\frac{R_3}{R_2}\right] = \frac{R_3}{R_s}v_i$$

$$\frac{v_o}{v_i} = \left(\frac{R_3}{R_s}\right)\Big/\left[1+(R_1+R_3)\left(\frac{1}{R_s}+\frac{1}{R_2}\right)-\frac{R_3}{R_2}\right]$$

or

$$\frac{v_o}{v_i} = \frac{R_2 R_3}{R_s(R_1+R_2)+R_2(R_1+R_3)} \quad \leftarrow$$

5.2.8.

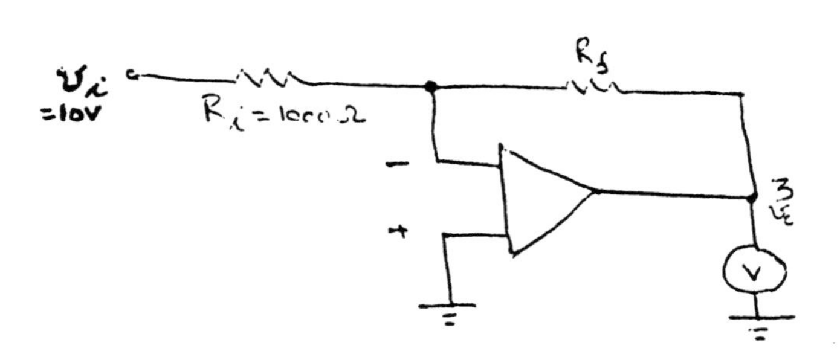

$$v_o = -\frac{R_f}{1000}v_i = -\frac{R_f}{100}$$

INDICATION $V_o = \frac{R_f}{100}$

$\therefore R_f = 100\,V_o$ ←

IF $V_o = 1$ V, THEN $R_f = 100\ \Omega$

IF $V_o = 10$ V, THEN $R_f = 1000\ \Omega = 1\ k\Omega$.

5.2.9.

$R_i = 1000\Omega$

$R_f = 2000\Omega$

i_1, i, v_o, i_2, R_L, v_i

(± 12V MAX @ ± 10mA)

(a) GAIN $= v_o/v_i = 1+\frac{R_f}{R_i} = 1+2 = 3$ WHICH IS INDEPENDENT OF R_L; $v_i = \frac{12}{3} = 4$V

$10\text{ mA} = i_{max} = i_{1\,max}+i_{2\,max} = \frac{v_{o\,max}-v_i}{2000}+\frac{v_{o\,max}}{R_{L\,min}} = \frac{12-4}{2000}+\frac{12}{R_{L\,min}}$ OR $R_{L\,min} = 2000\Omega$ ↖

(b) $v_{i\,max} = v_{o\,max}/\text{GAIN} = 12/3 = 4$ V ←

5.2.10.

(a) $A_o f_h = G_o f_H$

$A_o = 10^5$ NOTING THAT $20 \log_{10} A_o = A_{odB} = 100$

$10^5 \times 10 = 100\, f_H$

or $f_H = 10^4$ HZ ←

(b) $10^5 \times 10 = 1000\, f_H$

or $f_H = 10^3$ HZ ←

(c) GAIN-BANDWIDTH PRODUCT $= 10^6$ ←

5.2.11.

SUBSTITUTING THE OP-AMP MODEL WITH OPEN-CIRCUIT VOLTAGE GAIN $\bar{A}$ (BECAUSE OP AMP CAN INTRODUCE PHASE SHIFTS), WE HAVE

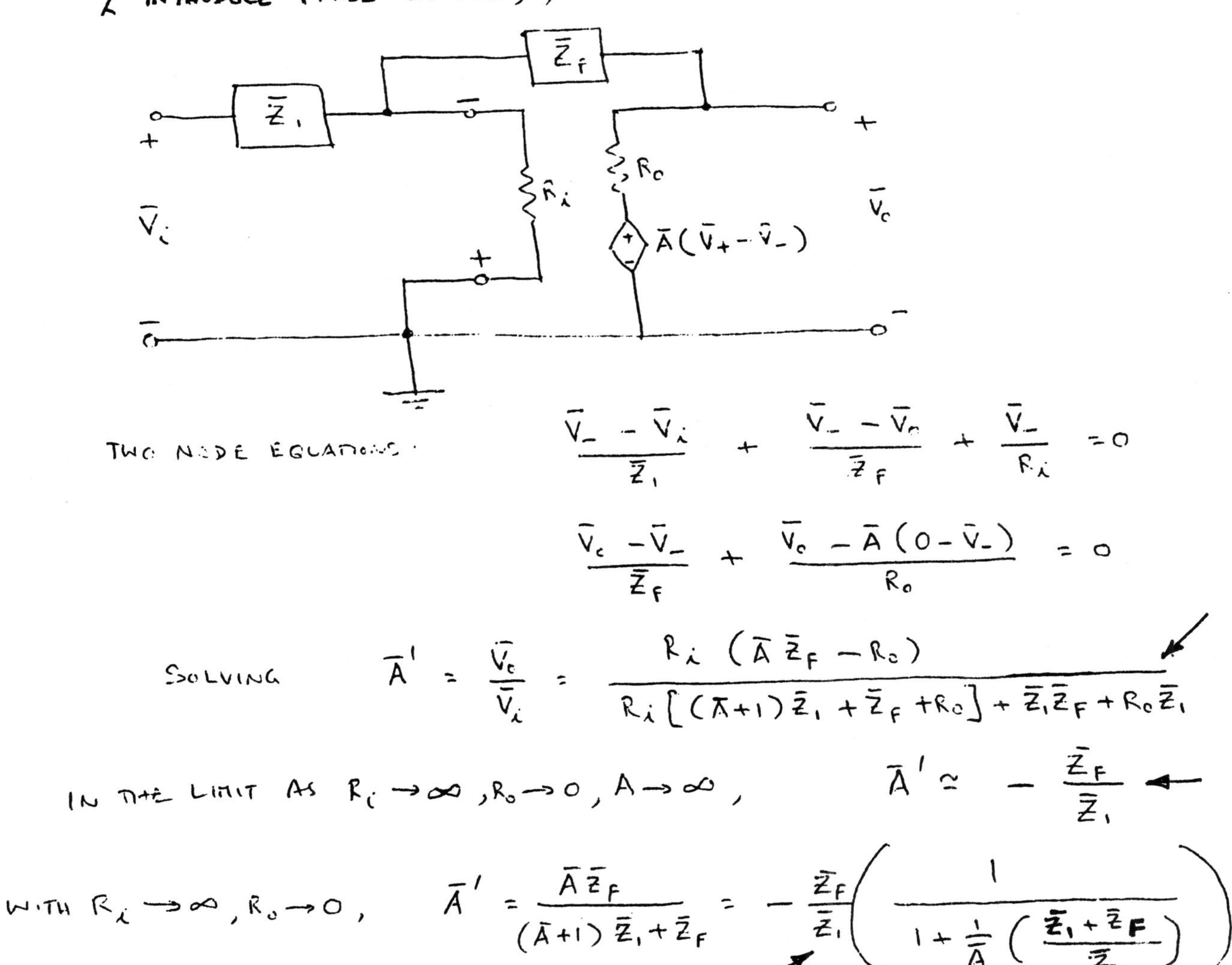

TWO NODE EQUATIONS:

$$\frac{\bar{V}_- - \bar{V}_i}{\bar{Z}_1} + \frac{\bar{V}_- - \bar{V}_o}{\bar{Z}_F} + \frac{\bar{V}_-}{R_i} = 0$$

$$\frac{\bar{V}_o - \bar{V}_-}{\bar{Z}_F} + \frac{\bar{V}_o - \bar{A}(0 - \bar{V}_-)}{R_o} = 0$$

SOLVING $\bar{A}' = \frac{\bar{V}_o}{\bar{V}_i} = \frac{R_i(\bar{A}\bar{Z}_F - R_o)}{R_i[(\bar{A}+1)\bar{Z}_1 + \bar{Z}_F + R_o] + \bar{Z}_1\bar{Z}_F + R_o\bar{Z}_1}$ ←

IN THE LIMIT AS $R_i \to \infty$, $R_o \to 0$, $A \to \infty$, $\bar{A}' \simeq -\frac{\bar{Z}_F}{\bar{Z}_1}$ ←

WITH $R_i \to \infty$, $R_o \to 0$, $\bar{A}' = \frac{\bar{A}\bar{Z}_F}{(\bar{A}+1)\bar{Z}_1 + \bar{Z}_F} = -\frac{\bar{Z}_F}{\bar{Z}_1}\left(\frac{1}{1 + \frac{1}{\bar{A}}\left(\frac{\bar{Z}_1 + \bar{Z}_F}{\bar{Z}_1}\right)}\right)$ ←

5.3.1. NOTING THAT $20 \log_{10} 100 = 20 \times 2 = 40$

$$C_1 = 100\,pF\ ;\ R_1 = 1.5\,k\Omega\ ;\ C_2 = 3\,pF. \leftarrow$$

5.3.2. FROM EX 5.3.2, SLEW RATE $= \omega V_m = \dfrac{0.5}{10^{-6}}$

$$\therefore V_m = \frac{0.5}{2\pi \times 10^5 \times 10^{-6}} = 0.8\,V \leftarrow$$

5.3.3. FROM EX. 5.3.2, SLEW RATE $= \omega V_m = \dfrac{0.7}{10^{-6}}$

$$\therefore V_m = \frac{0.7}{2\pi \times 50 \times 10^3 \times 10^{-6}} = 2.2\,V \leftarrow$$

FOR $V_m = 3\,V$, $\omega = \dfrac{0.7}{10^{-6}} \cdot \dfrac{1}{3}$

$$\text{or } f = \frac{1}{2\pi} \frac{0.7 \times 10^6}{3} = 37\,kHz \leftarrow$$

5.3.4.(a) $V_2 = \pm 12 \left(\dfrac{200}{500(10^3) + 200} \right) \simeq \pm 4.8\,mV \leftarrow$

RESISTANCE FROM TERMINAL 1 TO GROUND IS

$$R_1 \| R_2 = 1.5\,k\Omega \| 22\,k\Omega = 1404\,\Omega$$

RESISTANCE FROM TERMINAL 2 TO GROUND IS

$$R_3 + (R_4 \| R_5) = R_3 + \frac{200 \times 500 \times 10^3}{200 + 500 \times 10^3} \simeq R_3 + 200$$

$$\therefore R_3 \simeq 1404 - 200 = 1204\,\Omega \leftarrow$$

(b) $v_2 = -I_{b2} R_3$

ASSUMING $v_1 = v_2$ (ie ZERO DIFFERENTIAL INPUT VOLTAGE) BECAUSE OF LARGE OP-AMP'S GAIN

THEN $I_1 = -v_1/R_1 = -v_2/R_1 = I_{b2} R_3 / R_1$

$I_2 = I_{b1} - I_1 = I_{b1} - (I_{b2} R_3 / R_1)$ $\leftarrow$

5.3.4. (b) CONTINUED

AND $$v_o = I_2 R_2 + v_1 = R_2\left[I_{b1} - I_{b2} R_3 \left(\frac{R_1 + R_2}{R_1 R_2} \right) \right] \quad (A)$$

LET INPUT OFFSET CURRENT $I_{io} = I_{b2} - I_{b1}$

INPUT BIAS CURRENT $I_b = \dfrac{I_{b1} + I_{b2}}{2}$

THEN $I_{b1} = I_b - \dfrac{I_{io}}{2}$; $I_{b2} = I_b + \dfrac{I_{io}}{2}$

WHERE I_b IS THE COMMON COMPONENT OF I_{b1} AND I_{b2}, AND $I_{io}/2$ IS HALF OF THEIR DIFFERENCE.

USING THE ABOVE RELATIONSHIPS, ONE GETS

$$v_o = R_2 \left\{ I_b \left[1 - R_3 \left(\frac{R_1 + R_2}{R_1 R_2} \right) \right] - \frac{I_{io}}{2} \left[1 + R_3 \left(\frac{R_1 + R_2}{R_1 R_2} \right) \right] \right\} \quad (B)$$

IF $$R_3 = \frac{R_1 R_2}{R_1 + R_2} = R_1 \| R_2$$

THE COMPONENT OF v_o DUE TO I_b BECOMES ZERO. ←

THE REMAINING COMPONENT DUE TO I_{io} IS THEN

$$v_o = -I_{io} R_2$$

COMPARING THIS TO WHAT OCCURS IF R_3 WERE ZERO, ONE GETS

$$\left| \frac{v_o \text{ (FOR } R_3 = 0)}{v_o \text{ (FOR } R_3 = R_1 \| R_2)} \right| = \left| \frac{I_{b1}}{I_{io}} \right| \quad \leftarrow$$

SINCE $|I_{b1}/I_{io}|$ CAN EASILY BE 5 TO 10 OR MORE IN GENERAL-PURPOSE OP AMPS, THE REDUCTION IN OFFSET VOLTAGE IS SIGNIFICANT.

(C)

$$R_3 = R_1 \| R_2 = 5\,k\Omega \| 70\,k\Omega = \frac{5 \times 70}{75}\,k\Omega \simeq 4.7\,k\Omega \leftarrow$$

$I_{io} = I_{b2} - I_{b1} = 10\,nA$

FROM PART (b), $v_o = -I_{io} R_2 = -10 \times 10^{-9} \times 70 \times 10^{3}$

$= -0.7\,mV$ ←

5.3.5.

$$Y(\omega) = H_1(\omega)\, X_e(\omega)$$

$$X_e(\omega) = X(\omega) - X_f(\omega)$$

$$X_f(\omega) = H_2(\omega)\, Y(\omega)$$

SOLVING, ONE GETS, $H(\omega) = \dfrac{Y(\omega)}{X(\omega)} = \dfrac{H_1(\omega)}{1 + H_1(\omega)\, H_2(\omega)}$

$x(t)$ INPUT → $H(\omega) = \dfrac{H_1(\omega)}{1 + H_1(\omega)\, H_2(\omega)}$ → $y(t)$ OUTPUT

$$H_1(\omega) = A(\omega) = \frac{A_0}{1 + j(\omega/\omega_h)}$$

(NOTE THE FINITE GAIN AND FINITE BANDWIDTH OF THE OP AMP)

$$H_2(\omega) = \frac{R_1}{R_1 + R_2}$$

$$\therefore H(\omega) = \left(1 + \frac{R_2}{R_1}\right)\left[\frac{1}{1 + [(R_1 + R_2)/A_0 R_1]}\right]\left[\frac{1}{1 + j(\omega/\omega_h)/\{1 + [A_0 R_1/(R_1 + R_2)]\}}\right] \leftarrow$$

THE PRODUCT $H_1(\omega)\, H_2(\omega)$ IS CALLED THE LOOP GAIN.

FOR LARGE LOOP GAIN, $H(\omega) \simeq \dfrac{1}{H_2(\omega)}$

WHICH INDICATES THAT THE CLOSED-LOOP GAIN IS ALMOST ENTIRELY SET BY THE FEEDBACK NETWORK.

IF ONE ASSUMES $A_0 \to \infty$ FOR AN IDEAL OP AMP, THEN

$$H(\omega) = 1 + \frac{R_2}{R_1}$$

WHICH IS THE EXPECTED OP-AMP GAIN AS IN EXAMPLE 5.2.2.

5.4.1.

FROM EQ (5.4.6) $\frac{v_o}{v_i} = -15 = -\frac{20\times10^3}{R_1}\frac{1}{1+\frac{1}{50}+\frac{20\times10^3}{50R_1}}$

$$\frac{15R_1}{20(10^3)}\cdot\left[\left(1+\frac{1}{50}\right)+\frac{20(10^3)}{50R_1}\right] = 1$$

$$\frac{15R_1}{20(10^3)}\left(\frac{51}{50}\right) = 1-\frac{15}{50} = \frac{35}{50}$$

$$R_1 = \frac{35}{51}\frac{20(10^3)}{15} = \frac{140(10^3)}{153} = 0.91\ k\Omega \longleftarrow$$

WITH NEW $A_o = 60$,

$$\frac{v_o}{v_i} = \frac{-20(10^3)153}{140(10^3)}\cdot\left[\frac{1}{1+\frac{20(10^3)+\frac{140}{153}(10^3)}{60\times\frac{140}{153}(10^3)}}\right] = \frac{-20(153)60}{11600}$$

$$\simeq -15.83 \longleftarrow$$

5.4.2.

$$v_B = v_A = +1$$

$$i = \frac{v_S - v_B}{R_1} = \frac{3-1}{10\times10^3} = 0.2\ mA$$

$$v_o = -iR_2 + v_B = (-0.2\times10^{-3}\times30\times10^3)+1 = -5V. \longleftarrow$$

NOTE: HAD THE RESULT OF SUCH A CALCULATION GIVEN A VALUE OF v_o **OUTSIDE** THE RAIL VOLTAGES, THEN v_o WOULD BE LIMITED TO THE NEAREST RAIL VOLTAGE.

SEE PR 5.4.3 AND ITS SOLUTION

5.4.3.

FOR $v_S = +3V$, $v_o = -5V$ FROM PR. 5.4.2 SOLUTION.

WHEN $v_S = 0$ V, i IS REVERSED BECAUSE v_B IS POSITIVE WRT v_S

$$\therefore\ i = \frac{-1}{10(10^3)}$$

$$v_o = -iR_2 + v_B = -\left(\frac{-1}{10^4}\right)30\times10^3 + 1 = 4V$$

NOTE THAT A -3V CHANGE IN v_S PRODUCES A $+9$V CHANGE IN v_o, WHICH CORRESPONDS TO THE INVERTING GAIN OF $-R_2/R_1 = -3$.

WHEN $v_S = -3V$, $i = \frac{-3-1}{10\times10^3} = \frac{-4}{10^4}$

AND $v_o = -iR_2 + v_B = 12+1 = 13V$

WHICH IMPLIES A GAIN OF -3 AS BEFORE

v_S AND v_o ARE PLOTTED AS SHOWN

FOR $R_2 = 40k\Omega$, THE PLOT OF v_o IS SKETCHED BELOW. NOTE THAT THE POSITIVE EXCURSION OF v_o ATTEMPTS TO GO TO $+16V$; BUT IS LIMITED BY THE POSITIVE SUPPLY RAIL TO $+15V$.

5.4.3. CONTINUED

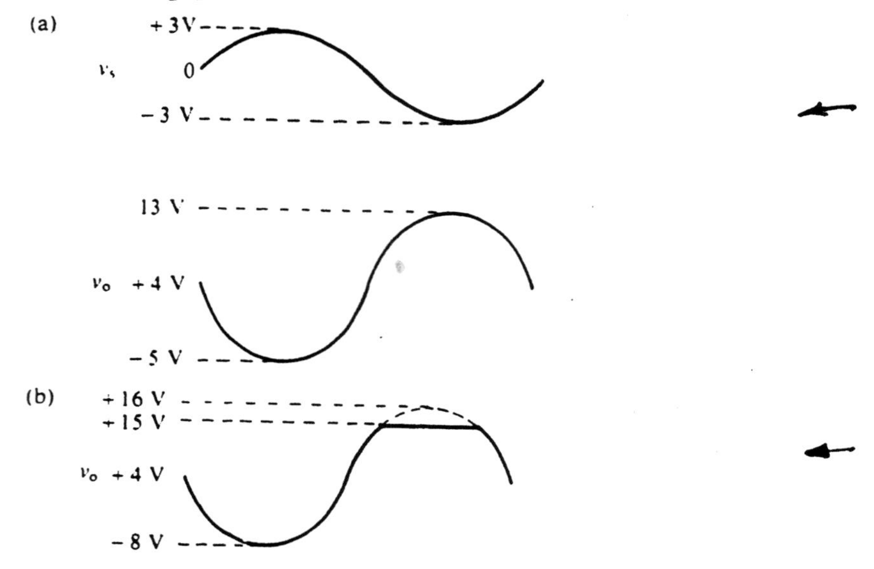

5.4.4.

NOTE THERE IS A **CONSTANT** VOLTAGE ON THE NONINVERTING INPUT AND AN ALTERNATING VOLTAGE ON THE **INVERTING** INPUT

WHEN $v_S = 0$, $v_O = \left(1 + \frac{R_2}{R_1}\right)(-1.5) = -4.5$ V

WHICH IS THE QUIESCENT VALUE OF v_O

[ALTERNATIVELY: $v_B = -1.5$ V; FIND CURRENT IN R_1; FIND v_O AS IN PR. 5.4.2.]

SINCE v_S IS CONNECTED TO THE INVERTING INPUT, IT IS AMPLIFIED (−2) TIMES.

∴ AT THE OUTPUT, THE AC COMPONENT WILL HAVE A PEAK VALUE OF 4 V BUT BE PHASE INVERTED WRT v_S, AND BECAUSE OF THE DC COMPONENT, IT WILL VARY AS SKETCHED IN FIGURE BELOW.

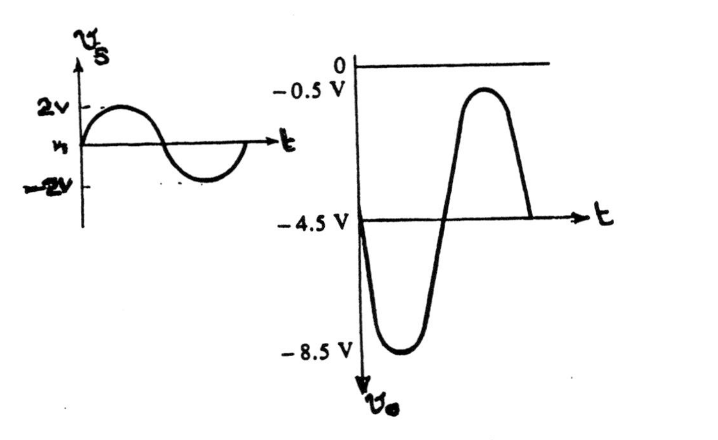

5.4.5. FROM EQ.(5.4.16)

$$v_o = -10^4\left(\frac{v_{i1}}{R_1} + \frac{v_{i2}}{R_2}\right)$$ WHICH MUST BE

EQUAL TO $(-10v_{i1} - 5v_{i2})$ SO THAT

$$-10 = -\frac{10^4}{R_1} \text{ or } R_1 = \frac{10^4}{10} = 10^3 = 1\,k\Omega \leftarrow$$

AND $$-5 = -\frac{10^4}{R_2} \text{ or } R_2 = +\frac{10^4}{5} = 2\times 10^3 = 2\,k\Omega \leftarrow$$

5.4.6.

$$(R_1 \| R_2 \| R_3 \| R_4 \| R_5) = \frac{1}{\frac{1}{\frac{3}{2}\times 50} + \frac{1}{\frac{5}{2}\times 50} + \frac{1}{\frac{7}{2}\times 50} + \frac{1}{\frac{9}{2}\times 50} + \frac{1}{\frac{11}{2}\times 50}}$$

$$= \frac{86625}{3043} \simeq 28.5\,\Omega$$

5.4.7.

FROM EQ (5.4.21)

$$v_o = \left(1 + \frac{3000}{500}\right)(28.5)\left[\frac{v_{i1}}{75} + \frac{v_{i2}}{125} + \frac{v_{i3}}{175} + \frac{v_{i4}}{225} + \frac{v_{i5}}{275}\right]$$

or $$v_o \simeq 2.66\,v_{i1} + 1.596\,v_{i2} + 1.14\,v_{i3} + 0.887\,v_{i4} + 0.725\,v_{i5} \leftarrow$$

FROM EQ (5.4.22)

$$v_o = \left(1 + \frac{R_f}{R_d}\right)\frac{1}{M}\sum_{m=1}^{M} v_{im}$$

$$= \left(1 + \frac{R_f}{1000}\right)\frac{1}{6}\sum_{m=1}^{6} v_{im}$$ WHICH MUST BE EQUAL TO $\sum_{m=1}^{6} v_{im}$.

$$\therefore \left(1 + \frac{R_f}{1000}\right)\frac{1}{6} = 1 \text{ or } R_f = (6-1)\,1000 = 5000 = 5\,k\Omega \leftarrow$$

5.4.8. $v_- = v_+ = 0$; KCL AT NODE X: $\frac{v_1}{R_1} + \frac{v_2}{R_2} + \frac{v_3}{R_3} + \frac{v_o}{R_F} = 0$

$$\therefore v_o = -R_F\left[\frac{v_1}{R_1} + \frac{v_2}{R_2} + \frac{v_3}{R_3}\right] \leftarrow$$

BY CHOOSING $R_1 = R_2 = R_3 = R_F$, $v_o = -[v_1 + v_2 + v_3]$; BY USING A UNITY-GAIN INVERTER TO REMOVE THE NEGATIVE SIGN, THE OUTPUT SIGNAL CAN BE SEEN TO BE THE SUM OF THE THREE INPUT SIGNALS. ←

5.4.9. FROM EQ. (5.4.31), $R_{in} = -R(R_1/R_1) = -R$; $i_{in} = i_L + i_R$

$$i_L = i_{in} - i_R = \frac{v_{in}}{Z_{in}} - \frac{v}{-R} = \frac{v_{in}}{Z_{in}} + \frac{1}{R}\,\frac{(-RZ_L)/(-R+Z_L)}{Z_{in}}\,v_{in}$$

$$= \frac{v_{in}}{Z_{in}}\left(1+\frac{Z_L}{R-Z_L}\right) = \frac{v_{in}}{Z_{in}}\left(\frac{R}{R-Z_L}\right)$$

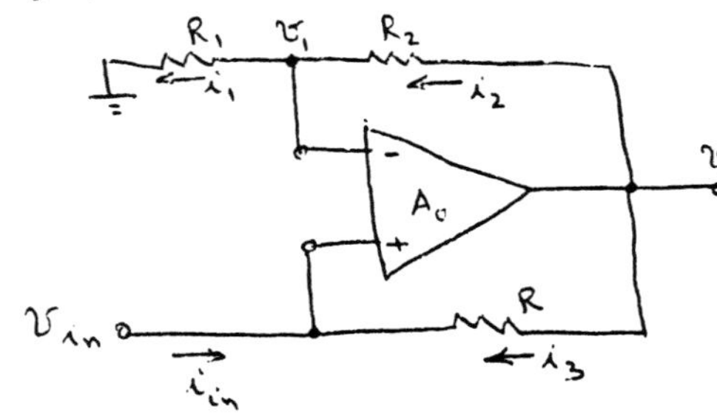

$$i_L = v_{in}\,\frac{1}{R + \frac{-RZ_L}{-R+Z_L}}\left(\frac{R}{R-Z_L}\right)$$

$$= v_{in}\,\frac{(Z_L-R)(-R)}{-R^2(Z_L-R)} = \frac{v_{in}}{R} \quad \leftarrow$$

5.4.10.

$$v_o = A_o(v_{in} - v_1)$$

$$\therefore\ v_1 = v_{in} - \frac{v_o}{A_o}$$

$$i_1 = \frac{v_1}{R_1} = \frac{v_{in}}{R_1} - \frac{v_o}{A_oR_1}\ ; \quad i_2 = i_1$$

$$v_o = i_2R_2 + i_1R_1 = i_1(R_1+R_2) = (R_1+R_2)\left[\frac{v_{in}}{R_1} - \frac{v_o}{A_oR_1}\right]$$

$$v_o\left[1 + \left(1+\frac{R_2}{R_1}\right)\frac{1}{A_o}\right] = \frac{R_1+R_2}{R_1}\,v_{in}$$

$$R_{in} = \frac{v_{in}}{i_{in}} = -\frac{v_{in}}{i_3} = -v_{in}\,\frac{R}{v_o - v_{in}} = -\frac{R}{\frac{v_o}{v_{in}}-1} = \frac{-R}{-1+\frac{(R_1+R_2)/R_1}{1+\frac{(R_1+R_2)}{R_1}\frac{1}{A_o}}}$$

$$= -R\,\frac{R_1}{R_2}\left[\frac{A_o+1+\frac{R_2}{R_1}}{A_o-1-\frac{R_1}{R_2}}\right] \quad \leftarrow$$

↑ CHANGE FACTOR DUE TO FINITE A_o

5.4.11.

$$v_2 = v_{in}\ ; \quad v_4 = v_2 = v_{in}$$

$$i_5 = \frac{v_4}{Z_5} = \frac{v_{in}}{Z_5}\ ; \quad i_4 = i_5 = \frac{v_{in}}{Z_5}\ ; \quad v_3 = v_4 + i_4Z_4 = v_{in} + \frac{v_{in}}{Z_5}Z_4 = v_{in}\left(1+\frac{Z_4}{Z_5}\right)$$

$$i_3 = \frac{v_3 - v_2}{Z_3} = \frac{v_3 - v_{in}}{Z_3} = \frac{v_{in}\left(1+\frac{Z_4}{Z_5}\right) - v_{in}}{Z_3} = v_{in}\,\frac{Z_4}{Z_3Z_5}$$

$$i_2 = i_3\ ; \quad v_1 = v_2 - i_2Z_2 = v_{in} - i_2Z_2 = v_{in} - i_3Z_2 = v_{in} - \frac{v_{in}Z_4}{Z_3Z_5}\cdot Z_2$$

$$\therefore\ v_1 = v_{in}\left(1 - \frac{Z_2Z_4}{Z_3Z_5}\right)\ ; \quad i_1 = (v_{in}-v_1)/Z_1 = \frac{v_{in} - v_{in}\left(1-\frac{Z_2Z_4}{Z_3Z_5}\right)}{Z_1}$$

$$i_1 = v_{in}\,\frac{Z_2Z_4}{Z_1Z_3Z_5} = i_{in}\ , \quad Z_{in} = \frac{v_{in}}{i_{in}} = \frac{v_{in}}{i_1} = \frac{Z_1Z_3Z_5}{Z_2Z_4} \quad \leftarrow$$

5.4.12.

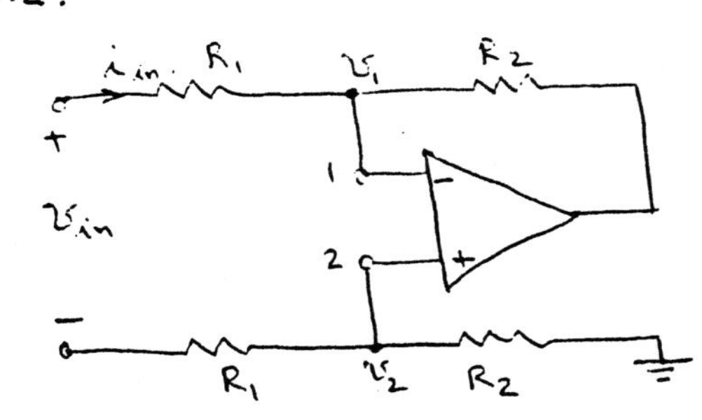

$v_2 - v_1 = 0$ (∵ IDEAL OP AMP)

NO POTENTIAL DIFF. BETWEEN 1 & 2.

$R_{in} = 2R_1$ ←

5.4.13.

$$i_1 = \frac{v_b - v}{100} \quad ; \quad i_2 = \frac{v_a - v}{100}$$

$$-i_1 = \frac{v_c - v}{200} \quad ; \quad -i_2 = \frac{v_c - v}{500}$$

$$v_b - v = \frac{100}{200}(v - v_c) \quad ; \quad v_a - v = \frac{100}{500}(v - v_c)$$

$$v_b = \left(1 + \frac{100}{200}\right)v - \frac{100}{200}v_c$$

$$v = \left(v_a + \frac{100}{500}v_c\right)\frac{1}{\left(1 + \frac{100}{500}\right)}$$

$$\therefore v_b = \left(1 + \frac{100}{200}\right)\frac{\left(v_a + \frac{100}{500}v_c\right)}{\left(1 + \frac{100}{500}\right)} - \frac{100}{200}v_c$$

$$= \frac{(200 + 100)\,500}{200\,(500 + 100)}\left(v_a + \frac{100}{500}v_c\right) - \frac{100}{200}v_o$$

$$v_b - \frac{(100 + 200)\,500}{(100 + 500)\,200}v_a = v_o\left[\frac{(100 + 200)\,100}{(100 + 500)\,200} - \frac{100}{200}\right]$$

$$v_b - \frac{30(50)}{60(20)}v_a = v_o\left[\frac{30(10)}{60(20)} - \frac{10}{20}\right]$$

$$= v_o\,\frac{10(-30)}{60(20)}$$

$$20(60)\,v_b - 30(50)\,v_a = -10(30)\,v_o$$

$$v_o = \frac{30(50)\,v_a - 20(60)\,v_b}{10(30)} = (5v_a - 4v_b) \leftarrow$$

5.4.14.

From Eq. (5.4.42) $\frac{v_o}{v_i} = -\frac{R_f}{R_i}\frac{1}{1+j\omega R_f C_f}$

$\omega_l R_f C_f = 1$; $R_f = 10^6\ \Omega$

$\therefore C_f = \frac{1}{\omega_l R_f} = \frac{1}{2\pi(10^3)10^6} = 1.59\times 10^{-10} = 159\ pF$ ←

5.4.15.

$\frac{R_f}{R_i} \simeq 2.5$; $\frac{1}{2\pi R_i C_i} = f_h = 1\ MHz = 10^6$

$\therefore R_i = \frac{1}{2\pi(10^6)(10^{-10})} = 1592\ \Omega$ ←

AND $R_f = 2.5 R_i = 2.5(1592) = 3980\ \Omega$ ←

5.4.16.

THE PROBLEM IS CONCERNED WITH THE RELATIVE VALUES OF THE INDIVIDUAL IMPEDANCES IN THE INPUT LINE AND IN THE FEEDBACK LOOP. AS THE IMPEDANCE OF C_1 IS INFINITELY LARGE COMPARED WITH R_1 AT LOW FREQUENCIES, THE OP AMP IS ISOLATED FROM DC. HENCE IT IS POSSIBLE TO DEDUCE IMMEDIATELY THAT AT LOW FREQUENCIES THE OVERALL IMPEDANCE IN THE INPUT LINE IS GOVERNED BY THE CAPACITOR. BY THE SAME REASONING, MOST OF THE CURRENT IN THE FEEDBACK LOOP MUST FLOW THROUGH THE RESISTOR R_F. AS A RESULT, THE CIRCUIT BEHAVES AS A DIFFERENTIATOR AT LOW FREQUENCIES AND AS AN INTEGRATOR AT HIGH FREQUENCIES. ←

IF ONE THINKS OF THIS CIRCUIT TO BE THAT OF A BANDPASS ACTIVE FILTER (FIGS. 5.4.13 AND 5.4.17), WITH $f_l \ll f_h$, AT LOW FREQUENCIES IT HAS HIGH-PASS FILTER CHARACTERISTICS AND LOW-PASS FILTER CHARACTERISTICS AT HIGH FREQUENCIES. (SEE FIG. 5.4.16)

5.4.17.

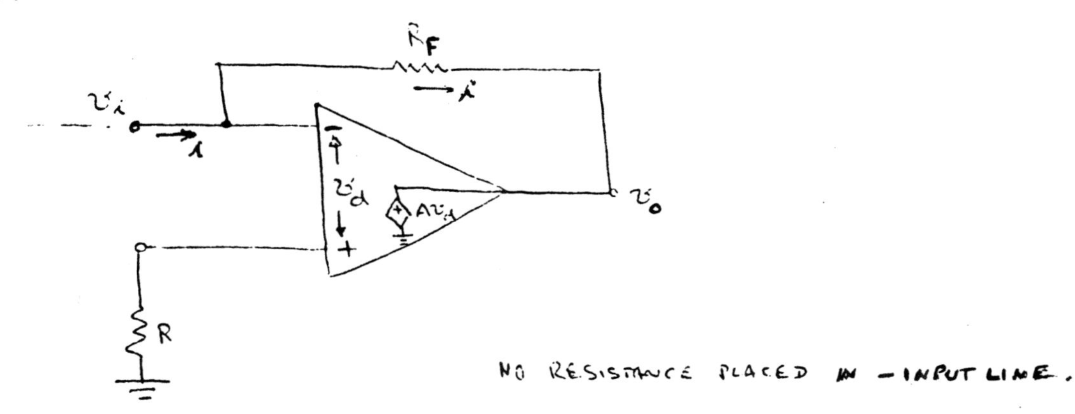

NO RESISTANCE PLACED IN − INPUT LINE.

KVL: $v_i - Av_d - R_F i = 0$

ALSO $v_i + v_d = 0$ or $v_i = -v_d$

THEN $v_i + Av_i = R_F i$

$$R_i = \frac{v_i}{i} = \frac{R_F}{(A+1)}$$

THIS IS SOMETIMES REFERRED TO AS THE MILLER INPUT RESISTANCE (IMPEDANCE

WHEN A IS LARGE BUT FINITE, $R_i = R_F / A$.

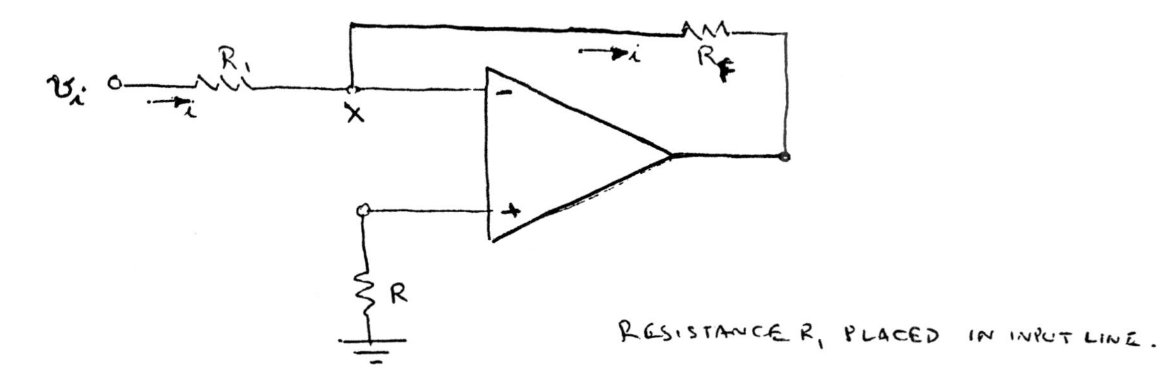

RESISTANCE R_1 PLACED IN INPUT LINE.

BECAUSE NODE X IS A VIRTUAL GROUND, FOR LARGE A, $R_i = R_1$

5.4.18.

USING EQ. (5.4.39) $v_c = -\frac{1}{CR}\int_0^{0.5} v_s\, dt = -50\,[t]_0^{0.5}$

SINCE v_c IS PROPORTIONAL TO t, IT WILL BE A LINEAR RAMP, NEGATIVE GOING AS INDICATED BY THE MINUS SIGN.

EVALUATING v_o AT $t = 0.3$ s, $v_o = -15$ V

EVALUATING v_o AT $t = 0.5$ s, $v_c = -25$ V

SINCE THE NEGATIVE SUPPLY RAIL IS ONLY -15 V, THE v_o RAMPS DOWN TO THIS VALUE OF -15V, WHERE IT WILL REMAIN AS SHOWN IN SKETCH.

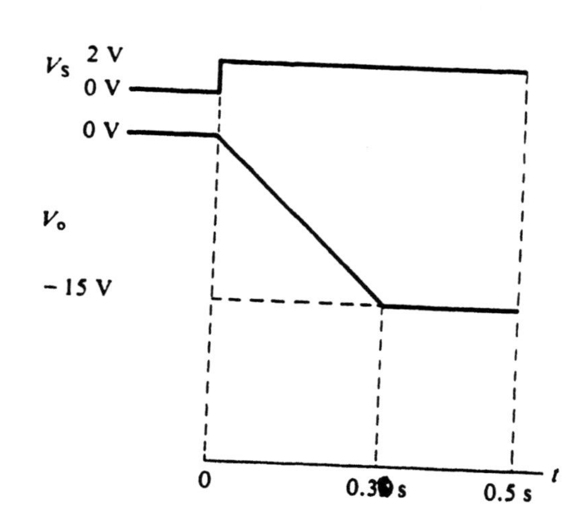

5.4.19.

BEFORE S IS OPENED, $v_o = v_B = v_A = +3\text{V}$

WHEN S IS OPENED, $v_o = -\frac{1}{C}\int i\,dt + 3$

$i = \frac{5-3}{40(10^3)} = 0.05 \text{ mA}$

$$v_o = \frac{-1}{0.2\times10^{-6}}\left[0.5\times10^{-4}\, t\right]_0^{60\text{ms}} + 3$$

WHICH IS A NEGATIVE - GOING LINEAR RAMP BEGINNING AT 3V.

AT 60 ms, $v_o = -5\times10^6\left[0.5\times10^{-4}\times60\times10^{-3}\right] + 3$
$= -12\text{V}$

THE WAVEFORM IS SKETCHED BELOW.

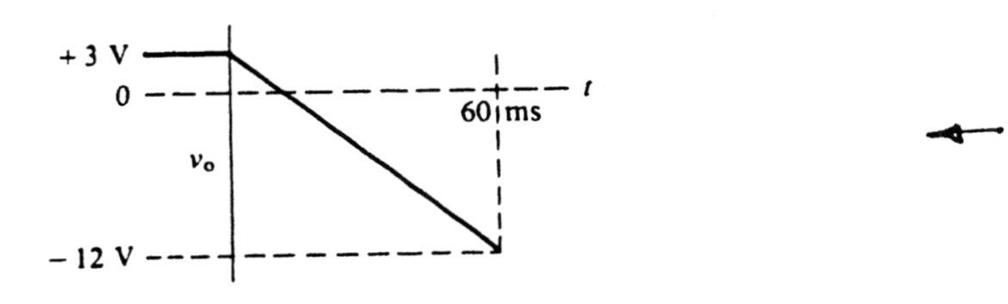

5.4.20.

IN THIS CIRCUIT $i = i_1 + i_2$

$$v_c = -\frac{1}{C}\int i\,dt + k$$

WHERE k IS THE INITIAL VALUE OF v_o

At $t = 0$

$$i = i_1 + i_2 = \frac{v_1 - V_A}{R_1} + \frac{v_2 - V_A}{R_2} = \frac{5-(-2)}{5\text{ k}\Omega} + \frac{0-(-2)}{1\text{ k}\Omega}$$

$$= 1.4\text{ mA} + 2\text{ mA} = 3.4\text{ mA}$$

Then from $t = 0$ to 1 ms,

$$v_o = \frac{-1}{C}\int_0^{1\text{ ms}} i\,dt - 2$$

$$= -2\times 10^{-6}[3.4\times 10^{-3}t]_0^{10^{-3}} - 2\text{ V}$$

Which is a negative-going linear sweep beginning at -2 V as shown for the period 0 to 1 ms in SKETCH. After 1 ms,

$$v_o = -6.8 - 2 = -8.8\text{ V}.$$

From $t = 1$ ms to 2 ms

$$= i_1 + i_2 = 1.4\text{ mA} + \frac{(-5-[-2])}{1\text{ k}\Omega} = 1.4\text{ mA} - 3\text{ mA}$$

$$= -1.6\text{ mA}$$

$$v_o = \frac{-1}{C}\int_{1\text{ ms}}^{2\text{ ms}} i\,dt - 8.8\text{ V}$$

as -8.8 V was the value of v_o at $t = 1$ ms.

$$\text{Then } v_o = -2\times 10^{-6}\int_{1\text{ ms}}^{2\text{ ms}} -1.6\times 10^{-3}\,dt - 8.8$$

$$= 2\times 10^{-6}[1.6t]_{1\text{ ms}}^{2\text{ ms}} - 8.8\text{ V}$$

which is a positive going linear sweep beginning at -8.8 V

At $t = 2$ ms,

$$v_o = 3.2 - 8.8 = -5.6\text{ V}$$

For $t = 2$ ms to 3 ms

$$i = i_1 + i_2 = \frac{0-(-2)}{5\text{ k}} - 3\text{ mA} = -2.6\text{ mA}$$

and v_o is a positive going linear sweep beginning at -5.6 V and reaching $5.2 - 5.6 = -0.4$ V at $t = 3$ ms. The complete waveform for v_o is SHOWN BELOW.

(SEE NEXT PAGE)

5.4.20. CONTINUED

SKETCH:

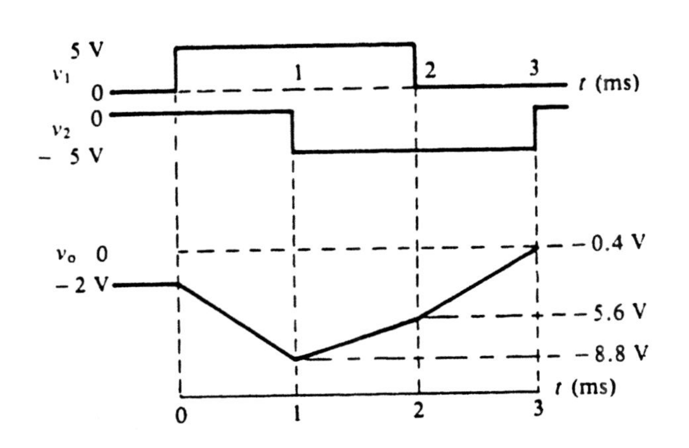

5.4.21.

WITH $R_2 = 30\,k\Omega$ AND $R_1 = 10\,k\Omega$

$$\frac{v_o}{v_i} = 1 + \frac{R_2}{R_1} = 1 + \frac{30}{10} = 4$$

WAVEFORMS OF v_i AND v_o ARE SKETCHED BELOW.

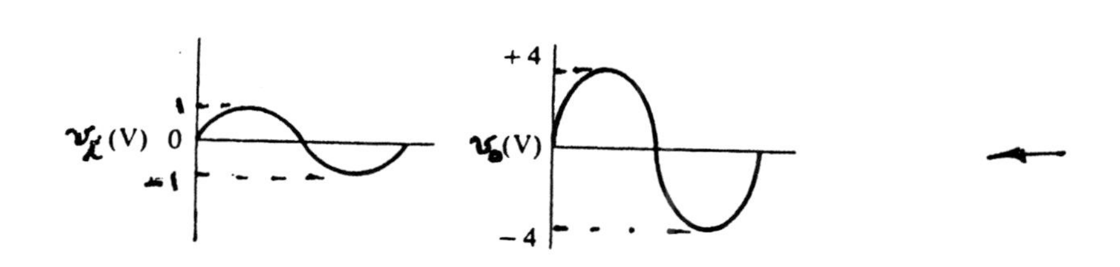

5.4.22. $\ddot{y} = -12\dot{y} - 5y + 10$

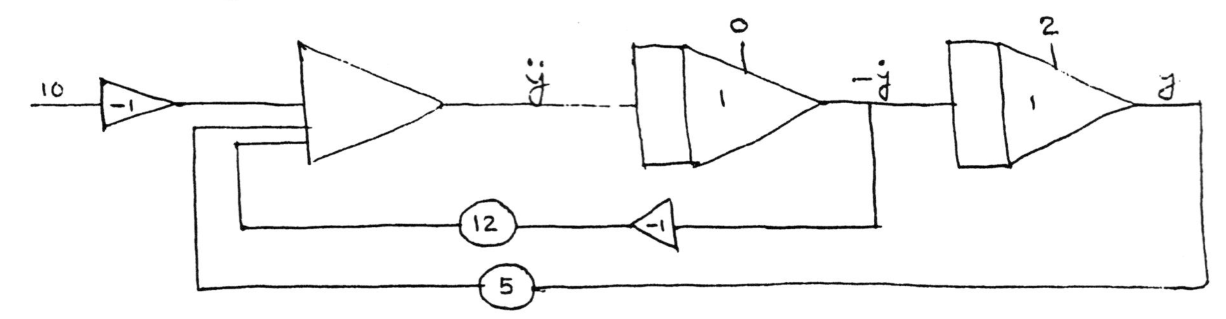

5.4.23. $t = \alpha\tau$

$t = 1\text{ ms}$, $\tau = 1\text{ s}$ $\therefore \alpha = \frac{t}{\tau} = 10^{-3}$

$$\therefore \frac{d^2y(\tau)}{d\tau^2} = -10^{-3} \times 12 \frac{dy(\tau)}{d\tau} - 10^{-6} \times 5\, y(\tau) + 10 \times 10^{-6}$$

$$\text{or } \ddot{y} = -(12\times10^{-3})\,\dot{y} - 5\times10^{-6}\, y + 10^{-5}$$

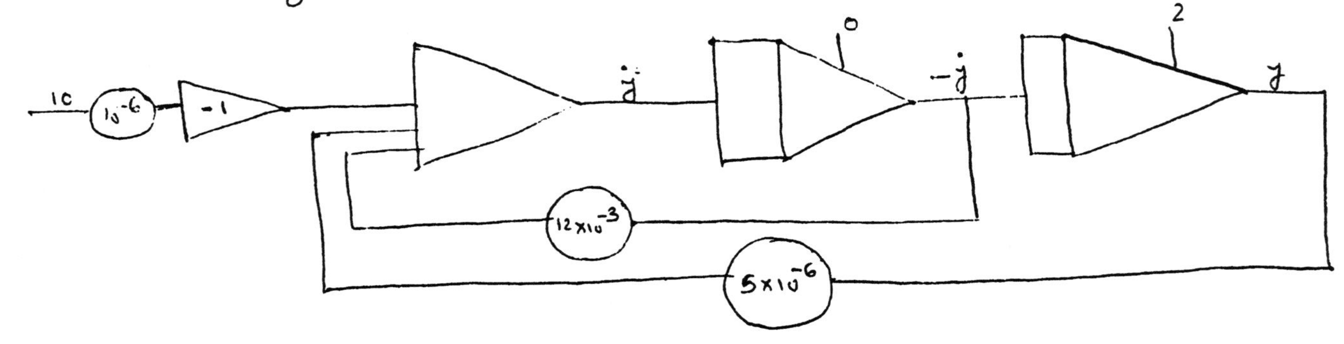

5.4.24.

$$\frac{dy(t)}{dt} = -2000\, y(t) \;;\; y(t) = -2000 \int_0^t y(\tau)\,d\tau + 5$$

INITIAL VOLTAGE = 5V ;

$R_i C_f = \frac{1}{2000}$; WITH $R_i = 10^4\,\Omega$, $C_f = \frac{1}{2000} 10^{-4} = 0.05 \times 10^{-6}$

$= 0.05\,\mu F$ ←

5.4.25.

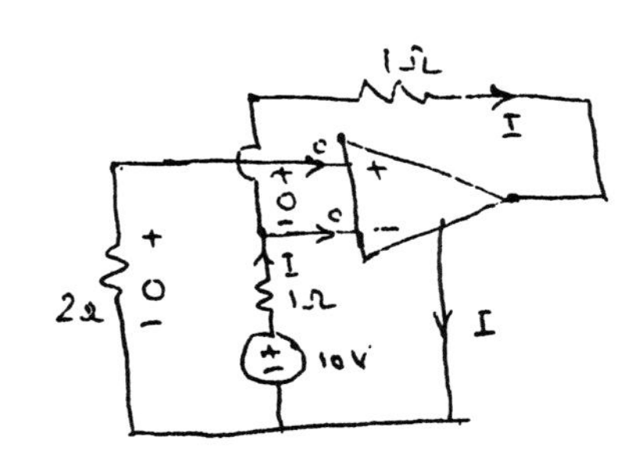

$I = \frac{10}{1} = 10A$ ←

CHAPTER 6

6.1.1.

(a) $(10110)_2 = 1\times2^4 + 0\times2^3 + 1\times2^2 + 1\times2^1 + 0\times2^0 = 16+0+4+2+0 = (22)_{10}$ ←

(b) $(101100)_2 = 1\times2^5 + 0\times2^4 + 1\times2^3 + 1\times2^2 + 0\times2^1 + 0\times2^0 = 32+0+8+4+0+0 = (44)_{10}$ ←

(c) $(11010101)_2 = 1\times2^7 + 1\times2^6 + 0\times2^5 + 1\times2^4 + 0\times2^3 + 1\times2^2 + 0\times2^1 + 1\times2^0 =$

$= 128 + 64 + 0 + 16 + 0 + 4 + 0 + 1 = (213)_{10}$ ←

(d) $(11101.101)_2 = 1\times2^4 + 1\times2^3 + 1\times2^2 + 0\times2^1 + 1\times2^0 + 1\times2^{-1} + 0\times2^{-2} + 1\times2^{-3}$

$= 16 + 8 + 4 + 0 + 1 + 0.5 + 0 + 0.125 = (29.625)_{10}$ ←

(e) $(.00101)_2 = 0\times2^{-1} + 0\times2^{-2} + 1\times2^{-3} + 0\times2^{-4} + 1\times2^{-5}$

$= 0 + 0 + 0.125 + 0 + 0.03125 = (0.15625)_{10}$ ←

6.1.2.

(a) $(255)_{10} = (?)_2$

	QUOTIENT	REMAINDER	
$255 \div 2 =$	127	1	(LSB)
$127 \div 2 =$	63	1	
$63 \div 2 =$	31	1	
$31 \div 2 =$	15	1	
$15 \div 2 =$	7	1	
$7 \div 2 =$	3	1	
$3 \div 2 =$	1	1	
$1 \div 2 =$	0 ← STOP	1	(MSB)

$(255)_{10} = (11111111)_2$ ← (MSB at left, LSB at right)

(b) $(999)_{10} = (?)_2 = (1111100111)_2$ ←

(c) $(1066)_{10} = (?)_2 = (10000101010)_2$ ←

(d) $(0.375)_{10} = (.011)_2$ ←

(e) $(1259.00125)_{10} = (10011101011.00000000010100011 1)_2$ ←

6.1.3.

(a) $(257)_8 = 2\times8^2 + 5\times8 + 7\times8^0 = 128 + 40 + 7 = (175)_{10}$ ←

(b) $(367)_8 = 3\times8^2 + 6\times8 + 7\times8^0 = 192 + 48 + 7 = (247)_{10}$ ←

(c) $(0.321)_8 = 3\times8^{-1} + 2\times8^{-2} + 1\times8^{-3} = 0.375 + 0.03125 + 0.0019531$
$= (0.4082031)_{10}$ ←

(d) $(367.240)_8 = (\ ?\ )_{10}$

FROM PART (b) $(367)_8 = (247)_{10}$

$(0.240)_8 = 2\times8^{-1} + 4\times8^{-2} + 0\times8^{-3} = 0.250 + 0.0625 + 0 = (.3125)_{10}$

$\therefore\ (367.240)_8 = (247.3125)_{10}$ ←

(e) $(2103.45)_8 = (1091.578125)_{10}$ ←

6.1.4.

(a) $(175)_{10} = (\ ?\ )_8$

	QUOTIENT	REMAINDER	
$175 \div 8 =$	21	7	(LSB)
$21 \div 8 =$	2	5	
$2 \div 8 =$	0 ↑STOP	2	(MSB)

$(175)_{10} = (257)_8$ ← (See Pr. 6.1.3a)

(b) $(247)_{10} = (367)_8$ ← (See Pr. 6.1.3b)

(c) $(65{,}535)_{10} = (177777)_8$ ←

(d) $(0.125)_{10} = (.1)_8$ ←

(e) $(379.25)_{10} = (573.2)_8$ ←

6.1.5.

(a) $(3425)_8 = (011\ 100\ 010\ 101)_2$ ←

(b) $(3651)_8 = (011\ 110\ 101\ 001)_2$ ←

(c) $(0.214)_8 = (.010\ 001\ 100)_2$ ←

(d) $(4125.016)_8 = (100\ 001\ 010\ 101.000\ 001\ 110)_2$ ←

(e) $(4573.26)_8 = (100\ 101\ 111\ 011.010\ 110)_2$ ←

6.1.6.

(a) $(011\ 100\ 010\ 101)_2 = (3425)_8$ ← See Pr. 6.1.5a

(b) $(1011010)_2 = (001\ 011\ 010)_2 = (132)_8$ ← (ADDED)

(c) $(.110101)_2 = (.65)_8$ ←

(d) $(100\ 101\ 111\ 011.\ 010\ 110)_2 = (4573.26)_8$ ← (ADDED)

(e) $(001\ 110\ 110\ 111.\ 101\ 100)_2 = (1667.54)_8$ ← (ADDED) (ADDED)

6.1.7.

(a) $(6B)_{16} = 6 \times 16^1 + B \times 16^0 = 6 \times 16^1 \times 11 \times 16^0 = 96 + 11 = (107)_{10}$ ←

(b) $(1F4)_{16} = (500)_{10}$ ←

(c) $(C59)_{16} = (3{,}161)_{10}$ ←

(d) $(256.72)_{16} = (598.4453125)_{10}$ ←

(e) $(.0E3)_{16} = (.05541992)_{10}$ ←

6.1.8.

(a) $(97)_{10} = (61)_{16}$ ←

(b) $(864)_{10} = (360)_{16}$ ←

(c) $(5{,}321)_{10} = (14C9)_{16}$ ←

(d) $(0.00125)_{10} = (.0051EB851E\underline{B851E})_{16}$ ←

(e) $(449.375)_{10} = (1C1.6)_{16}$ ←

6.1.9.

(a) $(0011\ 1011\ 1010\ 0100.\ 1001\ 1100)_2 = (3BA4.9C)_{16}$ ← (ADDED) (ADDED)

(b) $(0010\ 1101\ 1101)_2 = (2DD)_{16}$ ← (ADDED)

(c) $(.1110\ 1000)_2 = (.E8)_{16}$ ← (ADDED)

(d) $(0011\ 0111\ 0001.\ 1101\ 1110)_2 = (371.DE)_{16}$ (ADDED)

(e) $(.0000\ 1101\ 1100\ 0101)_2 = (0.0DC5)_{16}$ ←

6.1.10.

(a) $(2ABF5)_{16} = (0010\ 1010\ 1011\ 1111\ 0101)_2$ ←

(b) $(3BA4.9C)_{16} = (0011\ 1011\ 1010\ 0100.\ 1001\ 1100)_2$ ←

(c) $(0.0DC5)_{16} = (.0000\ 1101\ 1100\ 0101)_2$ ←

(d) $(15CE.FB3)_{16} = (0001\ 0101\ 1100\ 1110.\ 1111\ 1011\ 0011)_2$ ←

(e) $(2AB.F8)_{16} = (0010\ 1010\ 1011.\ 1111\ 1000)_2$ ←

6.1.11.

(a) $(567)_{10} = (0101\ 0110\ 0111)_{BCD}$ ←

(b) $(1975)_{10} = (0001\ 1001\ 0111\ 1000)_{BCD}$ ←

(c) $(163.25)_{10} = (0001\ 0110\ 0011.\ 0010\ 0101)_{BCD}$ ←

(d) $(0.659)_{10} = (.0110\ 0101\ 1001)_{BCD}$ ←

(e) $(2153.436)_{10} = (0010\ 0001\ 0101\ 0011.\ 0100\ 0011\ 0110)_{BCD}$ ←

6.1.12.

(a) $(0101\ 0110\ 0111)_{BCD} = (567)_{10}$ ←

(b) $(.0110\ 0101\ 1001)_{BCD} = (.659)_{10}$ ←

(c) $(.1001\ 1000\ 0100)_{BCD} = (.984)_{10}$ ←

(d) $(1001\ 0010.\ 0000\ 0001)_{BCD} = (92.01)_{10}$ ←

(e) $(0010\ 0001\ 0101\ 0011.\ 0100\ 0011\ 0110)_{BCD} = (2153.436)_{10}$ ←

6.1.13.

NOTE THAT INPUT A IS COMPLEMENTED BEFORE IT IS APPLIED TO THE AND GATE. F IS FOUND BY PERFORMING THE AND OPERATION ON B AND $\overline{A}$.

A	B	$\overline{A}$	F
0	0	1	0
0	1	1	1
1	0	0	0
1	1	0	0

←

6.1.14.

(a) $Y = A\bar{B} + \bar{A}B$ ← (b) $Y = (\bar{A} + \bar{B})(A + B) = A\bar{B} + \bar{A}B$ ←

6.1.15.

(a) $Y = (\overline{AB + CD})$ (b) $Y = (\overline{AB})(\overline{CD})$

BY DeMORGAN'S THEOREM, $(\overline{AB+CD}) = (\overline{AB})(\overline{CD})$

THUS BOTH ARE EQUIVALENT. ←

6.1.16.

A BCD-CODE USING FOUR BINARY DIGITS A, B, C, AND D WITH D BEING THE MOST SIGNIFICANT BIT IS GIVEN BELOW:

D	C	B	A	DECIMAL DIGIT
0	0	0	0	0
0	0	0	1	1
0	0	1	0	2
0	0	1	1	3
0	1	0	0	4
0	1	0	1	5
0	1	1	0	6
0	1	1	1	7
1	0	0	0	8
1	0	0	1	9

(a) Y_1 IS ON FOR DIGITS 0, 1, 2, 3, 4, 7, 8, AND 9.

NOTE THAT 5 AND 6 ARE THE MISSING DIGITS

$\therefore Y_1 = \bar{C} + \bar{A}\bar{B} + AB$ ←

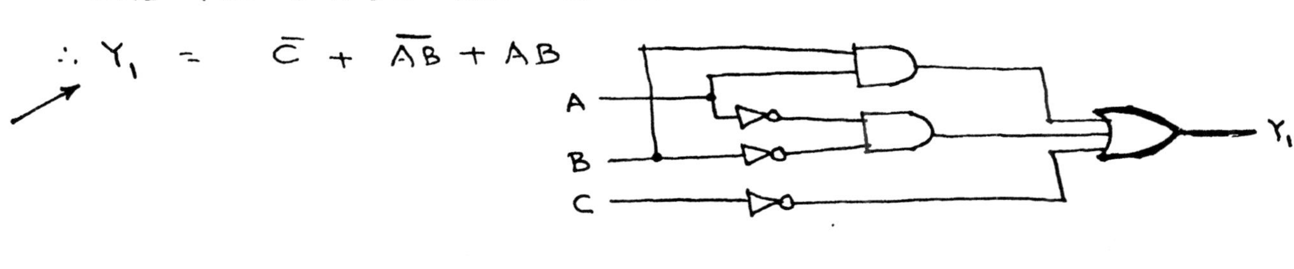

6.1.16. CONTD. (b)

Y_2 IS ON FOR DIGITS 0, 1, 3, 4, 5, 6, 7, 8, AND 9.

NOTE THAT 2 IS THE MISSING DIGIT.

$\therefore Y_2 = \overline{A B C D}$ OR BY DE MORGAN'S THEOREM $Y_2 = A + \bar{B} + C + D$

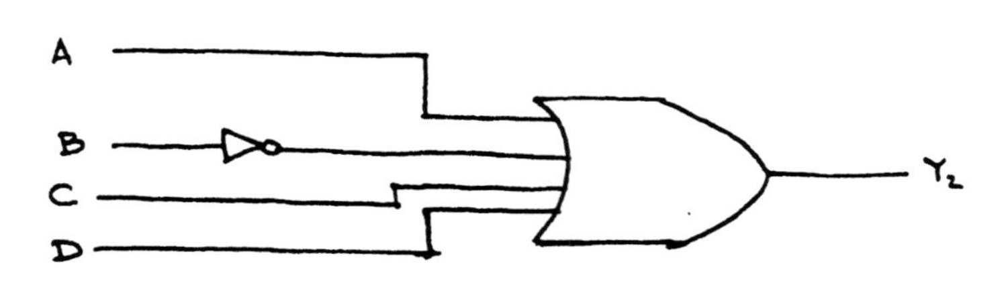

(c)

Y_3 IS ON FOR DIGITS 0, 2, 3, 5, 6, AND 8.

NOTE THAT 1, 4, 7, AND 9 ARE THE MISSING DIGITS.

$\therefore Y_3 = \overline{A}\overline{B}\overline{C} + A\bar{B}C + B\bar{C} + \bar{A}B$

$= \bar{B}(AC + \bar{A}\bar{C}) + B(\bar{A} + \bar{C})$

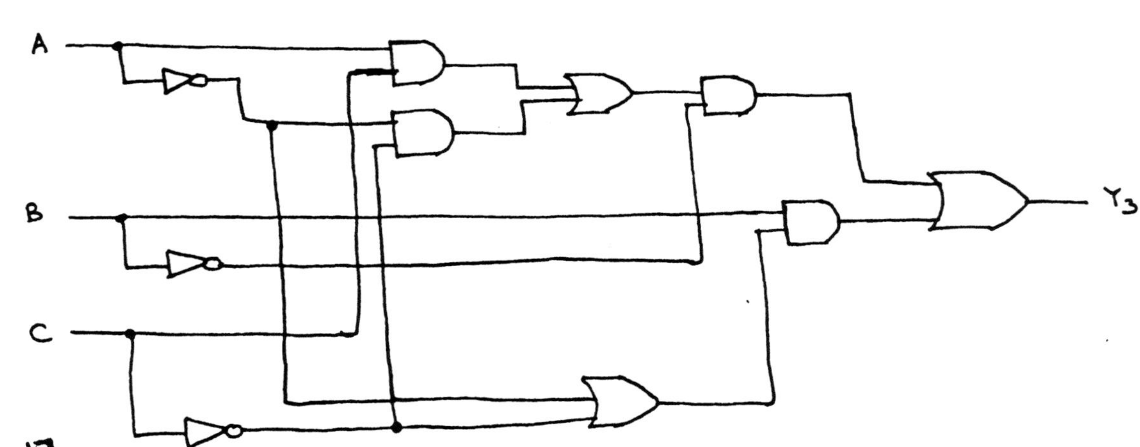

6.1.17.

(a)

A	B	Y
0	0	0
0	1	0
1	0	0
1	1	1

AND

$Y = A \cdot B$

(b)

A	B	Y
0	0	0
0	1	1
1	0	1
1	1	1

OR

$Y = A + B$

(c)

A	B	Y	$\bar{Y}$
0	0	0	1
0	1	1	0
1	0	1	0
1	1	1	0

NOR

$Y = A + B$; $\bar{Y} = \overline{A + B}$

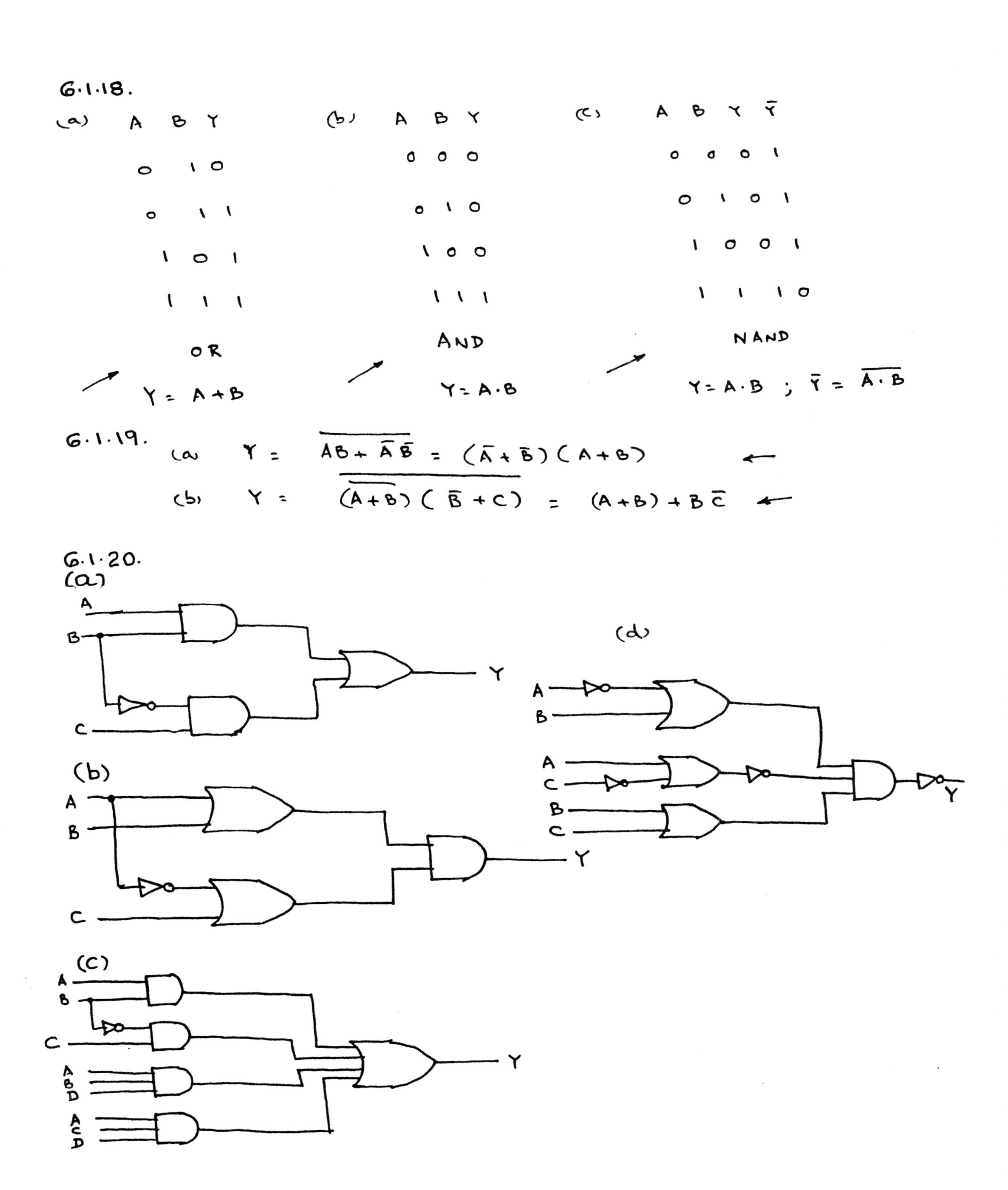

6.1.18.
(a)
A B Y
0 1 0
0 1 1
1 0 1
1 1 1
OR
Y = A + B
(b)
A B Y
0 0 0
0 1 0
1 0 0
1 1 1
AND
Y = A · B
(c)
A B Y Ȳ
0 0 0 1
0 1 0 1
1 0 0 1
1 1 1 0
NAND
Y = A · B ; Ȳ = $\overline{A \cdot B}$
6.1.19.
(a) Y = $\overline{AB + \bar{A}\bar{B}}$ = $(\bar{A} + \bar{B})(A + B)$
(b) Y = $\overline{(\overline{A+B})(\bar{B} + C)}$ = $(A + B) + B\bar{C}$
6.1.20.
(a)
A
B
C
Y
(b)
A
B
C
Y
(c)
A
B
C
A
B
D
A
C
D
Y
(d)
A
B
A
C
B
C
Y

6.1.21.

(a) $Y = \bar{A} + B\bar{C} + \bar{B}\bar{C}$ $\leftarrow$

(b) $Y = (A+B)(B+C)$ $\leftarrow$

(c) $Y = (A+\bar{B})\ (\bar{C})(C+D)$ $\leftarrow$

6.1.22.

(a)

$Y = A + \bar{A}\cdot B$

$= A\cdot 1 + \bar{A}\cdot B$ BY USING IDENTITY 6 OF TABLE 6.1.3 OF TEXT

$= A\cdot 1 + \bar{A}\cdot B + 1\cdot B$ " " 18

$= A + \bar{A}\cdot B + B$ " " 6

$= A + B$ " " 16 $\leftarrow$

(b)

$Y = A\cdot B + \bar{B}\cdot C + A\cdot C\cdot D + A\cdot B\cdot D$

$= A\cdot B + \bar{B}\cdot C + A\cdot C\cdot D$ BY USING IDENTITY 16

$= A\cdot B + \bar{B}\cdot C + A\cdot C + A\cdot C\cdot D$ " " 18

$= A\cdot B + \bar{B}\cdot C + A\cdot C$ " " 16

$= A\cdot B + \bar{B}\cdot C$ " " 18 $\leftarrow$

(c)

$Y = \overline{(\bar{A}+B+C)\cdot(\bar{A}+B+C)\cdot\bar{C}}$

$= \overline{(\bar{A}+B+C)} + \overline{(\bar{A}+B+C)} + \bar{\bar{C}}$ BY USING IDENTITY 20

$= \bar{\bar{A}}\cdot\bar{B}\cdot\bar{C} + \bar{\bar{A}}\cdot\bar{B}\cdot\bar{C} + C$ " " 19

$= A\cdot\bar{B}\cdot\bar{C} + A\cdot\bar{B}\cdot\bar{C} + C$ " " 9

$= A\cdot\bar{B}\cdot\bar{C} + C$ " " 3

$= A\cdot\bar{B}\cdot\bar{C} + C + A\cdot\bar{B}$ " " 6 & 18

$= C + A\cdot\bar{B}$ " " 16 $\leftarrow$

6.1.22.(d)

$F = B \cdot C + \bar{B} \cdot \bar{C} + A \cdot \bar{B} \cdot C$

$F = B \cdot C + \bar{B} \cdot \bar{C} + A \cdot \bar{B} \cdot C + A \cdot \bar{B}$ BY USING IDENTITY 18 (BETN. 2nd & 3rd TERMS)

$= B \cdot C + \bar{B} \cdot \bar{C} + A \cdot \bar{B}$ " " 16 ←

OR $F = B \cdot C + \bar{B} \cdot \bar{C} + A \cdot \bar{B} \cdot C + A \cdot C$ BY USING IDENTITY 18 (BETN. 1st & LAST TERMS)

$= B \cdot C + \bar{B} \cdot \bar{C} + A \cdot C$ " " 16 ←

THE STUDENT IS ENCOURAGED TO VERIFY THAT THESE TWO EXPRESSIONS ARE EQUIVALENT.

NOTE: A BOOLEAN FUNCTION MAY HAVE MORE THAN ONE MINIMUM REPRESENTATION.

6.1.23. (a) $F(A,B,C) = m_2 + m_3 + m_4 + m_5$

$= \bar{A} \cdot B \cdot \bar{C} + \bar{A} \cdot B \cdot C + A \cdot \bar{B} \cdot \bar{C} + A \cdot \bar{B} \cdot C$ ←

(b) Apply consensus identity betn the 1st & 2nd Terms and betn 3rd & 4th terms of the function.

$F(A,B,C) = \bar{A} \cdot B \cdot \bar{C} + \bar{A} \cdot B \cdot C + \bar{A} \cdot B + A \cdot \bar{B} \cdot \bar{C} + A \cdot \bar{B} \cdot C + A \cdot \bar{B}$

FINALLY, USING ABSORPTION IDENTITY

$F(A,B,C) = \bar{A} \cdot B + A \cdot \bar{B}$

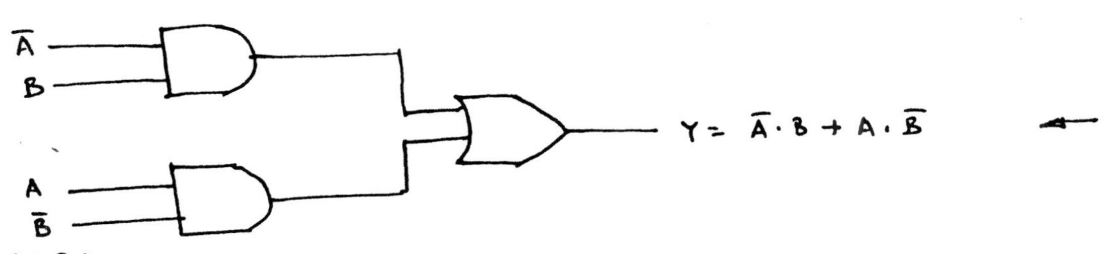

6.1.24. (a) $F(A,B,C) = (A+B+C) \cdot (A+B+\bar{C}) \cdot (\bar{A}+\bar{B}+C) \cdot (\bar{A}+\bar{B}+\bar{C})$ ←

NOTE: THE MAX TERM M_i IS THE COMPLEMENT OF THE MINTERM m_i.

(b) $F(A,B,C) = (A+B) \cdot (\bar{A}+\bar{B})$ FROM IDENTITY 15 OF TABLE 6.1.3 OF TEXT.

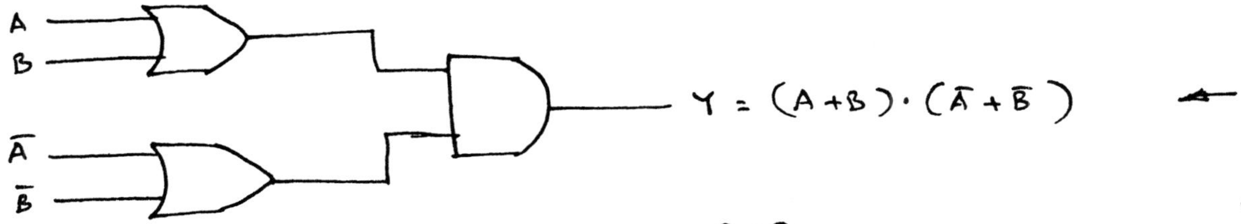

A, B, Ā, B̄ → $Y = (A+B) \cdot (\bar{A}+\bar{B})$ ←

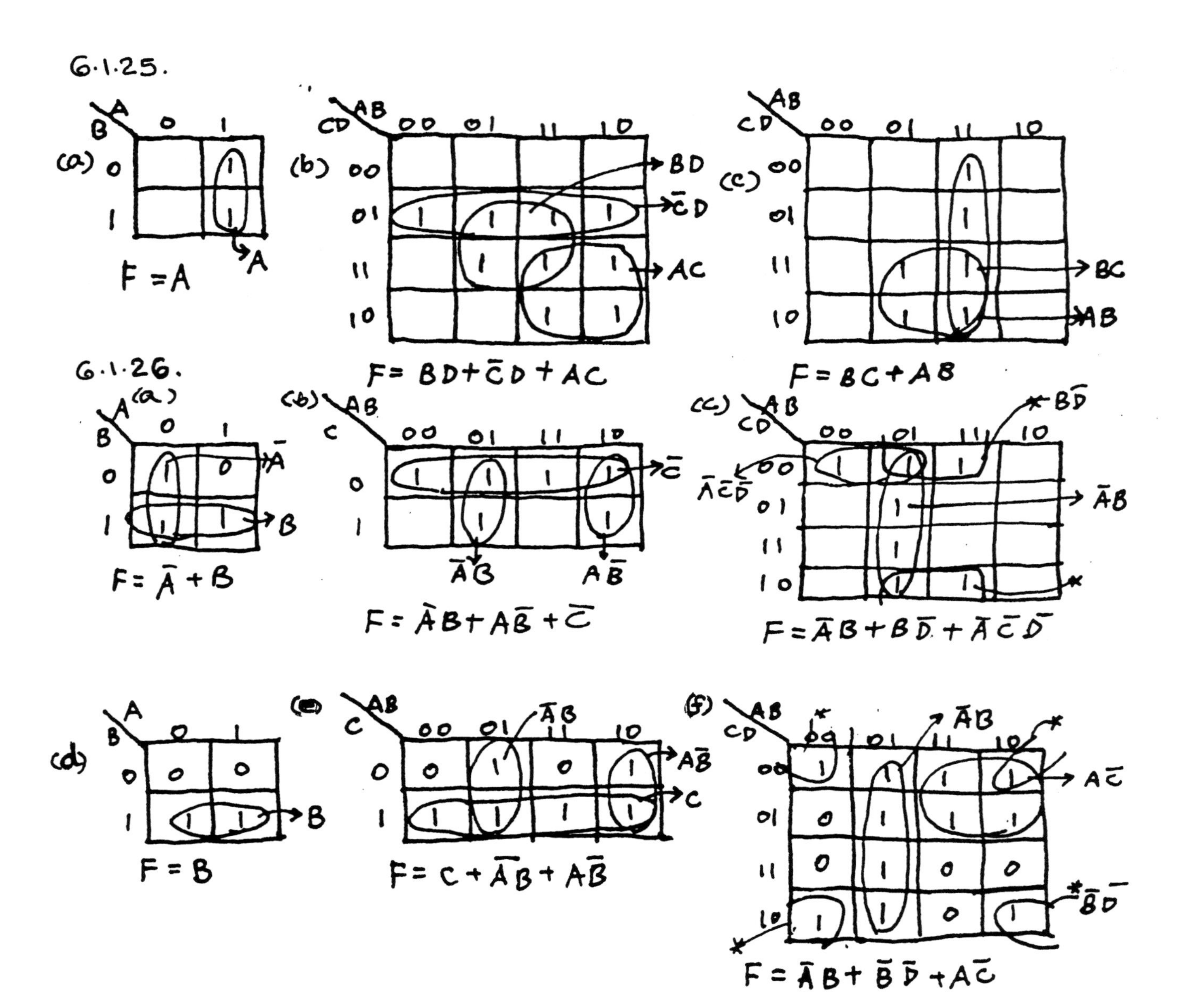

6.1.25.
(a)
F = A
(b)
F = BD + C̄D + AC
(c)
F = BC + AB
6.1.26.
(a)
F = Ā + B
(b)
F = ĀB + AB̄ + C̄
(c)
F = ĀB + BD̄ + ĀC̄D̄
(d)
F = B
(e)
F = C + ĀB + AB̄
(f)
F = ĀB + B̄D̄ + AC̄

6.1.27.

THE K-MAP IS SHOWN BELOW:

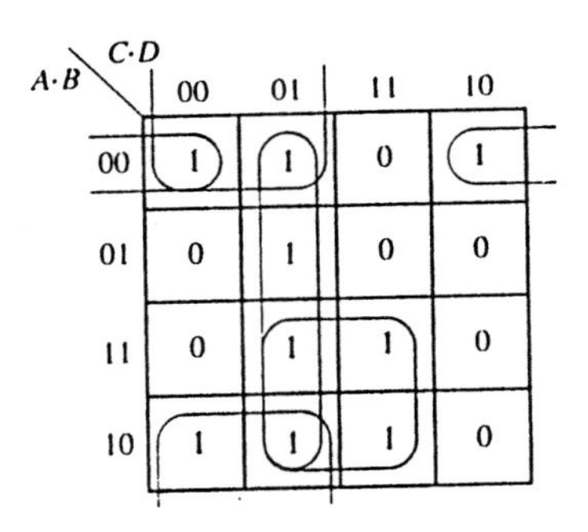

$\therefore\ y = A\cdot D + \bar{C}\cdot D + \bar{A}\cdot\bar{B}\cdot\bar{D} + \bar{B}\cdot\bar{C}$

THE IMPLEMENTATION USING TWO-INPUT GATES IS SHOWN BELOW:

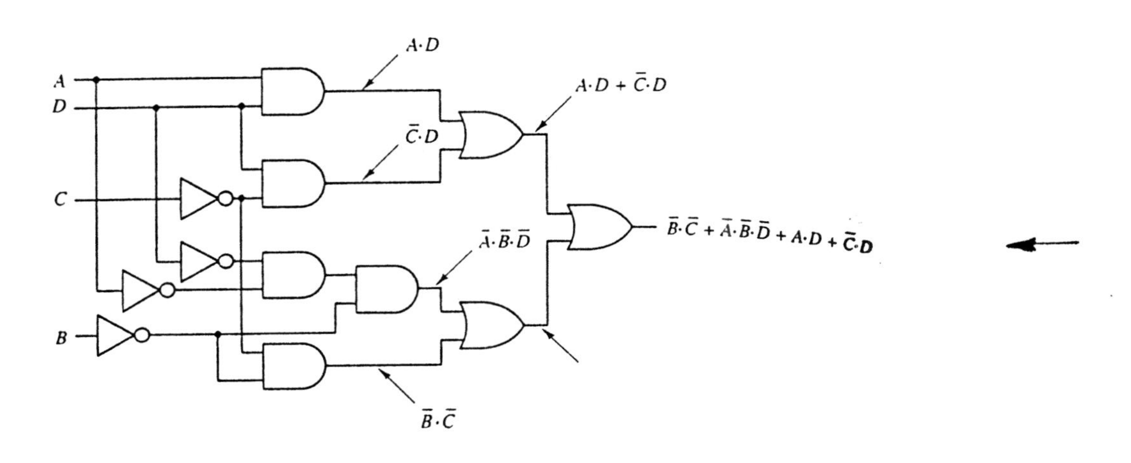

6.1.28. THE FIRST STEP IS TO MAKE A TRUTH TABLE. THE SECOND STEP IS TO USE A K-MAP TO SIMPLIFY AND IMPLEMENT THE FUNCTION. THE TRUTH TABLE MAY BE OBTAINED BY WRITING THE FUNCTION CORRESPONDING TO THE CIRCUIT:

$f = \bar{x}\cdot\bar{y} + y\cdot z$ AND FILLING THE TABLE ACCORDINGLY, AS FOLLOWS:

x	y	z	f
0	0	0	1
0	0	1	1
0	1	0	0
0	1	1	1
1	0	0	0
1	0	1	0
1	1	0	0
1	1	1	1

6.1.28. CONTINUED

CORRESPONDINGLY, WE CAN SELECT THE APPROPRIATE ENTRIES IN A THREE-VARIABLE K-MAP.

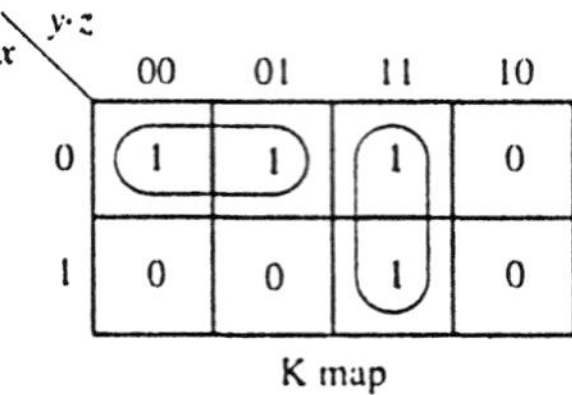

THE K-MAP VERIFIES THAT THE MINIMUM SUM-OF-PRODUCTS EXPRESSION IS:

$$f = \bar{x} \cdot \bar{y} + y \cdot z$$

6.1.29. THE LOGIC FUNCTION CORRESPONDING TO THE GIVEN LOGIC CIRCUIT IS

$$f = \bar{y} \cdot \bar{z} + x \cdot z + y \cdot \bar{z}$$

THE CORRESPONDING K-MAP IS SHOWN BELOW:

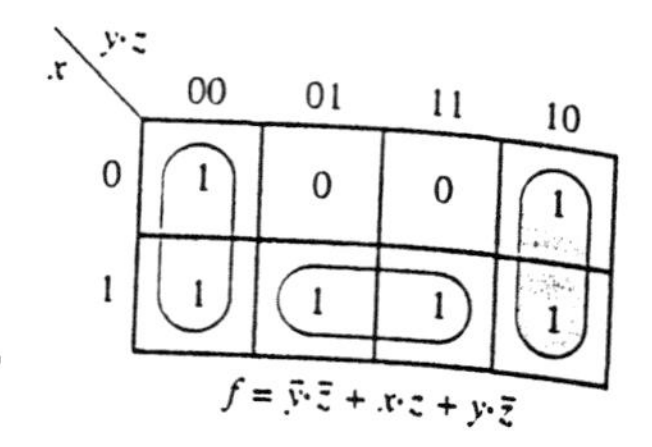

INSPECTION OF THE K-MAP, THOUGH, REVEALS THAT THE MAP HAS BEEN COVERED INEFFICIENTLY, SINCE TWO FOUR-CELL SUBCUBES COULD HAVE BEEN USED IN PLACE OF THREE TWO-CELL SUBCUBES. FIGURE BELOW DEPICTS THE ALTERNATIVE COVERING, WHICH LEADS TO THE MUCH SIMPLER FUNCTION

$$f = x + \bar{z}$$

THE CORRESPONDING GATE IMPLEMENTATION IS ALSO SHOWN IN FIGURE BELOW:

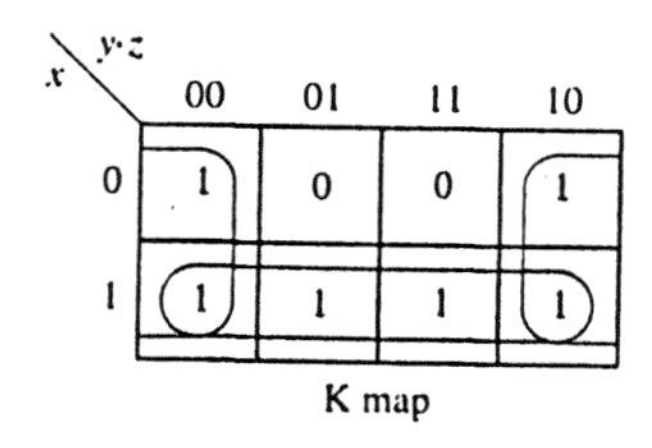

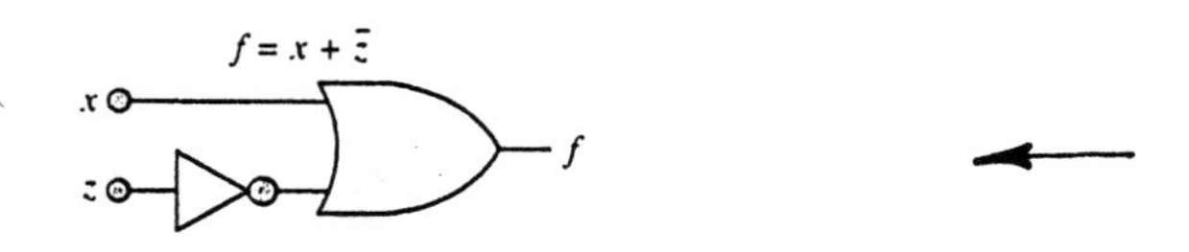

6.1.30. (a) USING 0s, WE OBTAIN THE K-MAP SHOWN BELOW LEADING TO THE PRODUCT-OF-SUMS EXPRESSION:

$$f = (x + y + z) \cdot (\bar{x} + \bar{y})$$

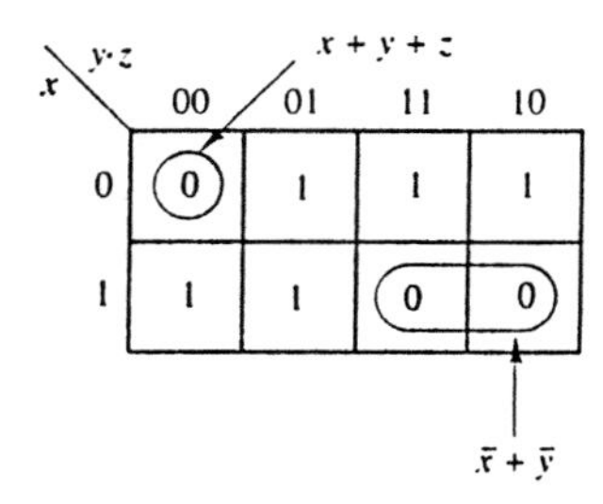

(b) IF 1s ARE USED, AS SHOWN BELOW, A SUM-OF-PRODUCTS EXPRESSION IS OBTAINED, OF THE FORM

$$f = \bar{x} \cdot y + x \cdot \bar{y} + \bar{y} \cdot z$$

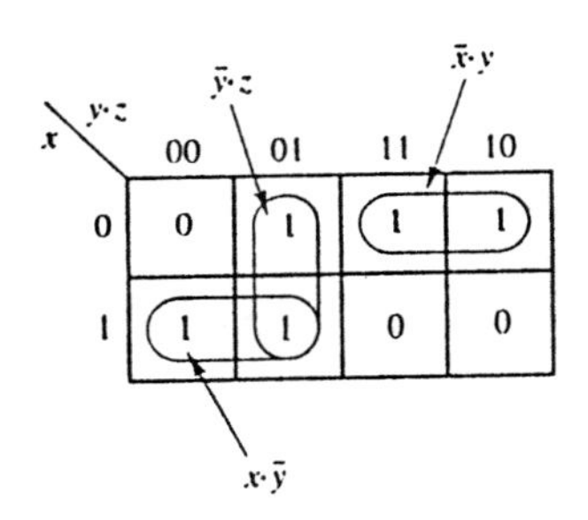

6.1.31. TO REALIZE THIS FUNCTION IN PRODUCT-OF-SUMS FORM, THE 0s ARE USED IN THE K-MAP SHOWN BELOW. THE RESULTING EXPRESSION IS THEREFORE GIVEN BY

$$f = \bar{z} \cdot (\bar{x} + \bar{y})$$

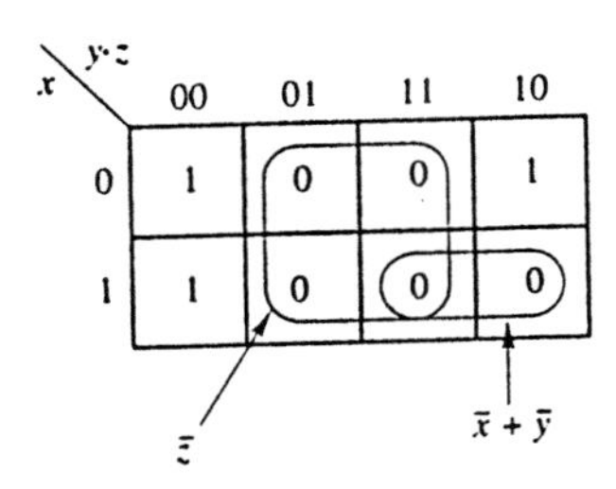

6.1.32.

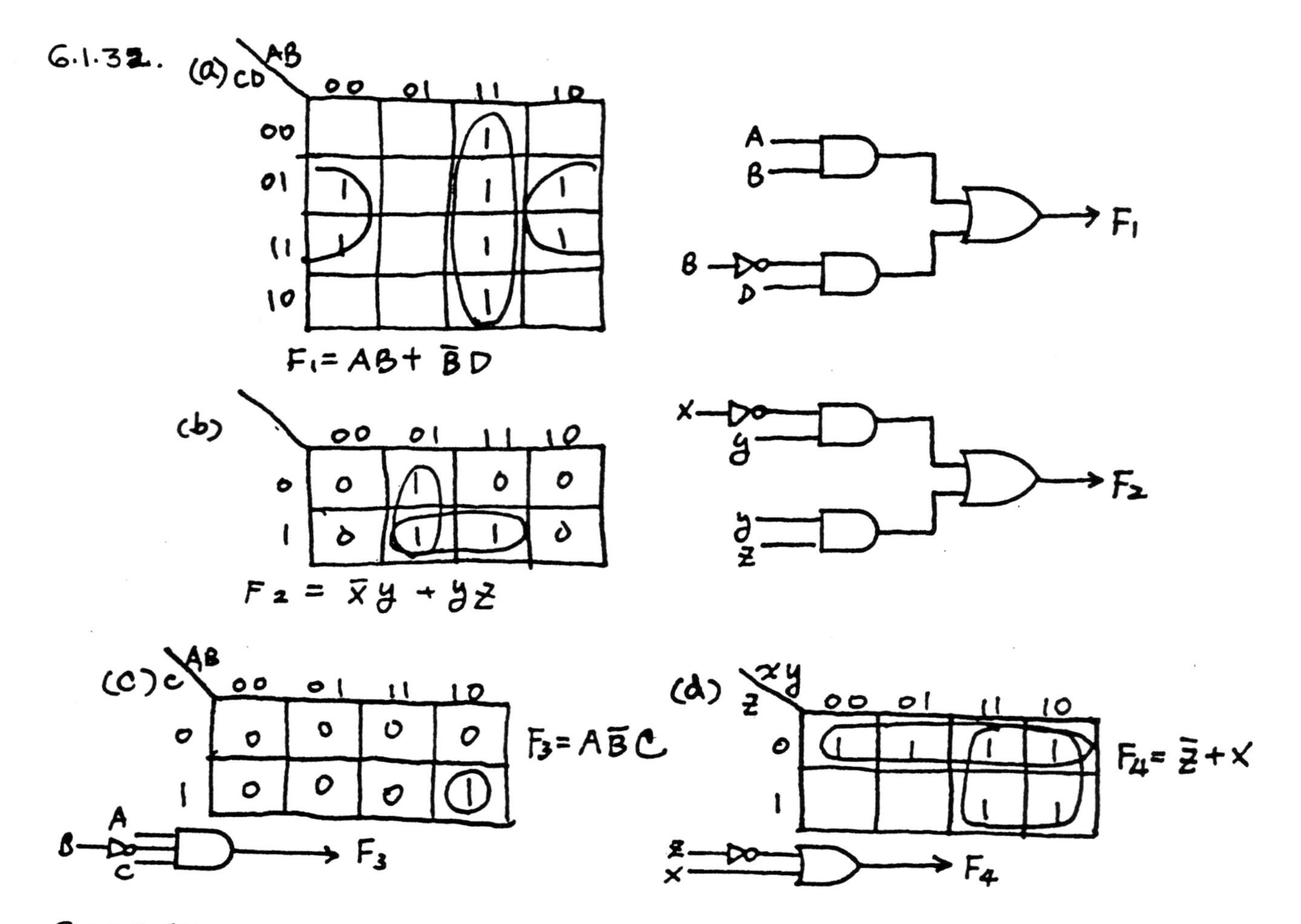

6.1.33. (a)

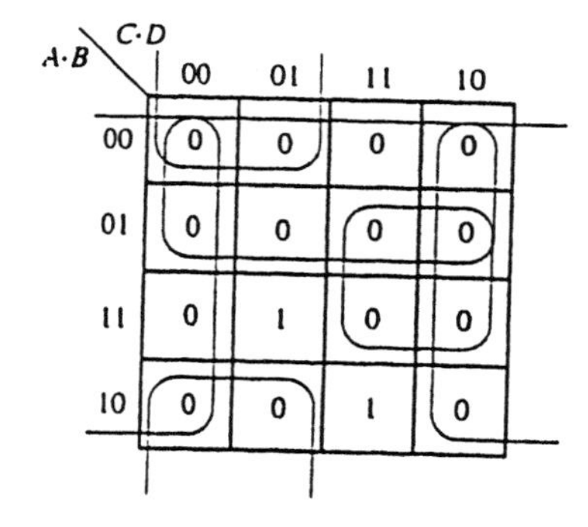

THE FOUR SUBCUBES SHOWN YIELD THE FOLLOWING EQUATION (CORRESPONDING TO THE POS REPRESENTATION):

$$y = A \cdot D \cdot (C + B) \cdot (\bar{C} + \bar{B})$$

NOTE THAT THERE ARE MANY MORE 0s THAN 1s IN THE TABLE; HENCE, THE USE OF 0s IN COVERING THE MAP WILL LEAD TO A GREATER SIMPLIFICATION.

BY DEMORGAN'S LAW,

$$y = A \cdot D \cdot (C + B) \cdot (\overline{C \cdot B}) \leftarrow$$

WHICH CAN BE REALIZED BY THE FOLLOWING CIRCUIT.

6.1.33.(a) CONTD.

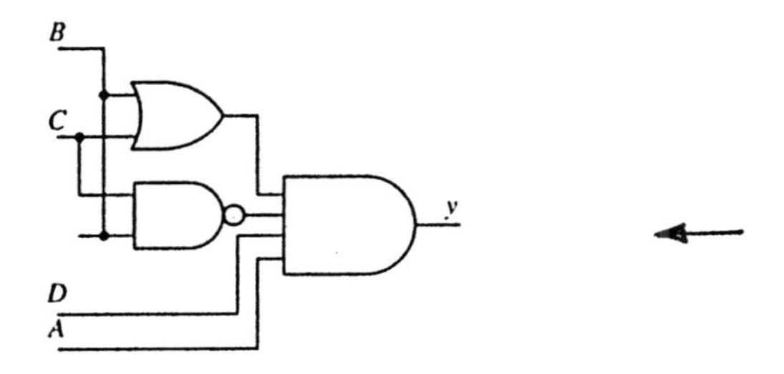

(b) THE SOP CIRCUIT IS SHOWN BELOW:

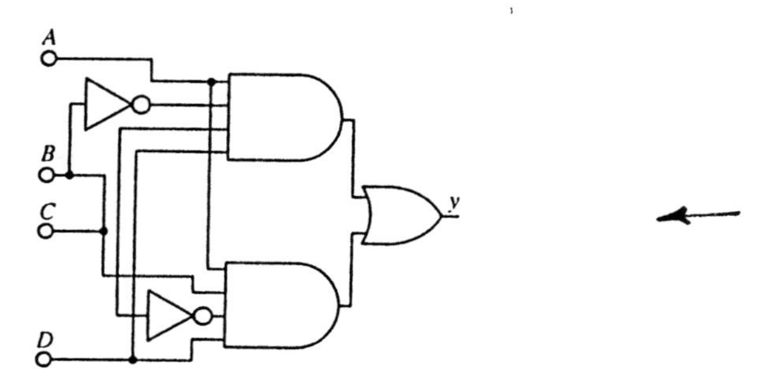

THIS CIRCUIT EMPLOYS A GREATER NUMBER OF GATES AND WILL THEREFORE LEAD TO A MORE EXPENSIVE DESIGN.

6.1.34. GIVING ALL d's A VALUE OF 1, THE FOLLOWING EXPRESSION CANBE OBTAINED WITH THE AID OF SUBCUBES FORMED, AS SHOWN BELOW:

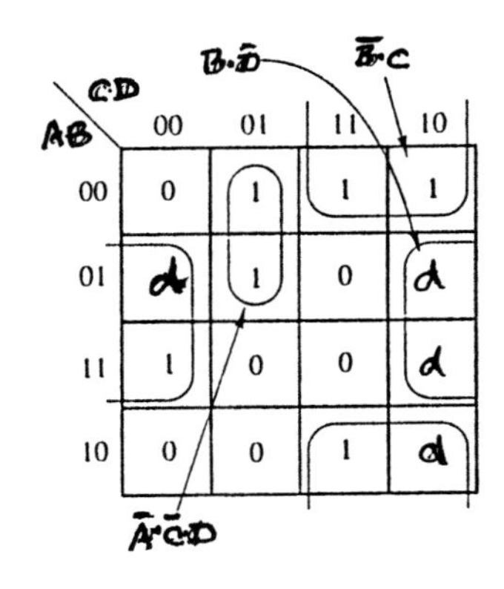

$$F(A,B,C,D) = B\cdot\bar{D} + \bar{B}\cdot C + \bar{A}\cdot\bar{C}\cdot D$$

6.1.35.

For the minimum SOP, all the 1s have to be covered. The subcubes are shown below:

AB \ CD	00	01	11	10
00	1	0	1	0
01	0	0	0	1
11	0	d	d	d
10	0	1	d	d

From the above figure, it can be seen that the function is given by

$$F = \bar{A}\cdot\bar{B}\cdot\bar{C}\cdot\bar{D} + B\cdot C\cdot\bar{D} + A\cdot D + \bar{B}\cdot C\cdot D \leftarrow$$

6.1.36.

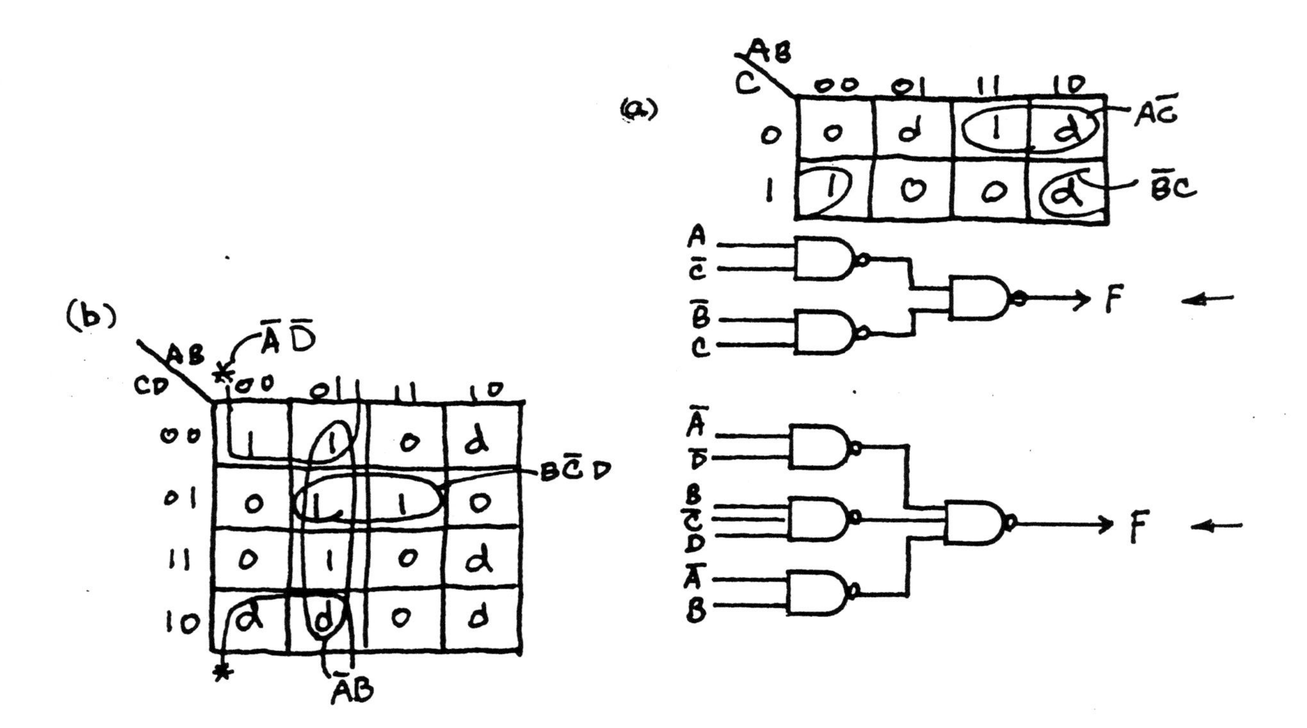

6.1.37.

(a)

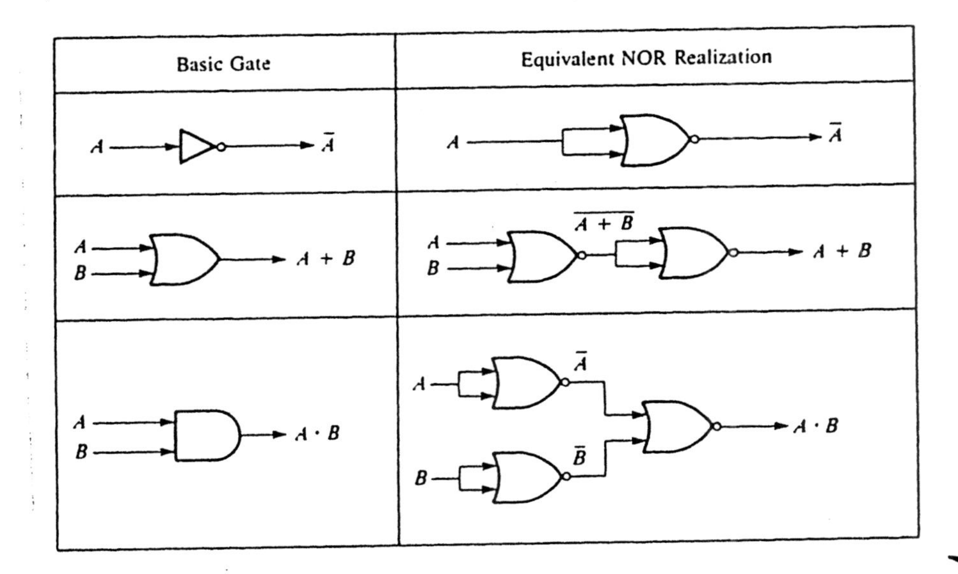

(b)

Basic Gate	Fquivalent NAND Realization
A → $\bar{A}$	A → $\bar{A}$
A, B → F = A · B	A, B → $\overline{A \cdot B}$ → F = A · B
A, B → F = A + B	A → $\bar{A}$, B → $\bar{B}$ → $\overline{\bar{A} \cdot \bar{B}}$ → F = A + B

6.1.38.

THE PROCEDURE FOR OBTAINING A MINIMUM NAND - GATE'S REALIZATION FOR A BOOLEAN EXPRESSION IS AS FOLLOWS:

(i) OBTAIN A MINIMUM 2-LEVEL AND-OR REALIZATION, WHERE NO INPUTS ARE DIRECTLY CONNECTED TO THE OR GATE; i.e., EVERY INPUT MUST PASS THROUGH AN AND GATE.

(ii) REDRAW THE ABOVE CIRCUIT BY SIMPLY REPLACING ALL THE AND AND OR GATES BY NAND GATES.

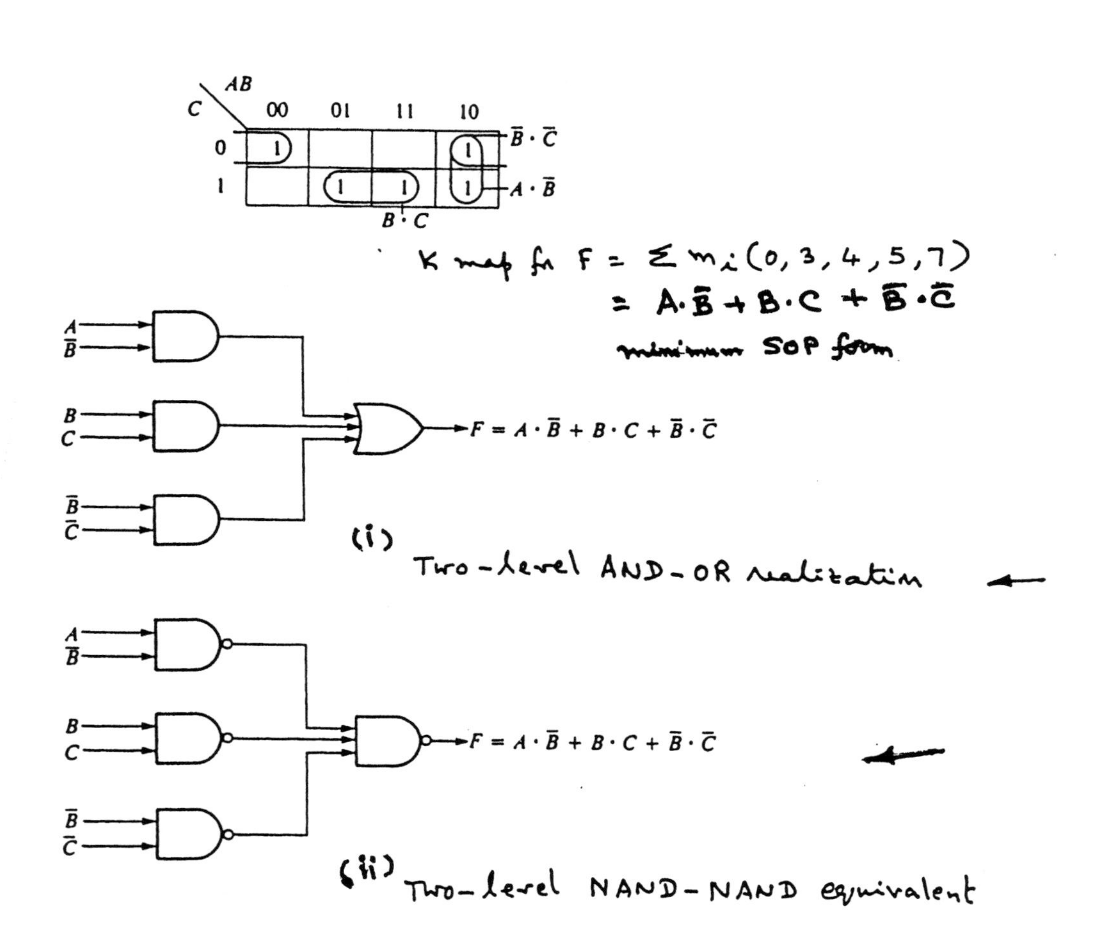

K map for $F = \sum m_i(0,3,4,5,7)$

$= A\cdot\bar{B} + B\cdot C + \bar{B}\cdot\bar{C}$

~~minimum~~ SOP form

(i) Two-level AND-OR realization

(ii) Two-level NAND-NAND equivalent

6.1.39.

(a) FULL-ADDER TRUTH TABLE IS SHOWN BELOW:

A	B	C_i	S'	C_o'	C_o''	C_o	S
0	0	0	0	0	0	0	0
0	0	1	0	0	0	0	1
0	1	0	1	0	0	0	1
0	1	1	1	0	1	1	0
1	0	0	1	0	0	0	1
1	0	1	1	0	1	1	0
1	1	0	0	1	0	1	0
1	1	1	0	1	0	1	1

(b)

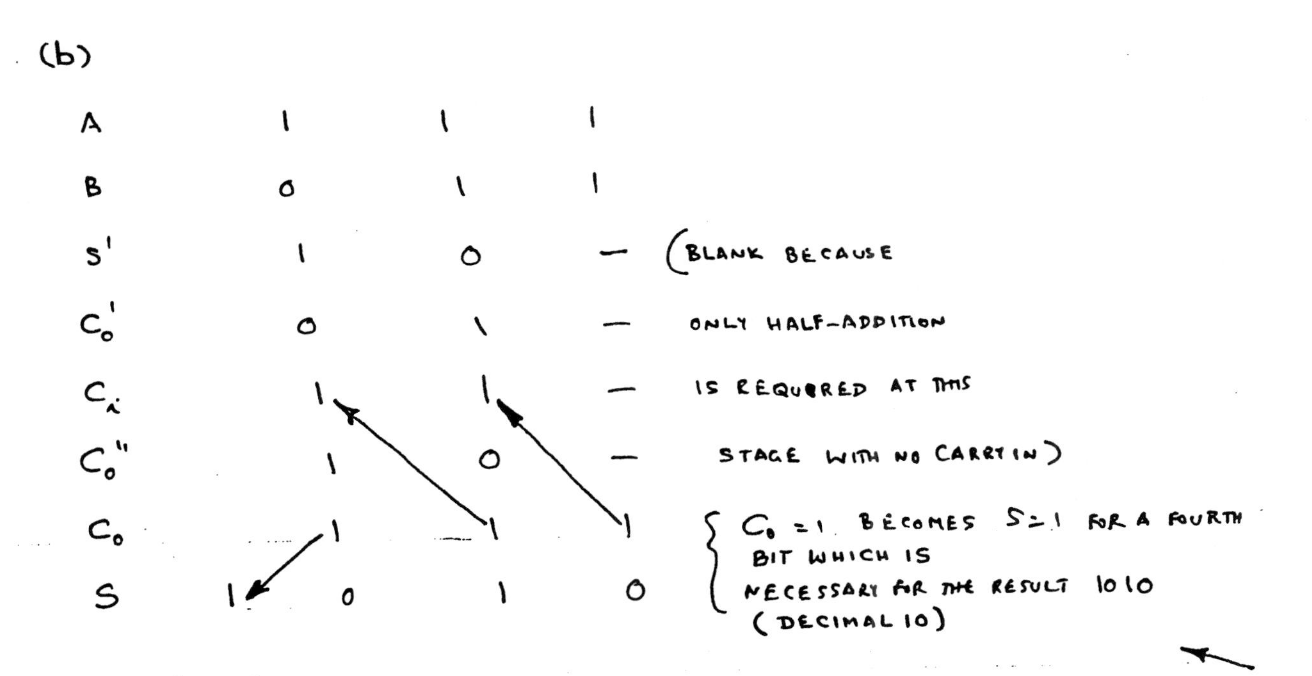

A 1 1 1
B 0 1 1
S' 1 0 —
C₀' 0 1 —
Cᵢ 1 1 —
C₀'' 1 0 —
C₀ 1 1 1
S 1 0 1 0
(BLANK BECAUSE
ONLY HALF-ADDITION
IS REQUIRED AT THIS
STAGE WITH NO CARRY IN)
C₀ = 1 BECOMES S = 1 FOR A FOURTH
BIT WHICH IS
NECESSARY FOR THE RESULT 1010
(DECIMAL 10)

6.1.40. THE GIVEN CIRCUIT IS EQUIVALENT TO THE FOLLOWING:

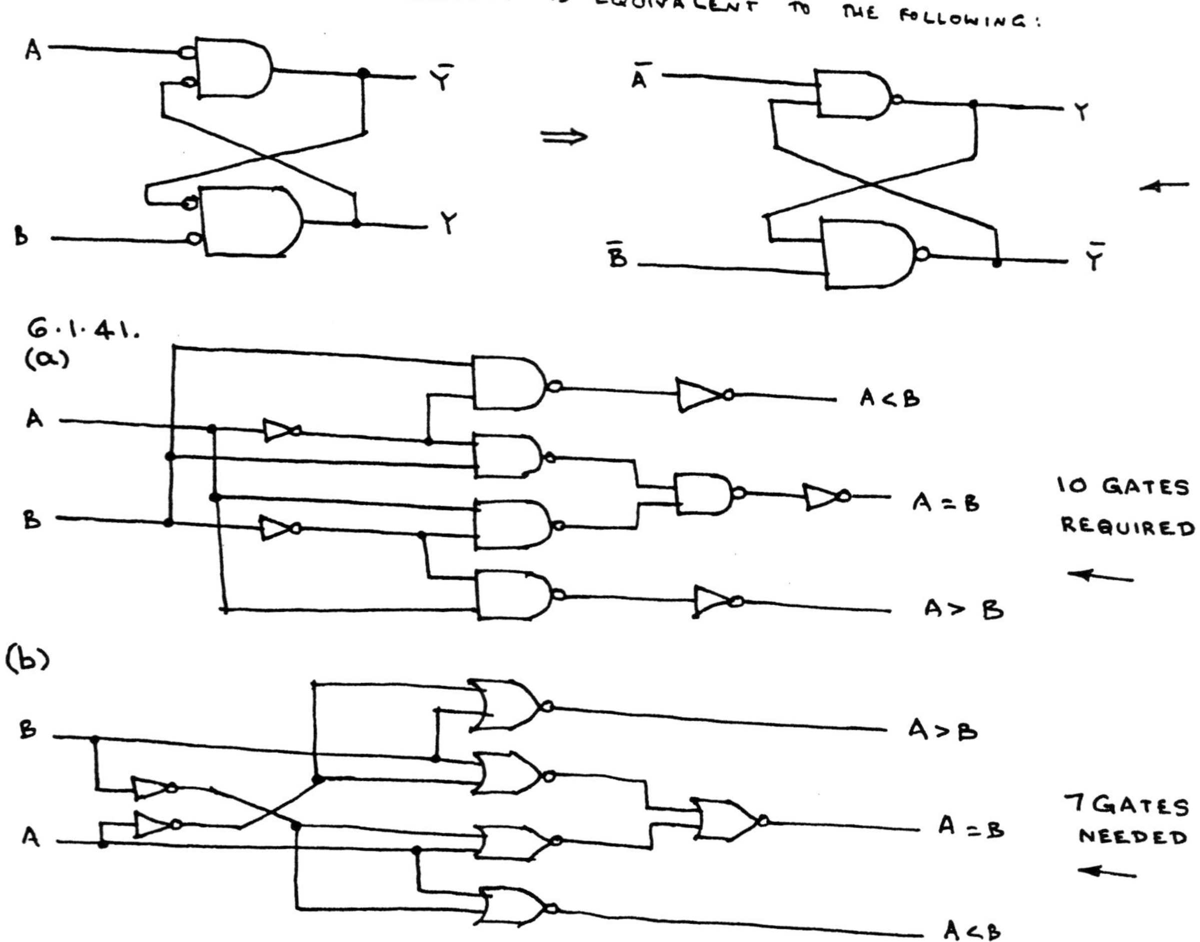

(c) THE NOR-INVERTER GATE REALIZATION REQUIRES FEWER GATES.

6.1.42.

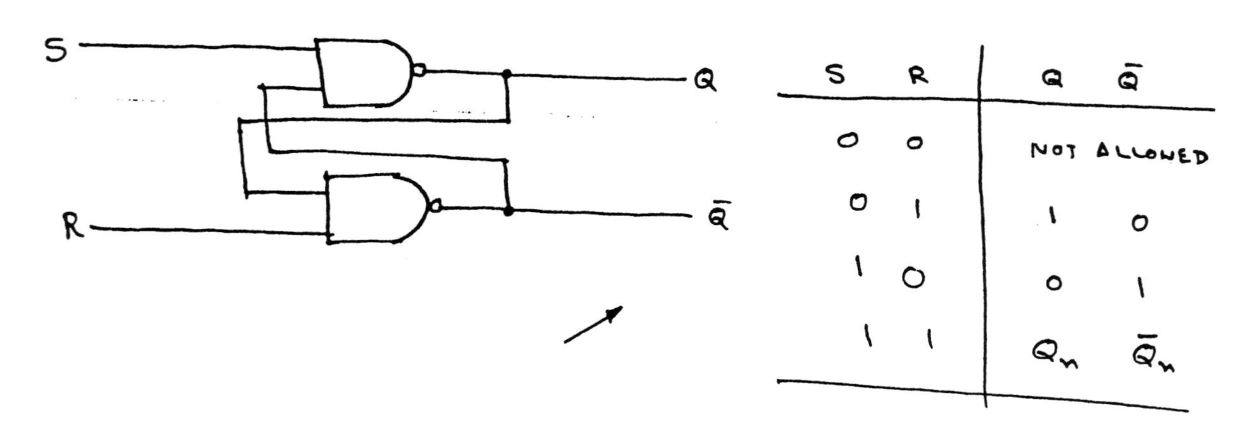

S	R	Q	$\bar{Q}$
0	0	NOT ALLOWED	
0	1	1	0
1	0	0	1
1	1	Q_n	$\bar{Q}_n$

6.1.43.

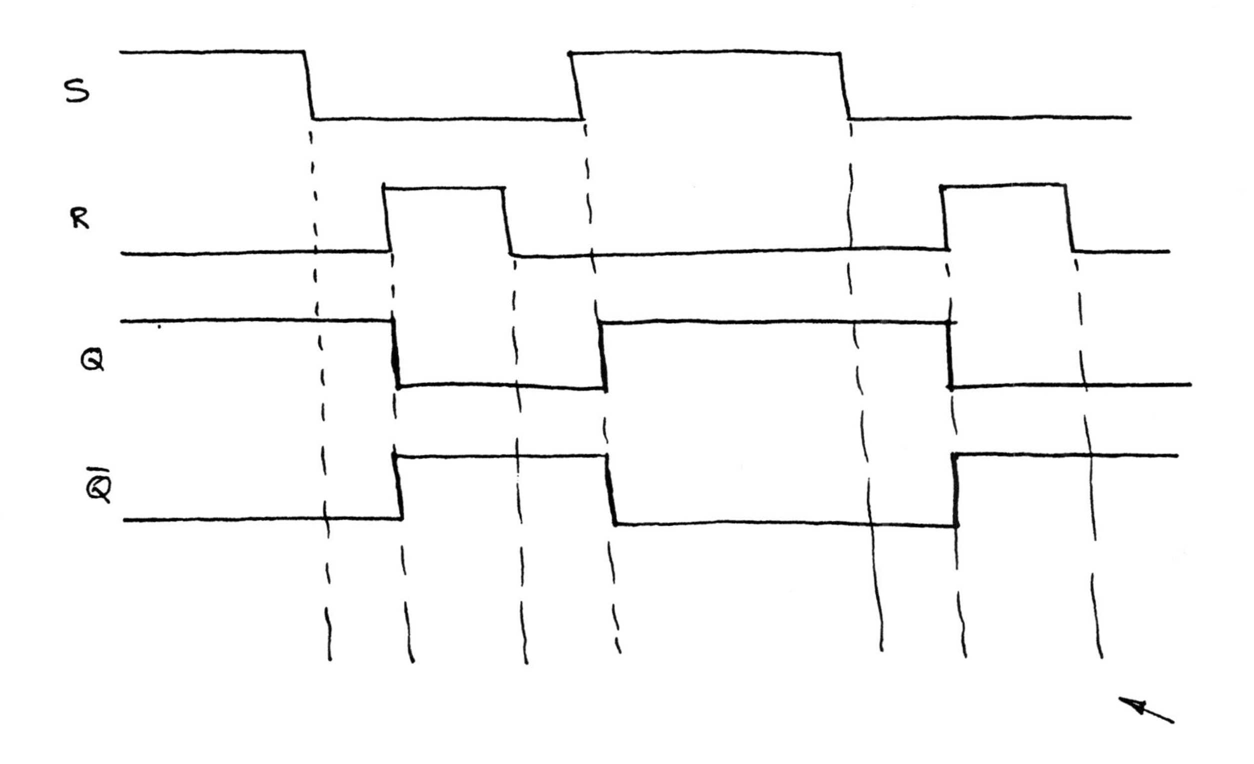

6.1.44.

ROW (a): WITH OUR GUESS THAT $Q=1$, THE OUTPUT OF THE LOWER NAND GATE (WHOSE INPUTS ARE 1 AND 1) MUST BE ZERO, AGREEING WITH OUR GUESS THAT $\bar{Q}=0$. THE OUTPUT OF THE UPPER NAND GATE (WHOSE INPUTS ARE 1 AND 0) IS 1, AGREEING WITH OUR GUESS THAT $Q=1$. THUS THE LAST TWO COLUMNS OF THE TABLE ($Q=1$ AND $\bar{Q}=0$) AGREE WITH THE "GUESS" COLUMNS. THIS MEANS THAT OUR GUESSES ARE SELF-CONSISTENT, AND DO THUS REPRESENT A POSSIBLE STATE OF THE SYSTEM.

ROW (b): PROCEEDING AS BEFORE, WE FIND $Q=0$ AND $\bar{Q}=1$; THE GUESS IN THIS CASE IS ALSO SELF-CONSISTENT. THUS WHEN BOTH INPUTS ARE ZERO, THE FLIP FLOP CAN BE IN EITHER OF ITS TWO STATES. ACTUALLY WE MIGHT HAVE GUESSED THE RESULT FROM THE SYMMETRY OF THE CIRCUIT.

ROW (c): FOR THIS CASE WE FIND $Q=1$ AND $\bar{Q}=1$ WHICH DO NOT AGREE WITH THE GUESS COLUMNS; A WRONG GUESS WAS OBVIOUSLY MADE. WE THROW THIS GUESS AWAY BY CROSSING OUT A LINE IN THE TRUTH TABLE.

ROW (d) AND ROW (e): GUESSES IN THESE CASES CAN BE SEEN TO BE SELF-CONSISTENT.

ROW (f): THIS FINAL ROW SHOWS THE RESULT OF THE "FORBIDDEN" INPUTS $S=1$ AND $R=1$. NOTE THAT IT IS POSSIBLE TO APPLY THIS INPUT, NOTHING DISASTROUS HAPPENS ELECTRICALLY. HOWEVER, YOU ARE NOT USING IT PROPERLY AS AN SRFF. EVEN THOUGH THE INSTRUCTION IS IMPROPER, IT CAN STILL BE USED, PROVIDED THAT ONE KNOWS WHAT THE RESULT WILL BE.

6.1.44. CONTD. THE COMPLETED MODIFIED TRUTH TABLE IS GIVEN BELOW:

S	R	Q (GUESS)	$\bar{Q}$ (GUESS)	Q	$\bar{Q}$
0	0	1	0	1	0
0	0	0	1	0	1
1	0	0	1	1	1
1	0	1	0	1	0
0	1	0	1	0	1
1	1	1	1	1	1

6.1.45.

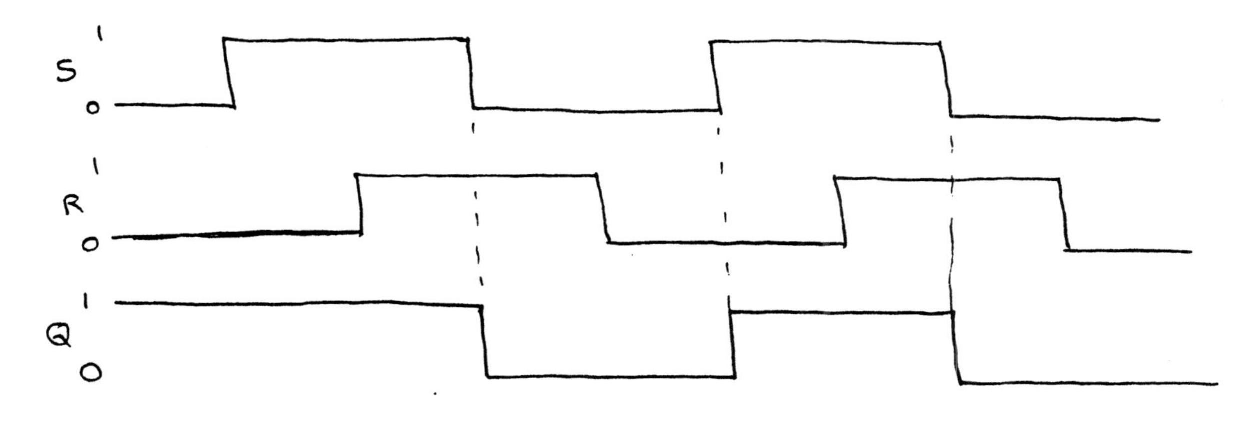

6.1.46.

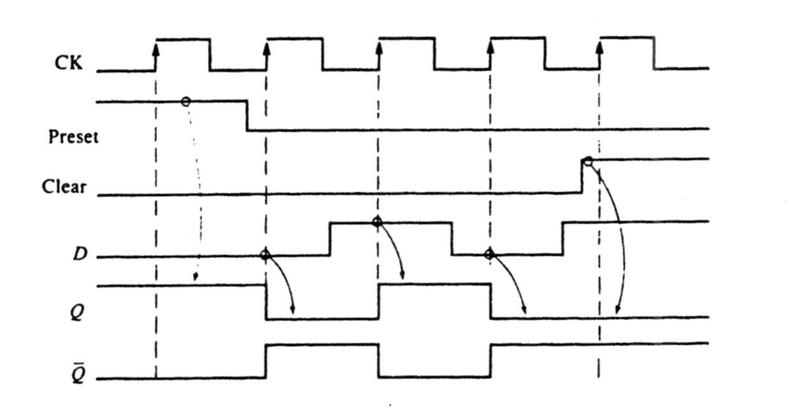

6.1.47. (a)

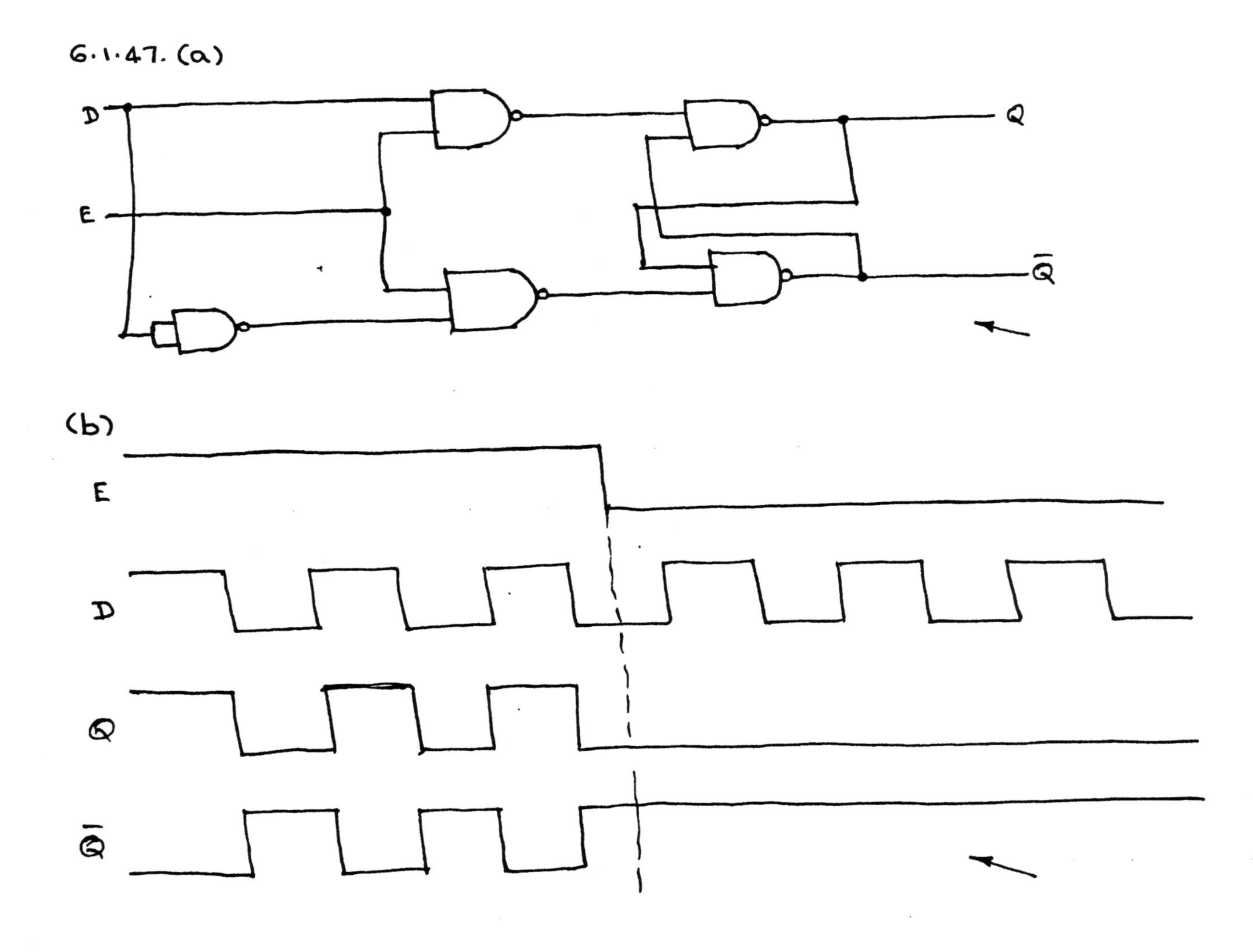

6.1.48.

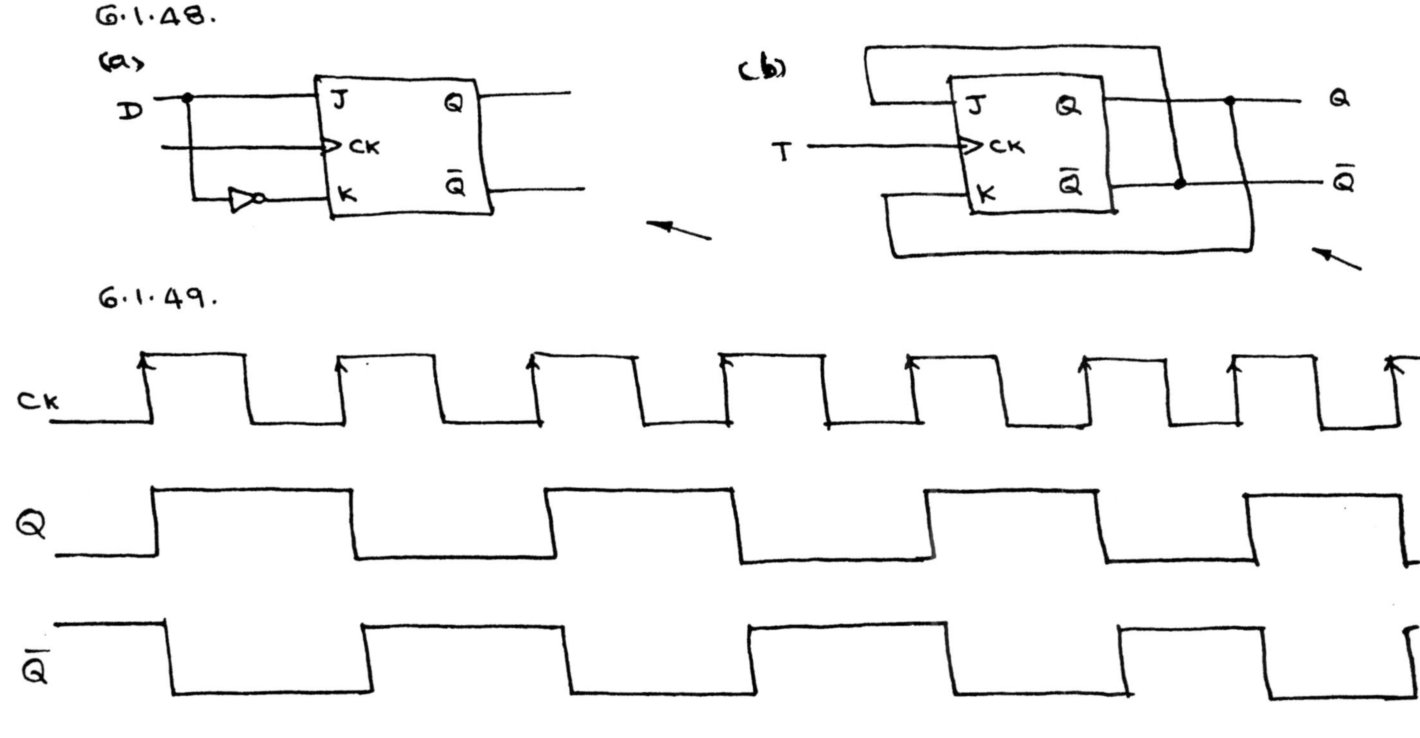

6.1.50.

WHEN THE SWITCH MAKES, S BECOMES 'HIGH' AND SR = 10. THE FF SETS WITH Q = 1. IF THE SWITCH NOW BREAKS, S GOES 'LOW' AND SR = 00; Q STAYS AT 1. WHEN THE SWITCH MAKES AGAIN, SR = 10, AND THE FF SETS. BUT BECAUSE Q IS ALREADY AT 1, IT RETAINS THIS STATE. THUS THE SWITCH CAN NO LONGER INFLUENCE THE LOGIC STATE OF THE CIRCUIT IN AN ADVERSE WAY.

6.1.51.

(a) JK = 00

IF $Q\bar{Q} = 10$, AB = 00 = SR AND Q STAYS AT 1

IF $Q\bar{Q} = 01$, AB = 00 = SR AND Q STAYS AT 1

(b) JK = 10

IF $Q\bar{Q} = 01$, AB = 10 = SR AND THE FLIP FLOP SETS i.e. Q GOES TO 1

FOR $Q\bar{Q} = 10$, AB = 00 = SR AND Q STAYS AT 1.

6.1.51. CONTD. (c)

JK = 01

IF $Q\bar{Q} = 10$, AB = 01 = SR AND THE FLIP FLOP RESETS i.e. Q GOES TO 0 (ZERO)

IF $Q\bar{Q} = 01$, AB = 00 = SR AND Q STAYS AT 0 (ZERO) ←

(d)

JK = 11

IF $Q\bar{Q} = 10$, AB = 01 = SR AND THE FLIP FLOP RESETS.

FOR $Q\bar{Q} = 01$, AB = 10 = SR AND THE FLIP FLOP SETS.

THIS MEANS THAT WE RETURN TO $Q\bar{Q} = 10$ ... i.e. THE OUTPUT OSCILLATES CONTINUOUSLY BETWEEN 1 AND 0 SO LONG AS THE CLOCK IS HIGH ←

NOTE THAT CASE (d) PRESENTS A PROBLEM, WHICH MAY BE SOLVED BY MAKING THE CLOCK PULSES OF EXTREMELY SHORT DURATION ... IN WHICH CASE THE CIRCUIT LOSES ITS VERSATALITY. SO THE JKFF IS MODIFIED BY USING A SECOND SRFF, WHEN IT CALLED THE MASTER-SLAVE JKFF.

6.1.52.

OPERATION:

(i) WHEN THE CLOCK IS 'HIGH', GATES 3 AND 4 ARE DISABLED OWING TO THE PRESENCE OF THE INVERTER. SO $S_2R_2 = 00$ AND Q_2 STAYS THE SAME. GATES 1 AND 2 ARE ENABLED SO DATA CAN ENTER THE MASTER AND Q_1 IS FREE TO CHANGE ITS STATE.

(ii) WHEN THE CLOCK IS 'LOW', GATES 1 AND 2 ARE DISABLED; i.e. $S_1R_1 = 00$ AND Q_1 STAYS THE SAME. GATES 3 AND 4 ARE ENABLED; DATA ENTERS THE SLAVE FROM THE MASTER AND Q_2 MAY CHANGE ITS VALUE.

THUS DATA ENTER JKFF WHEN THE CLOCK IS 'HIGH'; BUT THE OUTPUT MAY ALTER ONLY WHEN THE CLOCK GOES LOW...

G.1.52. CONTD.

TRUTH TABLE FOR JKFF

J	K	Q_t	Q_{t+1}	
0	0	0	0	
0	0	1	1	
1	0	0	1	SET
1	0	1	1	
0	1	0	0	
0	1	1	0	RESET
1	1	0	1	TOGGLE
1	1	1	0	TOGGLE

THE ARROWS INDICATE THOSE INPUT COMBINATIONS WHICH GENERATE THE SAME SET OF OUTPUTS. FOR EXAMPLE, $Q_t \rightarrow Q_{t+1} = 0 \rightarrow 0$ FOR JK = 00 AND 01. THIS MEANS THAT IT IS ESSENTIAL THAT J STAYS AT 0 WHEREAS WE 'DON'T CARE' WHAT VALUE K HAS. SIMILARLY, $Q_t \rightarrow Q_{t+1} = 0 \rightarrow 1$ OCCURS FOR JK = 10 AND 11. J MUST STAY AT 1; BUT WE 'DON'T CARE' WHAT VALUE K HAS.

A REDUCED TRUTH TABLE (OR TRANSITION TABLE) CAN BE SET UP WHICH SHOULD BE USED FOR DETERMINING THE OUTPUT STATE AFTER EACH CLOCK PULSE. NOTE THAT THE LETTER 'd' IS USED TO INDICATE A 'DON'T CARE' CONDITION.

TRANSITION TABLE FOR JKFF

J	K	$Q_t \rightarrow Q_{t+1}$
0	d	$0 \rightarrow 0$
d	0	$1 \rightarrow 1$
1	d	$0 \rightarrow 1$
d	1	$1 \rightarrow 0$

6.2.1.

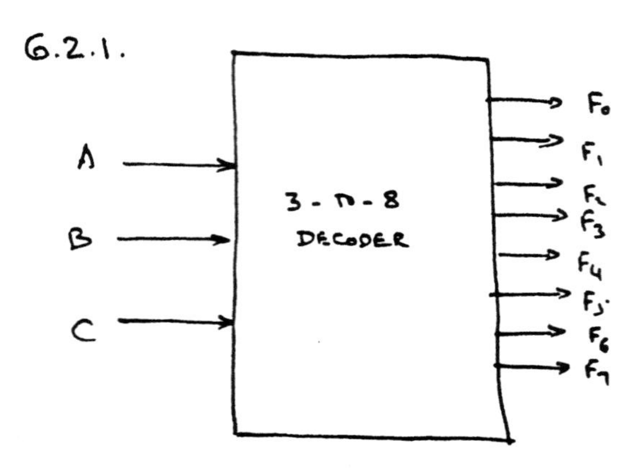

THE DECODER LOGIC MUST BE SUCH THAT, FOR EXAMPLE, $F_3 = 1$ WHEN ABC = 011, AND THEREFORE THE DECODER LOGIC FOR OUTPUT F_3 WILL BE $F_3 = \bar{A}BC$. THE COMPLETE DECODER COULD BE MADE UP OF EIGHT 3-INPUT AND GATES AND THREE INVERTERS.

6.2.2.

(a) EXCESS-3 CODE

DECIMAL	A	B	C	D
0	0	0	1	1
1	0	1	0	0
2	0	1	0	1
3	0	1	1	0
4	0	1	1	1
5	1	0	0	0
6	1	0	0	1
7	1	0	1	0
8	1	0	1	1
9	1	1	0	0

(b) EXCESS-3 DECODING

A	B	C	D	OUTPUT = 1
0	0	1	1	F_0
0	1	0	0	F_1
0	1	0	1	F_2
0	1	1	0	F_3
0	1	1	1	F_4
1	0	0	0	F_5
1	0	0	1	F_6
1	0	1	0	F_7
1	0	1	1	F_8
1	1	0	0	F_9

G.2.3.

DICIMAL	BCD				EXCESS-3			
DIGIT	A	B	C	D	P	Q	R	S
0	0	0	0	0	0	0	1	1
1	0	0	0	1	0	1	0	0
2	0	0	1	0	0	1	0	1
3	0	0	1	1	0	1	1	0
4	0	1	0	0	0	1	1	1
5	0	1	0	1	1	0	0	0
6	0	1	1	0	1	0	0	1
7	0	1	1	1	1	0	1	0
8	1	0	0	0	1	0	1	1
9	1	0	0	1	1	1	0	0

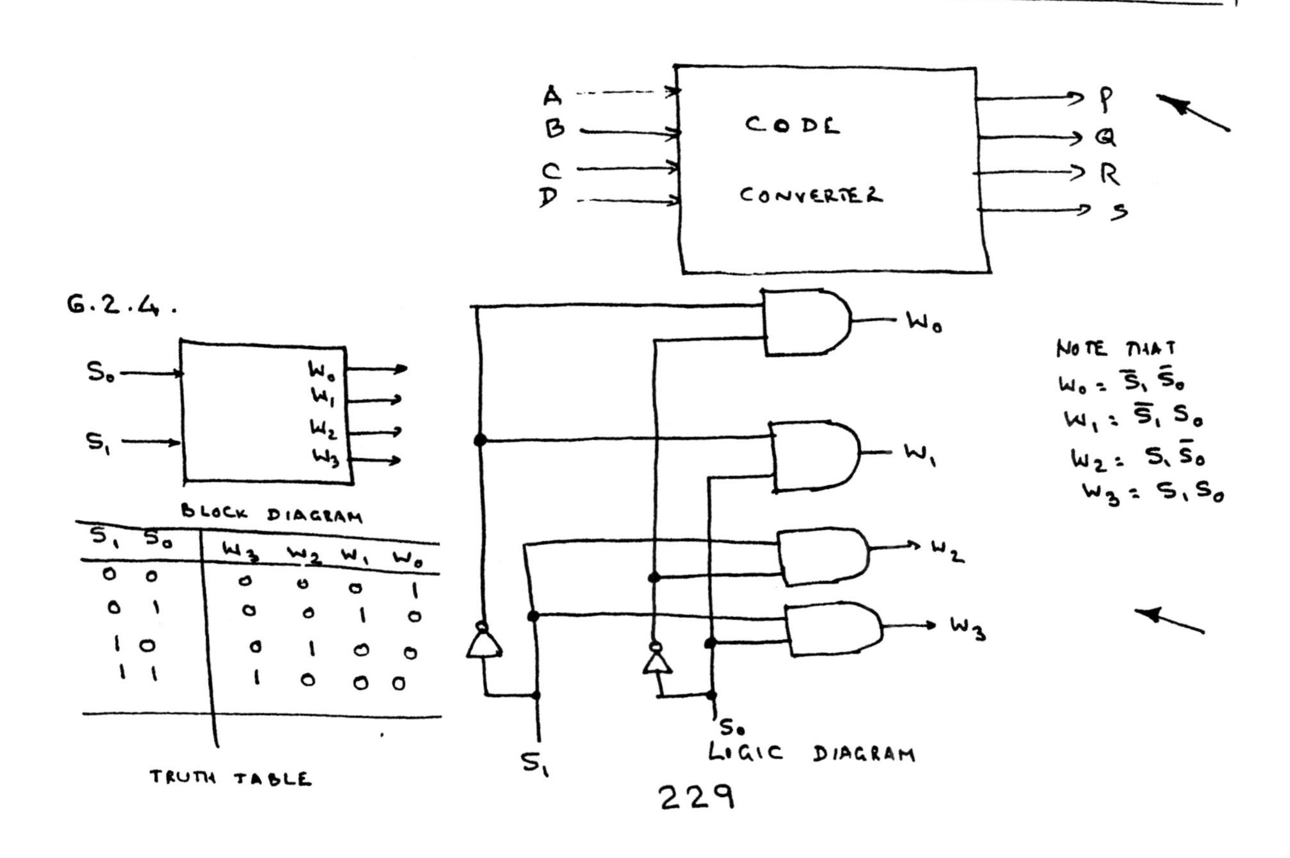

S_1	S_0	W_3	W_2	W_1	W_0
0	0	0	0	0	1
0	1	0	0	1	0
1	0	0	1	0	0
1	1	1	0	0	0

NOTE THAT

$W_0 = \bar{S}_1 \bar{S}_0$

$W_1 = \bar{S}_1 S_0$

$W_2 = S_1 \bar{S}_0$

$W_3 = S_1 S_0$

6.2.5.

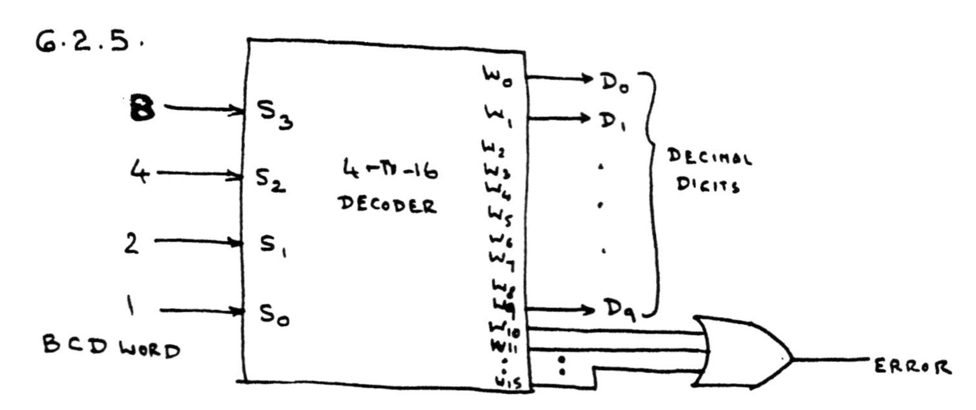

TRUTH TABLE

8421 WORD S_3	S_2	S_1	S_0	ACTIVE OUTPUT	ASSIGNMENT
0	0	0	0	W_0	D_0
0	0	0	1	W_1	D_1
⋮				⋮	⋮
1	0	0	1	W_9	D_9
1	0	1	0	W_{10}	ERROR
⋮				⋮	⋮
1	1	1	1	W_{15}	ERROR

AN 8421 BCD-CODE WORD APPLIED AT THE CONTROL TERMINALS SELECTS ONE OF THE TEN DECIMAL DIGIT OUTPUTS, D_0 THROUGH D_9.

THUS, FOR EXAMPLE, THE INPUT WORD $[S_3 S_2 S_1 S_0] = 0001$ ACTIVATES W_1 CORRESPONDING TO DIGIT D_1.

AN INVALID BCD-CODE WORD IS DETECTED BY THE OR GATE CONNECTED TO OUTPUTS W_{10} THROUGH W_{15}.

6.2.6.

AN ENCODER IS A DEVICE WHICH HAS A NUMBER OF INPUT LINES EQUAL TO THE NUMBER OF CODE COMBINATIONS TO BE GENERATED. THIS IS TEN FOR A BCD-ENCODER. THE NUMBER OF OUTPUT LINES IS EQUAL TO THE NUMBER OF BITS IN THE CODE i.e. FOUR FOR BCD.

DECIMAL	8421 BCD CODE A	B	C	D
0	0	0	0	0
1	0	0	0	1
2	0	0	1	0
3	0	0	1	1
4	0	1	0	0
5	0	1	0	1
6	0	1	1	0
7	0	1	1	1
8	1	0	0	0
9	1	0	0	1

THE LOGIC NEEDED FOR THE ENCODER CAN BE OBTAINED FROM THE BCD-CODE. FOR EXAMPLE, A MUST BE 1 IF EITHER K_8 OR K_9 IS PRESSED, WHICH MAKES K_8 AND K_9 INPUTS 1. i.e., $A = K_8 + K_9$

SIMILARLY $B = K_4 + K_5 + K_6 + K_7$

$C = K_2 + K_3 + K_6 + K_7$

$D = K_1 + K_3 + K_5 + K_7 + K_9$

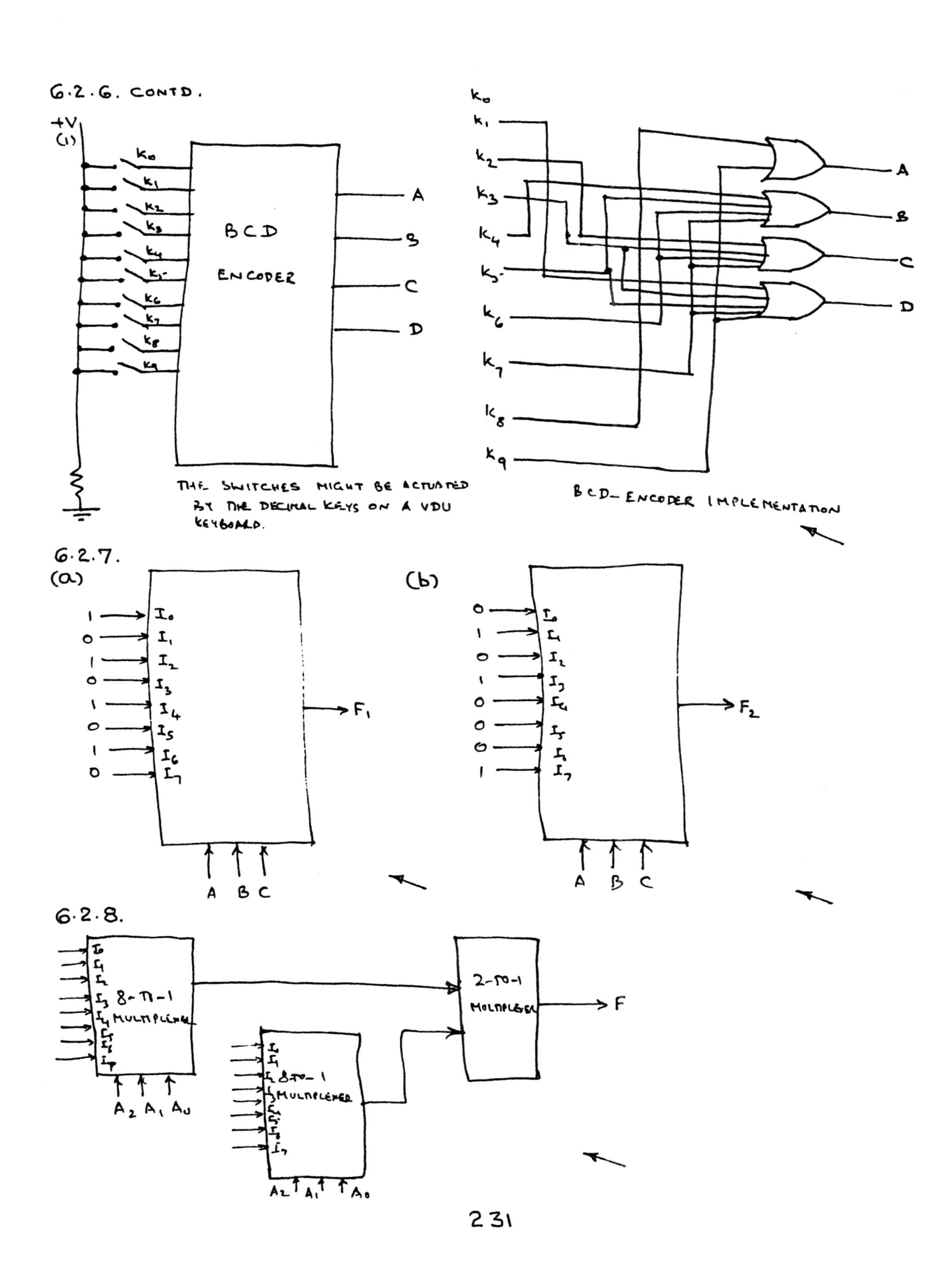
6.2.6. CONTD.
+V
(1)
k0
k1
k2
k3
k4
k5
k6
k7
k8
k9
BCD
ENCODER
A
B
C
D
THE SWITCHES MIGHT BE ACTUATED BY THE DECIMAL KEYS ON A VDU KEYBOARD.
BCD-ENCODER IMPLEMENTATION
6.2.7.
(a)
(b)
I0
I1
I2
I3
I4
I5
I6
I7
F1
F2
A B C
6.2.8.
8-TO-1
MULTIPLEXER
2-TO-1
MULTIPLEXER
F
A2 A1 A0

6.2.9.

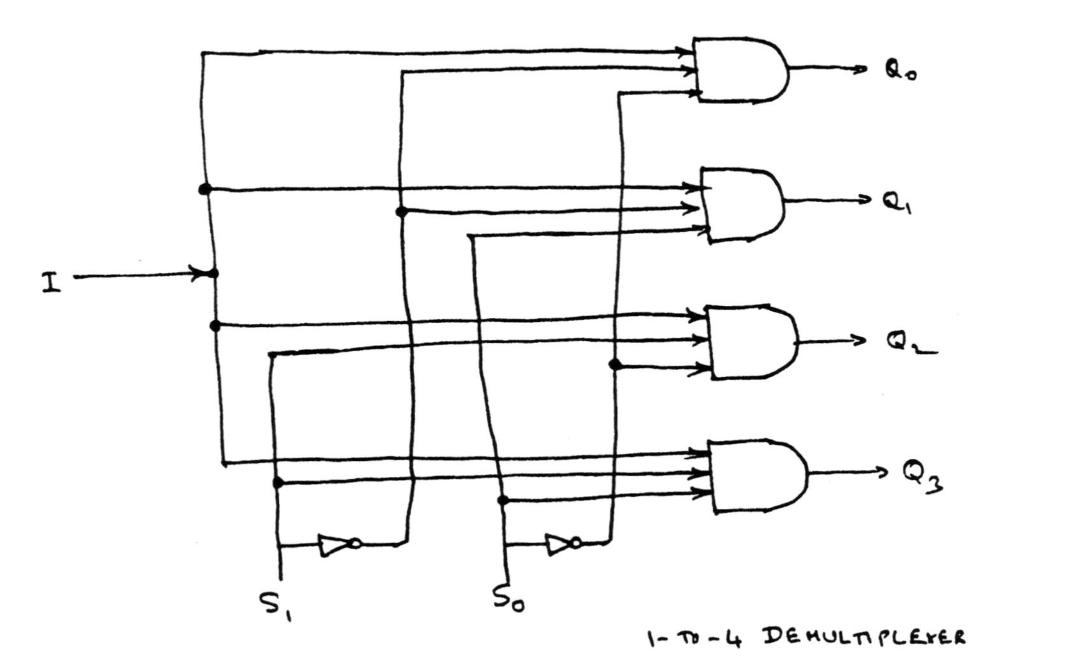

1-TO-4 DEMULTIPLEXER

6.2.10.

(a) TRUTH TABLE

S_1	S_0	Q
0	0	1
0	1	1
1	0	1
1	1	0

THIS CAN BE RELATED TO FIG. 6.2.3 C.

WITH THE CONTROL INPUTS S_1 AND S_0, THE OUTPUT Q WILL BE 'HIGH' IF I_0, I_1, AND I_2 ARE 'HIGH' AND I_3 LOW.

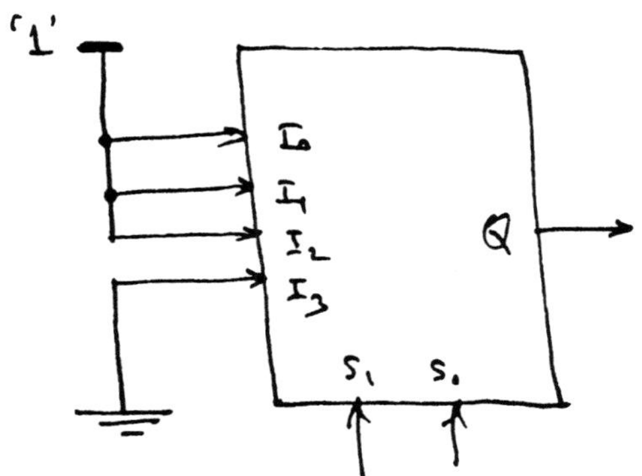

6.2.10. CONTD.

(b)

TRUTH TABLE

S_1	S_0	Q
0	0	0
0	1	1
1	0	1
1	1	0

I_1 AND I_2 NEED TO BE 'HIGH'; AND I_0 AND I_3 'LOW'.

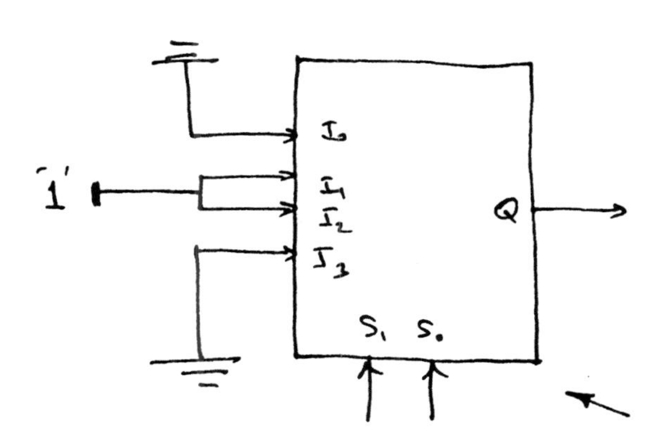

(c)

DECIMAL	BINARY	MINTERM
4	100	$A\cdot\bar{B}\cdot\bar{C}$
2	010	$\bar{A}\cdot B\cdot\bar{C}$
1	001	$\bar{A}\cdot\bar{B}\cdot C$

$$Q = A\cdot\bar{B}\cdot\bar{C} + \bar{A}\cdot B\cdot\bar{C} + \bar{A}\cdot\bar{B}\cdot C$$

MINTERMS INCLUDE $\bar{B}\cdot\bar{C}$, $\bar{B}\cdot C$, AND $B\cdot\bar{C}$ LOOKING AT THE BINARY TERMS INVOLVING B AND C. THE ONLY OMISSION FROM THE SET IS $B\cdot C$; BUT IT CAN BE INCLUDED BY REWRITING IN THE FOLLOWING WAY:

$$Q = A\cdot\bar{B}\cdot\bar{C} + \bar{A}\cdot\bar{B}\cdot C + \bar{A}\cdot B\cdot\bar{C} + 0\cdot B\cdot C$$

SO LET B AND C BE THE CONTROL VARIABLES AND WIRE UP THE MULTIPLEXER WITH I_0 PERMANENTLY CONNECTED TO A; I_1 AND I_2 CONNECTED TO $\bar{A}$; I_3 EARTHED.

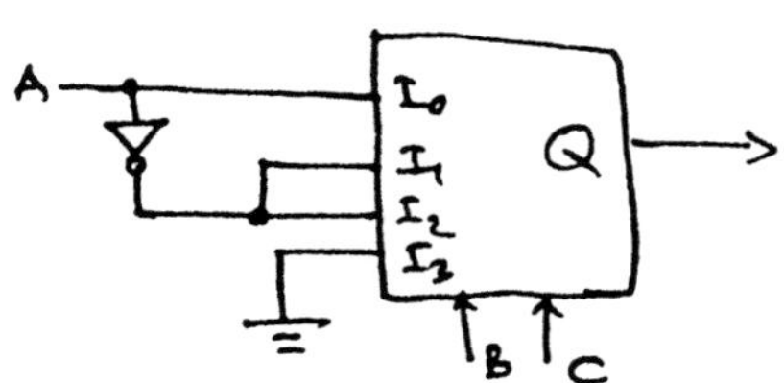

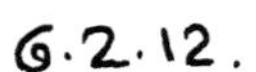
6.2.11.

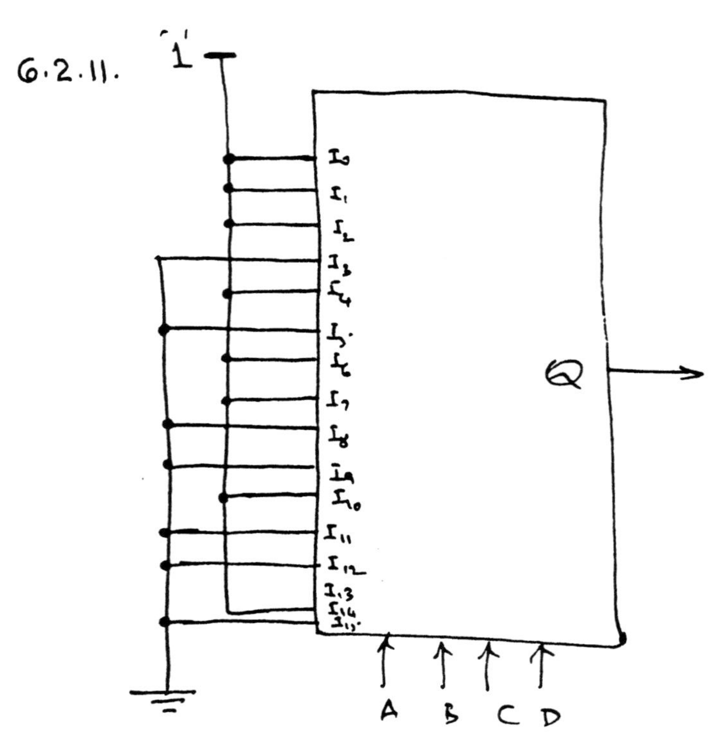

ALL 'HIGH' DATA INPUTS ARE CONNECTED TOGETHER; ALL 'LOW' INPUTS ARE CONNECTED TOGETHER. A, B, C, AND D ACT AS THE CONTROL INPUTS.

6.2.12.

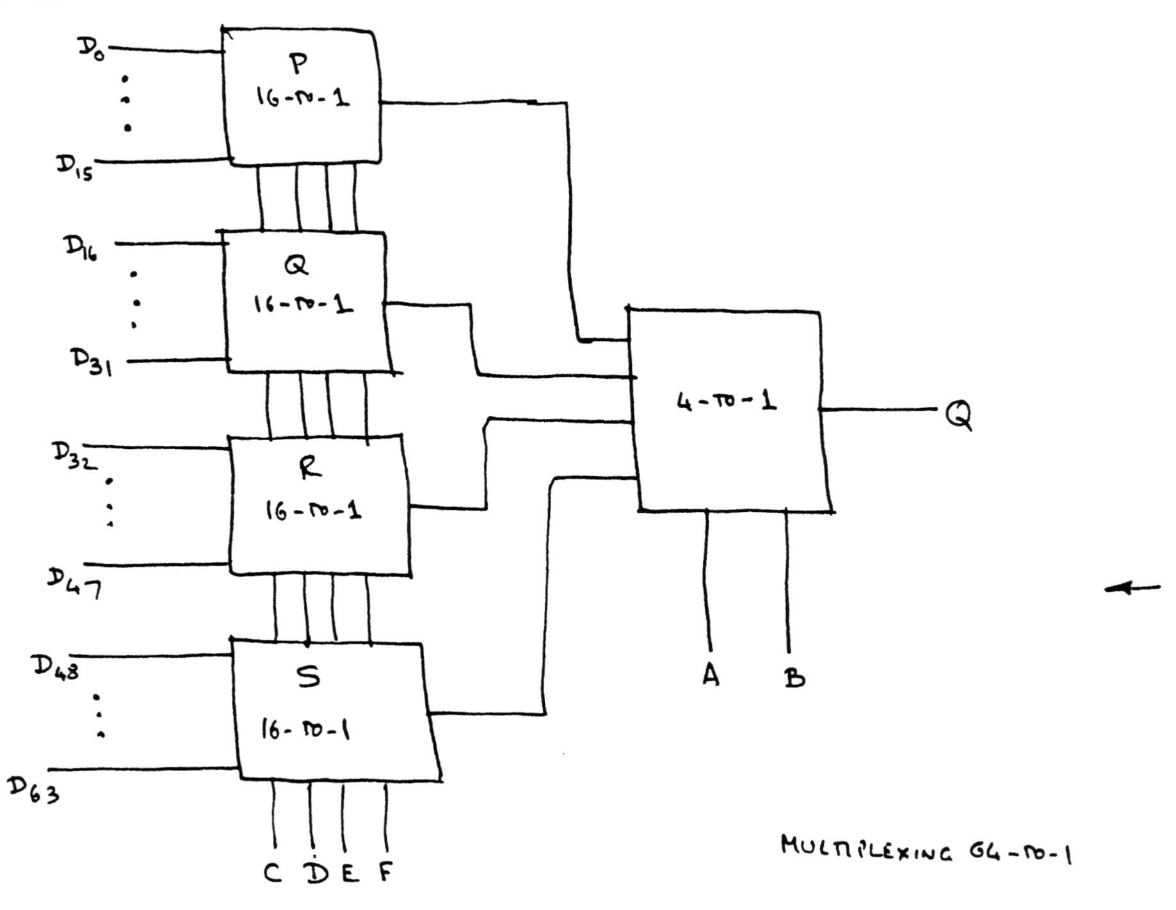

MULTIPLEXING 64-TO-1

6.2.13.

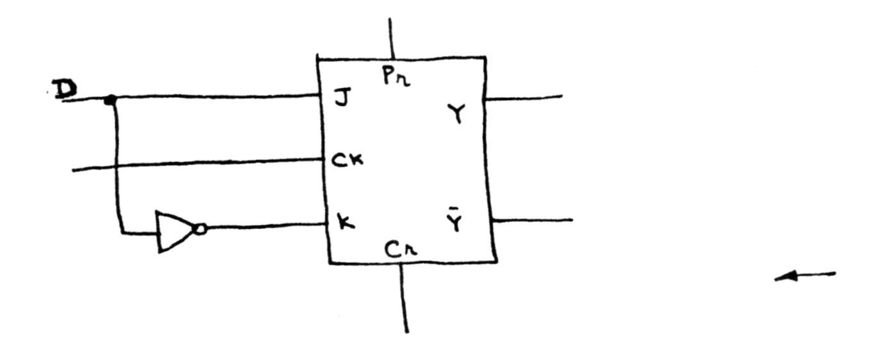

USING THIS FIG. FOR WIRING CONFIGURATION, THE OUTPUT WAVEFORMS WILL BE THE SAME AS FIG. 6.2.4b.

6.2.14.

WHEN THE CONTROL SIGNAL LOAD/$\overline{\text{SHIFT}}$ IS AT LOGIC 1, THE VALUES OF INPUT LINES I_0 THROUGH I_3 WILL BE TRANSFERRED TO Q_0 THROUGH Q_3 AT EVERY CLOCK PULSE OF THE MASTER CLOCK. BUT WHEN THE CONTROL SIGNAL LOAD/$\overline{\text{SHIFT}}$ IS AT LOGIC 0, THEN A SHIFT-RIGHT OPERATION ON THE DATA (next MSB, Q_3, Q_2, Q_1, AND Q_0) WILL OCCUR AT EVERY CLOCK PULSE OF THE MASTER CLOCK, i.e. FOR EVERY CLOCK PULSE, THE VALUE OF NEXT MSB WILL TRANSFER TO Q_3, THE VALUE OF Q_3 TO Q_2, THE VALUE OF Q_2 TO Q_1, AND THE VALUE OF Q_1 TO Q_0.

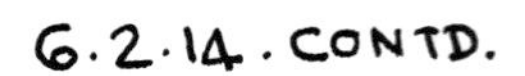
6.2.14. CONTD.

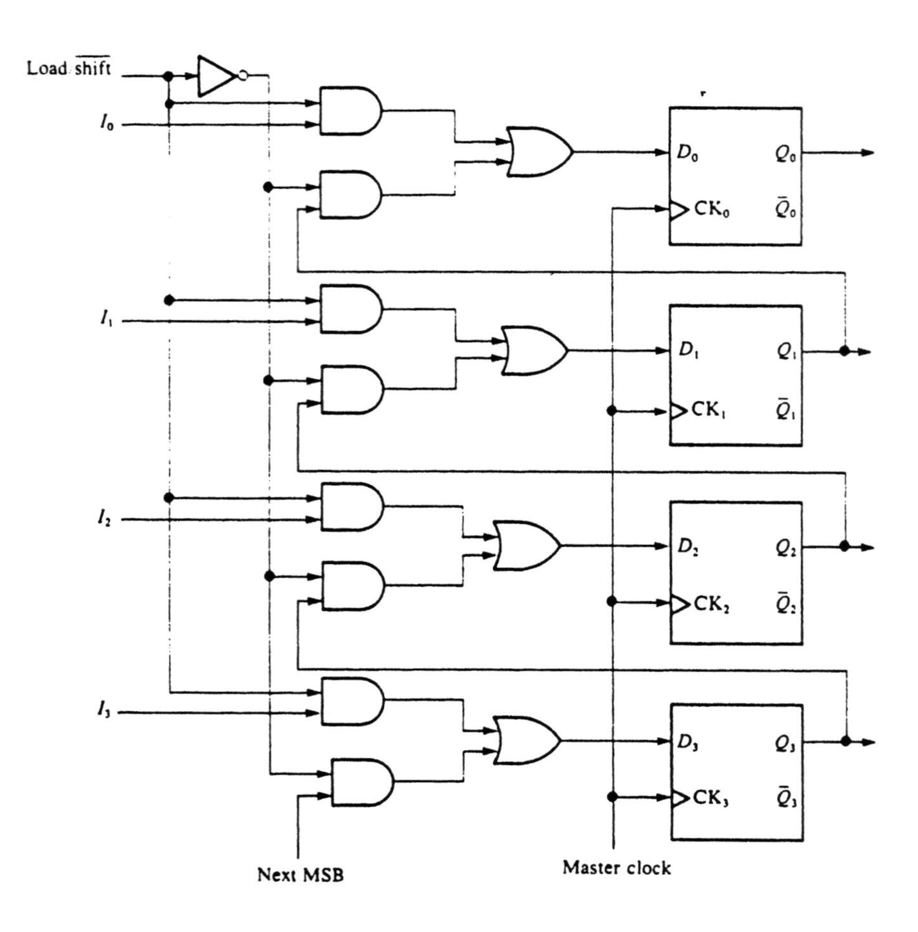

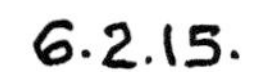
6.2.15.

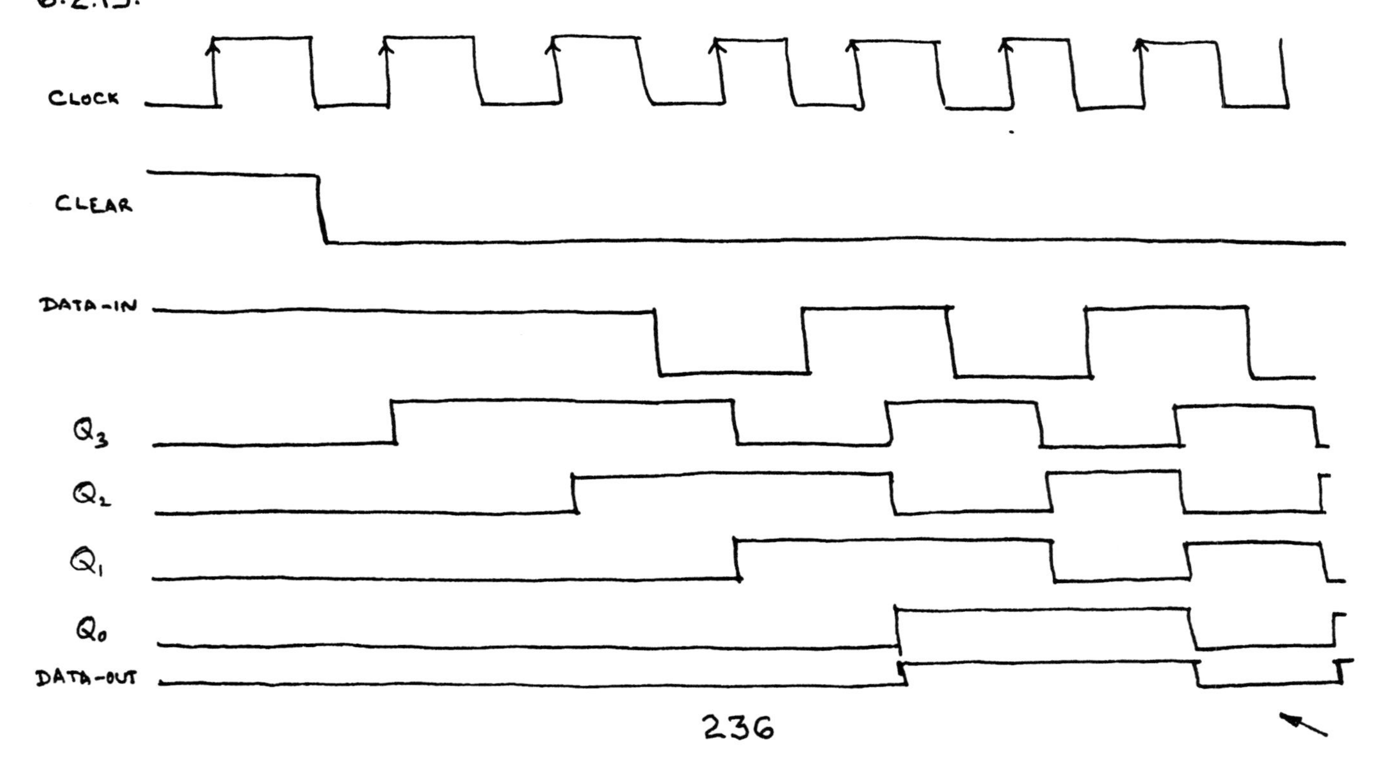

6.2.16. 1101

6.2.17.

(i) TO DIVIDE BY 2, SHIFT LEFT THE CONTENT OF THE REGISTER, ONE TIME WITH DATA-IN BEING Q.

(ii) TO MULTIPLY BY 2, SHIFT RIGHT THE CONTENT OF THE REGISTER, ONE TIME WITH DATA-IN BEING THE SAME AS THE SIGN BIT.

6.2.18.

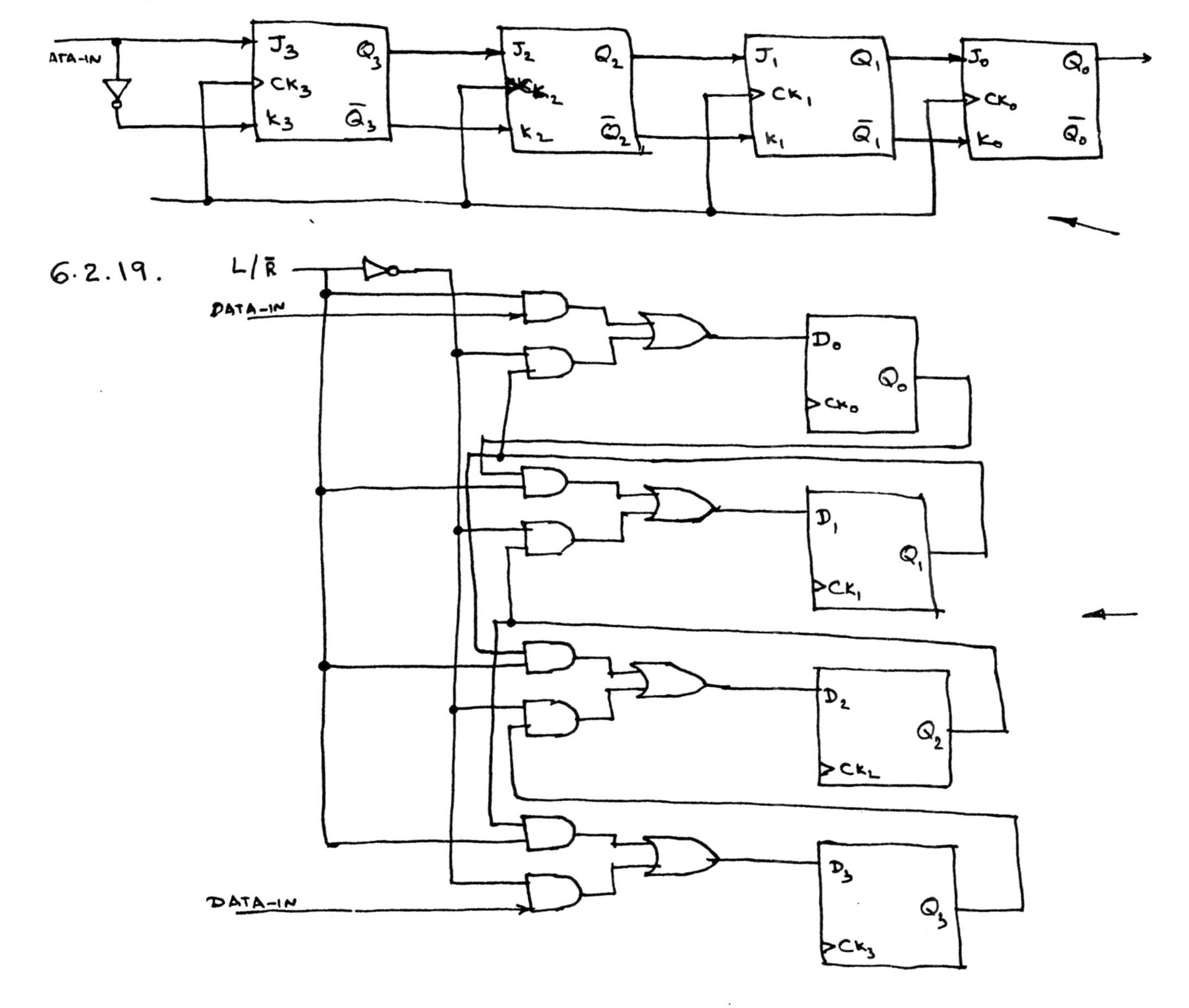

6.2.20. REFER TO 74178 IN A TTL DATA BOOK.

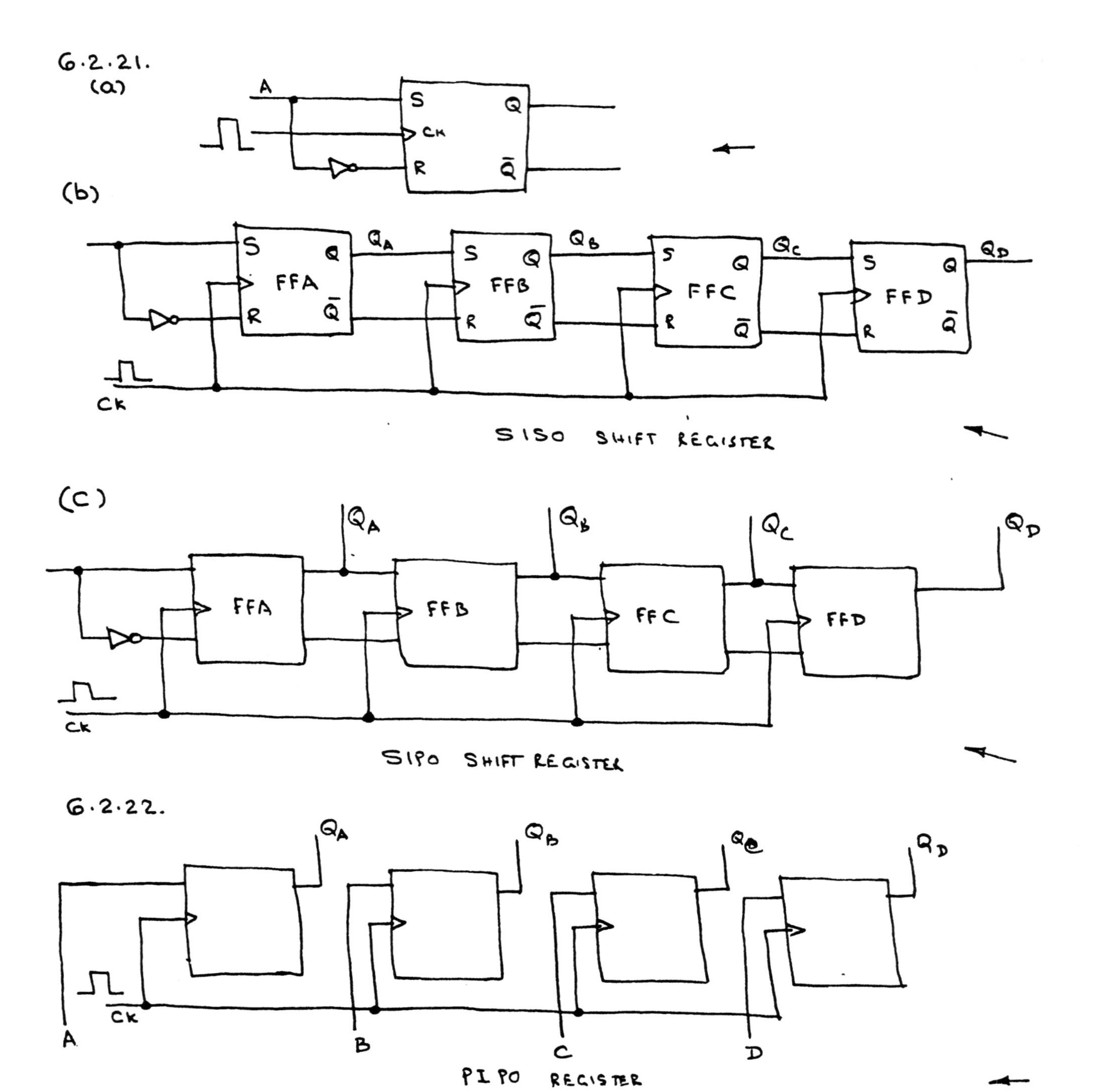

4 BITS OF THE BINARY WORD ABCD ARE READ IN TO THE REGISTER SIMULTANEOUSLY WHEN CK GOES POSITIVE, AND 4 BITS OF THE RESULTING STORED OUTPUT WORD $Q_A\ Q_B\ Q_C\ Q_D$ ARE ALL AVAILABLE TO BE READ OUT AFTER THE CLOCK PULSE.

WHATEVER MAY BECOME TO THE INPUT LOGIC LEVELS IN BETWEEN CLOCK PULSES, IT DOES NOT AFFECT THE OUTPUT.

6.2.23.

WHEN THE INPUT DATA ARE AVAILABLE, MAKING S/L (SHIFT/LOAD) = 1 ENABLES G_1 TO G_4 SO THAT THE FOUR FLIP FLOPS ARE LOADED ON THE NEXT CLOCK PULSE. ALSO, S/L = 1 DISABLES G_5 TO G_7 SO THE SHIFT MECHANISM DOES NOT OPERATE. WHEN S/L = 0, G_1 TO G_4 ARE DISABLED AND G_5 TO G_7 ENABLED SO THAT CONSECUTIVE CLOCK PULSES SHIFT THE DATA OUT SERIALLY AT Q_D

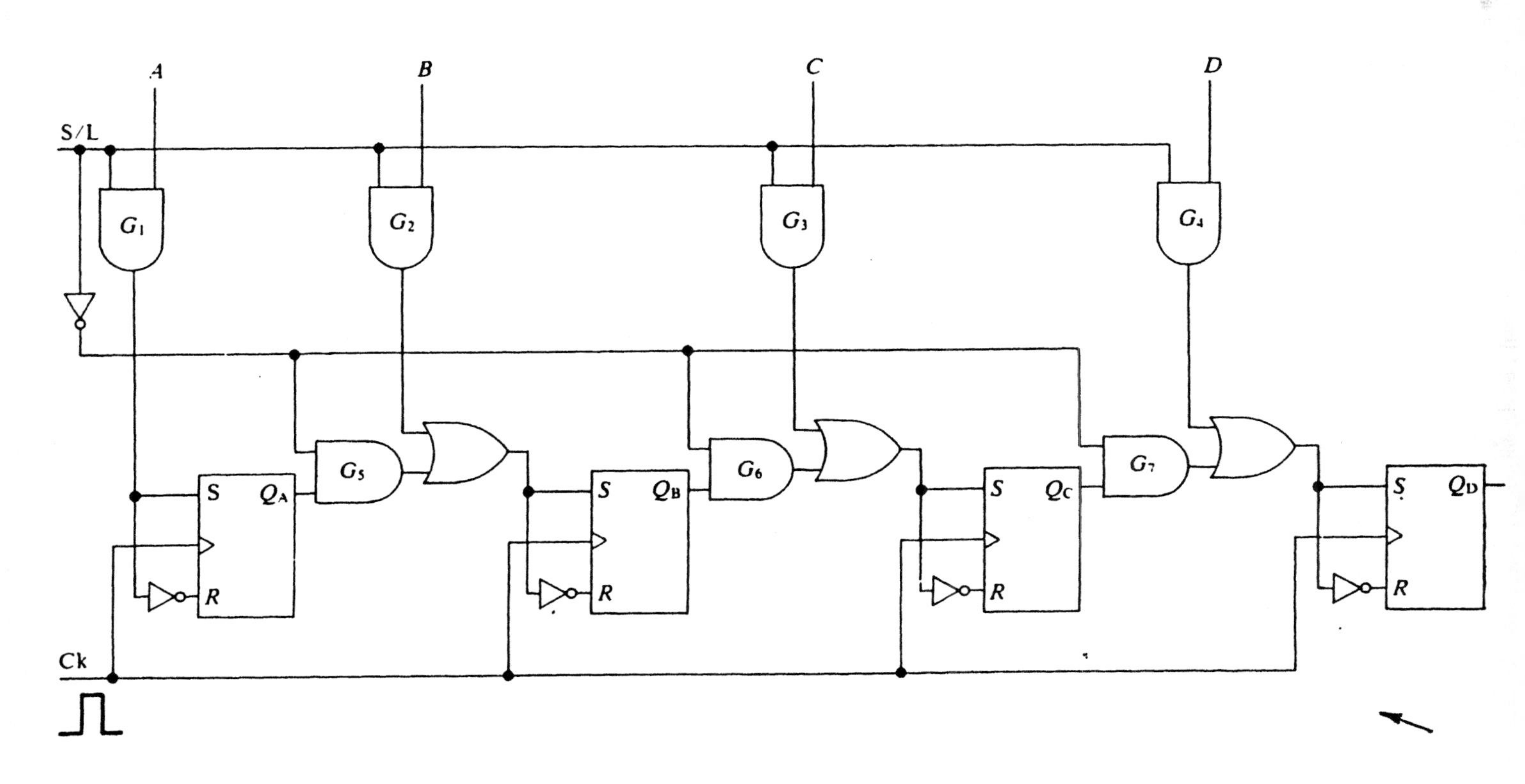

6.2.24.

A MODULO-5 COUNTER COUNTS UP FROM 0 TO 4 AND THEN CLEARS ALL FLIP FLOPS ON THE FIFTH PULSE. SINCE WE NEED TO COUNT UP TO 4 i.e. $(100)_2$, THREE FLIPFLOPS ARE NEEDED. BLOCK DIAGRAM IS SHOWN BELOW; NOTE THE AND GATE IS USED TO DETECT THE STATE 101 AND CLEAR ALL FLIPFLOPS.

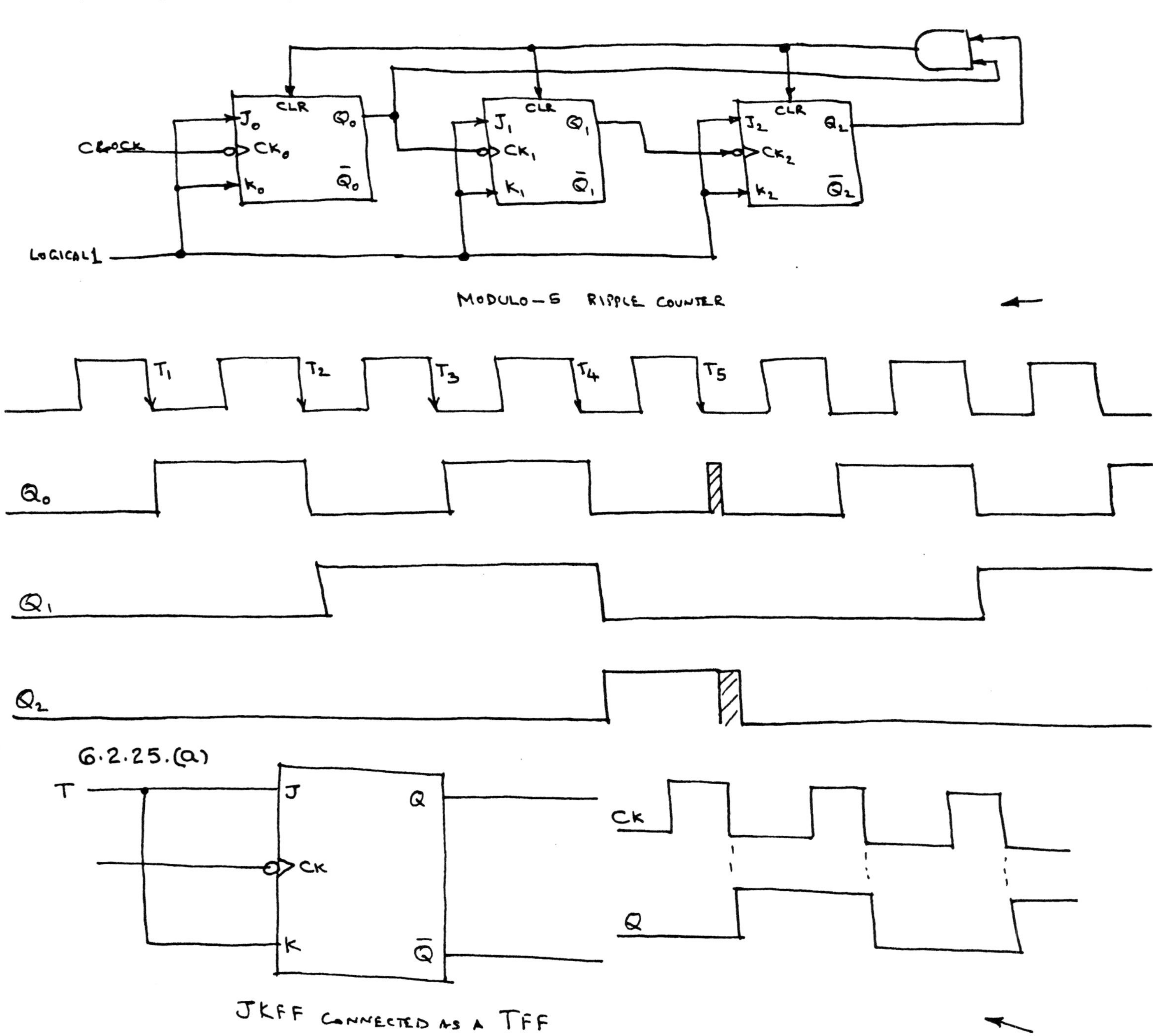

MODULO-5 RIPPLE COUNTER

6.2.25.(a)

JKFF CONNECTED AS A TFF

6.2.25.(b)

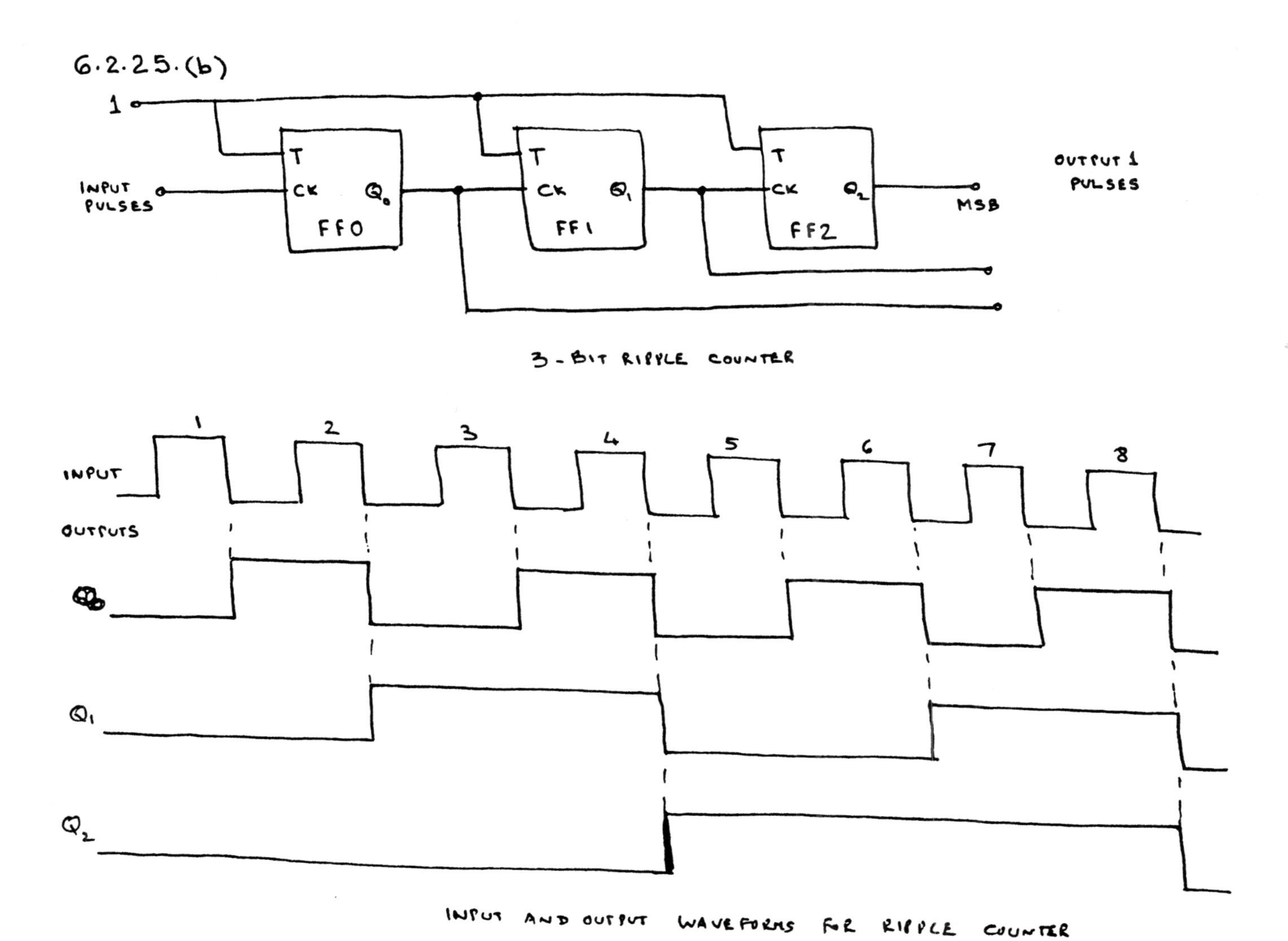

3-BIT RIPPLE COUNTER

INPUT AND OUTPUT WAVEFORMS FOR RIPPLE COUNTER

WITH THE INPUT HELD AT LOGICAL 1, EACH PULSE APPLIED AT THE CLOCK INPUT CAUSES A CHANGE IN STATE WHICH IS PROPAGATED (OR RIPPLED) THROUGH THE REGISTER. THE INPUT FF CHANGES STATE ON THE FALLING EDGE OF THE CLOCK PULSES AND EACH SUCCESSIVE TRANSITION IN THE OTHER FFs OCCURS ONLY WHEN THE PREVIOUS FF CHANGES ITS STATE FROM 1 TO 0.

6.2.26.

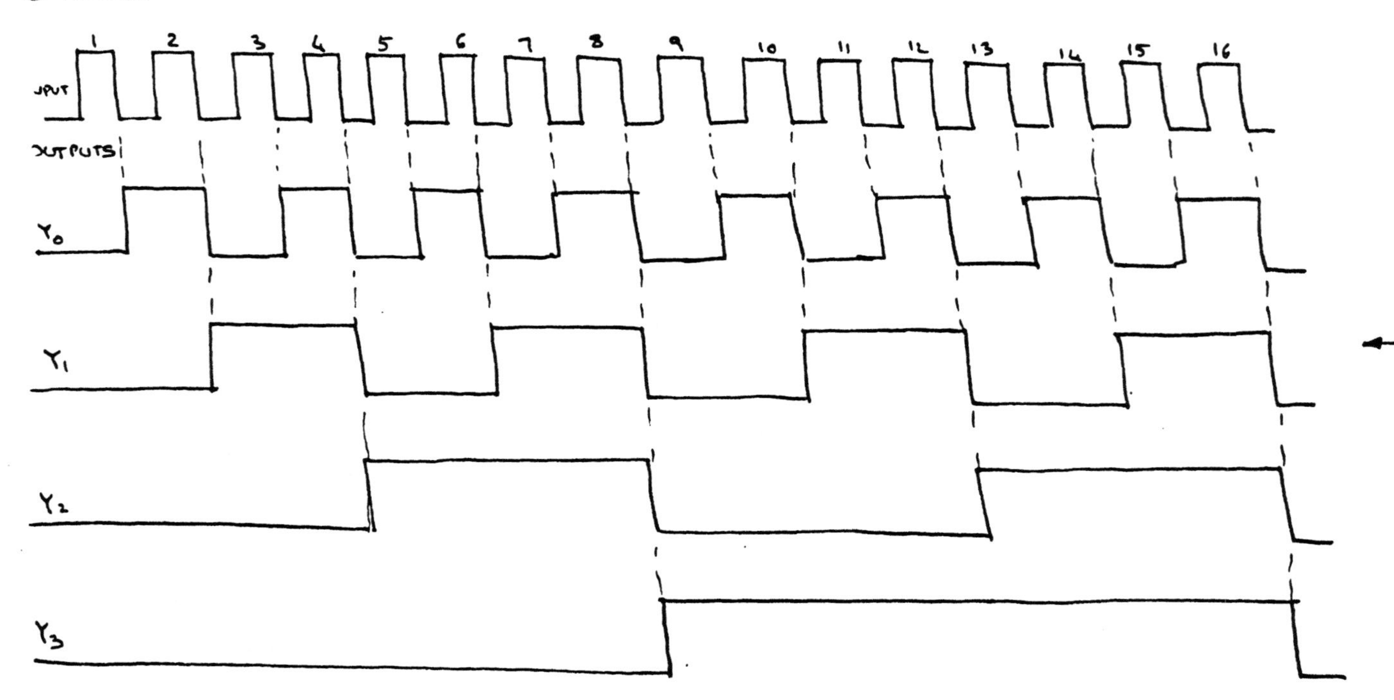

6.2.27. ON THE TENTH CLOCK PULSE, ALLOW THE COUNTER TO GO INTO THE STATE 1010. A NAND GATE WITH INPUTS $Q_3 \bar{Q}_2 Q_1 \bar{Q}_0$ WILL GIVE AN OUTPUT 0 WHEN THE COUNTER REACHES 1010 AND THIS CAN BE USED TO CLEAR THE FFs.

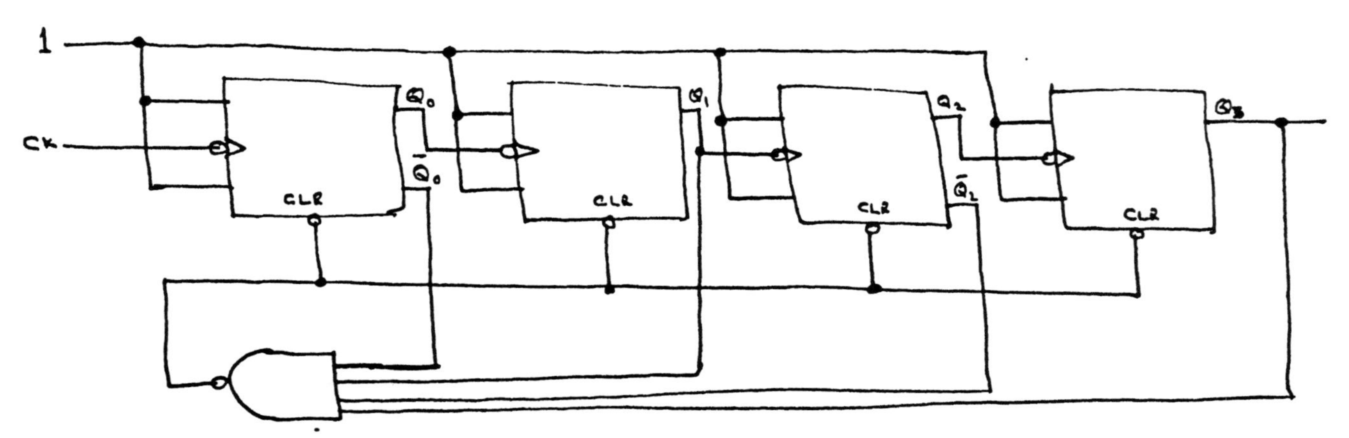

ASYNCHRONOUS DECADE COUNTER

6.2.27. CONTD.

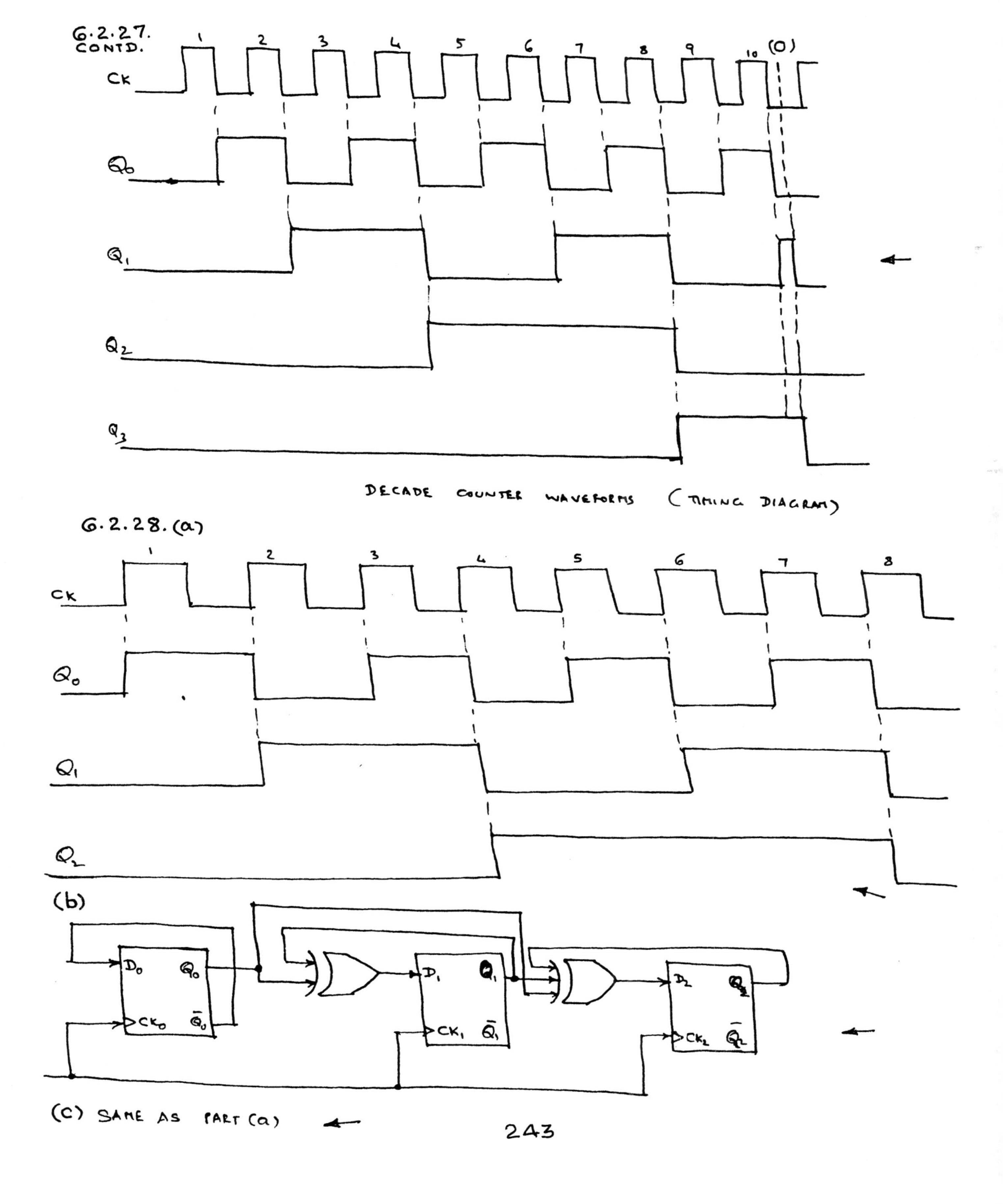

(c) SAME AS PART (a)

6.2.29.

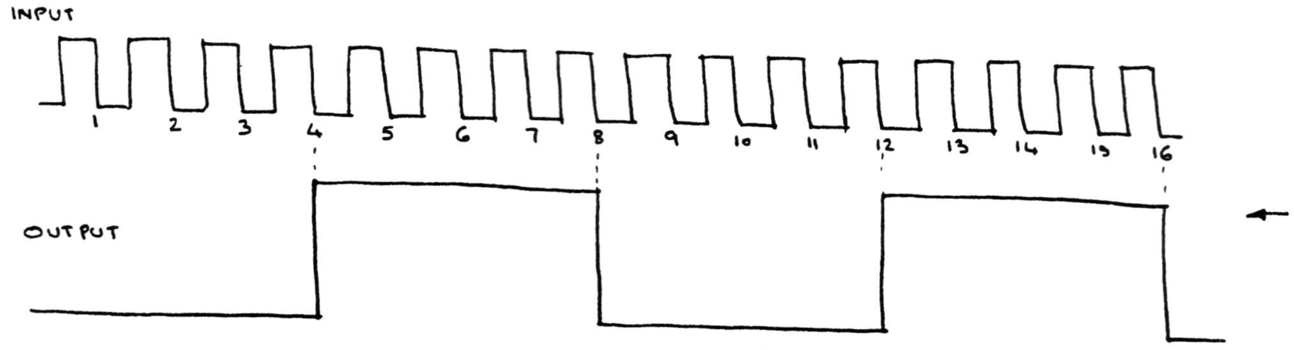

INPUT

1 2 3 4 5 6 7 8 9 10 11 12 13 14 15 16

OUTPUT

6.2.30.

THE OBJECT IS TO COUNT THE CLOCK PULSES. OUTPUT A IS USED AS THE LEAST SIGNIFICANT BIT.

THE OPERATION OF THE COUNTER CAN BE BEST EXPLAINED BY TABULATING THE STATE OF THE OUTPUTS A, B, AND C AFTER SUCCESSIVE CLOCK PULSES.

COUNTS	FF 1	FF 2	FF 3	COMMENTS
0	A = 0	B = 0	C = 0	INITIAL COUNT
1	A = 1	B = 0	C = 0	
		JK = 11 ↓		
2	A = 0	B = 1	C = 0	
		JK = 00	JK = 00 ↓	AND GATE DISABLED
3	A = 1	B = 1	C = 0	
		JK = 11 ↓	JK = 11 ↓	AND GATE ENABLED
4	A = 0	B = 0	C = 1	
	. .	. .	. .	

THUS WE SEE THAT AFTER 4 CLOCK PULSES THE COUNTER INDICATES THE BINARY NUMBER 100.

THE PROCESS CAN BE CONTINUED UP TO 7 COUNTS. THE DIAGONAL ARROWS INDICATE THE WAY IN WHICH THE INPUTS OF ONE FLIPFLOP ARE INFLUENCED BY THE OUTPUT OF THE PRECEDING FLIPFLOP.

6.2.30. CONTD.

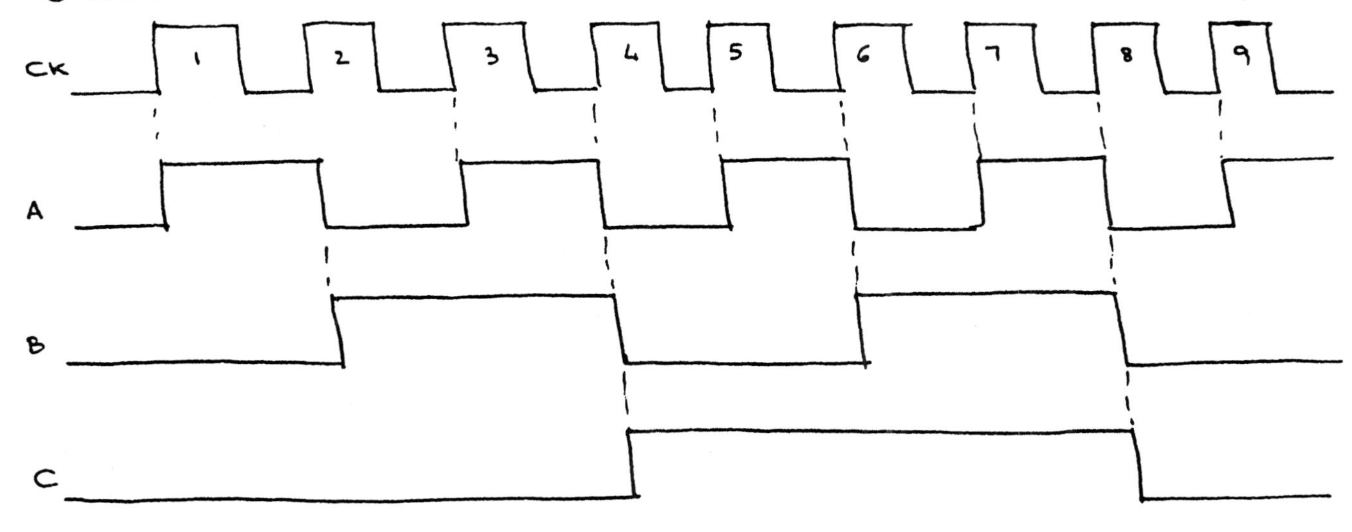

WAVEFORMS WITH POSITIVE EDGE TRIGGERING

6.2.31.
(a) $\frac{6\text{kHz}}{100} = 60\text{ Hz}$

$\frac{60\text{Hz}}{60} = 1\text{ Hz} = 1\text{ SEC} \Rightarrow Y_0$ DISPLAYS SECONDS

$\frac{1\text{Hz}}{10} = \frac{1}{10}\text{ Hz} = 10\text{ SEC} \Rightarrow Y_1$ DISPLAYS 10's OF SECONDS

$\frac{(1/10)\text{Hz}}{6} = \frac{1}{60}\text{ Hz} = 60\text{ SEC} = 1\text{ MIN} \Rightarrow Y_2$ DISPLAYS MINUTES

$\frac{(1/60)\text{Hz}}{10} = \frac{1}{600}\text{ Hz} = 600\text{ SEC} = 10\text{ MIN} \Rightarrow Y_3$ DISPLAYS 10's OF MINUTES

$\frac{(1/600)\text{Hz}}{6} = \frac{1}{3600}\text{ Hz} = 3600\text{ SEC} = 60\text{ MIN} = 1\text{ HOUR} \Rightarrow Y_4$ DISPLAYS HOURS

$\frac{(1/3600)\text{Hz}}{10} = \frac{1}{36000}\text{ Hz} = 36000\text{ SEC} = 600\text{ MIN} = 10\text{ HOURS} \Rightarrow Y_5$ DISPLAYS 10's OF HOURS.

(b)

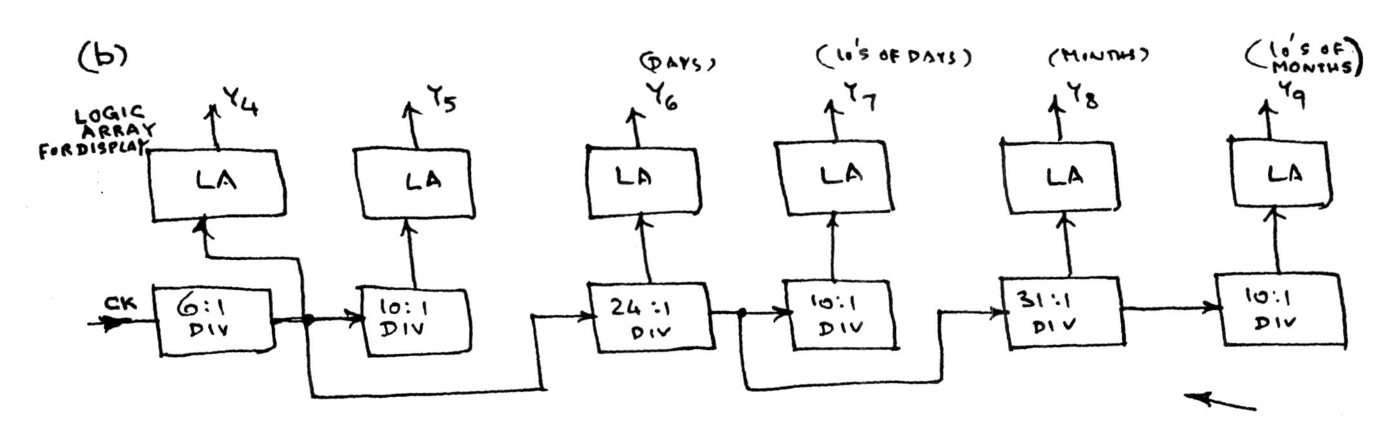

6.2.32.

$$\Delta V = \frac{V_{ref}}{2^{n-1}} \quad \text{or} \quad 2^{n-1} = \frac{V_{ref}}{\Delta V} = \frac{15}{1} = 15$$

$$n-1 = \log_2 15 = 3.91 \quad \text{or} \quad n = 4.91$$

So USE $n = 5$ ←

6.2.33.

(a) $V_{out} = 10\left(1 + 1\times 2^{-1} + 1\times 2^{-2} + 1\times 2^{-3}\right) = 10 \times 1.875 = 18.75\,V$ ←

(b) $V_{out} = 0$ ←

(c) $\Delta V = \frac{10}{8} = 1.25\ V$ ←

6.2.34.

DECIMAL	BINARY EQUIVALENT				I_{in}, pu
0	0	0	0	0	0
1	0	0	0	1	0.125
2	0	0	1	0	0.25
3	0	0	1	1	0.375
4	0	1	0	0	0.5
5	0	1	0	1	0.625
6	0	1	1	0	0.75
7	0	1	1	1	0.875
8	1	0	0	0	1.000
⋮	⋮	⋮	⋮	⋮	⋮
15	1	1	1	1	1.875

6.2.34. CONTD.

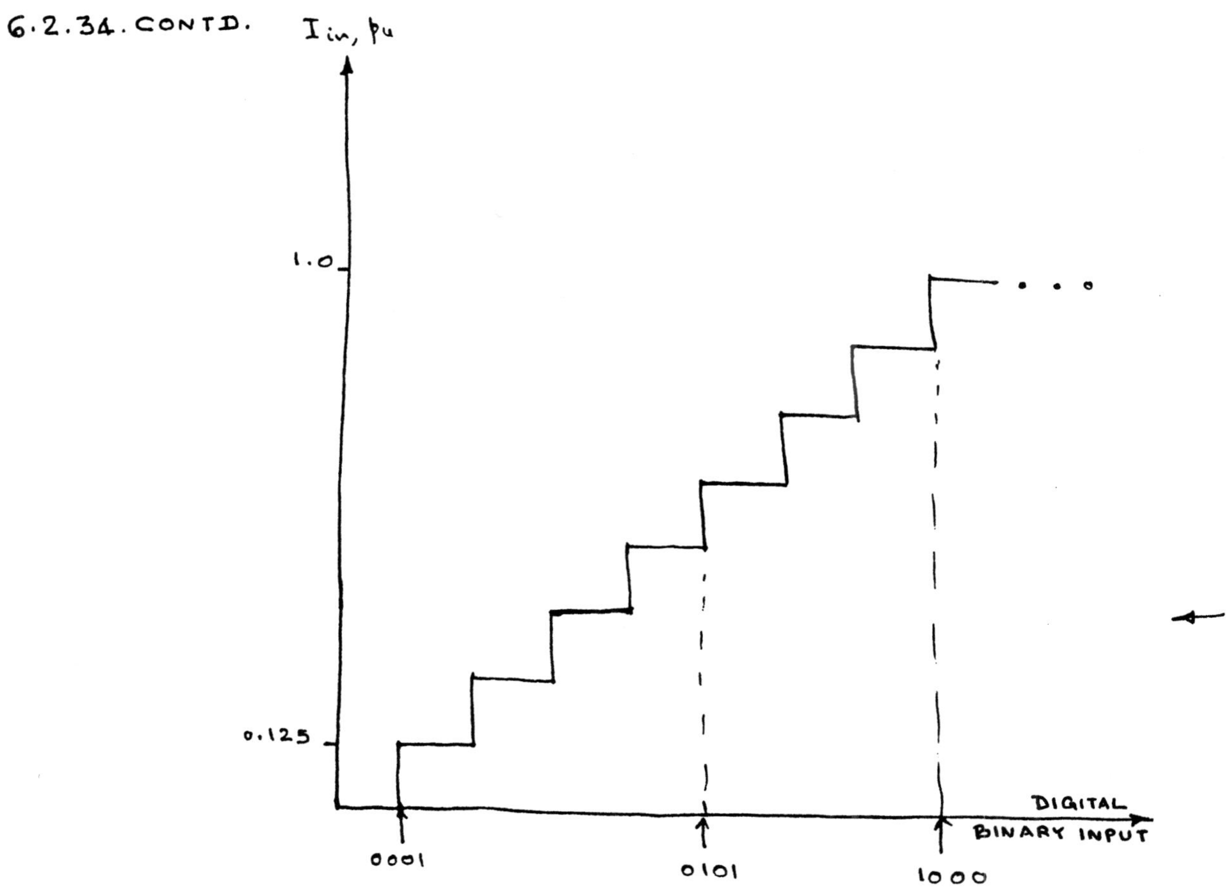

6.2.35.

R, 2R, 4R, 8R, 16R, 32R

↑ MSB ... ↑ LSB

MAXIMUM ANALOG-OUTPUT VOLTAGE $= 15\left(1 + 1\times 2^{-1} + 1\times 2^{-2} + 1\times 2^{-3} + 1\times 2^{-4} + 1\times 2^{-5}\right)$

$= 15 \times 1.96875 = 29.53125\ V$

MINIMUM ANALOG-OUTPUT VOLTAGE = 0

$\Delta V = \frac{15}{32} = 0.46875\ V$

6.2.36.

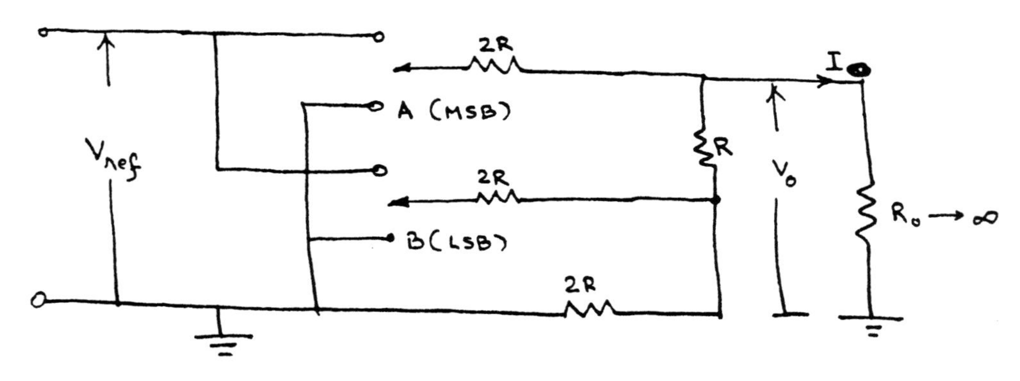

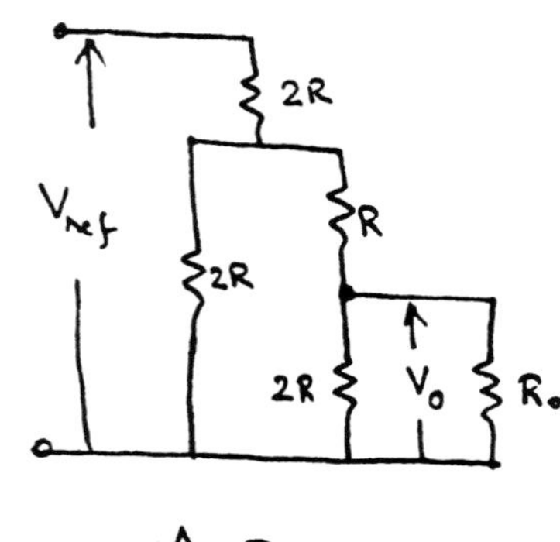

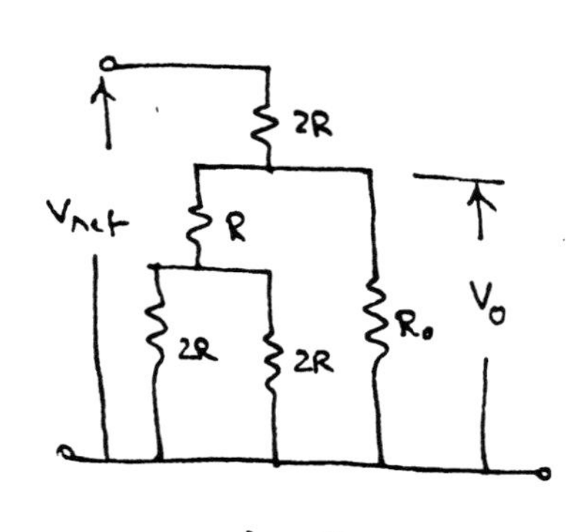

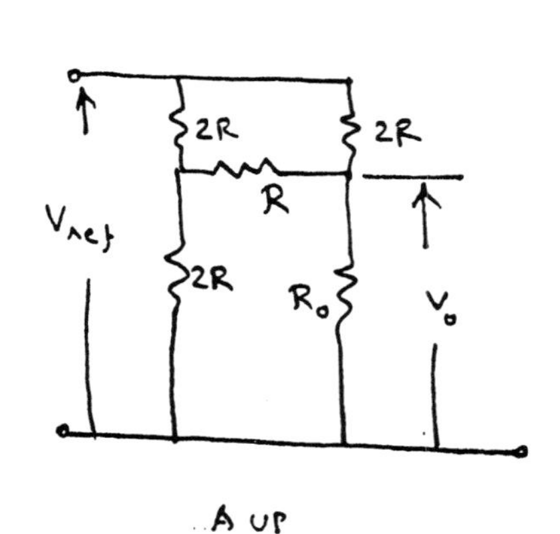

A DOWN
B UP
BINARY 01

$V_o = V_{ref}/4$

A UP
B DOWN
BINARY 10

$V_o = V_{ref}/2$

A UP
B UP
BINARY 11

$V_o = 3V_{ref}/4$

THE OUTPUTS ARE IN THE PROPER DECIMAL RELATIONSHIP OF 1:2:3, AND REPRESENT THE FIRST THREE STEPS OF THE STAIRCASE ANALOG OUTPUT SIGNAL.

6.2.37.

$$V_{out} = 10\left(1\times 2^{-1} + 1\times 2^{-2} + 0\times 2^{-3} + 0\times 2^{-4}\right)$$ AS PER EQ. (6.2.3)

$$= 10\,(0.75) = 7.5\,V$$

FOR A DECIMAL-OUTPUT VOLTAGE, $V_{out} = 12\,V$ WHEN THE BINARY INPUT CODE IS 1100.

$$\therefore\ 12 = -V_{ref}\left(1\times 2^{-1} + 1\times 2^{-2} + 0\times 2^{-3} + 0\times 2^{-4}\right) = -V_{ref}\,(0.75)$$

OR $$V_{ref} = -\frac{12}{0.75} = -16\,V$$

6.2.38.

FOR A R-2R LADDER NETWORK, THE VALUE OF THE LSB RESISTOR IS SAME AS THAT OF THE MSB RESISTOR. i.e. $10\,k\Omega$.

6.2.39.

WEIGHTED-RESISTOR DAC REQUIRES n RESISTORS OF VALUES $R, 2R, 4R, \ldots, 2^{n-1}R$;

THE R-2R LADDER DAC NEEDS $(n-1)$ RESISTORS OF VALUE R AND $(n+1)$ RESISTORS OF VALUE 2R. ←

6.2.40.
(a)

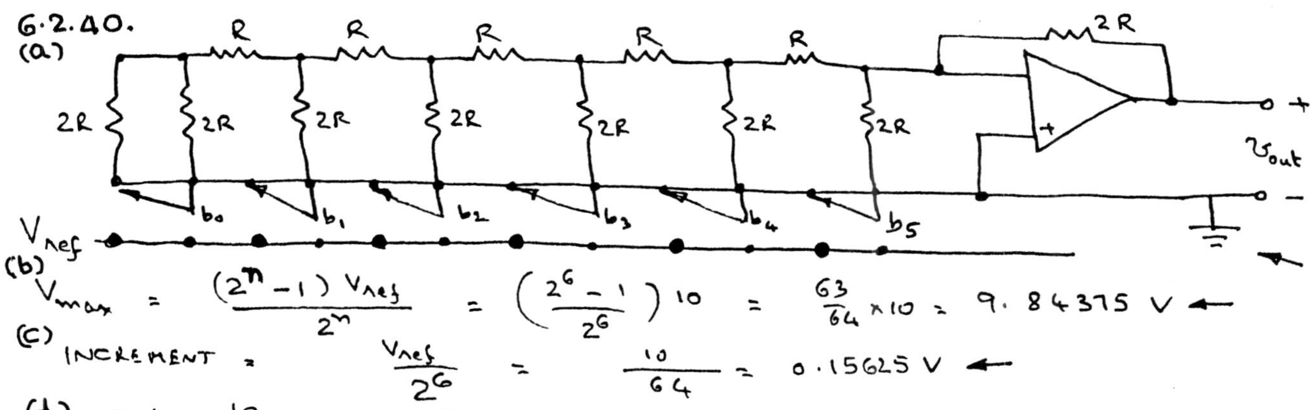

(b)

$$V_{max} = \frac{(2^n - 1)V_{ref}}{2^n} = \left(\frac{2^6-1}{2^6}\right)10 = \frac{63}{64} \times 10 = 9.84375 \text{ V} \leftarrow$$

(c)

$$\text{INCREMENT} = \frac{V_{ref}}{2^6} = \frac{10}{64} = 0.15625 \text{ V} \leftarrow$$

(d) $0.1 = \frac{10}{2^n}$ or $2^n = \frac{10}{0.1} = 100 \Rightarrow n = 7$ FOR A REF. VOLTAGE OF 10V.

ONE CAN ALSO CHOOSE A REF. VOLTAGE APPROPRIATELY FOR A CHOSEN NUMBER OF BITS IN ORDER TO GET THE DESIRED OUTPUT-VOLTAGE INCREMENT. ←

6.2.41.

THE COUNTER-CONTROLLED ADC HAS A MAXIMUM CONVERSION TIME OF 2^n CLOCK PERIODS;

THE SUCCESSIVE-APPROXIMATION ADC HAS A MAXIMUM CONVERSION TIME OF n CLOCK PERIODS. ←

6.2.42.

(a) $$Q_c = \int_0^T i(t)\,dt = \int_0^T \frac{V_{in}}{R}\,dt \;;\; Q_c = \frac{V_{in}}{R}T \leftarrow$$

(b) $$Q_d = \frac{V_{ref}}{R} t_d$$

SINCE $Q_c = Q_d$, $\frac{V_{in}}{R}T = \frac{V_{ref}}{R}t_d$; $\therefore\ t_d = \frac{V_{in}}{V_{ref}}T$ ←

6.2.43.

(a) $T = 1/f = 1/10^6 = 1\,\mu s.$

SINCE $n = 8$, NUMBER OF REQUIRED CLOCK CYCLES $= 2^n = 2^8 = 256$

$$t_{max} = 256 \times 1 = 256\,\mu s. \leftarrow$$

(b) FOR $n = 8$, NUMBER OF REQUIRED CLOCK CYCLES $= n = 8$.

$$\therefore t_{max} = 8 \times 1 = 8\,\mu s. \leftarrow$$

6.2.44. BYTES/BOOK = 2500 × 500 = 1.25×10^6

BYTES/TAPE = $2400 \frac{ft}{tape} \times 12 \frac{in}{ft} \times 1600 \frac{\text{BYTES}}{\text{IN}} = 41.6 \times 10^6$ BYTES/TAPE

$\therefore \frac{\text{BOOKS}}{\text{TAPE}} = \frac{41.6 \times 10^6}{1.25 \times 10^6} = 36.8$ ←

6.2.45.

(a) $8192 = 2^{13}$ BITS

SO INDIVIDUAL ADDRESSES WOULD HAVE TO BE 13 BITS LONG. ←

(b) $2^{13}/8 = 2^{13}/2^3 = 2^{10} = 1024$ WORDS; 10 ADDRESS BITS ←

(c) ROM STORES 1K × 8 BITS OR 1K-BYTE, WHERE 1K STANDS FOR 1024. ←

(d) 1024 GATES ←

(e) 2^{13} BITS ARE STORED IN A 64-WORD × 128 BIT MATRIX;

THE FIRST 6 ADDRESS BITS ARE APPLIED TO A DECODER HAVING 64 GATES AND WORDLINES. THE WORDLINES INTERSECT 128 BIT LINES THAT ARE DIVIDED INTO 8 GROUPS AND CONNECTED TO MULTIPLEXERS WITH 128/8 = 16 INPUTS. THE REMAINING 4 ADDRESS BITS APPLIED TO THE MULTIPLEXERS SELECT FROM THE 16 AVAILABLE WORDS THE DESIRED 8-BIT WORD TO BE READ AT THE OUTPUT.

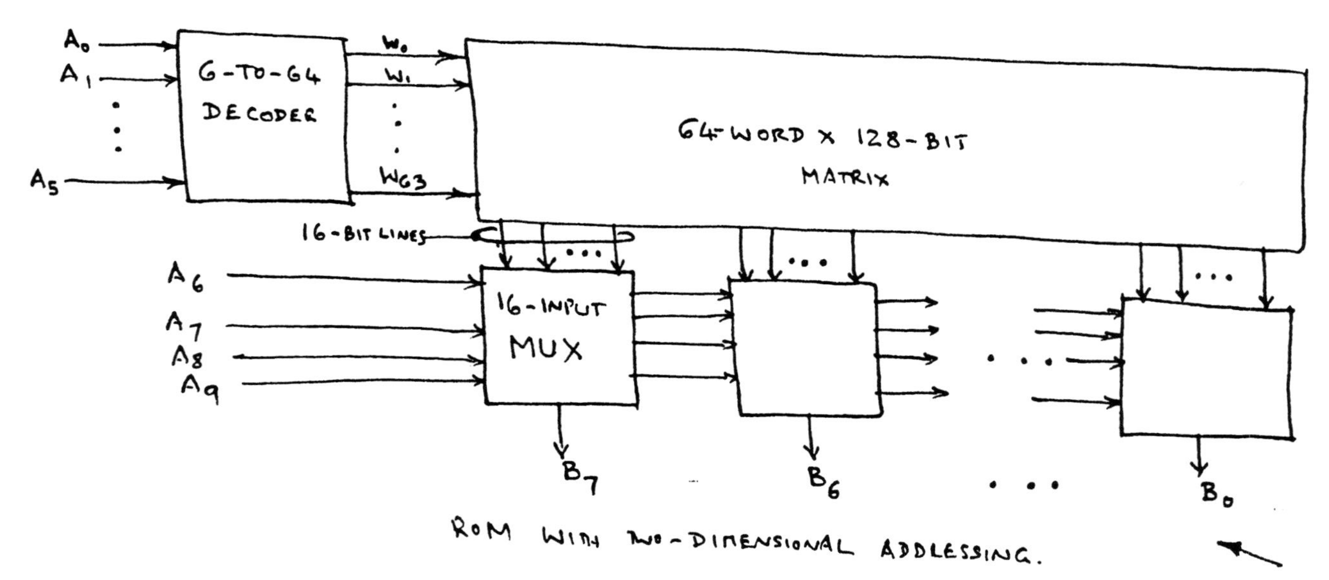

ROM WITH TWO-DIMENSIONAL ADDRESSING. ←

SINCE EACH MUX CONTAINS 16 AND GATES PLUS ONE OR GATE, THIS ADDRESSING SCHEME REDUCES THE GATE COUNT FROM 1024 TO $64 + 8 \times (16+1) = 200$.

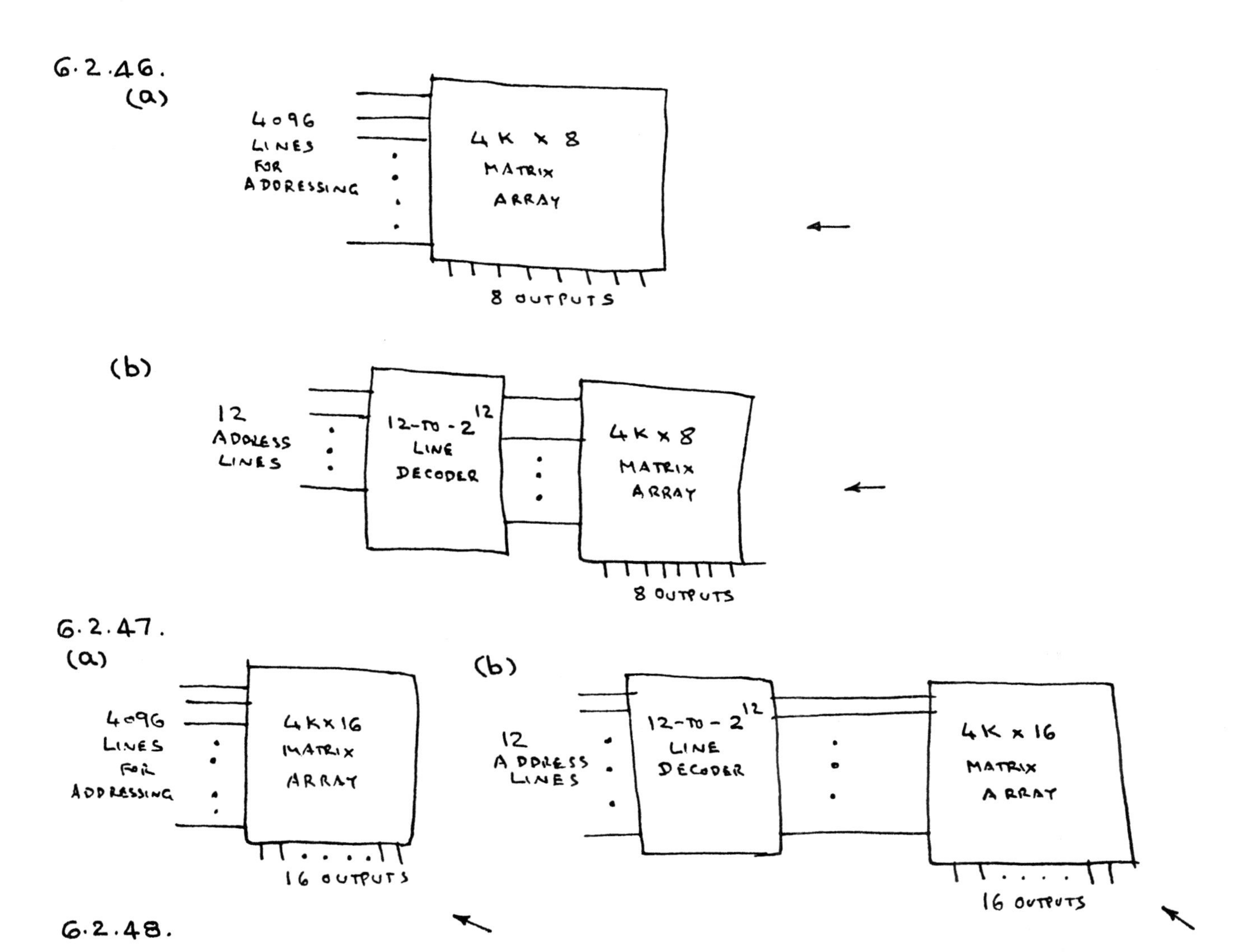

A 2-BIT INPUT WORD DESCRIBES 4 STATES, EACH OF WHICH SELECTS ONE LINE IN THE OUTPUT.

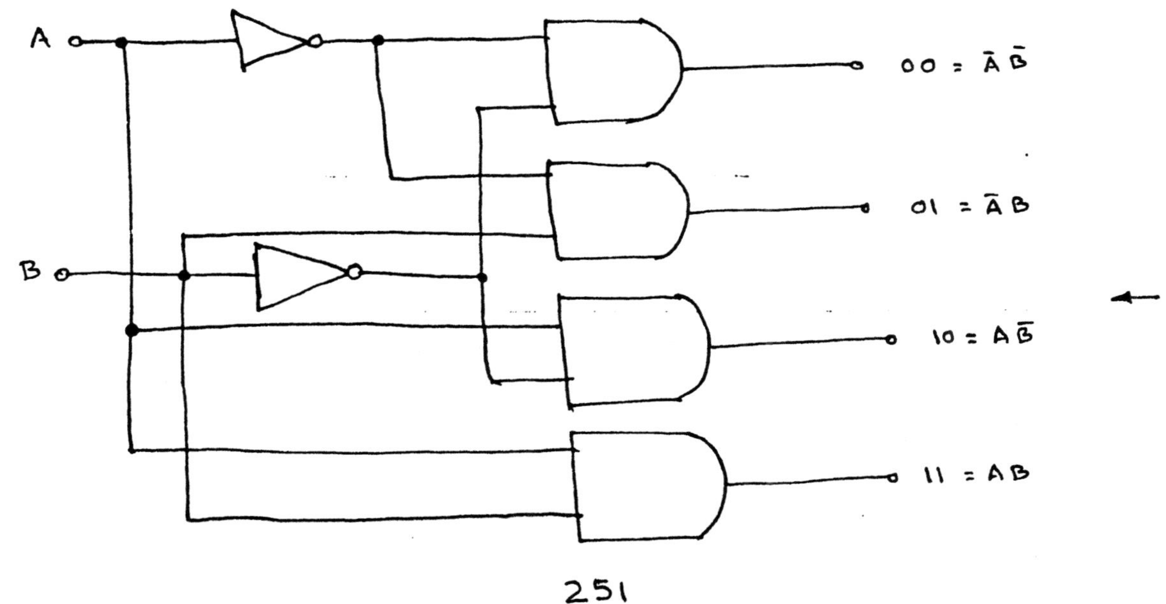

6.2.49.

(a)

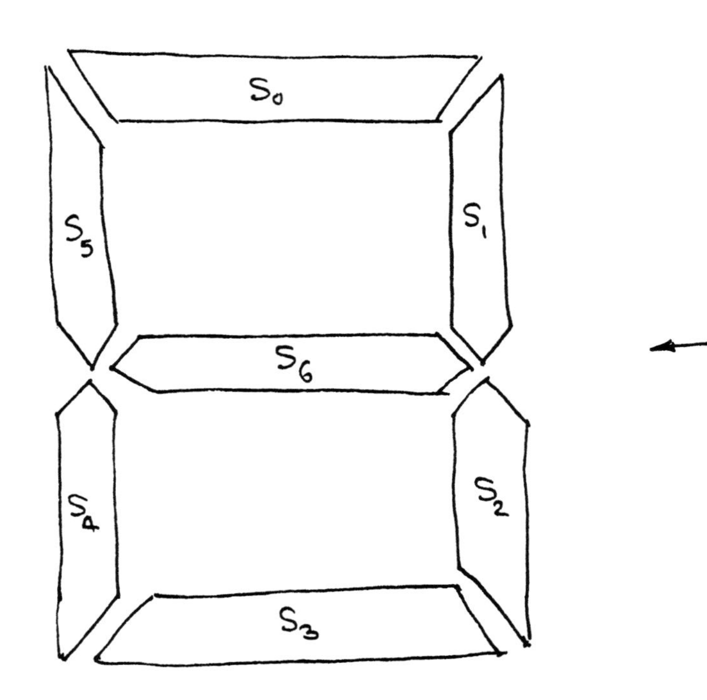

(b) NOTE: $S_1 = 0$, SEGMENT 1 IS LIT; AND WHEN $S_1 = 1$, SEGMENT 1 IS DARK; AND SO FORTH.

DECIMAL INTEGER	W	X	Y	Z	S_0	S_1	S_2	S_3	S_4	S_5	S_6
0	0	0	0	0	0	0	0	0	0	0	1
1	0	0	0	1	1	0	0	1	1	1	1
2	0	0	1	0	0	0	1	0	0	1	0
3	0	0	1	1	0	0	0	0	1	1	0
4	0	1	0	0	1	0	0	1	1	0	0
5	0	1	0	1	0	1	0	0	1	0	0
6	0	1	1	0	1	1	0	0	0	0	0
7	0	1	1	1	0	0	0	1	1	1	1
8	1	0	0	0	0	0	0	0	0	0	0
9	1	0	0	1	0	0	0	1	1	0	0

NOTE THAT, SINCE BINARY CODES ABOVE 1001 WILL NEVER OCCUR IN BCD, THE OUTPUTS FOR SUCH CODES SHOULD BE DON'T-CARE.

6.2.49. CONTD.

(C) S_0 IS EQUAL TO 1 IF WXYZ IS 0001, 0100, OR 0110.

A CIRCUIT TO REALIZE THIS IS SHOWN BELOW:

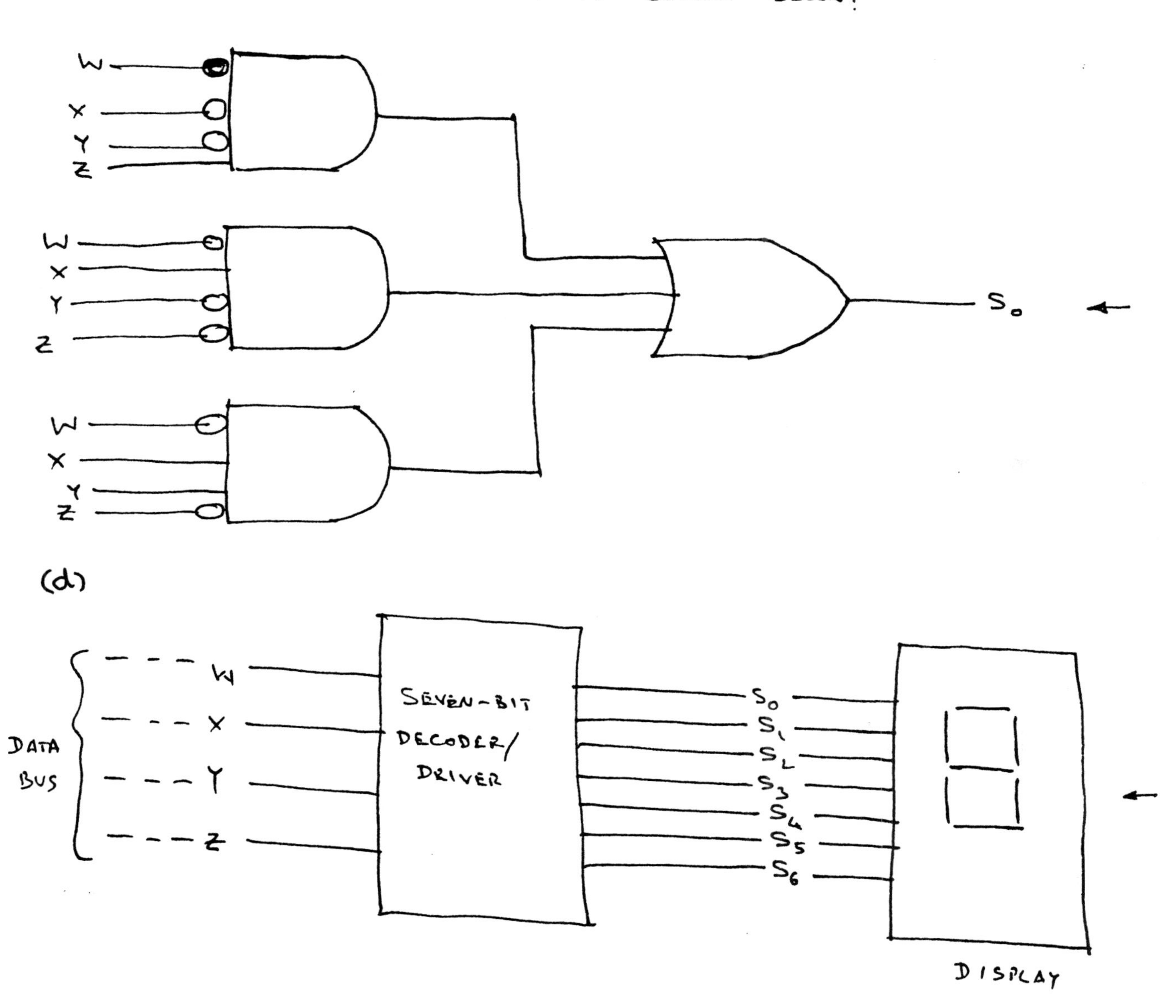

6.2.50.

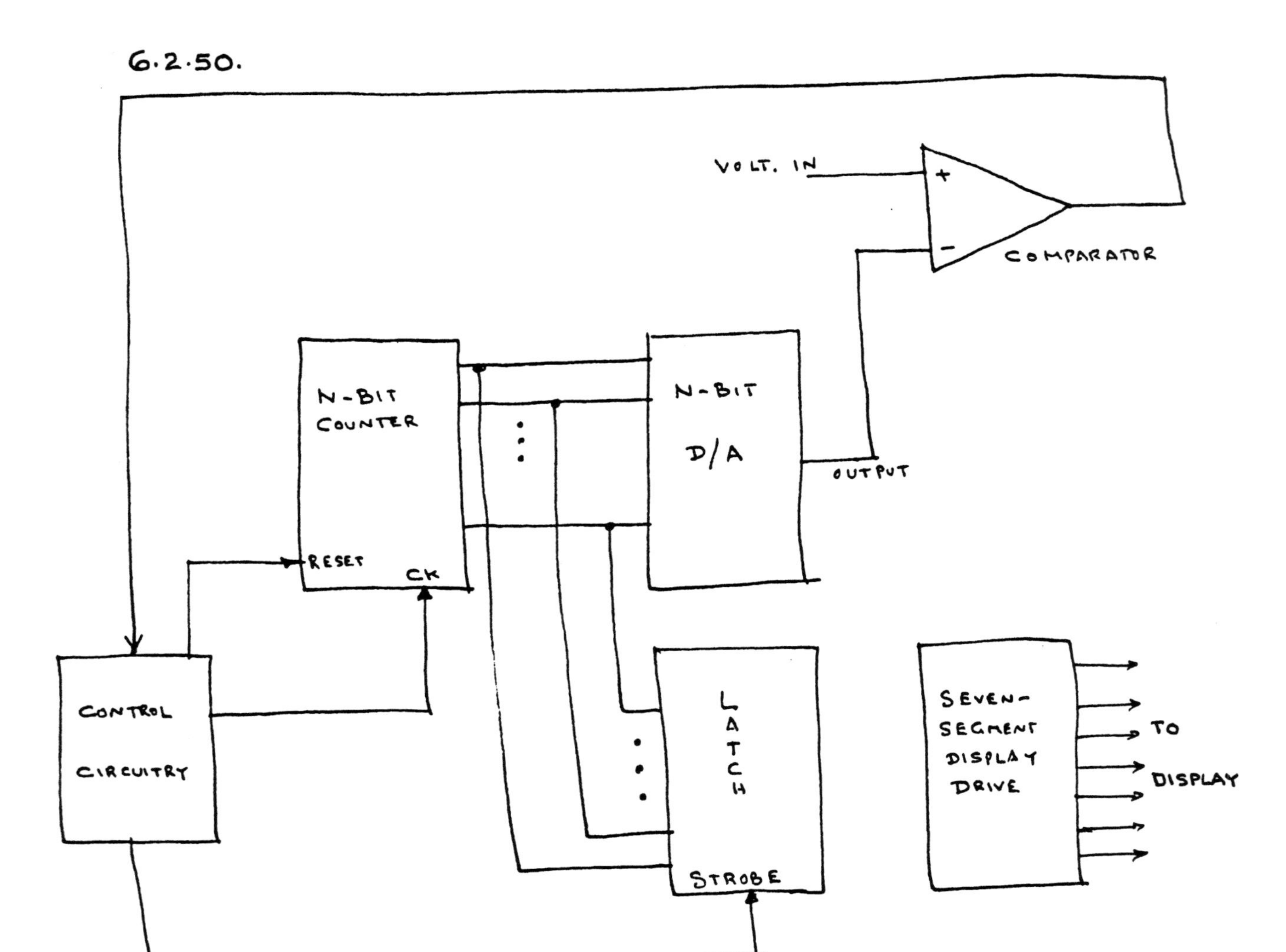

VOLT. IN
+
−
COMPARATOR
N-BIT
COUNTER
N-BIT
D/A
OUTPUT
RESET
CK
CONTROL
CIRCUITRY
LATCH
STROBE
SEVEN-
SEGMENT
DISPLAY
DRIVE
TO
DISPLAY

CHAPTER 7

7.2.1.

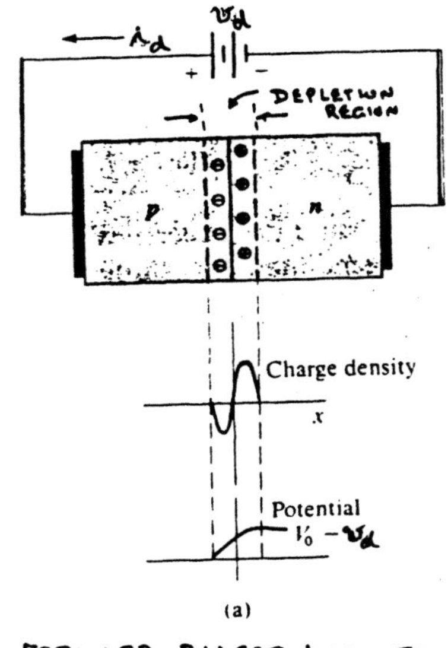

(a)

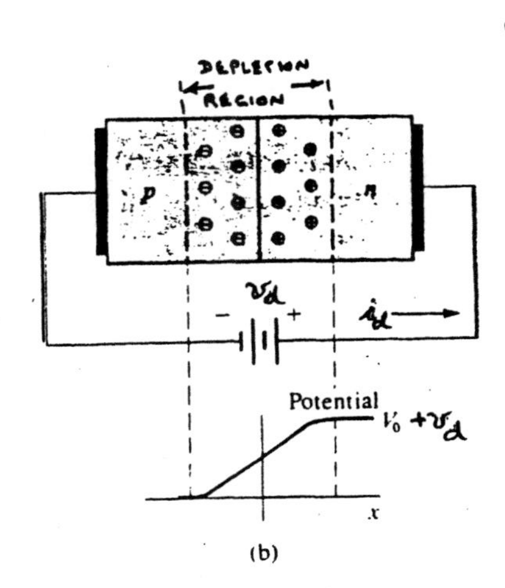

(b)

A FORWARD-BIASED pn-JUNCTION WITH AN EXTERNAL VOLTAGE v_d IS SHOWN IN FIG. (a). THE BARRIER VOLTAGE V_0 IS REDUCED BY THE APPLIED VOLTAGE v_d. THE EFFECT IS TO CAUSE A DIFFUSION CURRENT OF MAJORITY CARRIERS (HOLES FROM p TO n REGION AND ELECTRONS FROM n TO p REGION) TO FLOW. THIS CURRENT INCREASES EXPONENTIALLY WITH v_d AND CAN BE QUITE SIGNIFICANT FOR v_d ONLY SLIGHTLY ABOVE ABOUT 0.6V FOR SILICON.

FORWARD-BIASED pn-JUNCTION   REVERSE-BIASED pn-JUNCTION

THE OPPOSITE OCCURS WHEN THE pn-JUNCTION IS REVERSE-BIASED, AS SHOWN IN FIG. (b). THE BARRIER VOLTAGE IS RAISED; MAJORITY CARRIER CURRENT EXPONENTIALLY BECOMES NEGLIGIBLE QUICKLY; THE PRINCIPAL CURRENT IS DUE TO MINORITY CARRIERS (ELECTRONS GOING FROM p TO n REGION, AND HOLES FROM n TO p REGION). THIS DRIFT CURRENT QUICKLY SATURATES AND BECOMES CONSTANT AT A VERY LOW LEVEL, BECAUSE MINORITY CARRIERS ARE GENERATED ONLY THROUGH THERMAL AGITATION AND AT THE USUAL OPERATING TEMPERATURES ARE IN SMALL NUMBERS. THIS SMALL CURRENT IS KNOWN AS THE REVERSE SATURATION CURRENT I_S.

7.2.2.

V	I	V/I
−2	-10^{-13}	2×10^{13}
−0.5	-10^{-13}	5×10^{12}
0.3	1.0×10^{-8}	3×10^{7}
0.5	2.2×10^{-5}	2.3×10^{4}
0.7	0.05	14
1.0	5.1×10^{3}	2.0×10^{-4}
1.5	1.1×10^{12}	1.4×10^{-12}

NOTE:

1. V/I IS A FUNCTION OF APPLIED VOLTAGE.
2. ONE-WAY CONDUCTION PROPERTY OF THE DIODE IS ILLUSTRATED.
3. REVERSE VOLTAGES YIELD VERY SMALL REVERSE CURRENTS.
4. THE LAST TWO LINES OF THE TABLE ARE UNREALISTIC.

7.2.3.

$$10^{-5}\left(e^{40V}-1\right)=30\times10^{-3}$$

$$e^{40V}-1=3000 \quad \text{or} \quad V=(\ln 3001)/40=0.2\text{ V}$$

$\therefore$ SOURCE VOLTAGE $V_s = (30\times10^{-3}\times10^{3})+0.2 = 30.2$ V ←

$$10^{-5}\left(e^{40V}-1\right)=-8\times10^{-6}$$

$$e^{40V}-1=-8\times10^{-6}\times10^{5}=-0.8$$

$$V=(\ln 0.2)/40=-1.6/40=-0.0402\text{ V}$$

$\therefore$ SOURCE VOLTAGE $V_s = (-8\times10^{-6}\times10^{3})-0.0402$

$= -0.0482$ V ←

7.2.4.

FROM EQ. (7.2.1), WITH $V_T = 0.025$ V

$$I_s=\frac{10\times10^{-3}}{e^{0.5/(2\times0.025)}-1}=4.54\times10^{-7}=0.454\,\mu\text{A}$$ ←

7.2.5.

$$I_s=\frac{16\times10^{-6}}{(e^{40\times0.05}-1)}=2.5\,\mu\text{A}$$

DIFFUSION CURRENT $= 16+2.5 = 18.5\,\mu$A ←

7.2.6.

LOAD-LINE EQUATION IS GIVEN BY $I=\dfrac{5-V}{1000}$

THE INTERSECTION OF THE DEVICE CHARACTERISTIC WITH THE LOAD LINE YIELDS

$I = 4.5$ mA ←

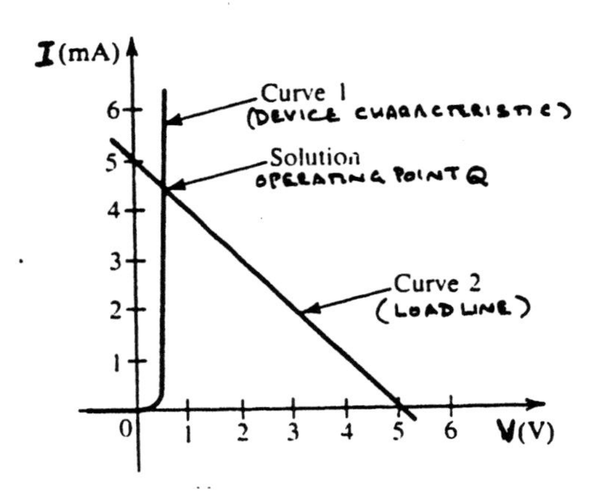

7.2.7.

FOR DETERMINING THE R_{TH} AS SEEN BY THE DIODE:

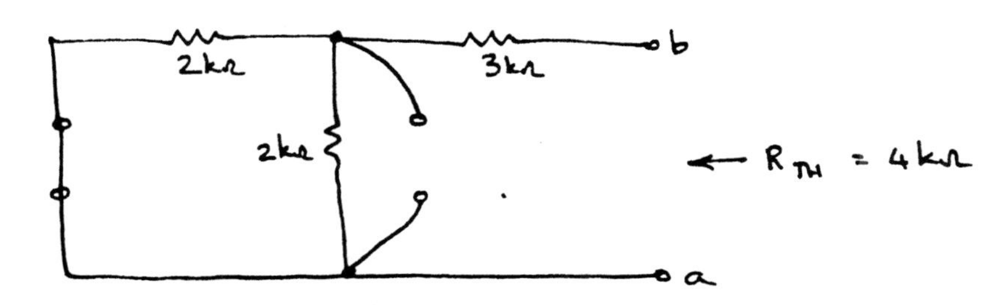

$\leftarrow R_{TH} = 4\,k\Omega$

FOR DETERMINING THE V_{OC} AT TERMINALS a AND b:

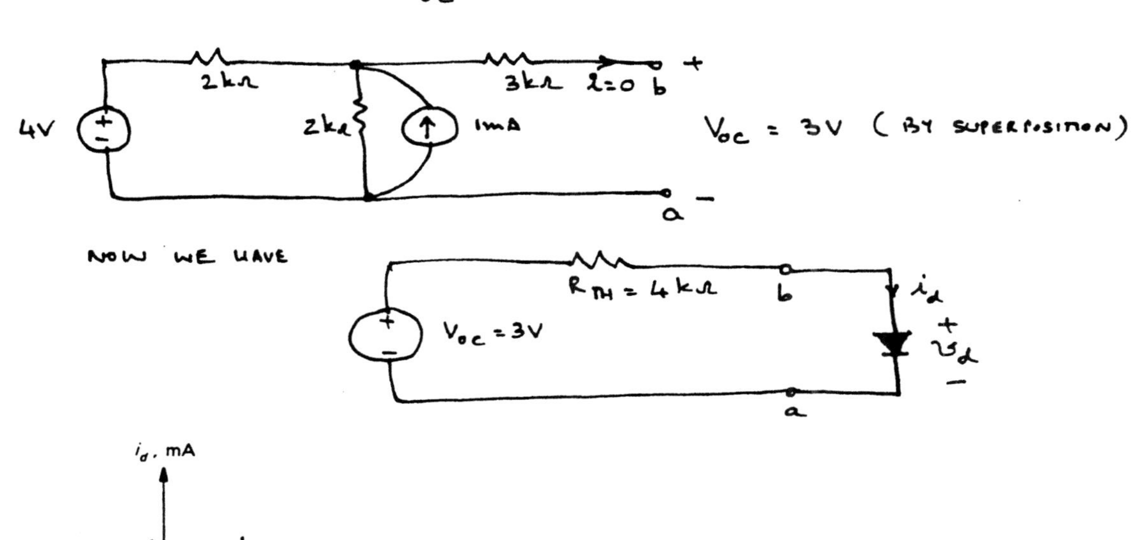

$V_{oc} = 3V$ (BY SUPERPOSITION)

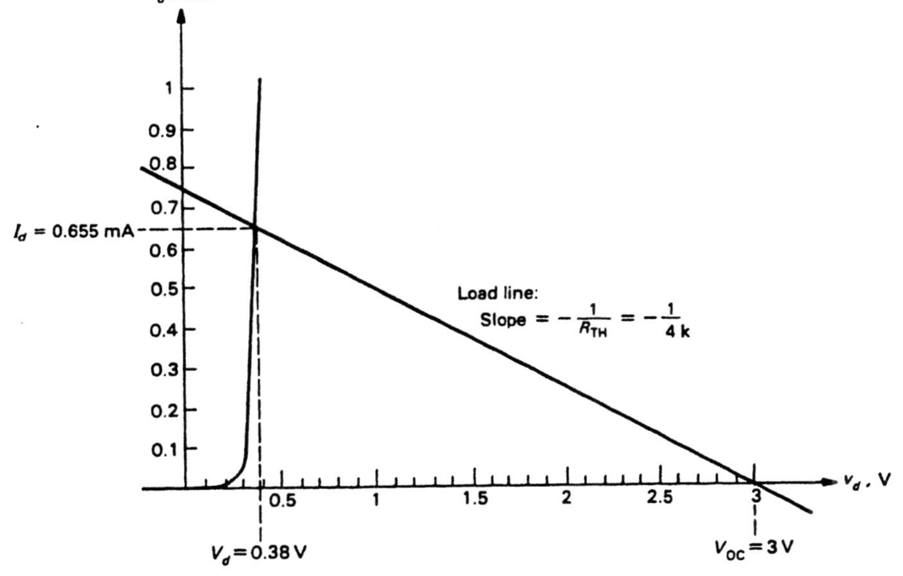

$V_d = 0.38\text{ V}\ ;\ I_d = 0.655\text{ mA}$

$$i = I_d - 0.001 + \frac{V_d + (3000)\,I_d}{2000}$$

$$= 0.8275\text{ mA} \leftarrow$$

7.2.8.

(a)

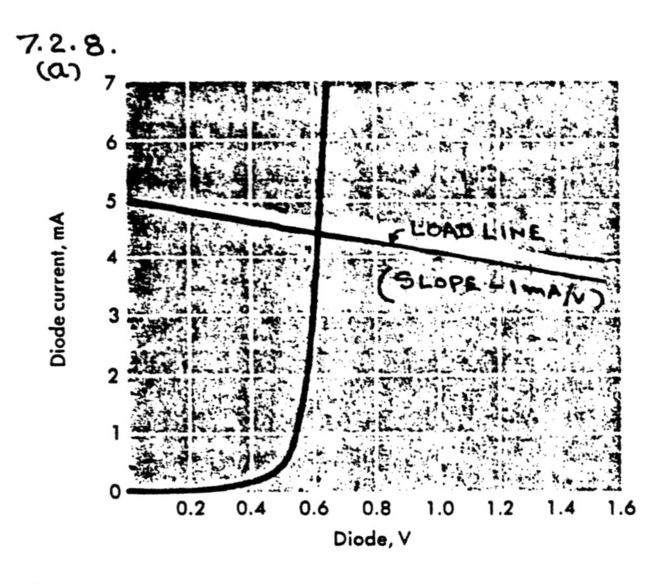

THE LOAD LINE HAS A SLOPE OF $-1/10^3$ S OR -1 mA/V.

THE y - INTERCEPT IS 5 mA.

OPERATING POINT : 4.5 mA ; 0.6 V ;

VOLTAGES ACROSS LOAD RESISTANCE $= 5 - 0.6 = 4.4\text{V}$ ←

(b) $P = I_{DQ} V_{DQ} = 4.5 \times 6 = 2.7$ mW ←

(c) LOAD-LINE SLOPE CHANGES TO -0.5 mA/V, -0.2 mA/V, -2 mA/V, AND -5 mA/V, RESPECTIVELY. CURRENT IS 4.75 mA, 4.9 mA, 3.75 mA, AND 2.1 mA, RESPECTIVELY.

7.2.9. (a)

LOAD LINE : y - INTERCEPT $10/2\text{ k}\Omega = 5$ mA ; SLOPE $= -5\text{mA}/10\text{V} = -0.5$ mA/V

OPERATING POINT : 4.7 mA ; 6.2 V ; VOLTAGE ACROSS $R_L = 10 - 6.2 = 3.8\text{V}$ ←

(b) $P_D = 6.2 \times 4.7 \times 10^{-3} = 29.1$ mW ←

(C) CONSTRUCTING LOAD LINES, CORRESPONDING CURRENTS ARE 1.0, 2.2, 7.2, AND 9.7 mA, RESPECTIVELY.

NOTE : LOAD-LINE SLOPE DOES NOT CHANGE AS R_L IS HELD CONSTANT. DIODE-CURVE MAY HAVE TO BE EXTENDED FOR 15 & 20 V CASES.

7.2.10.

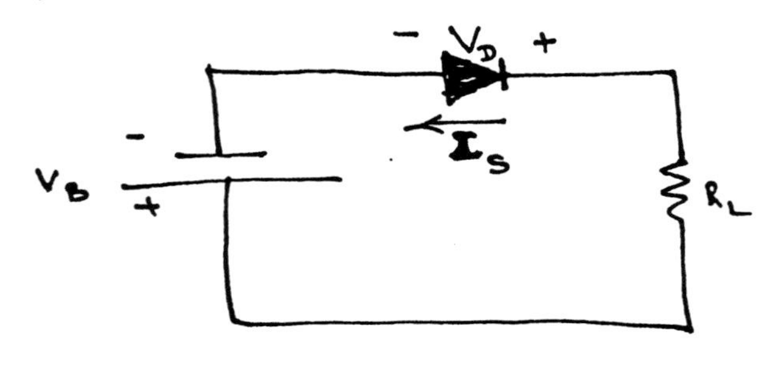

$V_B = I_S R_L + V_D$

WHEN THE DIODE IS REVERSE BIASED, THE USE OF LOAD LINE IS UNNECESSARY. THE CURRENT FLOWING IS JUST THE DIODE SATURATION CURRENT I_S. ←

7.2.11.

DIODE A IS FORWARD-BIASED AND DIODE B IS REVERSE-BIASED. SINCE THEY ARE IN SERIES, THE CURRENT IS THAT IN THE REVERSE-BIASED DIODE, WHICH IS 0.1 μA.

FOR THE FORWARD-BIASED DIODE $10^{-7} = 10^{-7}\left(e^{V_A/\eta V_T} - 1\right)$

OR $V_A = \eta V_T \ln 2 = 2 \times 25 \times 10^{-3} \ln 2 = 34.7$ mV ←

THE DROP ACROSS R_L IS $IR_L = 10^{-7} \times 10^5 = 10$ mV

AND THE VOLTAGE ACROSS THE REVERSE-BIASED DIODE B IS

$$-15 + (34.7 + 10)\,10^{-3} \simeq -14.955 \text{ V} \quad \leftarrow$$

7.2.12.

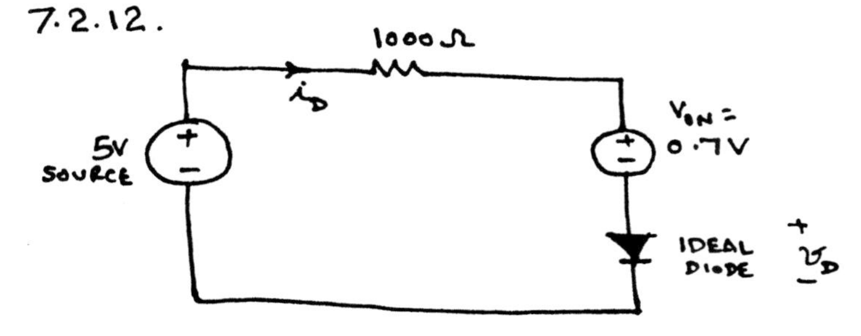

KVL: $-5 + 1000\, i_D + 0.7 + v_D = 0$

4.3 V MUST BE DROPPED ACROSS THE COMBINATION OF RESISTOR AND THE PERFECT DIODE.

THE DIODE IS OBVIOUSLY FORWARD-BIASED, THERE IS NO VOLTAGE DROP ACROSS AN IDEAL DIODE WITH FORWARD-BIAS.

∴ 4.3 V MUST BE DROPPED ACROSS THE RESISTOR.

THUS $i_D = \dfrac{4.3}{1000} = 4.3$ mA. ←

THE ANSWER IN PR. 7.2.6 IS 4.5 mA.

7.2.13.

WITH $V_{ON} = 0.7$V, $i_1 = \dfrac{10 - 0.7}{300} \simeq 31$ mA ←

$i_2 = \dfrac{0.7}{500} \simeq 1.4$ mA ←

THE DIODE IS FORWARD-BIASED AND IT IS CARRYING A CURRENT OF (31 − 1.4) = 29.6 mA ←

7.2.14.

W.R.T. THE DIODE TERMINALS, REPLACE THE CIRCUIT BY ITS THEVENIN'S EQUIVALENT.

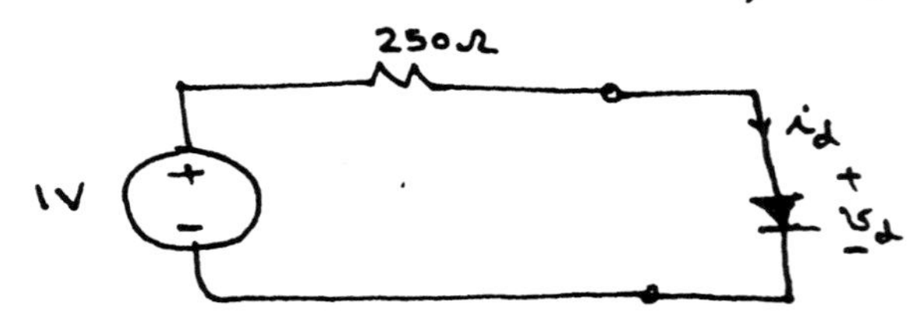

$$v_d = 1 - 250\, i_d \quad \text{or} \quad i_d = \frac{1 - v_d}{250}$$

LOAD LINE IS DRAWN ON DIODE'S v-i CURVE : STRAIGHT LINE JOINING $v_d = 1$ V ON X-AXIS AND $i_d = 1/250 = 4$ mA ON Y-AXIS.

THE INTERSECTION YIELDS $V_Q = 0.55$ V AND $i_Q = 1.8$ mA ←

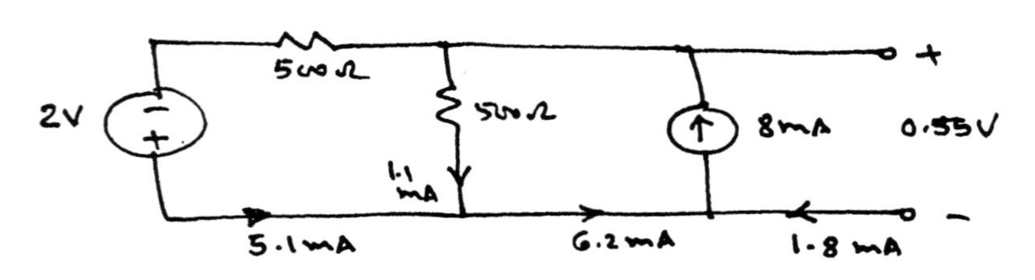

$$P_{\text{VOLTAGE-SOURCE}} = 2 \times 5.1 \text{ mA} = 10.2 \text{ mW} \leftarrow$$

7.2.15.

$I_d = 2$mA

BY DRAWING A TANGENT TO THE v-i CURVE AT THE OPERATING POINT

$V_{ON} = 0.55$V ←

$$V_{ON} = V_d - R_f I_d$$

$$R_f = \frac{-(V_{ON} - V_d)}{I_d} = \frac{0.05}{2\text{mA}} = 25\,\Omega \leftarrow$$

R_f, $V_D = 0.6$V, V_{ON}

7.2.16.

DIODE IS FORWARD-BIASED.

9V, 600Ω, 300Ω ⟹ 3V, 200Ω

(a)

3V, 200Ω, 100Ω

$$I = \frac{3}{200 + 100} = 10 \text{ mA} \leftarrow$$

7.2.16. CONTD.

(b)

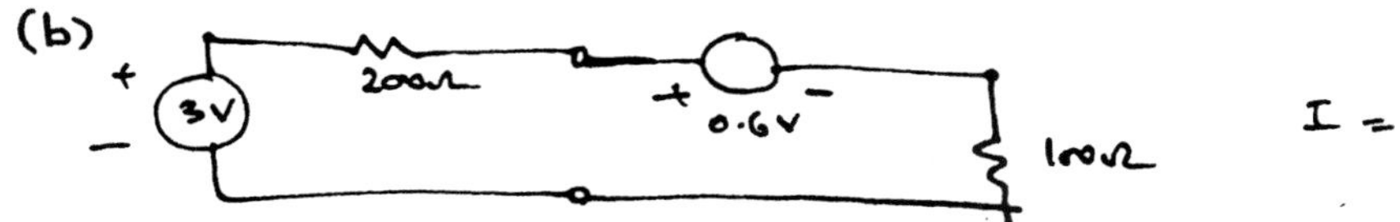

$$I = \frac{3-0.6}{200+100} = 8\text{ mA}$$

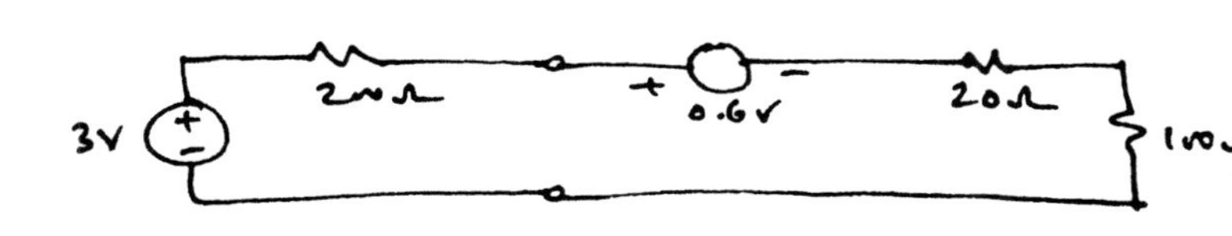

$$I = \frac{3-0.6}{200+100+20} = 7.5\text{ mA}$$

7.2.17. DIODE CONDUCTS WHEN $v(t) \geq 6V$ OR FOR $6 \leq t \leq 10ms$.

IN THIS INTERVAL

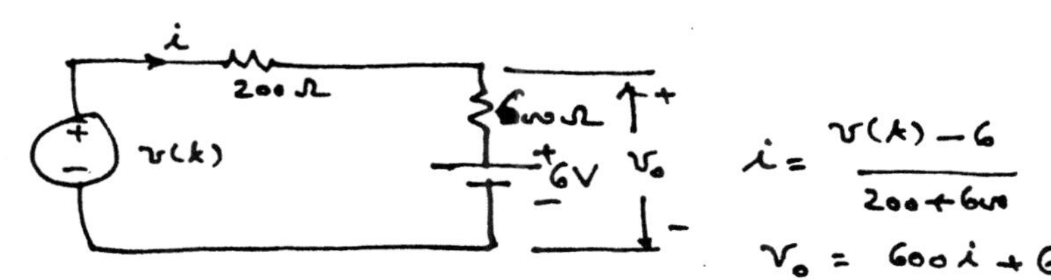

$$i = \frac{v(t)-6}{200+600}$$

$$v_o = 600i + 6$$

$$v_o(t) = \frac{3}{4}v(t) - 4.5 + 6 = \frac{3}{4}v(t) + 1.5$$

WHEN $v(t) = 10V$, $v_o(t) = 9V$

WHEN THE DIODE IS OPENCIRCUITED, $v_o = 6V$

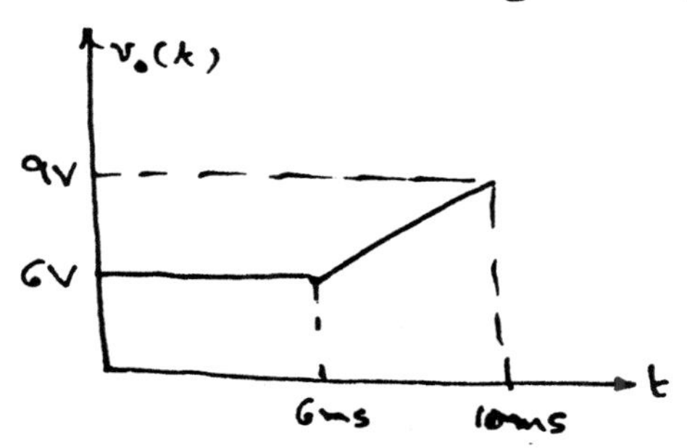

7.2.18.

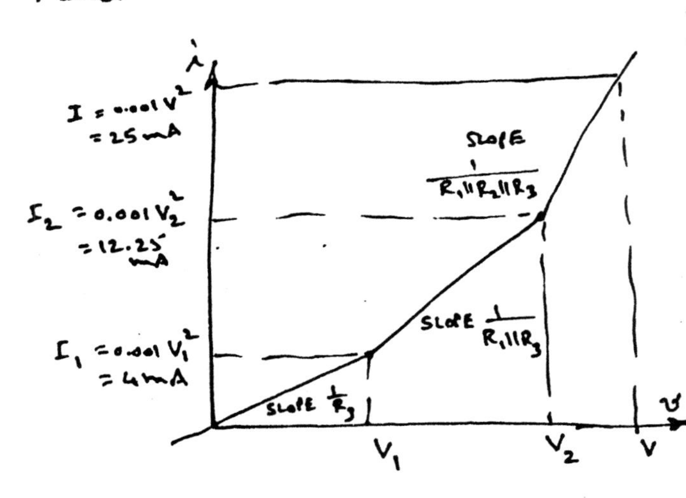

THE RESISTORS ARE SO CHOSEN THAT THE BREAK POINTS AT V_1 AND V_2 AND THE POINT AT 5V FALL ON THE DESIRED IDEAL CURVE.

$$R_3 = V_1/I_1 = 500\,\Omega$$

$$R_1 \| R_3 = \frac{V_2-V_1}{I_2-I_1} = 182\,\Omega \Rightarrow R_1 = 286\,\Omega$$

$$R_1 \| R_2 \| R_3 = \frac{V-V_2}{I-I_2} = 118\,\Omega \Rightarrow R_2 = 333\,\Omega$$

REVISED VALUES: $R_1 = 276\,\Omega$; $R_2 = 323\,\Omega$;

$R_3 = 500\,\Omega$; $V_1 = 1.5V$; $V_2 = 3V$

7.2.19.

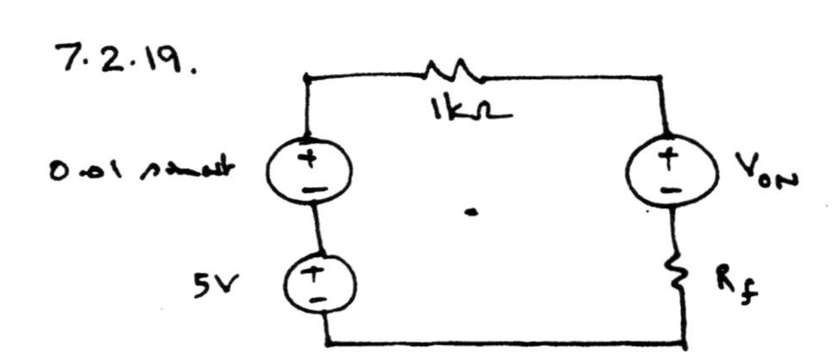

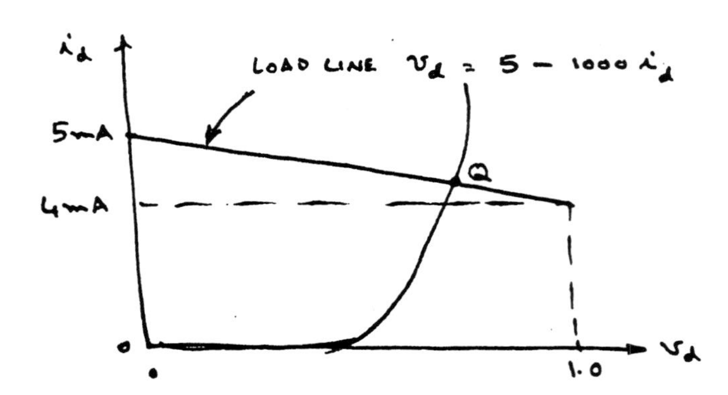

$v_Q = 0.61\text{V}$; $i_Q = 4.3\text{ mA}$

$$i_d = \frac{0.01}{1000 + R_f} \sin\omega t + \frac{5 - V_{ON}}{1000 + R_f}$$

$$R_f = 1/\text{SLOPE AT OPERATING POINT Q} = \frac{50}{i_Q\,(\text{mA})} = \frac{50}{4.3} = 11.63\,\Omega$$

$$V_{ON} = v_d - R_f\, i_d = 0.61 - 0.05 = 0.56\text{V}$$

$$\therefore i_d = \frac{0.01}{1000 + 11.63} \sin\omega t + \frac{5 - 0.56}{1000 + 11.63} = (9.885 \times 10^{-6} \sin\omega t + 4.389)\text{ mA} \leftarrow$$

7.2.20. (a) IN ORDER TO OPERATE THE ZENER IN ITS REVERSE BREAKDOWN REGION, THE CURRENT THROUGH IT MUST ALWAYS LIE BETWEEN (i) $i_d = 0$ AND (ii) $i_d = -P_{max}/V_Z$

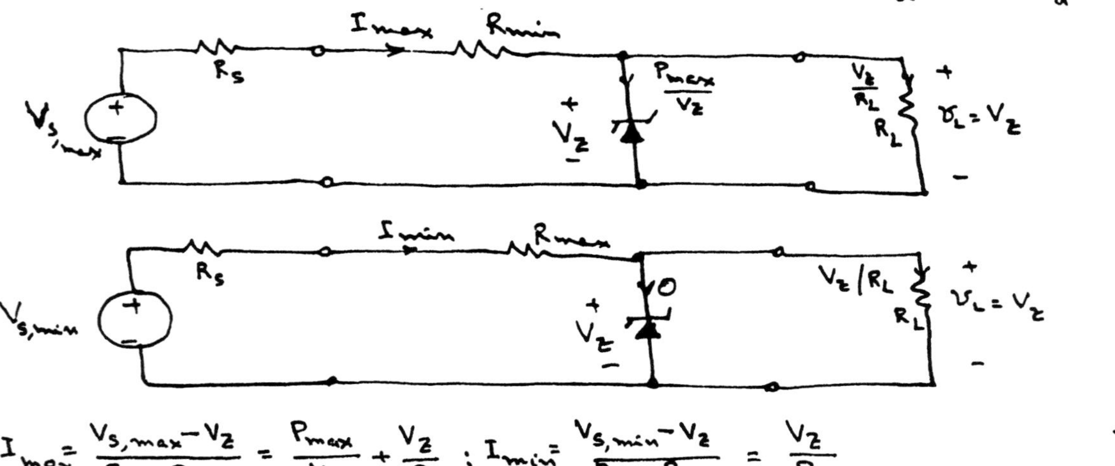

$$I_{max} = \frac{V_{s,max} - V_Z}{R_s + R_{min}} = \frac{P_{max}}{V_Z} + \frac{V_Z}{R_L}; \quad I_{min} = \frac{V_{s,min} - V_Z}{R_s + R_{max}} = \frac{V_Z}{R_L} \leftarrow$$

(b)

SELECTING A ZENER VOLTAGE OF 60V, $R_{max} = \dfrac{V_{s,min} - V_Z}{I_{min}} = \dfrac{75 - 60}{0.06} = 250\,\Omega$

$\left(\because I_{min} = \dfrac{V_Z}{R_L} = \dfrac{60}{1000} = 0.06\right)$

POWER RATING OF THE ZENER IS DETERMINED

WHEN $V_S = V_{S,max}$ AND THE ZENER DRAWS THE MAXIMUM CURRENT OF P_{max}/V_Z:

$$\frac{P_{max}}{V_Z} = \frac{V_{s,max} - V_Z}{R} - \frac{V_Z}{R_L} = \frac{120 - 60}{250} - \frac{60}{1000} = 0.24 - 0.06 = 0.18\text{A}$$

SO THAT $P_{max} = 0.18 \times 60 = 10.8\text{ W} \leftarrow$

7.2.21.

$R_s = 100\,\Omega$

$40 \le V_s \le 60$, I, R, $V_Z = 30\text{V} = V_L$, $R_L = 1000\,\Omega$, $30/1000$

AT $V_S = 40\text{V}$, $I_{min} = \dfrac{40-30}{100+R} = \dfrac{30}{1000}$ $(\because i_d = 0)$

$= 30\text{ mA}$

$100 + R = \dfrac{40-30}{30\text{ mA}} \Rightarrow R_{max} = 233\,\Omega \leftarrow$

AT $V_S = 60\text{V}$, $I_{max} = \dfrac{60-30}{100+R}$

WITH $R = 233\,\Omega$

$I_{max} = 90\text{ mA}$

$\therefore I_d = 90 - 30 = 60\text{ mA}$

$\therefore P_{ZENER} = 30\text{V} \times 60\text{ mA} = 1.8\text{ W} \leftarrow$

7.2.22.

$V_{S,max} = 30\text{V}$; $V_{S,min} = 20\text{V}$

THE LARGEST DIODE CURRENT OCCURS WHEN THE LOAD IS OPENCIRCUITED AND THE VOLTAGE IS MAXIMUM; $I = I_Z = \dfrac{V_{Smax} - V_Z}{R} \le I_{Zmax}$ WHERE I_Z IS THE CURRENT THROUGH THE ZENER.

$\therefore R \ge \dfrac{V_{Smax} - V_Z}{I_{Zmax}}$

IN OUR CASE $R \ge \dfrac{30-12}{20 \times 10^{-3}} = 900\,\Omega \leftarrow$

SMALLEST DIODE CURRENT OCCURS FOR THE LARGEST LOAD CURRENT AND SMALLEST SOURCE VOLTAGE; $I_Z = I - I_L = \dfrac{V_{Smin} - V_Z}{R} - \dfrac{V_Z}{R_{Lmin}} \ge I_{Zmin}$

$\therefore R_{Lmin} \ge \dfrac{V_Z R}{V_{Smin} - V_Z - R I_{Zmin}}$

IN OUR CASE $R_{Lmin} \ge \dfrac{12(900)}{20 - 12 - (900 \times 10^{-3})} = 1521\,\Omega \leftarrow$

IN PRACTICE, ONE MAY CHOOSE THE NEAREST RESISTOR OF $910\,\Omega$, CORRESPONDING TO WHICH $R_{Lmin} \ge 1540\,\Omega$.

7.2.23.

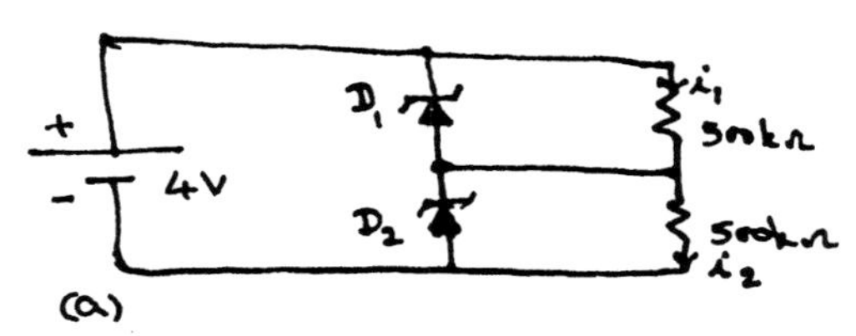

(a)

ASSUMING D_1 AND D_2 TO BE REVERSE BIASED BUT NEITHER IN BREAKDOWN

$$\left.\begin{aligned} V_s = 4 &= i_1 \times \frac{10^6}{2} + i_2 \times \frac{10^6}{2} \\ 2(10^{-6}) + i_1 &= 4(10^{-6}) + i_2 \end{aligned}\right\} \Rightarrow$$

$i_1 = 5\,\mu A$; $i_2 = 3\,\mu A$

$v_1 = \frac{1}{2} 10^6 (5)(10^{-6}) = 2.5V$

$v_2 = \frac{1}{2} 10^6 (3) 10^{-6} = 1.5V$

(b)

ASSUMING AGAIN THAT BOTH DIODES ARE NOT IN BREAKDOWN

$i_1 = 9\,\mu A$; $i_2 = 7\,\mu A$

$v_1 = 4.5V$; $v_2 = 3.5V$ (BOTH OF WHICH ARE LESS THAN 5V; SO NO BREAKDOWN OCCURS.)

7.2.24.

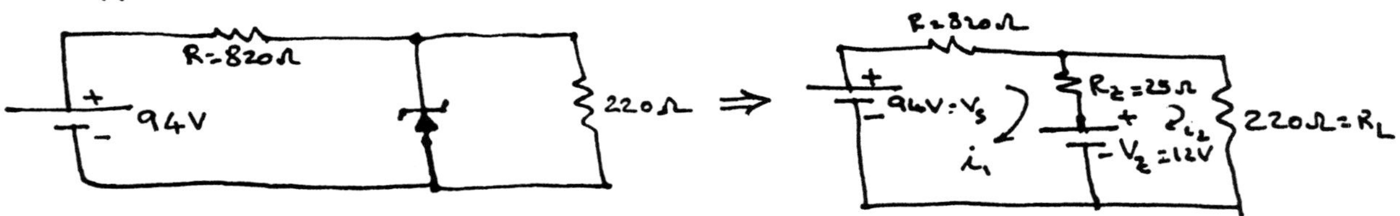

$V_S = i_1(R+R_Z) - i_2 R_Z + V_Z$; $V_Z = -i_1 R_Z + i_2(R_L+R_Z) \Rightarrow i_1 = \dfrac{i_2(R_L+R_Z) - V_Z}{R_Z}$

$V_S = \dfrac{R+R_Z}{R_Z}\left[i_2(R_L+R_Z) - V_Z\right] - i_2 R_Z + V_Z = \left[\dfrac{(R+R_Z)(R_L+R_Z)}{R_Z} - R_Z\right] i_2 - \dfrac{R}{R_Z} V_Z$

$\therefore i_2 = \dfrac{V_S R_Z + R V_Z}{(R+R_Z)(R_L+R_Z) - R_Z^2} = \dfrac{(94\times25)+(820\times12)}{(845\times245) - 25^2} = \dfrac{12{,}190}{206{,}400} = 0.05906\,A$

$V_L = i_2 R_L = 0.05906(220) = 12.99322V$; $P_L = i_L^2 R_L = 0.76738\,W$

(b) $P_R = (94 - 12.99322)^2 / 820 = 8.00256\,W$

$P_Z = \left[\left(\dfrac{94-12.99322}{820}\right) - 0.05906\right] 12.99322 = 0.51620\,W$

7.2.25.

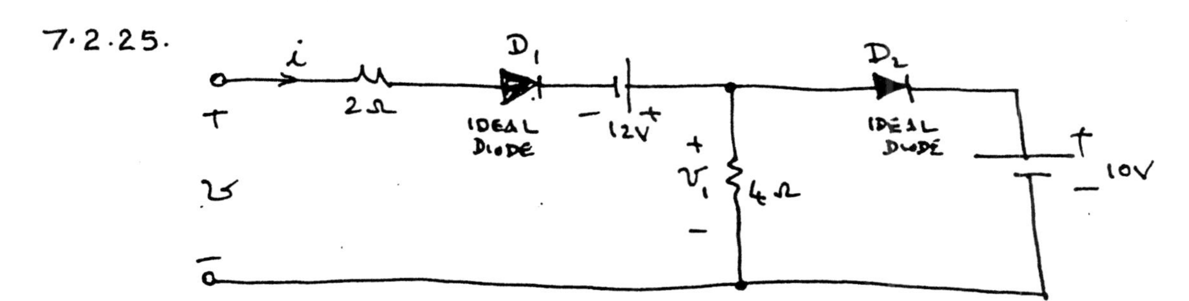

FOR $v \to \infty$, D_1 AND D_2 WILL BE FORWARD BIASED.

$v = 2i - 12 + 10$ OR $2i = v + 2$ OR $i = 0.5v + 1$; FOR $v \to -\infty$, $i = 0$

D_1 AT BREAKPOINT

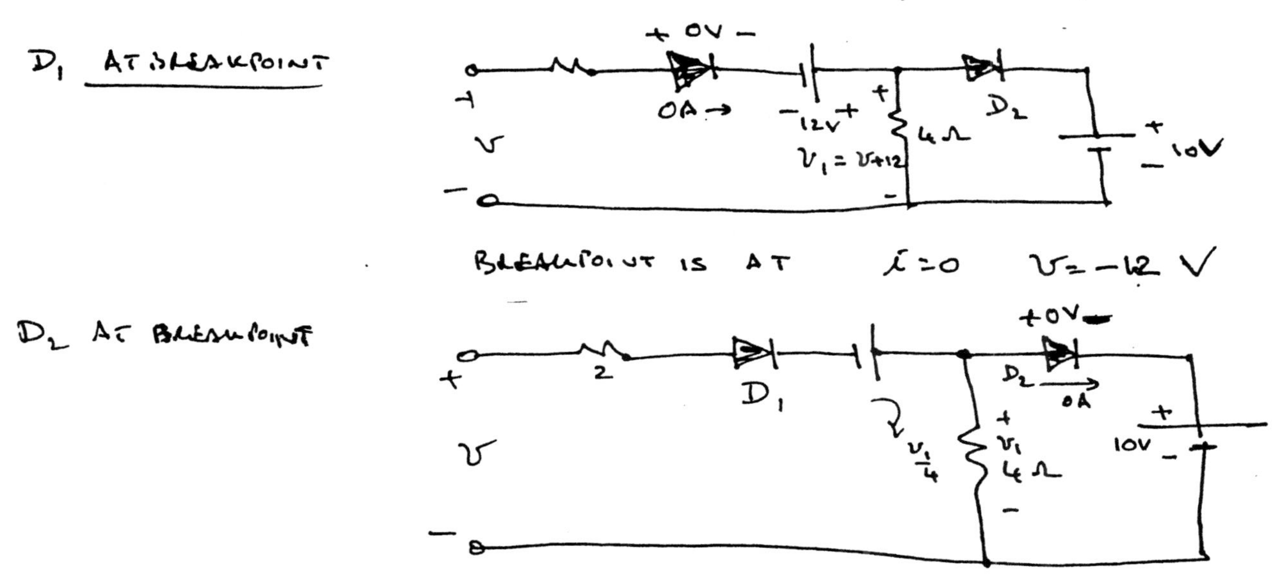

$v = 2i - 12 + v_1$; $v_1 = 10$V ; $i = \frac{10}{4} = 2.5$A ; WITH $i = 2.5$A, $v = 5 - 12 + 10 = 3$V

BREAKPOINT IS AT $i = 2.5$A , $v = 3$V.

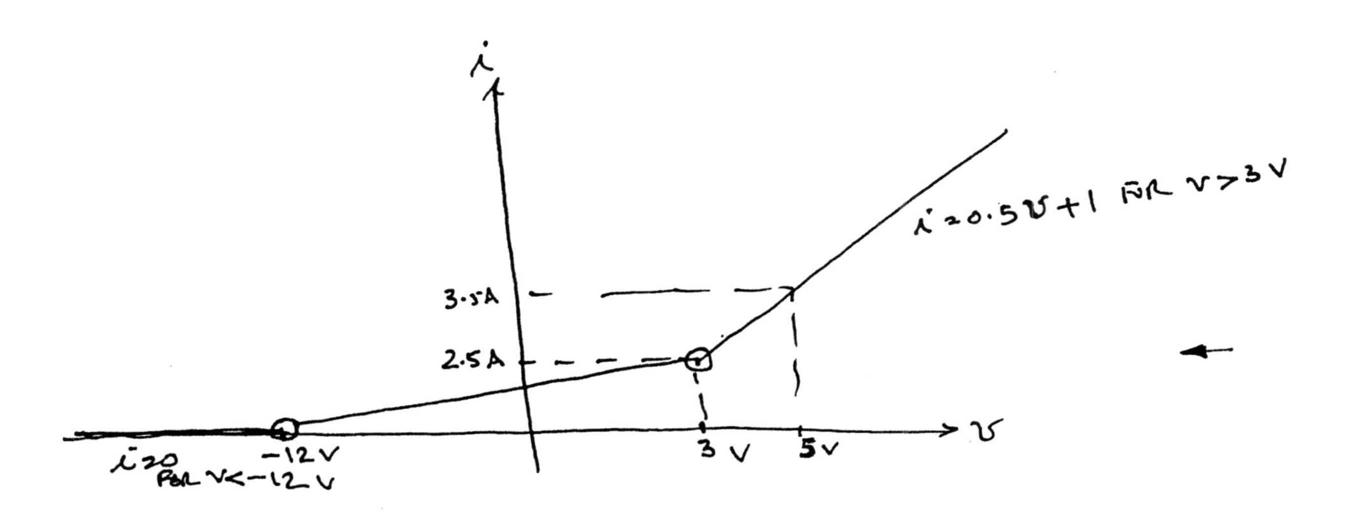

7.2.26. $v_L(t) = V_L + a_1 \sin\omega t + a_2 \sin 2\omega t + \cdots$
$+ b_1 \cos\omega t + b_2 \cos 2\omega t + \cdots$

$$V_L = \frac{1}{2\pi}\int_0^{\pi} V_s \sin\omega t \, d(\omega t) = \frac{V_s}{\pi}$$ WHICH IS THE AVERAGE DC VALUE

$$a_n = \frac{2}{T}\int_0^T v_L(t) \sin n\omega t \, dt$$
$$b_n = \frac{2}{T}\int_0^T v_L(t) \cos n\omega t \, dt$$

$$a_1 = \frac{2}{T}\int_0^T v_L(t) \sin\omega t \, dt = \frac{1}{\pi}\int_0^{\pi} V_s \sin\omega t \sin\omega t \, d(\omega t)$$
$$= \frac{V_s}{\pi}\int_0^{\pi} \frac{1}{2}(1-\cos 2\omega t)\, d(\omega t) = \frac{V_s}{2}$$

$$b_1 = 0 \; ; \; b_2 = -\frac{2V_s}{3\pi} \; ; \; b_3 = 0 \; ; \; b_4 = -\frac{2V_s}{15\pi}, \; b_5 = 0$$

for $n > 1$, $a_n = 0$ $\because$ $a_n = \frac{2}{T}\int_0^T v_L(t) \sin n\omega t \, dt$
$$= \frac{1}{\pi}\int_0^{\pi} V_s \sin\omega t \sin n\omega t \, d(\omega t)$$
$$= \frac{V_s}{\pi}\int_0^{\pi} \frac{1}{2}\left[\cos(n-1)\omega t - \cos(n+1)\omega t\right] d(\omega t)$$
$$= 0$$

$$\therefore \; v_L(t) = \frac{V_s}{\pi} + \frac{V_s}{2}\sin\omega t - \frac{2V_s}{3\pi}\cos 2\omega t - \frac{2V_s}{15\pi}\cos 4\omega t + \cdots$$

7.2.27.

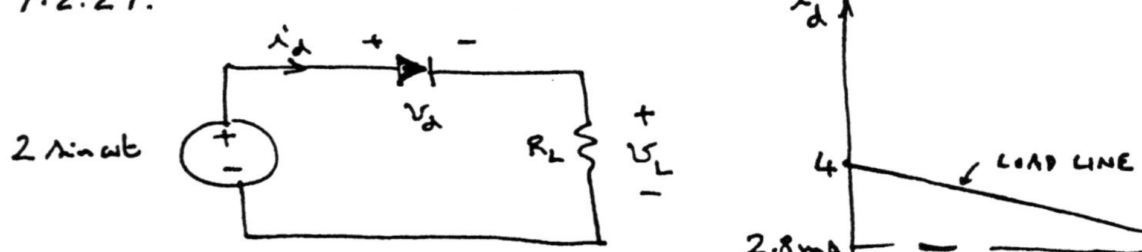

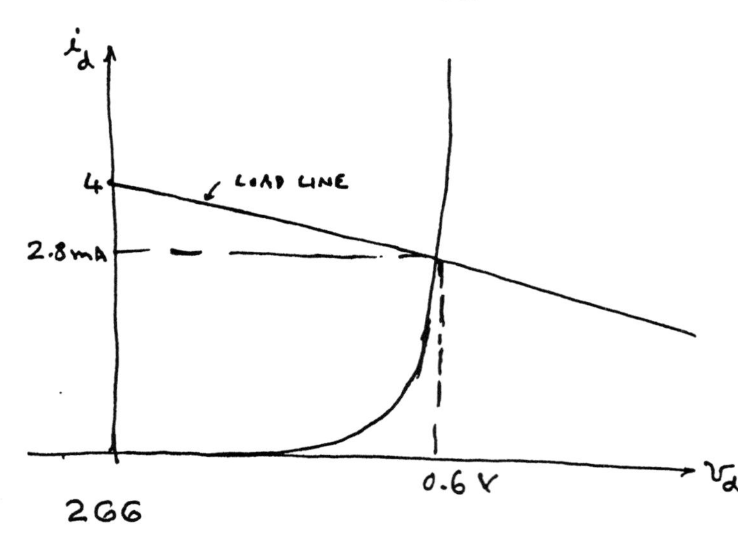

$V_s = 2\,V$; $R_L = 500\,\Omega$

$v_d = 2\sin\omega t - R_L i_d$

$V_{L\,max} = 2.8 \times 500 = 1.4\,V$

7.2.27. CONTD.

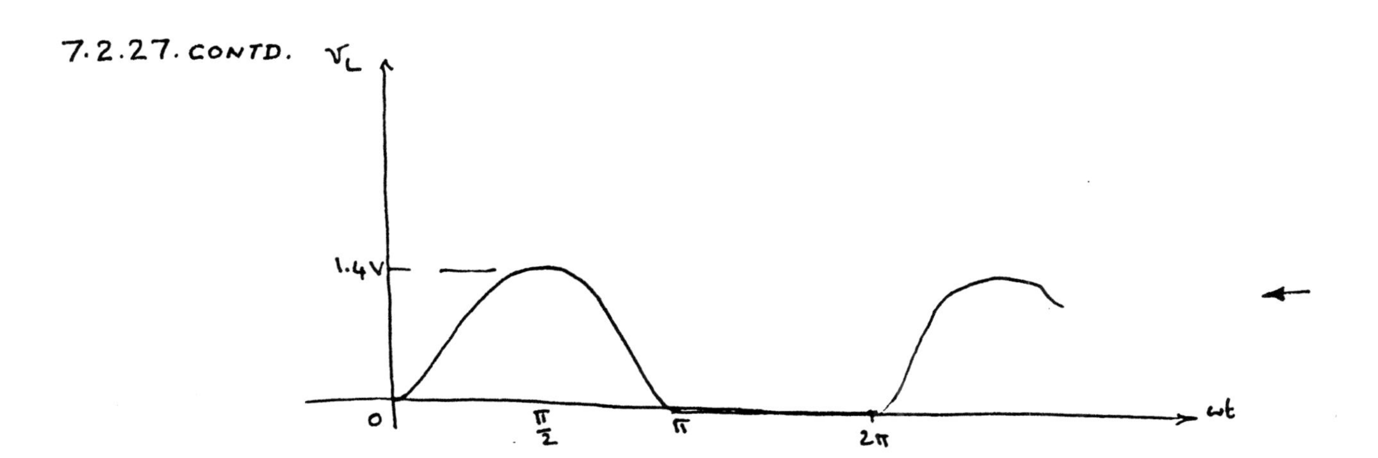

7.2.28.

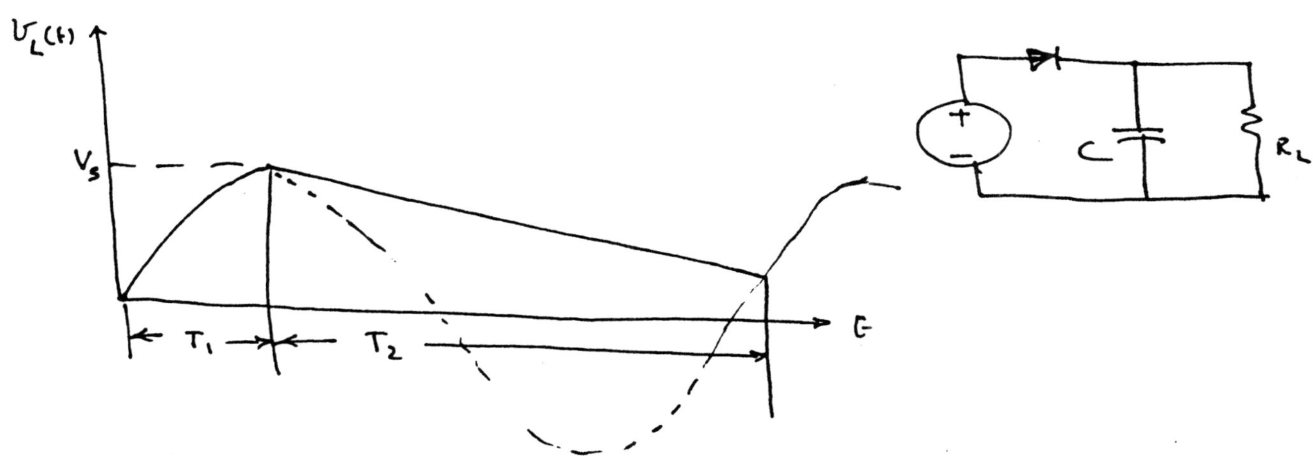

$V_s = 10\text{V}$; $f = 1\text{kHz}$; $C = 10\mu\text{f}$; $R_L = 1\text{k}\Omega$

$$v_L(t) = A\, e^{-(t-T_1)/R_L C} \qquad T_1 \le t \le T_1 + T_2$$

AT $t = T_1$, $v_L = V_s$

$$\therefore \quad v_L(t) = V_s\, e^{-(t-T_1)/R_L C}$$

ASSUMING $T_1 < T_2$ WHICH MEANS $R_L C >> T$ WHERE $T = 1/f = 10^{-3}$

$$V_s\, e^{-T_2/R_L C} \simeq V_s\, e^{T/R_L C} \simeq V_s \left(1 - \frac{T}{R_L C}\right)$$

$$v_{L\,min} \simeq V_s \left(1 - \frac{T}{R_L C}\right) \simeq 10\left(1 - \frac{10^{-3}}{10^{-2}}\right)$$

$$= 90 \leftarrow$$

7.2.29.

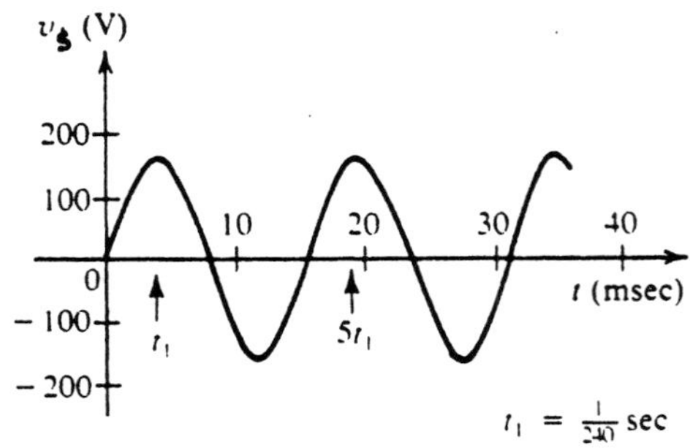

(a)

v_L, C, R, i_0

for $t_1 \le t \le t_c$

$R = 1000\ \Omega$

$C = 50 \times 10^{-6}$ F

(b)

v_L(V)

$165\, e^{-(t-t_1)/R_L C}$

t_1, t_C, t (msec)

(c)

v_L(V)

t (msec)

(d)

THE LOAD CURRENT HAS THE SAME FORM AS THE OUTPUT VOLTAGE

SINCE $i_L(t) = v_L(t)/R_L$.

7.2.30.

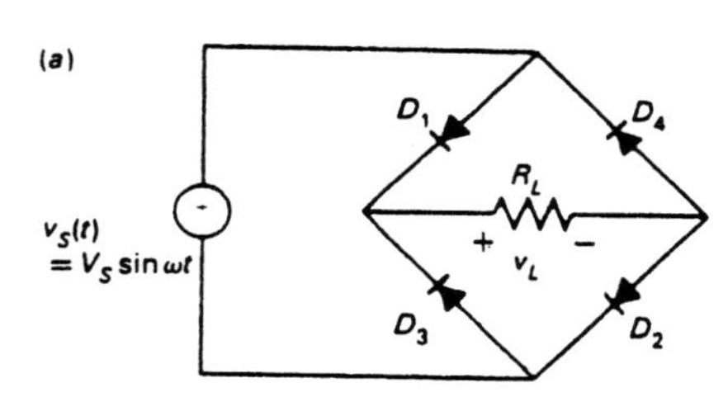

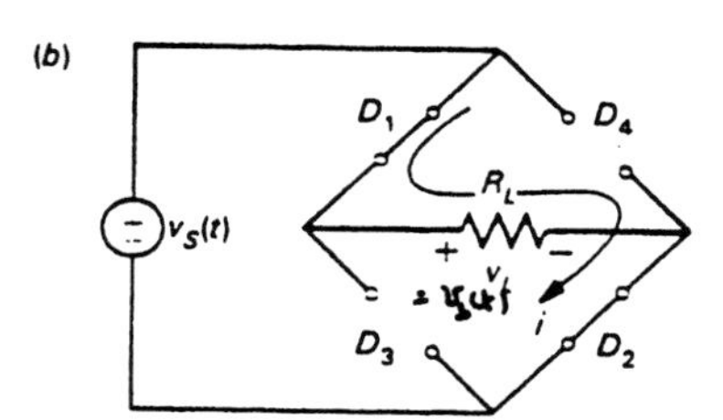

$v_S(t) > 0$

D_1 & D_2 FORWARD BIASED AND CLOSED

D_3 & D_4 REVERSE BIASED AND OPEN

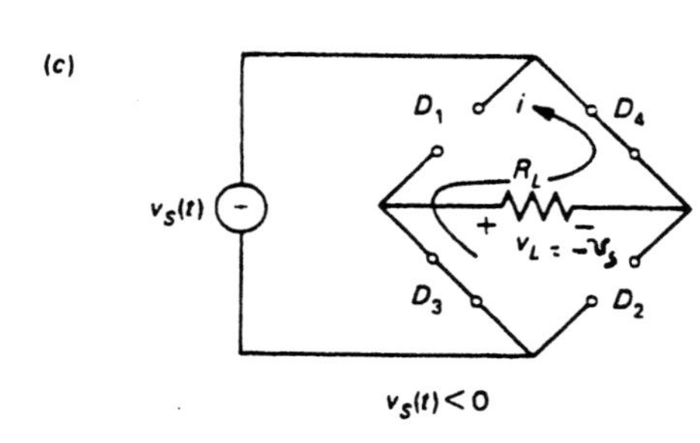

$v_S(t) < 0$

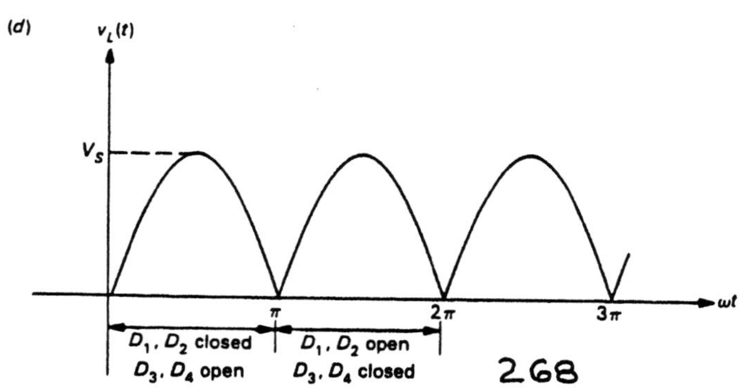

7.2.31.

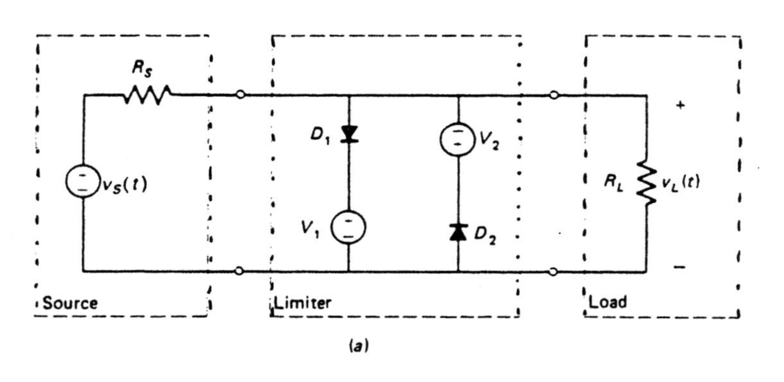

(a)

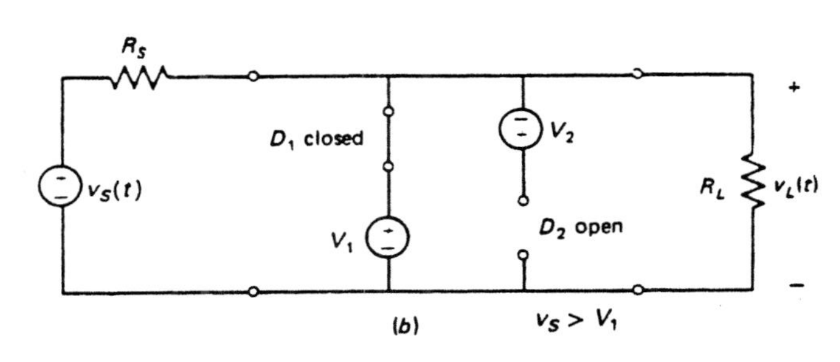

(b)

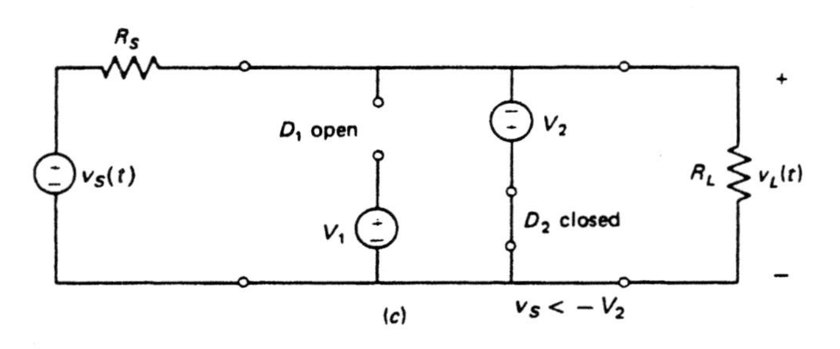

(c)

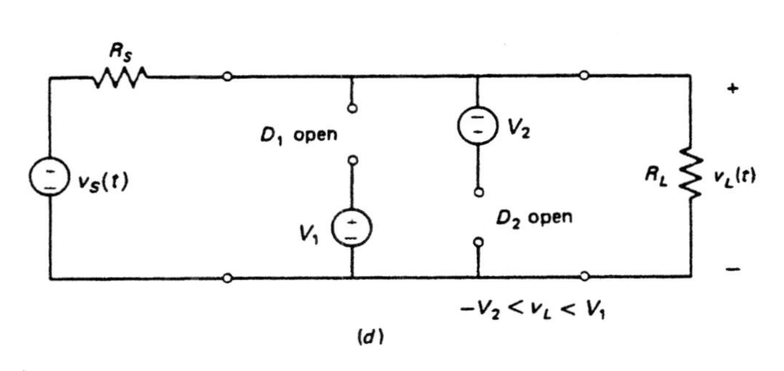

(d)

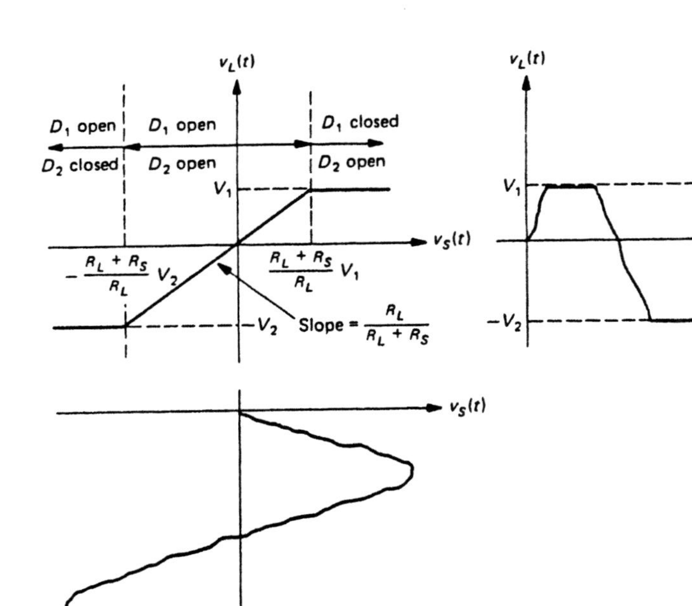

(e)

$$-\frac{R_L + R_S}{R_L} V_2 < v_s(t) < \frac{R_L + R_S}{R_L} V_1$$

7.3.1. $i_B = 25\mu A$; $\alpha = 0.985$; $I_{CBO} \simeq 0$

$$\beta = \frac{\alpha}{1-\alpha} = \frac{0.985}{0.015} = 65.67 \leftarrow$$

$$i_E = \frac{i_B}{1-\alpha} = \frac{25}{0.015}\,\mu A = 1666.67\,\mu A = 1.67\,mA \leftarrow$$

$$i_C = \alpha\, i_E = 0.985 \times 1.67 = 1.645\ mA \leftarrow$$

7.3.2.

NOMINAL $\alpha = 0.99$

$$\text{NOMINAL } \beta = \frac{\alpha}{1-\alpha} = \frac{0.99}{0.01} = 99$$

$$\text{FOR } \alpha = 0.99\,(1.01), \quad \beta = \frac{0.99\,(1.01)}{1-(0.99\times 1.01)} = 9{,}999$$

$$\text{FOR } \alpha = 0.99\,(0.99), \quad \beta = \frac{0.99\,(0.99)}{1-(0.99\times 0.99)} = 49.2313$$

% CHANGE IN β = $+10{,}000\%$ TO -50.2512% ←

7.3.3. $i_E = 5\,mA$; 300K ; Si ; $v_{BE} = 0.7\,V$; $\alpha = 0.99$

FROM EQ. (7.3.1) $\quad I_{SE}$ = REV. SAT. CURRENT OF BEJ $= \dfrac{i_E}{e^{v_{BE}/V_T}}$

$$= \frac{5\times 10^{-3}}{e^{0.7/(25.861\times 10^{-3})}} = \frac{5}{5.6936}\times 10^{-14}$$

$$= 0.878\times 10^{-14}\ A \leftarrow$$

$$i_C = \alpha\, i_E = 0.99 \times 5\ mA = 4.95\ mA \leftarrow$$

$$\beta = \frac{\alpha}{1-\alpha} = \frac{0.99}{0.01} = 99 \leftarrow$$

$$i_B = (1-\alpha)\, i_E = 0.01 \times 5\ mA = 0.05\ mA = 50\,\mu A \leftarrow$$

7.3.4. $\alpha = 0.98$; $I_{CBO} = 90\times 10^{-9}\,A$; $i_C = 7.5\,mA$

$$\beta = \frac{\alpha}{1-\alpha} = \frac{0.98}{0.02} = 49 \leftarrow$$

$$i_B = \frac{i_C}{\beta} - \frac{\beta+1}{\beta} I_{CBO} = \frac{7.5\times 10^{-3}}{49} - \frac{50}{49}(90\times 10^{-9})$$

$$= 0.153\ mA - 91.8\ nA$$

$$= (0.153 - 0.0000918)\ mA = 0.1529\ mA \leftarrow$$

$$i_E = \frac{1+\beta}{\beta} i_C - \frac{1+\beta}{\beta} I_{CBO}$$

$$= \frac{50}{49}\,7.5\times 10^{-3} - \frac{50}{49}\times 90\times 10^{-9}$$

$$= 7.653\ mA - 91.8\ nA = (7.653 - 0.0000918)\ mA$$

$$= 7.6529\ mA \leftarrow$$

7.3.5. $I_{CBO} = 4\ nA = 4\times10^{-9}A$; $i_E = 1mA = 1\times10^{-3}A$; $i_C = 0.9\ mA = 0.9\times10^{-3}A$

$v_{BE} = 0.7V$

$$\alpha = \frac{i_C - I_{CBO}}{i_E} = \frac{(0.9\times10^{-3}) - (4\times10^{-9})}{1\times10^{-3}} = 0.9 \leftarrow$$

$$\beta = \frac{\alpha}{1-\alpha} = \frac{0.9}{0.1} = 9 \leftarrow$$

$$i_B = (1-\alpha)\, i_E - I_{CBO} = 0.1\times1\times10^{-3} - 4\times10^{-9}$$

$$= 0.1\ mA \leftarrow$$

$$I_{SE} \approx i_E\, e^{-v_{BE}/V_T} = 1\times10^{-3}\, e^{-0.7/(25.861\times10^{-3})}$$

$$= 1\times10^{-3} \times \frac{1}{5.6936\times10^{11}} = 17.56\times10^{-16}A \leftarrow$$

7.3.6. (a) ASSUMING ACTIVE STATE OF OPERATION, $v_{BE} = 0.7\ V$

Then $i_B = \frac{12-0.7}{430\times10^3} = \frac{11.3}{430\times10^3} = 26.28\ \mu A$

$i_C = \beta i_B = 2.234\ mA$; $i_E = (1+\beta)\, i_B = 2.26\ mA$

FOR THIS i_C $v_{CE} = 12 - i_C(4700) = 1.5016\ V$

SO THAT $v_{CB} = 1.5016 - 0.7 = 0.8016\ V$

WHICH IS ENOUGH TO REVERSE BIAS THE CBJ SO THAT OPERATION IS IN THE ACTIVE STATE $\leftarrow$

(b) FOR $\beta = 85 - 8.5 = 76.5$

$i_C = 2.0103\ mA$; $v_{CE} = 12 - 2.0103(4.7) = 2.5514\ V$

SO THAT $v_{CB} = 1.8514\ V$ AND THE DEVICE IS IN THE ACTIVE STATE OF OPERATION. $\leftarrow$

(c) FOR $\beta = 85 + 17 = 102$, REPEATING THE PROCEDURE,

$i_B = 26.2791\ \mu A$; $i_C = 2.6305\ mA$ SO THAT

$v_{CE} = 12 - 2.6305(4.7) = -0.5932\ V$

AND $v_{CB} = -1.2982\ V$ WHICH IS MORE THAN ENOUGH TO FORWARD BIAS THE CBJ. THUS THE OPERATION IS NOT IN THE ACTIVE REGION BUT IS IN THE SATURATION REGION. $\leftarrow$

7.3.7. (a) $\beta = 75$; $V_A = 65V$; $g_m = 0.03\,S$; $i_C = 6\,mA$

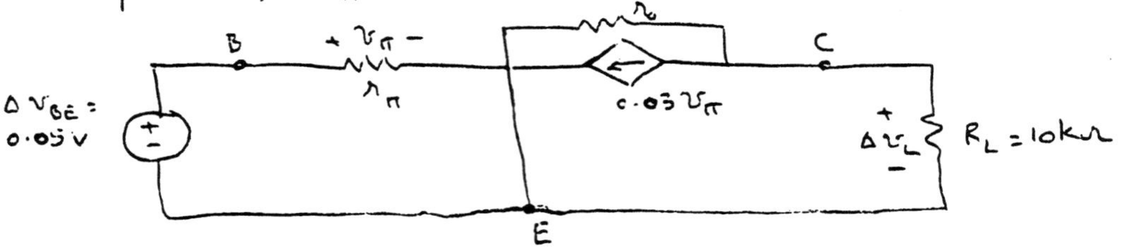

$r_\pi = \beta/g_m = 75/0.03 = 2500\,\Omega$; $\frac{1}{r_o} = I_{CQ}/V_A = 6\times10^{-3}/65$ OR $r_o = 10{,}833.3\,\Omega$

$v_\pi = \Delta v_{BE}$; $g_m v_\pi = -\frac{\Delta v_L}{R_L} - \frac{\Delta v_L}{r_o}$ OR $\Delta v_L = \frac{-g_m v_\pi R_L r_o}{R_L + r_o} = \frac{-g_m R_L r_o}{(R_L + r_o)} \Delta v_{BE}$

$$= \frac{-0.03\times10^4 \times 1.03333(10^4)\,0.05}{10^4\,(1+1.03333)} = -7.8V \leftarrow$$

(b) $\Delta i_B = \frac{\Delta v_{BE}}{r_\pi} = \frac{0.05}{2500} = 20\,\mu A \leftarrow$

7.3.8.

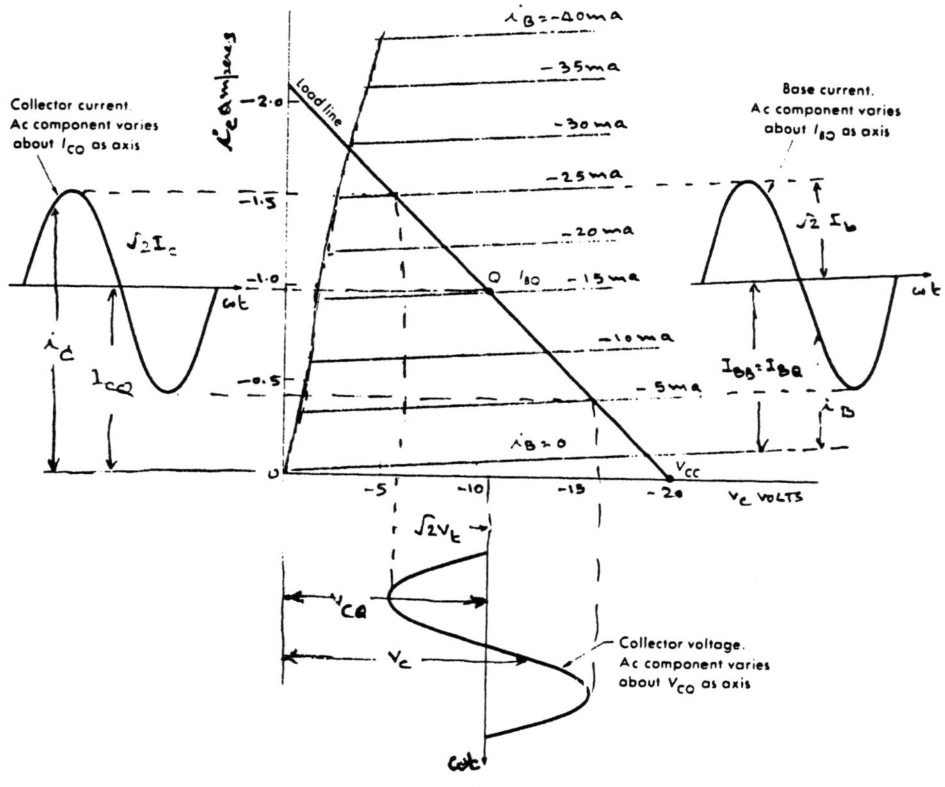

Sinusoidal variations of collector current i_c, collector voltage v_c, and base current i_b superimposed on the direct values I_{CO}, V_{CO}, and I_{BO}, respectively, for the circuit of Fig. P7.3.8b. The collector characteristics of the *pnp* transistor are idealized.

7.3.8. CONTD.

KVL: $v_C + i_C R_L + V_{CC} = 0$ DEFINES THE LOAD LINE

SINCE KIRCHHOFF'S LAWS MUST BE SATISFIED AT EACH INSTANT OF TIME, THE INSTANTANEOUS BEHAVIOR OF THE CIRCUIT IS CONSTRAINED TO BE ALONG THE LOAD LINE, RESULTING IN WAVEFORMS SHOWN ABOVE.

FOR A CHANGE IN BASE CURRENT OF 10 mA (PEAK VALUE), CORRESPONDING CHANGE IN COLLECTOR CURRENT IS APPROXIMATELY 0.6 A.

$$\therefore \text{ CURRENT GAIN} = \frac{0.6}{10 \times 10^{-3}} = 60 \leftarrow$$

7.3.9.
(a) THE CIRCUIT IS REDRAWN BELOW:

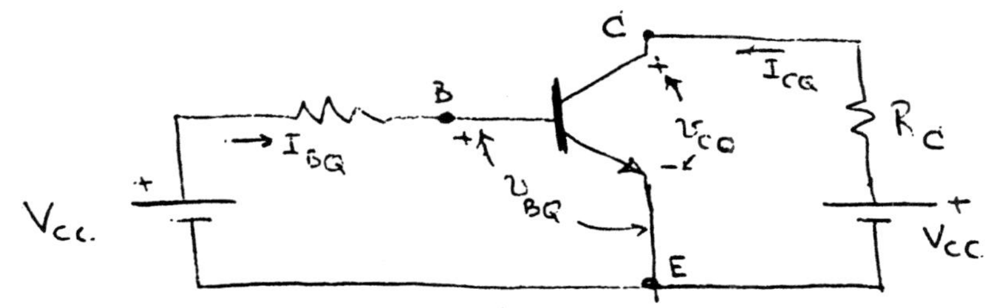

$V_C + I_C R_C = V_{CC}$ LOAD LINE

$V_B + I_B R_B = V_{CC}$ WHERE $V_B \cong 0.7\,V$ FOR SILICON BJT
$\simeq 0.3\,V$ FOR GERMANIUM TRANSISTORS

$$\therefore I_{BQ} = \frac{V_{CC} - V_B}{R_B}$$

IF $V_B << V_{CC}$, $I_{BQ} \simeq \dfrac{V_{CC}}{R_B}$

THE VALUE OF I_{BQ} IN CONJUNCTION WITH THE LOAD LINE DETERMINES THE OPERATING POINT. ←

(b)
LOAD LINE IS DRAWN (AS SHOWN BELOW) AS A STRAIGHT LINE THROUGH Q INTERSECTING THE HORIZONTAL AXIS AT V_{CC}. SLOPE OF LOAD LINE IS $-1/R_C$.

$$-\frac{1}{R_C} = \frac{I_{CQ} - 0}{V_{CQ} - V_{CC}} = \frac{16 \times 10^{-3}}{10 - 18} = -2 \times 10^{-3}\,S \quad \text{OR} \quad R_C = 500\,\Omega \leftarrow$$

$I_{BQ} = 100\,\mu A$; FROM $V_{CC}/R_B \simeq I_{BQ}$, $R_B = \dfrac{V_{CC}}{I_{BQ}} = \dfrac{18}{100 \times 10^{-6}} = 180{,}000\,\Omega$ ←

7.3.9.

(b) CONTD.

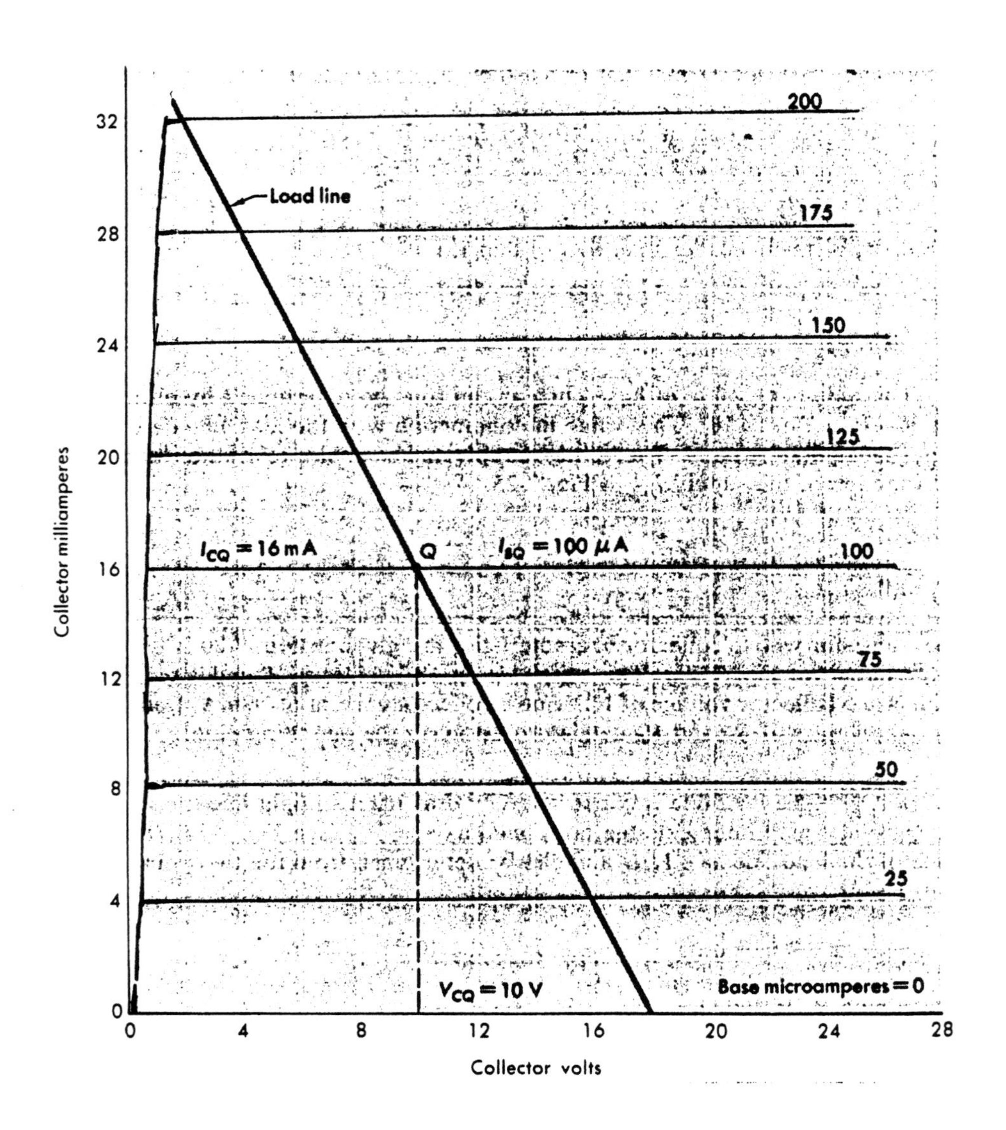

Load line
200
175
150
125
$I_{CQ} = 16\,mA$
Q
$I_{BQ} = 100\,\mu A$
100
75
50
25
$V_{CQ} = 10\,V$
Base microamperes = 0
Collector milliamperes
Collector volts
0
4
8
12
16
20
24
28
32

7.3.10.

$$I_{C1} = -\alpha I_{E1} \; ; \quad I_{B2} = -I_{E1} \; ; \quad I_{E2} = -\frac{I_{B2}}{1-\alpha} \; , \quad I_{C2} = \frac{\alpha}{1-\alpha} I_{B2}$$

$$I_{CC} = I_{C1} + I_{C2} =$$

EMITTER CURRENT OF THE COMBINATION IS I_{E2}

$$\therefore \quad \alpha_C = \frac{I_{CC}}{|I_{E2}|} = 1-(1-\alpha)^2 \quad \leftarrow$$

BASE CURRENT OF THE COMBINATION $I_{BC} = I_{B1}$

$$I_{C1} = \beta I_{B1} \; ; \quad I_{E1} = -I_{B2} = -(\beta+1) I_{B1}$$

$$I_{C2} = \beta I_{B2}$$

USING $I_{CC} = I_{C1} + I_{C2}$ AND NOTING $\beta_C = \frac{I_{CC}}{I_{BC}}$

ONE GETS $\beta_C = \beta(\beta+2) \quad \leftarrow$

7.3.11. (a)

DRAW A LOAD THROUGH $(0, -2A)$ $(-60V, 0)$ WITH A SLOPE OF $-1/30$...

$$I_{BQ} = \frac{-60-(-0.7)}{6\times10^3} = -9.88\,mA \simeq -10\,mA$$

FROM THE LOAD LINE, $V_{CQ} = -25V$; $I_{CQ} = -1.15A \quad \leftarrow$

(b) POWER $= (-60)(I_C + I_B) = -60(-1.15-0.01)$

$$= 69.6\,W \quad \leftarrow$$

7.3.12.

CURRENT THRO' R_C IS $I_C + I_B$

KVL FOR COLLECTOR-EMITTER LOOP: $-V_{CC} + R_C(I_C+I_B) + V_{CE} = 0$ — (1)

FOR BASE LOOP: $-V_{CC} + (I_C+I_B)R_C + I_B R_F + V_{BE} = 0$ — (2)

USING $I_C = \beta I_B$ AND SOLVING FOR I_C YIELDS

$$I_C = \frac{V_{CC}-V_{BE}}{R_C\left(1+\frac{1}{\beta}\right) + R_F/\beta}$$

SUBSTITUTING VALUES $I_{CQ} = \dfrac{10-0.6}{2.7\times10^3\left(1+\frac{1}{99}\right) + 2\times10^5/99} = 1.98\,mA \quad \leftarrow$

SUBSTITUTING IN (1) $\quad -10 + 2.7\times10^3(1.98\times10^{-3} + 1.98\times10^{-3}/99) + V_{CE} = 0$

$$\Rightarrow V_{CEQ} = 4.6\,V \quad \leftarrow$$

7.3.13.

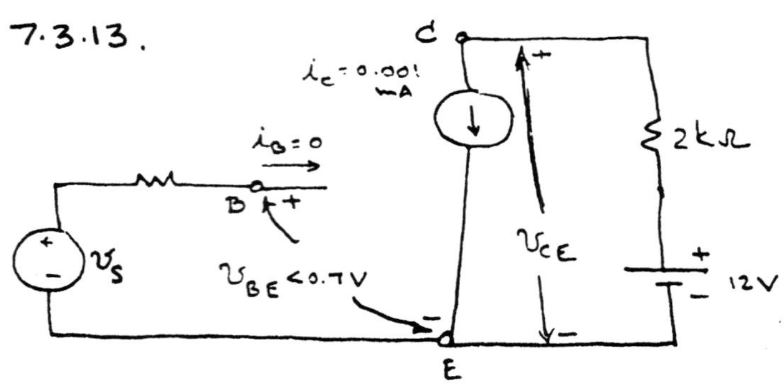

MODEL FOR CUTOFF CONDITIONS

$\because i_B = 0,\ v_{BE} = v_S$

SO $v_S < 0.7$V FOR CUTOFF ←

$i_C = I_{CEO} \simeq 0.001$ mA

$v_{CE} = 12 - 2000 \times 0.001 \times 10^{-3} = 11.998\text{V} \simeq 12\text{V}$

I_{CEO} COULD BE AS WELL OMITTED.

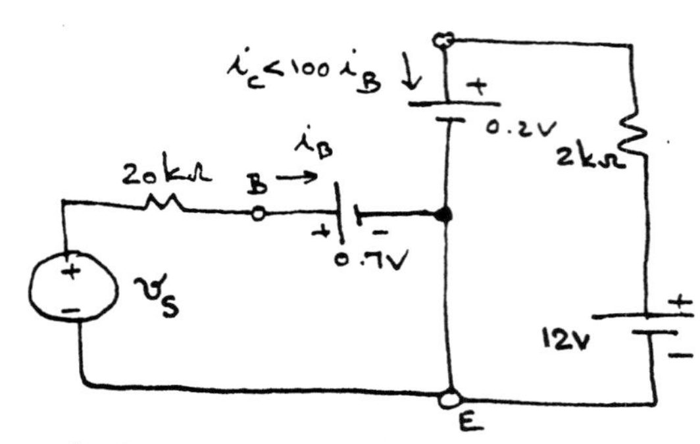

MODEL FOR SATURATED CONDITIONS

$i_B = \frac{v_S - 0.7}{20000}$; $i_C = \frac{12 - 0.2}{2000} = 5.9$ mA

FOR SAT. COND. $i_C < 100\, i_B$ REQUIRES

$$v_S > 0.7 + \frac{20000 \times 5.9 \times 10^{-3}}{100} = 1.88\text{V} \leftarrow$$

HAD WE ASSUMED $V_{sat} \simeq 0$, $v_S > 1.9$ V ← FOR SATURATION

7.3.14.

$$i_E = 50\, i_B + i_B = 51\, i_B$$

OUTER LOOP EQ : $20 - 62{,}000\, i_B - 0.7 = 0$

or $i_B = \frac{19.3}{62000} = 0.3113$ mA

$\therefore i_E = 51 \times 0.3113 \text{ mA} = 15.9 \text{ mA} \leftarrow$

7.3.15.

(a) ACTIVE STATE

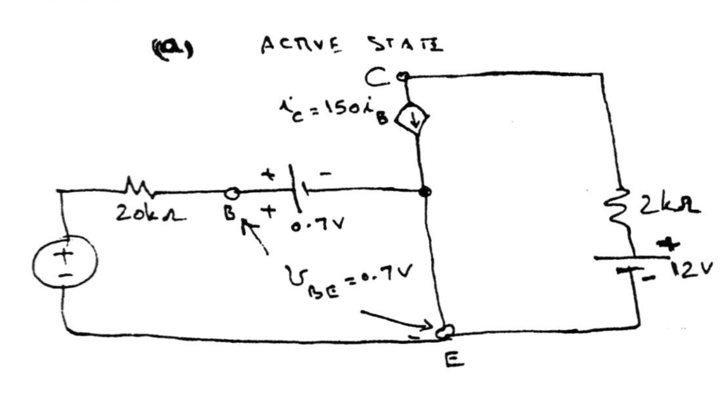

$i_C = 150 \times 20\mu\text{A} = 3\text{mA} \leftarrow$

$v_{CE} = 12 - 2 \times 3 = 6\text{V} \leftarrow$

(b) SATURATED STATE

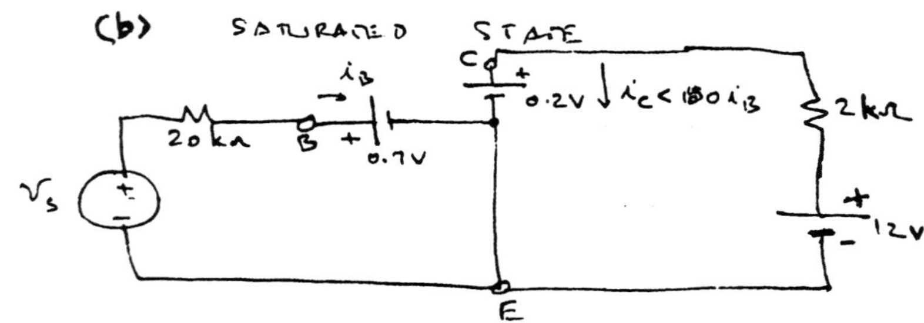

$i_C = \frac{12 - 0.2}{2000} = 5.9\text{mA} \leftarrow$
$< 150 \times 60\mu\text{A}$ i.e. 9mA

$v_{CE} = 12 - 2000 \times 5.9 \times 10^{-3}$
$= 12 - 11.8 = 0.2 = V_{sat.} \leftarrow$

7.3.16.

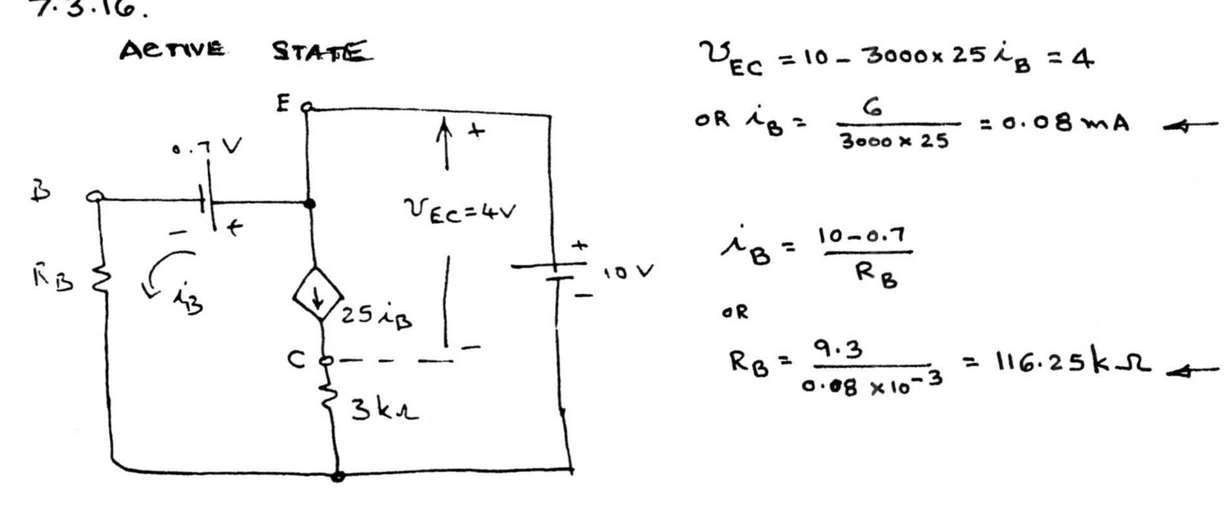

$$v_{EC} = 10 - 3000 \times 25 i_B = 4$$

$$\text{OR } i_B = \frac{6}{3000 \times 25} = 0.08 \text{ mA} \leftarrow$$

$$i_B = \frac{10 - 0.7}{R_B}$$

OR

$$R_B = \frac{9.3}{0.08 \times 10^{-3}} = 116.25 \text{ k}\Omega \leftarrow$$

7.3.17.

SATURATED CONDITION

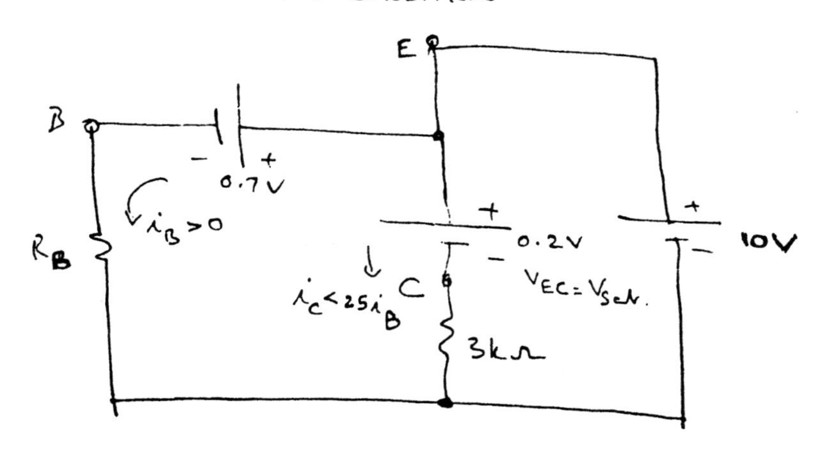

$$i_C = \frac{10 - 0.2}{3 \text{ k}\Omega} = 3.267 \text{ mA} < 25 i_B$$

$$\therefore i_B > \frac{3.267}{25} \text{ mA}$$

$$i_B > 0.131 \text{ mA} \leftarrow$$

$$i_B = \frac{10 - 0.7}{R_B}$$

$$\therefore R_B = \frac{9.3}{i_B}$$

$$R_B < \frac{9.3}{0.131 \text{ mA}}$$

$$R_B < 71 \text{ k}\Omega \leftarrow$$

7.4.1.

(a) $v_{DS} > v_{GS} + V_p \quad ; \quad v_{GS} > -V_p$ ←

(b) $$i_D = \frac{I_{DSS}}{V_p^2}(v_{GS} + V_p)^2 = I_{DSS}\left(1 + \frac{v_{GS}}{V_p}\right)^2$$

WHERE I_{DSS} IS THE VALUE OF i_D WHEN $v_{GS} = 0$

$$i_D = I_{DSS}\left(\frac{v_{DS}}{V_p}\right)^2 \quad \text{AT BOUNDARY}$$ ←

(c) LINEAR OHMIC OPERATION REQUIRES

$$|v_{DS}| \leq \tfrac{1}{4}(v_{GS} + V_p) \quad ; \quad v_{GS} > -V_p$$

EQUIVALENT DRAIN-TO-SOURCE RESISTANCE

$$R_{DS} = \frac{V_p^2}{2 I_{DSS}(v_{GS} + V_p)}$$ ←

(d) CUTOFF REGION OCCURS WHEN $v_{GS} \leq -V_p$, SO $i_D = 0$ ←

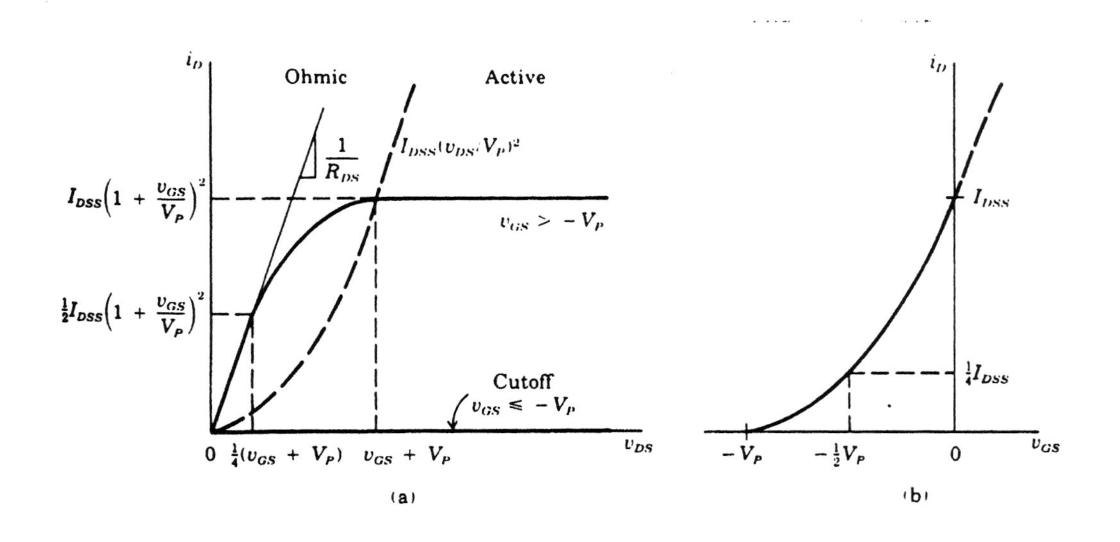

7.4.2.

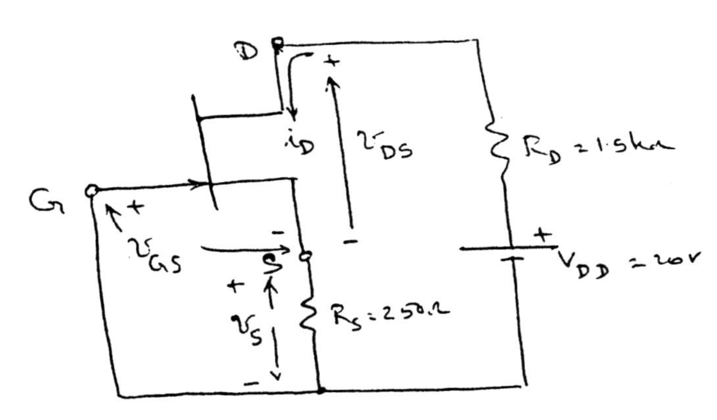

$$v_{GS} + v_S = 0 \quad ; \quad v_S = R_S\, i_D$$

$$v_{GS} = -v_S = -R_S\, i_D < 0$$

ASSUME ACTIVE OPERATION

$$\therefore \quad R_S i_D = -v_{GS}$$

MULTIPLY EQ. (7.4.1) BY R_S BOTH SIDES; ELIMINATE i_D.

$$-v_{GS} = \frac{R_S I_{DSS}}{V_p^2}\,(v_{GS} + V_p)^2$$

SUBSTITUTING & REARRANGING

$$v_{GS}^2 + 20 v_{GS} + 36 = 0$$

$$\therefore v_{GS} = \tfrac{1}{2}\left(-20 \pm \sqrt{20^2 - 4\times 36}\right) = -10 \pm 8$$

BUT ONE MUST HAVE $v_{GS} > -V_p = -6\,V$ FOR ACTIVE OPERATION

$\therefore$ CHOOSE $v_{GS} = -10 + 8 = -2\,V$ ←

USING THIS BACK INTO EQ 7.4.1 YIELDS $i_D = 8\,mA$ ←

KVL: $v_{DS} = V_{DD} - R_D i_D - R_S i_D = 6\,V$ ←

WHICH SATISFIES THE ACTIVE CONDITION $v_{DS} > v_{GS} + V_p = +4\,V.$

7.4.3.

$$-v_{GS} = \frac{R_S I_{DSS}}{V_p^2}(v_{GS} + V_p)^2$$

$$-v_{GS} = \frac{R_S(18\times10^{-3})}{36}(v_{GS}+6)^2$$

$$v_{GS} = -v_S = -R_S i_D < 0$$

$$v_{GS} = -R_S(2\times10^{-3})$$

SUBSTITUTING $R_S = -v_{GS}/(2\times10^{-3})$

$$-v_{GS} = -\frac{v_{GS}}{2\times10^{-3}}\frac{18\times10^{-3}}{36}(v_{GS}+6)^2$$

OR $v_{GS}^2 + 12v_{GS} + 32 = 0$

OR $(v_{GS}+4)(v_{GS}+8) = 0$

OR $v_{GS} = -4$ OR -8 V

WE MUST HAVE $v_{GS} > -V_p = -6$V

SO CHOOSE $v_{GS} = -4$ V ←

$$R_S = -v_{GS}/i_D = 4/(2\times10^{-3}) = 2\,k\Omega \leftarrow$$

∵ FOR ACTIVE CONDITION $v_{DS} > v_{GS} + V_p$

$$v_{DS} = V_{DD} - R_D i_D - R_S i_D = 20-3-4 = 13V \leftarrow$$

(WHICH IS $> v_{GS}+V_p = -4+6 = 2$; HENCE OK)

7.4.4. AT THE BOUNDARY $i_D = I_{DSS}\left(\frac{v_{DS}}{V_P}\right)^2$

$$i_D = 32\times10^{-3}\left(\frac{4}{5}\right)^2 = 20.48 \text{ mA}$$

SINCE i_D IS LARGER THAN 20.48 mA WHEN $v_{DS} = 4V$, THE JFET IS OPERATING IN THE **OHMIC REGION**. ←

7.4.5.

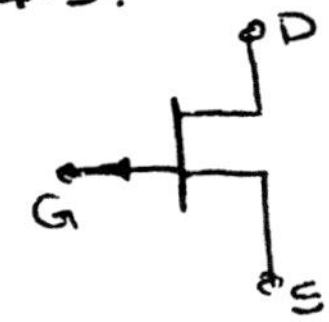

$v_{DS} < 0$; $v_{GS} > 0$ ←

CURRENTS OUT OF D ; INTO S ; NONE AT G. ←

7.4.6.
LOAD LINE IS THE KVL EQ. IN THE DRAIN LOOP:

$$V_{DD} = i_D R_D + v_D$$

$i_G = 0$; Gate voltage $v_G = -V_{GG} + \sqrt{2}\, V_g \sin\omega t$

THE OPERATING QUIESCENT POINT OCCURS WHEN $\sin\omega t = 0$

THE FIG. BELOW SHOWS THE RESULTS NEEDED

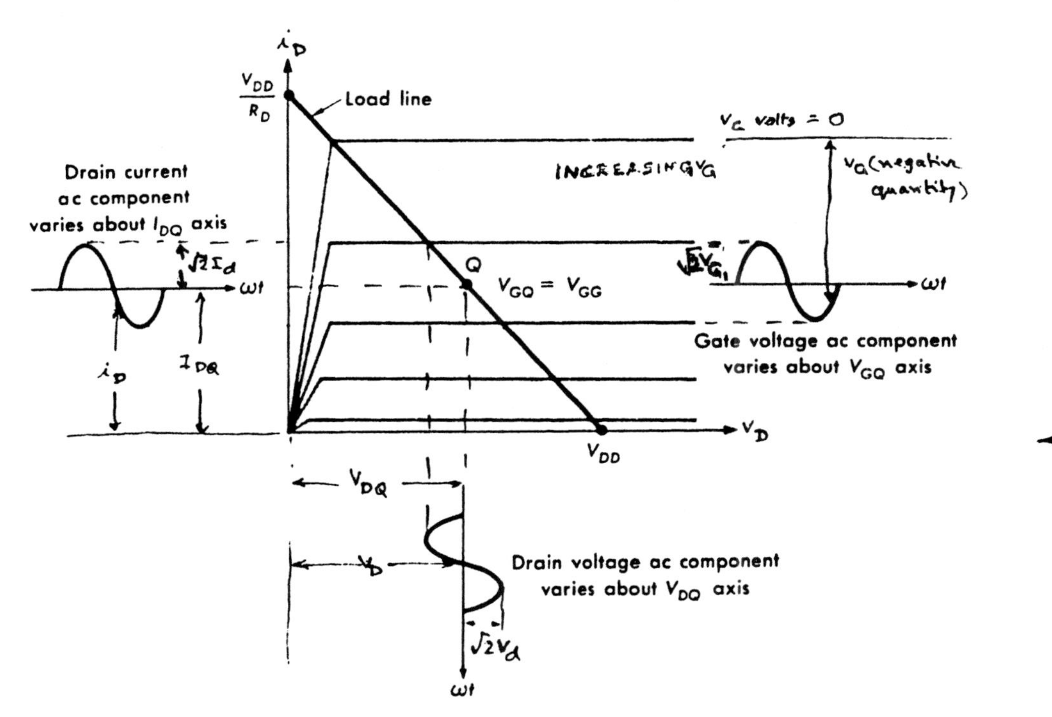

Sinusoidal variations of drain current i_d, drain voltage v_d, and gate voltage v_g superimposed on the direct values I_{DQ}, V_{DQ}, and V_{GQ}, respectively, for the circuit of Fig. P 7.4.6. ←

7.4.7. GATE LOOP: $V_{GS} = -I_D R_S$

DRAW THE BIAS LINE (INDICATED BY ab) ON THE TRANSFER CHARACTERISTIC.

AT INTERSECTION Q: $V_{GSQ} = -2.5\text{V}$ ← ; $I_{DQ} = 2.5\text{ mA}$ ←

KVL DRAIN LOOP: $V_{DD} = I_D(R_S + R_D) + V_{DS}$

PLOT AS THE LOAD LINE ON THE CHARACTERISTICS. $V_{DSQ} = 11\text{V}$ ←

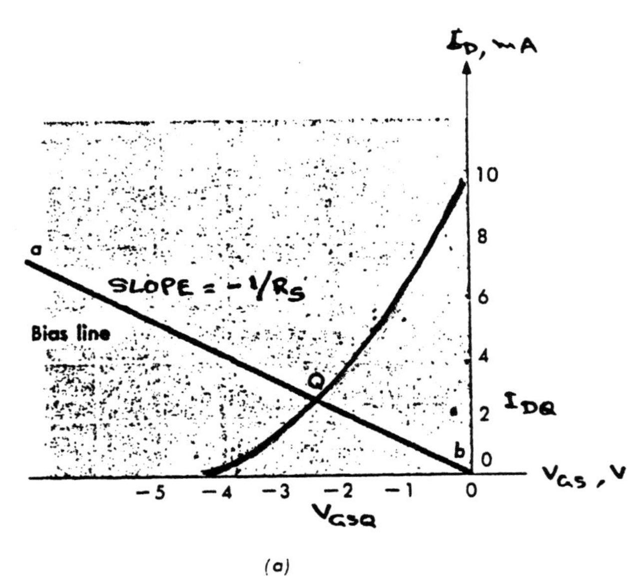

(a)

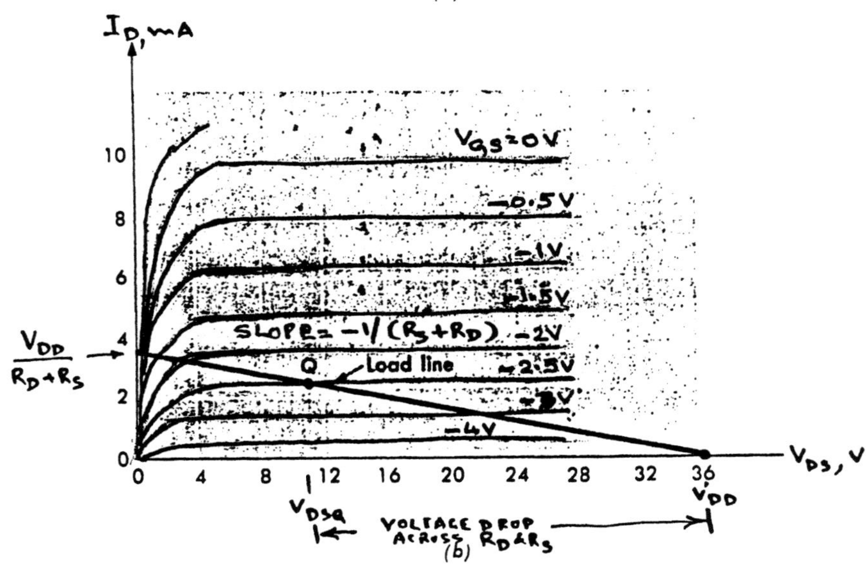

(b)

JFET (a) transfer characteristic and (b) output characteristics

7.4.8.

USING EQ (7.4.1) $\quad -2\times10^{-3} = -5\times10^{-3}\left(1+\frac{V_{GS}}{-4}\right)^2$

$\therefore V_{GS} = +0.422\ V$

$|V_{GSQ}| = |I_{DQ}|R_S \quad$ or $\quad R_S = \frac{0.422}{2\times10^{-3}} = 211\ \Omega \leftarrow$

KVL FOR THE DRAIN LOOP: $\quad I_D R_D = V_{DD} - I_D R_S - V_{DSQ}$

$\therefore R_D = \{-12 + (2\times10^{-3}\times 211) + 4\} / (-2\times10^{-3})$

$= 3.79\ k\Omega \leftarrow$

7.4.9.

$-R_S I_{DQ} = V_{GS}$

$R_S = \frac{-V_{GS}}{I_{DQ}} = \frac{1}{6\times10^{-3}} = 166.7\ \Omega \leftarrow$

(a) $\quad V_{DD} = I_{DQ}R_D + I_{DQ}R_S + V_{DS}$

$= 6\times10^{-3}(2000 + 166.7) + 5 = 12 + 1 + 5 = 18\ V \leftarrow$

FOR $R_D = 2\ k\Omega$

(b) $\quad$ FOR $R_D = 4\ k\Omega$, NO CHANGE OCCURS IN GATE BIASING.

$\therefore R_S$ & I_{DQ} REMAIN AT SAME VALUES AS IN (a)

$V_{DD} = I_{DQ}(R_D + R_S) + V_{DS}$

$= 6\times10^{-3}(4000 + 166.7) + 5 = 24 + 1 + 5 = 30V \leftarrow$

7.4.10.

$V_p = 3V$, $\quad I_{DSS} = 6\ mA$; $\quad$ n-CHANNEL JFET

(a) $\quad v_{GS} = -2V$; ACTIVE REGION

OHMIC REGION ENDS WHEN

$v_{DS} = V_{GS} + V_p = -2 + 3 = 1V \leftarrow$

(b) $\quad$ FROM EQ. (7.4.1) $\quad i_D = 6\times10^{-3}\left(1+\frac{-2}{3}\right)^2 = 6\times\left(\frac{1}{3}\right)^2\times10^{-3}$

$= \frac{2}{3}\ mA \leftarrow$

7.4.11.

IF OPERATING IN ACTIVE REGION $\quad i_D = I_{DSS}\left(1+\frac{V_{GS}}{V_p}\right)^2$

$= 12\times10^{-3}\left(1+\frac{-3.2}{5}\right)^2$

$= 1.62\ mA$ WOULD OCCUR

SINCE $i_D < 1.62\ mA$, OPERATION IS IN THE OHMIC REGION. $\leftarrow$

7.4.12.

SINCE REQUIRED $i_D > I_{DSS}$, $\quad$ JFET IS IN ACTIVE MODE

EQ. (7.4.2) APPLIES: $\quad 11\times10^{-3} = 10\times10^{-3}\left(1+\frac{0}{+3}\right)^2\left(1+\frac{v_{DS}}{350}\right)$

$V_{DS} = \left(\frac{11}{10}-1\right)350 = 35\ V \leftarrow$

7.4.13. FIG. 7.4.2. APPLIES.

$$i_D = 10\times10^{-3}\left(1+\frac{-0.5}{2}\right)^2\left(1+\frac{10}{300}\right) = 10\times10^{-3}\times 0.75\times\frac{310}{300} = 7.75\text{ mA}$$

7.4.14.(a)

$$g_m = \frac{2 I_{DSS}}{V_P}\left(1+\frac{v_{GS}}{V_P}\right)\left(1+\frac{v_{DS}}{V_A}\right) \quad \text{EQ.(7.4.4)}$$

$$= \frac{2\times10\times10^{-3}}{3}\left(1+\frac{v_{GS}}{3}\right)\left(1+\frac{10}{100}\right) = \frac{22}{3}\times10^{-3}\left(1+\frac{v_{GS}}{3}\right)$$

WHEN $V_A \to \infty$, $\frac{20}{3}\times10^{-3}\left(1+\frac{v_{GS}}{3}\right)$

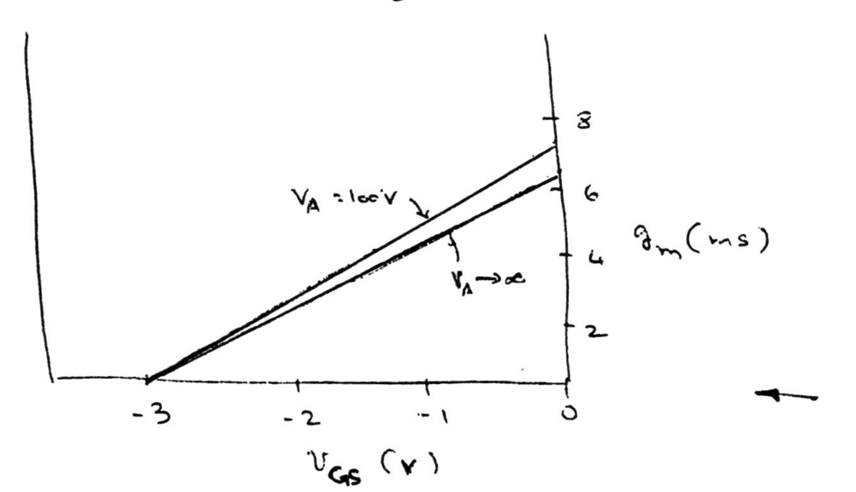

7.4.14.(b)

EQ.(7.4.5.)
$$r_o = \frac{V_A}{I_{DSS}\left(1+\frac{v_{GS}}{V_P}\right)^2} = \frac{V_A}{10\times10^{-3}\left(1+\frac{v_{GS}}{3}\right)^2}$$

$$r_o = \frac{100\,V_A}{\left(1+\frac{v_{GS}}{3}\right)^2}$$

FOR $V_A = 100$ V, $r_o = \frac{10^4}{\left(1+\frac{v_{GS}}{3}\right)^2}$

FOR $V_A \to \infty$, $r_o \to \infty$

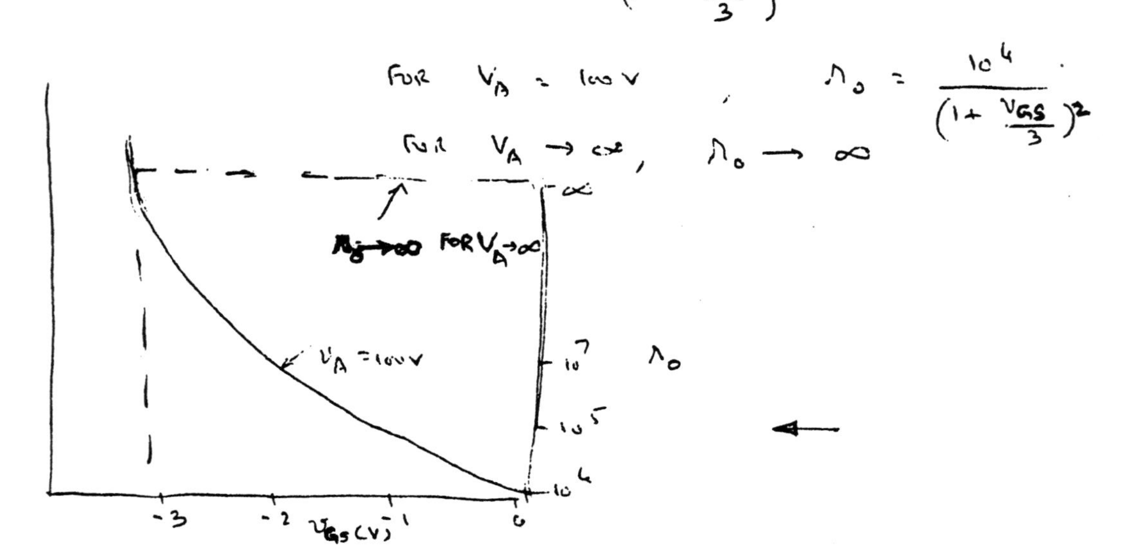

7.4.15.

For small v_{DS} :

$$i_D \simeq \frac{2\,I_{DSS}}{V_P}\left(1+\frac{v_{GS}}{V_P}\right)(v_{DS})$$

$$r_{DS} = \frac{v_{DS}}{i_D} = \frac{V_P}{2\,I_{DSS}\left(1+\frac{v_{GS}}{V_P}\right)} = \frac{V_P^{\,2}}{2\,I_{DSS}\,(V_P+v_{GS})}$$

$$= \frac{3^2 \times 10^3}{2\times 25\,(3-2)} = 9\times 20 = 180\ \Omega \leftarrow$$

7.4.16.

APPLYING EQ. (7.4.2)

$$k \simeq \frac{i_D}{(v_{GS}-V_T)^2} = \frac{1\times 10^{-3}}{(6-4)^2} = 0.25\times 10^{-3}\ \mathrm{A/V^2} \text{ OR } 0.25\ \mathrm{mA/V^2}. \leftarrow$$

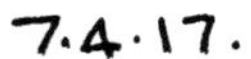

7.4.17.

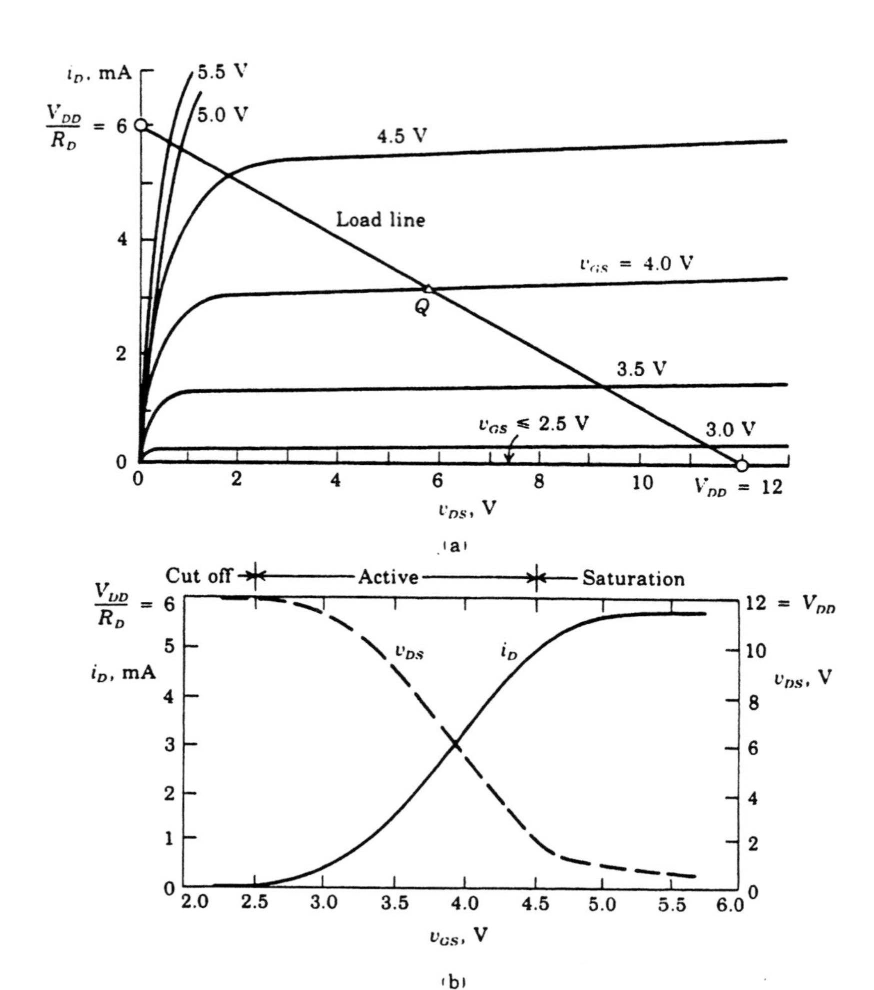

FIG. S 7.4.17.(A)

(a)

$v_{DS} = V_{DD} - R_D i_D$, $i_D = \dfrac{V_{DD} - v_{DS}}{R_D}$ LOAD-LINE EQ. ←

FOR $v_{GS} = 4$ V, $i_D \simeq 3$ mA AND $v_{DS} \simeq 6$ V. ←

(b)

CUT OFF REGION: $i_D \ll V_{DD}/R_D$; $v_{DS} \simeq V_{DD}$ MOSFET'S NORMALLY-OFF STATE

SATURATION REGION: $i_D \simeq V_{DD}/R_D$, $v_{DS} \ll V_{DD}$ MOSFET'S OHMIC STATE

ACTIVE REGION: MOSFET'S CONSTANT-CURRENT STATE ←

7.4.17. CONTD.

(C)

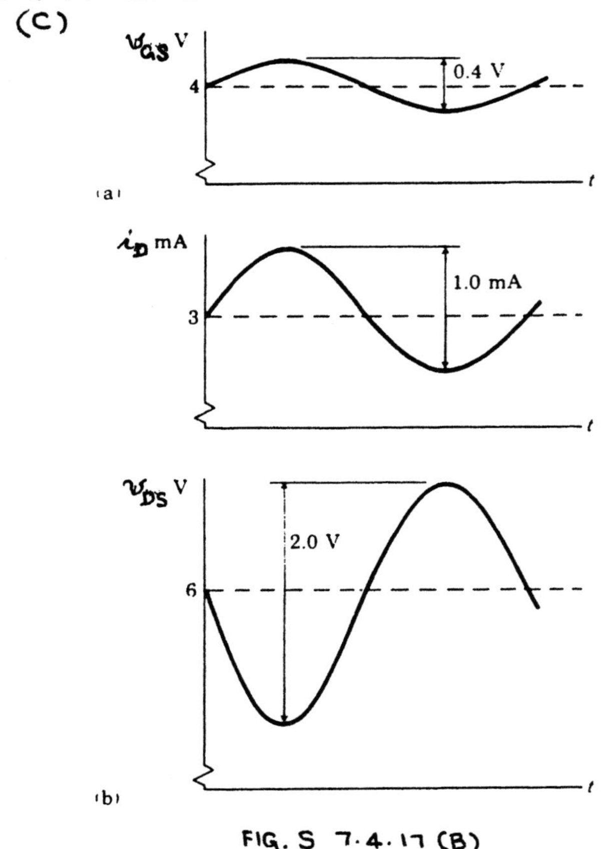

$i_D(t) \approx (3 + 0.5 \sin\omega t)$ mA

$v_{DS}(t) \simeq 6 - 1.0 \sin\omega t$

0.4V VARIATION OF v_{GS} PRODUCES A 2-V VARIATION OF v_{DS} IN THE OPPOSITE DIRECTION.

$\therefore$ VOLTAGE AMPLIFICATION $A_v = \dfrac{\Delta v_{DS}}{\Delta v_{GS}}$

$\simeq -\dfrac{2}{0.4} = -5$ ←

WHERE THE MINUS SIGN REFLECTS THE VOLTAGE INVERSION.

FIG. S 7.4.17 (B)

(d) SEVERE DISTORTION IN THE FORM OF TOP AND BOTTOM "CLIPPING" APPEARS IN THE OUTPUT WAVEFORMS BECAUSE THE CIRCUIT IS CUT OFF WHENEVER $v_{GS} \leq 2.5$V AND SATURATED WHENEVER $v_{GS} > 4.5$V.

IT IS A SWITCHING CIRCUIT THAT GOES FROM 'OFF' TO 'ON' WHEN THE INPUT CHANGES BY ABOUT 2V. THUS THE MOSFET FUNCTIONS AS A SWITCH, TURNING POWER OFF AND ON TO THE LOAD RESISTOR R_D AND IT DOES SO WITHOUT DRAWING POWER FROM THE CONTROLLING INPUT.

7.4.17 (d) CONTD.

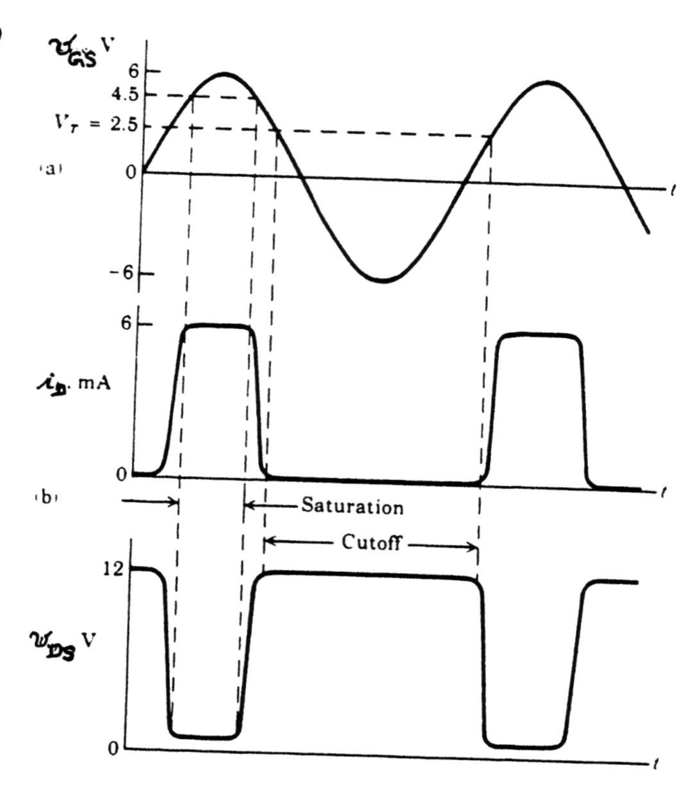

FIG. S 7.4.17(C)

7.4.18.

ACTIVE STATE

$$v_{DS} > v_{GS} - V_T \quad ; \quad v_{GS} > V_T$$

$$i_D = \frac{I_{DSS}}{V_T^2}(v_{GS} - V_T)^2 = I_{DSS}\left(\frac{v_{GS}}{V_T} - 1\right)^2$$

WHERE I_{DSS} IS THE VALUE OF i_D WHEN $v_{GS} = 2V_T$

$$\therefore \; i_D = K(v_{GS} - V_T)^2 \quad \text{WHERE } K = I_{DSS}/V_T^2$$

MOSFET BEHAVES LIKE A NON-LINEAR VOLTAGE-CONTROLLED CURRENT SOURCE.

LINEAR OHMIC STATE : MOSFET ACTS AS A VOLTAGE-CONTROLLED RESISTOR

$$|v_{DS}| \leq \frac{1}{4}(v_{GS} - V_T) \quad ; \quad v_{GS} > V_T$$

$$i_D \simeq v_{DS}/R_{DS}$$

WHERE $R_{DS} = \dfrac{V_T^2}{2 I_{DSS}(v_{GS} - V_T)}$ IS KNOWN AS THE EQUIVALENT DRAIN-TO-SOURCE RESISTANCE.

7.4.18. EQUATIONS SUGGEST THAT OHMIC OPERATION ALLOWS BIDIRECTIONAL CURRENT FLOW.
CONTD.

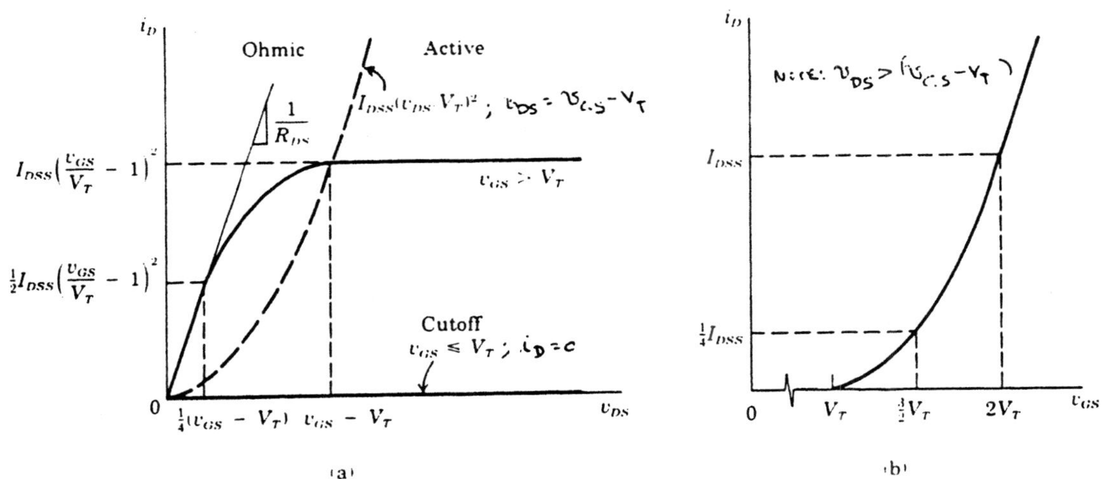

(a) STATIC CHARACTERISTICS (b) TRANSFER CHARACTERISTIC

FIG. S 7.4.18 UNIVERSAL IDEALIZED ENHANCEMENT MOSFET CURVES

7.4.19.

(a)

$$v_{DS} = V_{DD} - R_D i_D = 12 - (2\times10^{3})(4\times10^{-3}) = 4V \leftarrow$$

WHICH IS LARGE ENOUGH TO SUGGEST ACTIVE-REGION OPERATION.

Eq (7.4.3):
$$i_D = \frac{8.3\times10^{-3}}{(2.5)^2}(v_{GS} - 2.5)^2 = 4\times10^{-3}$$

$$\therefore v_{GS} = 2.5\left(1 \pm \sqrt{4/8.3}\right) \simeq 4.24V \leftarrow$$

NOTE THAT POSITIVE VALUE IS TAKEN HERE AS OTHERWISE THE NEG. VALUE WOULD YIELD $v_{GS} < 2.5V$, CORRESPONDING TO CUTOFF RATHER THAN ACTIVE OPERATION.

ALSO NOTE $v_{DS} > v_{GS} - V_T \simeq 1.7V$ WHICH IS WELL SATISFIED BY $v_{DS} = 4V$, FOR ACTIVE OPERATION.

(b)

i_D INCREASES WHILE v_{DS} DECREASES AND PRESUMABLY PUTS THE MOSFET IN ITS LINEAR OHMIC REGION. (SEE EQ. 7.4.2)

$$R_{DS} = \frac{V_T^2}{2I_{DSS}(v_{GS} - V_T)} = \frac{(2.5)^2}{2(8.3\times10^{-3})(6-2.5)} = 0.108\,k\Omega$$

$$i_D = \frac{V_{DD}}{R_D + R_{DS}} = 5.69mA\ ;\ v_{DS} = R_{DS}\, i_D = 0.615\,V \leftarrow$$

NOTE $v_{DS} \leq \frac{1}{4}(v_{GS} - V_T) = 0.875V$ HENCE LINEAR OHMIC OPERATION.

7.4.20. BECAUSE OF THE CONNECTION, $v_{GS} = v_{DS} = v$

$i = i_D$ SINCE THE GATE DRAWS THE CURRENT.

OHMIC OPERATION IS IMPOSSIBLE HERE BECAUSE $v_{DS} = v_{GS} > v_{GS} - V_T$.

HENCE, THE i–v CURVE SHOWN BELOW SHOWS CUTOFF OPERATION FOR $v \le V_T$ ←

AND ACTIVE OPERATION FOR $v > V_T$.

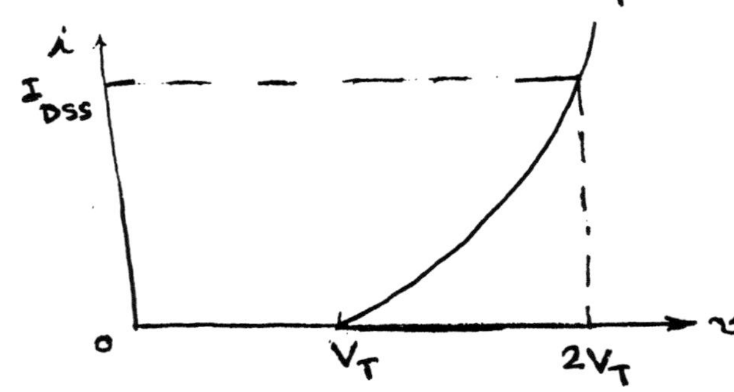

FOR ACTIVE OPERATION,

$$v = V_T\left(1 + \sqrt{i/I_{DSS}}\right)$$

WHICH FOLLOWS FROM

$$i_D = \frac{I_{DSS}}{V_T^2}(v_{GS} - V_T)^2 = I_{DSS}\left(\frac{v_{GS}}{V_T} - 1\right)^2$$

7.4.21. $i_D = \frac{I_{DSS}}{V_T^2}(v_{SG} - V_T)^2 = I_{DSS}\left(\frac{v_{SG}}{V_T} - 1\right)^2$

$i_D = 0 = I_{DSS}\left(\frac{3}{V_T} - 1\right)^2$ — ① $\Rightarrow V_T = 3V$ ←

$i_D = 5\times10^{-3} = I_{DSS}\left(\frac{8}{V_T} - 1\right)^2$ — ②

$= I_{DSS}\left(\frac{8}{3} - 1\right)^2 = I_{DSS} \times 2.78$ OR $I_{DSS} = \frac{5\times10^{-3}}{2.78} = 1.8\,mA$ ←

7.4.22.

$v_{DS} = v_{GS} + V_p$ SINCE v_{DS} IS THE LARGEST FOR OHMIC REGION.

EQ.(7.4.7) APPLIES SINCE i_D IS AT THE HIGHEST END OF OHMIC REGION.

NEGLECTING THE EFFECT OF V_A, $i_D = I_{DSS}\left(1 + \frac{v_{GS}}{V_p}\right)^2$

$$\left(1 + \frac{v_{GS}}{V_p}\right) = \sqrt{i_D/I_{DSS}}$$

$$v_{GS} = V_p\left[\sqrt{i_D/I_{DSS}} - 1\right] = 3\left[\sqrt{\frac{3\times10^{-3}}{11\times10^{-3}}} - 1\right]$$

$$= -1.43V$$ ←

7.4.23.

(a) SINCE $v_{DS} = 4.5 \geq v_{GS} + V_p = 1.2 + 2.8 = 4V$

OPERATION IS IN THE ACTIVE REGION. ←

(b) EQ. (7.4.7) : $i_D \simeq 4.3 \times 10^{-3} \left(1 + \frac{1.2}{2.8}\right)^2 = 8.78$ mA ←

(c) SINCE $v_{GS} > 0$, OPERATION IS IN THE ENHANCEMENT MODE. ←

CHAPTER 8

8.1.1.

$$I_{EQ} = I_{CQ} + I_{BQ} = 11.11 \times 10^{-3}$$

$$V_{R_E} = I_{EQ} R_E = 11.11 \times 10^{-3} \times 240 = 2.67\ V$$

$$V_{R_C} = V_{CC} - V_{CEQ} - V_{R_E} = 24 - 14 - 2.67 = 7.33 V$$

$$R_C = V_{R_C} / I_{CQ} = \frac{7.33}{11 \times 10^{-3}} = 666.36\ \Omega \leftarrow$$

$$I_{R_2} = (0.8 + 2.67)/3000 = 1.157\ mA$$

$$I_{R_1} = 0.11 + 1.157 = 1.267\ mA$$

$$V_{R_1} = 24 - 3.47 = 20.53\ V$$

$$R_1 = 20.53/(1.267 \times 10^{-3}) = 16{,}203.6\ \Omega \leftarrow$$

8.1.2.

FROM EQ. (8.1.1) : $R_C = 3(10)/\ 8\ (5 \times 10^{-3}) = 750\ \Omega \leftarrow$

(8.1.2) : $R_E = 10(100)/[\ 8(1+100)\ 5 \times 10^{-3}] = 247.5\ \Omega \leftarrow$

(8.1.3) : $R_2 = [0.7 + (10/8)]/(5 \times 5 \times 10^{-3}/100)$
$= 1.95/2.5 \times 10^{-4} = 0.78 \times 10^4 = 7800\ \Omega \leftarrow$

(8.1.4) : $R_1 = \frac{(7 \times 10/8) - 0.7}{6 \times 5 \times 10^{-3}/100} = \frac{8.05 \times 100}{30 \times 10^{-3}}$
$= \frac{8.05}{3} \times 10^4 = 26{,}833\ \Omega \leftarrow$

8.1.3.

$$R_B = \frac{V_{CC} - V_{BEQ}}{I_{BQ}} ;\quad I_{BQ} = \frac{I_{CQ}}{\beta} = \frac{14 \times 10^{-3}}{70} = 0.2\ mA$$

$$R_B = \frac{12 - 0.7}{(14/70)\ 10^{-3}} = 56{,}500\ \Omega \leftarrow$$

$$R_C = \frac{V_{CC} - V_{CEQ}}{I_{CQ}} = \frac{12 - 7}{14 \times 10^{-3}} = \frac{5000}{14} = 357\ \Omega \leftarrow$$

8.1.4.

$$R_B = \frac{V_{CEQ} - V_{BEQ}}{I_{BQ}} = \frac{7 - 0.7}{(14/70)\ 10^{-3}} = \frac{6.3 \times 10^3 \times 5}{} = 31{,}500\ \Omega \leftarrow$$

$$V_{CEQ} = V_{CC} - (I_{CQ} + I_{BQ}) R_C ;$$

$$\therefore R_C = \frac{-V_{CEQ} + V_{CC}}{I_{CQ} + \frac{I_{CQ}}{\beta}} = \frac{-7 + 12}{14 \times 10^{-3} + \frac{14 \times 10^{-3}}{70}} = \frac{5}{14.2 \times 10^{-3}} = 352\ \Omega \leftarrow$$

8.1.5. $I_{CQ} = 5\text{ mA}$; $\beta = 100$;

So $I_{BQ} = \frac{5}{100}\text{ mA} = 0.05\text{ mA}$

$$I_{EQ} = \left(5 + \frac{5}{100}\right) = 5.05\text{ mA} \leftarrow$$

$$\therefore V_{RE} = \left[(5\times10^{-3}) + \frac{5}{100}(10^{-3})\right]750 = 3.79\text{ V}$$

$$V_B = V_{BEQ} + V_{RE} = 0.7 + 3.79 = 4.49\text{ V}$$

$$I_2 = \frac{4.49}{9000} = 0.5\text{ mA} \leftarrow$$

$$I_1 = \left(\frac{5}{100} + \frac{4.49}{9}\right)10^{-3} = 0.55\text{ mA} \leftarrow$$

8.2.1. From Eqs. (8.2.6) through (8.2.12)

$$R_S = \frac{5.768\,V_p}{I_{DSS}} = \frac{5.768\times3.5}{5\times10^{-3}} = 4{,}037.6\ \Omega \leftarrow$$

$$R_2 = 100\,R_S = 403{,}760\ \Omega \leftarrow$$

$$R_1 = 100\,R_S\left(\frac{V_{DD}}{1.5V_p} - 1\right) = (403{,}760)\left(\frac{24}{1.5\times3.5} - 1\right) = 1.442\text{ M}\Omega \leftarrow$$

$$R_D = \frac{3\,(V_{DD} - 2.923\,V_p)}{2\,I_{DSS}} = \frac{3\left[24 - (2.923\times3.5)\right]}{2\times5\times10^{-3}} = 4130.85\ \Omega \leftarrow$$

$$V_G = 1.5\,V_p = 1.5\times3.5 = 5.25\text{ V} \leftarrow$$

8.2.2.

28 V

2 mA

100 kΩ

3000 Ω

(a) From Eqs. (8.2.2) & (8.2.3)

$$\sqrt{\frac{2}{5}} - 1 = \frac{v_{GS}}{V_p} = \frac{V_G - i_D R_S}{V_p} = \frac{V_G - 2(10^{-3})3(10^3)}{3.5}$$

$$\therefore V_G = \left(\sqrt{\frac{2}{5}} - 1\right)3.5 + 6 = 4.71\text{ V} \leftarrow$$

(b)

$$V_{R_D} = 28 - 12 - 6 = 10\text{ V}$$

$$\therefore R_D = \frac{10}{2(10^{-3})} = 5000\ \Omega \leftarrow$$

(c) From Eq. (8.2.10):

$$R_1 = \frac{100(10^3)\,(28 - 4.71)}{4.71} = \frac{100(10^3)\,23.29}{4.71} = 494.5\text{ k}\Omega \leftarrow$$

8.2.3.

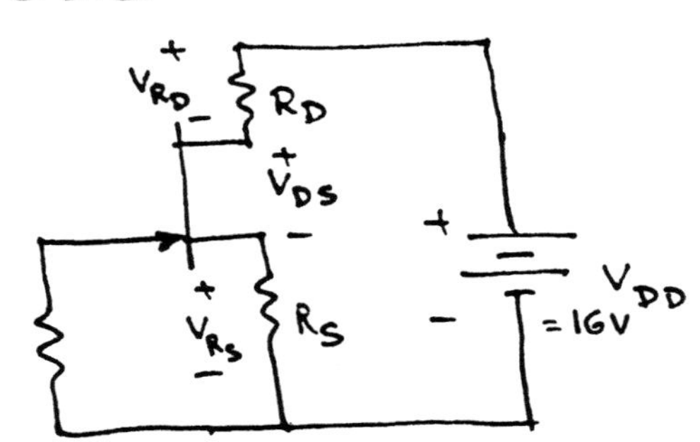

$$I_{DQ} = I_{DSS}\left(1 + \frac{V_{GSQ}}{V_p}\right)^2$$

$$5 = 24\left(1 + \frac{V_{GSQ}}{3}\right)^2$$

$$\therefore V_{GSQ} = \left[\sqrt{\frac{5}{24}} - 1\right]3 = -1.63\text{ V} = -V_{RS}$$

$$\therefore R_S = \frac{V_{RS}}{I_{DQ}} = \frac{1.63}{5\times10^{-3}} = 326\ \Omega \leftarrow$$

$\because$ $V_{DS} \geq V_p (=3V)$ for active-region operation, then $V_{DS} \geq 3$ is required. This is easily realized with V_{DD} as large as 16V.

$$V_{DD} = I_{DQ}(R_D + R_S) + V_{DSQ}$$

$$\therefore 16 = 5(10^{-3})(R_D + 326) + 8$$

$$\therefore R_D = \frac{8\times10^3}{5} - 326 = 1274\ \Omega \leftarrow$$

8.2.4.

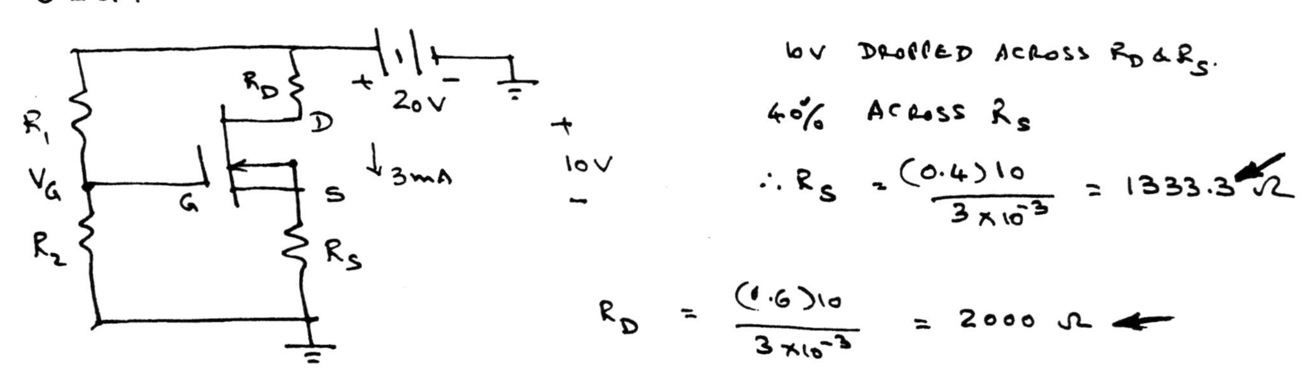

6V dropped across R_D & R_S.

40% across R_S

$$\therefore R_S = \frac{(0.4)10}{3\times10^{-3}} = 1333.3\ \Omega \leftarrow$$

$$R_D = \frac{(1.6)10}{3\times10^{-3}} = 2000\ \Omega \leftarrow$$

Since $I_{DQ} = I_{DSS} = I_{DSS}\left(1 + \frac{V_{GS}}{V_p}\right)^2$, $V_{GS} = 0$

$$V_G = V_{GS} + I_{DQ}R_S = 0 + 4 = 4V \leftarrow$$

$$V_G = V_{DD}\frac{R_2}{R_1+R_2}\ ; \quad R_1 = (V_{DD}R_2 - V_G R_2)/V_G = \frac{V_{DD}}{V_G}R_2 - R_2$$

$$R_1 = R_2\left(\frac{V_{DD}}{V_G} - 1\right) = 10^6\left(\frac{20}{4} - 1\right) = 4\ M\Omega \leftarrow$$

8.2.5.

FROM EQ. (8.2.16) $\quad V_{GSQ} = V_T + \sqrt{\frac{I_{DQ}}{K}} = 1.5 + \sqrt{\frac{10}{5}} = 2.914 \text{ V}$

SINCE $V_{R_S} = 3\text{V}$, $\quad R_S = \frac{3}{10\times10^{-3}} = 300\ \Omega \leftarrow$

DROP ACROSS R_D IS $24 - 3 - 15 = 6\text{ V}$; SO $R_D = \frac{6}{10(10^{-3})} = 600\ \Omega \leftarrow$

FROM EQ. (8.2.18) $\quad V_G = 2.914 + 10(10^{-3})(300) = 2.914 + 3 = 5.914\text{ V} \leftarrow$

NEXT, $\quad R_1 = \frac{(V_{DD} - V_G)R_2}{V_G} = \frac{(24 - 5.914)\,1\times10^6}{5.914} = 3.058\times10^6\ \Omega = 3.058\ \text{M}\Omega \leftarrow$

POWER $= 10(10^{-3})15 = 0.15\text{ W} \leftarrow$

8.3.1.

$$v_L = -g_m v_\pi (r_o \| R_C \| R_L)$$

$$v_\pi = \frac{(r_\pi \| R_B)v_s}{R_s + (r_\pi \| R_B)}$$

$$v_s = i_s[R_s + (r_\pi \| R_B)]$$

$$i_L = \frac{v_L}{R_L} = -g_m[i_s(r_\pi \| R_B)](r_o \| R_C \| R_L)/R_L$$

$$\frac{i_L}{i_s} = \frac{-g_m(r_\pi \| R_B)(r_o \| R_C \| R_L)}{R_L} = \frac{-g_m(r_o \| R_C)R_L(r_\pi \| R_B)}{R_L[(r_o \| R_C) + R_L)}$$

$$= \frac{-g_m(r_o \| R_C)(r_\pi \| R_B)}{R_L + (R_C \| r_o)} = A_i \quad (9.E.1D) \leftarrow$$

8.3.2.

AC EQUIVALENT CIRCUIT FOR SMALL SIGNALS WITH $r_o \to \infty$

8.3.2
CONTD.

WITH $r_o \to \infty$

$$R_i = r_\pi + R_{E1}(1 + g_m r_\pi) \qquad \text{——— (i)}$$

$$A_{v1} = \frac{-g_m R_L R_C}{R_L + R_C}\left(\frac{r_\pi}{R_i}\right) \qquad \text{——— (ii)}$$

$$A_i = -\frac{g_m R_C R_{in}}{R_L + R_C}\left(\frac{r_\pi}{R_i}\right) \qquad \text{——— (iii)}$$

WHERE $R_{in} = R_1 \parallel R_2 \parallel R_i$ ——— (iv)

SINCE $\beta \approx g_m r_\pi$ FROM EQ (7.3.9)

EQ (i) INDICATES THAT THE INPUT RESISTANCE HAS INCREASED BY $R_{E1}(1+\beta)$ OVER ITS VALUE r_π, (SEE EQ. 8.3.2), WHICH OCCURS WHEN THE EMITTER RESISTANCE IS FULLY BYPASSED.

THUS THE INPUT RESISTANCE HAS INCREASED BY A FACTOR $1 + [R_{E1}(1+\beta)/r_\pi] = R_i/r_\pi$ COMPARED WITH THE INPUT RESISTANCE WHEN THE FULL EMITTER'S RESISTOR IS BYPASSED.

ON COMPARING EQ. (ii) WITH EQ. (8.3.5), THE VOLTAGE GAIN IS REDUCED BY THE RECIPROCAL OF THIS FACTOR. CURRENT GAIN IS ALSO REDUCED; NOT BY THE SAME FACTOR THOUGH! A_i WITH R_{E1} UNBYPASSED IS REDUCED BY A FACTOR EQUAL TO THE RECIPROCAL OF $1 + [R_{E1}(1+\beta)/(r_\pi + R_B)]$ RELATIVE TO A_i WITH R_{E1} BYPASSED.

WHILE REDUCED CURRENT AND VOLTAGE GAINS ARE DISADVANTAGES, THE ADVANTAGES ARE: INCREASE OF AMPLIFIER'S BANDWIDTH; THE BASE-EMITTER VOLTAGE IS REDUCED BY THE SAME FACTOR AS IS THE VOLTAGE GAIN. THE SIGNAL LEVEL THAT CAN BE ACCOMMODATED AT THE INPUT IS LARGER BECAUSE A LARGE FRACTION OF THE INPUT v_1 APPEARS ACROSS R_{E1} AND NOT ACROSS THE BASE-EMITTER JUNCTION.

8.3.3. $g_m = \frac{I_{CQ}}{V_T} = \frac{80(10^{-3})}{25.861 \times 10^{-3}} = 3.093$ S

$r_o = \frac{V_A}{I_{CQ}} = \frac{50}{80(10^{-3})} = 625\ \Omega$

$r_o \| R_C = 625 \| 70 = \frac{625 \times 70}{625 + 70} = 62.95\ \Omega$

$A_{v1} = \frac{-3.093(150)\,62.95}{150 + 62.95} = -137.15 \leftarrow$

$R_B = R_1 \| R_2 = \frac{R_1 R_2}{R_1 + R_2} = \frac{1600 \times 400}{2000} = 320\ \Omega$

$r_\pi = \frac{70}{3.093} = 22.632\ \Omega = R_i$

$R_{in} = R_B \| R_i = \frac{R_B r_\pi}{R_B + r_\pi} = \frac{(320)(22.632)}{342.632} = 21.137\ \Omega$

$A_i = -\frac{3.093(62.95)\,21.137}{150 + 62.95} = -19.326 \leftarrow$

8.3.4.

TO A GOOD APPROXIMATION A_{v1} IS ALTERED BY THE FACTOR (r_π / R_i) WHERE

$r_\pi = 22.632\Omega$ AND $R_i = r_\pi + R_{E1}(1 + g_m r_\pi)$

$= 22.632 + 20(1 + 70) = 1442.632\ \Omega$

$\therefore\ r_\pi / R_i = 22.632 / 1442.632 = 0.0157$

SO V_{v1}(UNBYPASSED) $\simeq 0.0157 \times (-137.15) = -2.153 \leftarrow$

CURRENT GAIN IS REDUCED BY FACTOR $1 \Big/ \left(1 + \frac{R_{E1}(1+\beta)}{r_\pi + R_B}\right)$

$= 1 \Big/ \left(1 + \frac{20(1+70)}{22.632 + 320}\right) = \frac{1}{1 + 4.144} = 0.1944$

SO $A_i = 0.1944(-19.326) = -3.76 \leftarrow$

8.3.5. $R_{E1} = 10\Omega$

VOLTAGE GAIN REDUCTION FACTOR: $\frac{r_\pi}{r_\pi + R_{E1}(1+\beta)} = \frac{22.632}{22.632 + 10(1+70)} = 0.0309$

SO $A_{v1} = 0.0309(-137.15) = -4.24 \leftarrow$

CURRENT GAIN REDUCTION FACTOR $= \frac{1}{1 + \frac{R_{E1}(1+\beta)}{r_\pi + R_B}} = \frac{1}{1 + \frac{10(1+70)}{22.632 + 320}}$

$= \frac{1}{1 + \frac{710}{342.632}} = 0.3255$

$\therefore$ So $A_i = 0.3255(-19.326) = -6.29 \leftarrow$

8.3.6.

$$g_m = \frac{I_{CQ}}{V_T} = \frac{5(10^{-3})}{25.861(10^{-3})} = 0.1933 \text{ S}$$

$$r_o = \frac{V_A}{I_{CQ}} = \frac{100}{5(10^{-3})} = 20{,}000\ \Omega \quad ; \quad r_\pi = \frac{\beta}{g_m} = \frac{100}{0.1933} = 517.33\ \Omega$$

$$r_o \| R_C = \frac{1}{\frac{1}{20000} + \frac{1}{1400}} = \frac{1}{(5\times10^{-5}) + (71.43\times10^{-5})} = \frac{10^5}{76.43} = 1308\ \Omega$$

$$A_{v1} = \frac{-g_m R_L (r_o \| R_C)}{R_L + (r_o \| R_C)} = \frac{-0.1933\,(2000)\,(1308)}{2000 + 1308} = -152.86 \leftarrow$$

$$r_\pi \| R_B = \frac{1}{\frac{1}{28000} + \frac{1}{8000} + \frac{1}{517.33}} = \frac{1}{(3.57\times10^{-5}) + (12.5\times10^{-5}) + (193.3\times10^{-5})}$$

$$= \frac{10^5}{209.37} = 477.62\ \Omega$$

$$A_i = \frac{-g_m (r_o \| R_C)(r_\pi \| R_B)}{R_L + (r_o \| R_C)} = -152.86\ \frac{477.62}{2000} = -36.5 \leftarrow$$

8.3.7.

$$\text{VOLTAGE GAIN REDUCTION FACTOR} = \frac{r_\pi}{r_\pi + R_{E1}(1+\beta)} = \frac{517.33}{517.33 + 200(101)} = \frac{517.33}{20717.33}$$

$$= 0.025$$

$$\therefore A_{v1} = 0.025(-152.86) = -3.82 \leftarrow$$

$$\text{CURRENT GAIN REDUCTION FACTOR} = \frac{1}{1 + \frac{R_{E1}(1+\beta)}{r_\pi + R_B}} = \frac{1}{1 + \frac{200(1+100)}{517.33 + 6222.2}}$$

$$\left[\text{Note: } R_B = R_1 \| R_2 = \frac{R_1 R_2}{R_1 + R_2} = \frac{(28000)(8000)}{36000} = 6{,}222.2\ \Omega\right]$$

$$= \frac{1}{1 + 2.997} = 0.25$$

$$\therefore A_i = 0.25(-36.5) = -9.125 \leftarrow$$

8.3.8.

$$g_m = \frac{I_{CQ}}{V_T} = \frac{5(10^{-3})}{25.861(10^{-3})} = 0.1933 \text{ S}$$

$$r_o = \frac{V_A}{I_{CQ}} = \frac{50}{5(10^{-3})} = 10{,}000\,\Omega$$

$$r_o \| R_d = \frac{1}{\frac{1}{10{,}000} + \frac{1}{750}} = \frac{1}{(1\times10^{-4}) + (13.33\times10^{-4})} = \frac{10^4}{14.33} = 697.84\,\Omega$$

$$A_v = \frac{-0.1933(1000)(697.84)}{1000 + 697.84} = -79.45 \leftarrow$$

$$R_B = \frac{(30{,}000)(9000)}{39{,}000} = 6923\,\Omega$$

$$r_\pi = \frac{\beta}{g_m} = \frac{150}{0.1933} = 776\,\Omega$$

$$R_{in} = r_\pi \| R_B = \frac{(776)(6923)}{776 + 6923} = \frac{(776)(6923)}{7699} = 697.8\,\Omega$$

$$A_i = -\frac{0.1933(697.84)(697.8)}{1000 + 697.84} = -55.44 \leftarrow$$

8.3.9. TO A GOOD APPROXIMATION A_{v1} IS ALTERED BY THE FACTOR $\frac{r_\pi}{R_i}$

WHERE $r_\pi = 776\,\Omega$ AND $R_i = r_\pi + R_{E1}(1 + g_m r_\pi)$

$$= 776 + 250(1 + 150)$$
$$= 38{,}526\,\Omega$$

$$\frac{r_\pi}{R_i} = \frac{776}{38{,}526} = 0.02$$

$$\therefore A_{v1}(\text{UNBYPASSED}) \approx 0.02(-79.45) = -1.6 \leftarrow$$

CURRENT GAIN IS REDUCED BY FACTOR

$$\frac{1}{1 + \frac{R_{E1}(1+\beta)}{r_\pi + R_B}} = \frac{1}{1 + \frac{250(151)}{776 + 6923}} = \frac{1}{1 + 4.903} = \frac{1}{5.903} = 0.1694$$

$$\therefore A_i = 0.1694(-55.44) = -9.39 \leftarrow$$

8.3.10

CE STAGE

$$A_p = \frac{v_L i_L}{v_1 i_s} = A_{v1} \cdot A_i = A_{v1}^2 \frac{R_{in}}{R_L} = \left[\frac{-g_m R_L (r_o \| R_C)}{R_L + (r_o \| R_C)}\right]^2 \frac{(r_\pi \| R_B)}{R_L} \leftarrow$$

8.3.11. $R_W = r_o \| R_E \| R_L$

IF $r_o \to \infty$, $R_W = R_E \| R_L$

$\therefore R_i = r_\pi + R_W(1 + g_m r_\pi)$ WILL BECOME

$R_i = r_\pi + (R_E \| R_L)(1 + g_m r_\pi)$ WHICH IS EQ. (8.3.13) ←

$$A_{v1} = \frac{R_W(1+g_m r_\pi)}{r_\pi + R_W(1+g_m r_\pi)} \xrightarrow[r_o \to \infty]{IF} \frac{(1+g_m r_\pi)(R_E \| R_L)}{r_\pi + (1+g_m r_\pi)(R_E \| R_L)} \simeq 1 \leftarrow$$

WHICH IS EQ (8.3.14)

$$R_{in} = R_B \| R_i \rightarrow R_B \| [r_\pi + (R_E \| R_L)(1 + g_m r_\pi)]$$

$$\therefore A_i = \frac{R_{in} R_W(1+g_m r_\pi)}{R_L[r_\pi + R_W(1+g_m r_\pi)]} \rightarrow \frac{R_B[r_\pi + (R_E\|R_L)(1+g_m r_\pi)](R_E\|R_L)(1+g_m r_\pi)}{R_L[R_B + r_\pi + (1+g_m r_\pi)(R_E\|R_L)][r_\pi + (R_E\|R_L)(1+g_m r_\pi)]}$$

$$= \frac{(1+g_m r_\pi)(R_E \| R_L) R_B}{R_L[R_B + r_\pi + (1+g_m r_\pi)(R_E \| R_L)]}$$ WHICH IS EQ. (8.3.15) ←

8.3.12.

$$g_m = \frac{I_{CQ}}{V_T} = \frac{1.2(10^{-3})}{25.861(10^{-3})} = 0.0464$$

$$r_o = \frac{V_A}{I_{CQ}} = \frac{75}{1.2(10^{-3})} = 62.5\ k\Omega$$

$$r_\pi = \frac{\beta}{g_m} = \frac{100}{0.0464} = 2155.2\ \Omega$$

$$R_W = \frac{1}{\frac{1}{62500} + \frac{1}{2000} + \frac{1}{3000}} = \frac{1}{(1.6\times10^{-5}) + (50\times10^{-5}) + (33.33\times10^{-5})}$$

$$= \frac{10^5}{84.93} = 1177.44\ \Omega$$

$$R_i = r_\pi + R_W(1+\beta)$$

$$= 2155.2 + 1177.44\ (101) = 121{,}076.6\ \Omega \leftarrow$$

$$R_{in} = \frac{1}{\frac{1}{121{,}076.6} + \frac{1}{150{,}000} + \frac{1}{150{,}000}} = \frac{1}{(8.26\times10^{-6}) + (6.66\times10^{-6}) + (6.66\times10^{-6})}$$

$$= \frac{10^6}{21.58} = 46{,}339.2\ \Omega \leftarrow$$

8.3.12. CONTD.

$$A_{v1} = \frac{R_W(1+\beta)}{R_i} = \frac{1177.44(101)}{121076.6} = 0.9822 \leftarrow$$

$$A_i = \frac{R_{in}}{R_L} A_{v1} = \frac{46339.2}{3000}(0.9822) = 15.17 \leftarrow$$

8.3.13.

COMMON-COLLECTOR STAGE

$$P_L = v_L i_L \; ; \; P_s = v_1 i_s \; ; \; A_p = \frac{v_L i_L}{v_1 i_s} = A_{v1} A_i$$

$$\therefore A_p = A_{v1} \cdot A_{v1} \frac{R_{in}}{R_L} = \left[\frac{R_W(1+\beta)}{r_\pi + R_W(1+\beta)}\right]^2 \frac{R_B \| R_i}{R_L}$$

$$= \frac{R_W^2(1+\beta)^2 R_B R_i}{R_i^2(R_B+R_i)R_L} = \frac{R_W^2(1+\beta)^2 R_B}{R_i R_L (R_B+R_i)} \leftarrow$$

8.3.14.

$$g_m = \frac{I_{CQ}}{V_T} = \frac{4\times10^{-3}}{25.861(10^{-3})} = 0.1547$$

$$r_o = \frac{V_A}{I_{CQ}} = \frac{70}{4(10^{-3})} = 17{,}500\ \Omega$$

$$r_\pi = \frac{\beta}{g_m} = \frac{50}{0.1547} = 323.2$$

$$R_B = \frac{1}{\frac{1}{32000}+\frac{1}{22000}} = \frac{1}{(3.125\times10^{-5})+(4.545\times10^{-5})} = \frac{10^5}{7.67} = 13{,}038\ \Omega$$

$$R_W = r_o \| R_E \| R_L = \frac{1}{\frac{1}{17500}+\frac{1}{400}+\frac{1}{250}} = 152.51\ \Omega$$

$$R_i = r_\pi + R_W(1+\beta) = 323.2 + 152.51(51) = 8101.2\ \Omega \leftarrow$$

$$R_{in} = R_B \| R_i = \frac{1}{\frac{1}{8101.2}+\frac{1}{13038}} = 4997.5\ \Omega \leftarrow$$

$$A_{v1} = R_W(1+\beta)/R_i = \frac{152.51\times51}{8101.2} = 0.96 \leftarrow$$

$$A_i = A_{v1} R_{in}/R_L = 0.96 \times 4997.5/250 = 19.19 \leftarrow$$

8.3.15. CB STAGE

BY USING EQS. (8.3.16) THROUGH (8.3.19)

$$A_p = \frac{v_L i_L}{v_1 i_s} = A_{v1}^2 \frac{R_{in}}{R_L} = \left[\frac{R_H(1+g_m r_o)}{R_H + r_o}\right]^2 \frac{(r_\pi \| R_E)(r_o + R_H)}{[r_o + R_H + (r_\pi \| R_E)(1+g_m r_o)]R_L}$$

$$= \frac{R_H^2 (1+g_m r_o)^2 (r_\pi \| R_E)}{(r_o + R_H)[r_o + R_H + (r_\pi \| R_E)(1+g_m r_o)]R_L} \quad \leftarrow$$

8.3.16. REDRAW FIG. 8.3.3b.

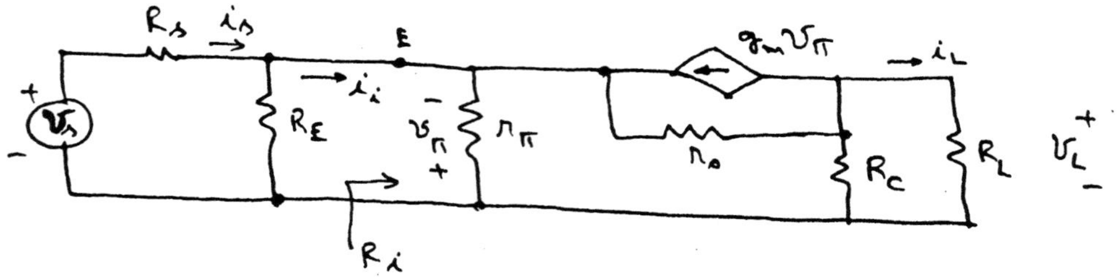

$$i_i' + g_m v_\pi = -\frac{v_\pi}{r_\pi} + \frac{(-v_\pi - v_L)}{r_o}$$

$$\text{or } i_i = -v_\pi\left[\frac{1}{r_o} + \frac{1}{r_\pi} + g_m\right] - \frac{v_L}{r_o} \quad \text{—①}$$

$$g_m v_\pi + \frac{v_L - (-v_\pi)}{r_o} + \frac{v_L}{R_C} + \frac{v_L}{R_L} = 0$$

OR $v_L\left(\frac{1}{r_o} + \frac{1}{R_C} + \frac{1}{R_L}\right) = -v_\pi\left(\frac{1}{r_o} + g_m\right)$ OR $v_L = \dfrac{-v_\pi\left(\frac{1}{r_o} + g_m\right)}{\left(\frac{1}{r_o} + \frac{1}{R_C} + \frac{1}{R_L}\right)}$ —②

USING ② IN ①

$$i_i = -v_\pi\left(\frac{1}{r_o} + \frac{1}{r_\pi} + g_m\right) - \frac{-v_\pi\left(\frac{1}{r_o} + g_m\right)}{r_o\left(\frac{1}{r_o} + \frac{1}{R_C} + \frac{1}{R_L}\right)}$$

$$= -v_\pi\left[\frac{1}{r_o} + \frac{1}{r_\pi} + g_m - \frac{\frac{1}{r_o} + g_m}{r_o\left(\frac{1}{r_o} + \frac{1}{R_C} + \frac{1}{R_L}\right)}\right]$$

$$R_i = \frac{-v_\pi}{i_i} = \frac{1}{\left(\frac{1}{r_o} + \frac{1}{r_\pi} + g_m\right) - \dfrac{\frac{1}{r_o} + g_m}{r_o\left(\frac{1}{r_o} + \frac{1}{R_C} + \frac{1}{R_L}\right)}}$$

$$= \frac{r_\pi(r_o + R_H)}{r_\pi + R_H + r_o(1+\beta)}$$ WHICH IS EQ. (8.3.17) $\leftarrow$

8.3.17. $r_\pi = \dfrac{\beta}{g_m} = \dfrac{60}{0.03} = 2000\,\Omega$; $g_m = \dfrac{I_{CQ}}{V_T}$ or $I_{CQ} = g_m V_T$

$$r_o = \frac{V_A}{I_{CQ}} = \frac{V_A}{g_m V_T} = \frac{70}{0.03(25.861)10^{-3}} = 90{,}226\,\Omega$$

$$R_H = R_C \| R_L = \frac{1}{\frac{1}{1000} + \frac{1}{6000}} = \frac{1}{(1\times10^{-3}) + (0.167\times10^{-3})}$$

$$= 10^3/1.167 \simeq 857\,\Omega$$

$$R_i = \frac{r_\pi(r_o + R_H)}{r_\pi + R_H + r_o(1+\beta)} = \frac{2000(90226 + 857)}{2000 + 857 + 90226(61)} = \frac{182.166\times10^6}{5.5066\times10^6}$$

$$= 33.1\,\Omega \leftarrow$$

$$R_{in} = \frac{R_E R_i}{R_E + R_i}\;;\quad R_E R_{in} + R_i R_{in} = R_E R_i\;;\quad R_E = -\frac{R_{in} R_i}{R_{in} - R_i}$$

$$\therefore R_E = -\frac{20(33.1)}{20 - 33.1} = 50.53\,\Omega \leftarrow$$

$$A_{v1} = \frac{R_H(1 + g_m r_o)}{r_o + R_H} = \frac{857(1 + (0.03\times 90226))}{90226 + 857} = 25.48 \leftarrow$$

$$A_i = A_{v1}\frac{R_{in}}{R_L} = 25.48\,\frac{20}{6000} = 0.085 \leftarrow$$

8.3.18. SEE PR. 8.3.8.

$g_m = 0.1933\,S$; $r_o = 10\,k\Omega$; $\beta = 150$; $r_\pi = 776\,\Omega$

$$R_H = R_C \| R_L = \frac{1}{\frac{1}{750} + \frac{1}{1000}} = 428.63\,\Omega$$

$$R_i = \frac{r_\pi(r_o + R_H)}{r_\pi + R_H + r_o(1+\beta)} = \frac{776(10000 + 429)}{776 + 429 + 10000(151)} = 5.364\,\Omega \leftarrow$$

$$R_{in} = R_E \| R_i = \frac{1}{\frac{1}{250} + \frac{1}{5.364}} = 5.252\,\Omega \leftarrow$$

8.3.18. CONTD.

$$A_{v1} = \frac{R_4(1+g_m r_o)}{R_4 + r_o} = \frac{429(1+(0.1933 \times 10{,}000))}{429 + 10{,}000}$$

$$= \frac{829686}{10429} = 79.56 \leftarrow$$

$$A_i = A_{v1}\frac{R_{in}}{R_L} = 79.56\,\frac{5.252}{1000} = 0.42 \leftarrow$$

8.4.1.

$$r_o \| R_D \| R_L = \frac{1}{\frac{1}{15000} + \frac{1}{2000} + \frac{1}{3000}} = \frac{1}{(6.667\times10^{-5}) + (50\times10^{-5}) + (33.333\times10^{-5})}$$

$$= \frac{10^5}{90} = 1111\ \Omega$$

$$A_{v1} = -4.5 = -g_m(r_o \| R_D \| R_L)$$

$$g_m = \frac{4.5}{1111} = 4.05\times10^{-3}\ \text{S} \leftarrow$$

8.4.2.

$$A_{v1} = \frac{-g_m r_o R_F}{r_o + R_F + R_{SS1}(1+g_m r_o)}$$

$$A_i = \frac{-g_m r_o R_F(R_1 \| R_2)}{R_L[r_o + R_F + R_{SS1}(1+g_m r_o)]}$$

NOTE THAT THESE WILL BE THE SAME AS EQS. (8.4.3) AND (8.4.4), AS THEY SHOULD, WHEN $R_{SS1} = 0$. THE RESULT IS LOWER AMPLIFICATION GAIN IN MAGNITUDE FOR BOTH THE VOLTAGE AND CURRENT. WHILE THAT IS A DISADVANTAGE, HOWEVER, ADVANTAGES OF INCREASED BANDWIDTH, GAIN STABILITY, AND LARGER ALLOWABLE SIGNAL AMPLITUDES CAN BE REALIZED.

8.4.3. (a) BECAUSE $r_o = \frac{V_A}{I_{DQ}}$, $I_{DQ} = \frac{V_A}{r_o} = \frac{80}{2000} = 0.04\ \text{A}$

$$r_o \| R_L \| R_D = \frac{1}{\frac{1}{2000} + \frac{1}{300} + \frac{1}{130}} = 95.24\ \Omega$$

$$A_{v1} = -g_m(r_o \| R_D \| R_L)$$

$$-2.8 = -g_m(95.24) \Rightarrow g_m = \frac{2.8}{95.24} = 0.03\text{S} \leftarrow$$

$$g_m = 0.03 = \frac{2 I_{DSS}}{V_p}\sqrt{\frac{I_{DQ}}{I_{DSS}}} = \frac{2\sqrt{I_{DSS}}\sqrt{I_{DQ}}}{V_p};\ \sqrt{I_{DSS}} = \frac{0.03\times4}{2\sqrt{0.04}} = 0.3 \text{ OR } I_{DSS} = 0.09\text{A} \leftarrow$$

8.4.3. CONTD.
(b)

$$A_{v1} = -1.4 = \frac{-g_m r_o R_F}{r_o + R_F + R_{SS1}(1+g_m r_o)}$$ (See Pr. 8.4.2 & ITS SOLN)

$$R_F = R_D \| R_L = \frac{1}{\frac{1}{150}+\frac{1}{300}} = \frac{1}{(6.667\times10^{-3})+(3.333\times10^{-3})}$$

$$= \frac{10^{+3}}{10} = 100\ \Omega$$

$$R_{SS1} = \left[\frac{-g_m r_o R_F}{-1.4} - r_o - R_F\right]\frac{1}{1+g_m r_o}$$

$$= \frac{[-0.03(2000)(100)/-1.4] - 2000 - 100}{1+(0.03)(2000)}$$

$$= \frac{4285.7 - 2100}{1+60} = \frac{2185.7}{61} = 35.83\ \Omega \leftarrow$$

8.4.4.

$$R_{in} = R_1 \| R_2 = \frac{330(110)}{330+110}\ k\Omega = 82.5\ k\Omega$$

$$r_o \simeq \frac{V_A}{I_{DQ}} = \frac{90}{6\times10^{-3}} = 15\ k\Omega$$

$$g_m = \frac{2\ I_{DSS}}{V_p}\sqrt{\frac{I_{DQ}}{I_{DSS}}} = \frac{2(20\times10^{-3})}{4}\sqrt{\frac{6}{20}} = 5.477\times10^{-3}\ S$$

FROM EQS. (8.4.3) THROUGH (8.4.5)

$$R_F = R_D \| R_L = \frac{(1000)(1000)}{2000} = 500\Omega;\quad A_{v1} = \frac{-g_m r_o R_F}{r_o + R_F} = \frac{-(5.477\times10^{-3})(15000)(500)}{15000+500}$$

$$= -2.65 \leftarrow$$

$$A_i = \frac{R_{in}}{R_L}A_{v1} = \frac{82.5}{1}(-2.65) = -218.625 \leftarrow$$

RELATIVELY SMALL VOLTAGE GAIN, BUT VERY LARGE CURRENT GAIN BECAUSE R_1 AND R_2 WERE SELECTED SO LARGE.

8.4.5.

$$A_{v1} = \frac{g_m r_o (R_{SS} \| R_L)}{r_o + (R_{SS} \| R_L)(1 + g_m r_o)} \quad (8.4.8)$$

$$= \frac{g_m r_o (R_{SS} \| R_L)}{r_o + (R_{SS} \| R_L) + g_m r_o (R_{SS} \| R_L)}$$

$$= \frac{g_m r_o (R_{SS} \| R_L)}{[r_o + (R_{SS} \| R_L)]\left[1 + \dfrac{g_m r_o (R_{SS} \| R_L)}{r_o + (R_{SS} \| R_L)}\right]} = \frac{g_m r_o (r_o \| R_{SS} \| R_L)}{1 + g_m (r_o \| R_{SS} \| R_L)} \leftarrow$$

$$A_i = \frac{g_m r_o (R_{SS} \| R_L)(R_1 \| R_2)}{R_L[r_o + (R_{SS} \| R_L)(1 + g_m r_o)]} \quad (8.4.9)$$

$$= \frac{g_m r_o (R_{SS} \| R_L)(R_1 \| R_2)}{R_L [r_o + (R_{SS} \| R_L)]\left[1 + \dfrac{g_m r_o (R_{SS} \| R_L)}{r_o + (R_{SS} \| R_L)}\right]}$$

$$= \frac{g_m (r_o \| R_{SS} \| R_L)(R_1 \| R_2)}{R_L [1 + g_m (r_o \| R_{SS} \| R_L)]} \leftarrow$$

8.4.6.

$$A_{v1} = 0.7 = \frac{g_m r_o (R_{SS} \| R_L)}{r_o + (R_{SS} \| R_L)(1 + g_m r_o)}$$

$$0.7 r_o + 0.7 (R_{SS} \| R_L)(1 + g_m r_o) = g_m r_o (R_{SS} \| R_L)$$

$$(R_{SS} \| R_L) = 0.7 r_o / (g_m r_o - 0.7 - 0.7 g_m r_o)$$

$$= 0.7 r_o / (0.3 g_m r_o - 0.7)$$

$$(R_{SS} \| R_L) = \frac{0.7 \times 5000}{0.3(0.025)(5000) - 0.7} = \frac{3500}{36.8} = 95.11$$

$$= \frac{R_{SS} R_L}{R_{SS} + R_L}$$

$$95.11 R_{SS} + 95.11 R_L = R_{SS} R_L$$

$$R_L = \frac{95.11 R_{SS}}{R_{SS} - 95.11} = \frac{95.11(300)}{300 - 95.11} = \frac{28533}{204.89}$$

$$= 139.26\,\Omega \leftarrow$$

8.4.7.

R_{OUT} AS RESISTANCE OF THÉVENIN'S SOURCE

$$R_{OUT} = \frac{V_{OC}}{I_{SS}}$$ WHERE I_{SS} IS THE SHORT-CIRCUIT CURRENT.

8.4.7. CONTD.

$$A_{v1} = \frac{v_L}{v_1} = \frac{g_m r_o (R_{SS} \| R_L)}{r_o + (R_{SS} \| R_L)(1 + g_m r_o)} \quad \text{FOR ANY } R_L$$

$$V_{OC} = v_L \Big|_{R_L \to \infty} = A_{v1} v_1 \Big|_{R_L \to \infty} = \frac{g_m r_o (R_{SS} \| R_L) v_1}{r_o + (R_{SS} \| R_L)(1 + g_m r_o)} \Bigg|_{R_L \to \infty} = \frac{g_m r_o R_{SS} v_1}{r_o + R_{SS}(1 + g_m r_o)}$$

$$I_{SS} = \frac{v_L}{R_L} \Big|_{R_L \to 0} = \frac{A_{v1} v_1}{R_L} \Big|_{R_L \to 0} = \frac{g_m r_o R_{SS} R_L v_1}{R_2 (R_{SS} + R_L)[r_o + (R_{SS} \| R_L)(1 + g_m r_o)]} \Bigg|_{R_L \to 0} = g_m v_1$$

$$R_{OUT} = \frac{V_{OC}}{I_{SS}} = \frac{r_o R_{SS}}{r_o + R_{SS}(1 + g_m r_o)} = \frac{r_o R_{SS}}{r_o + R_{SS} + g_m r_o R_{SS}}$$

$$= \frac{1}{\frac{1}{R_{SS}} + \frac{1}{r_o} + g_m} = \left(r_o \| R_{SS} \| \frac{1}{g_m} \right) \leftarrow$$

8.4.8.

FROM EX. 8.4.1 : $r_o = 26{,}666.7\ \Omega$; $g_m = 2.7386 \times 10^{-3}$ S ; $R_F = 750\ \Omega$

$$A_{v1} = \frac{R_F (1 + g_m r_o)}{r_o + R_F} = \frac{750 [1 + (2.7386 \times 10^{-3} \times 26{,}666.7)]}{26{,}666.7 + 750}$$

$$= \frac{750 (74.03)}{27416.7} = 2.025 \leftarrow$$

$$A_i = \frac{R_F (1 + g_m r_o) R_{SS}}{R_L [r_o + R_F + R_{SS}(1 + g_m r_o)]} = \frac{750 (1 + \{2.7386 \times 10^{-3} \times 26666.7\}) 1000}{3000 [26666.7 + 750 + 1000 (1 + 2.7386 \times 10^{-3} \times 26666.7)]}$$

$$= \frac{750 (74.03) 1000}{3000 \times 101446.7} = 0.182 \leftarrow$$

$$R_{in} = \frac{R_{SS}(r_o + R_F)}{r_o + R_F + R_{SS}(1 + g_m r_o)} = \frac{1000 (26666.7 + 750)}{26666.7 + 750 + 1000 [1 + \{2.7386 \times 10^{-3} \times 26666.7\}]}$$

$$= \frac{1000 \times 27416.7}{101446.7} = 270.26\ \Omega \leftarrow$$

8.4.9.

$$R_{in} = \frac{R_i R_{SS}}{R_i + R_{SS}} \rightarrow R_{in} R_i + R_{in} R_{SS} = R_i R_{SS}, \quad R_i = \frac{R_{in} R_{SS}}{R_{SS} - R_{in}}$$

$$R_{SS} - R_{in} = R_{SS} - \frac{R_{SS}(r_o + R_F)}{r_o + R_F + R_{SS}(1+g_m r_o)} = \frac{R_{SS}^2(1+g_m r_o)}{r_o + R_F + R_{SS}(1+g_m r_o)}$$

$$R_i = \frac{\dfrac{R_{SS}^2(r_o + R_F)}{[r_o + R_F + R_{SS}(1+g_m r_o)]}}{\dfrac{R_{SS}^2(1+g_m r_o)}{[r_o + R_F + R_{SS}(1+g_m r_o)]}}$$

$$\therefore R_i = \frac{r_o + R_F}{1+g_m r_o} \leftarrow$$

8.4.10.

$$R_F = R_D \| R_L = \frac{1}{\frac{1}{15000} + \frac{1}{7500}} = 5000 \ \Omega$$

$$A_{v1} = \frac{R_F(1+g_m r_o)}{R_F + r_o} = \frac{5000(1 + 5\times10^{-3} \times 100\times10^3)}{5000 + 100{,}000} = \frac{5000 \times 501}{105{,}000}$$

$$= 23.86 \leftarrow$$

$$A_i = \frac{R_F(1+g_m r_o) R_{SS}}{R_L[r_o + R_F + R_{SS}(1+g_m r_o)]} = \frac{5000(1+500)5000}{15000[100{,}000 + 5000 + 5000\times 501]}$$

$$= \frac{5000 \times 501 \times 5000}{15000 \times 2610000} = 0.32 \leftarrow$$

$$R_{in} = \frac{R_{SS}(r_o + R_F)}{r_o + R_F + R_{SS}(1+g_m r_o)} = \frac{5000(100000 + 5000)}{100000 + 5000 + 5000(501)}$$

$$= \frac{5000 \times 105000}{2610000} = 201.15 \ \Omega \leftarrow$$

8.5.1.

BY INSPECTION $\omega_H = 5\pi(10^4)$ rad/s $\leftarrow$ AND $\omega_L = 120\pi$ rad/s $\leftarrow$

FOR $\omega > 120\pi$ AND $\omega < 5\pi(10^4)$

$$A_v \simeq \frac{20 j\omega}{j\omega} = 20 \leftarrow$$

8.5.2.

$$R_G = R_1 \| R_2 = \frac{1}{\frac{1}{520000} + \frac{1}{140000}} = \frac{1}{(1.923\times10^{-6}) + (7.143\times10^{-6})}$$

$$= 110{,}302.23 \ \Omega$$

$$A_{vo} = \frac{-g_m R_G R_D R_L}{(R_S + R_G)(R_D + R_L)} = \frac{-5\times10^{-3}(110302)(1000)(2000)}{(500 + 110302)(1000 + 2000)} = -3.32 \leftarrow$$

8.5.2. CONTD.

$$R_{L1} = \frac{R_S}{1+g_m R_{SS}} = \frac{1400}{1+5(1.4)} = \frac{1400}{8} = 175\ \Omega$$

$$R_{L2} = R_D + R_L = 1000 + 2000 = 3000\ \Omega$$

$$R_{L3} = R_S + R_G = 500 + 110{,}302 = 110{,}802\ \Omega$$

$$C_S = \frac{1}{R_{L1}\,\omega_L} = \frac{1}{175\,(40\pi)} = 45.45\ \mu F \leftarrow$$

$$C_D = \frac{1}{R_{L2}\,\omega_L/10} = \frac{10}{3000\,(40\pi)} = 26.52\ \mu F \leftarrow$$

$$C_G = \frac{1}{R_{L3}\,\omega_L/10} = \frac{10}{110802\,(40\pi)} = 0.72\ \mu F \leftarrow$$

8.5.3.

$$R_{DL} = r_o \| R_D \| R_L = \frac{1}{\frac{1}{20000} + \frac{1}{1000} + \frac{1}{2000}}$$

$$= \frac{1}{(5\times10^{-5}) + (100\times10^{-5}) + (50\times10^{-5})}$$

$$= \frac{10^5}{155} = 645.16\ \Omega$$

$$\omega_H = \frac{1}{R_A\left[C_{gs} + C_{gd}\left(1 + g_m R_{DL} + \frac{R_{DL}}{R_A}\right)\right]}$$

WHERE $R_A = R_S \| R_G = \frac{R_S R_G}{R_S + R_G} = \frac{500 \times 110302}{110802}$

$= 497.7\ \Omega$

$$\omega_H = \frac{1}{497.7\left[2\times10^{-12} + 2\times10^{-12}\left(1 + 5\times10^{-3}\times 645.16 + \frac{645.16}{497.7}\right)\right]}$$

$$= \frac{1}{497.7\left[2\times10^{-12} + 2\times10^{-12}(5.53)\right]}$$

$$= \frac{10^{12}}{497.7 \times 13.06} = 153.8\times10^{6}\ \text{rad/s} \leftarrow$$

8.5.3. CONTD.

$$R_{DL} = r_o \| R_D \| R_L \simeq R_D \| R_L = R_L/2$$

$$R_A = R_S \| (R_1 \| R_2) \simeq R_S$$

$$\omega_H = \frac{1}{R_A\left[C_{gs} + C_{gd}(1+g_m R_{DL}) + \frac{R_{DL}}{R_A}\right]} \simeq \frac{1}{R_S C_{gd}\left[2 + g_m \frac{R_L}{2} + \frac{R_L}{2R_S}\right]} \quad (1)$$

$$\omega_H = \frac{\omega_Z}{75} = \frac{g_m}{75\, C_{gd}} \quad — (2)$$

SOLVE (1) & (2) TOGETHER:

$$g_m = \frac{75}{R_S\left[2 + \frac{g_m R_L}{2} + \frac{R_L}{2R_S}\right]} \quad ; \quad 2g_m + \left(\frac{g_m^2}{2} + \frac{g_m}{2R_S}\right)R_L = \frac{75}{R_S}$$

$$\therefore R_L = \left(\frac{75}{R_S} - 2g_m\right)\frac{2}{g_m\left(g_m + \frac{1}{R_S}\right)} = \frac{2(75 - 2g_m R_S)}{g_m(1 + g_m R_S)}$$

$$= \frac{2\left[75 - 2(5)10^{-3}(1\times10^3)\right]}{5\times10^{-3}\left\{1 + (5\times10^{-3}\times1\times10^3)\right\}} = \frac{2(75-10)}{5\times10^{-3}\times 6} = \frac{2\times65\times10^3}{30}$$

$$R_L = 4333\ \Omega \leftarrow$$

$$C_{gs} = \frac{g_m}{\omega_Z} = \frac{g_m}{75\,\omega_H} \simeq \frac{g_m}{75\,\omega_{3dB}} \simeq \frac{5\times10^{-3}}{75\,(25)\,10^6} = 2.67\ pF.$$

$$A_{VO} = \frac{-g_m R_G R_{DL}}{R_S + R_G} \simeq -g_m R_{DL} \simeq -g_m \frac{R_L}{2} = -5(10^{-3})\frac{4333}{2}$$

$$= -10.83 \leftarrow$$

8.5.4.

(a) FOR $|v_d| \le 0.02V$, $v_{out} \simeq 50 v_d$

THE UNDISTORTED OUTPUT IS LIMITED TO $|v_{out}|_{max} \simeq 1V$. (See Fig. a below)

(b) $v_{in} = v_d + B v_{out}$. FIG. b Shows curves for $B = 0.03$ & 0.08.

FROM Fig b: $|v_{out}|_{max} \simeq 2V$ WITH $B = 0.03$

$|v_{out}|_{max} \simeq 3V$ WITH $B = 0.08$

REDUCED GAIN TO GET IMPROVED LINEARITY.

$\because v_{out}/v_d \simeq 50$ FOR SMALL SIGNALS, EQ. (8.6.1) PREDICTS THAT THE GAIN WITH FEEDBACK WILL BE

$$A_f \simeq \frac{50}{1+50B} = \begin{cases} 20 & \text{for } B = 0.03 \\ 10 & \text{for } B = 0.08. \end{cases}$$

8.5.4. CONTD.

(a) Nonlinear transfer characteristic. (b) Improved linearity with feedback.

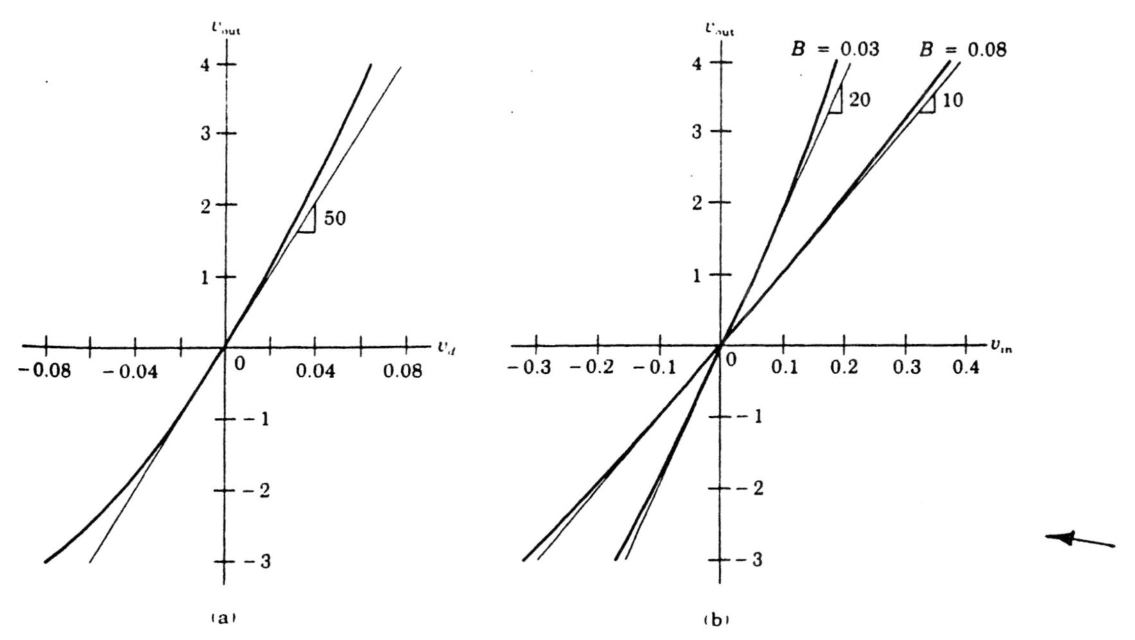

FIG. S 8.5.4.

8.5.5. (a) $A_f = \dfrac{A}{1+AB}$

$$-20 = \frac{-100}{1-100B} \Rightarrow B = -0.04 \leftarrow$$

(b) $A_f = \dfrac{-50}{1-50(-0.04)} = -16.7$ FOR $A = -50$

$A_f = \dfrac{-200}{1-200(-0.04)} = -22.2$ FOR $A = -200$

THUS $-22.2 \le A_f \le -16.7$ FOR $-200 \le A \le -50 \leftarrow$

8.5.6. (a)

B, i_b, hie, v_b, $h_{re} v_c$, $h_{fe} i_b$, i_c, C, $h_{oe} v_c$, hoe, v_c, E, E

(b)

B, i_b, v_b, hie, $h_{fe} i_b$, i_c, C, v_c, E, E

$$I_L = -I_c = \frac{-1/h_{oe}}{(1/h_{oe}) + R_L} h_{fe} I_b = \frac{-h_{fe}}{1 + h_{oe} R_L} I_b \quad (1)$$

$$A_{I1} = \frac{I_L}{I_b} = \frac{-I_c}{I_b} = -\frac{h_{fe}}{1 + h_{oe} R_L} \quad (2) \leftarrow$$

$$A_{I2} = \frac{I_L}{I_s} = \frac{I_L}{I_b} \cdot \frac{I_b}{I_s} \quad ; \quad \frac{I_b}{I_s} = \frac{R_B}{R_B + h_{ie}} = \frac{1}{1 + h_{ie}/R_B}$$

$$\therefore A_{I2} = \frac{-h_{fe}}{(1 + h_{oe} R_L)(1 + h_{ie}/R_B)} \quad (3) \leftarrow$$

8.5.6. CONTD.

(c)

$$A_{v1} = \frac{V_c}{V_b} = \frac{-I_c R_L}{V_b} = \frac{A_{I1} I_b R_L}{V_b} = \frac{A_{I1} R_L}{V_b / I_b}$$

$$= A_{I1} R_L / h_{ie}$$

$$A_{v1} = \frac{-h_{fe}}{1 + h_{oe} R_L} \frac{R_L}{h_{ie}} \leftarrow (4)$$

$$A_{v2} = \frac{V_c}{V_s} = \frac{V_c}{V_b} \cdot \frac{V_b}{V_s} = A_{v1} \frac{V_b}{V_s}$$

$$R_i = (R_B \| h_{ie}) = \frac{R_B h_{ie}}{R_B + h_{ie}} = \frac{h_{ie}}{1 + h_{ie}/R_B}$$

$$\frac{V_b}{V_s} = \frac{R_i}{R_i + r_s} = \frac{1}{1 + r_s / R_i}$$

$$\therefore A_{v2} = \frac{-h_{fe}}{1 + h_{oe} R_L} \left(\frac{R_L}{h_{ie}} \right) \frac{1}{1 + r_s / R_i} \leftarrow (5)$$

(d)

THE h-PARAMETER EQUIVALENT CIRCUIT OF THE COMMON-EMITTER AMPLIFIER FOR THE SMALL MAGNITUDE OF THE INPUT SIGNAL IS SHOWN BELOW: h_{re} IS ASSUMED EQUAL TO ZERO; $R_B = (R_1 \| R_2)$

(NOTE: DEALING WITH AN AUDIO AMPLIFIER, THE DIFFUSION AND TRANSITION CAPACITORS IN THE COMPLETE EQUIVALENT CIRCUIT OF THE TRANSISTOR MAY BE NEGLECTED.

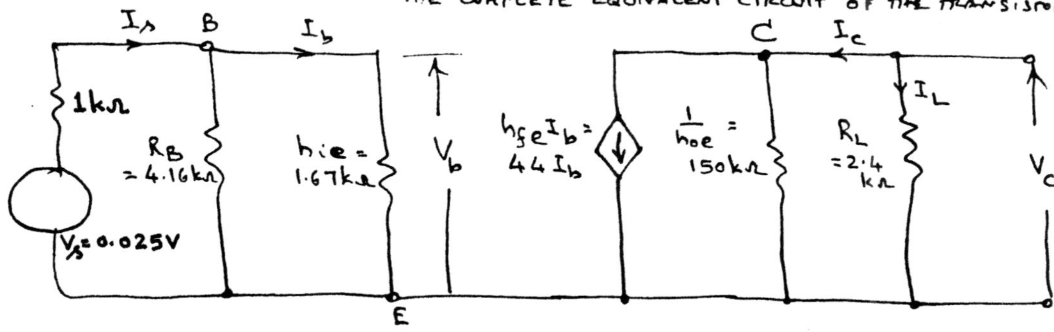

$$R_i = (R_B \| h_{ie}) = \frac{4.16 \times 1.67}{4.16 + 1.67} \text{ k}\Omega = 1.19 \text{ k}\Omega$$

$$I_s = \frac{V_s}{1000 + R_i} = \frac{0.025}{2.19 \times 10^3} = 11.4 \ \mu A$$

$$I_b = \frac{4.16}{5.83} I_s = 8.15 \ \mu A \text{ (RMS)}$$

$$h_{fe} I_b = 44(8.15) = 0.358 \text{ mA}$$

8.5.6.(d) CONTD.

$$I_c = \frac{150}{152.4}(0.358) = 0.352 \text{ mA}$$

$$A_{I2} = \frac{I_L}{I_s} = \frac{-I_c}{I_s} = -\frac{0.352\times10^{-3}}{11.4\times10^{-6}} = -30.9 \leftarrow$$

[ALSO CHECK BY EQ. (3) OF PART(C) SOLUTION:

$$A_{I2} = \frac{I_L}{I_s} = \frac{-44}{(1+2.4/150)(1+1.67/4.16)} \approx -30.9 \text{]}$$

$$A_{I1} = \frac{I_L}{I_b} \approx \frac{-I_c}{I_b} = -\frac{352}{8.15} \approx -43.2 \leftarrow$$

{OR FROM EQ.(2) OF PART (C) SOLUTION: $A_{I1} = \frac{-44}{1+2.4/150} \approx -43.2$]

$$A_{v2} = \frac{V_L}{V_s} = \frac{V_c}{V_s} = \frac{-I_c R_L}{V_s} = -\frac{0.352(2.4)}{0.025} = -33.8 \leftarrow$$

[CAN ALSO APPLY EQ. 5 OF PART(C) SOLUTION]

$$\text{POWER GAIN} = |A_{v2}|\,|A_{I2}| = (33.8)(30.9) = 1042 \leftarrow$$

$$R_i = 1.19\text{ k}\Omega \leftarrow$$

$$R_o = \left(R_L \,\|\, \frac{1}{h_{oe}}\right) = \frac{2.4\times150}{152.4}\text{ k}\Omega = 2.36\text{ k}\Omega \leftarrow$$

(e) THE h-PARAMETER EQUIVALENT CIRCUIT OF THE TWO-STAGE AMPLIFIER IS SHOWN BELOW:

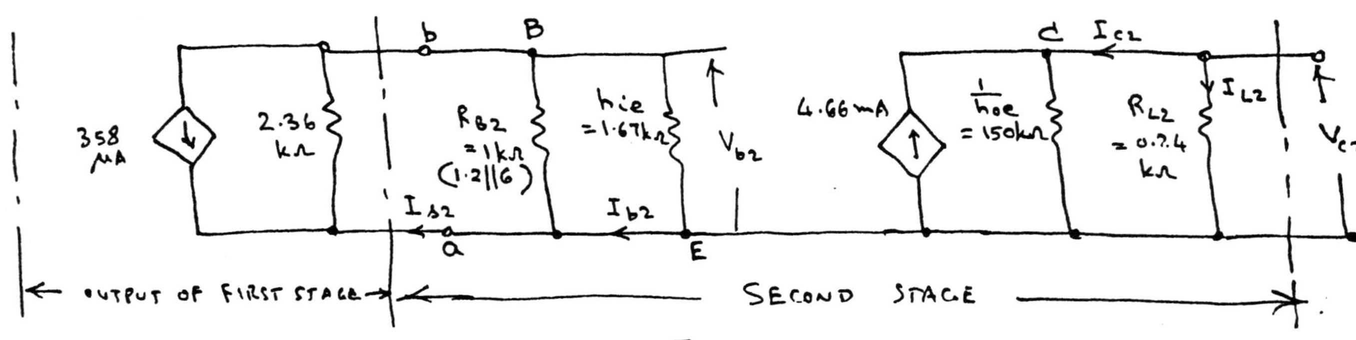

8.5.6.(e) CONTD.

$$R_{i2} = \frac{1(1.67)}{1+1.67} = 0.625\,k\Omega$$

$$I_{s2} = \frac{2.36}{2.36+0.625}(358) = 283\,\mu A$$

$$I_{b2} = \frac{1}{1+1.67} I_{s2} = \frac{1}{2.67}(283) = 106\,\mu A$$

MAXM. INPUT CURRENT SWING IS $\sqrt{2}\,(106) = 150\,\mu A$ WHICH IS SUFFICIENTLY SMALL (FOR 2N104 TRANSISTOR) TO ALLOW COMPUTATIONS TO BE MADE BY MEANS OF THE EQUIV. CKT RATHER THAN BY GRAPHICAL ANALYSIS. HOWEVER, SUCH VALIDITY MAY NOT HOLD GOOD FOR A THIRD STAGE!

$$h_{fe} I_{b2} = 44(106) = 4.66\,mA$$

OUTPUT RESISTANCE OF THE SECOND STAGE AS SEEN BETN. C & E $\simeq 240\,\Omega$

$$I_{c2} \approx h_{fe} I_{b2} = -4.66\,mA$$

$$A_{I1} = -\frac{I_{c1}}{I_{s1}} = -\frac{352}{11.4} = -30.9$$ FOR FIRST STAGE

$$A_{I1} = \frac{I_{s2}}{I_{s1}} = -\frac{283}{11.4} = -24.8$$

BECAUSE OF THE LOADING EFFECT WHEN THE SECOND STAGE IS MADE TO FOLLOW THE FIRST STAGE. (REDUCED CURRENT GAIN MAGNITUDE)

$$A_{I2} = \frac{I_{L2}}{I_{s2}} = -\frac{I_{c2}}{I_{s2}} = \frac{4.66}{-0.283} = -16.5$$

(UNLOADED CURREN GAIN BECAUSE THERE IS NO THIRD STAGE HERE!)

TOTAL CURRENT GAIN OF THE COMPLETE 2-STAGE AMPLIFIER

$$\frac{I_{L2}}{I_{s1}} = -\frac{I_{c2}}{I_{s1}} = \frac{4.66}{0.0114} = 408 \leftarrow$$

$$\left[\text{OR } \frac{I_{L2}}{I_{s1}} = -\frac{I_{c2}}{I_{s1}} = \frac{I_{s2}}{I_{s1}}\left(-\frac{I_{c2}}{I_{s2}}\right) = 24.8\,(16.5) = 408\right]$$

$$\frac{V_{c2}}{V_s} = -\frac{I_{c2} R_{L2}}{V_s} = \frac{4.66\,(0.24)}{0.025} = 44.74 \leftarrow$$

8.5.6.(e) CONTD. POWER GAIN FOR BOTH STAGES = $G = 408 \times 44.74 = 18{,}254$ ←

R_{i2} = INPUT RESISTANCE TO SECOND STAGE = $\dfrac{1 \times 1.67}{1 + 1.67} = 0.625\,k\Omega = R_{p2}$ ←

OUTPUT RESISTANCE OF THE SECOND STAGE = $R_{o2} = \dfrac{(150\,k\Omega)(0.24\,k\Omega)}{150.24\,k\Omega} \simeq 0.24\,k\Omega$ ←

(f)

$$I_2 = \frac{-R_L}{R_L + R_2} h_{fe} I_b = \frac{-h_{fe}}{1 + R_2/R_L} I_b$$

$$I_b = \frac{R_B}{R_B + h_{ie}} I_s = \frac{1}{1 + h_{ie}/R_B} I_s$$

$$\therefore I_2 = \frac{-h_{fe}}{(1 + R_2/R_L)(1 + h_{ie}/R_B)} I_s$$

∴ CURRENT GAIN OF THE COMPLETE STAGE INCLUDING THE LOADING EFFECT OF THE NEXT STAGE IS THEN

$$A_{Im} = \frac{I_2}{I_s} = \frac{-h_{fe}}{(1 + R_2/R_L)(1 + h_{ie}/R_B)} \leftarrow$$

[Note: Subscript m to denote reference to the midband freq. range]

A_{Im} IS INVARIANT WITH FREQUENCY.

(g)

$$I_2 = \frac{-R_L}{R_L + R_2 + 1/j\omega C_c} h_{fe} I_b = \frac{-R_L}{R_L + R_2}\left[\frac{1}{1 + \dfrac{1}{j\omega C_c (R_L + R_2)}}\right]$$

BUT $I_b = \dfrac{R_B}{R_B + h_{ie}} I_s = \dfrac{1}{1 + h_{ie}/R_B} I_s$

$$\therefore I_2 = \frac{-h_{fe}}{(1 + R_2/R_L)(1 + h_{ie}/R_B)}\left[\frac{1}{1 - j\dfrac{1}{\omega (R_L + R_2) C_c}}\right] I_s$$

FOR $\omega = \omega_\ell$, $\dfrac{1}{\omega_\ell (R_L + R_2) C_c} = 1$

$$\text{or } \omega_\ell = \frac{1}{(R_L + R_2) C_c} \text{ rad/s} \leftarrow$$

8.5.6.(g) CONTD.

$$f_\ell = \frac{1}{2\pi (R_L + R_2) C_c} \text{ Hz} \leftarrow$$

CURRENT GAIN AT LOW FREQUENCY FOR THE ENTIRE STAGE UNDER LOADED CONDITION IS GIVEN BY

$$A_{I\ell} = A_{Im} \frac{1}{1 - j(\omega_\ell/\omega)} \leftarrow$$

AT $\omega = \omega_\ell$, 30% DECREASE IN GAIN FROM THE MIDBAND VALUE.

ω_ℓ IDENTIFIES THE LOWER END OF THE USEFUL BANDWIDTH OF THE AMPLIFIER.

(h)

$$I_1 = I_D + I_T \quad \text{—(i)}$$

$$I_D = j\omega C_D v_{B'E} \quad \text{—(ii)} \; ; \quad I_T = j\omega C_T (v_{B'E} - V_c)$$

WHERE V_c IS THE RMS VALUE OF THE COLLECTOR VOLTA.

ALSO $I_T - g_m v_{B'E} + I_c = 0$ WHERE I_c IS GOING THROUGH R_p, WHICH IS THE EQUIV. RES OF || COMBN of R_L & R_2.

$$V_c = -I_c R_p$$

$$\frac{1}{R_p} = \frac{1}{R_L} + \frac{1}{R_2}$$

IT FOLLOWS THEN

$$V_C = -g_m v_{B'E} R_p + I_T R_p$$

$$\because I_T << g_m v_{B'E}$$

$$V_c \simeq -g_m v_{B'E} R_p \quad \text{—(iii)}$$

THEN $I_T = j\omega C_T (1 + g_m R_p) v_{B'E}$

SUBSTITUTING (iii) AND (ii) INTO (i), ONE GETS

$$I_1 = j\omega [C_D + C_T (1 + g_m R_p)] v_{B'E} = j\omega C v_{B'E} \quad \text{—(iv)}$$

WHERE C IS AN EQUIV. CAP AS SEEN BETN. TERMINALS B' AND E.

AND DEFINED BY $C = C_D + C_T (1 + g_m R_p)$

THE CIRCUIT MAY NOW BE REPLACED BY THE FOLLOWING BY MEANS OF EQ (iv)

8.5.6.(h) CONTD.

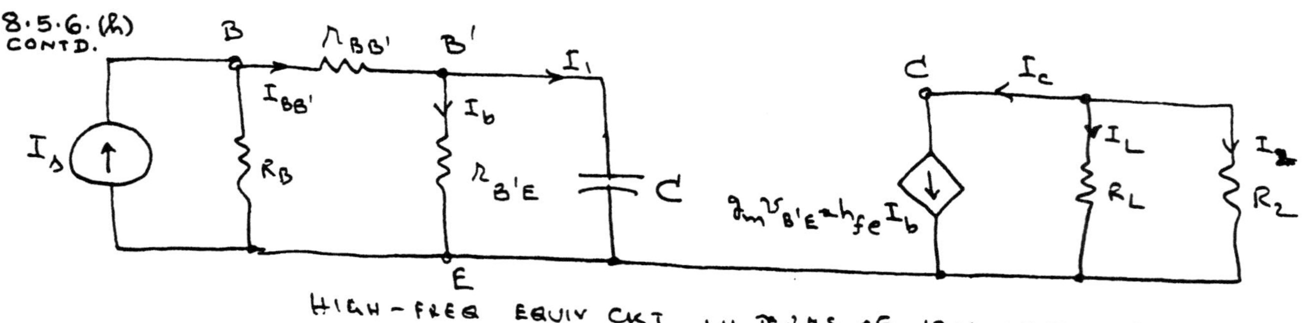

HIGH-FREQ EQUIV CKT IN TERMS OF ISOLATED INPUT & OUTPUT SECTIONS

ONE-LOOP EQUIV. OF THE INPUT SECTION OF THE ABOVE IS SHOWN BELOW:

B — $r_{BB'}$ — B'; R_B; $+$ $I_s R_B$; $\dfrac{r_{B'E}}{1+j\omega r_{B'E} C}$; E; C; I_c; $h_{fe} I_b$; I_L; R_L; I_2; R_2

$$I_2 = \frac{-R_L}{R_L + R_2} h_{fe} I_b = \frac{-h_{fe}}{1 + R_2/R_L} I_b$$

$$\left(r_{B'E} \,\|\, \frac{1}{j\omega C}\right) = \frac{r_{B'E}}{1 + j\omega r_{B'E} C}$$

$$v_{B'E} = \frac{\dfrac{r_{B'E}}{1 + j\omega r_{B'E} C}}{R_B + r_{BB'} + \dfrac{r_{B'E}}{1 + j\omega r_{B'E} C}} R_B I_s$$

BY VOLTAGE-DIVISION PRINCIPLE

$$= \frac{r_{B'E} R_B I_s}{R_B + h_{ie} + j\omega r_{B'E}(R_B + r_{BB'}) C}$$

WHERE $h_{ie} = r_{BB'} + r_{B'E}$

DIFFUSION BASE CURRENT $I_b = \dfrac{v_{B'E}}{r_{B'E}} = \dfrac{1}{1 + h_{ie}/R_B} \dfrac{1}{1 + j(\omega/\omega_h)} I_s$

WHERE $\omega_h = \dfrac{R_B + h_{ie}}{r_{B'E}(R_B + r_{BB'}) C}$ rad/s $\leftarrow$

8.5.6. (h) CONTD.

Substituting for I_b into EQ for I_2 and finding ratio of I_2 to I_s gives the desired expression for the complete stage current gain under loaded conditions for high-band frequencies

$$A_{Ih} = \frac{I_2}{I_s} = \frac{-h_{fe}}{1+R_2/R_L}\;\frac{1}{1+h_{ie}/R_B}\;\frac{1}{1+j(\omega/\omega_h)}$$

$$= A_{Im}\,\frac{1}{1+j(\omega/\omega_h)} \leftarrow$$

(i)

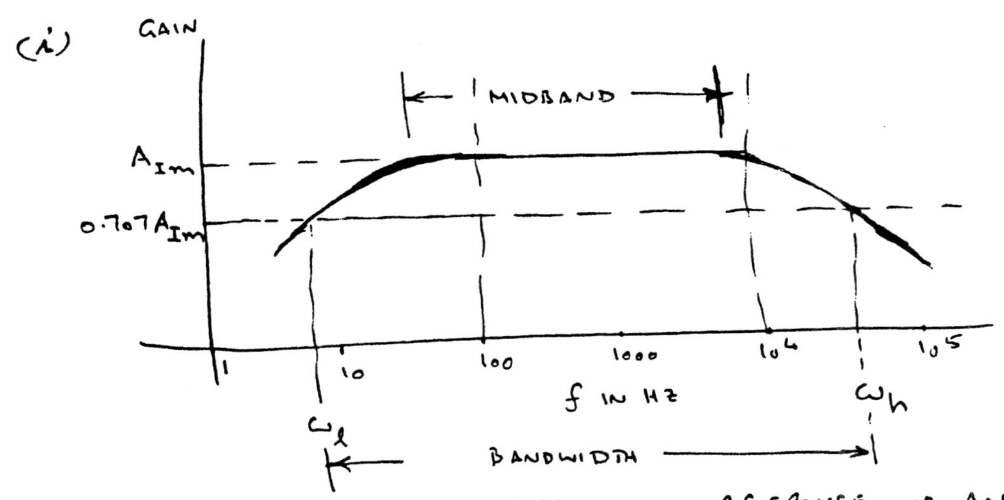

FREQUENCY RESPONSE OF AN RC COUPLED AMPLIFIER

CHAPTER 9

9.1.1.

CONSIDERING THE ANALYSIS OF THE BJT SWITCH OF FIG. 9.1.2, WE ASSUMED THAT R_B WAS SMALL ENOUGH SUCH THAT $i_B \geq 50\,\mu A$ SO AS TO DRIVE THE TRANSISTOR INTO SATURATION. THERE IS NO POINT IN INCREASING i_B ANY FURTHER. THE BJT OPERATING POINT REMAINS AT ② IN FIG. 9.1.2b. BUT ONE MUST BE SURE THAT i_B IS SUFFICIENTLY LARGE SINCE $i_C \simeq \beta i_B + I_{CEO}$, ONE MUST HAVE (NEGLECTING I_{CEO}), $I_{B\,SAT} > I_{C,SAT}/\beta$.

BUT i_C AT SATURATION IS $I_{C,SAT} = (V_{CC} - V_{SAT})/R_C$.

THUS $I_{B\,SAT} > (V_{CC} - V_{SAT})/(\beta R_C)$

ALSO $I_{B\,SAT} = (V_i - V_T)/R_B$

$\therefore$ SATURATION WILL OCCUR WHEN

$$\frac{V_i - V_T}{R_B} > \frac{V_{CC} - V_{SAT}}{\beta R_C}$$

OR

$$V_i > (V_{CC} - V_{SAT})\frac{R_B}{\beta R_C} + V_T \quad \longleftarrow$$

9.1.2.

A NODAL EQUATION AT THE COLLECTOR TERMINAL YIELDS

$$i_C = \frac{V_{CC} - v_{CE}}{R_C} + i_{OUT}$$

IF THE OUTPUT IS "HIGH", THE TRANSISTOR IS CUT OFF, AND $i_C = 0$

$$\therefore \quad \frac{V_{CC} - v_{OUT}}{R_C} = -i_{OUT} \quad \& \quad v_{OUT} = V_{CC} + R_C i_{OUT}$$

SINCE i_{OUT} HAS A NEGATIVE VALUE, ITS EFFECT IS TO REDUCE v_{OUT}.

SINCE v_{OUT} CAN NOT BE ALLOWED TO BECOME LESS THAN 4V AS IT MUST REMAIN IN THE "HIGH" RANGE, THE LARGEST $|i_{OUT}|$ OCCURS WHEN

$$4 = V_{CC} + i_{OUT,MAX} R_C$$

FROM WHICH ONE GETS

$$|i_{OUT,MAX}| = \left|\frac{4 - V_{CC}}{R_C}\right| = 1\,mA \quad \longleftarrow$$

9.1.3.

WHEN THE SIGNAL IS ZERO, THE TRANSISTOR IS CUT OFF AND $I_{C,\,CUTOFF} = I_{CEO} = 0.1\,mA$

THEN $V_o = V_{CC} - R_C I_{CEO} = 4.95V \simeq 5V$ WHICH IS V_{CC}.

WITH THE TRANSISTOR IN SATURATION,

$$I_{C,\,SAT} = \frac{V_{CC} - V_{SAT}}{R_C} = \frac{5-0.2}{500} = 9.6\,mA$$

BUT $i_C \simeq \beta i_B + I_{CEO}$

SO THAT $I_{B,\,SAT} \simeq \dfrac{I_{C,\,SAT} - I_{CEO}}{\beta} = 95\,\mu A$

FOR THE TRANSISTOR TO BE IN SATURATION, ONE MUST HAVE

$$I_B > I_{B,\,SAT},$$

R_B CAN BE FOUND FROM $I_B = \dfrac{V_i - V_T}{R_B} \geq I_{B,\,SAT}$

OR $R_B \leq \dfrac{V_i - V_T}{I_{B,\,SAT}} = 45.26\,k\Omega$ ←

THE RESULTING OUTPUT-VOLTAGE WAVEFORM IS SKETCHED BELOW:

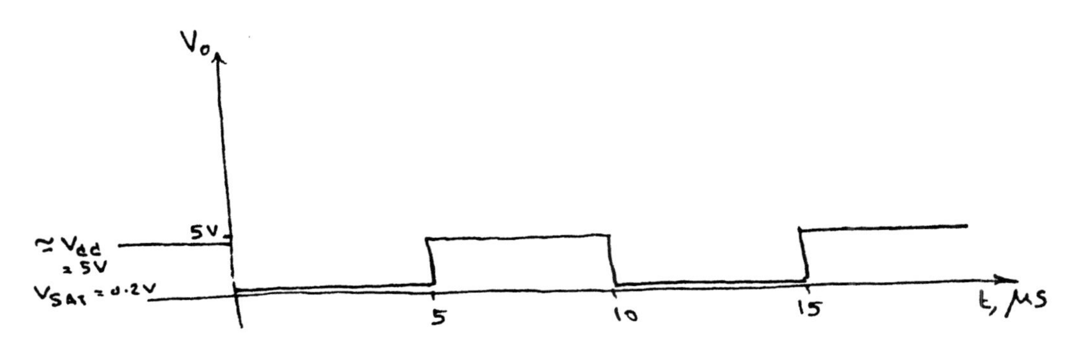

9.1.4.

$$V_i > V_T + \frac{R_B}{\beta R_C}(V_{CC} - V_{SAT})$$

$$5 > 0.7 + \frac{R_B}{R_C}\left(\frac{5-0.2}{25}\right) \Rightarrow \frac{R_B}{R_C} \leq \frac{4.3 \times 25}{4.8}$$

$$\leq 22.396$$

$$i_B \leq 0.1\,mA = \frac{5-0.7}{R_B} \Rightarrow R_B \geq \frac{4.3}{0.1\,mA} = 43\,k\Omega$$ ←

$$R_C \geq \frac{R_B}{22.396} = 1920\,\Omega$$ ←

9.1.5. SATURATION WILL OCCUR WHEN

$$V_i > (V_{cc} - V_{SAT}) \frac{R_B}{\beta R_c} + V_T \qquad (9.1.6)$$

$$V_i > (5 - 0.2) \frac{10}{100(0.5)} + 0.7$$

$$> 1.66V$$

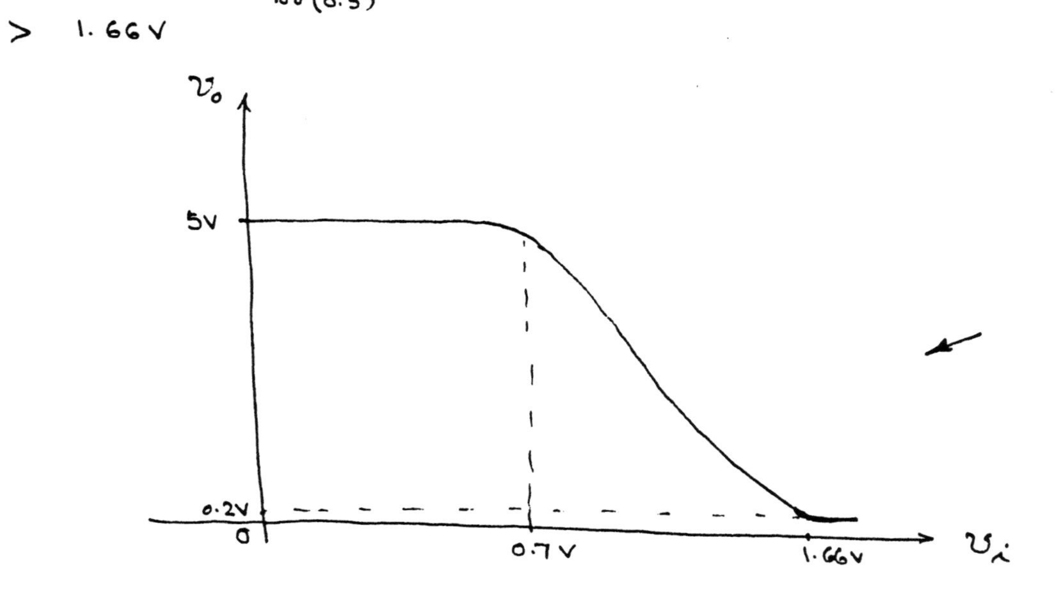

9.1.6.

$$\text{SLOPE} = -\frac{1}{R_B} = -\frac{1}{10,000} = -0.1 \times 10^{-3};$$

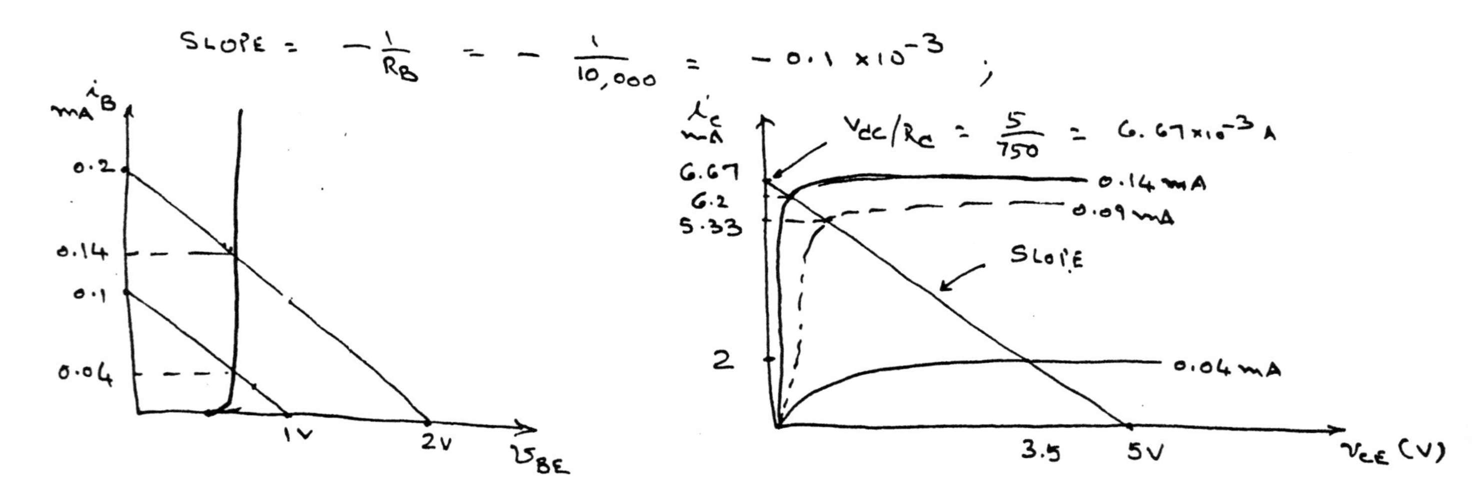

LOOKING AT THE CHARACTERISTICS OF FIG. P. 9.1.6a

AT ABOUT $I_B = 0.09$ mA, IT COMES OUT OF SATURATION.

THIS OCCURS AT ABOUT $\frac{V_i - 0.6}{R_B = 10K} = 0.09$ mA or $V_i = 1.5V$

9.1.6. CONTD.

CORRESPONDING TO WHICH THE TIME $t = 0.75$ ms FROM FIG. P9.1.6b

AT WHICH POINT $V_o = 5 - 750\left(5.33 \times 10^{-3}\right) = 5 - 4 = 1\text{V}$

THUS WE HAVE THE FOLLOWING OUTPUT SKETCH:

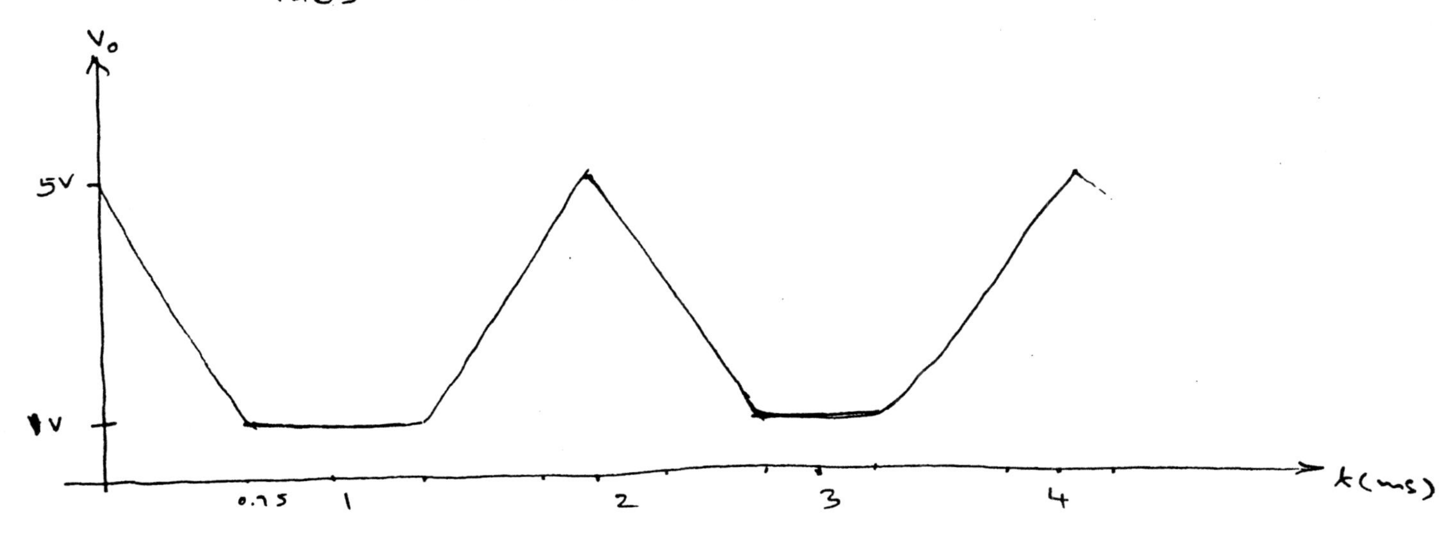

9.2.1.

2ND LINE: $v_A = 0$ (LOW) ; $v_B = 0.2$ V (LOW)

ONE MIGHT GUESS THAT ONLY D_A CONDUCTS, OR THAT ONLY D_B CONDUCTS, OR BOTH. ONLY ONE OF THESE GUESSES WILL BE CORRECT. WE NEED TO GUESS A VALUE OF v_X AND CHECK IT FOR SELF-CONSISTENCY.

IF DIODE B IS CONDUCTING, $v_X = v_B + 0.7 = 0.2 + 0.7 = 0.9V$

BUT THIS GUESS IMPLIES A DROP OF 0.9 V ACROSS D_A AND HENCE CAN NOT BE CORRECT.

ON THE OTHER HAND, IF WE GUESS $v_X = 0.7V$, D_A WILL BE CONDUCTING; BUT D_B WILL NOT SINCE IT HAS ONLY 0.5 V ACROSS IT. SINCE THIS GUESS DOES NOT VIOLATE ANY RULES, IT IS CORRECT. NOTE ALSO THAT $v_X = 0.7V$ IS INSUFFICIENT TO PRODUCE BASE CURRENT. $\therefore$ $V_F = V_{CC} = 5V$. (HIGH)

3rd LINE: $v_A = 0.3V$ (LOW); $v_B = 4.6V$ (HIGH)

SINCE THE AVAILABLE VOLTAGE ACROSS THE PATH $V_{CC} - R_A - D_B$ IS ONLY 0.4 V, D_B WOULD NOT CONDUCT.

WITH A GUESS OF $v_X = 1.0$ V, D_A CONDUCTS AND D_B DOES NOT.

THE BASE CURRENT STILL BEING ZERO, $v_F = V_{CC} = 5V$.(HIGH)

4th line: $v_A = 4.6$ V (HIGH); $v_B = 0.3$ V (LOW)

SINCE THE AVAILABLE VOLTAGE ACROSS THE PATH $V_{CC} - R_A - D_A$ IS ONLY 0.4V, D_A WOULD NOT CONDUCT.

WITH A GUESS OF $v_X = 1.0V$, D_B CONDUCTS AND D_A DOES NOT.

THE BASE CURRENT STILL BEING ZERO, $v_F = V_{CC} = 5V$ (HIGH)

5th LINE: $v_A = 4.8$ V (HIGH); $v_B = 4.1$ V (HIGH)

D_B CANNOT CONDUCT UNLESS $v_X = 4.8V$. BUT THIS WOULD IMPLY TOO MUCH VOLTAGE ACROSS THE CHAIN $D_1 D_2 T_1$. V_X CANNOT BE ABOVE 2.1V WITHOUT OVERBIASING THIS CHAIN. THUS WITH A GUESS OF $v_X = 2.1V$, NEITHER D_A NOR D_B WILL CONDUCT, BUT BASE CURRENT FLOWS IN THE TRANSISTOR. HENCE THE TRANSISTOR WILL BE SATURATED. $\therefore$ $V_F = V_{CE\,SAT} \approx 0.2$ V (LOW).

9.2.2.

IF $v_X = 2.1V$, D_1, D_2, AND T WOULD ALL BE FORWARD BIASED, WHICH IS POSSIBLE. HOWEVER, D_A AND D_B WOULD EACH HAVE 2.1V OF FORWARD BIAS ACROSS THEM. THIS WOULD BE INCONSISTENT WITH THE RULE THAT A CURRENT-CARRYING DIODE HAS 0.7V ACROSS IT.

9.2.3.

$v_X = 0.7 + 0.3 = 1.0\ V$

$v_F = 5\ V$ (HIGH)

9.2.4.

WITH $v_X = 0.7V$, NEITHER D_A, NOR D_B, NOR THE COMBINATION $D_1D_2T_1$ CAN BE CARRYING ANY CURRENT. IN SUCH A CASE THE CURRENT THROUGH R_A WOULD HAVE TO BE ZERO. BUT THIS WOULD IMPLY $v_X = V_{CC}$ WHICH IS INCONSISTENT WITH THE GUESS $v_X = 0.7V$.

9.2.5.

IF D_1 AND D_2 WERE OMITTED, v_X COULD RISE TO ONLY 0.7V BEFORE i_B WOULD BEGIN TO FLOW. WITH THESE INPUTS THE TRANSISTOR WOULD SATURATE AND v_F WOULD BE 'LOW' INSTEAD OF 'HIGH' AS REQUIRED FOR THE NAND GATE. HOWEVER, WITH D_1 AND D_2 PRESENT, NO i_B CAN FLOW UNTIL v_X REACHES 2.1V, WHICH CAN NEVER OCCUR WHEN EITHER v_A or v_B IS IN THE 'LOW' RANGE.

9.2.6. (a)

WITH $V_A = 5V$ (HIGH) AND $V_B = 0$ (LOW), D_A WILL BE CLOSED AND D_B WILL BE OPEN.

$\therefore$ BASE CURRENT $i_B = \dfrac{V_A - 0.7 - 0.7}{R_B} = \dfrac{5 - 1.4}{12{,}000} = 0.3\ mA$

FOR THE TRANSISTOR TO BE IN SATURATION,

$$i_C > I_{C\,SAT} = \frac{V_{CC} - V_{SAT}}{R_C} = \frac{5 - 0.2}{500} = 9.6\ mA$$

BUT $i_C \simeq \beta i_B = 35(0.3) = 10.5\ mA$

$\therefore$ THE TRANSISTOR IS IN SATURATION.

THEN $V_o = V_{SAT} = 0.2V$ (LOW)

(b)

WITH $V_A = V_B = 0$ (LOW), BOTH DIODES WILL BE OPEN AND $i_B = 0$

THUS, THE TRANSISTOR IS CUT OFF AND $V_o \simeq V_{CC} = 5V$ (HIGH)

9.2.6.(C)

THUS WE HAVE

V_A	V_B	V_O
HIGH	LOW	LOW
LOW	LOW	HIGH

OR WITH POSITIVE LOGIC

V_A	V_B	V_O
1	0	0
0	0	1

THE DEVICE FUNCTIONS AS A NOR GATE.

9.2.7.

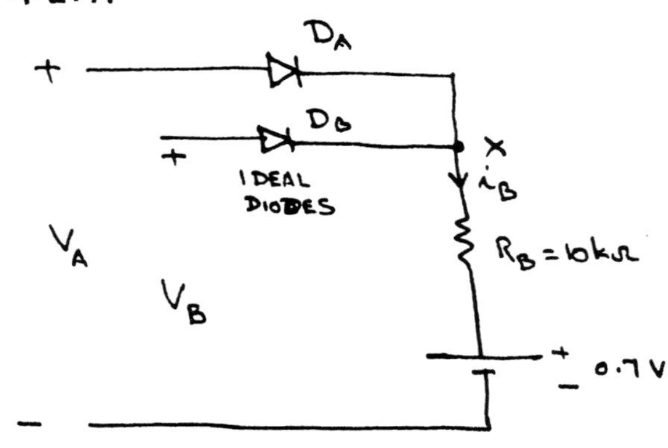

$$i_{C\,MAX} = \frac{V_{CC} - V_{SAT}}{R_C} = \frac{5 - 0.2}{500} = 9.6\,mA$$

$$i_{B\,MAX} = \frac{i_{C\,MAX}}{\beta} = \frac{9.6}{20} = 0.48\,mA$$

WITH $V_A = 10V$ AND $V_B = 6V$

$$i_B = \frac{10 - 0.7}{10\,k\Omega} = 0.93\,mA$$

SINCE D_B IS OPEN BECAUSE $V_X = 10\,V$

WITH $V_A \simeq 0$ AND $V_B = 6V$

$$i_B = \frac{6 - 0.7}{10\,k\Omega} = 0.53\,mA$$

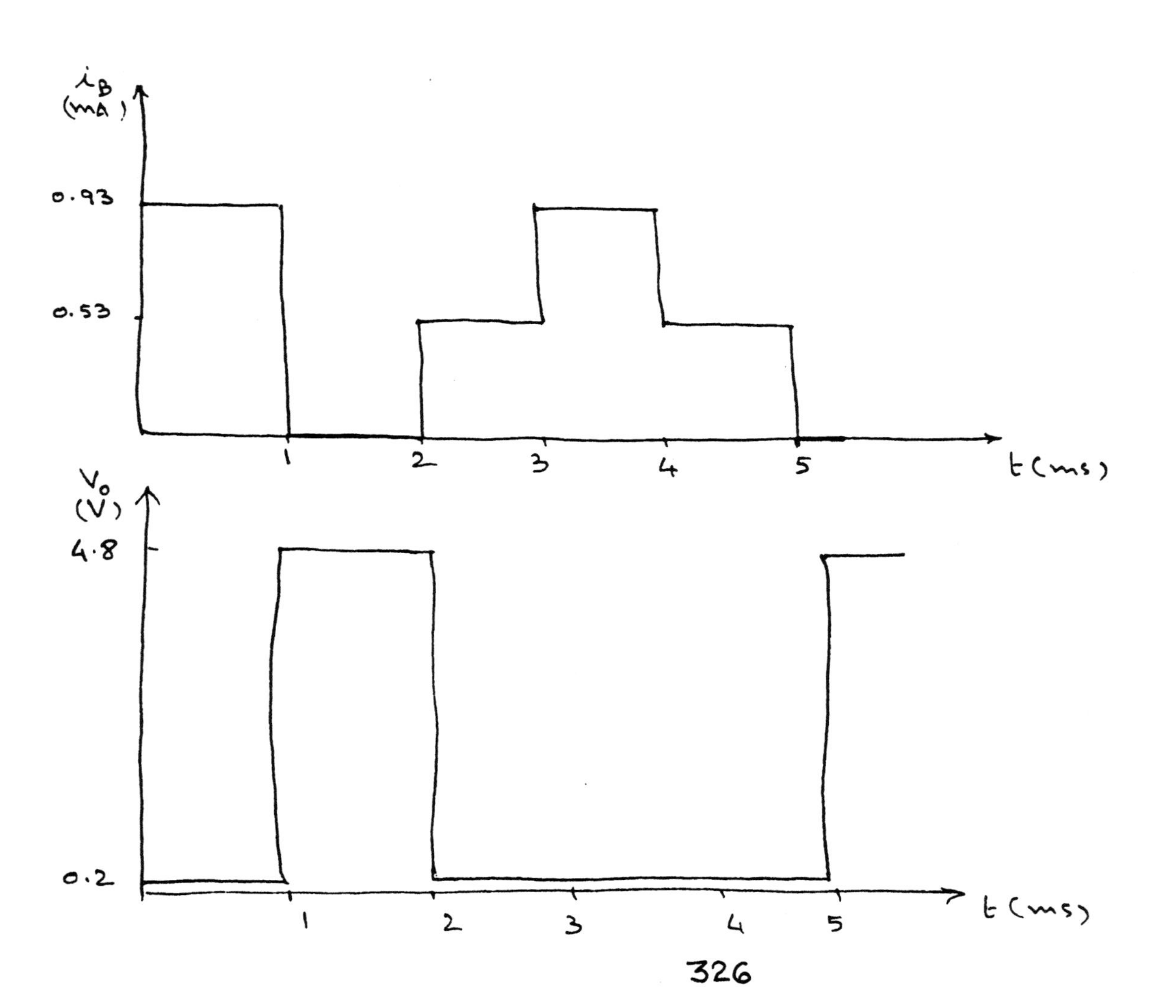

9.2.8. SUPPOSE V_A IS HIGH BUT V_B AND V_C ARE LOW.

IN THIS CASE, T_1 IS ON AND T_2 & T_3 ARE OFF, SO THAT V_O IS LOW.

IF BOTH V_A AND V_B ARE HIGH, V_o IS ALSO LOW.

V_o IS HIGH ONLY IF V_A, V_B, AND V_C ARE ALL LOW.

THUS THE GATE PERFORMS THE NOR FUNCTION.

TRUTH TABLE OF NOR GATE:

INPUTS A	B	C	OUTPUT F
0	0	0	1
0	0	1	0
0	1	0	0
0	1	1	0
1	0	0	0
1	0	1	0
1	1	0	0
1	1	1	0

9.2.9.

THE CIRCUIT IS OBTAINED BY CONNECTING THE AND DIODE GATE TO A TRANSISTOR SWITCH WHICH IS AN INVERTER.

THE GATE BEHAVES LIKE A NAND GATE.

9.2.10.

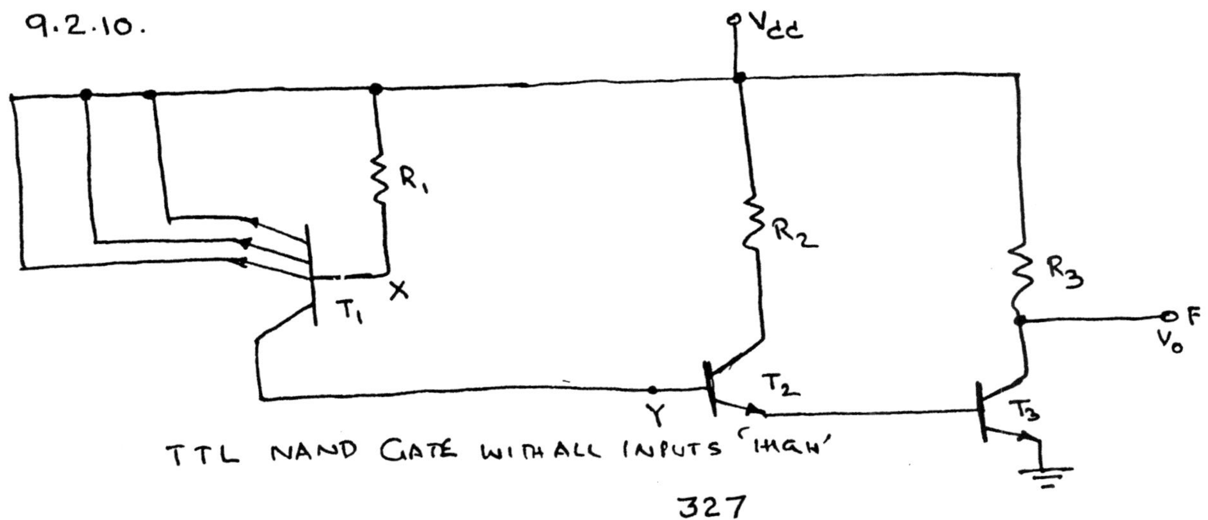

TTL NAND GATE WITH ALL INPUTS 'HIGH'

9.2.10. CONTD. A CURRENT PATH NOW EXISTS FROM V_{CC} DOWN THROUGH R_1 AND THROUGH THE BASE-COLLECTOR JUNCTION OF T_1 (WHICH IS FORWARD BIASED) AND THE BASE-EMITTER JUNCTIONS OF T_2 AND T_3. T_1 IS NOW EFFECTIVELY OPERATING IN THE REVERSE MODE WITH THE COLLECTOR JUNCTION ACTING AS THE EMITTER. WE NOW EXPECT THAT $v_X \simeq 2.1V$, $v_Y \simeq 1.4V$, AND T_3 WILL BE SATURATED, THEREBY MAKING THE OUTPUT 'LOW'.

9.2.11.

$$V_X \simeq 0.7 + 0.1 = 0.8V$$

$$V_Y \simeq 0.3V$$

$$V_o \simeq V_{CC}$$

9.2.12.(a)

REPLACING THE BASE-EMITTER AND BASE COLLECTOR JUNCTIONS WITH DIODES, USING FIG. 9.2.4, THE CIRCUIT MAY BE REDRAWN FOR CONVENIENCE. WITH V_A, V_B, AND V_C ALL 'HIGH', ALL THREE DIODES D_A, D_B, AND D_C ARE REVERSE-BIASED AND OPEN. CURRENT FLOWS THROUGH R_B AND D TO SATURATE T_2, WHICH SATURATES T_3. BY ADDING THE BASE-EMITTER DROPS OF T_2 AND T_3, THE BASE VOLTAGE OF T_2 BECOMES $2V_T = 1.4V$. ADDING THE 0.7V DROP OF T_1 ACROSS THE BASE-COLLECTOR DIODE D GIVES THE BASE CURRENT TO T_2 OF $\{(V_{CC} - 2.1)/R_B\}$.
WITH T_2 ON, THE BASE VOLTAGE OF T_4 IS TOO LOW TO CAUSE IT TO SATURATE, AND T_4 IS OFF. THUS WITH ALL INPUTS 'HIGH', V_o WILL BE 'LOW'.

(b)
LET V_A BE 'LOW', SAY $V_A = 0$. THE ASSOCIATED BASE-EMITTER DIODE D_A WILL BE CLOSED, AND THE BASE VOLTAGE OF T_1 WILL BE $V_A + 0.7V = 0.7V$. COMBINED WITH 0.7V DROP ACROSS D, THE BASE VOLTAGE OF T_2 WILL BE $-0.7 + 0.7 = 0$, SO THAT T_2 IS CUT OFF. THIS SERVES TO CUT OFF T_3, AND THE OUTPUT IS THEN 'HIGH'. SINCE T_2 IS CUT OFF, ITS COLLECTOR VOLTAGE RISES, TURNING ON T_4.

9.2.13.

THE COMBINATION OF R_4 AND T_4 ACTS AS A VARIABLE RESISTANCE. WHEN T_3 IS ON, T_4 IS OFF, LOWERING THE POWER CONSUMPTION; BUT WHEN T_3 IS OFF, T_4 IS ON, WHICH PROVIDES A LOW RESISTANCE AS SEEN BY THE SUCCEEDING GATE, THEREBY IMPROVING THE SWITCHING TIME. THE DIODE BETWEEN T_4 AND T_3 SERVES TO ADD A VOLTAGE DROP TO HELP ENSURE THAT T_4 IS OFF WHEN T_3 IS ON.

9.2.14.

LET V_A, V_B, AND V_C BE ALL 'HIGH', AND THEIR DIODES OPEN.

THEN TO SATURATE T_2

$$\frac{V_{CC} - 2.1}{R_B} > 3.8 \text{ mA} \qquad \text{or} \qquad R_B \leq \frac{V_{CC} - 2.1}{3.8 \text{ mA}} = 763\ \Omega$$

MAXIMUM VALUE OF R_B IS THEN $763\ \Omega$ ←

9.3.1.

(a) FOR $v_{GS} < -V_T$, NO DRAIN CURRENT FLOWS AND THE DEVICE IS CUTOFF. WITH $V_i = 0$, THE FET IS IN SATURATION; WITH V_i MORE NEGATIVE THAN THE THRESHOLD VOLTAGE V_T, THE FET IS CUT OFF.

WITH POSITIVE LOGIC, $V_i = 0$ MAY REPRESENT A LOGICAL 1, AND $V_i = -V_p = -V_T$ MAY REPRESENT A LOGICAL 0 (ZERO).

(b)

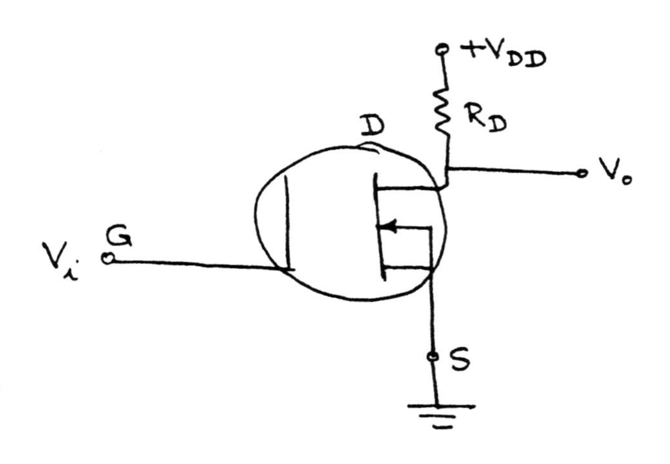

(c)

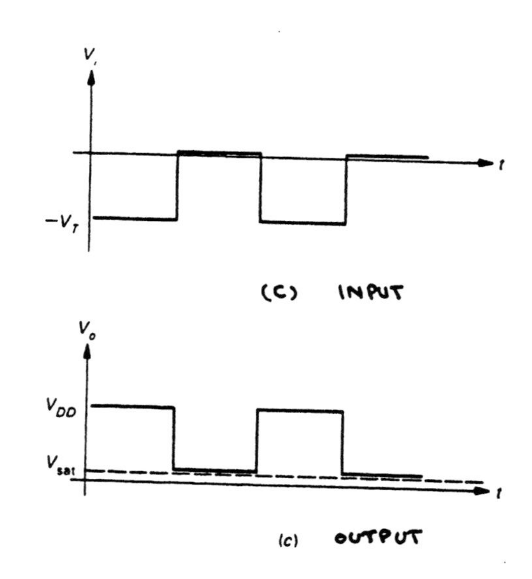

9.3.2.

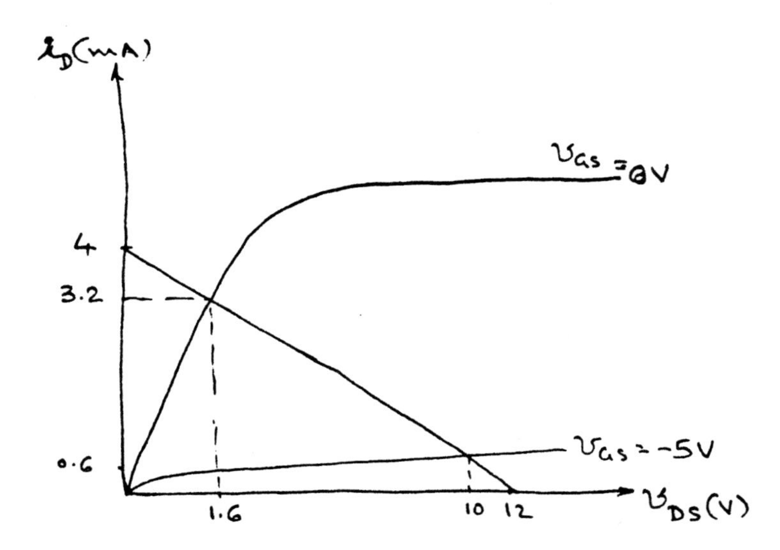

i_D(mA)
v_{GS} = 0V
4
3.2
0.6
v_{GS} = −5V
1.6
10 12
v_{DS}(V)

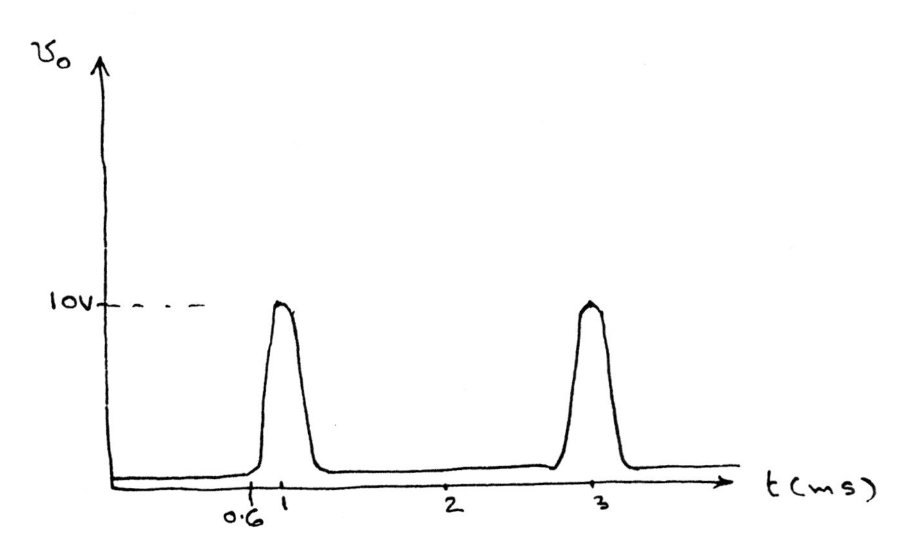

v_o
10V
0.6
1
2
3
t (ms)

9.3.3.

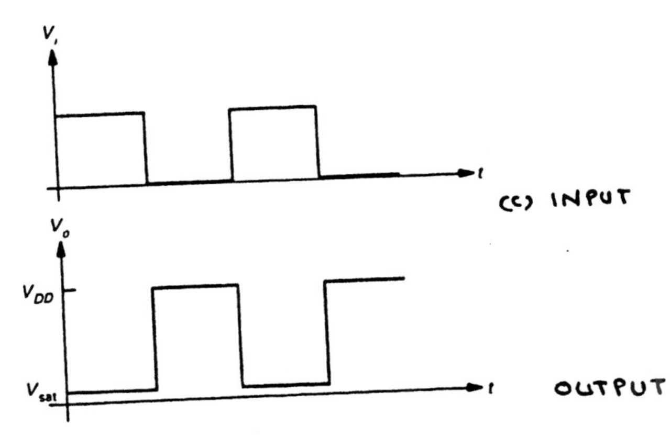

V_i
t
(c) INPUT
V_o
V_{DD}
V_{sat}
t
OUTPUT

9.3.4. $v_{GS\ NMOS} = v_{IN}$; $v_{GS\ PMOS} = v_{IN} - V_{SS}$

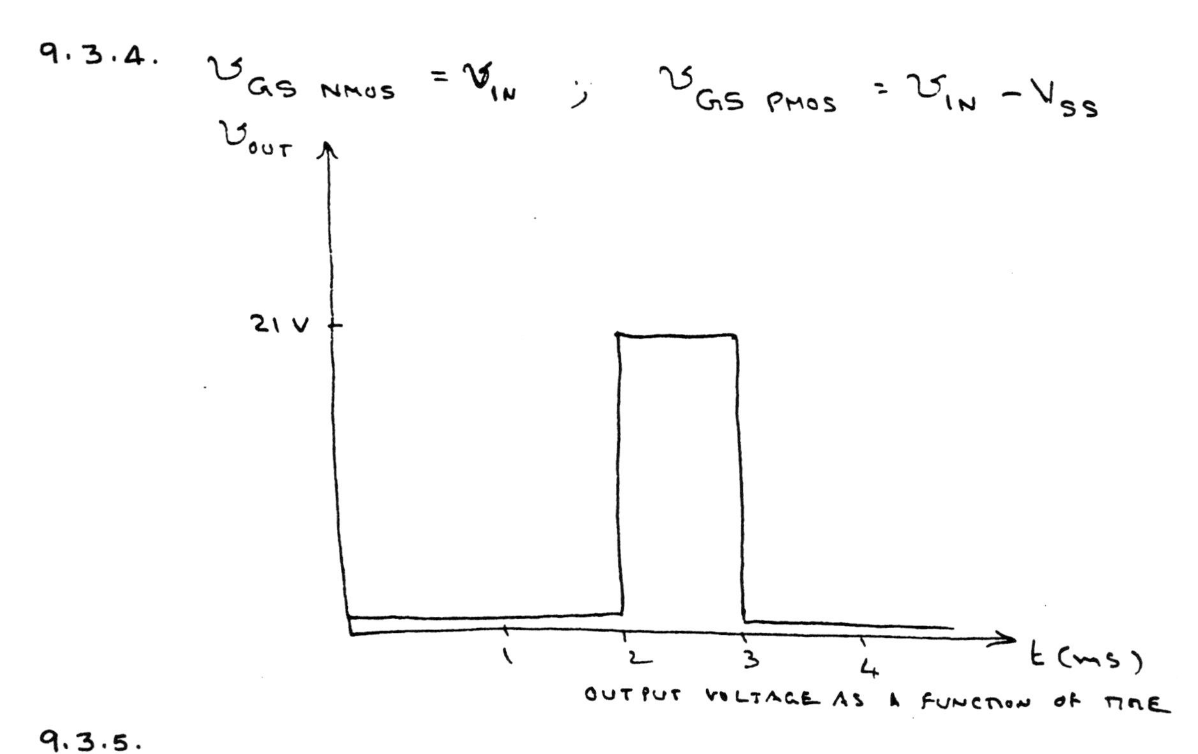

OUTPUT VOLTAGE AS A FUNCTION OF TIME

9.3.5.

(a) THE I-V CHARACTERISTICS OF THE p- CHANNEL MOSFET T_2 ARE SHOWN BELOW:

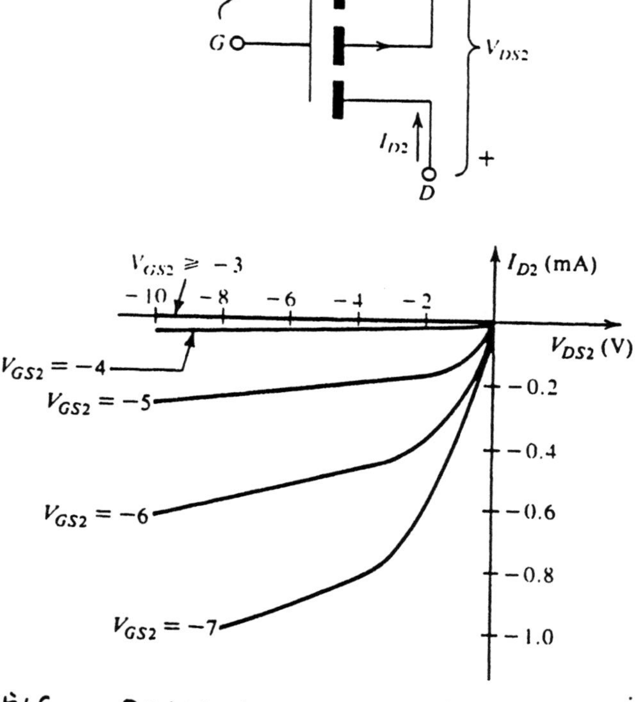

FIG. CHARACTERISTICS OF T_2

9.3.5. CONTD. THE CMOS INVERTER CIRCUIT IS REDRAWN TO ILLUSTRATE THE GRAPHICAL ANALYSIS.

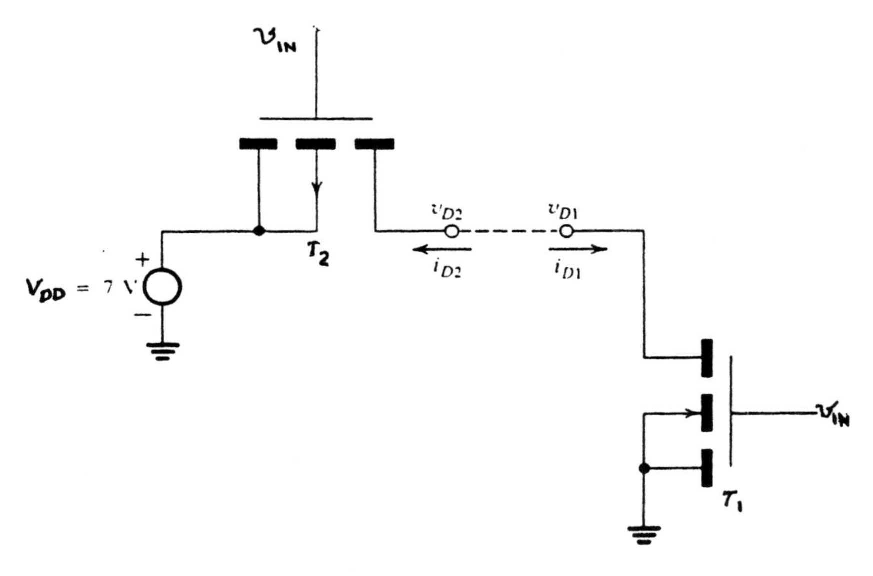

CMOS INVERTER CIRCUIT REDRAWN

THE OPERATING POINT OF THE CIRCUIT WILL BE THAT AT WHICH $V_{D1} = V_{D2}$ AND $-I_{D2} = I_{D1}$. TO FIND THIS POINT, WE NEED A GRAPH OF $-I_{D2}$ VS V_{D2}. TOWARDS THAT END, FIRST PLOT $+I_{D2}$ VS V_{DS2} (WHICH IS NOT THE SAME AS V_{D2}). CHANGING I_{D2} TO $-I_{D2}$ SIMPLY INVERTS THE GRAPH. THIS IS SHOWN BELOW IN FIG. S 9.3.5 (a). NOTE THAT $V_{GS2} = v_{IN} - V_{DD}$ IN OUR CIRCUIT. CHOOSE $V_{DD} = 7V$. DIFFERENT CURVES IN FIG. S 9.3.5 (a) ARE ACCORDINGLY RELABELED IN TERMS OF v_{IN}. NOTING THAT $V_{DS2} = V_{D2} - V_{DD} = V_{D2} - 7$, GRAPH $-I_{D2}$ VS V_{D2} AS SHOWN IN FIG. S 9.3.5 (b). THE FINAL STEP IS TO SUPERIMPOSE FIG. S 9.3.5 (b) ON TO $I_{D1} - V_{D1}$ CHARACTERISTIC OF T_1. THE RESULT IS SHOWN IN FIG. S 9.3.5 (c).

9.3.5. CONTD.

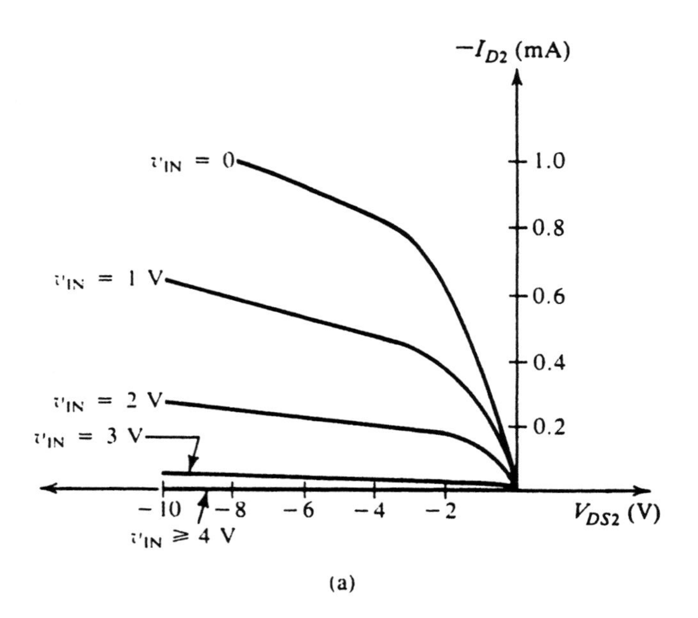

(a)

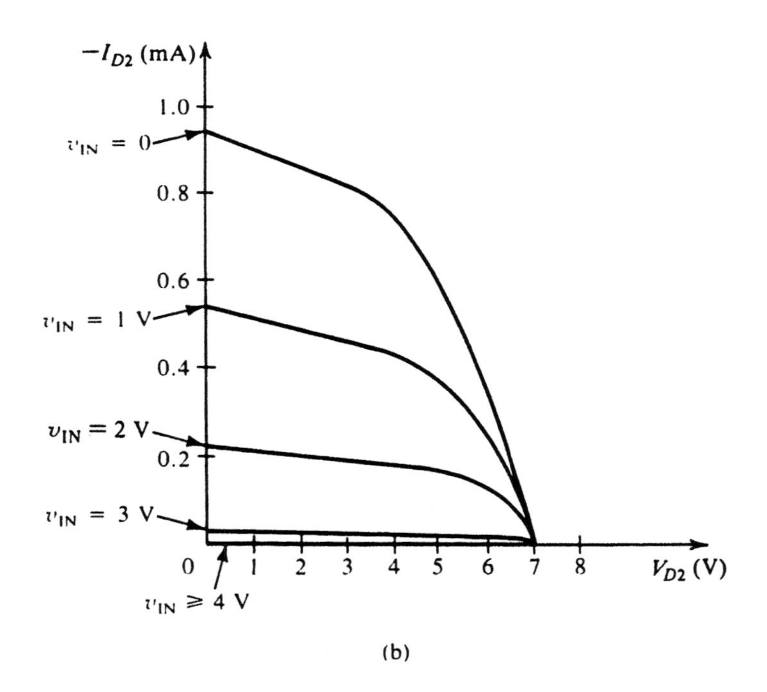

(b)

FIGURE S9.3.5 (a) $-I_{D2}$ vs V_{DS2} (b) $-I_{D2}$ vs V_{D2}

9.3.5. CONTD.

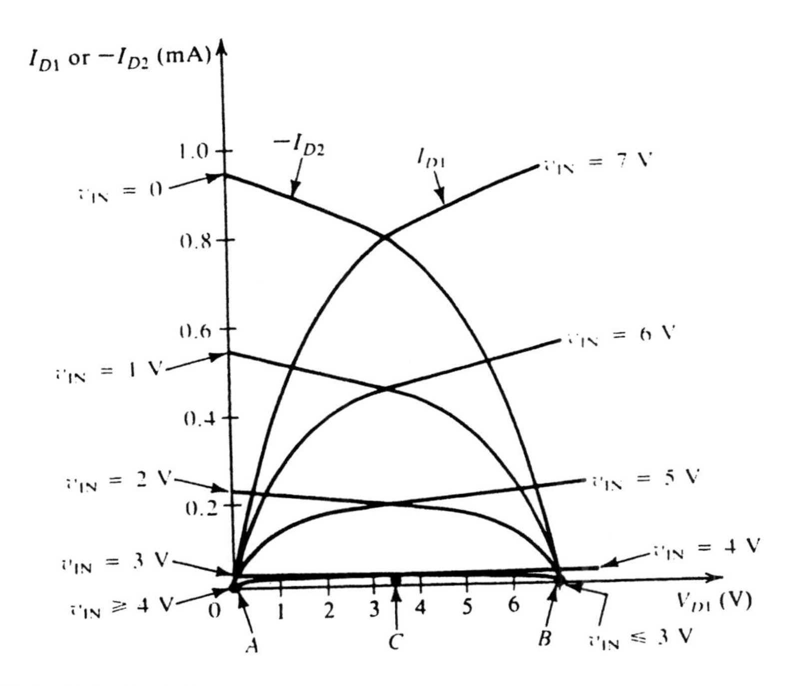

FIG. S9.3.5(c) SUPERPOSITION

TO FIND THE OPERATING POINT, SIMPLY LOCATE THE POINT AT WHICH THE TWO CURVES CORRESPONDING TO A GIVEN VALUE OF v_{IN} INTERSECT.

FOR $v_{IN} \leq 3V$, THE INTERSECTION IS SEEN TO LIE AT POINT B;

FOR $v_{IN} \geq 4V$, THE INTERSECTION IS SEEN TO LIE AT POINT A;

INTERPOLATING CURVES FOR $v_{IN} = 3.5V$, C IS APPROXIMATELY THE OPERATING POINT.

(b)

VOLTAGE-TRANSFER CHARACTERISTIC (v_{OUT} VS v_{IN}) IS SHOWN BELOW IN FIGURE S9.3.5(d). ALSO, USING FIGURE S9.3.5(c), GRAPH I_{D1} $(=-I_{D2})$ VS v_{IN} IS DRAWN IN FIGURE S9.3.5(e).

MAXIMUM VALUE OF I_{D1} OCCURS AT POINT C IN FIGURE S9.3.5(C), ALTHOUGH THE MAXIMUM ITSELF IS VERY SMALL.

9.3.5 CONTD.

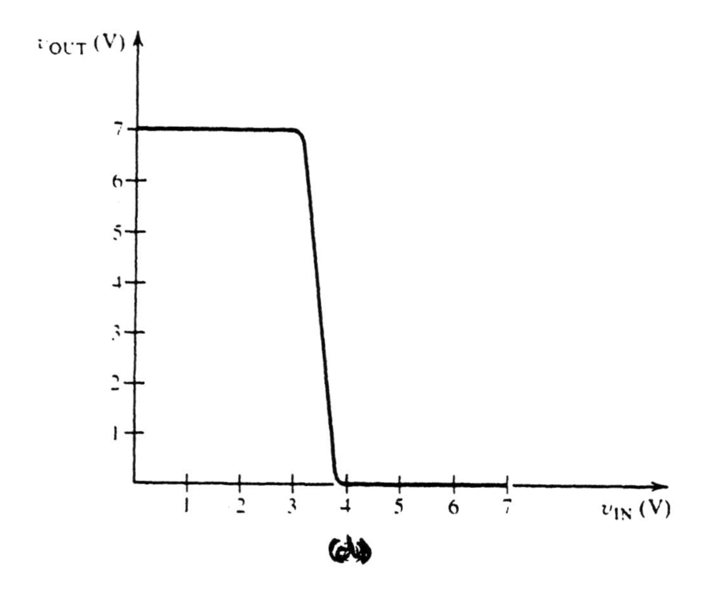

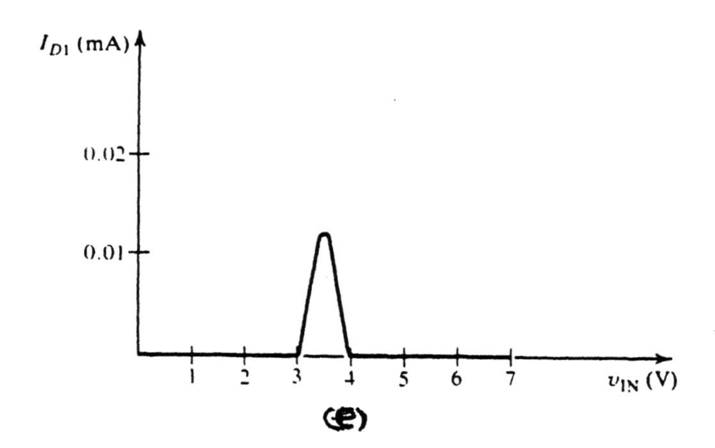

FIG. S9.3.5 (d) OUTPUT VOLTAGE AS A FUNCTION OF INPUT VOLTAGE
(e) DRAIN CURRENT VS v_{IN}

9.3.6. LET $V_{DD} = 7V$; 'LOW' = 0 V; 'HIGH' = 7V; AND $V_T = 3V$ WHICH IS TURN-ON VOLTAGE. WHEN C IS 'LOW' (AND $\bar{C}$ IS 'HIGH') THE GATES OF T_1 AND T_2 ARE UNBIASED WITH RESPECT TO BULKS. FURTHERMORE, v_{IN} IS UNBIASED OR POSITIVE (DEPENDING ON THE VALUE OF v_{IN}) WITH RESPECT TO G_1 AND UNBIASED OR NEGATIVE WITH RESPECT TO G_2. THUS NEITHER GATE CAN INDUCE A CHANNEL, AND THE CIRCUIT ACTS AS A HIGH RESISTANCE BETWEEN v_{IN} AND v_{OUT}. THE TRANSMISSION GATE IS EFFECTIVELY AN OPEN CIRCUIT.

ON THE OTHER HAND, WHEN C IS 'HIGH' (AND $\bar{C}$ IS 'LOW'), LET v_{IN} BE ZERO. THEN T_1 HAS A LARGE POSITIVE V_{GS} AND PROVIDES A LOW-RESISTANCE PATH BETWEEN v_{IN} AND v_{OUT}. WHEN $v_{IN} = 7V$, T_2 HAS A LARGE NEGATIVE V_{GS} AND PROVIDES A LOW-RESISTANCE PATH BETWEEN v_{IN} AND v_{OUT}.

FOR SOME VALUES OF v_{IN} BETWEEN 0 (ZERO) AND 7 V, IT IS POSSIBLE FOR BOTH T_1 AND T_2 TO CONDUCT. THUS THE TRANSMISSION GATE PROVIDES A LOW-RESISTANCE PATH BETWEEN INPUT AND OUTPUT WHEN C IS 'HIGH' FOR ALL ALLOWED VALUES OF v_{IN}.

9.3.7.

A 'HIGH' GATE VOLTAGE APPLIED TO AN n-CHANNEL DEVICE CREATES A LOW-RESISTANCE CHANNEL THAT ACTS NEARLY LIKE A SHORT CIRCUIT, WHILE A 'LOW' GATE VOLTAGE APPLIED TO AN n-CHANNEL DEVICE RESULTS IN A NONEXISTANT CHANNEL, WHICH IS NEARLY AN OPEN CIRCUIT. FOR THE TWO p-CHANNEL DEVICES, WITH THEIR SOURCES CONNECTED TO V_{DD}, OPPOSITE STATEMENTS APPLY. THE APPROXIMATE BEHAVIOR OF TRANSISTORS IN CMOS LOGIC IS SHOWN BELOW:

TYPE	GATE 'HIGH'	GATE 'LOW'
n-CHANNEL	SHORT	OPEN
p-CHANNEL	OPEN	SHORT

IT IS NOW EASILY VERIFIED THAT THE CIRCUIT FUNCTIONS AS A NAND GATE WITH POSITIVE LOGIC.

CHAPTER 10

10.2.1.

$\bar{E}_1 = 120\angle 0°$ V ; $\bar{E}_2 = 110\angle 45°$ V ; $\bar{Z} = (1.5 + j6)\ \Omega$

(a)

$$\bar{I} = \frac{\bar{E}_1 - \bar{E}_2}{\bar{Z}} = \frac{120\angle 0° - 110\angle 45°}{(1.5 + j6)} = 14.3\angle -137.5°\ \text{A}$$

$$\bar{S}_1 = \bar{E}_1 \bar{I}_1^* = (120\angle 0°)(14.3\angle 137.5°) = (-1265 + j1159)\ \text{VA}$$

$$\bar{S}_2 = \bar{E}_2 \bar{I}_2^* = (110\angle 45°)(14.3\angle 137.5°) = (-1572 - j68.6)\ \text{VA}$$

$P_1 = 1265$ W ABSORBED ; $P_2 = 1572$ DELIVERED ←

(b) $Q_1 = 1159$ VAR DELIVERED ; $Q_2 = 68.6$ VA DELIVERED ←

(c)

$$\bar{S}_Z = (120\angle 0° - 110\angle 45°)(14.3\angle 137.5°) = (306 + j1228)\ \text{VA}$$

$P_Z = 306$ W ABSORBED ; $Q_Z = 1228$ VAR ABSORBED ←

10.2.2.

$$\bar{I} = \frac{\bar{E}_1 - \bar{E}_2}{\bar{Z}} = \frac{120\angle 0° - 110\angle 45°}{1.5 - j6} = 14.3\angle 14.4°$$

(a)

$$\bar{S}_1 = (120\angle 0°)(14.3\angle -14.4°) = (1662 - j427)\ \text{VA}$$

$$\bar{S}_2 = (110\angle 45°)(14.3\angle -14.4°) = (1354 + j800)\ \text{VA}$$

$P_1 = 1662$ W DELIVERED ; $P_2 = 1354$ W ABSORBED ←

(b) $Q_1 = 427$ VAR ABSORBED ; $Q_2 = 800$ VAR ABSORBED ←

(c)

$$\bar{S}_Z = (120\angle 0° - 110\angle 45°)(14.3\angle -14.4°) = (308 - j1227)\ \text{VA}$$

$P_Z = 308$ W ABSORBED ; $Q_Z = 1227$ VAR DELIVERED ←

10.2.3.

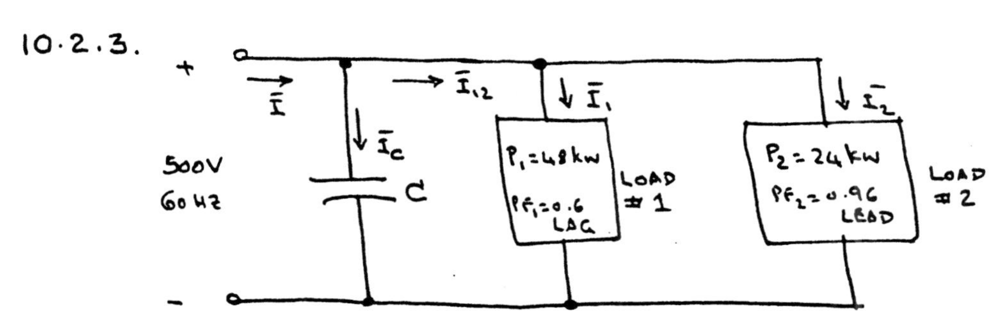

LOAD # 1: $S_1 = 48/0.6 = 80$ kVA ; $Q_1 = \sqrt{80^2 - 48^2} = 64$ kVAR (POSITIVE SINCE PF_1 IS LAGGING)

$I_1 = 80 \times 10^3 / 500 = 160$ A

LOAD # 2: $S_2 = 24/0.96 = 25$ kVA; $|Q_2| = \sqrt{25^2 - 24^2} = 7$ kVAR ; $Q_2 = -7$ kVAR

(Q_2 IS NEGATIVE SINCE PF_2 IS LEADING) ; $I_2 = 25 \times 10^3 / 500 = 50$ A

COMBINATION OF LOADS # 1 & # 2: $P_{12} = P_1 + P_2 = 48 + 24 = 72$ kW; $Q_{12} = Q_1 + Q_2 = 64 - 7 = 57$ kVAR

$S_{12} = \sqrt{72^2 + 57^2} = 91.8$ kVA ; $I_{12} = 91.8 \times 10^3 / 500 = 184$ A

$PF_{12} = 72/91.8 = 0.784$ LAGGING (SINCE $Q_{12} > 0$)

USING AN IDEAL LOSSLESS CAPACITOR C, $P_c = 0$; $Q_c = -57$ kVAR

IN ORDER TO BRING THE OVERALL PF TO UNITY.

$$\therefore C = -Q_c / \omega V_c^2 = \frac{57 \times 10^3}{2\pi (60)(500)^2} = 605\ \mu F \leftarrow$$

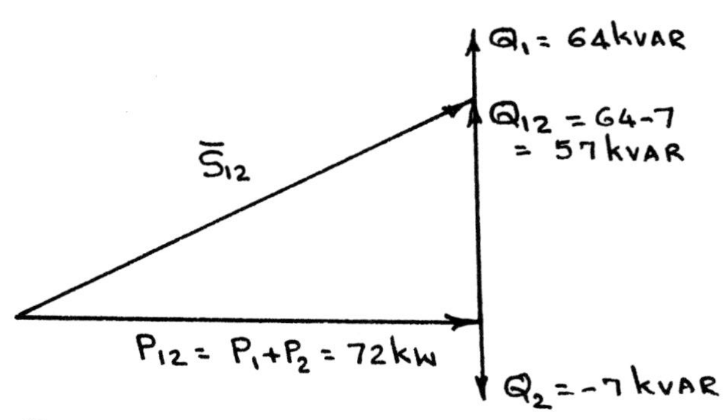

POWER TRIANGLE OF THE COMBINATION OF LOADS # 1 & # 2

$P = P_{12} + P_c = 72$ kW

$Q = Q_{12} + Q_c = 0$

POWER TRIANGLE OF THE OVERALL PLANT WITH CAPACITOR

OVERALL CURRENT DRAWN $= \dfrac{72 \text{ kVA}}{500 \text{ V}} = 144$ A $\leftarrow$

10.2.4.

$S_1 = 10\text{kVA}$; $PF_1 = 0.80$ LAGGING ; $\bar{V} = 230\angle 0°$ V

$P_2 = 6\text{kW}$; $PF_2 = 0.90$ LAGGING

$$\bar{I}_1 = \frac{10\times 10^3}{230}\angle -\cos^{-1}0.8 = 43.48\angle -36.9° \text{ A}$$

$$\bar{I}_2 = \frac{6000}{230(0.9)}\angle -\cos^{-1}0.9 = 28.99\angle -25.8° \text{ A}$$

$$\bar{I}_S = \bar{I}_1 + \bar{I}_2 = 43.48\angle -36.9° + 28.99\angle -25.8°$$

$$= 43.48(0.8 - j0.6) + 28.99(0.9 - j0.43523)$$

$$= (34.78 + 26.1) - j(26.1 + 12.6)$$

$$= 60.88 - j38.7 = 72.1\angle -32.4° \text{ A}$$

$$I_S = 72.1 \text{ A} \longleftarrow$$

10.2.5.

THE PER-PHASE EQUIVALENT CIRCUIT IS SHOWN BELOW:

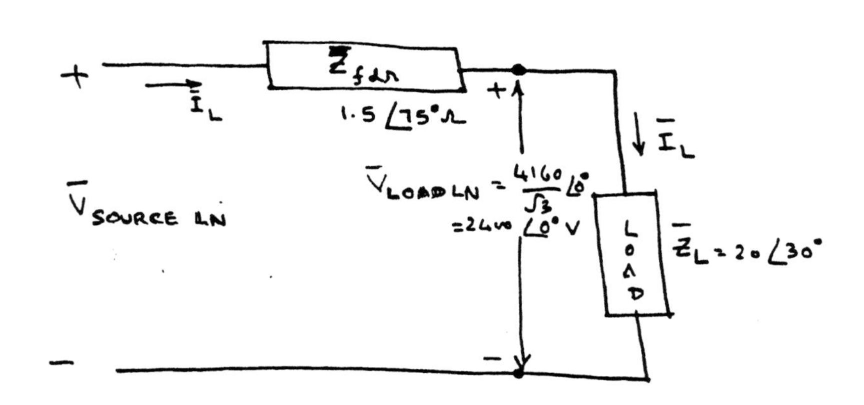

$$\bar{I}_L = \frac{2400\angle 0°}{20\angle 30°} = 120\angle -30° \text{ A}$$

$$\bar{V}_{SOURCE\,LN} = 2400\angle 0° + (120\angle -30°)(1.5\angle 75°)$$

$$= 2527 + j127 = 2530\angle 3° \text{ V}$$

(a) $V_{SOURCE\,LL} = 2530\sqrt{3} = 4382 \text{ V} \longleftarrow$

(b) $I_L = 120 \text{ A} \longleftarrow$

10.2.6. $S = 600$ MVA ; $V_{LL} = 345$ kV ; WYE LOAD ; PF = 0.866 LAGGING

(b) $$I_{ph} = I_L = \frac{600 \times 10^6}{\sqrt{3}\,(345 \times 10^3)} = 1004 \text{ A} \leftarrow$$

(a) $$\bar{Z}_{Y,ph} = \frac{10^3 (345/\sqrt{3}) \angle 0°}{1004 \angle -\cos^{-1} 0.866} = \frac{10^3 \times 199.2 \angle 0°}{1004 \angle -30°} = 198.4 \angle 30° \ \Omega \leftarrow$$

(c) $$P_{ph} = \frac{345}{\sqrt{3}} \times 10^3 \times 1004 \times 0.866 = 173.18 \text{ MW} \leftarrow$$

$$Q_{ph} = \frac{345}{\sqrt{3}} \times 10^3 \times 1004 \times 0.5 = 99.99 \text{ MVAR} \leftarrow$$

(d) $$P_{3\phi} = \sqrt{3}\,(345)\,10^3 \times 1004 \times 0.866 = 519.54 \text{ MW} = 3P_{ph} \leftarrow$$

$$Q_{3\phi} = \sqrt{3}\,(345)\,10^3 \times 1004 \times 0.5 = 299.97 \text{ MVAR} = 3Q_{ph} \leftarrow$$

10.2.7.

$P_L = 120$ kW ; $V_L = 440$ V ; $PF_L = 0.85$ LAGGING

$Q_c = 50$ kVAR

$$\bar{I}_L = \frac{120000}{\sqrt{3}(440)(0.85)} \angle -\cos^{-1} 0.85 = 185.2 \angle -31.8° \text{ A}$$

$$\bar{I}_c = \frac{50000}{\sqrt{3}\,(440)} \angle 90° = 65.6 \angle 90° \text{ A}$$

$$\bar{I}_T = \bar{I}_L + \bar{I}_c = 185.2 \angle -31.8° + 65.6 \angle 90° = 160.6 \angle -11.49° \text{ A}$$

$$I_{RESULTANT} = 160.6 \text{ A} \leftarrow$$

$$PF_{RESULTANT} = \cos 11.49° = 0.98 \text{ LAGGING} \leftarrow$$

10.2.8.

$S_1 = 15$ kVA ; $V_L = 2400$; $PF_1 = 0.8$ LAGGING

$P_2 = 20$ kW ; $V_L = 2400$; $PF_2 = 0.9$ LEADING

(a)
$$\bar{I}_1 = \frac{15000}{\sqrt{3}(2400)} \angle -\cos^{-1} 0.8 = 3.61 \angle -36.9° \text{ A}$$

$$\bar{I}_2 = \frac{20000}{\sqrt{3}(2400)(0.9)} \angle \cos^{-1} 0.9 = 5.34 \angle 25.8° \text{ A}$$

$$\bar{I}_T = \bar{I}_1 + \bar{I}_2 = 7.7 \angle 1.2° \text{ A}$$

$$I_T = 7.7 \text{ A} \leftarrow$$

10.2.8. CONTD.

(b) $P_1 = 15(0.8) = 12\,kW$; $P_2 = 20\,kW$

$Q_1 = 15 \sin 36.9^\circ = 15 \times 0.6 = 9\,kVAR$; $Q_2 = -20 \tan 25.8^\circ = -9.7\,kVAR$

$P_T = P_1 + P_2 = 12 + 20 = 32\,kW \leftarrow$

$Q_T = Q_1 + Q_2 = 9 - 9.7 = -0.7\,kVAR \leftarrow$

(c) $PF = \cos 1.2^\circ = 0.999$ LEADING $\leftarrow$

10.2.9.

(a) $P_1 = 50\,kW$; $V_{LOAD\,LL} = 460\,V$; $PF_1 = 0.866$ LAGGING

$S_2 = 36\,kVA$; $V_{LOAD\,LL} = 460\,V$; $PF_2 = 0.9$ LEADING

$\bar{Z}_{fdr} = (0.5 + j2)\,\Omega/ph$; $V_{LOAD\,L-N} = \dfrac{460}{\sqrt{3}} = 265.6\,V$

$$\bar{I}_1 = \frac{50000}{\sqrt{3}(460)(0.866)} \angle -\cos^{-1} 0.866 = 72.46\angle -30^\circ\,A$$

$$\bar{I}_2 = \frac{36000}{\sqrt{3}(460)} \angle \cos^{-1} 0.9 = 45.18\angle 25.8^\circ\,A$$

$$\bar{I}_T = \bar{I}_1 + \bar{I}_2 = 104.7\angle -9.1^\circ\,A$$

$$I_T = 104.7\,A \leftarrow$$

(b) $$\bar{Z}_1 = \frac{265.6\angle 0^\circ}{72.46\angle -30^\circ} = 3.66\angle 30^\circ\,\Omega/ph \leftarrow$$

$$\bar{Z}_2 = \frac{265.6\angle 0^\circ}{45.18\angle 25.8^\circ} = 5.88\angle -25.8^\circ\,\Omega/ph \leftarrow$$

(c) $$\bar{V}_{BUS\,LN} = \bar{V}_{LOAD\,LN} + \bar{Z}_{fdr}\,\bar{I}_T$$

$$= 265.6\angle 0^\circ + (0.5 + j2.0)(104.7\angle -9.1^\circ)$$

$$= 402.7\angle 29.5^\circ\,V$$

$$V_{BUS\,LL} = 402.7\sqrt{3} = 697.5\,V \leftarrow$$

(d) $$\bar{S}_{BUS}/ph = (402.7\angle 29.5^\circ)(104.7\angle -9.1^\circ)^* = 42.2\angle 38.6^\circ\,kVA = (32.95 + j26.3)\,kVA$$

$P_{BUS\,3\phi} = 3 \times 32.95 = 98.85\,kW \leftarrow$

$Q_{BUS\,3\phi} = 3 \times 26.3 = 78.9\,kVAR \leftarrow$

10.2.10.(a)

$\bar{Z}_{\Delta} = 45\angle 60^\circ \ \Omega/ph$; $\bar{Z}_{fdr} = (1.2 + j1.6)\ \Omega$

$\bar{Z}_Y = \frac{1}{3}\bar{Z}_{\Delta} = 15\angle 60^\circ \ \Omega/ph = (7.5 + j13)\ \Omega/ph$ ← EQUIV. Y CONVERSION

$\bar{Z}_T = \bar{Z}_{fdr} + \bar{Z}_Y = 7.5 + j13 + 1.2 + j1.6 = (8.7 + j14.6)\ \Omega/ph.$

$$\bar{I} = \frac{(208/\sqrt{3})\angle 0^\circ}{(8.7 + j14.6)} = 7.06\angle -59.2^\circ \ A$$

$\bar{V}_{LOAD\ LN} = 120\angle 0^\circ - (1.2 + j1.6)(7.06\angle -59.2^\circ) = 106\angle 0.81^\circ$

THEREFORE

$V_{LOAD\ LL} = \sqrt{3}(106) = 183.6V$ ←

$(1.2 + j1.6)\ \Omega$ $\bar{Z}_{fdr}$ + + $\frac{208}{\sqrt{3}}\angle 0^\circ = 120\angle 0^\circ V$ $\bar{V}_{LOAD\ LN}$ LOAD $\bar{Z}_Y/ph = (7.5 + j13)\ \Omega$ − −

(b)

$\bar{Z}_{c,\Delta} = 60\angle -90^\circ = -j60\ \Omega/ph$

$\bar{Z}_{c,Y} = \frac{1}{3}\bar{Z}_{c,\Delta} = 20\angle -90^\circ = -j20\ \Omega/ph$

$$\bar{Z}_{eq} = \frac{(7.5 + j13)(-j20)}{7.5 + j13 - j20} = (28.5 + j6.6)\ \Omega/ph$$

$$\bar{I} = \frac{120\angle 0^\circ}{28.5 + j6.6 + 1.2 + j1.6} = 3.89\angle -15.4^\circ \ A$$

$\bar{V}_{LOAD\ LN} = 120\angle 0^\circ - (1.2 + j1.6)(3.89\angle -15.4^\circ)$
$= 113.9\angle -2.4^\circ$

$\therefore \ V_{LOAD\ LL} = \sqrt{3}(113.9) = 197.3V$ ←

10.3.1.

(a)

$$\bar{I}_R = \bar{I}_S = \frac{\bar{V}_S - \bar{V}_R}{jX} = \frac{V_S e^{j\delta} - V_R}{jX}$$

$$\bar{S}_R = \bar{V}_R \bar{I}_R^* = V_R \left(\frac{V_S e^{-j\delta} - V_R}{-jX} \right)$$

$$= \frac{V_R V_S \sin\delta + j V_R V_S \cos\delta - j V_R^2}{X}$$

THE REAL POWER DELIVERED IS

$$P_R = P_S = P = \mathrm{Re}(\bar{S}_R) = \frac{V_R V_S}{X'} \sin\delta \quad \leftarrow$$

(b) $$P_{MAX} = V_R V_S / X \quad \leftarrow$$

(c)

IF V_R & V_S ARE LINE-TO-LINE VOLTAGES IN VOLTS, AND X IS THE PER-PHASE REACTANCE IN OHMS, THE EXPRESSION FOR P IN WATTS APPLIES FOR THREE-PHASE CASE

ALSO, IF V_R & V_S ARE LINE-TO-LINE VOLTAGES IN kV, AND X IS THE PER-PHASE REACTANCE IN OHMS, THE EXPRESSION FOR P IN MW APPLIES FOR THREE-PHASE CASE.

10.3.2.

(a)

$$\bar{V}_R = \frac{33}{\sqrt{3}} \angle 0° = 19.05 \angle 0° \text{ kV}_{LN}$$

$$\bar{I}_R = \frac{\bar{S}_R}{\sqrt{3} V_{RLL}} \angle -\cos^{-1}(PF) = \frac{10}{\sqrt{3}(33)} \angle -\cos^{-1} 0.9 = 0.175 \angle -25.84° \text{ kA}$$

$$\bar{Z} = (0.19 + j0.34)20 = 7.79 \angle 60.8° \; \Omega/\text{ph.}$$

$$\bar{V}_S = \bar{V}_R + \bar{Z}\bar{I}_R = 19.05 + (7.79 \angle 60.8°)(0.175 \angle -25.84°)$$

$$= 19.05 + 1.363 \angle 34.96° = 20.17 + j0.7809$$

$$\bar{V}_S = 20.19 \angle 2.22° \text{ kV}_{LN}$$

(b) $$V_S = 20.19\sqrt{3} = 34.96 \text{ kV}_{LL} \quad \leftarrow$$

$$\bar{I}_R = 0.175 \angle +25.84° \text{ kA}$$

$$\bar{V}_S = 19.05 + (7.79 \angle 60.8°)(0.175 \angle 25.84°) = 19.18 \angle 4.07° \text{ kV}_{LN}$$

$$V_S = 19.18\sqrt{3} = 33.22 \text{ kV}_{LL} \quad \leftarrow$$

10.3.2.(C)(i) CONTD. 0.9 PF LAGGING ; NOTE $\bar{I}_S = \bar{I}_R$

$$P_R = 10 \times 0.9 = 9 \text{ MW}$$

$$P_S = \sqrt{3}\,(34.96)\,0.175 \cos(25.84 + 2.22)^\circ \text{ MW} = 9.35 \text{ MW}$$

$$\therefore \eta_{0.9 \text{ PF LAGGING}} = \frac{9}{9.35} \times 100 = 96.26\% \leftarrow$$

(ii) 0.9 PF LEADING; $\bar{I}_S = \bar{I}_R$

$$P_R = 10 \times 0.9 = 9 \text{ MW}$$

$$P_S = \sqrt{3}\,(33.22)(0.175) \cos(25.84 - 4.07)^\circ \text{ MW}$$

$$= 9.35 \text{ MW}$$

$$\eta_{0.9 \text{ PF LEADING}} = \frac{9}{9.35} \times 100 = 96.26\% \leftarrow$$

(d)
(i)

$$\text{TLVR} = \frac{|V_{R\,NL}| - |V_{R\,FL}|}{|V_{R\,FL}|} \times 100$$

$$= \frac{34.96 - 33}{33} \times 100 = 5.94\% \leftarrow$$ FOR 0.9 PF LAGGING CASE

(ii) TLVR FOR 0.9 PF LEADING CASE $= \dfrac{33.22 - 33}{33} \times 100$

$$= 0.67\% \leftarrow$$

10.3.3.

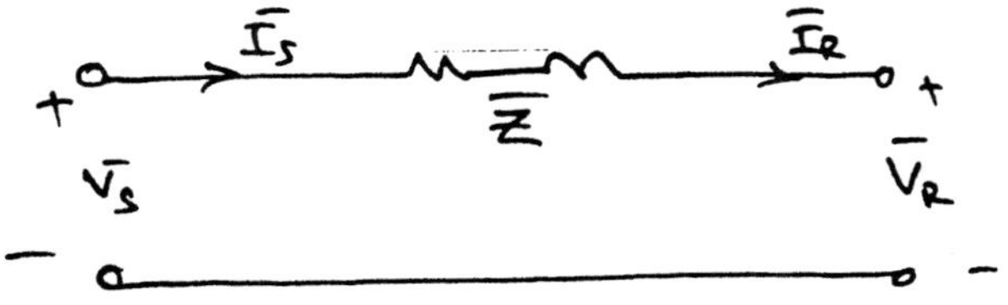

$\bar{I}_S = \bar{I}_R$; $\bar{V}_S = \bar{V}_R + \bar{Z}\,\bar{I}_R$

$\therefore$ $\bar{A} = 1$; $\bar{B} = \bar{Z}\ \Omega$

$\bar{C} = 0\ S$; $\bar{D} = 1$

$\bar{A}$ AND $\bar{D}$ ARE DIMENSIONLESS; $\bar{B}$ HAS UNITS OF OHMS ;

$\bar{C}$ HAS UNITS OF SIEMENS.

$$\bar{A}\bar{D} - \bar{B}\bar{C} = 1 - 0 = 1$$

NOTE THAT THIS RELATION ALWAYS HOLDS GOOD FOR LINEAR, PASSIVE, AND BILATERAL TWO-PORT NETWORKS.

10.3.4.

+ $\bar{I}_S$ $\bar{Z} = R + jX$ $\bar{I}_R = \bar{I}_S$ +

$\bar{V}_S$ $\bar{V}_R$

− −

MODEL

PHASOR DIAGRAMS ARE SHOWN BELOW:

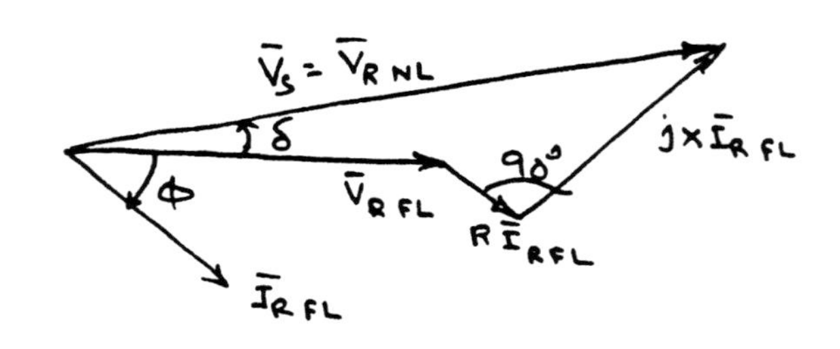

(a) LAGGING − PF − LOAD

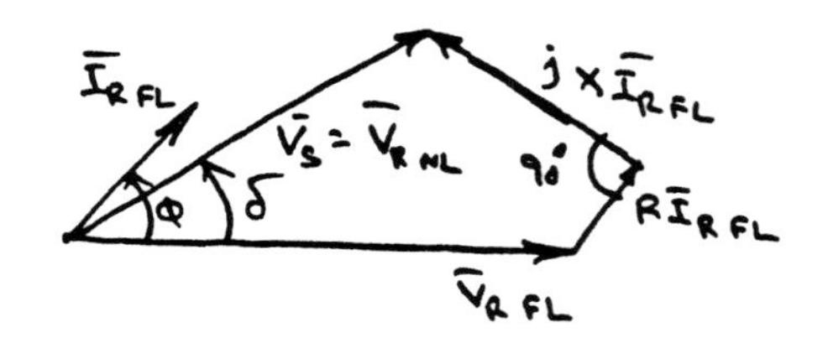

(b) LEADING − PF − LOAD

10.3.5. $\bar{V}_S = \bar{A}\bar{V}_R + \bar{B}\bar{I}_R$; AT NO LOAD $\bar{I}_R = 0$; $\therefore \bar{V}_{R\,NL} = \bar{V}_S / \bar{A}$ ←

10.3.6.

$$S_1 = \sqrt{3}\, V_{L1} I_{L1}$$

WHEN V_{L1} IS DOUBLED, $V_{L2} = 2V_{L1}$

I_{L1} IS HALVED ; $I_{L2} = I_{L1}/2$

$$\therefore S_2 = \sqrt{3}\, V_{L2} I_{L2} = \sqrt{3}\,(2V_{L1})(I_{L1}/2)$$

$$= \sqrt{3}\, V_{L1} I_{L1} = S_1$$

PER-PHASE $P_{LOSS} = \left(\frac{I_{L1}}{2}\right)^2 R = \frac{1}{4} I_{L1}^2 R$

PER-PHASE $Q_{LOSS} = \left(\frac{I_{L1}}{2}\right)^2 X = \frac{1}{4} I_{L1}^2 X$ } THESE ARE ONE-FOURTH OF THE PREVIOUS VALUE. ←

PER-PHASE $V_{DROP} = \left(\frac{I_{L1}}{2}\right) Z = \frac{1}{2} I_{L1} Z$ WHICH IS ONE-HALF OF THE PREVIOUS VALUE ←

10.3.7. TABLE 10.3.1 ENTRIES

$S = 200$ MVA (GIVEN); $V = 115$ kV (GIVEN); $I = \dfrac{200}{\sqrt{3}\,115} = 1$ kA

$PF = \cos\theta = 0.8$ LAGGING (GIVEN) $\Rightarrow \theta = \cos^{-1}0.8 = 36.9^\circ$

$P = 200\,(0.8) = 160$ MW

$Q = 200\,(0.6) = 120$ MVAR

LOAD IMPEDANCE $Z = \dfrac{V^2}{S} = \dfrac{115^2}{200} = 66.1\,\Omega$

$R = Z\cos\theta = 66.1 \times 0.8 = 52.9\,\Omega$

$X_L = Z\sin\theta = 66.1 \times 0.6 = 39.7\,\Omega$

TABLE 10.3.2 ENTRIES

$R = 3\,\Omega$ GIVEN

$X_L = 8\,\Omega$ GIVEN

$Z = \sqrt{3^2+8^2} = \sqrt{73} = 8.5\,\Omega$

$\theta = \tan^{-1}\dfrac{8}{3} = 69.4^\circ$

$P_{LOSS\,3\phi} = 3\,I^2R = 3\,(1)^2\,3 = 9$ MW (1 kA)

$Q_{LOSS\,3\phi} = 3\,I^2X_L = 3\,(1)^2\,8 = 24$ MVAR

$V_{DROP} = IZ = 1\,(8.5) = 8.5$ kV

TABLE 10.3.3 ENTRIES

$\bar{V}_{S\,LN} = \bar{V}_{R\,LN} + \bar{I}\bar{Z} = \dfrac{115}{\sqrt{3}} + (1\angle -36.9^\circ)(8.5\angle 69.4^\circ)$

$= 66.4 + 8.5\angle 32.5^\circ = 73.57 + j\,4.57$

$= 73.71\angle 3.55^\circ$

$\therefore V_{S\,LL} = \sqrt{3}\,(73.71) = 127.7$ kV ←

REAL POWER $= 160 + 9 = 169$ MW ;

REACTIVE POWER $= 120 + 24 = 144$ MVAR ;

10.3.7. CONTD.

APPARENT POWER = $\sqrt{(169)^2+(144)^2} = 222$ MVA

$\theta = \tan^{-1}(Q/P) = \tan^{-1}(144/169) = 40.4^\circ$

OR θ = ANGLE BETWEEN $\bar{V}_{S\,LN}$ AND $\bar{I}_S = \bar{I}_R = 36.9 + 3.5 = 40.4^\circ$

10.3.8. (a) TABLE 10.3.4 ENTRIES (WITH PF CORRECTION)

V = 115 kV (GIVEN); REAL POWER = 160 MW; REACTIVE POWER = 0 MVAR

$\therefore$ S = 160 MVA ; $I = 160/\sqrt{3}(115) = 0.8$ kA ; PF = $\cos\theta = 1.0$ (BECAUSE OF 100% COMPENSATION)

$\theta = \cos^{-1} 1.0 = 0^\circ$

$$Z = \frac{115/\sqrt{3}}{0.8} \text{ OR } \frac{115^2}{160} = 82.7\ \Omega$$

$R = (82.7)(1) = 82.7\ \Omega$; $X_L = 0\ \Omega$

(b)

$P_{LOSS\ NEW\ 3\phi} = 3(0.8)^2 3 = 5.76$ MW

WHICH IS LESS THAN $P_{LOSS\ OLD} = 9$ MW

$Q_{LOSS\ NEW\ 3\phi} = 3(0.8)^2(8) = 15.36$ MVAR

WHICH IS LESS THAN $Q_{LOSS\ OLD} = 24$ MVAR

NEW $V_{DROP}/_{PH} = (0.8)\,8.5 = 6.8$ kV

WHICH IS LESS THAN OLD $V_{DROP}/_{PH} = 8.5$ kV

NOTE:
1. CURRENT FLOW IS REDUCED FROM 1 kA TO 0.8 kA
2. THE ENTIRE REACTIVE POWER REQUIREMENT OF 120 MVA HAS BEEN SATISFIED BY THE APPLICATION OF THE SHUNT-CAPACITOR BANK.

10.3.9.

WITHOUT PF CORRECTION:

$$\eta = \frac{P_{LOAD}}{P_{SOURCE}} \times 100 = \frac{160}{169} \times 100 = 94.7\%$$

WITH PF CORRECTION

$$\eta = \frac{160}{160+5.76} \times 100 = \frac{160}{165.8} \times 100 = 96.5\%$$

WITHOUT PF CORRECTION:

$$TLVR = \frac{V_{DROP}}{V_{NOMINAL}(\text{OR } V_{R\,FL})} \times 100\%$$

$$= \frac{8.5}{115/\sqrt{3}} \times 100\% = 12.8\%$$

WITH PF CORRECTION

$$TLVR = \frac{(0.8)(8.5)}{115/\sqrt{3}} \times 100\% = 10.2\%$$

10.3.10.

(a) $s(t) = vi = (\sqrt{2}\, V \cos\omega t)(\sqrt{2}\, I \cos(\omega t - \theta))$

USING TRIG. IDENTITIES

$$s(t) = VI \cos\theta\,(1 + \cos 2\omega t) + VI \sin\theta \sin 2\omega t$$

$$s(t) = P(1 + \cos 2\omega t) + Q \sin 2\omega t \quad \longleftarrow$$

(b) $E = \int_{t_1}^{t_2} s\, dt = \int_{t_1}^{t_2} \{P(1 + \cos 2\omega t) + Q \sin 2\omega t\}\, dt$

$$= P\int_{t_1}^{t_2} dt + \frac{P}{2\omega}\int_{t_1}^{t_2} \cos 2\omega t\,(2\omega\, dt) + \frac{Q}{2\omega}\int_{t_1}^{t_2} \sin 2\omega t\,(2\omega\, dt)$$

$$= Pt\Big|_{t_1}^{t_2} + \underbrace{\frac{P}{2\omega} \sin 2\omega t\Big|_{t_1}^{t_2}}_{\Downarrow\; 0} - \underbrace{\frac{Q}{2\omega} \cos(2\omega t)\Big|_{t_1}^{t_2}}_{\Downarrow\; 0}$$

$$= P(t_2 - t_1) = P\, t_{interval} \quad \longleftarrow$$

10.3.11.
(a)

LOAD: 40 MW, 220 kV, 0.9 PF LAGGING

$$\bar{I}_R = \frac{40{,}000}{\sqrt{3}\,(220)(0.9)} \angle -\cos^{-1} 0.9 = 116.6 \angle -25.8^\circ \text{ A}$$

$$\bar{V}_S = \bar{V}_R + \bar{Z}\,\bar{I}_R$$

$$= \frac{220}{\sqrt{3}} \times 10^3 \angle 0 + (35 + j140)(116.6 \angle -25.8^\circ)$$

$$= 138.4 \angle 5.4^\circ \text{ kV LN}$$

$$V_{S\,LL} = 138.4\sqrt{3} = 239.7 \text{ kV} \leftarrow$$

Circuit: $\bar{I}_S$, $\bar{Z} = (35 + j140)\,\Omega = 144.3\angle 76^\circ$, $\bar{I}_R$, $\bar{V}_{S\,LN}$, $\frac{220}{\sqrt{3}}\angle 0^\circ$ kV, LOAD

$$\bar{I}_S = \bar{I}_R; \qquad I_S = 116.6 \text{ A} \leftarrow$$

$$PF_{\text{SENDING END}} = \cos[5.4 - (-25.8)]^\circ = 0.86 \text{ LAGGING} \leftarrow$$

(b)

$$\% \, VR = \frac{V_{RNL} - V_{RFL}}{V_{RFL}} \times 100$$

$$= \frac{239.7 - 220}{220} \times 100 = 8.95\% \leftarrow$$

$$P_S = \sqrt{3}\,(239.7)\,10^3\,(116.6)(0.86) = 41.6 \text{ MW}$$

$$\therefore \eta = \frac{P_R}{P_S} \times 100 = \frac{40}{41.6} \times 100 = 96.2\% \leftarrow$$

10.3.12.

Circuit: $\bar{V}_S$, $\bar{I}_S$, $22.86\angle 62.3^\circ\,\Omega$, $\bar{V}_R$

(a) 0.8 PF LAGGING

$$\bar{V}_R = \frac{13.8}{\sqrt{3}} \angle 0^\circ \text{ kV} = 7967.4 \angle 0^\circ \text{ V}$$

$$\bar{I}_R = \frac{2.5 \times 10^3}{\sqrt{3}\,(13.8)(0.8)} \angle -\cos^{-1} 0.8 = 130.7 \angle -26.9^\circ \text{ A}; \quad I_S = I_R = 130.7 \text{ A} \leftarrow$$

$$\bar{V}_{S\,LN} = \bar{V}_{R\,LN} + \bar{Z}\,\bar{I}_R = 7967.4 \angle 0^\circ + (22.86 \angle 62.3^\circ)(130.7 \angle -36.9^\circ)$$

$$= 10743 \angle 6.85^\circ$$

$$V_{S\,LL} = 10743\sqrt{3} = 18.6 \text{ kV}_{LL} \leftarrow$$

10.3.12. CONTD. (a)

$$\bar{S}_s = 3\bar{V}_{S\,LN}\bar{I}_S^* = 3(10743\angle 6.85°)(130.7\angle 36.9°)$$
$$= 4.21\times 10^6 \angle 43.8°$$
$$= (3.04 + j2.91)\,10^6 \text{ VA}$$

$P_S = 3.04$ MW ; $Q_S = 2.91$ MVAR ←

(b) UNITY PF

$$\bar{I}_R = \frac{2500}{\sqrt{3}(13.8)}\angle 0° = 104.6\angle 0° \text{ A}$$

$I_S = 104.6$ A ←

$$\bar{V}_{S\,LN} = 7967.4\angle 0° + (22.86\angle 62.3°)(104.6\angle 0°)$$
$$= 9323\angle 13.5° \; V_{LN}$$

$V_{S\,LL} = 16.15$ kV ←

$$\bar{S}_s = 3(9323\angle 13.1°)(104.6\angle 0°)^* \text{ VA}$$
$$= 2.92\angle 13.1° \text{ MVA} = (2.85 + j0.66) \text{ MVA}$$

$P_S = 2.85$ MW ; $Q_S = 0.66$ MVAR ←

(C) 0.9 PF LEADING

$$\bar{I}_R = \frac{2500}{\sqrt{3}(13.8)(0.9)}\angle \cos^{-1}0.9 = 116.2\angle 25.8° \text{ A}$$

$I_S = 116.2$ A ←

$$\bar{V}_{S\,LN} = 7967.4\angle 0° + (22.86\angle 62.3°)(116.2\angle 25.8°)$$
$$= 8482\angle 18.2° \text{ LN V}$$

$V_{S\,LL} = 14.7$ kV ←

$$\bar{S}_s = 3(8482\angle 18.2°)(116.2\angle 25.8°)^*$$
$$= 2.96\times 10^6\angle -7.6° \text{ VA}$$
$$= (2.93 - j0.39) \text{ MVA}$$

$P_S = 2.93$ MW ←

$Q_S = -0.39$ MVAR ←

CHAPTER 11

11.1.1.

M-19: $\mu_r = \frac{1}{\mu_0}\frac{1}{90} = \frac{10^7}{4\pi(90)} = 8842.$ ←

ANSI 1020: $\mu_r = \frac{1}{\mu_0}\frac{1}{1600} = \frac{10^7}{4\pi(1600)} = 497.$ ←

11.1.2. UNITS OF $(HB) = \frac{\text{AMPERES}}{\text{METER}} \times \frac{\text{WEBERS}}{(\text{METER})^2} = \frac{\text{AMPS}}{\text{METER}} \times \frac{\text{VOLTS} \times \text{SECONDS}}{(\text{METER})^2}$

$= \frac{\text{JOULES}}{(\text{METER})^3} = J/m^3$ ←

11.1.3. 1 cm^2 OF THE AREA OF HYSTERESIS LOOP REPRESENTS

$$400 \frac{At}{m} \times 0.3\ T = 120\ J/m^3$$

HYSTERESIS LOSS PER CYCLE =

$$= (400 \times 10^{-6})\ (6.2 \times 120)\ J$$

$$= 0.3\ J$$ ←

11.1.4. 1 cm^2 OF THE AREA OF LOOP REPRESENTS

$$200 \frac{At}{m} \times 0.03\,T = 6\ J/m^3$$

HYSTERESIS LOSS = $(20 \times 10^{-6}) \times 400 \times (80 \times 6)$

$$= 3.84\ W$$ ←

11.1.5. $P_e \propto f^2 B_m^2$ FROM EQ. (11.1.5)

$P_{e1} = k_1 (500)^2 (1)^2 = 15 \Rightarrow k_1 = \frac{15}{(500)^2}$

$P_{e2} = k_1 (750)^2 (0.8)^2$

$= \frac{15}{(500)^2}(750)^2(0.8)^2 = 21.6\ W$ ←

11.1.6. $P_h \propto f$; $P_e \propto f^2$

FOR $f = 60\,HZ$, $k_1(60) + k_2(3600) = 1800$
FOR $f = 90\,HZ$, $k_1(90) + k_2(8100) = 3000$ $\Rightarrow k_1 = 23.33$; $k_2 = 0.111$

∴ AT $f = 60\,HZ$, $P_h = 23.33 \times 60 = 1400\,W$; $P_e = 3600 \times 0.111 = 400\,W$ ←
AT $f = 90\,HZ$, $P_h = 23.33 \times 90 = 2100\,W$; $P_e = 8100 \times 0.111 = 900\,W$

11.1.7. $P_h \propto B_m^{1.5} \propto I_m^{1.5}$

OR $P_{h1} = k_1 2^{1.5} = 10$ OR $k_1 = \frac{10}{2^{1.5}} = \frac{10}{2.83} = 3.53$

FOR $I_m = 0.5A$, $P_{h2} = 3.53(0.5)^{1.5} = 3.53 \times 0.3536 = 1.25$ W ←

FOR $I_m = 8A$, $P_{h3} = 3.53(8)^{1.5} = 79.87$ W ←

11.1.8. $P_h \propto f$; $P_e \propto f^2$

FOR $f = 50$ Hz, $k_1(50) + k_2(2500) = 10$

FOR $f = 60$ Hz, $k_1(60) + k_2(3600) = 13$

$\therefore k_1 = \frac{7}{60}$ $k_2 = \frac{1}{600}$

THEN FOR $f = 400$ Hz

TOTAL LOSS $= k_1(400) + k_2(400)^2$

$= \left(\frac{7}{60} \times 400\right) + \left(\frac{1}{600} \times 400 \times 400\right)$

$= 46.67 + 266.67$

$= 313.3$ W ←

11.2.1.

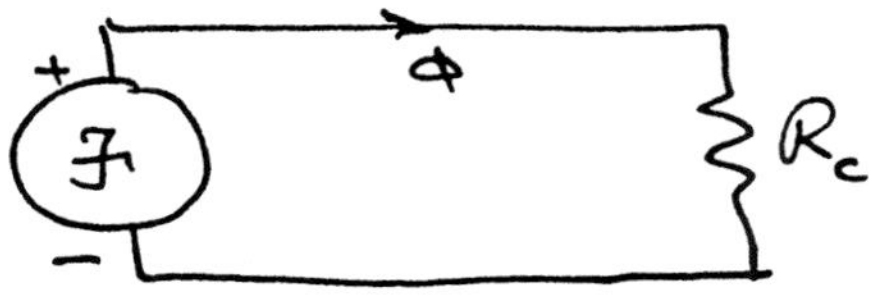

(a) $\Phi = BA = 1.5(16 \times 10^{-4}) = 2.4$ mWb ←

(b) $\lambda = N\Phi = (100)(2.4 \times 10^{-3}) = 0.24$ Wb-t ←

(c) $\mathcal{R}_c = \frac{l_c}{\mu_r \mu_0 A_c} = \frac{40 \times 10^{-2}}{(50000)(4\pi \times 10^{-7})(16 \times 10^{-4})} = 3979 \frac{1}{H}$

$\mathcal{F} = NI = \Phi \mathcal{R}_c$ OR $I = \frac{\Phi \mathcal{R}_c}{N} = \frac{2.4 \times 10^{-3} \times 3979}{100}$

$= 0.095A \simeq 0.1$ A ←

11.2.2.

Circuit: $\mathcal{F}$, ϕ, $\mathcal{R}_c$, $\mathcal{R}_g$

(a) $\phi = BA = 1.5(16\times10^{-4}) = 2.4$ mWb ←

(b) $\lambda = N\phi = 100(2.4\times10^{-3}) = 0.24$ Wb-t ←

(c)

$$\mathcal{R}_c = \frac{l_c - l_g}{\mu_c A_c} \simeq \frac{l_c}{\mu_c A_c} = 3979 \ \text{1/H} \quad \text{(FROM Sol. of Pr. 11.2.1)}$$

$$\mathcal{R}_g = \frac{l_g}{\mu_o A_c} = \frac{0.1\times10^{-3}}{4\pi\times10^{-7}\times16\times10^{-4}} = 49{,}736 \ \text{1/H}$$

$$\mathcal{R}_{total} = \mathcal{R}_c + \mathcal{R}_g = 3979 + 49736 = 53715 \ \text{1/H}$$

SERIES CIRCUIT

$$\therefore I = \frac{\mathcal{R}_{total}\,\phi}{N} = \frac{(53715)(2.4\times10^{-3})}{100} = 1.3\ \text{A} \leftarrow$$

IN PR. 11.2.1 WITHOUT AIRGAP I = 0.1 A WHEREAS NOW IT IS 1.3 A WHICH IS 13 TIMES THE PREVIOUS VALUE.

11.2.3.

$\mu_r = 2500$; $r_{av} = 10$ cm ; $B = 1.25$ T

(a) $$H = \frac{B}{\mu_r\mu_o} = \frac{1.25}{(2500)(4\pi\times10^{-7})} = 397.9 \ \text{At/m}$$

$$NI = Hl = (397.9)(2\pi\times0.1) = 250$$

$$\therefore I = \frac{250}{2500} = 0.1\ \text{A} \leftarrow$$

(b)
$$\phi = BA = 1.25\left[\frac{\pi d^2}{4}\right] = \frac{1.25\times\pi(0.04)^2}{4} = 1.57\ \text{mWb} \leftarrow$$

CIRCULAR CROSSSECTION

4cm dia.

(c)

$$\mathcal{R}_c = \frac{l_c - l_g}{\mu A_c} \simeq \frac{l_c}{\mu A_c} = \frac{2\pi(0.1)}{(2500)(4\pi\times10^{-7})\left[\frac{\pi}{4}(0.04)^2\right]} = 159{,}155 \ \text{1/H}$$

$$\mathcal{R}_g = \frac{l_g}{\mu_o A} = \frac{0.01}{(4\pi\times10^{-7})(\frac{\pi}{4}\times0.04^2)} = 6{,}332{,}574 \ \frac{1}{H}$$

$$\mathcal{R}_{Total} = \mathcal{R}_c + \mathcal{R}_g = 6{,}491{,}729 \ \text{1/H}$$

$$NI = \mathcal{R}_{Total}\,\phi = \mathcal{R}_{Total} BA = (6491729)(1.25)\ \frac{\pi}{4}(0.04^2) = 10{,}197 \ \text{AT}$$

$$\therefore I = \frac{10197}{2500} = 4.08\text{A} \leftarrow$$

11.2.4.

$$H_g = \frac{1.6}{\mu_0} = 0.127 \times 10^7 \text{ At/m};$$

$$\mathcal{F}_g = H_g l_g = 0.127 \times 10^7 \times 0.001 = 1270 \text{ At}$$

SECTION 1: $B_1 = \frac{16}{24} \times 1.6 = 1.067 \text{ T}$

FROM FIG. 11.1.2 (M-19) $H_1 \approx 95 \text{ At/m}$

$$\mathcal{F}_1 = 95 \times 0.6 = 48 \text{ At}$$

SECTION 2: $B_2 = 1.6 \text{ T}$; $H_2 \simeq 3000 \text{ At/m}$ FROM FIG. 11.1.2 (M-19)

$$\mathcal{F}_2 = 2 \times 3000 \times 0.1 = 600 \text{ At}$$

TOTAL MMF (OF THE SERIES CIRCUIT) $= \mathcal{F}_g + \mathcal{F}_1 + \mathcal{F}_2$

$= 1270 + 48 + 600$

$= 1918 \text{ At}$ ←

11.2.5.

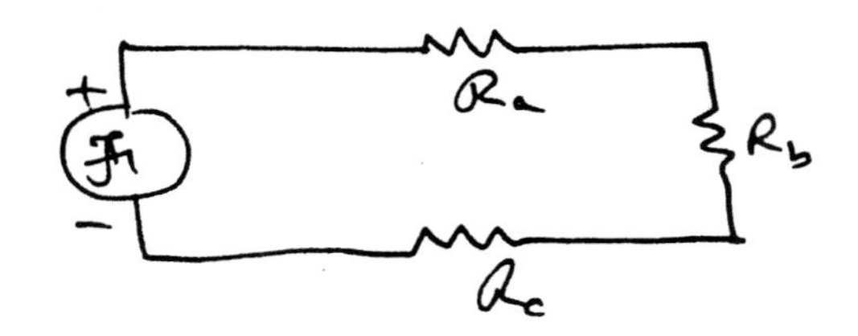

$$\mathcal{F}_{TOTAL} = \mathcal{F}_a + \mathcal{F}_b + \mathcal{F}_c = H_a l_a + H_b l_b + H_c l_c$$

$$B_a = B_b = B_c = \frac{\Phi}{A} = \frac{0.0012}{0.002} = 0.6 \text{T}$$

FROM FIG. P.11.2.5b: $H_a = 10 \text{ At/m}$; $H_b = 77 \text{ At/m}$; $H_c = 270 \text{ At/m}$.

$$\therefore \mathcal{F}_{TOTAL} = 10(0.6) + 77(0.4) + 270(0.2)$$

$= 6 + 30.8 + 54 = 90.8 \text{ At}$ ←

11.2.6.

$$\mathcal{F}_1 = N_1 I_1 = 1000 \times 5 = 5000 \text{ At} \; ; \; \mathcal{F}_2 = N_2 I_2 = 1000 \times 5 = 5000 \text{ At}$$

$$\mathcal{R}_{leg} = l/\mu A = (3 \times 50 \times 10^{-2})/(1000 \times 4\pi \times 10^{-7} \times 4 \times 10^{-4}) = 3 \times 10^6 \text{ At/Wb}$$

$$\mathcal{R}_g = l_g/\mu_0 A_g = (5 \times 10^{-3})/\left[4\pi \times 10^{-7}\{(2+0.5)10^{-2} \times (2+0.5)10^{-2}\}\right] = 6.37 \times 10^6$$

$$\mathcal{R}_{center\ leg} = \mathcal{R}_g + \frac{49.5 \times 10^{-2}}{1000 \times 4\pi \times 10^{-7} \times 4 \times 10^{-4}} = 7.37 \times 10^6 \text{ At/Wb}$$

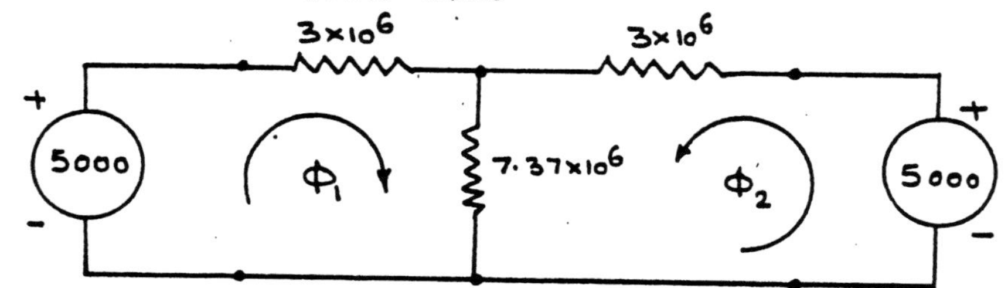

$$\left.\begin{aligned} \Phi_1(3\times10^6 + 7.37\times10^6) + \Phi_2(7.37\times10^6) &= 5000 \\ \Phi_1(7.37\times10^6) + \Phi_2(7.37\times10^6 + 3\times10^6) &= 5000 \end{aligned}\right\} \Rightarrow \Phi_1 = \Phi_2 = 0.3\times10^{-3} \text{ Wb}$$

(Note the symmetry)

$$\Phi_g = \Phi_1 + \Phi_2 = 0.6 \times 10^{-3} \text{ Wb} \leftarrow \; ; \; B_g = \frac{\Phi_g}{A_g} = \frac{0.6\times10^{-3}}{2.5\times2.5\times10^{-4}} = 0.96 \text{ T} \nearrow \; ; \; H_g = \frac{B_g}{\mu_0} = 0.764\times10^6 \text{ At/m} \nearrow$$

11.2.7.

$$l_0 = l_2 = 0.8 \text{ m} \; ; \; l_1 = 0.3 \text{ m}$$

$\mathcal{R}_0$ $\mathcal{R}_2$ $\mathcal{F}$ $\mathcal{R}_1$

$$H_1 l_1 = H_2 l_2$$

SINCE μ IS A CONSTANT, $B_1 l_1 = B_2 l_2$

$$\therefore B_2 = 1 \times 0.3/0.8 = 0.375 \text{ T}$$

$$\Phi_0 = B_0 A = \Phi_1 + \Phi_2 = B_1 A + B_2 A$$

$$\therefore B_0 = B_1 + B_2 = 1.375 \text{ T}$$

$$\mathcal{F} = 500\, I = H_0 l_0 + H_1 l_1 = \frac{1.375 \times 0.8}{1000\,\mu_0} + \frac{1 \times 0.3}{1000\,\mu_0} = \frac{1.4 \times 10^4}{4\pi}$$

OR $I = 2.23$ A $\leftarrow$

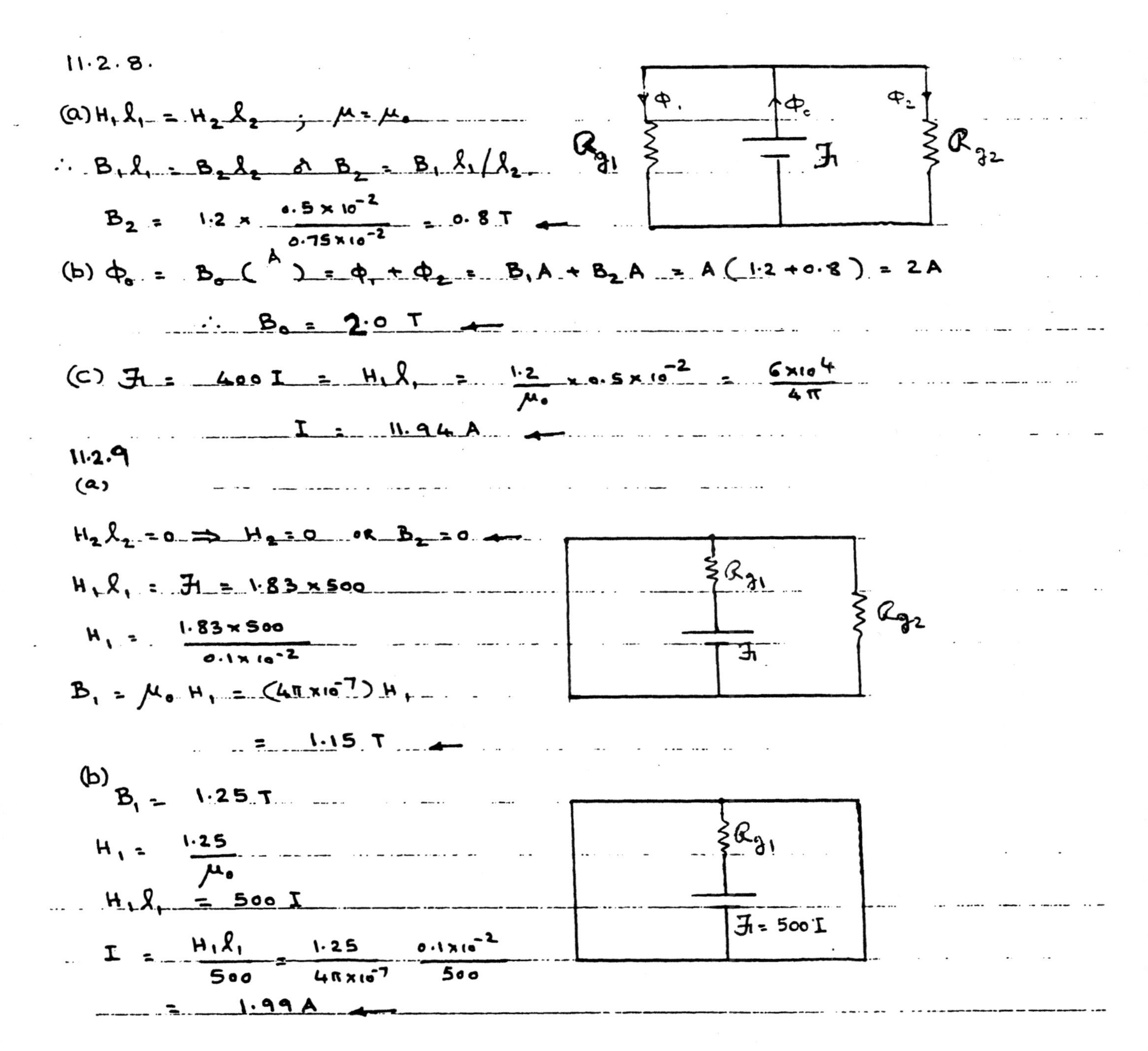

11.2.8.

(a) $H_1 l_1 = H_2 l_2$; $\mu = \mu_0$

$\therefore B_1 l_1 = B_2 l_2$ or $B_2 = B_1 l_1 / l_2$

$$B_2 = 1.2 \times \frac{0.5 \times 10^{-2}}{0.75 \times 10^{-2}} = 0.8\ T \leftarrow$$

(b) $\Phi_0 = B_0 (A) = \Phi_1 + \Phi_2 = B_1 A + B_2 A = A(1.2 + 0.8) = 2A$

$$\therefore B_0 = 2.0\ T \leftarrow$$

(c) $\mathcal{F} = 400 I = H_1 l_1 = \frac{1.2}{\mu_0} \times 0.5 \times 10^{-2} = \frac{6 \times 10^4}{4\pi}$

$$I = 11.94\ A \leftarrow$$

11.2.9

(a)

$H_2 l_2 = 0 \Rightarrow H_2 = 0$ or $B_2 = 0 \leftarrow$

$H_1 l_1 = \mathcal{F} = 1.83 \times 500$

$$H_1 = \frac{1.83 \times 500}{0.1 \times 10^{-2}}$$

$B_1 = \mu_0 H_1 = (4\pi \times 10^{-7}) H_1$

$$= 1.15\ T \leftarrow$$

(b)

$B_1 = 1.25\ T$

$$H_1 = \frac{1.25}{\mu_0}$$

$H_1 l_1 = 500 I$

$$I = \frac{H_1 l_1}{500} = \frac{1.25}{4\pi \times 10^{-7}} \cdot \frac{0.1 \times 10^{-2}}{500}$$

$$= 1.99\ A \leftarrow$$

11.2.10.

(a)

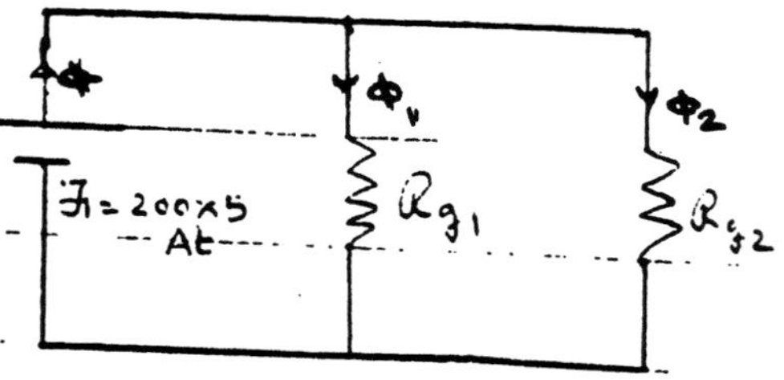

$H_1 l_1 = 1000$

$$H_1 = \frac{1000}{0.4 \times 10^{-2}} = 0.25 \times 10^6$$

$$B_1 = \mu_0 H_1 = 4\pi \times 10^{-7} \times 0.25 \times 10^6$$

$$= 0.1\pi = 0.3142 \text{ T} \leftarrow$$

(b)

$H_2 l_2 = 1000$

$$H_2 = \frac{1000}{0.5 \times 10^{-2}} = 0.2 \times 10^6$$

$$B_2 = \mu_0 H_2 = 4\pi \times 10^{-7} \times 0.2 \times 10^6 = 0.2513 \text{ T} \leftarrow$$

(c)

$\phi = \phi_1 + \phi_2$; WITH SAME CROSSSECTION $B = B_1 + B_2 = 0.5655$ T $\leftarrow$

11.2.11.

(a)

$$\mathcal{R}_g = \frac{g}{\mu_0 A} = \frac{2 \times 10^{-3}}{(4\pi \times 10^{-7})(20 \times 10^{-4})} = 795{,}770 \quad \text{/H}$$

$$\mathcal{F} = NI = 500 \times 4 = 2000 \text{ At}$$

$$\phi = \frac{NI}{\mathcal{R}_g} = \frac{2000}{795770} = 0.0025 \text{ Wb or } 2.5 \text{ mWb} \leftarrow$$

(b) $\lambda = N\phi = (500)(2.5 \times 10^{-3}) = 1.25$ Wb-t $\leftarrow$

(c) $L = \dfrac{\lambda}{i} = \dfrac{1.25}{4} = 0.3125$ H $\leftarrow$

(d) ENERGY STORED $= \frac{1}{2} L i^2 = \frac{1}{2}(0.3125)\, 4^2$

$= 2.5$ J $\leftarrow$

11.2.12.

(a) $\mathcal{R}_g = 795\,770 \quad 1/\text{H}$

$$\mathcal{R}_c = \frac{l_c}{\mu A} = \frac{1}{(2000)(4\pi \times 10^{-7})(20 \times 10^{-4})} = 198944 \quad 1/\text{H}$$

$$\mathcal{R}_t = \mathcal{R}_g + \mathcal{R}_c = 994714 \quad 1/\text{H}$$

$$\mathcal{F} = NI = (500)\,4 = 2000$$

$$\Phi = \frac{NI}{\mathcal{R}_t} = \frac{2000}{994714} = 0.002 \approx 2\text{ mWb} \leftarrow$$

(b) $\lambda = N\phi = 500\,(2 \times 10^{-3}) = 1 \text{ Wb-t} \leftarrow$

(c) $L = \dfrac{\lambda}{i} = \dfrac{1}{4} = 0.25 \text{ H} \leftarrow$

(d) ENERGY STORED $= \frac{1}{2} L i^2 = \frac{1}{2}(0.25)\,4^2 = 2 \text{ J} \leftarrow$

11.2.13.

BEFORE AIR GAP IS CUT:

$$\lambda = N\Phi = 2500\,(1.57 \times 10^{-3}) = 3.925 \text{ Wb-t}$$

$$L = \frac{\lambda}{i} = \frac{3.925}{0.1} = 39.25 \text{ H} \leftarrow$$

ENERGY STORED: $\frac{1}{2} L i^2 = \frac{1}{2}(39.25)(0.1)^2$

$$= 0.2 \text{ J} \leftarrow$$

AFTER AIR GAP IS CUT:

$$\lambda = N\phi = 2500\,(1.57 \times 10^{-3}) = 3.925 \text{ Wb-t}$$

$$L = \frac{\lambda}{i} = \frac{3.925}{4.08} = 0.962 \text{ H} \leftarrow$$

ENERGY STORED: $\frac{1}{2} L i^2 = \frac{1}{2}(0.962)(4.08)^2$

$$= 8 \text{ J} \leftarrow$$

11.3.1.

(a). $a = \frac{4800}{240} = 20$

$$\bar{Z}_{eq\,LV} = \frac{1}{a^2}\bar{Z}_{eq\,HV} = \frac{1}{400}(120 + j300) = (0.3 + j0.75)\,\Omega$$

(b).

$+$ $\bar{V}_1$ $-$; $\bar{I}_1$; $(120 + j300)\,\Omega$; $+$ $a\bar{V}_2$ $-$; $\bar{I}_2/a = 2.08\angle 0°$ A

$\bar{V}_2 = 230\angle 0°$ V ; $a\bar{V}_2 = 4600\angle 0°$ V

$\bar{I}_{2\,RATED} = \frac{10 \times 10^3}{240}\angle -\cos^{-1} 1 = 41.67\angle 0°$ A ;

$\bar{I}_{2\,RATED}/a = \frac{41.67}{20}\angle 0°$ A $= 2.08\angle 0°$ A

KVL: $\bar{V}_1 = a\bar{V}_2 + \frac{\bar{I}_2}{a}(120 + j300)$

$= 4600\angle 0° + (2.08\angle 0°)(120 + j300) = 4890\angle 7.3°$ V

$\therefore V_1 = 4890$ V ←

11.3.2.

$a = 2300/230 = 10$

(a) $\bar{Z}_{eq\,HV} = \{1.5 + (10)^2\,0.015\} + j\,[2.4 + (10)^2\,0.024]$

$= (3 + j4.8)\,\Omega$ ←

(b) $\bar{Z}_{eq\,LV} = \frac{1}{100}(3 + j4.8) = (0.03 + j0.048)\,\Omega$ ←

(c)

$+$ $\bar{V}_1$ $-$; $\bar{I}_1$; $(3 + j4.8)\,\Omega$; $+$ $a\bar{V}_2$ $-$; $\bar{I}_2/a = 10.87\angle -30°$ A

$\bar{V}_2 = 230\angle 0°$ V ; $a\bar{V}_2 = 2300\angle 0°$ V

$\bar{I}_{2\,RATED\,(FL)} = \frac{25000}{230}\angle -\cos^{-1} 0.866 = 108.7\angle -30°$ A

$\bar{I}_2/a = 10.87\angle -30°$ A

KVL: $\bar{V}_1 = 2300\angle 0° + (10.87\angle -30°)(3 + j4.8)$

$= 2354.5\angle 0.7°$ V

$\therefore V_1 = 2354.5$ V ←

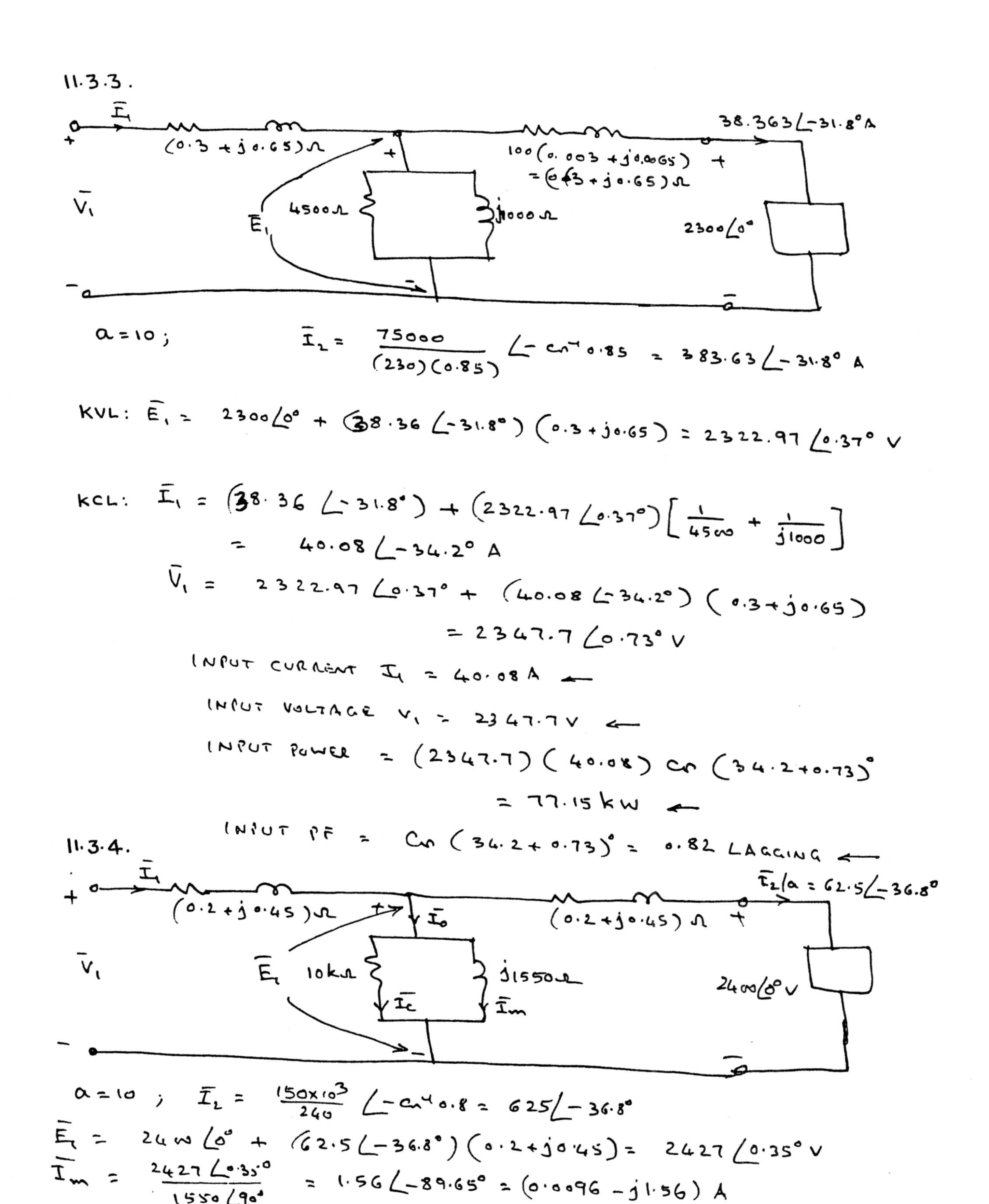

11.3.3.

$a = 10$; $\bar{I}_2 = \dfrac{75000}{(230)(0.85)} \angle -\cos^{-1} 0.85 = 383.63 \angle -31.8^\circ$ A

$100(0.003 + j0.0065) = (0.3 + j0.65)\,\Omega$

KVL: $\bar{E}_1 = 2300\angle 0^\circ + (38.36\angle -31.8^\circ)(0.3 + j0.65) = 2322.97\angle 0.37^\circ$ V

KCL: $\bar{I}_1 = (38.36\angle -31.8^\circ) + (2322.97\angle 0.37^\circ)\left[\dfrac{1}{4500} + \dfrac{1}{j1000}\right]$
$= 40.08\angle -34.2^\circ$ A

$\bar{V}_1 = 2322.97\angle 0.37^\circ + (40.08\angle -34.2^\circ)(0.3 + j0.65)$
$= 2347.7\angle 0.73^\circ$ V

INPUT CURRENT $I_1 = 40.08$ A ←

INPUT VOLTAGE $V_1 = 2347.7$ V ←

INPUT POWER $= (2347.7)(40.08)\cos(34.2 + 0.73)^\circ$
$= 77.15$ kW ←

INPUT PF $= \cos(34.2 + 0.73)^\circ = 0.82$ LAGGING ←

11.3.4.

$a = 10$; $\bar{I}_2 = \dfrac{150 \times 10^3}{240} \angle -\cos^{-1} 0.8 = 625\angle -36.8^\circ$

$\bar{I}_2/a = 62.5\angle -36.8^\circ$

$\bar{E}_1 = 2400\angle 0^\circ + (62.5\angle -36.8^\circ)(0.2 + j0.45) = 2427\angle 0.35^\circ$ V

$\bar{I}_m = \dfrac{2427\angle 0.35^\circ}{1550\angle 90^\circ} = 1.56\angle -89.65^\circ = (0.0096 - j1.56)$ A

11.3.4. CONTD.

$$\bar{I}_c = \frac{2427\angle 0.35^\circ}{10{,}000} \simeq (0.2427 + j0)\text{ A}$$

$$\bar{I}_0 = \bar{I}_c + \bar{I}_m = (0.25 - j1.56)\text{ A}$$

$$\bar{I}_1 = \bar{I}_0 + (\bar{I}_2/a)$$
$$= (0.25 - j1.56) + (50 - j37.5)$$
$$= 50.25 - j39.06 = 63.65\angle -37.85^\circ\text{ A}$$

$$\bar{V}_1 = (2427\angle 0.35^\circ) + (63.65\angle -37.85^\circ)(0.2 + j0.45)$$
$$= 2455\angle 0.7^\circ\text{ V}$$

SUPPLY VOLTAGE = 2455 V ←

11.3.5.

(a)

$0.06\,\Omega$ $j0.3\,\Omega$ $(0.003)\left(\frac{2400}{480}\right)^2 = 0.075\,\Omega$ $j0.012\left(\frac{2400}{480}\right)^2 = j0.3\,\Omega$ $\frac{500\times 10^3}{2400}\angle -\cos^{-1}0.8 = 208\angle -36.8^\circ$

$\bar{I}_1$ $\bar{V}_1$ $\bar{E}_1$ $\bar{I}_0$ $\bar{I}_c$ $\bar{I}_m$ $2000\,\Omega$ $j500\,\Omega$ $2400\angle 0^\circ$

$$\bar{E}_1 = 2400\angle 0^\circ + 208\angle -36.8^\circ(0.075 + j0.3) = 2449.92 + j40.56$$

$$\bar{I}_0 = \frac{\bar{E}_1}{2000} + \frac{\bar{E}_1}{j500} = 1.306 - j4.88\ ;\ I_c = \frac{\bar{E}_1}{2000} = 1.225 + j0.02 = 1.225\angle$$

$$\bar{I}_1 = 208\angle -36.8^\circ + \bar{I}_0 = 167.706 - j129.68 = 212\angle -37.71^\circ \leftarrow$$

$$\bar{V}_1 = \bar{E}_1 + 212\angle -37.71^\circ(0.06 + j0.3) = 2498.9 + j83.09 = 2500\angle 1.9^\circ \leftarrow$$

power factor $= \cos\theta_1 = \cos(37.71^\circ + 1.9^\circ) = 0.77$ ←

(b) $(R_{eq} + jX_{eq})_{HV} = (0.06 + 0.075) + j(0.3 + 0.3) = (0.135 + j0.6)\,\Omega$ ←

$(R_{eq} + jX_{eq})_{LV} = (0.135 + j0.6)\frac{1}{25} = (0.0054 + j0.024)\,\Omega$ ←

(c) $Z_{Th} = \bar{V}_1/\bar{I}_1 = 2500\angle 1.9^\circ / 212\angle -37.71^\circ = 11.79\angle 39.61^\circ\,\Omega$ ←

11.3.6.

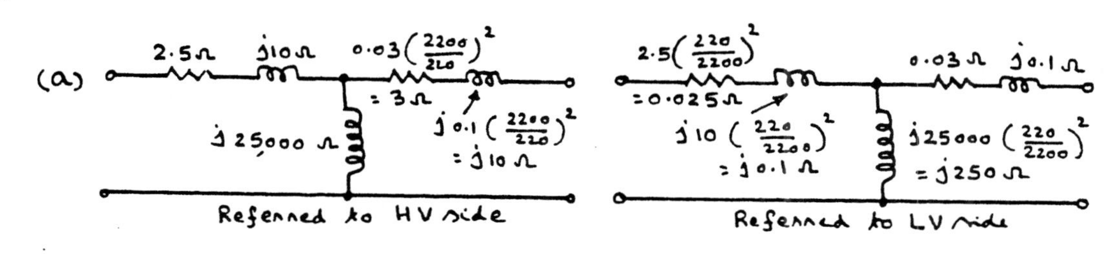

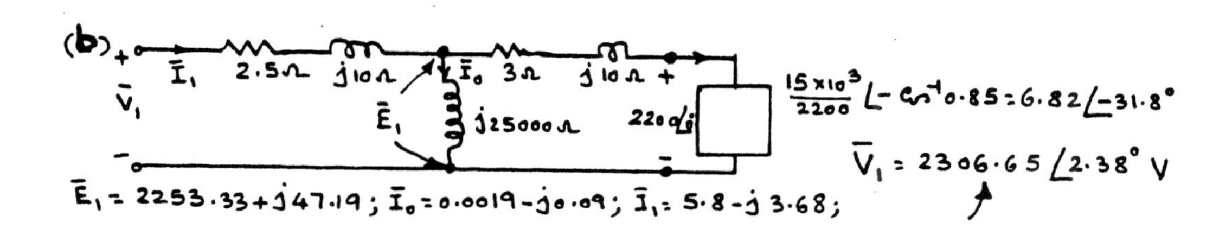

$\frac{15\times10^3}{2200}\angle -\cos^{-1}0.85 = 6.82\angle -31.8°$

$\bar{V}_1 = 2306.65\angle 2.38°$ V

$\bar{E}_1 = 2253.33 + j47.19$; $\bar{I}_0 = 0.0019 - j0.09$; $\bar{I}_1 = 5.8 - j3.68$;

11.4.1.

(a) $\cos\theta_{02} = \frac{P_{oc}}{V_{oc}I_{oc}} = \frac{70}{230\times0.45} = 0.676$; $\theta_{02} = 47.4°$; $\sin\theta_{02} = 0.737$.

$R_{c\,LV} = \frac{230}{0.45\times0.676} = 756\,\Omega$; $X_{m\,LV} = \frac{230}{0.45\times0.737} = 694\,\Omega$

$R_{c\,HV} = 756\,(2300/230)^2 = 75{,}600\,\Omega$; $X_{m\,HV} = 694\,(2300/230)^2 = 69{,}400\,\Omega$

$R_{eq\,HV} = 224/(4.35)^2 = 11.84\,\Omega$; $Z_{sc\,HV} = 120/4.35 = 27.6\,\Omega$

$X_{eq\,HV} = \sqrt{27.6^2 - 11.84^2} = 24.93\,\Omega$; $R_{eq\,LV} = 11.84\left(\frac{230}{2300}\right)^2 = 0.1184\,\Omega$

$X_{eq\,LV} = 0.2493\,\Omega$

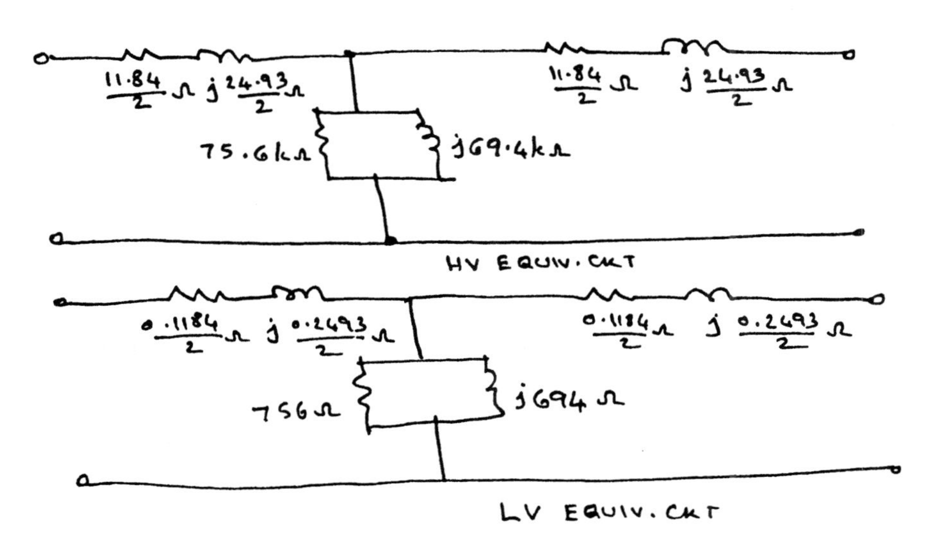

11.4.1. CONTD.

(b) $I_0 = 0.45\,A = 0.45/43.48 = 0.01$ or 1% ← Note that $I_{FL\,LV} = \frac{10\times10^3}{230} = 43.48$ Amps.

(c) output $= 10\times10^3 \times 0.8 = 8000\,W$; $P_{oc} = 70\,W$; $P_{sc} = 224\,W$ corresponding to a current of 4.35 A.

Since $I_{FL\,HV} = \frac{10\times10^3}{2300} = 4.35\,A$, Copper loss at rated FL current $= P_{sc} = 224\,W$.

$\therefore$ Effcy $= \frac{8000}{8000+70+224} = 0.965$ ← or 96.5%

(d) output $= 7.5\times10^3 \times 0.85 = 6375\,W$; $P_{oc} = 70\,W$; Copperloss: $224\left(\frac{7.5}{10}\right)^2 = 126\,W$

$\therefore$ Effcy $= \frac{6375}{6375+70+126} = \frac{6375}{6571} = 0.97$ ← or 97%

(e) $k = \sqrt{\frac{P_{oc}}{P_{sc}}} = \sqrt{\frac{70}{224}} = 0.56$ ←

Max^m. Effcy $= \frac{10\times10^3 \times 0.56 \times 0.85}{(10\times10^3 \times 0.56\times0.85)+70+70} = 0.971$ or 97.1%

(f) Energy output in watt-hours during 24 hours =

$(0.85\times10\times10^3\times0.85\times8) + (0.60\times10\times10^3\times0.85\times12) + 0 = 119{,}000\,Wh$

Core loss in 24 hours $= 24\times70 = 1680\,Wh$; Since Copperloss corresponding to rated FL current is 224 W, Copper loss during 24 hours =

$(0.85^2\times224\times8) + (0.6^2\times224\times12) = 2262.4\,Wh.$

$\therefore$ All-day or energy efficiency $= \frac{119000}{119000+1680+2262.4} = 0.968$ ← or 96.8%

(g) $\bar{E}_1 = 2351.85 + j26.18$

$\bar{I}_0 = 0.0315 - j0.0336$

$\bar{I}_1 = 4.4\angle -36.97°$

$\bar{V}_1 = 2404.89\angle +1.25°$

VOL. REG. $= \frac{2404.89-2300}{2300} = 0.046$ ← or 4.6%

pf at HV terminals: $\cos(36.97° + 1.25°) = 0.786$ ← LAGGING

11.4.2. OUTPUT $= 2\times10^3 \times 0.85 = 1700\,W$; core loss $= P_{oc} = 50\,W$

copper loss $= 30\left(\frac{2}{3}\right)^2 = 13.33\,W$

NOTE: FL CURRENT $= \frac{3\times10^3}{220} = 13.64\,A$ WHICH IS SAME AS THAT IN SC TEST DATA.

$\eta = \frac{1700}{1700+50+13.33} = 0.964$ or 96.4% ←

11.4.3.

(a)

$$Z_s = \frac{9.5}{326} = 0.029\ \Omega \leftarrow$$

(b)

$$R_{eq1} = P_{sc} / I_{sc}^2 = 1200/(326)^2 = 0.0113\ \Omega$$

$$X_{eq1} = \sqrt{0.029^2 - 0.0113^2} = 0.027\ \Omega$$

$$\bar{V}_1 = 230 + (326\angle -\cos^{-1}0.8)(0.0113 + j0.027)$$

$$= 238.28 \angle$$

$$\therefore V_1 = 238.28\ V$$

THEN, VOLTAGE REGULATION $= \dfrac{V_1 - aV_2}{aV_2} = \dfrac{238.28 - 230}{230} = 0.036$ OR 3.6% ←

(c)

$$\eta_{\text{rated load}} = \frac{(75 \times 10^3)(0.8)}{(60 \times 10^3) + 1200 + 750} = 96.85\% \leftarrow$$

$$\eta_{\frac{1}{2}\text{rated load}} = \frac{(37.5 \times 10^3)(1)}{(37.5 \times 10^3) + 300 + 750} = 97.27\% \leftarrow$$

(d) FOR MAXIMUM EFFICIENCY, COPPER LOSS = CORE LOSS = 750 W

HENCE $I_1 = 326\sqrt{\dfrac{750}{1200}} = 326 \times 0.79 = 257.72\text{A} \leftarrow$

THE POWER OUTPUT IS

$$\frac{I_1}{326}(75 \times 10^3) = 0.79(75 \times 10^3)\ W$$

AND SO

$$\eta_{max} = \frac{0.79(75 \times 10^3)}{[0.79(75 \times 10^3)] + 750 + 750} = 97.53\% \leftarrow$$

11.4.4.(a) $P_c = 1500\,W$; $P_{cu} = 4500\,W$ AT FL

UNITY PF

% OF LOAD	OUTPUT (W)	COPPER LOSS (W)	CORE LOSS (W)	$\eta = \frac{OUTPUT}{OUTPUT+LOSSES} \times 100$
25	$300\times10^3 \times \frac{1}{4} \times 1 = 75000$	$4500 \times \frac{1}{16} = 281.25$	1500	97.7%
50	$300\times10^3 \times \frac{1}{2} \times 1 = 150000$	$4500 \times \frac{1}{4} = 1125$	1500	98.3
75	$300\times10^3 \times \frac{3}{4} \times 1 = 225000$	$4500 \times (\frac{3}{4})^2 = 2531.25$	1500	98.2
100	$300\times10^3 \times 1 \times 1 = 300000$	4500	1500	98.2
125	$300\times10^3 \times 1.25 \times 1 = 375000$	$4500 (\frac{5}{4})^2 = 7031.25$	1500	98. / 97.8

(b)
AT 25% FL, 0.8 pf, OUTPUT $= 300\times10^3 \times \frac{1}{4} \times 0.8 = 60{,}000\,W$

Copper loss $= 4500 \times \frac{1}{16} = 281.25\,W$; core loss $= 1500\,W$; $\eta = 97.1\%$ ←

AT 25% FL, 0.6 pf, OUTPUT $= 300\times10^3 \times \frac{1}{4} \times 0.6 = 45000\,W$

Copper loss $= 4500 \times \frac{1}{16} = 281.25\,W$; core loss $= 1500\,W$; $\eta = 96.2\%$ ←

(c)
$k = \sqrt{\frac{1500}{4500}} = 0.577$ or 57.7% of FL

AT 57.7% FL & UNITY PF

OUTPUT $= 300\times10^3 \times 0.577 \times 1 = 173100\,W$

Copper loss $= 4500 \times \frac{1}{3} = 1500$; core loss $= 1500\,W$

$$\eta_{max} = \frac{173100}{173100+1500+1500} \times 100 = 98.3\% \leftarrow$$

AT 57.7% FL, 0.8 pf: OUTPUT $= 300\times10^3 \times 0.577 \times 0.8 = 138480\,W$

$$\eta_{max} = \frac{138480}{138480+1500+1500} \times 100 = 97.9\% \leftarrow$$

AT 57.7% FL, 0.6 pf: OUTPUT $= 300\times10^3 \times 0.577 \times 0.6 = 103860\,W$

$$\eta_{max} = \frac{103860}{103860+1500+1500} \times 100 = 97.2\% \leftarrow$$

11.4.5.

$a\bar{V}_2$ $\bar{V}_1$ θ_L

PHASOR DIAG AT THE PF FOR GREATEST (POOREST) REGULATION

$$\cos\theta_L = \cos\theta_{sc} = \frac{650}{(52)(20.8)} = 0.6 \text{ Lagging}$$

$$\%\,Reg = \frac{I_{sc} Z_{sc}}{aV_2} \times 100 = \frac{V_{sc}}{aV_2} \times 100 = \frac{52}{2400} \times 100 = 2.17\% \leftarrow$$

11.4.6.

NOTE: $I_{FL\,HV} = \frac{10\times10^3}{400} = 25\,A$ (SAME AS THAT IN SC TEST DATA)

$$\eta = \frac{0.5\times 10{,}000\times 0.85}{(0.5\times 10{,}000\times 0.85) + 450 + [600\times(0.5)^2]}$$

$$= \frac{4250}{4250+450+150} = 0.8763 \text{ or } 87.63\% \leftarrow$$

11.4.7.

(a) $\frac{4890-4600}{4600}\times 100 = 6.3\% \leftarrow$

$P_{out} = 10\times10^3 = 10{,}000\,W$; $P_{in} = (4890)(2.08)\cos(7.3)^\circ = 10089$

$\therefore \eta = \frac{10000}{10089}\times 100 = 99.12\% \leftarrow$

(b) $\frac{2354.5-2300}{2300}\times 100 = 2.37\% \leftarrow$

$P_{out} = 25\times10^3\times 0.866 = 21650\,W$

$P_{in} = (2354.5)(10.87)\cos(30+0.7)^\circ = 22006$

$\eta = \frac{21650}{22006}\times 100 = 98.4\% \leftarrow$

11.4.8.

$a = \frac{2300}{230} = 10$

$\bar{Z}_{eq1} = (R_1 + a^2R_2) + j(x_1 + a^2x_2) = 0.6 + j1.3$

$\bar{V}_2 = 230\angle 0^\circ\,V$

(a) $\bar{I}_2 = \frac{100\times10^3}{230}\angle -\cos^{-1}0.8 = 434.78\angle -36.9^\circ\,A$

$\bar{V}_1 = 2300\angle 0^\circ + (43.48\angle -36.9^\circ)(0.6+j1.3) = 2355\angle 0.72^\circ\,V$

$\%\,VR = \frac{2355-2300}{2300}\times 100 = 2.39\% \leftarrow$

(b) $\bar{I}_2 = 434.78\angle +36.9^\circ\,A$

$\bar{V}_1 = 2300\angle 0^\circ + (43.48\angle +36.9^\circ)(0.6+j1.3) = 2287.7\angle 1.52^\circ\,V$

$\%\,VR = \frac{2287.7-2300}{2300}\times 100 = -0.53\% \leftarrow$

11.4.9.

(a) $a = 10$; $\bar{V}_2 = 240\angle 0^\circ\,V$; $\bar{I}_{2FL} = \frac{25\times10^3}{240}\angle -\cos^{-1}0.85 = 104.17\angle -31.8^\circ\,A$

$\bar{V}_1 = 2400\angle 0^\circ + (10.42\angle -31.8^\circ)(3.45+j5.75) = 2462.3\angle 0.74^\circ\,V$

(b) $\%\,VR = \frac{2462.3-2400}{2400}\times 100 = 2.6\% \leftarrow$

$P_{core} = 120\,W$; $P_{cu} = (10.42)^2(3.45) = 374.3\,W$; $P_{out} = 25\times10^3\times0.85 = 21{,}250\,W$

$\eta = \frac{21250}{21250+120+374.3}\times 100 = 97.7\% \leftarrow$

11.4.10.

$$\bar{Z}_{eq1} = 3.5 + j4 = 5.315 \angle 48.8^\circ$$

2200∠0°
$\bar{V}_1$
48.8° = θ
θ
$\bar{I}_1 = 11.364 \angle -48.8^\circ$ A

$$\theta = 48.8^\circ \qquad \cos\theta = 0.66$$

$$\bar{I}_2 = \frac{25000}{220} \angle -\theta = 113.64 \angle -48.8^\circ \text{ A}$$

$$\bar{V}_1 = 2200 + (11.364 \angle -48.8^\circ)(5.315 \angle 48.8^\circ)$$

$$= 2260.4 \angle 0^\circ$$

$$\% \text{ VR} = \frac{2260.4 - 2200}{2200} \times 100 = 2.74\% \leftarrow$$

$$pf = 0.66 \text{ LAGGING} \leftarrow$$

11.4.11.

$$Z_{eq1} = \frac{V_{sc}}{I_{sc}} = \frac{55}{10.4} = 5.29\,\Omega$$

$$R_{eq1} = P_{sc}/I_{sc}^2 = \frac{375}{(10.4)^2} = 3.47\,\Omega$$

$$X_{eq1} = \sqrt{5.29^2 - 3.47^2} = 3.99\,\Omega$$

$$\bar{Z}_{eq1} = 3.47 + j3.99 = 5.29 \angle 49^\circ$$

$$\bar{I}_{2FL} = \frac{25000}{240} = 104 \text{ A}$$

$$\bar{I}_1 = 10.4 \angle -49^\circ$$

$$\bar{V}_1 = 2400 \angle 0^\circ + (10.4 \angle -49^\circ)(5.29 \angle 49^\circ)$$

$$= 2455 \angle 0^\circ$$

$$\% \text{ VR} = \frac{2455 - 2400}{2400} \times 100 = 2.29\% \leftarrow$$

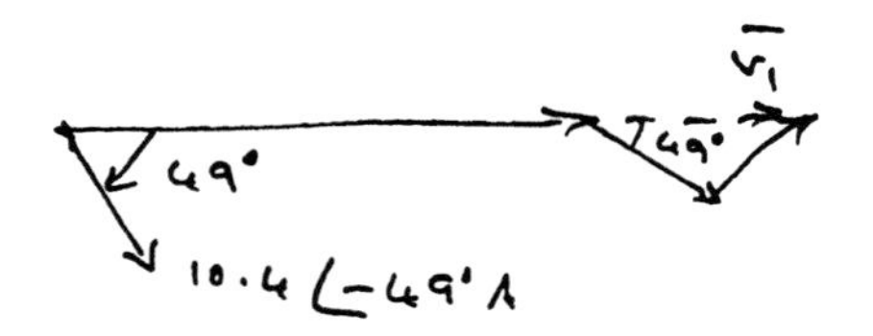

11.4.12.

$$0.9877 = \frac{400\times10^3\times0.8}{(400\times10^3\times0.8) + P_{cu} + P_{oc}}$$

$$0.9913 = \frac{200\times10^3\times1}{(200\times10^3\times1) + P_{oc} + P_{cu}\left(\frac{1}{2}\right)^2}$$

SOLVING TWO EQUATIONS WITH TWO UNKNOWNS, ONE GETS

(a) $P_{oc} = 1012$ W; (b) P_{cu} AT FULL LOAD $= 2973$ W

(c) $$\eta_{\frac{3}{4}\text{ LOAD}} = \frac{300\times10^3\times0.9}{(300\times10^3\times0.9) + 1012 + 2973\left(\frac{3}{4}\right)^2} = 0.9902$$

or 99.02%

11.4.13.

ENERGY OUTPUT IN kW-HRS DURING 24 HOURS IS

$$(12\times2) + (6\times12) + (6\times18\times0.9) = 193.2$$

$$\eta_{MAX} = 0.98 = \frac{15\times1}{(15\times1) + 2P} \quad \text{or } P = 153\text{ W}$$

$P_{oc} = 153$ W ; P_{cu} CORRESPONDING TO 15 kVA $= 153$ W

CORE LOSS DURING 24 HOURS IS $24\times153 = 3672$ W-hrs.

COPPER LOSS DURING 24 HOURS IS GIVEN BY

$$153\left(\frac{2}{0.5}\cdot\frac{1}{15}\right)^2 12 + 153\left(\frac{12}{0.8}\cdot\frac{1}{15}\right)^2 6 + 153\left(\frac{18}{15}\right)^2 6$$

$$= 2370.6\text{ W-hrs.}$$

$$\eta_{AD} = \frac{193.2}{193.2 + 3.672 + 2.371} = 97\%$$

11.4.14. (a) $a = 220/110 = 2$

$R_{eq\,HV} = 0.3 + 2^2(0.06) = 0.54\ \Omega$

$X_{eq\,HV} = 0.8 + 2^2(0.2) = 1.6\ \Omega$

$I_{HV\ Rated} = \frac{3 \times 10^3}{220} = 13.63\ A$

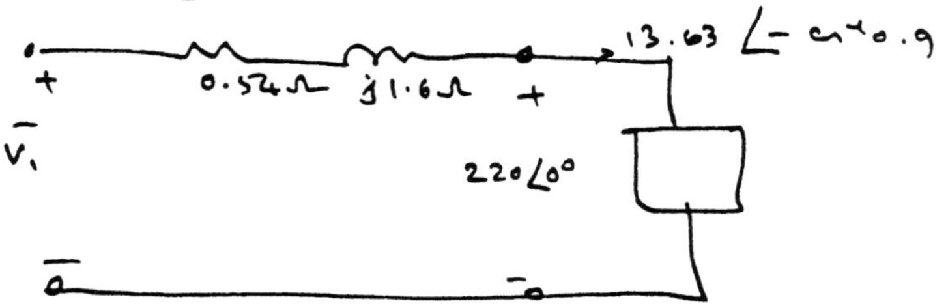

$\bar{V}_1 = 220\angle 0° + (13.63\angle -25.8°)(0.54 + j1.6)$

$= 236.68\angle 3.98°\ V$

$\%\,VR = \frac{236.68 - 220}{220} \times 100 = 7.6\%$ ←

(b) output = 3kW; $kVA = \frac{3}{0.95} = 3.16$; $I_{HV} = \frac{3.16 \times 10^3}{220} = 14.36\,A$;

Copper loss $= (14.36)^2\, 0.54 = 111.35\,W$; Effcy $= \frac{3}{3 + 0.045 + 0.111} = 0.95$ ←

(c) output in kWhrs $= (3.5 \times 0.95 \times 3) + (3 \times 0.9 \times 8) + (2.4 \times 0.8 \times 4) + (1 \times 0.95 \times 9)$
$= 47.805$

Copper loss in Whrs $= \left\{\left(\frac{3.5}{3}\right)^2 100 \times 3\right\} + \left\{\left(\frac{3}{3}\right)^2 100 \times 8\right\} + \left\{\left(\frac{2.4}{3}\right)^2 100 \times 4\right\} + \left\{\left(\frac{1}{3}\right)^2 100 \times 9\right\}$
$= 1564$

Core loss in Whrs: $45 \times 24 = 1080$; All-day or Energy Effcy $= \frac{47805}{47805 + 1080 + 1564} = 0.948$ ←

11.4.15.

OUTPUT FOR 24h $= (150 \times 0.8 \times 12) + (150 \times 0.5 \times 1 \times 8) = 2040$ kWh

CORE LOSS: $(0.2427)^2\ 10{,}000 \times 24 = 14.14$ kWh

COPPER LOSS: FULL LOAD, 12h: $12\left[63.65^2(0.2) + 625^2(0.002)\right] = 19.1$ kWh

$\frac{1}{2}$ FULL LOAD, 8h: $8\left[\left(\frac{63.63}{2}\right)^2(0.2) + \left(\frac{625}{2}\right)^2(0.002)\right] = 3.18$ kWh

TOTAL LOSSES $= 14.14 + 19.1 + 3.18 = 36.42$ kWh

INPUT FOR 24 hrs $= 2040 + 36.42 = 2076.42$ kWh

$\eta_{AD} = \frac{2040}{2076.42} \times 100 = 98.2\%$ ←

11.4.16.

$$\eta_{AD} = \frac{(75\times 8) + 0 + (37.5\times 8)}{(75\times 8) + (37.5\times 8) + (1\times 24) + (1\times 8) + (\frac{1}{4}\times 8)} \times 100 = 96.4\% \leftarrow$$

11.4.17.

$$W_{core} = (24)(150) = 3.6 \text{ kwh}$$

$$W_{cu} = 4(250) + 8(0.75)^2 250 + 12(0.5)^2 250 = 2.875 \text{ kwh}$$

$$W_{OUT} = 4(10)(0.8) + 8(0.75)(10)(1.0) + 12(0.5)(10)(0.6)$$

$$= 128 \text{ kwh}$$

$$\eta_{AD} = \frac{128}{128 + 3.6 + 2.875} \times 100 = 95.2\% \leftarrow$$

11.4.18.

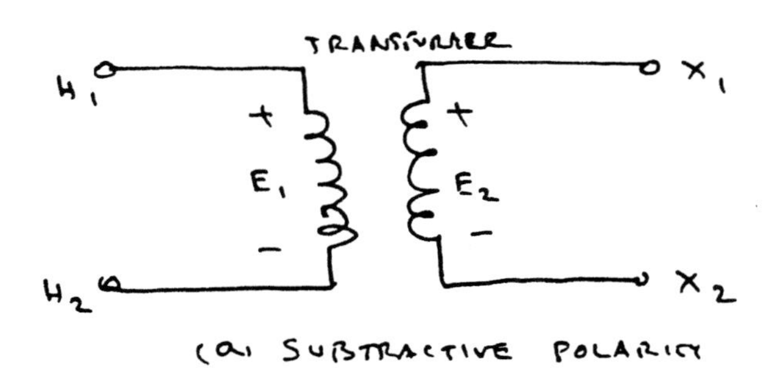

(a) SUBTRACTIVE POLARITY

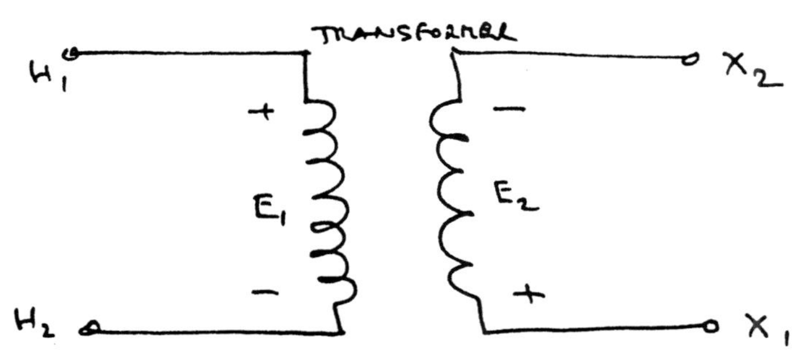

(b) ADDITIVE POLARITY

11.4.19.
(a)

- 4800 V +

2400 2400

H2 H1 H2 H1

X2 X1 X2 X1

120 120

240

b b

- +

11.4.19. (b)

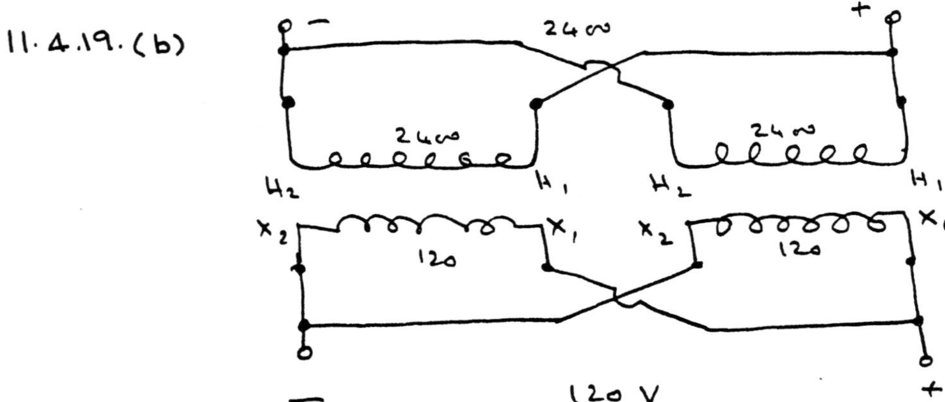

11.4.20.

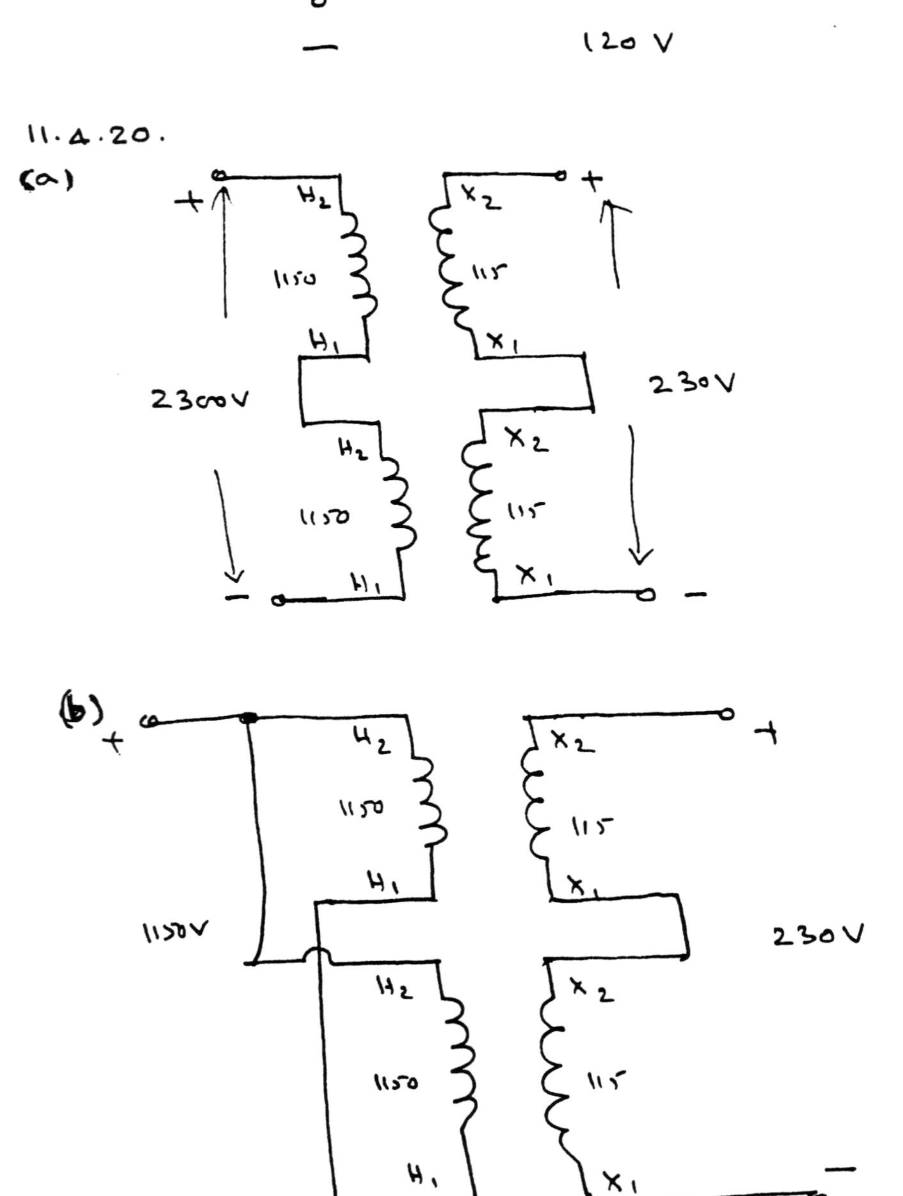

11.5.1.

The analysis is carried out on a per-phase-of-Y basis by referring all quantities to the high-voltage Y-side of the transformer bank. The impedance of the low-voltage feeder—referred to the high-voltage side by means of the square of the ideal line-to-line voltage ratio of the transformer bank—is

$$(0.00083 + j0.0033)\left(\frac{2{,}400\sqrt{3}}{240}\right)^2 = (0.25 + j1.0)\ \Omega$$

Thus we have, referred to the high-voltage side, the combined series impedance of the high-voltage and low-voltage feeders as

$$Z_{f\,HV} = (0.25 + j1.0) + (0.25 + j1.0)$$
$$= (0.5 + j2.0)\ \Omega \text{ per phase of Y}$$

The equivalent series impedance of the transformer bank referred to its high-voltage Y-side is given from Example 11.4.1 as

$$Z_{eq\,HV} = (1.5 + j2.0)\ \Omega \text{ per phase of Y}$$

The equivalent circuit per phase of Y referred to the Y-connected primary side is then given by exactly the same circuit shown in FIG. E 11.A.2 c. For the conditions given in the problem, the solution on a per-phase basis is exactly the same as the solution of Example 11.A.2, from which it follows that the voltage at the sending end of the high-voltage feeder is 2,483.5 volts. The actual line-to-line voltage, because of the primary Y-connection, is given by

$$2{,}483.5\sqrt{3} = 4{,}301.4 \text{ V} \longleftarrow$$

11.5.2.

(a) $I = \dfrac{4500}{\sqrt{3} \times 13.2} = 197\,A$; $R_{LV} = \dfrac{4500 \times 10^3}{3} \times \dfrac{1}{(197)^2} = 38.65\,\Omega$

$$R_{HV} = \left(\frac{66}{13.2}\right)^2 (38.65) = 966.25\,\Omega \leftarrow$$

(b) L-L low-tension Voltage $= \dfrac{13.2}{\sqrt{3}}$ kV; $R_{HV} = \left(\dfrac{66}{13.2/\sqrt{3}}\right)^2 \times 38.65 = 2898.75\,\Omega \leftarrow$

11.5.3.

(a)

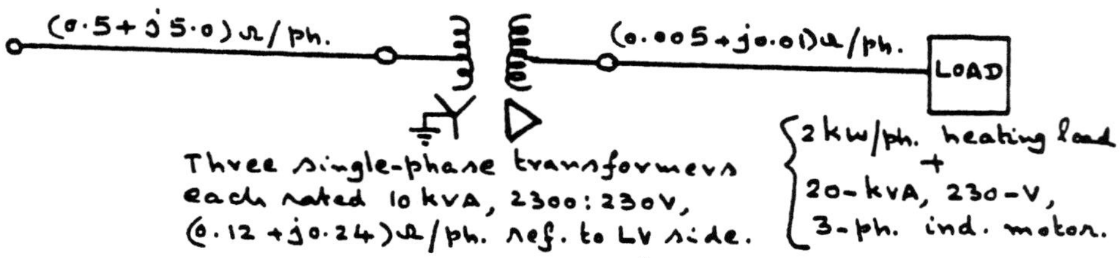

(b) Impedance of LV feeder ref. to HV side $= (0.005 + j0.01)\left(\dfrac{2300\sqrt{3}}{230}\right)^2 = (1.5 + j3.0)\,\Omega$

$Z_{f\,HV} = (0.5 + j5.0) + (1.5 + j3.0) = (2.0 + j8.0)\,\Omega$ per phase of Y.

Transf. $Z_{eq\,HV} = (0.12 + j0.24)(2300/230)^2 = 12 + j24$

Induction motor: 20 kVA; 0.8 pf lagging; $P_m = 16$ kW; $Q_m = 12$ kVAR

Heating load: $P_h = 3 \times 2 = 6$ kW; $S = \sqrt{22^2 + 12^2} = 25.06$ kVA;

$I_1 = I_2' = \dfrac{25.06 \times 10^3}{\sqrt{3} \times 2300 \times \sqrt{3}} = 3.63\,A$, at angle $\theta = -\cos^{-1}\dfrac{22}{25.06} = -28.6°$

(c) $\bar{V}_S = 2401.6 \angle 1.85°$

L-L Voltage $= 2401.6\sqrt{3} = 4159.6\,V$ ↑

Line current $= 3.63\,A \leftarrow$

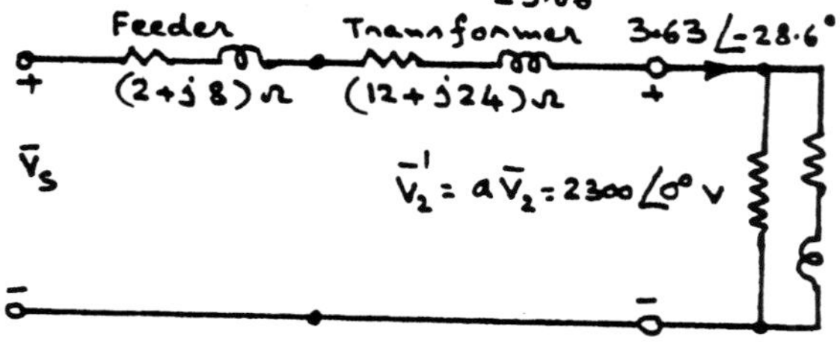

11.5.4.

$$V_{L\text{-}L\,HV} = V_{ph\,HV} = 2400\,V; \quad V_{L\text{-}L\,LV} = 120\sqrt{3} = 208\,V$$

$$V_{ph\,LV} = 120\,V$$

$I_{L\,LV} = \dfrac{30 \times 10^3}{\sqrt{3}(208)} = 83.27\,A = I_{ph\,LV}$; RATIO $= \dfrac{V_{ph\,HV}}{V_{ph\,LV}} = \dfrac{2400}{120} = 20$

$I_{ph\,HV} = \dfrac{83.27}{20} = 4.16\,A$; $I_{L\,HV} = 4.16\sqrt{3} = 7.21\,A$

CHECK: $\sqrt{3} \times 2400 \times 7.21 = 30$ kVA

$$\frac{V_{L\text{-}L\,HV}}{V_{L\text{-}L\,LV}} = \frac{2400}{208} = 11.54 \leftarrow$$

$$\frac{I_{L\,HV}}{I_{L\,LV}} = \frac{7.21}{83.27} = \frac{1}{11.54} = 0.087 \leftarrow$$

11.5.5.

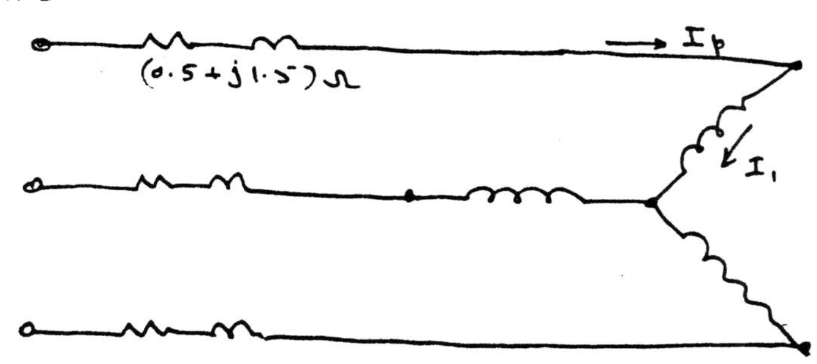

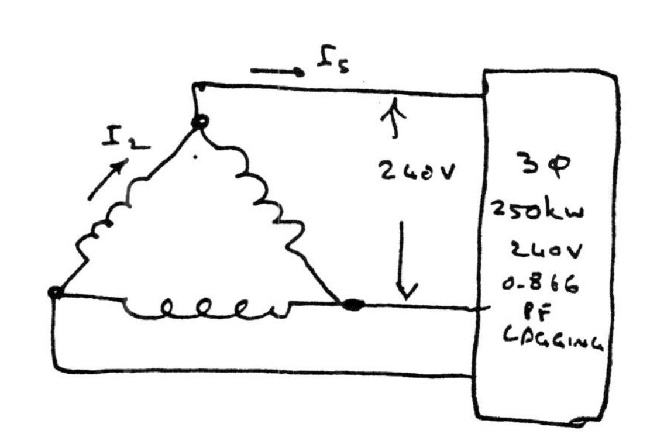

(a)

$2400\sqrt{3} \approx 4160$ V ; $\therefore$ Y – Δ CONNECTION

$$I_S = \frac{250{,}000}{\sqrt{3}(240)(0.866)} = 694.5 \text{ A} \leftarrow$$

$$I_2 = I_S/\sqrt{3} = \frac{694.5}{\sqrt{3}} = 400 \text{ A} \leftarrow$$

TR. WDG RATIO = 2400/240 = 10

$$\therefore \; I_p = I_1 = \frac{I_2}{a} = \frac{400}{10} = 40 \text{ A} \leftarrow$$

(b)

$$\bar{Z}_{eq1} = a^2 \bar{Z}_{eq2} = (10)^2(0.045 + j0.16) = (4.5 + j16)\,\Omega/\text{ph}$$

+ $(0.5 + j1.5)\,\Omega$ → $\bar{I}_1$ $(4.5 + j16)\,\Omega$ + → $\bar{I}_2/a = 40\angle{-30^\circ}$ A

$\bar{V}_{p,\text{SOURCE}}$ $a\bar{V}_2$

– –

PER-PHASE EQUIV. CKT

$$\bar{V}_{p,\text{ SOURCE}} = 2400\angle 0^\circ + (40\angle{-30^\circ})(5 + j17.5)$$

$$= 2966.7\angle 9.8^\circ \text{ V} \quad (\text{L-N})$$

$$V_{L,\text{ SOURCE}} = \sqrt{3}\, V_{p,\text{ SOURCE}} = \sqrt{3}\,(2966.7)$$

$$= 5138.5 \text{ V (L-L)} \leftarrow$$

11.5.6.

(a)

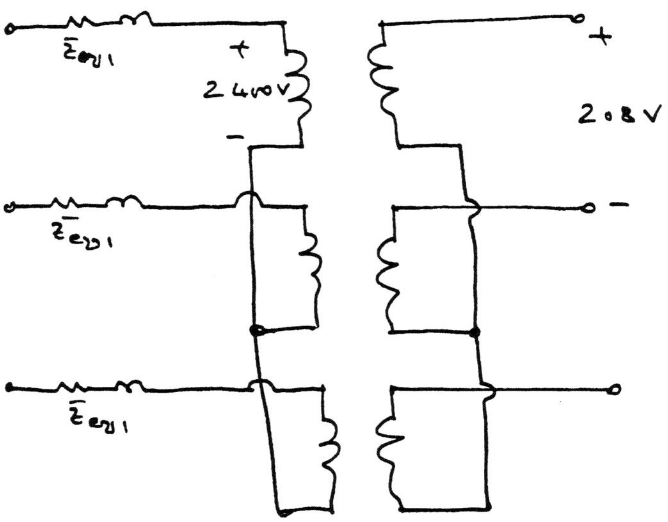

$\bar{Z}_{eq1}$ + 2400V −   + 208V −

$\bar{Z}_{eq1}$

$\bar{Z}_{eq1}$

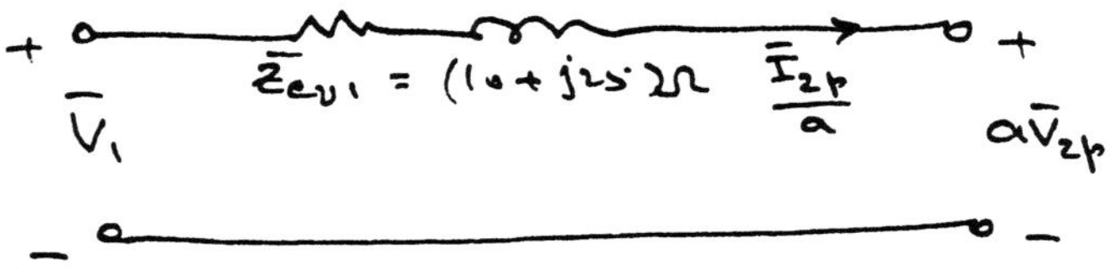

$+ \; \bar{V}_1 \; -$ $\bar{Z}_{eq1} = (10 + j25)\,\Omega$ $\frac{\bar{I}_{2p}}{a}$ $+ \; a\bar{V}_{2p} \; -$

$a = \frac{2400}{120} = 20$

PER-PHASE EQUIV. CKT

(b)

$$\bar{I}_{2p} = \bar{I}_{2L} = \frac{27000}{\sqrt{3}\,(208)(0.9)} \angle +\cos^{-1} 0.9 = 83.27 \angle +25.8^\circ \text{ A}$$

$$\bar{I}_{1L} = \bar{I}_{1p} = \frac{\bar{I}_{2p}}{a} = 4.16 \angle 25.8^\circ \text{ A} \leftarrow$$

$$\bar{V}_{1p} = a\bar{V}_{2p} + \frac{\bar{I}_{2p}}{a}\bar{Z}_{eq1}$$

$$= 20\,(120\angle 0^\circ) + (4.16\angle 25.8^\circ)(10 + j25)$$

$$= 2394.8 \angle 2.67^\circ \text{ V}$$

$$V_{1L} = \sqrt{3}\,(2394.8) = 4148 \text{ V} \leftarrow$$

$$\text{PF} = \cos\,(2.67 - 25.8)^\circ = 0.92 \text{ LEADING} \leftarrow$$

(c)

$$\%\,\text{VR} = \frac{2394.8 - 2400}{2400} \times 100 = -0.2\,\% \leftarrow$$

11.5.7.

$P_{out} = (0.7)(600)(0.85) = 357$ kW

$P_{core} = 4.4$ kW

$P_{cu} = (0.7)^2 (7.6) = 3.724$ kW

$$\eta = \frac{357}{357 + 4.4 + 3.724} \times 100 = \frac{357}{365.124} \times 100 = 97.78\% \leftarrow$$

11.5.8.

$S = 300$ MVA

$V_{2L} = 34.5$ kV

$$I_{2L} = \frac{300 \times 10^3}{\sqrt{3}(34.5)} = 5 \text{ kA}$$

$V_{1L} = 230$ kV

$$I_{1L} = \frac{300 \times 10^3}{\sqrt{3}(230)} = 753 \text{ A}$$

CONNECTION	V_{1p}(kV)	I_{1p}(A)	MVA_{ph}	V_{2p}(kV)	I_{2p}(kA)
Δ – Δ	230	435	100	34.5	2.9
Y – Δ	133	753	100	34.5	2.9
Y – Y	133	753	100	20	5
Δ – Y	230	435	100	20	5

11.6.1.

(a) $I_H = \dfrac{10000}{230} = 43.5\,\text{A}$

$I_c = 43.5\left(\dfrac{230}{2300}\right) = 4.35\,\text{A}$

$I_x = I_H + I_c = 47.85\,\text{A}$

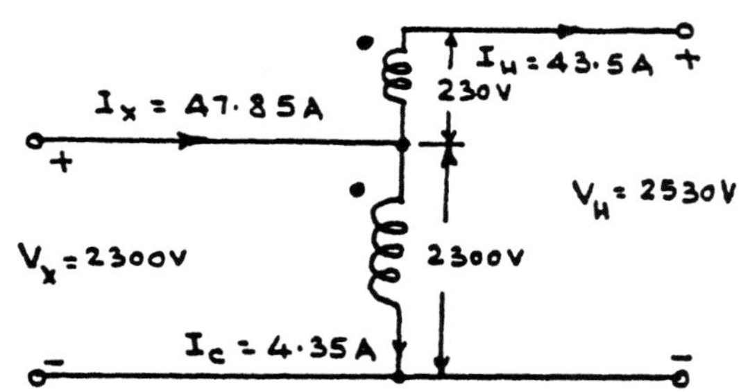

(b) Autotransformer kVA rating $= \dfrac{V_H I_H}{1000} = \dfrac{V_x I_x}{1000} = \dfrac{2530 \times 43.5}{1000} = 110$ ←

kVA transformed by em induction $= \dfrac{230 \times 43.5}{1000} = 10$ ←

(C) From the solution of Problem 11.4.1, part (C),

Copper loss at rated FL current = 224 W; No-load core loss = 70W

Output as an autotransformer = 110 × 1000 × 0.8 = 88,000 W

$\therefore \eta = \dfrac{88000}{88294} = 0.997$ ← Very high because the losses are only those due to transforming 10 kVA.

11.6.2.

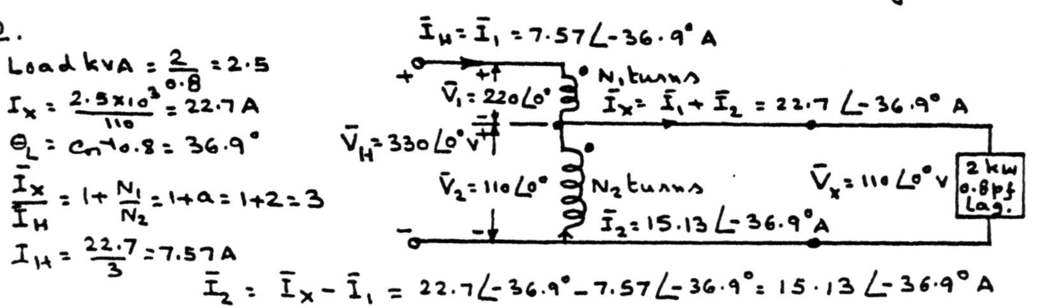

Load kVA $= \dfrac{2}{0.8} = 2.5$

$I_x = \dfrac{2.5 \times 10^3}{110} = 22.7\,\text{A}$

$\theta_L = \cos^{-1} 0.8 = 36.9°$

$\dfrac{I_x}{I_H} = 1 + \dfrac{N_1}{N_2} = 1 + a = 1 + 2 = 3$

$I_H = \dfrac{22.7}{3} = 7.57\,\text{A}$

$\bar{I}_2 = \bar{I}_x - \bar{I}_1 = 22.7\angle -36.9° - 7.57\angle -36.9° = 15.13\angle -36.9°$ A

11.6.3.

130.43 A
+
2415 V
+
2300 V
6.52 A
−
−

RATED CURRENT OF 2300-V WDG

$= \dfrac{15000}{2300} = 6.52\,\text{A}$

RATED CURRENT OF 115-V WDG

$= \dfrac{15000}{115} = 130.43\,\text{A}$

AUTO-TRAN. kVA $= 2415 \times 130.43 \times 10^{-3} = 315\,\text{kVA}$ ←

$\eta = \dfrac{315000 \times 0.8}{(315000 \times 0.8) + 75 + 250} \times 100 = 99.87\%$ ←

11.6.4.

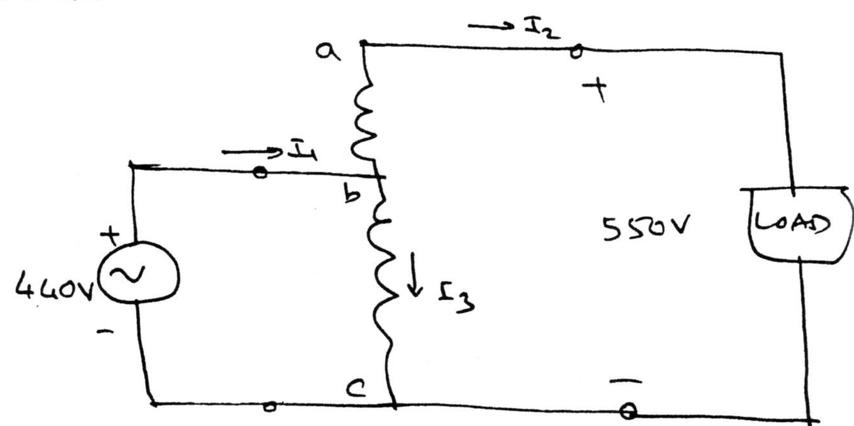

$$I_2 = \frac{10{,}000}{110} = 90.9\text{ A} \quad ; \quad I_3 = \frac{10{,}000}{440} = 22.7\text{ A}$$

$$I_1 = I_2 + I_3 = 90.9 + 22.7 = 113.6\text{ A}$$

$$\text{kVA} = (440)(113.6)10^{-3} = 50\text{ kVA}$$

$$\text{OR } (550)(90.9)10^{-3} = 50\text{ kVA} \leftarrow$$

$$S_{IND} = V_1 I_3 = (440)(22.7)10^{-3} = 10\text{ kVA} \leftarrow$$

$$S_{COND} = V_1 I_2 = (440)(90.9)10^{-3} = 40\text{ kVA} \leftarrow$$

11.6.5.

+ 75A 2200V − ; 68.18A + 2420 − ; 6.82A

$$I_p = \frac{15000}{2200} = 6.82\text{A} \; ; \; V_p = 2200\text{V}$$

$$I_s = \frac{15000}{220} = 68.18\text{A} \; ; \; V_s = 220\text{V}$$

$$I_2 = I_s = 68.18\text{A} \; ; \; I_1 = I_p + I_s = 75\text{A} \; ;$$

$$V_1 = V_p = 2200\text{V}$$

$$\text{kVA} = 2200 \times 75 \times 10^{-3} \text{ OR } 2420 \times 68.18 \times 10^{-3} = 165\text{ kVA} \leftarrow$$

$$\text{kVA}_{TRANS.\ ACTION} = V_p I_p = 2200(6.82)10^{-3} = 15\text{ kVA} \leftarrow$$

$$\text{kVA}_{COND} = V_1 I_2 = 2200(68.18)10^{-3} = 150\text{ kVA} \leftarrow$$

11.6.6.

(a)

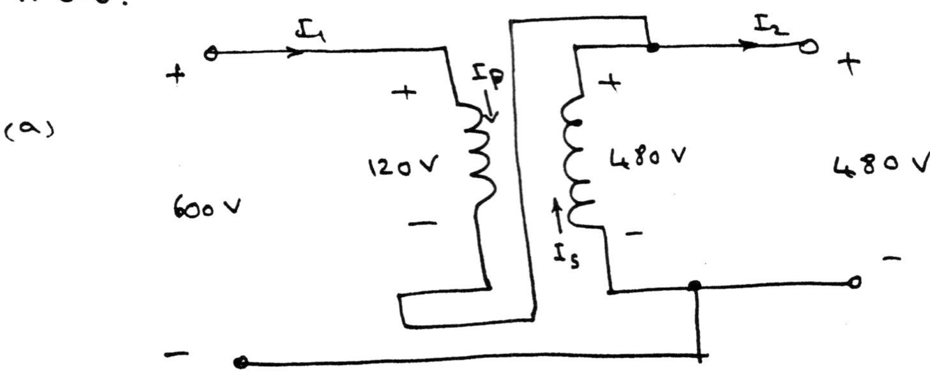

$$I_p = \frac{5000}{120} = 41.67\,A \quad ; \quad I_s = \frac{5000}{480} = 10.42\,A$$

(b) $I_1 = I_p = 41.67\,A$

$V_1 = 600\,V$

MAXM. KVA RATING = $V_1 I_1 = 600\,(41.67)\,10^{-3} = 25\,kVA$ ←

(c) $P_{out} = 5000 \times 0.8 = 4000\,W$

$$\eta_T = 0.95 = \frac{4000}{4000 + \text{LOSSES}}$$

$$\therefore \text{LOSSES} = \frac{4000}{0.95} - 4000 = 210\,W$$

$$\eta_{AUTO} = \frac{(25000)(0.8)}{(25000 \times 0.8) + 210} \times 100 = 99\,\%$$ ←

CHAPTER 12

12.1.1.

(a) FROM THE SYMMETRY OF THE GEOMETRY ONLY THE ϕ - COMPONENT OF THE MAGNETIC FIELD EXISTS AND THAT IS ONLY A FUNCTION OF RADIUS r. CHOOSING A CONSTANT r- LOOP, ACCORDING TO AMPERE'S CIRCUITAL LAW, ONE HAS

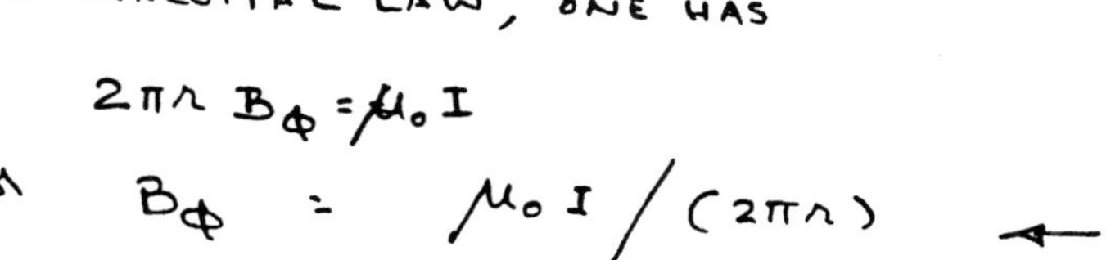

$$2\pi r B_\phi = \mu_0 I$$

$$\text{or} \quad B_\phi = \mu_0 I / (2\pi r) \quad \longleftarrow$$

(b) ACCORDING TO $B l I$ FORMULA, THE FORCE PER UNIT LENGTH ALONG THE WIRE IS GIVEN BY $F_{12} = F_{21} = \dfrac{\mu_0}{2\pi} \dfrac{I_1 I_2}{r}$

THE FORCE $\bar{F}_{12}$ (i.e THE FORCE ON WIRE 1 DUE TO THE FIELD OF I_2) AND THE FORCE $\bar{F}_{21}$ (i.e. THE FORCE ON WIRE 2 DUE TO THE FIELD OF I_1) ARE IN THE DIRECTIONS SHOWN IN THE FIGURE, WHEN I_1 AND I_2 ARE IN THE SAME DIRECTION; THE FORCES, BEING ATTRACTIVE IN NATURE, PULL THE WIRES TOGETHER.

WHEN THE WIRES CARRY CURRENTS IN OPPOSITE DIRECTIONS, THE WIRES ARE REPELLED BY THE MAGNETIC FORCES.

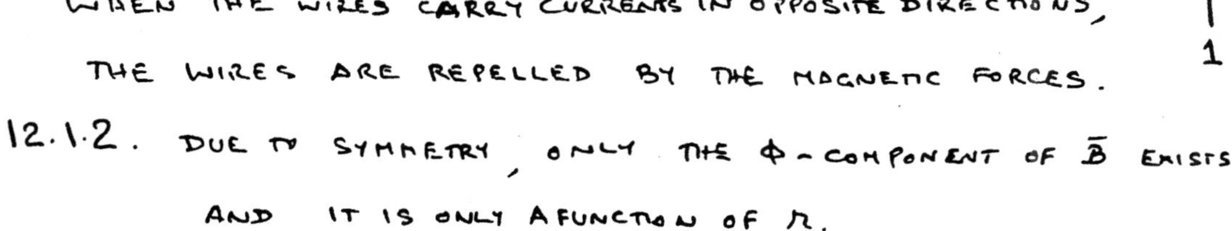

12.1.2. DUE TO SYMMETRY, ONLY THE ϕ - COMPONENT OF $\bar{B}$ EXISTS AND IT IS ONLY A FUNCTION OF r.

$$B_\phi = \mu H_\phi = \mu I / (2\pi r)$$

FLUX IN ELEMENTAL AREA ($dA = l\, dr$) IS GIVEN BY

$$d\phi = B_\phi \, dA = \frac{\mu I l}{2\pi} \frac{dr}{r}$$

$$\therefore \phi = \frac{\mu I l}{2\pi} \int_{r_1}^{r_2} \frac{dr}{r} = \frac{\mu I l}{2\pi} \ln \frac{r_2}{r_1} \quad \longleftarrow$$

12.1.3.

(a) THE MAGNETIC FLUX $\phi_t = B w l \cos \theta_m$ LINKING THE CONDUCTING LOOP, WHICH VARIES FROM A MINIMUM VALUE OF ZERO AND A MAXIMUM VALUE OF $B w l$, AS THE LOOP ROTATES.

SINCE THE LOOP HAS 1 TURN, $N = 1$; $e_{aa'} = \dfrac{d\lambda}{dt} = N \dfrac{d\phi}{dt} = \dfrac{d\phi}{dt}$ in accordance with Faraday's Laws

$$\therefore e_{aa'} = -B w l \omega_m \sin \omega_m t = B w l \omega_m \cos(\omega_m t + 90^\circ) \quad \longleftarrow$$

12.1.3.(b) CONTD. A CURRENT $i = e_{aa'}/R$ WILL FLOW THROUGH THE RESISTOR AND THE CONDUCTING LOOP.

ELECTRICAL POWER $e_{aa'}\, i = i^2 R$ IS DELIVERED TO THE RESISTOR.

THIS ELECTRICAL POWER ORIGINATES FROM THE MECHANICAL POWER REQUIRED TO KEEP THE LOOP ROTATING AT A SPEED OF ω_m.

THE MECHANICAL POWER IS GIVEN BY $P_m = \omega_m T_m = e_{aa'}\, i$

THUS THE PRINCIPLE OF CONSERVATION OF ENERGY IS SATISFIED.

12.1.4. (a) FOLLOWING THE EXPRESSION DEVELOPED IN PR. 12.1.3a,

$$e_{TURN} = 2 \times 0.1 \times 2.5 \times (2\pi)(30) \cos(60\pi t + 90^\circ)$$

WHERE $\omega_m = 2\pi(30) = 60\pi$ rad/s IS USED.

∴ THE INDUCED VOLTAGE ACROSS THE COIL OF 15 TURNS IS GIVEN BY

$$e_{COIL} = 15\, e_{TURN} = 1413.7 \cos(188.5t + 90^\circ)\ V \leftarrow$$

(b) RMS VALUE $E_{COIL} = 1413.7/\sqrt{2} \simeq 1{,}000\ V = 1\ kV$

RMS CURRENT FLOWING THROUGH THE RESISTOR AND COIL IS

$$I = E_{COIL}/R = 1000/500 = 2\ A$$

THE AVERAGE POWER DELIVERED TO THE RESISTOR $= P = I^2 R = (2)^2\, 500 = 2\ kW \leftarrow$

(c) THE AVERAGE MECHANICAL TORQUE REQUIRED TO ROTATE THE COIL OF PART b IS GIVEN BY $T_m = P/\omega_m = 2000/188.5 = 10.6\ N{\cdot}m \leftarrow$

THE ACTION OF THE DEVICE CORRESPONDS TO THAT OF A <u>GENERATOR</u> ←

12.1.5.

$$\omega_m = \frac{2\pi(1800)}{60} = 60\pi = 188.5\ rad/s$$

$$T = \frac{VI}{\omega_m} = \frac{(1000)(2)}{188.5} = 10.6\ N{\cdot}m \leftarrow$$

12.1.6.

(a) $\phi_m = (0.1)(0.1)(1) = 0.01\ Wb = 10\ mWb \leftarrow$

(b) $\omega_m = \dfrac{2\pi(300)}{60} = 10\pi = 31.42\ rad/s$

$\lambda(t) = N\phi(t) = N\phi_m \cos\omega_m t = (50)(0.01)\cos 31.42t$
$= 0.5 \cos 31.42t$ Wb-t ←

(c) $e(t) = \dfrac{d\lambda(t)}{dt} = -0.5(31.42)\sin 31.42t = -15.7 \sin 31.42t\ V \leftarrow$

(d) $e_{av} = \frac{1}{T}\int_0^T e(t)\,dt = 0 \leftarrow$

(e) $e_{30^\circ} = -15.7 \sin 30^\circ = -15.7\ V \leftarrow$

12.1.7.

$$E_{MAX} = \sqrt{2}\, E_{RMS} = \sqrt{2}\,(1000) = 1414\ V$$

$$e_{COIL} = \frac{E_{MAX}}{N} = \frac{1414}{100} = 14.14\ V$$

$$\omega_m = \frac{2\pi(1200)}{60} = 125.66\ rad/s$$

$$B = \frac{e}{\omega l \omega_m} = \frac{14.14}{(0.1)(1.125)(125.66)} \approx 1\ T \leftarrow$$

12.1.8.

CIRCUMFERENCE OF THE PLUNGER: $2\pi(2.5) = 15.71\ cm$

AREA OF GAP AT SIDES OF THE PLUNGER = $2\pi(2.5)(1.25) = 19.63\ cm^2$

AREA OF MAIN GAP: $\pi(2.5)^2 = 19.63\ cm^2$

BECAUSE THE TWO GAPS HAVE THE SAME AREA, THE GAP AT THE SIDES OF THE PLUNGER CAN BE INCLUDED BY ADDING 0.25 mm TO THE MAIN GAP LENGTH g.

$$\therefore\ H_{AIR} = \frac{3 \times 1000}{(1.25 + 0.025)10^{-2}} = \frac{3000}{1.275 \times 10^{-2}}\ At/m$$

$$B_{AIR} = \mu_0 H_{AIR} = \frac{4\pi \times 10^{-7} \times 3 \times 10^5}{1.275} = \frac{0.12\pi}{1.275} = 0.296\ T$$

FORCE ACTING ON TWO PLANE PARALLEL IRON FACES $= F = \frac{B^2 A}{2\mu_0} = \frac{0.296^2 \times 19.63 \times 10^{-4}}{2 \times 4\pi \times 10^{-7}} = 68.4\ N \leftarrow$

12.1.9.

$$H = \frac{1800 \times 1}{2 \times 0.125 \times 10^{-2}} = \frac{1800}{0.25 \times 10^{-2}}\ At/m$$

$$B = \mu_0 H = \frac{4\pi \times 10^{-7} \times 1800}{0.25 \times 10^{-2}} = 16\pi \times 18 \times 10^{-3} = 0.905\ T$$

$$A = \frac{\pi(1.25)^2 \times 10^{-4}}{4} = \frac{0.0005}{4}\ m^2$$

$$F_{TOTAL} = 2 \times F_{\substack{PER\ POLE \\ (PER\ GAP)}} = \frac{B^2}{\mu_0} A = \frac{(0.905)^2}{4\pi \times 10^{-7}} \left(\frac{0.0005}{4}\right)$$

$$= 81.5\ N \leftarrow$$

12.1.10.

FOLLOWING THE EXPRESSION DEVELOPED IN PR. 12.1.3a

$$\text{MAX. } E_{TURN} = 2\,(\text{FLUX PER POLE})\,\omega_m = 2 \times 0.02 \times \frac{2\pi(1800)}{60} = 7.54\ V \leftarrow$$

12.2.1.

(a)

$$\omega_m = \frac{2\pi n}{60} = \frac{2\pi(3600)}{60} = 120\pi = 377 \text{ rad/s}$$

$$\omega = \frac{P}{2}\omega_m = \frac{2}{2} \times 377 = 377 \text{ rad/s}$$

$$f = \frac{\omega}{2\pi} = \frac{377}{2\pi} = 60 \text{ Hz} \leftarrow$$

(b)

$$\Phi_p = \frac{2}{P}(2 B_m l r) = \frac{2}{2}\left(2 \mu_0 \frac{2}{\pi} \frac{N_f I_f}{g}\right) l r$$

$$= \frac{4}{\pi} \times 4\pi \times 10^{-7} \times \frac{400 \times 1}{1 \times 10^{-3}} \times 1.5 \times 0.25$$

$$= 0.24 \text{ Wb}$$

$$E_{MAX} = \omega N_a \Phi_p = 2\pi \times 60 \times 50 \times 0.24 = 4523.9 \text{ V}$$

$$E_{RMS} = \frac{4523.9}{\sqrt{2}} = 3199 \text{ V} \leftarrow$$

12.2.2.

$$\lambda = 2N\ell r B_m \sin \omega_1 t \cos \omega_2 t$$

Thus,

$$v = \frac{d\lambda}{dt} = 2N\ell r B_m (\omega_1 \cos \omega_1 t \cos \omega_2 t - \omega_2 \sin \omega_1 t \sin \omega_2 t)$$

When $\omega_1 = \omega_2 = \omega$, this reduces to

$$v = 2N\ell r B_m \omega \cos 2\omega t \leftarrow$$

i.e. a double-frequency generator.

12.2.3.

$v = 2NB_m \ell r \omega \sin \omega t$. Substituting the numerical values,

$$v = 2(150)(0.6)(0.100)(0.050)(377) \sin 377t = 339.3 \sin 377t \quad \text{(V)} \leftarrow$$

AND

$$V = \frac{339.3}{\sqrt{2}} = 240 \text{ V rms} \leftarrow$$

12.2.4.

(a) RMS phase voltage $= \dfrac{\omega N \Phi}{\sqrt{2}} = \dfrac{377(2\times3)(0.1)}{\sqrt{2}} = 160\,\text{V}$ ←

(b) RMS line-to-line voltage $= \sqrt{3} \times 160 = 277\,\text{V}$ ←

(c) $e_{an} = 160\sqrt{2}\sin 377t$; $e_{bn} = 160\sqrt{2}\sin(377t - 120°)$; $e_{cn} = 160\sqrt{2}\sin(377t + 120°)$

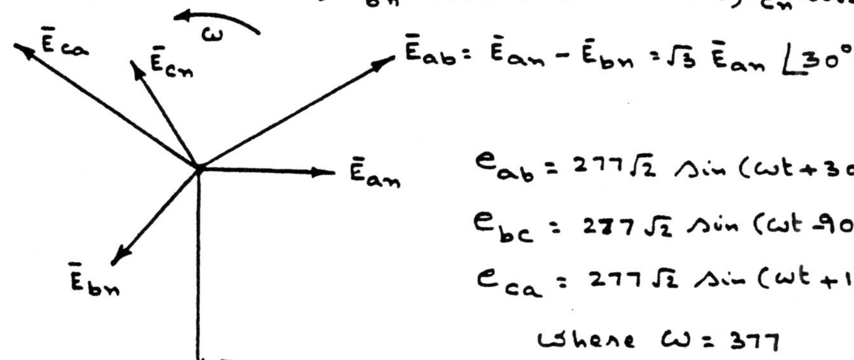

$$\left.\begin{aligned} e_{ab} &= 277\sqrt{2}\sin(\omega t + 30°) \\ e_{bc} &= 277\sqrt{2}\sin(\omega t - 90°) \\ e_{ca} &= 277\sqrt{2}\sin(\omega t + 150°) \\ &\text{where } \omega = 377 \end{aligned}\right\}$$ ←

12.2.5. New synchronous speed $= \dfrac{120 f}{P} = \dfrac{120 \times 60}{6} = 1200\,\text{rpm}$ ←

Terminal voltage $\propto f \Rightarrow 1000 \times \dfrac{60}{50} = 1200\,\text{V L-L RMS}$ ←

12.2.6

Syn. Speed $= \dfrac{120 \times 60}{6} = 1200\,\text{rpm}$; $\omega_m = \dfrac{1200}{60} \times 2\pi = 40\pi\,\text{rad/s}$.

$200\,\text{hp} = 200 \times 746\,\text{W}$; $T_e \omega_m = 200 \times 746$ or $T_e = \dfrac{200 \times 746}{40\pi} = 1187.3\,\text{N·m}$ ←

12.2.7.

$$\frac{120 f_m}{P_m} = \frac{120 f_g}{P_g}; \; f_m = 60\,\text{Hz}; \; f_g = 50\,\text{Hz}; \; \frac{P_m}{P_g} = \frac{f_m}{f_g} = \frac{60}{50} = \frac{6}{5} = \frac{12}{10}$$

(a) min. $P_m = 12$ (b) min. $P_g = 10$ (c) $\dfrac{120 \times 60}{12} = \dfrac{120 \times 50}{10} = 600\,\text{rpm}$ ←

12.2.8.

$$\frac{120 f_1}{P_1} = \frac{120 f_2}{P_2} \Rightarrow \frac{f_1}{f_2} = \frac{P_1}{P_2} = \frac{25}{60} = \frac{5}{12} = \frac{10}{24} = \frac{20}{48} = \frac{30}{72}$$

$P_1/P_2 = 10/24$ CORRESPONDS TO 300 RPM.
" 20/48 " " 150 "
" 30/72 " " 100 "

} 300, 150, 100 RPM ←

12.2.9. $E_{max} = BNA\omega = (1.1)(10)(0.200)^2(2\pi \times 1800)/60 = 82.94\,\text{V}$ ←

12.2.10.

(a) $E_a = \dfrac{P\phi n Z}{60\alpha}$ (5.3.23) ; $\alpha = P = 4$

$\therefore E_a = \dfrac{4 \times 30 \times 10^{-3} \times 1800 \times 728}{60 \times 4} = 655.2\text{ V}$ ←

(b) $\alpha = 2$; $P = 4$

$\therefore E_a = 655.2 \times 2 = 1310.4\text{ V}$ ←

(c) PER-PATH CURRENT = 100/4 = 25 A MAXIMUM ; $I_a = 100\text{A}$

MAX. EM POWER DEVELOPED = $E_a I_a = 655.2 \times 100 = 65.52\text{ kW}$ ←

MAX. EM TORQUE DEVELOPED = $(65.52/\omega_m)10^3 = \dfrac{65.52 \times 10^3}{(1800/60)\,2\pi} = 347.6\text{ N·m.}$ ←

(d) $I_a = 2 \times 25 = 50\text{ A}$; $E_a = 1310.4\text{ V}$

$\therefore E_a I_a = 1310.4 \times 50 = 65.52\text{ kW}$, SAME AS IN PART (C).

$\therefore$ MAX. DEVELOPED POWER OR TORQUE REMAINS UNCHANGED. ←

12.2.11. $P = \alpha = 4$; $n = 720$; $\phi = 0.02$; $Z = 144 \times 2 \times 2 = 576$

SUBSTITUTING INTO EQ. (5.3.23), ONE GETS

$$E = \frac{0.02 \times 720 \times 576}{60}\left(\frac{4}{4}\right) = 138.24\text{ V} \leftarrow$$

12.2.12. A PORTION OF THE MACHINE IS ILLUSTRATED IN THE FIGURE BELOW:

FROM $E_a = 252 = \dfrac{\phi n Z}{60}\left(\dfrac{P}{\alpha}\right)$, $\phi = 71.35\text{ mWb}$.

The pole surface area is

$$A = r\theta\ell = (0.21)(\pi/3)(0.28) = 0.0616\text{ m}^2$$

Hence

$$B = \frac{\phi}{A} = \frac{71.35 \times 10^{-3}}{0.0616} = 1.16\text{ T} \leftarrow$$

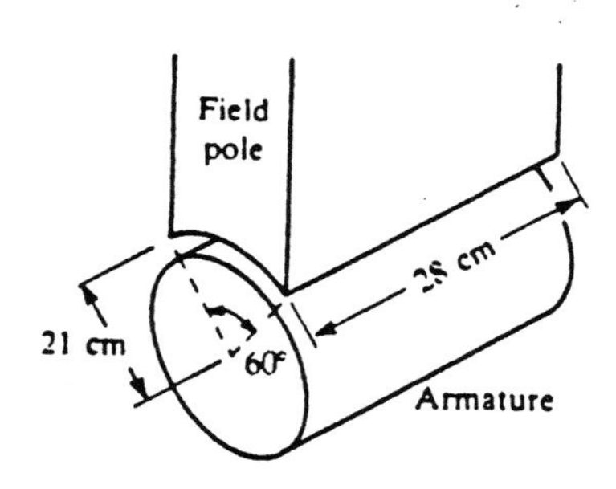

12.2.13.

$$E_a = \frac{P\phi n Z}{a \times 60} \quad (5.3.23); \quad P = 6; \ \phi = \underset{\text{Wb}}{0.025}; \ n = \underset{\text{rpm}}{1200};$$

Z = no. of conductors = $28 \times 2 \times 5 = 280$; $a = 6$ for case (a)
$a = 2$ for case (b)

(a) $E_a = \dfrac{6 \times 0.025 \times 1200 \times 280}{6 \times 60} = 140\text{V} \leftarrow$ (b) $E_a = 3 \times 140 = 420\text{V} \leftarrow$

12.2.14.

$$E_a = \frac{P\phi n Z}{a \times 60} \quad (5.3.23); \quad P = 4; \ \phi = \underset{\text{Wb}}{0.02}; \ a = 4; \ Z = 400;$$

$$E_a = \frac{4 \times 0.02 \times n \times 400}{4 \times 60} = \frac{2n}{15}, \text{ where } n \text{ is rpm}; \quad \omega = \frac{n}{60} \times 2\pi$$

$$T = \frac{E_a I_a}{\omega} = \frac{E_a I_a (60)}{2\pi n} = \frac{2n}{15} \cdot 50 \cdot \frac{60}{2\pi n} = 63.64 \text{ N·m} \leftarrow$$

12.3.1.

$$i_a = I \cos \omega_s t \ ; \quad i_b = I \cos(\omega_s t - 90°)$$

$$F_\theta = F_a \cos\theta + F_b \cos(\theta - 90°)$$

$$F_a = F_m \cos \omega_s t \ ; \quad F_b = F_m \cos(\omega_s t - 90°)$$

$$\therefore F(\theta, t) = F_m \cos\theta \cos\omega_s t + F_m \cos(\theta - 90°) \cos(\omega_s t - 90°)$$

$$= F_m \cos(\theta - \omega_s t) \leftarrow$$

12.3.2.

$$i_a = i_b = i_c = i = I\cos\omega t$$

$$F_T = F_a + F_b + F_c$$

$$= Ni\cos\theta_m + Ni\cos(\theta_m - 120°) + Ni\cos(\theta_m - 240°)$$

$$= Ni\left[\cos\theta_m + \cos(\theta_m - 120°) + \cos(\theta_m - 240°)\right]$$

$$= 0 \leftarrow$$

$$i_a = I\cos\omega t \;;\quad i_b = I\cos(\omega t - 120°) \;;\quad i_c = I\cos(\omega t - 240°)$$

WHERE $\omega = 2\pi f$

$$F_T = F_a + F_b + F_c = NI\left[\cos\omega t\cos\theta_m + \cos(\omega t - 120°)\cos(\theta_m - 120°) + \cos(\omega t - 240°)\cos(\theta_m - 240°)\right]$$

$$= NI\left[\tfrac{1}{2}\cos(\omega t + \theta_m) + \tfrac{1}{2}\cos(\omega t - \theta_m) + \tfrac{1}{2}\cos(\omega t + \theta_m - 240°) + \tfrac{1}{2}\cos(\omega t - \theta_m) + \tfrac{1}{2}\cos(\omega t + \theta_m - 120°) + \tfrac{1}{2}\cos(\omega t - \theta_m)\right]$$

$$= \frac{3}{2}NI\cos(\omega t - \theta_m) \leftarrow$$

$$F_T = \frac{3}{2}NI\cos(\omega t - \omega_m t - \alpha)$$

FOR $\omega = \omega_m = 2\pi f$ AND $\alpha = 0$

$$F_T = \frac{3}{2}NI \quad \leftarrow$$

12.3.3.

(a) PEAK F_a & F_b & $F_c = F_m = \frac{2}{\pi}NI = \frac{2}{\pi}(15)(100) = 955$ At $\leftarrow$

(b) $F(\theta_m, t) = \frac{3}{2}F_m\cos(\theta_m - 377t)$

$$= 1432.5\cos(377t - \theta_m) \leftarrow$$

12.3.4.

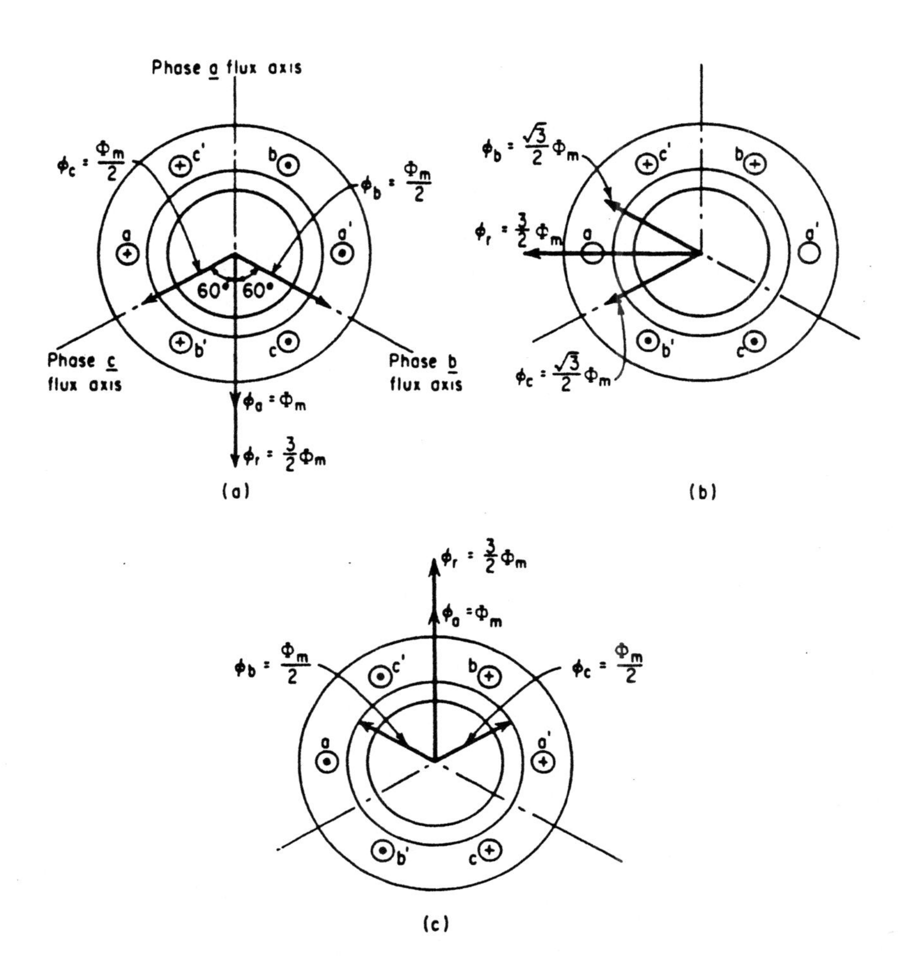

Representing the rotating magnetic field at three different instants of time: (a) time t_1 in Fig. (b) time t_3; (c) time t_5.
P 12.4.4

12.4.1.

$$\lambda = \frac{k\, i^{2/3}}{x+t} \quad \text{or} \quad i = \left\{\frac{\lambda (x+t)}{k}\right\}^{3/2};$$ Note that λ-i relationship is nonlinear.

$$W_m'(i, x) = \lambda i - W_m = \frac{k\, i^{5/3}}{(x+t)} - \frac{2}{5}\,\frac{k}{(x+t)}\, i^{5/3} = \frac{3}{5}\,\frac{k}{(x+t)}\, i^{5/3}$$

$$F_e = \frac{\partial W_m'(i, x)}{\partial x} = -\frac{3}{5}\, i^{5/3}\, \frac{k}{(x+t)^2} \leftarrow \text{Eq. (12.5.19)}$$

$$F_e = -\frac{\partial W_m(\lambda, x)}{\partial x} = -\frac{\partial}{\partial x}\left\{\frac{2}{5}\,\frac{(x+t)^{3/2}}{k^{3/2}}\,\lambda^{5/2}\right\} = -\frac{3}{5}\, i^{5/3}\, \frac{k}{(x+t)^2} \leftarrow \text{Eq. (12.5.20)}$$

The negative sign indicates that the force F_e acts in such a direction as to decrease x; i.e., the plunger is attracted towards the coil. ←

12.4.2.

Applying Ampere's laws: $Ni = H_x x + H_t t$, where H_x and H_t are the mag. intensities in the airgaps of lengths x and t, respectively.

$Ni = (B_x/\mu_0)x + (B_t/\mu_0)t = (\Phi/\mu_0 A)x + (\Phi/2 / \mu_0 A/2)t$; or $\Phi = \dfrac{N i A \mu_0}{x+t}$

$L = \lambda/i = N\Phi/i = \dfrac{N^2 A \mu_0}{x+t}$ ← Note that λ-i relationship is linear. $F_e = \frac{1}{2} i^2 \dfrac{\partial L}{\partial x} = -\frac{1}{2} i^2 \dfrac{N^2 A \mu_0}{(x+t)^2}$ ←

12.4.3.

$$i = a\lambda^2 + b\lambda (x-d)^2; \quad W_m = \int_0^\lambda i\, d\lambda = a\frac{\lambda^3}{3} + b\frac{\lambda^2}{2}(x-d)^2$$

$$F_e = -\frac{\partial W_m(\lambda, x)}{\partial x} = -b\frac{\lambda^2}{2}\cdot 2(x-d) = -b\lambda^2 (x-d) \leftarrow$$

12.4.4.

For the magnetic circuit, the reluctance is

$$\mathcal{R} = \frac{g}{\mu_0 \pi c^2} + \frac{b}{\mu_0 2\pi a \ell} \qquad \text{where} \qquad c = a - \frac{b}{2}$$

The inductance L is then given by

$$L = \frac{N^2}{\mathcal{R}} = \frac{2\pi\mu_0 a\ell c^2 N^2}{2a\ell g + bc^2} = \frac{k_1}{k_2 g + k_3}$$

where $k_1 = 2\pi\mu_0 a\ell c^2 N^2$, $k_2 = 2a\ell$, and $k_3 = bc^2$.

12.4.4. CONTD.

(a) THE FORCE IS GIVEN BY

$$F_e = \frac{1}{2} I^2 \frac{\partial L}{\partial g} = -\frac{I^2 k_1 k_2}{2(k_2 g + k_3)^2} \leftarrow$$

where the minus sign indicates that the force tends to decrease the air gap.

(b) Substituting the numerical values in the force expression of (a) yields 600 N as the magnitude of F_e.

12.4.5.

(a) The instantaneous force is given by

$$F_e = -\frac{(10\sqrt{2}\cos 120\pi t)^2 k_1 k_2}{2(k_2 g + k_3)^2} = -\frac{100 k_1 k_2}{(k_2 g + k_3)^2}\cos^2 120\pi t \quad \text{(N)} \leftarrow$$

(b) Because the $\cos^2$ has average value 1/2, the average force is the same as the force due to 10 A dc; namely, 600 N. ←

12.4.6.

(a)

reluctance: $\mathcal{R} = \dfrac{b + g}{2\mu_0 a^2}$

∴ INDUCTANCE : $L = N^2/\mathcal{R} = \dfrac{2\mu_0 a^2 N^2}{b + g}$

electrical force: $F_e = \dfrac{1}{2} I^2 \dfrac{\partial L}{\partial g} = -\dfrac{\mu_0 a^2 N^2 I^2}{(b + g)^2}$ ←

(b) 256.4 N. ←

12.4.7.

(a) The electrical equation of motion of the system has the form

$$Ri + \frac{d}{dt}(Li) = v$$

in which, from Pr. 12.5.6 (a),

$$L = \frac{2\mu_0 a^2 N^2}{b + g_0} = \text{constant}$$

Thus, we have to do with the familiar

$$L\frac{di}{dt} + Ri = V \sin \omega t$$

The desired steady-state solution is simply $\mathbf{I} = \mathbf{V}/\mathbf{Z}$, or

$$i = \frac{V_m}{\sqrt{R^2 + (\omega L)^2}} \sin(\omega t - \psi) \qquad \text{where} \qquad \psi = \arctan \frac{\omega L}{R} \leftarrow$$

(b) Because the magnetic circuit is linear, we may determine the electrical force by

$$F_e = \frac{1}{2} i^2 \frac{\partial L}{\partial g_0} = -\frac{\mu_0 a^2 N^2}{(b + g_0)^2} i^2 \leftarrow$$

where $i(t)$ is as found in (a) above.

12.4.8.

$$T_e = \frac{\partial W_m'(i_s,\theta)}{\partial\theta} = \frac{1}{2} i_s^2 \frac{\partial L(\theta)}{\partial\theta} = \frac{1}{2} i_s^2 [0.08 \sin 2\theta + 0.12 \sin 4\theta]$$

$i_s = I_s \sin\omega_s t = 5\sqrt{2} \sin(2\pi \times 60 t)$; $\theta = \omega_m t - \delta$; Substituting & rearranging

$$T_e = \frac{I_s^2}{4}[0.08 \sin 2(\omega_m t - \delta) + 0.12 \sin 4(\omega_m t - \delta)]$$

$$- \frac{I_s^2}{4}\Big[0.04 \sin 2\{(\omega_m + \omega_s)t - \delta\} + 0.04 \sin 2\{(\omega_m - \omega_s)t - \delta\}$$
$$+ 0.06 \sin 4\{(\omega_m + \frac{\omega_s}{2})t - \delta\} + 0.06 \sin 4\{(\omega_m - \frac{\omega_s}{2})t - \delta\}\Big]$$

(a) for $\omega_m \neq 0$, for average torque, $\omega_m = \pm\omega_s = \pm 377$ rad/s or $\omega_m = \pm \frac{\omega_s}{2} = \pm 188.5$ rad/s

(b) for $\omega_m = 377$ rad/s and $I_s = 5\sqrt{2}$ A,

$|T_{av}|_{max} = \frac{I_s^2}{4} \times 0.04 = 0.5$ N·m ; $P_m = 0.5 \times 377 = 188.5$ W ←

for $\omega_m = 188.5$ rad/s and $I_s = 5\sqrt{2}$ A,

$|T_{av}|_{max} = \frac{I_s^2}{4} \times 0.06 = 0.75$ N·m ; $P_m = 0.75 \times 188.5 = 141.4$ W ←

12.4.9.

The flux-linkage relations are given by

$$\lambda_s = L_{ss}i_s + L_{sr}(\theta)i_r = L_{ss}i_s + Li_r\cos\theta \qquad \lambda_r = L_{sr}(\theta)i_s + L_{rr}i_r = Li_s\cos\theta + L_{rr}i_r$$

in which L_{ss}, L_{rr}, and L are constants. The volt-ampere equations are then given by

$$v_s = R_s i_s + p\lambda_s; \qquad v_r = R_r i_r + p\lambda_r$$

where R_s and R_r are the winding resistances, and p is the time-derivative operator d/dt.

Substituting for λ_s and λ_r, and recognizing that θ is a varible and a function of time, we obtain

$$v_s = R_s i_s + L_{ss}pi_s + L\cos\theta(pi_r) - Li_r\sin\theta(p\theta)$$
$$v_r = R_r i_r + L_{rr}pi_r + L\cos\theta(pi_s) - Li_s\sin\theta(p\theta)$$

where $p\theta$ is the instantaneous speed ω_m. The fourth term on the right-hand side of each equation is caused by mechanical motion and is proportional to the instantaneous speed. These are the speed-voltage terms, which are the coupling terms relating the interchange of power between electrical and mechanical systems.

The voltages induced in the stator and rotor windings will now be found for each of the parts in (c).

(1) $i_s = I_s;\quad i_r = I_r;\quad \omega_m = 0$ so that

$$e_s = 0 \text{ and } e_r = 0.$$

(2) $i_s = I_s\cos\omega_s t;\quad i_r = I_r\cos(\omega_s t + \alpha);\quad \omega_m = 0$ so that

$$e_s = -\omega_s L_{ss}I_s\sin\omega_s t - \omega_s LI_r\cos\theta_0\sin(\omega_s t + \alpha)$$
$$e_r = -\omega_s L_{rr}I_r\sin(\omega_s t + \alpha) - \omega_s LI_s\sin\omega_s t\cos\theta_0$$

(3) $i_s = I_s\cos\omega_s t;\quad i_r = I_r;\quad \omega_m = \omega_s$ so that

$$e_s = -\omega_s L_{ss}I_s\sin\omega_s t - \omega_s LI_r\sin(\omega_s t + \theta_0)$$
$$e_r = -\omega_s LI_s\sin\omega_s t\cos(\omega_s t + \theta_0) - \omega_s LI_s\cos\omega_s t\sin(\omega_s t + \theta_0)$$

or

$$e_r = -\omega LI_s\sin(2\omega_s t + \theta_0)$$

Note that the stator current induces a double-frequency voltage in the rotor circuit.

(4) $i_s = I_s\cos\omega_s t;\quad i_r = I_r\cos(\omega_r t + \alpha);\quad \omega_m = \omega_s - \omega_r$ so that

$$e_s = -\omega_s L_{ss}I_s\sin\omega_s t - \omega_r LI_r\sin(\omega_r t + \alpha)\cdot\cos[(\omega_s - \omega_r)t + \theta_0]$$
$$-(\omega_s - \omega_r)LI_r\cos(\omega_r t + \alpha)\cdot\sin[(\omega_s - \omega_r)t + \theta_0]$$

or

$$e_s = -\omega_s L_{ss}I_s\sin\omega_s t$$
$$-\frac{\omega_s LI_r}{2}\{\sin(\omega_s t + \alpha + \theta_0) - [\sin(-\omega_s + 2\omega_r)t + \alpha - \theta_0]\}$$
$$-\omega_r LI_r\sin[(-\omega_s + 2\omega_r)t + \alpha - \theta_0]$$
$$e_r = -\omega_r L_{rr}I_r\sin(\omega_r t + \alpha)$$
$$-\omega_s LI_s\sin\omega_s t\cdot\cos[(\omega_s - \omega_r)t + \theta_0]$$
$$-(\omega_s - \omega_r)LI_s\cos\omega_s t\sin[(\omega_s - \omega_r)t + \theta_0]$$

12.4.9. CONTD.

or

$$
\begin{aligned}
e_r = & - \omega_r L_{rr} I_r \sin(\omega_r t + \alpha) \\
& + \frac{\omega_r L I_s}{2} \{\sin[(2\omega_s - \omega_r)t + \theta_0] - \sin(\omega_r t - \theta_0)\} \\
& - \omega_s L I_s \sin[(2\omega_s - \omega_r)t + \theta_0]
\end{aligned}
$$

12.4.10.

Because $L_{11} = L_{22}$ and $L_{12} = L_{21}$, $i_1 = i_2 = i$. Hence the stored magnetic energy is

$$W'_m = \frac{1}{2}L_{11}i_1^2 + \frac{1}{2}L_{22}i_2^2 + L_{12}i_1i_2 = (L_{11} + L_{12})i^2$$

and the electrical force is

$$F_e = \frac{\partial W'_m(i, x)}{\partial x} = i^2\frac{\partial}{\partial x}(L_{11} + L_{12}) = -\frac{(k_1 + k_2)i^2}{x^2}$$

The current i is related to the voltage $v_1 = v_2 = v$ through

$$v = \frac{d}{dt}[(L_{11} + L_{12})i]$$

or

$$i = \frac{1}{L_{11} + L_{12}}\int v\,dt = \frac{x}{k_1 + k_2}\left(-\frac{V_m}{\omega}\cos\omega t\right)$$

Then

$$F_e = -\frac{V_m^2\cos^2\omega t}{(k_1 + k_2)\omega^2}$$

It is seen that F_e is apparently independent of x, a result which arises from having ignored the leakage flux.

12.4.11.

(a)

$$W'_m = \frac{1}{2} L_{11} I_0^2 \qquad T_e = \frac{\partial W'_m}{\partial \theta} = 0 \quad \longleftarrow$$

(b)

$$W'_m = \frac{1}{2}(A + B) I_0^2 + C I_0^2 \cos\theta \qquad T_e = -C I_0^2 \sin\theta \quad \longleftarrow$$

(c)

$$W'_m = \frac{1}{2} A I_m^2 \sin^2 \omega t + \frac{1}{2} B I_0^2 + C I_0 I_m \sin\omega t \cos\theta$$

$$T_e = -C I_0 I_m \sin\omega t \sin\theta \quad \longleftarrow$$

(d)

$$T_e = -C I_m^2 \sin^2 \omega t \sin\theta \quad \longleftarrow$$

(e) For coil 1:

$$\frac{d}{dt}(L_{11} i_1 + L_{12} i_2) = 0 \qquad \text{or} \qquad L_{11} i_1 + L_{12} i_2 = k = \text{constant}$$

Therefore, for given i_2 and L_{11},

$$i_1 = \frac{k - L_{12} I_0}{A}$$

and

$$W_m = \frac{A}{2}\left(\frac{k - L_{12} I_0}{A}\right)^2 + \frac{B}{2} I_0^2 + L_{12} I_0 \left(\frac{k - L_{12} I_0}{A}\right)$$

$$= \frac{k^2}{2A} - \frac{L_{12}^2 I_0^2}{2A} - \frac{B}{2} I_0^2 = \frac{k^2}{2A} - \frac{B}{2} I_0^2 - \frac{I_0^2}{2A} C^2 \cos^2\theta$$

Hence,

$$T_e = \frac{\partial W'_m}{\partial \theta} = \frac{I_0^2}{A} C^2 \cos\theta \sin\theta \quad \longleftarrow$$

12.4.12.

(a) $T_e = i_a i_f \dfrac{dL_{af}}{d\theta} + i_b i_f \dfrac{dL_{bf}}{d\theta} = -i_a i_f L \sin\theta + i_b i_f L \cos\theta$ ←

(b) $T_e = I_a I_f L[-\cos\omega t \sin(\omega t + \delta) + \sin\omega t \cos(\omega t + \delta)] = -I_a I_f L \sin\delta$ ←

A steady torque.

(c) $v_{ta} = \dfrac{d}{dt}(L_{aa} i_a + L_{af} i_f) = L_{aa} \dfrac{di_a}{dt} + e_{af}$

where speed voltage $e_{af} = -\omega L I_f \sin(\omega t + \delta)$ ←

$v_{tb} = L_{aa} \dfrac{di_b}{dt} + e_{bf}$; $e_{bf} = +\omega L I_f \cos(\omega t + \delta)$

12.4.13.

With $I_b = I_a$, from part (a) of Prob. 12.5.12,

$$T_e = I_a I_f L (-\sin\theta + \cos\theta)$$

The rotor comes to rest at $\theta = 45°$, where magnetic fields of stator and rotor line up with $T_e = 0$; This is stable equilibrium because $\dfrac{dT_e}{d\theta}$ is negative. ←

12.4.14.

(a) $T_e = \frac{1}{2} i_a^2 \frac{dL_{aa}}{d\theta} + \frac{1}{2} i_b^2 \frac{dL_{bb}}{d\theta} + i_a i_b \frac{dL_{ab}}{d\theta} + I_f \left(i_a \frac{dL_{af}}{d\theta} + i_b \frac{dL_{bf}}{d\theta} \right)$

Substituting the expressions for inductances and time variation of currents, and simplifying, $T_e = I_a^2 L_2 \sin(2\omega t - 2\theta) + I_a I_f L \sin(\omega t - \theta)$

With $\theta = \omega t + \delta$, $T_e = - I_a^2 L_2 \sin 2\delta - I_a I_f L \sin\delta$ (a steady torque) ←

(b) part (b) of Prob. 12.5.12: $T_e = - I_a I_f L \sin\delta$; here, in Prob. 12.5.14, we have an additional term proportional to $\sin 2\delta$ that does not depend on I_f.

(c) With a negative value of δ and a positive value of T_e, the machine will run as a synchronous motor; it will run as a generator, if driven mechanically, with a positive value of δ and a negative value of T_e.

(d) With $I_f = 0$, $T_e = - I_a^2 L_2 \sin 2\delta$; The machine will run as a 2-phase reluctance motor or generator.

12.4.15.

$T_e = - \frac{P}{2} K F_S F_r \sin\delta$ Eq. (5.2.33); $K = \pi D l \mu_0 / (2g)$

(a) $T_e = - \frac{P}{2} K F F_S \sin\delta_S$, since $F_r \sin\delta = F \sin\delta_S$ ←

(b) $T_e = - \frac{P}{2} K F F_r \sin\delta_r$, since $F_S \sin\delta = F \sin\delta_r$ ←

(c) $B = \mu_0 F / g$, where g is the uniform airgap; or $F = Bg/\mu_0$

$T_e = - \frac{P}{2} K_1 B F_r \sin\delta_r$, where $K_1 = \pi D l / 2$ ←

(d) Φ = (average value of flux density over a pole) (pole area)

$= \frac{2}{\pi} B \frac{\pi D l}{P} = \frac{2}{P} K_2 B$, where D is the diameter and l is the length; the average value of a sinusoid over a half wavelength is $2/\pi$ times its peak value.

$T = - \left(\frac{P}{2}\right)^2 K_3 \Phi F_r \sin\delta_r$, where $K_3 = \pi/2$ ←

12.4.16.

$v_1 = \frac{d}{dt}(L_{11} i_1 + L_{12} i_2) = v_2 = \frac{d}{dt}(L_{21} i_1 + L_{22} i_2)$

$i_1 = i_2 = \frac{1}{L_{11} + L_{12}} \int v_1 \, dt = - \frac{V_m \cos\omega t}{\omega (L_{11} + L_{12})}$

$W_m'(i_1, i_2, \theta) = \frac{1}{2} L_{11} i_1^2 + L_{12} i_1 i_2 + \frac{1}{2} L_{22} i_2^2 = (L_{11} + L_{12}) i_1^2 = \frac{V_m^2 \cos^2\omega t}{\omega^2 (L_{11} + L_{12})}$

$T_e = + \frac{\partial W_m'(i_1, i_2, \theta)}{\partial\theta} = \frac{V_m^2 \cos^2\omega t}{\omega_m^2} \frac{\partial}{\partial\theta} \left\{ \frac{1}{L_{11} + L_{12}} \right\}$; Substituting values,

$T_e = \frac{(220\sqrt{2})^2}{314^2} \cos^2 314t \left[\frac{\sin\theta + 4\sin 2\theta}{(6 + \cos\theta + 2\cos 2\theta)^2} \right] = 0.98 \cos^2 314t \left[\frac{\sin\theta + 4\sin 2\theta}{(6 + \cos\theta + 2\cos 2\theta)^2} \right]$ N.m ←

12.4.17.

2 poles on stator; 4 poles on rotor.

Let the axes of the rotor field be at an arbitrary angle with the axis of the stator field and let the two pairs of poles on the rotor have equal strengths in order to avoid unbalanced radial magnetic pull. On the N_1N_2-axis, pole N_1 is repelled by pole N and attracted by pole S, causing a counterclockwise torque; Pole N_2 is likewise repelled by N and attracted by S, causing an equal clockwise torque. Hence the net torque is zero. A similar situation exists on S_1S_2-axis. Hence no net em torque. ←

Axis of stator field

℄ of rotor poles

N

S

N_1 S_1 S_2 N_2

℄ of rotor poles

12.4.18. No reluctance torque is produced if the movement of the rotor does not alter the reluctance or inductance: parts c, d, e, f ←

If the movement of the rotor alters the flux distribution and hence the reluctance, the reluctance torque is produced: parts a, b, g, h ←

12.4.19.

$|\omega_m| = |\omega_s \pm \omega_r|$; $\omega_s = 120\pi$ rad/s; $\omega_r = 50\pi$ rad/s;

$\omega_m = 170\pi$ or 70π rad/s i.e. 5100 or 2100 rpm in either direction. ←

12.4.20. $M = 2\cos\theta$; $dM/d\theta = -2\sin\theta$

$$T_e = i_1 i_2 \frac{dM}{d\theta} = -i_1 i_2 (2\sin\theta) = -(\sqrt{2}\,I_1 \sin\omega t)(\sqrt{2}\,I_2 \sin(\omega t - \phi))(2\sin\theta)$$

$$= -2 I_1 I_2 \sin\theta\,[\cos\phi - \cos(2\omega t - \phi)]$$

Where ϕ is the phase angle between the two currents $\bar{I}_1$ and $\bar{I}_2$.

$$(T_e)_{av} = -2 I_1 I_2 \sin\theta \cos\phi$$

$\bar{V}_1 = j\omega L_1 \bar{I}_1 + j\omega M \bar{I}_2$; $0 = j\omega L_2 \bar{I}_2 + j\omega M \bar{I}_1$ or $\bar{I}_2 = -\bar{I}_1 M / L_2$

The angle ϕ is 180°; $(T_e)_{av} = 2 I_1^2 \frac{M}{L_2} \sin\theta$; for $\theta = 45°$,

$$(T_e)_{av} = 2(10)^2 \left(\frac{2\cos 45°}{1}\right)\sin 45° = 200 \text{ N·m} \leftarrow$$

12.4.21.

$$W' = \frac{1}{2} L_{11} i_1^2 + M i_1 i_2 + \frac{1}{2} L_{22} i_2^2$$

$$= \frac{1}{2}(1+\sin\theta) i_1^2 + (1-\sin\theta) i_1 i_2 + (1+\sin\theta) i_2^2$$

$$T_e = \frac{\partial W'}{\partial \theta} = \left(\frac{1}{2}\cos\theta\right) i_1^2 - (\cos\theta) i_1 i_2 + \cos\theta\, i_2^2$$

(a)

$$T_e = \frac{1}{2}(\cos 45^\circ)(15)^2 - \cos 45^\circ (15)(-4) + \cos 45^\circ (-4)^2$$

$$= 133.3 \text{ N.m}, \text{ IN A DIRECTION OF INCREASING } \theta \leftarrow$$

(b)

$$W_1 = \frac{1}{2}\lambda_1 i_1 = \frac{1}{2} L_{11} i_1^2 + \frac{1}{2} M i_2 i_1$$

$$= \frac{1}{2}(1+\sin 45^\circ)(15)^2 + \frac{1}{2}(1-\sin 45^\circ)(-4)(15)$$

$$= 192 - 8.8 = 183.2 \text{ J} \leftarrow$$

$$W_2 = \frac{1}{2}\lambda_2 i_2 = \frac{1}{2} M i_1 i_2 + \frac{1}{2} L_{22} i_2^2$$

$$= \frac{1}{2}(1-\sin 45^\circ)(15)(-4) + \frac{1}{2}(2)(1+\sin 45^\circ)(-4)^2$$

$$= -8.8 + 27.3 = 18.5 \text{ J} \leftarrow$$

(c)

$$\lambda_2 = M i_1 + L_{22} i_2 = 0$$

$$i_2 = -\frac{M}{L_{22}} i_1$$

$$I_2 = -\frac{(1-\sin 45^\circ)}{2(1+\sin 45^\circ)} 10 = -0.858 \text{ A (RMS)} \leftarrow$$

(d)

$$i_1 = \sqrt{2}\, 10 \cos 377t;\quad i_2 = \sqrt{2}(-0.858)\cos 377t$$

$$T_e = \left(\frac{1}{2}\cos 45^\circ\right)(\sqrt{2}\, 10 \cos 377t)^2$$

$$-(\cos 45^\circ)(\sqrt{2}\, 10 \cos 377t)[-\sqrt{2}(0.858)\cos 377t]$$

$$+(\cos 45^\circ)[-\sqrt{2}(0.858)\cos 377t]^2$$

$$= 118.6 \cos 45^\circ \cos^2 377t$$

$$= 41.9(1+\cos 754t) \text{ N.m} \leftarrow$$

(e) $T_{e\,av} = 41.9 \text{ N.m} \leftarrow$

12.4.22.

(a)

$$L_{ss} = \frac{4\mu_0 N_s^2 l r}{\pi g} \quad ; \quad L_{rr} = \frac{4\mu_0 N_r^2 l r}{\pi g} \quad ; \quad L_{sr} = \frac{4\mu_0 N_s N_r l r}{\pi g}\cos\theta$$

$$= M\cos\theta \quad \longleftarrow$$

(b)

$$T_e = \frac{\partial W'(i_s, i_r, \theta)}{\partial \theta}$$

$$W' = \tfrac{1}{2} L_{ss} i_s^2 + \tfrac{1}{2} L_{rr} i_r^2 + L_{sr} i_s i_r$$

$$T_e = -M\, i_s\, i_r \sin\theta$$

$$\text{or} \quad M\, i_s\, i_r \cos(\theta + 90^\circ) \quad \longleftarrow$$

(c)

$$M = \frac{4\mu_0 N_s N_r l r}{\pi g}$$

$$T_e = -M\, i_s\, i_r \sin\theta$$

$$= -\frac{4\mu_0 N_s N_r l r}{\pi g} i_s i_r \sin\theta$$

$$= -\frac{\pi\mu_0}{g}\left(\frac{2N_s i_s}{\pi}\right)\left(\frac{2N_r i_r}{\pi}\right) l r \sin\theta$$

$$= -\frac{\pi\mu_0 l r}{g} F_s F_r \sin\delta \quad \longleftarrow$$

12.5.1.

(a) η at FL $= \dfrac{10\times 746}{(10\times 746)+600+350+350+50} = \dfrac{7460}{8810} = 0.85$ ←

(b) η at half-full load $= \dfrac{3730}{3730+600+50+(700\times\frac{1}{4})} = \dfrac{3730}{4555} = 0.82$ ←

12.5.2.

$$\text{rms hp} = \sqrt{\frac{50^2\times 8 + 100^2\times 8 + 150^2\times 10 + 120^2\times 20 + 0^2\times 14}{8+8+10+20+14}}$$

$$= \sqrt{\frac{613{,}000}{60}} = 101.1 \text{ hp}.$$

May choose 100 to 120-hp motor depending on other considerations, such as availability, cost, etc. ←

CHAPTER 13

13.1.1.

(a) The no-load speed will be quite close to syn. speed, but somewhat less; the number of poles has to be an even, whole integer.

Syn. Speed = $\frac{120f}{P}$ ⇒ P = 8 with syn. speed = 750 rpm for f = 50 Hz

(b) Slip at FL = $\frac{750-700}{750} = 0.067$; $f_r = Sf_s = \frac{1}{15} \times 50 = 3\frac{1}{3}$ Hz ←

(c) $120 \times \frac{10}{3} \times \frac{1}{8} = 50$ rpm w.r.t rotor; 700 + 50 = 750 rpm w.r.t. stator.

13.1.2. (a) $\frac{120 \times 60}{P} = 1800$ or P = 4 ←

(b) Slip at FL = $\frac{1800-1710}{1800} = 0.05$ ←

(c) $f_r = Sf_s = 0.05 \times 60 = 3$ Hz ←

(d) i) $120 \times 3 \times \frac{1}{4} = 90$ rpm w.r.t. rotor ←

ii) 1710 + 90 = 1800 rpm w.r.t. stator ←

iii) 1800 − 1800 = 0 rpm, i.e., stationary w.r.t. each other. ←

13.1.3.

SYNCHRONOUS SPEED = $\frac{120 \times 60}{4}$ = 1800 rpm

(a) ROTOR SPEED FOR THE LOAD CONDITION = (1 − 0.03) 1800 = 1746 rpm ←

(b) $f_r = Sf_s$ = (0.03) 60 = 1.8 Hz ←

(c) 1800 rpm i.e. SYNCHRONOUS SPEED ←

(d) ZERO (STATIONARY WITH RESPECT TO EACH OTHER) ←

13.1.4.

(i) V and f are halved; Syn. speed is halved; flux is the same; at the same slip, the rotor current and torque are unchanged. Hence the torque-speed characteristic is the previous one moved to the left as shown in (a).

(ii) With V halved, and f being the same, the flux is halved and rotor-current is halved; the torque is $\frac{1}{4}$ of its previous value, as shown in (b).

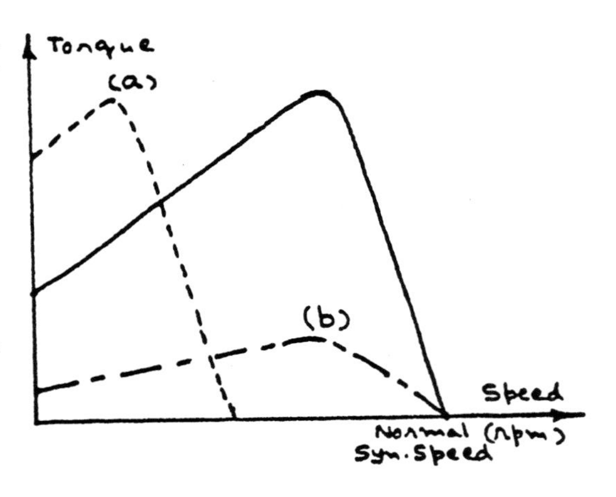

13.1.5.

(a) $S = \frac{1800 - (-1750)}{1800} = \frac{3550}{1800} = 1.97$ ←

(b) $f_r = S f_s = 1.97 \times 60 = 118.33$ Hz ←

13.1.6.

(a) Syn. Speed $= \frac{120 \times 60}{4} = 1800$ rpm

The slip range: $\frac{1800 - 3600}{1800}$ to $\frac{1800 + 3600}{1800}$, i.e., -1 to 3

∴ the frequency range available at the slip rings: 60 to 180 Hz;

20 V at 60 Hz and 60 V at 180 Hz. ←

(b) In order to give 150 Hz at the slip-ring terminals, the rotor must be driven such that the slip is $\pm \frac{150}{60} = \pm 2.5$, i.e. 2700 rpm in the opposite direction to that of the stator field, or 6300 rpm in the same direction to that of the stator field.

The slip-ring voltage is $400 \times \frac{150}{120} = 500$ V in either case. ←

13.1.7.

Syn. Speed of syn. motor $= \frac{120 \times 60}{4} = 1800$ rpm cw which is also the actual speed of the ind. m/c.

The rotating mag. field of the ind. m/c rotates at $\frac{120 \times 60}{8} = 900$ rpm ccw

∴ Slip $= \frac{900 - (-1800)}{900} = 3$; freq. of rotor voltages $= 3 \times 60 = 180$ Hz ←

13.1.8.

(a) $n = \frac{120f}{P} = \frac{120 \times 60}{6} = 1200$ rpm ←

~~(b)~~ $\omega_m = \frac{2\pi n}{60}$ can be computed, if needed, in rad/s.

~~(c)~~ (b) $n = \frac{120 \times 60}{12} = 600$ rpm ←

~~13.1.9.~~

~~(a)~~ (c) $n = \frac{120 \times 60}{2} = 3600$ rpm ←

13.1.9

~~(b)~~ a $n = \frac{120f}{P} = \frac{120 \times 50}{6} = 1000$ rpm ←

~~(c)~~ b $n = \frac{120 \times 50}{12} = 500$ rpm ←

(c) $\frac{120 \times 50}{2} = 3000$ rpm

13.1.10.

(a) $n = \frac{120 \times 50}{2} = 3000$ rpm ←

(b)

$n = \frac{120 \times f}{P}$ or $P = \frac{120f}{n} = \frac{120 \times 60}{1200} = 6$ poles

HAS TO BE EVEN WHOLE INTEGER

ALSO NL Speed $\lesssim$ Syn-Speed
1188 1200

SPEED NOT SYNCHRONOUS; INDUCTION MOTOR ←

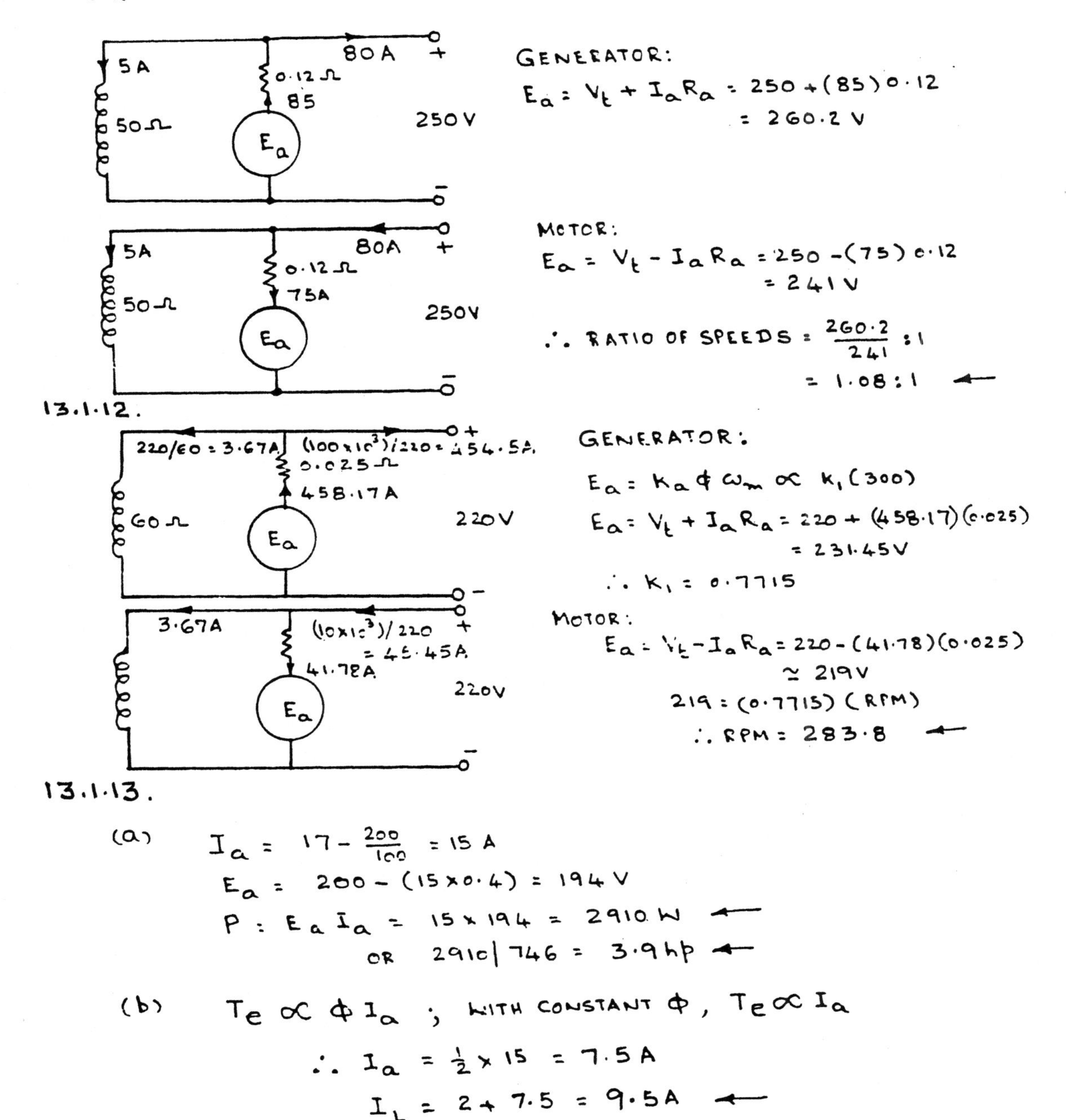

13.1.11.

GENERATOR:

$E_a = V_t + I_a R_a = 250 + (85)0.12$
$= 260.2\ V$

MOTOR:

$E_a = V_t - I_a R_a = 250 - (75)\ 0.12$
$= 241\ V$

$\therefore$ RATIO OF SPEEDS $= \frac{260.2}{241} : 1$
$= 1.08 : 1$ ←

13.1.12.

GENERATOR:

$E_a = K_a \phi\, \omega_m \propto k_1 (300)$

$E_a = V_t + I_a R_a = 220 + (458.17)(0.025)$
$= 231.45V$

$\therefore k_1 = 0.7715$

MOTOR:

$E_a = V_t - I_a R_a = 220 - (41.78)(0.025)$
$\simeq 219V$

$219 = (0.7715)$ (RPM)

$\therefore$ RPM $= 283.8$ ←

13.1.13.

(a) $I_a = 17 - \frac{200}{100} = 15\ A$

$E_a = 200 - (15 \times 0.4) = 194\ V$

$P = E_a I_a = 15 \times 194 = 2910\ W$ ←

OR $2910 / 746 = 3.9\ hp$ ←

(b) $T_e \propto \phi I_a$; WITH CONSTANT ϕ, $T_e \propto I_a$

$\therefore I_a = \frac{1}{2} \times 15 = 7.5\ A$

$I_L = 2 + 7.5 = 9.5A$ ←

13.1.14.

(a) $T_e \omega_m = E_a I_a = 20\times10^3$; $\omega_m = \frac{1200}{60}\times 2\pi = 40\pi$ rad/s.

$T_e = \frac{20\times10^3}{40\pi} = 159.15$ N·m; $I_a = \frac{20\times10^3}{400} = 50$ A ←

(b) 30 hp = 30×746 = 22,380 W; No losses

(i) $E_a I_a = 22380$ or $I_a = 22380/400 = 56$ A ←

(ii) Speed = 1200 rpm, since $E_a = K_a \Phi \omega_m = K_1 I_f \omega_m$ and the same values of $E_a = 400$ V and $I_f = 4$ A apply as in part (a)

(iii) $T_e \omega_m = 22380$ or $T_e = 22380/(40\pi) = 178$ N·m ←

(c) $E_a = K_a \Phi \omega_m = K_1 I_f \omega_m$; $400 = K_1 \times 4 \times 40\pi$ or $K_1 = 5/(2\pi)$

$440 = \frac{5}{2\pi}\times 4 \times \omega_m$ or $\omega_m = 44\pi$ rad/s or 1320 rpm ←

Since $E_a I_a = 0$, $I_a = 0$ ←

13.1.15.

DC Shunt motor

(i) Assuming $\Phi \propto I_f$ (Note that this is generally not true because of saturation)

since $E_a = K_a \Phi \omega_m$

if I_f is doubled, with the same E_a, ω_m gets halved. ←

Since $T_e = K_a \Phi I_a$ or $T_e \omega_m = E_a I_a$,

with the same T_e and E_a, I_a gets halved. ←

(ii) With I_f being the same and E_a halved, ω_m gets halved. ←

With T_e being the same and E_a halved, I_a remains to be the same because ω_m is halved. ←

(iii) With I_f and E_a halved, ω_m remains to be the same. ←

With hp-output to be the same, i.e. $T_e \omega_m$ or T_e being the same, with E_a halved, I_a is doubled. ←

(iv) With I_f being the same and E_a halved, ω_m gets halved. ←

With hp-output being the same, i.e. $T_e \omega_m$ being the same,

T_e gets doubled and I_a is doubled. ←

(v) With I_f being the same and E_a halved, ω_m gets halved. ←

With $T_e \propto \omega_m^2$, I_a is $\frac{1}{4}$ of its previous value. ←

13.2.1.

a) $120f/P = 120 \times 60/2 = 3600$ rpm or 60 rps. ←

b) $f_r = s f_s = 0.05 \times 60 = 3$ Hz ; 3 rps or 180 rpm ←

c) 3600 rpm or 60 rps (d) 0 (e) $f_r = 3$ Hz ←

f) Rotor emf at standstill $= \frac{100}{2} = 50$ V ; rotor emf at 0.05 slip $= 0.05 \times 50$
$= 2.5$ V ←

13.2.2.

$R_1 = 0.1\,\Omega$; $R_0 = \frac{1000}{3} \times \frac{1}{20^2} = 0.83\,\Omega$; $Z_0 = \frac{220}{\sqrt{3}} \times \frac{1}{20} = 6.35\,\Omega$; $X_0 = \sqrt{Z_0^2 - R_0^2}$
$= \sqrt{6.35^2 - 0.83^2} = 6.3\,\Omega$; $R_{eq} = \frac{1500}{3} \times \frac{1}{50^2} = 0.2\,\Omega$; $Z_{eq} = \frac{30}{\sqrt{3}} \times \frac{1}{50} = 0.346\,\Omega$
$X_{eq} = \sqrt{0.346^2 - 0.2^2} = 0.282\,\Omega$; $X_{l1} = X'_{l2} = 0.282/2 = 0.141\,\Omega$;
$X_m = X_0 - X_{l1} = 6.3 - 0.141 = 6.159\,\Omega$; $R'_2 = (0.2 - 0.1)\left(\frac{0.141 + 6.159}{6.159}\right)^2 = 0.105\,\Omega$
NL Rot. losses $= 3\left[\frac{1000}{3} - (20^2 \times 0.1)\right] = 880$ W which includes fric., wind., core losses.

13.2.3.

Mech. power output = P_o = 5 hp = 5 × 746 = 3730 W

Rot. + stray-load losses = $\frac{5}{100} \times 3730 = 187$ W

mech. power developed P_m = 3730 + 187 = 3917 W

Slip at FL = 0.025 ; P_g = Power across airgap = $\frac{3917}{1-S} = \frac{3917}{0.975} = 4017$ W ←

Rotor copper loss = $SP_g = 0.025 \times \frac{3917}{0.975} = 100.4$ W ←

$T = \frac{P_g}{\omega_s} = \left(\frac{3917}{0.975}\right)\frac{1}{40\pi} = 31.97$ N·m ←

Note: Syn. Sp. = $\frac{120 \times 60}{6} = 1200$ rpm ; $\omega_s = \frac{1200}{60} \times 2\pi = 40\pi$ rad/s.

13.2.4.

$P_g = 24 \times 10^3$ W ; $P_m = P_g(1-S) = 22 \times 10^3$ W

$\therefore 1 - S = 22/24$ or $S = 1 - \frac{22}{24} = 0.083$ ←

$P_o = P_m - P_{rot} = 22 \times 10^3 - 400 = 21600$ W $= T_o \omega_m = T_o \cdot \omega_s(1-S)$

Syn. Speed = $\frac{120 \times 60}{2} = 3600$ rpm ; $\omega_s = \frac{3600}{60} \times 2\pi = 120\pi$ rad/s.

$\therefore T_o = \frac{21600}{(0.917)(120\pi)} = 62.5$ N·m ←

13.2.5.

$R_2' = (2.5)^2\, 0.1 = 0.625\ \Omega$/ph ; $X_{\ell 2}' = (2.5)^2\, 0.2 = 1.25\ \Omega$/ph.

$R_2'/S = \frac{0.625}{0.05} = 12.5\ \Omega$; $Z_2' = \frac{R_2'}{S} + jX_{\ell 2}' = 12.5 + j1.25 = 12.56\angle 5.7^\circ\ \Omega$/ph.

$Z_1 = R_1 + jX_{\ell 1} = 0.5 + j1.25 = 1.35\angle 68.2^\circ\ \Omega$/ph.

Exciting impedance ref. to the stator $Z_M = \frac{R_c(jX_M)}{R_c + jX_M} = \frac{(360)(j40)}{360 + j40}$
$= 39.8\angle 83.65^\circ = (4.4 + j39.5)\ \Omega$/ph.

Z as seen from the input terminals $= Z_1 + \frac{Z_2' Z_M}{Z_2' + Z_M} = 11.01 + j5.48$
$= 12.3\angle 26.46^\circ\ \Omega$/ph.

$\bar{I}_1 = \frac{\bar{V}_1}{Z} = \frac{(440/\sqrt{3})\angle 0^\circ}{12.3\angle 26.46^\circ} = 20.65\angle -26.46^\circ$ A ; pf = Cos 26.46° = 0.895 lagging ; ←

$\bar{I}_2' = \frac{\bar{E}_2'}{Z_2'} = \left(\frac{\bar{I}_1 Z_2' Z_M}{Z_2' + Z_M}\right)\frac{1}{Z_2'} = \frac{\bar{I}_1 Z_M}{Z_2' + Z_M} = 18.63\angle -10.29^\circ$ A

ref. to the rotor, $\bar{I}_2 = 2.5\, \bar{I}_2' = 46.58\angle -10.29^\circ$ A ←

13.2.5. CONTD.

output power $= P_o = P_m - P_{rot}$; $P_m = m_1 (I_2')^2 R_2' \dfrac{1-s}{s}$

$P_m = 3(18.63)^2\, 0.625 \times \dfrac{1-0.05}{0.05} = 12365\,W$; $P_o = 12365 - (200 + 100)$
$= 12065\,W$ or $16.2\,hp$. ←

output torque $= \dfrac{P_o}{\omega_m}$; $\omega_m = \dfrac{2\pi n(1-s)}{60}$ where $n = \dfrac{120 \times 60}{8} = 900\,rpm$

$\omega_m = 2\pi \times \dfrac{900}{60}(1-0.05) = 89.54$; $T_o = \dfrac{P_o}{\omega_m} = \dfrac{12065}{89.54} = 134.74\,N{\cdot}m$ ←

Effcy. $= \dfrac{\text{output}}{\text{input}}$; Input $= mV_1 I_1 \cos\phi_1 = 3 \times \dfrac{440}{\sqrt{3}} \times 20.65 \times 0.895 = 14083\,W$

$\therefore\ \eta = 12065/14083 = 0.857$ ←

13.2.6.

a) $I_2 = \dfrac{E_2}{\sqrt{\left(\frac{R_2}{S}\right)^2 + X_{l2}^2}}$; $T = \dfrac{I_2^2 R_2}{S\,\omega_s}$ per phase;

$$\therefore T = \frac{E_2^2\,(R_2/S)}{\omega_s\left[\left(\frac{R_2}{S}\right)^2 + X_{l2}^2\right]} = \frac{E_2^2\, y}{\omega_s (y^2 + X_{l2}^2)}, \text{ where } y = R_2/S$$

For maximum T, $dT/dy = 0 \Rightarrow y = X_{l2}$ or $\dfrac{R_2}{S_{maxT}} = X_{l2}$ or $R_2 = S_{maxT} \cdot X_{l2}$

The same result can also be seen from Eq. (13.2.13) with $R_1'' = 0$; $X_1'' = 0$; $R_2' = R_2$ and $X_{l2}' = X_{l2}$.

b) $S_{maxT} = R_2 / X_{l2}$ ←

c) $T_{max} = \dfrac{1}{\omega_s}\,\dfrac{0.5\, m_1 E_2^2}{X_{l2}}$, where m_1 is the no. of phases. ←

d) at starting, $S = 1$; $\therefore$ R_2 for max. starting torque $= X_{l2}$ ←

13.2.7.

$S = 1$ at starting; $T_{st} = 1.6\,T_{FL} = 1.6 \times 0.5\,T_{max} = 0.8\,T_{max}$

$$\frac{T}{T_{max}} = \frac{2}{\left(\frac{S}{S_{maxT}} + \frac{S_{maxT}}{S}\right)}; \quad \frac{T_{st}}{T_{max}} = \frac{2}{\left(\frac{1}{S_{maxT}} + \frac{S_{maxT}}{1}\right)} = 0.8$$

$$(1/S_{maxT}) + S_{maxT} = 2/0.8 = 5/2 \text{ or } S_{maxT} = 0.5 \text{ or } 2$$

For the induction motor, $S_{maxT} = 0.5$ ←

$$\because T_{FL} = 0.5\,T_{max}, \quad \frac{T_{FL}}{T_{max}} = 0.5 = \frac{2}{\frac{S_{FL}}{0.5} + \frac{0.5}{S_{FL}}} = \frac{4 S_{FL}}{4 S_{FL}^2 + 1}$$

$$\text{or } 4 S_{FL}^2 - 8 S_{FL} + 1 = 0 \text{ or } S_{FL} = \frac{8 \pm \sqrt{64 - 16}}{8} = 0.134 \text{ or } 1.87$$

For the induction motor, $S_{FL} = 0.134$ ←

13.2.8.

Circuit: $\frac{400}{\sqrt{3}}\angle 0°$ V source, $\bar{I}_1$, $(1+j2)\,\Omega$, $\bar{I}_0$ through $400\,\Omega$ in parallel with $j40\,\Omega$, $j2\,\Omega$, $\bar{I}_2'$, $\frac{1}{S}$

OR

Circuit: $231\angle 0°$ V source, $\bar{I}_1$, $(1+j2)\,\Omega$, $\bar{I}_0$ through $1/(\frac{1}{400}+\frac{1}{j40}) = (4+j40)\,\Omega$, $(\frac{1}{S}+j2)\,\Omega$, $\bar{I}_2'$

a) at $S \simeq 0$, on no load, $Z_{in} = 5+j42 = 42.3\angle 83.2°$;

input current: $231\angle 0° / 42.3\angle 83.2° = 5.46\angle -83.2°$; ←

input power: $3\times 231\times 5.46\times \cos 83.2° = 447\,W$. ←

At standstill, $S=1$, $Z_{in} = (1+j2) + \frac{(4+j40)(1+j2)}{(5+j42)} = 4.36\angle 63.9°$;

input current: $231\angle 0° / 4.36\angle 63.9° = 53.0\angle -63.9°$; ←

input power: $3\times 231\times 53\times \cos 63.9° = 16{,}158\,W$. ←

Note that the blocked-rotor test will not, of course, be performed at full voltage.

b) at $S=0.05$, $\frac{1}{S}+j2 = 20+j2$;

$$\bar{I}_1 = \frac{231\angle 0°}{(1+j2) + \frac{(4+j40)(20+j2)}{(24+j42)}} = 12.5\angle -34° = 10.4 - j7 \quad ←$$

$$\bar{I}_2' = \bar{I}_1 \frac{(4+j40)}{(4+j40)+(20+j2)} = 10.4\angle -9.7° = 10.3 - j1.8$$

$$\bar{I}_0 = \bar{I}_1 \frac{(20+j2)}{(4+j40)+(20+j2)} = 5.2\angle -88.3° = 0.15 - j5.2$$

$T\omega_s = m_1 (I_2')^2 \frac{R_2'}{S} = 3(10.4)^2 \frac{1}{0.05} = 6489\,W$; $T = 6489$ syn. watts

or $\because \omega_s = 2\pi \times \frac{2\times 60}{4} = 60\pi$ rad/s., $T = \frac{6489}{60\pi} = 34.43$ N·m

Output power $= 6489(1-S) = 6489\times 0.95 = 6165\,W$ ←

input power: $3\times 231\times 12.5\times \cos 34° = 7182\,W$; $\eta = \frac{6165}{7182} = 0.858$ ←

c) $\bar{V}_{Th} = 231\angle 0° \frac{4+j40}{(1+j2)+(4+j40)} = 220\angle 1.1°$

$Z_{Th} = \frac{(4+j40)(1+j2)}{(4+j40)+(1+j2)} = 2.09\angle 64.5° = 0.9+j1.89$

Circuit: $\bar{V}_{Th} = 220\angle 1.1°$, $Z_{Th} = 0.9+j1.89$, $\bar{I}_2'$, $(\frac{1}{S}+j2)$

THEVENIN'S EQUIV. CKT.

At starting, $S=1$; $Z_{in} = 1.9+j3.89 = 4.33\angle 63.9°$

$\bar{I}_2' = \frac{220\angle 1.1°}{4.33\angle 63.9°} = 50.8\angle -62.8°$; $T_{st} = \frac{3\times(50.8)^2}{\omega_s}$ or 7742 Syn.W. or $\frac{7742}{60\pi} = 41.07$ N·m ←

13.2.8. CONTD.

c) CONTD.

$$S_{\max T} = \frac{R_2'}{\sqrt{(R_1'')^2 + (X_1'' + X_{l2}')^2}} = \frac{1}{\sqrt{(0.9)^2 + (1.89+2)^2}} = 0.25$$

Eq. (13.2.14)

$$T_{\max} = \frac{1}{\omega_s} \frac{0.5\, m_1 V_{1a}^2}{R_1'' + \sqrt{(R_1'')^2 + (X_1'' + X_{l2}')^2}} = \frac{1}{60\pi} \frac{0.5 \times 3 \times 220^2}{0.9 + \sqrt{0.9^2 + (1.89+2)^2}} = 78.72 \text{ N·m}$$

Eq. (13.2.15)

Output power is given by $m_1 (I_2')^2 R_2' \dfrac{1-S}{S}$

For maximum output power, $\dfrac{1}{S} - 1 = \sqrt{1.9^2 + 3.89^2} = 4.33$

or $S = 0.19$ ←

The total circuit impedance is now $0.9 + \dfrac{1}{0.19} + j3.89 = 6.16 + j3.89$

$$\therefore I_2' = \frac{220}{|6.16 + j3.89|} = \frac{220}{7.28} = 30.22;$$

Max^m. output power $= 3(30.22)^2\left(\dfrac{1}{0.19} - 1\right) = 11{,}671$ W ←

d) $S = 0.05$; $1 + \dfrac{1}{S} + j4 = 21 + j4 = 21.4\angle 10.8°$

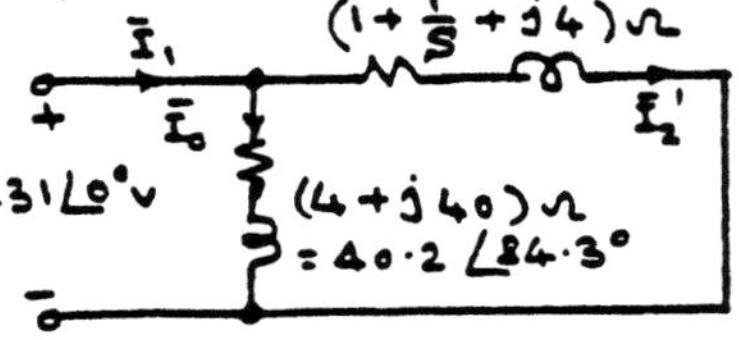

$$\therefore \bar{I}_2' = \frac{231\angle 0°}{21.4\angle 10.8°} = 10.8\angle -10.8° = 10.6 - j2$$

$$\bar{I}_0 = \frac{231\angle 0°}{40.2\angle 84.3°} = 5.7\angle -84.3° = 0.57 - j5.7$$

$$\bar{I}_1 = \bar{I}_0 + \bar{I}_2' = 11.2 - j7.7 = 13.6\angle -34.5° \leftarrow$$

$$T = \frac{1}{\omega_s} \times 3 \times (10.8)^2 \frac{1}{0.05} = 37.13 \text{ N·m} \leftarrow$$

output power $= (6998)(1-S) = 6998 \times 0.95 = 6648$ W ←

input power $= 3 \times 231 \times 13.6 \times \cos 34.5° = 7767$ W; $\eta = \dfrac{6648}{7767} = 0.856$ ←

e) $S = -0.05$; $\left(1 + \dfrac{1}{S} + j4\right) = -19 + j4 = 19.4\angle 168°$

$$\bar{I}_0 = \frac{231\angle 0°}{40.2\angle 84.3°} = 5.7\angle -84.3° = 0.57 - j5.72$$

$$\bar{I}_2' = 231\angle 0° / 19.4\angle 168° = 11.9\angle -168° = -11.6 - j2.47$$

$\bar{I}_1 = \bar{I}_0 + \bar{I}_2' = -11 - j8.19 = 13.6\angle 217°$; $T = \dfrac{1}{60\pi}(3)(11.9)^2 \dfrac{1}{0.05} = 45.08$ N·m ←

Mech. power input $= (1-S)(8497) = 1.05 \times 8497 = 8922$ W; Elec. power output $= 3(231)(13.6)\cos 217° = 7526$ W ←

13.2.8. CONTD.

f) If the machine is running with a slip of 0.05 and is then plugged the operating point moves from $S = 0.05$ to that of a slip $(2-S) = 1.95$. The slip will initially be 1.95 and the machine will then brake to a standstill. With the new value of the slip of 1.95, $(1 + \frac{1}{S} + j4) = 1.51 + j4 = 4.28 \angle 69.3°$;

$\bar{I}_0 = 231\angle 0° / 40.2 \angle 84.3° = 5.75 \angle -84.3° = 0.57 - j5.72$;

$\bar{I}_2' = 231\angle 0° / 4.28 \angle 69.3° = 54 \angle -69.3° = 19.1 - j50.5$;

$\bar{I}_1 = \bar{I}_0 + \bar{I}_2' = 19.7 - j56.6 = 60 \angle 70.7°$ ←

Braking torque $= \frac{1}{60\pi}(3)(54)^2 \frac{1}{1.95} = 23.8$ N.m ←

13.2.9. (a) If the rotor resistance is increased 5 times, the slip must increase 5 times for the same value of (R_2'/S) and therefore for the same torque. The new slip at F.L. torque is then $5 \times 0.02 = 0.1$ ←

b) $P_{g1\,FL} = \frac{P}{1-S} = \frac{(500)(746)}{(1-0.02)} = 380{,}612$ W

Rotor copper loss at full load $= S P_{g1} = (0.02)(380612) = 7612$ W

Since the effective value of the rotor current is the same as its F.L. value, the new rotor copper loss is 5 times the old F.L. value; i.e.,

$$5 \times 7612 = 38{,}060 \text{ W} \leftarrow$$

c) With added rotor resistance, the increased slip has caused the per-unit speed at F.L. torque to drop from $1 - 0.02 = 0.98$ to $1 - 0.1 = 0.9$; the torque is the same; the power output will therefore drop proportionally; i.e., $\frac{0.9}{0.98}(500) = 459.2$ hp ←

The decrease in output will be equal to the increase in rotor copper loss.

d) $S_{max\,T} \propto R_2'$; ∴ the new slip at maximum torque with the added rotor resistance is $5 \times 0.06 = 0.3$

e) The effective value of the rotor current at max^m. torque is independent of the rotor resistance. $I'_{2\,max\,T} = 3\, I'_{2\,FL}$ ←

13.2.9. CONTD.

f) With the rotor resistance increased by 5 times, the starting torque will be the same as the original running torque at a slip of 0.2 and so $T_{start} = 1.2\ T_{FL}$ ←

g) The rotor current at starting with the added rotor resistance will be the same as the old rotor current when running at a slip of 0.2 with the slip rings shortcircuited. $I'_{2start} = 4\ I'_{2FL}$ ←

13.2.10.

a) $S_{maxT} = R'_2 \Big/ \sqrt{(R''_1)^2 + (X''_1 + X'_{l2})^2}$ Eq. (13.2.14)

$$R''_1 + jX''_1 = \frac{jX_M (R_1 + jX_{l1})}{R_1 + j(X_{l1} + X_M)} = \frac{j50\,(0.75 + j2)}{0.75 + j52} = 0.72 + j1.92 = 2.05\angle 69.4^\circ$$

$$\therefore S_{maxT} = 0.8 \Big/ \sqrt{(0.72)^2 + (3.92)^2} = 0.2 \leftarrow$$

b) $$T_{max} = \frac{1}{\omega_s}\,\frac{0.5\,m_1 V_{1a}^2}{R''_1 + \sqrt{(R''_1)^2 + (X''_1 + X'_{l2})^2}}$$ Eq. (13.2.15)

4 pole; 60 Hz; Syn. Speed $= \frac{120 \times 60}{4} = 1800$ rpm; $\omega_s = \frac{1800}{60}(2\pi) = 60\pi$ rad/s

$$\bar{V}_{1a} = \bar{V}_1 \frac{jX_M}{R_1 + j(X_{l1} + X_M)} = \frac{600}{\sqrt{3}}\,\frac{j50}{0.75 + j52} = 333.05\ V$$

$$T_{max} = \frac{1}{60\pi}\,\frac{0.5 \times 3 \times 333.05^2}{0.72 + \sqrt{0.72^2 + 3.92^2}} = 187.6\ N\cdot m \leftarrow$$

c) $0 < S < 0.2$ or $1440 < \text{Speed} < 1800$ rpm

d) $$I'_{2start} = \frac{333.05}{\sqrt{(0.72 + 0.8)^2 + (3.92)^2}} = 79.55\ A$$

$$T_{st} = \frac{1}{60\pi} \times 3 \times 79.55^2 \times 0.8 = 80.5\ N\cdot m \leftarrow$$

e) $S_{maxT} = 1 = R'_{2\,total} \Big/ \sqrt{0.72^2 + 3.92^2}$; or $R'_{2\,total} = 3.98\ \Omega$;

∴ Additional rotor resistance to be inserted $= 3.18\ \Omega$ per phase (ref. to stator) ←

13.2.11.

$Z_f = R_f + jX_f$, WHICH IS $\left(\frac{R_2'}{S} + jX_{\ell 2}'\right)$ IN PARALLEL WITH (jX_M)

FOR $S = 0.02$, SUBSTITUTION OF NUMERICAL VALUES GIVES

$R_f + jX_f = 5.41 + j3.11$; $R_1 + jX_{\ell 1} = 0.29 + j0.5$; SUM: $5.7 + j3.61 = 6.75 \angle 32.4°\ \Omega$

STATOR CURRENT $I_1 = \frac{220/\sqrt{3}}{6.75} = 18.8\text{ A}$ ←

POWER FACTOR = $\cos 32.4° = 0.844$ LAGGING ←

SYNCHRONOUS SPEED = $\frac{120 \times 60}{6} = 1200$ rpm; $\omega_S = \frac{1200}{60} \times 2\pi = 40\pi$ rad/s

ROTOR SPEED = $(1 - 0.02)(1200) = 1176$ rpm ←

$P_g = m_1 (I_2')^2 R_2'/S = m_1 I_1^2 R_f = 3(18.8)^2\, 5.41 = 5740\text{ W}$

INTERNAL MECHANICAL POWER $P_m = (0.98)(5740) = 5630\text{ W}$

OUTPUT POWER = $5630 - 403 \simeq 5230\text{ W} \simeq 7\text{ hp}$ ←

OUTPUT TORQUE = $5230/(0.98 \times 125.6) = 42.5\text{ N·m}$ ←

STATOR COPPER LOSS = $3(18.8)^2\, 0.294 = 312\text{ W}$

ROTOR COPPER LOSS = $SP_g = 0.02 \times 5740 = 115\text{ W}$

FRICTION, WINDAGE, AND CORE LOSSES = 403 W

TOTAL LOSSES = 830 W

INPUT = 5230 + 830 = 6060 W

$\eta = 5230/6060 = 0.863$ or 86.3% ←

13.2.12.

$$T = \frac{1}{\omega_s} m_1 (I_2')^2 \frac{R_2'}{S} \quad (13.2.5); \quad \frac{T_{st}}{T_{FL}} = \frac{(I'_{2st})^2}{(I'_{2FL})^2} \frac{S_{FL}}{S_{st}} = 5^2 (0.05) = 1.25$$

$$I_2' = \frac{V_{1a}}{\sqrt{\left[R_1'' + \frac{R_2'}{S}\right]^2 + (x_1'' + x'_{l2})^2}} \quad (13.2.11); \text{ For our problem, } R_1'' = 0;$$

Let $(x_1'' + x'_{l2}) = x$; Then

$$I_2' = \frac{V_{1a}}{\sqrt{\left(\frac{R_2'}{S}\right)^2 + x^2}}; \quad I'_{2st} = \frac{V_{1a}}{\sqrt{(R_2')^2 + x^2}} = 5 I'_{2FL} = \frac{5 V_{1a}}{\sqrt{\left(\frac{R_2'}{0.05}\right)^2 + x^2}}$$

or $25\left[(R_2')^2 + x^2\right] = 400 (R_2')^2 + x^2$; $24 x^2 = 375 (R_2')^2$;

$$S_{maxT} = R_2'/x = \sqrt{\frac{24}{375}} = 0.253 \leftarrow$$

Applying Eq. (13.2.16),

$$\frac{T_{FL}}{T_{max}} = \frac{2}{\frac{S_{FL}}{S_{maxT}} + \frac{S_{maxT}}{S_{FL}}} = \frac{2}{\frac{0.05}{0.253} + \frac{0.253}{0.05}} = 0.38$$

$$\text{or } T_{max}/T_{FL} = 1/0.38 = 2.63 \leftarrow$$

13.2.13.

$$I_2' = \frac{V_{1a}}{\sqrt{\left(R_1'' + \frac{R_2'}{S}\right)^2 + (x_1'' + x'_{l2})^2}} \quad (13.2.11)$$

For small values of slip, $\frac{R_2'}{S} \gg (x_1'' + x'_{l2}) \gg R_1''$; $\therefore I_2' \simeq kS \leftarrow$

$T = \frac{1}{\omega_s} m_1 (I_2')^2 \frac{R_2'}{S}$ (13.2.5); for small values of slip $I_2' \simeq kS$

and $\therefore T \simeq k_1 S \leftarrow$

$P_m = T\omega_s (1 - S)$ (13.2.6); for small values of S, $P_m \simeq k_2 S \leftarrow$

13.2.14.

$S_{FL} = 0.03$; $T_{max} = 2T_{FL}$; $P_g = \frac{P_m}{1-S} = \frac{50 \times 746}{(1 - 0.03)}$

Rotor copper loss $= S P_g = \frac{0.03 \times 50 \times 746}{0.97} = 1150\,W \leftarrow$

$T_{FL}/T_{max} = 0.5 \Rightarrow S_{FL}/S_{maxT} = 0.275$ from Eq. (13.2.16); $S_{maxT} = \frac{0.03}{0.275} = 0.109$

$\therefore$ Speed at $T_{max} = (1 - 0.109)(120 \times 60/4) = 1600$ rpm $\leftarrow$

For max. starting torque, $R'_{2\,total} = 0.1/0.109 = 0.918\,\Omega$; i.e. per-phase rotor res. (ref. to stator) that must be added in series $= 0.918 - 0.1 = 0.818\,\Omega$/ph. of Y. $\leftarrow$

13.2.15.

BACKGROUND

PER-PHASE EQUIVALENT CIRCUITS OF A POLYPHASE INDUCTION MOTOR CORRESPONDING TO THE NO-LOAD TEST CONDITIONS ($S \simeq 0$) ARE GIVEN BELOW:

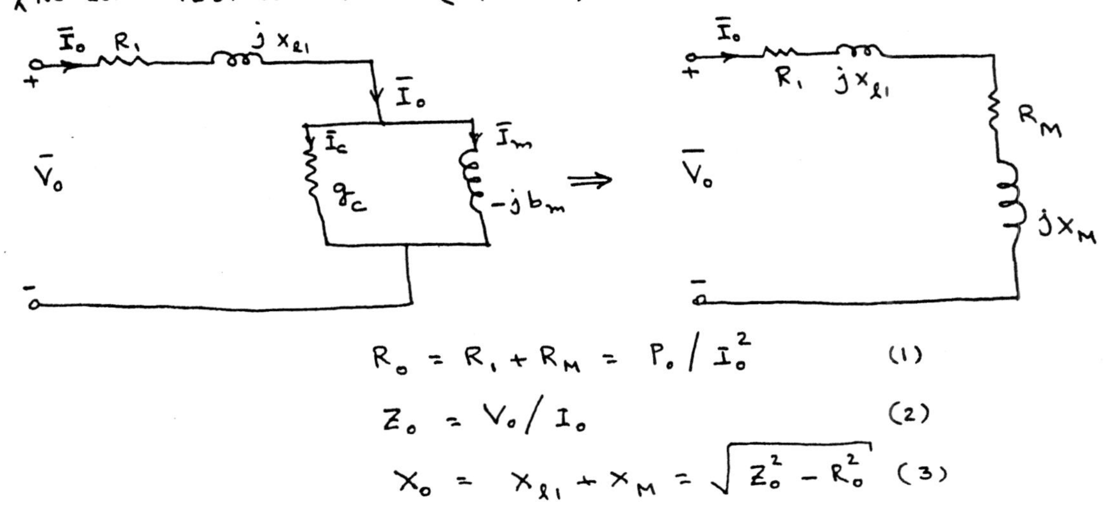

$$R_o = R_1 + R_M = P_o / I_o^2 \qquad (1)$$

$$Z_o = V_o / I_o \qquad (2)$$

$$X_o = X_{l1} + X_M = \sqrt{Z_o^2 - R_o^2} \qquad (3)$$

PER-PHASE EQUIVALENT CIRCUITS OF A POLYPHASE INDUCTION MOTOR CORRESPONDING TO THE BLOCKED-ROTOR TEST CONDITIONS ($S = 1$) ARE GIVEN BELOW:

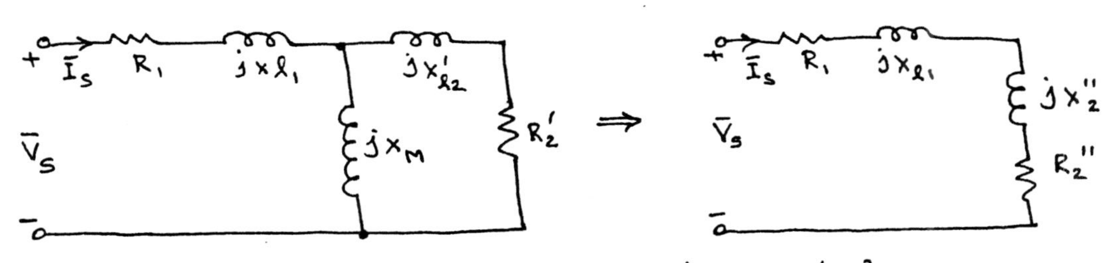

$$R_{eq} = R_1 + R''_2 = P_s / I_s^2 \qquad (4)$$

$$Z_{eq} = V_s / I_s \qquad (5)$$

$$X_{eq} = X_{l1} + X''_2 = \sqrt{Z_{eq}^2 - R_{eq}^2} \qquad (6)$$

ONE CAN ASSUME $X_{l1} = X'_{l2} = 0.5\, X_{eq}$ (7)

IT CAN BE SHOWN THAT $R'_2 = (R_{eq} - R_1) \dfrac{(X'_{l2} + X_M)^2}{X_M^2}$ (8)

13.2.15. CONTINUED

a) From the no-load test data and Eqs. (1) to (3) :

$R_0 = \frac{600}{3} \times \frac{1}{3^2} = 22.22\ \Omega$; $Z_0 = \frac{220}{\sqrt{3}} \times \frac{1}{3} = 42.34\ \Omega$; $X_0 = 36.04\ \Omega$

From the blocked-rotor test data and Eqs. (4) to (6) :

$R_{eq} = \frac{720}{3} \times \frac{1}{15^2} = 1.07\ \Omega$; $Z_{eq} = \frac{35}{\sqrt{3}} \times \frac{1}{15} = 1.35\ \Omega$; $X_{eq} = 0.823\ \Omega$

Taking $X_{\ell 1} = X'_{\ell 2} = 0.823/2 = 0.41\ \Omega$, ←

from Eq. (3), $X_M = X_0 - X_{\ell 1} = 36.04 - 0.41 = 35.63\ \Omega$; $R_1 = 0.5\ \Omega$;

from Eq. (8), $R'_2 = (1.07 - 0.5)\left(\frac{0.41 + 35.63}{35.63}\right)^2 = 0.583\ \Omega$. ←

b) $s = 0.05$; $P_g = m_1 (I'_2)^2 R'_2 / s = m_1 I_1^2 R_f$, where R_f is the real part of the parallel combination of jX_M and $\left(\frac{R'_2}{s} + jX'_{\ell 2}\right)$.

$$R_f = \text{real part of} \left[\frac{jX_M\left(\frac{R'_2}{s} + jX'_{\ell 2}\right)}{\frac{R'_2}{s} + j(X_M + X'_{\ell 2})}\right] = \text{real part of} \left[\frac{j35.63\left(\frac{0.583}{0.05} + j0.41\right)}{\frac{0.583}{0.05} + j(35.63 + 0.41)}\right]$$

$$= \text{real part of } [10.32 + j3.74] = 10.32$$

$Z_t = R_1 + jX_{\ell 1} + (10.32 + j3.74) = 10.82 + j4.15 = 11.6\angle 21°$; $\therefore I_1 = \frac{220/\sqrt{3}}{11.6} = 10.95$ A

$\therefore P_g = 3 \times 10.95^2 \times 10.32 = 3712.2$ W ; $P_m = P_g(1-s) = 0.95 \times 3712.2 = 3526.6$ W

P_0 = total mech. power output = $3526.6 - P_{rot}$; $P_{rot} = 3\left[\frac{600}{3} - (3^2 \times 0.5)\right] = 586.5$ W

$\therefore P_0 = 3526.6 - 586.5 = 2940$ W or $\frac{2940}{746} = 3.94$ hp ←

$$\text{Output torque} = \frac{\text{output power}}{(1-s)\ \omega_s} = \frac{2940}{(1-0.05)\left(\frac{4\pi \times 60}{4}\right)} = 16.42\ \text{N·m} \leftarrow$$

Total input power $= \sqrt{3}\,(220)\,10.95 \times \cos 21° = 3895$ W ; $\therefore \eta = \frac{2940}{3895} = 0.755$ ←

c) From Eq. (13.2.9), $\bar{V}_{1a} = \frac{220}{\sqrt{3}} \frac{j35.63}{0.5 + j(0.41 + 35.63)} \simeq 125.56$

From Eq. (13.2.10), $R''_1 + jX''_1 = \frac{(0.5 + j0.41)(j35.63)}{0.5 + j(0.41 + 35.63)} \simeq 0.49 + j0.41$

From Eq. (13.2.14), $s_{maxT} = \frac{0.583}{\sqrt{0.49^2 + (0.41 + 0.41)^2}} = 0.61$ ←

From Eq. (13.2.15), $T_{max} = \frac{1}{60\pi} \frac{0.5 \times 3 \times 125.56^2}{0.49 + \sqrt{(0.49)^2 + (0.41 + 0.41)^2}} = 86.8$ N·m ←

13.2.16. (SEE FOR BACKGROUND SOLUTION OF PR. 13.2.15)

(a)

$$R_2 - a^2 R_2 = \frac{P_s}{I_s^2} \quad \text{or} \quad 0.2 - 4R_2 = \frac{(45 \times 10^3)/3}{(193.6)^2}$$

whence $R_2 = 0.05\ \Omega$. ←

(b) Referred to the stator, the rotor resistance per phase is $R_2' = a^2 R_2 = 0.2\ \Omega$. Then

$$\text{starting torque} = \frac{3I_s^2 R_2'}{\omega_s} = \frac{3(193.6)^2(0.2)}{2\pi(900)/60} = 238.6\ \text{N}\cdot\text{m} \quad \leftarrow$$

13.2.17. (SEE FOR BACKGROUND SOLUTION OF PR. 13.2.15)

FROM THE NO-LOAD TEST DATA

$$V_0 = 400/\sqrt{3} = 231\ \text{V}\ ; \quad P_0 = \tfrac{1}{3}(1770 - 600) = 390\ \text{W}\ ; \quad I_0 = 18.5\ \text{A}$$

$$\therefore R_C = (231)^2/390 = 136.8\ \Omega \quad \leftarrow$$

$$X_M = \frac{(231)^2}{\sqrt{(231)^2(18.5)^2 - (390)^2}} = 12.5\ \Omega \quad \leftarrow$$

FROM BLOCKED-ROTOR TEST DATA

$$V_S = 45/\sqrt{3} = 25.98\ \text{V}\ ; \quad I_S = 63\ \text{A}\ ; \quad P_S = 2700/3 = 900\ \text{W}$$

THEN $R_{eq} = R_1 + a^2 R_2 = 900/(63)^2 = 0.23\ \Omega$

$$\therefore R_1 = R_2' = 0.115\ \Omega \quad \leftarrow$$

$$X_{eq} = X_{\ell 1} + a^2 X_{\ell 2} = \frac{\sqrt{(25.98)^2(63)^2 - (900)^2}}{(63)^2} = 0.34\ \Omega$$

$$\therefore X_{\ell 1} = X_{\ell 2}' = 0.17\ \Omega \quad \leftarrow$$

13.2.18.

DEVELOPED POWER PER PHASE $= P_d = (I_2')^2 \frac{R_2'}{S}(1-S)$

FROM THE CIRCUIT, $(I_2')^2 = V_1^2 / \left[\left(R_1 + \frac{R_2'}{S}\right)^2 + (X_{\ell 1} + X_{\ell 2}')^2\right]$

SUBSTITUTING THIS AND INSERTING NUMERICAL VALUES, ONE GETS

$$P_d = \text{CONSTANT} \times \frac{S(1-S)}{(S+1)^2 + 36S^2}$$

SETTING $dP_d/dS = 0$, A QUADRATIC EQUATION IN S IS OBTAINED,

THE SOLUTION OF WHICH YIELDS $S \simeq 0.14$ ←

13.2.19.

$$T = \frac{I_2'^2 R_2'}{s\omega_s} \qquad \text{where} \qquad I_2' \approx I_1 \qquad (1)$$

Let

$(I_2')_{SFV}$ ≡ rotor current at start if full voltage is applied

$(I_2')_S$ ≡ rotor current at start if reduced voltage is applied

k ≡ ratio of reduced voltage to full voltage

At a given slip—in particular, at $s = 1$—rotor current may be considered proportional to applied voltage. Hence,

$$\frac{(I_2')_S}{(I_2')_{SFV}} = k$$

and

$$(I_2')_S = k(I_2')_{SFV} \approx k5I_{2FL}' \qquad (2)$$

Applying (1) at reduced-voltage start and at full-load, and substituting (2), one obtains

$$\frac{T_S}{T_{FL}} \approx \left(\frac{k5I_{2FL}'}{I_{2FL}'}\right)^2 \frac{s_{FL}}{1} \qquad \text{or} \qquad \frac{1}{2} \approx (25k^2)(0.04)$$

from which $k \approx 0.707 = 70.7\%$. ←

13.2.20.

By the (approximate) current–voltage proportionality,

$$\frac{I_b}{I_{FL}} = \frac{400}{45} \qquad \text{and} \qquad \frac{I_S}{I_b} = k$$

where I_b is the full-voltage blocked-rotor current, I_{FL} is the full-load current, and I_S is the starting current. But it is given that

$$\frac{I_S}{I_{FL}} = 4$$

Therefore,

$$k = \frac{I_S/I_{FL}}{I_b/I_{FL}} = \frac{4}{400/45} = 45\%$$

Now, from (1) of the solution of Pr. 13.2.19,

$$\frac{T_S}{T_{FL}} = \left(\frac{I_S}{I_{FL}}\right)^2 s_{FL} = (4)^2(0.04) = 0.64 \qquad ←$$

13.2.21.

a) Ratio of the starting compensator $= \sqrt{693/300} = 1.52$ ←

Starting torque with the compensator $= 6250/(1.52)^2 = 2705$ N.m ←

b) With wye-starting connection, starting current per phase $= \frac{693}{3} = 231$ A ←

Since the starting torque is proportional to the square of the st. current,

$$T_{st} = \frac{6250}{3} = 2083 \text{ N.m} \leftarrow$$

13.2.22.

Stator copper loss $= 3 \times 12^2 \times 0.5 = 216$ W

Rotor copper loss $= S$ (measured output + stator copper loss + core loss)

$= 0.04\,[4000 + 216 + (220 - 70)] = 175$ W

Input = output + losses $= 4000 + 216 + 175 + 220 = 4611$ W; $\eta = \frac{4000}{4611} = 0.8675$ ←

13.2.23.

Syn. speed : $120 \times 60/16 = 450$ rpm; Slip $S = \frac{450-459}{450} = -0.02$

$R_1 + jx_{\ell 1} = 0.1 + j0.625$; $jx_m = j20$; $jx'_{\ell 2} = j0.625$; $\frac{R'_2}{S} = -\frac{0.1}{0.02} = -5$;

$$\bar{Z}_t = 0.1 + j0.625 + \frac{(j20)(-5 + j0.625)}{(-5 + j20.625)} = -4.34 + j2.325 = 4.92\angle 151.8^\circ$$

$$\bar{I}_1 = \frac{2200/\sqrt{3}}{4.92\angle 151.8^\circ} = 258.2\angle -151.8^\circ; \quad \bar{I}'_2 = \bar{I}_1 \frac{(j20)}{(-5 + j20.625)} = 243.36\angle -165.4^\circ \leftarrow$$

Output $S = P + jQ = -3\,\bar{V}_1 \bar{I}_1^* = -3 \times \frac{2200}{\sqrt{3}} \times 258.2\angle +151.8^\circ$

$= 867114 - j\,464943$

$\therefore$ P = 867 kW ; Q = − 465 kVAR ←

13.2.24.

For the same flux on 50 Hz, $V_1 = 440 \times 50/60 = 367$ V L-L ←

Slip speed $= (0.025)\,\frac{120 \times 60}{4} = 45$ rpm, which will remain to be the same.

New syn. speed on 50-Hz supply $= \frac{120 \times 50}{4} = 1500$ rpm

New slip $= 45/1500 = 0.03$ ←

13.2.25.

a) The volume of copper in the rotor is unchanged and the rotor-mmf at rated load is unchanged. With twice the number of turns, the rotor current is one-half the original value.

b) The length of the winding is doubled and the area of crosssection is reduced by one-half the original value. Therefore, the new resistance is 4 times the old value.

c) Since the number of turns is doubled, the rotor resistance referred to the stator is ($\frac{1}{4}$ × the new value of resistance) and hence is the same as before; i.e. unchanged.

Part 2:

a) The current-carrying capacity is reduced to one-half of the original value in the rotor winding. With the same number of turns, the full-load rotor current will be one-half the original value.

b) The new rotor resistance is 2 times the old value.

c) The rotor resistance ref. to the stator will be doubled with the unchanged turns ratio.

13.2.26.

a) $120 < f_r < 420$; $\because$ slip $= f_r / f_s$, $2 < S < 7$;

Speed in rpm $= (1-S)\,\frac{120 f_s}{P}$; for $S = 7$, speed in rpm is $(1-7)\,\frac{120 \times 60}{P}$, which must not exceed 3000 rpm; i.e. $\frac{6 \times 120 \times 60}{P} < 3000$

or $P > \frac{6 \times 120 \times 60}{3000} = 14.4$; So the induction machine must have 16 poles, corresponding to which the syn. rp. is 450 rpm.

b) Corresponding maximum speed $= 6 \times 450 = 2700$ rpm ←
and minimum speed $= 1 \times 450 = 450$ rpm ←

13.2.26.
(CONTD.)

c) Stator input $P_g = \frac{\text{Rotor output}}{S} = \frac{70}{7} = 10 \text{ kW}$ at 0.8 pf;

kVA rating of the induction-machine stator $= \frac{10}{0.8} = 12.5$ ←

d) $P_m = (1-S)\,P_g = (-6)(10) = -60$ kW

DC machine acts as a motor driving the induction machine against its rotating magnetic field. $hp = \frac{60}{0.746} = 80.4$ ←

13.2.27.

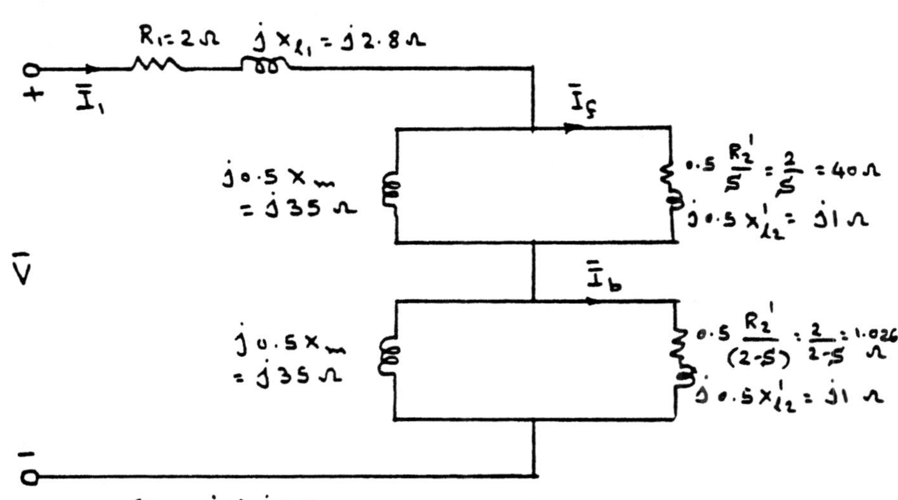

$$Z_f = R_f + jX_f = \frac{(40+j1)\,j35}{40+j36} = 16.92 + j19.77$$

$$Z_b = R_b + jX_b = \frac{(1.026+j1)\,j35}{1.026+j36} = 0.97 + j1$$

$$Z_e = Z_1 + Z_f + Z_b = (2+j2.8) + Z_f + Z_b = 19.89 + j23.57 = 30.84\angle 49.84^\circ$$

Input Current $= 110/30.84\angle 49.84^\circ = 3.57\angle -49.84^\circ = \bar{I}_1$

→ The input current is 3.57 A; pf $= \cos 49.84^\circ = 0.645$ lagging

Developed power $P_d = (I_1^2 R_f)(1-S) + (I_1^2 R_b)[1-(2-S)] = I_1^2 (R_f - R_b)(1-S)$
$= 3.57^2\,(16.92 - 0.97)(1-0.05) = 193$ W

Shaft output power $P_o = P_d - P_{rot} - P_{core} = 193 - 12 - 25 = 156\text{ W} = 0.21$ hp ←

Speed $= (1-S)$ Syn. Sp. $= 0.95 \times 120\times 60/4 = 1710$ rpm or 179 rad/s ←

Torque $= 156/179 = 0.87$ N·m ← $\eta = 156/(110 \times 3.57 \times 0.645) = 0.616$ ←

13.2.28.

$$Z_0 = \frac{110}{3.7} = 29.73; \quad I_0^2 (R_1 + 0.25R_2') = 3.7^2(2.0 + 0.25R_2') = 43$$

$\therefore \quad R_2' = 4.56\ \Omega$; Note $R_1 = 2.0\ \Omega$ (50 − 7) =

From the blocked-rotor test, $Z_{bl} = 50/5.6 = 8.93\ \Omega$

$Z_{bl} = (R_1 + jx_{\ell 1}) + (R_2' + jx_{\ell 2}') = (R_1 + R_2') + j(2x_{\ell 1})$ Eq. (7.7.5)

With $R_1 = 2$, $R_2' = 4.56$, $8.93^2 = (6.56)^2 + (2x_{\ell 1})^2$ or $x_{\ell 1} = x_{\ell 2}' = 3.03\ \Omega$

$Z_0 = (R_1 + jx_{\ell 1}) + j\,0.5x_m + (0.25R_2' + j0.5\,x_{\ell 2}')$ Eq. (7.7.4)

$29.73 = (2 + j3.03) + j\,0.5\,x_m + (0.25 \times 4.56 + j\,0.5 \times 3.03)$

$\left(\frac{x_m}{2} + 4.545\right) = \sqrt{29.73^2 - 3.14^2} = 29.56$ or $x_m = 50.04\ \Omega$

The double-revolving-field equivalent circuit can now be drawn.

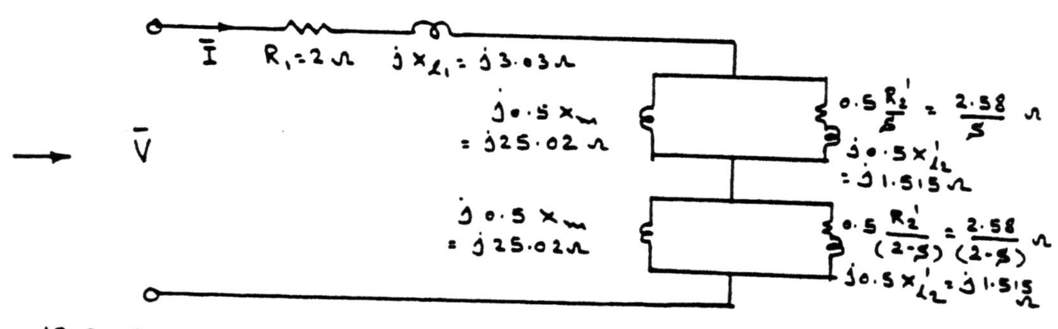

13.2.29.

SEE FIG. 13.2.13.

The impedance angle of the main winding is $\phi_m = \tan^{-1}\frac{3.8}{4.6} = 39.56°$

The impedance angle of the auxiliary winding must be

$$\phi_a = 39.56° - 90° = -50.44°$$

The reactance X_c of the cap. C must be such that

$$\tan^{-1}\frac{3.6 - X_c}{9.6} = -50.44° \text{ or } \frac{3.6 - X_c}{9.6} = -1.21 \text{ or } X_c = 15.22\ \Omega$$

The capacitance C is then given by $C = \frac{10^6}{15.22 \times 2\pi \times 60} = 174.3\,\mu f$ ←

13.3.1.

10 kVA ; 230 V ; $X_s = 1.5\,\Omega/\text{ph.}$; $R_a = 0.5\,\Omega/\text{ph.}$

a) 0.8 lagging pf ; full load ; $I_a = (10 \times 10^3)/(\sqrt{3} \times 230) = 25.1\,A$

$\bar{I}_a = 25.1 \angle -36.9°$; $\bar{V}_t = \frac{230}{\sqrt{3}} \angle 0°$; $Z_s = 0.5 + j1.5 = 1.58 \angle 72°\ \Omega/\text{ph.}$

$\bar{E}_f = \bar{V}_t + \bar{I}_a Z_s = \frac{230}{\sqrt{3}} \angle 0° + (25.1 \angle -36.9°)(0.5 + j1.5) = 166.8 \angle 7.86°\ V$

Per-unit voltage regulation $= \frac{E_f - V_t}{V_t} = \frac{166.8 - 132.8}{132.8} = 0.256$ ←

b) 0.8 leading pf; full load ; $\bar{E}_f = \frac{230}{\sqrt{3}} \angle 0° + (25.1 \angle +36.9°)(0.5 + j1.5)$

$= 119.95 + j37.52 = 125.7 \angle 17.4°$

Per-unit voltage regulation $= (125.7 - 132.8)/132.8 = -0.053$ ←

c) Let $\bar{I}_a = 25.1 \angle \phi$; $\bar{E}_f = 132.8 + (25.1 \angle \phi)(1.58 \angle 72°)$

$= 132.8 + 39.66 \cos(\phi + 72°) + j\,39.66 \sin(\phi + 72°)$

for zero voltage regulation, $E_f = V_t = 132.8$;

$\therefore \quad (132.8)^2 = [132.8 + 39.66 \cos(\phi + 72°)]^2 + [39.66 \sin(\phi + 72°)]^2$

or $\cos(\phi + 72°) = -\frac{39.66}{2(132.8)} = -0.15$ or $\phi = 26.6°$ ←

i.e., pf 0.894 leading

13.3.2.

(a) Rated kVA per phase $= \frac{1000}{3} = 333.3$ kVA

Rated voltage per phase $= \frac{2.3}{\sqrt{3}} = 1.328$ kV $= V_t$

Rated stator (or armature) Current $= I_a = \frac{333.3}{1.328} = 251$ A ←

(b)

(i) For the power factor of 0.866 lagging

$$\phi = \cos^{-1}(0.866) = 30°$$

$$\bar{E}_f = \bar{V}_t + j\,\bar{I}_a X_s = 1328\angle 0° + j\,5\,(251\angle -30°)$$

$$= 2495\angle 14.6°\ \text{V}$$

Noting that $I_f = E_f / K_{ag}$

$$I_f = \frac{2495}{200} = 12.48\ \text{A}$$

$\therefore$ $V_{ex} = I_f R_f = 12.48 \times 10 = 124.8$ V ←

(ii) For the power factor of 0.866 leading

$$\phi = \cos^{-1}(0.866) = 30°$$

$$\bar{E}_f = \bar{V}_t + j\,\bar{I}_a X_s = 1328\angle 0° + j5\,(251\angle +30°)$$

$$= 1293\angle 57.2°\ \text{V}$$

$$I_f = E_f / K_{ag} = 1293/200 = 6.465\ \text{A}$$

$$V_{ex} = I_f R_f = 6.465 \times 10 = 64.65\ \text{V}$$ ←

(c)

For part (b), the per-phase real power output $= V_t I_a \cos\phi$

$$= 1328 \times 251 \times 0.866 = 288.7\ \text{kW}$$

At unity power factor, $I_a = \frac{288.7}{1.328} = 217.4$ A

$$\phi = \cos^{-1}(1.0) = 0°$$

$$\bar{E}_f = \bar{V}_t + j\,\bar{I}_a X_s = 1.328\angle 0° + j\,5\,(217.4\angle 0°)$$

$$= 1716\angle 39.3°\ \text{V}$$

$$I_f = E_f / K_{ag} = \frac{1716}{200} = 8.58\ \text{A}$$

$$V_{ex} = I_f R_f = 8.58 \times 10 = 85.8\ \text{V}$$ ←

13.3.2. CONTD.

(d)

The complex power delivered by the generator to the system is given by $\bar{S}_{3\phi} = 3\,\bar{V}_t\,\bar{I}_a^*$

For part (b)

(i) $\bar{S}_{3\phi} = 3\,(1.328)\,(251\angle{+30^\circ})$

$= 866\text{ kW} + j\,500\text{ kVAR}$ ←

For lagging power factor, the generator delivers reactive power; in this case, 500 kVAR.

(ii) $\bar{S}_{3\phi} = 3\,(1.328)\,(251\angle{-30^\circ})$

$= 866\text{ kW} - j\,500\text{ kVAR}$ ←

For leading power factor, the generator absorbs reactive power; in this case, 500 kVAR.

For part (c), $\bar{S}_{3\phi} = 3\,(1.328)\,(217.4)$

$= 866\text{ kW} + j\,0\text{ kVAR}$ ←

13.3.3.

Output at rated voltage and 0.8 pf lagging $= 10\times10^6 \times 0.8 = 8\times10^6$ W

Losses = Armature copper loss + Field copper loss + Core loss + F&W loss + stray load loss

$= 50 + \frac{(342)^2\,0.3}{1000} + 70 + 80 + 20 = 255.1\text{ kW}.$

Note that 342 A is the field current calculated in Prob. 8-2.

$\eta = \frac{8\times10^6}{(8\times10^6)+(255.1\times10^3)} = \frac{8}{8.2551} = 0.969$ ←

13.3.4.

$$\bar{V}_t = \frac{550}{\sqrt{3}} \angle 0° \text{ V} = 317.54 \angle 0° \text{ V}$$

$$I_a = \frac{50 \times 10^3}{\sqrt{3}\,(550)} = 52.49 \text{A} \leftarrow$$

$$\bar{I}_a = 52.49 \angle -\cos^{-1} 0.95 = 52.49 \angle -18.19° \text{ A}$$

$$\bar{E}_f = \bar{V}_t + j\,\bar{I}_a\,X_s = 317.54 + j2(52.49 \angle -18.19°)$$

$$= 364.2 \angle 15.89°$$

$$E_f = 364.2 \text{V} \leftarrow$$

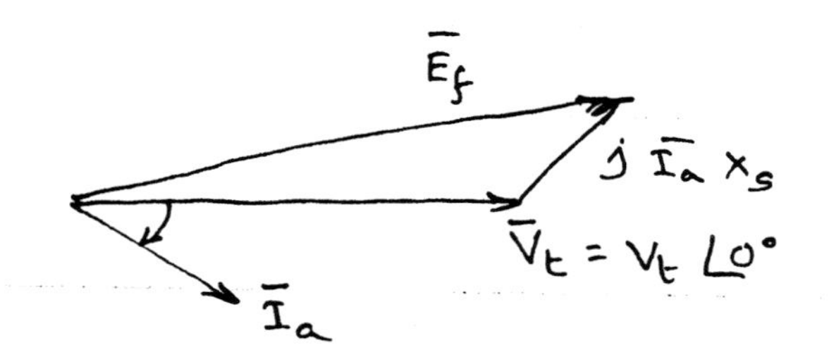

$$\text{Regulation} = \frac{E_f - V_t}{V_t} = \frac{364.2 - 317.54}{317.54} = 0.147 \leftarrow$$

13.3.5.
(a)

Rated kVA per phase = $\frac{1000}{3}$ = 333.3 kVA

Rated voltage per phase = $\frac{2.3}{\sqrt{3}}$ = 1.328 kV = V_t

Rated armature current = I_a = $\frac{333.3}{1.328}$ = 251 A

(i)
For the power factor 0.866 lagging,

$\phi = \cos^{-1}(0.866) = 30°$

$\bar{E}_f = \bar{V}_t - j\bar{I}_a X_s = 1328\angle 0° - j5(251\angle -30°)$

$= 1293\angle -57.2°$ V

Noting that $I_f = E_f / K_{ag}$

$= 1293/200 = 6.465$ A

$\therefore V_{ex} = I_f R_f = 6.465 \times 10 = 64.65$ V ←

(ii) For the power factor 0.866 leading,

$\phi = \cos^{-1}(0.866) = 30°$

$\bar{E}_f = \bar{V}_t - j\bar{I}_a X_s = 1328\angle 0° - j5(251\angle +30°)$

$= 2237\angle -29.1°$ V

$I_f = E_f / K_{ag} = 2237/200 = 11.19$ A

$V_{ex} = I_f R_f = 11.19 \times 10 = 111.9$ V ←

(b)

For part (a), the per-phase real power input = $V_t I_a \cos\phi$

$= 1328 \times 251 \times 0.866 = 288.7$ kW

At unity power factor, $I_a = \frac{288.7}{1.328} = 217.4$ A

$\phi = \cos^{-1}(1.0) = 0°$

$\bar{E}_f = \bar{V}_t - j\bar{I}_a X_s = 1328\angle 0° - j(5)(217.4\angle 0°)$

$= 1716\angle -39.3°$ V

$I_f = E_f/K_{ag} = 1716/200 = 8.58$ A

$V_{ex} = I_f R_f = 8.58 \times 10 = 85.8$ V ←

13.3.5. CONTD.

(c)

The complex power absorbed by the motor is given by

$$\bar{S}_{3\phi} = 3 \bar{V}_t \bar{I}_a^*$$

For part (a)

(i) $\bar{S}_{3\phi} = 3(1.328)(251 \angle +30°)$

$= 866 \text{ kW} + j\,500 \text{ kVAR}$ ←

For lagging power factor, the motor absorbs reactive power, in this case 500 kVAR.

(ii) $\bar{S}_{3\phi} = 3(1.328)(251 \angle -30°)$

$= 866 \text{ kW} - j\,500 \text{ kVAR}$ ←

For leading power factor, the motor delivers reactive power; in this case 500 kVAR.

FOR PART (b), $\bar{S}_{3\phi} = 3(1.328)(217.4)$

$= 866 \text{ kW} + j\,0 \text{ kVAR}$ ←

13.3.6.

In per unit the short-circuit load loss is

$$\frac{1.80}{45} = 0.040$$

at $I_a = 1.00$ per unit. Therefore,

$$R_{a\,eff} = \frac{0.040}{(1.00)^2} = 0.040 \text{ per unit}$$ ←

On a per phase basis the short-circuit load loss is

$$\frac{1800}{3} \text{ W/phase}$$

and consequently the effective resistance is

$$R_{a\,eff} = \frac{1800}{3(118)^2} = 0.043\ \Omega/\text{phase}$$ ←

The ratio of ac-to-dc resistance is

$$\frac{R_{a\,eff}}{R_{a\,dc}} = \frac{0.043}{0.0335} = 1.28$$ ←

NOTE: Because this is a small machine, its per unit resistance is relatively high. The armature resistance of machines with ratings above a few hundred kilovoltamperes usually is less than 0.01 per unit.

13.3.7.

a) $P_{ph} = \frac{5000 \times 746}{3} \times 10^{-3} = 1243\text{ kW}$; $V_t = \frac{4000}{\sqrt{3}} = 2310\text{ V/ph.}$; $I_a = \frac{1243 \times 10^3}{2310} = 538\text{ A}$

$\bar{E}_f = 2310\angle 0° + (538)(4)\angle -90° = 3157\angle -43°$

Check: $P = \frac{V_t E_f \sin\delta}{X_s} = \frac{2310 \times 3157 \times 0.682}{4} = 1243\text{ kW/ph.}$

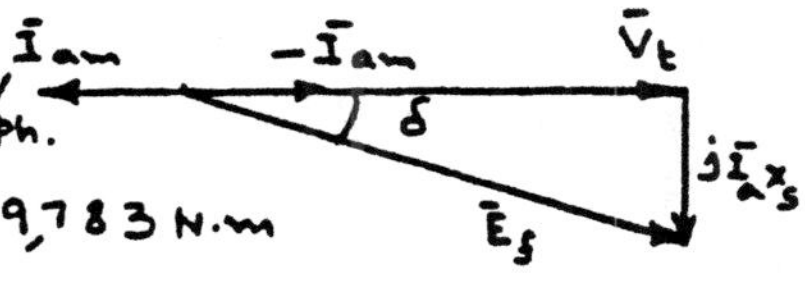

Rated torque per phase $= P/\omega = \frac{1243 \times 10^3}{2\pi\left(\frac{120 \times 60}{12} \times \frac{1}{60}\right)} = 19{,}783\text{ N.m}$

The machine torque is $3 \times 19783 = 59.35 \times 10^3$ N.m ←

For maximum torque, $\delta = 90°$;

$T_{max} = 3 \times \frac{2310 \times 3157}{4} \times \frac{1}{20\pi} = 87.05 \times 10^3$ N.m ←

13.3.7.
(CONTD.)

b) $I_{am} X_s = \sqrt{2310^2 + 3157^2} = 3912$

or $I_{am} = 3912/4 = 978\,A$ ←

$\phi = -\tan^{-1}\dfrac{2310}{3157} = -36.2°$

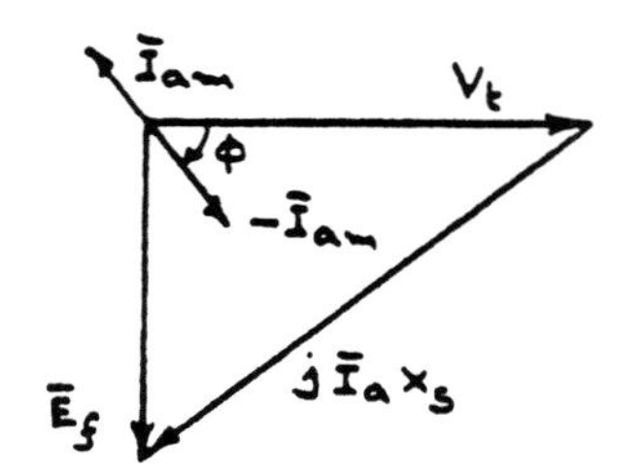

13.3.8

a) phase voltage $V_t = \dfrac{400}{\sqrt{3}} = 231V$; motor input per phase: $\dfrac{15 \times 746}{3} = 3730W$

Armature current $I_a = 3730/(231)(0.8) = 20.18A$

$I_a X_s = 20.18 \times 3 = 60.54\,V$; $\phi = \cos^{-1} 0.8 = 36.87°$

$E_f^2 = (231)^2 + (60.54)^2 - 2(231)(60.54)\cos 126.87°$

$E_f = 271.7V$; $|\sin\delta| = \dfrac{60.54}{271.7}\sin 126.87° = 0.178$

$\delta = -10.25°$ ← ; corresponding to $E_f = 271.7V$, $I_f = 9.1A$ ←

b) minimum $I_a = \dfrac{3730}{231} = 16.15A$ ←, corresponding to $\cos\phi = 1$ or $\phi = 0°$.

$E_f^2 = 231^2 + (3 \times 16.15)^2$ or $E_f = 236V$; Corresp. $I_f = 6.1A$ ←

c) With $I_f = 10A$, $E_f = 280V$; $280^2 = 231^2 + (3\times 25)^2 - 2(231)(75)\cos(90+\phi)$

$\cos\phi = 0.83$ leading ← ; Developed power = Input power $= \sqrt{3} \times 400 \times 25 \times 0.83 = 14,350W$ ←

d) $(231)^2 = 231^2 + (3\times 20)^2 + 2(231)(60)\sin\phi$; $\cos\phi = 0.99$ lagging

dev. power $= \sqrt{3} \times 400 \times 20 \times 0.99 = 13740W$; $\omega = \dfrac{120\times 60}{4} \times \dfrac{1}{60} \times 2\pi = 60\pi$ rad/s

∴ developed torque $= 13740/(60\pi) = 72.9\,N\cdot m$ ←

13.3.9.

SEE FIG. 13.3.4 (c)

a) $\bar{E}_f = \frac{2300}{\sqrt{3}} \angle 0° - j\,350 \angle \cos^{-1} 0.8 \,(2) = 1835.5 \angle -17.76°$

L-L excitation voltage $= \sqrt{3}\,(1835.5) = 3179$ V ; power angle $= -17.76°$ ←

b) $E_f = V_t$; $\delta = -20°$; $\bar{E}_f = \frac{2300}{\sqrt{3}} \angle 0° = 1328 \angle 0°$; $P = \frac{V_t E_f \sin\delta}{X_s} = V_t I_a \cos\phi$

Note that P in this case is the power absorbed by the machine.

$I_a \cos\phi = \frac{E_f \sin\delta}{X_s} = \frac{1328\,(-0.342)}{2} = -227$;

$Q = \frac{V_t E_f \cos\delta - V_t^2}{X_s} = \frac{1328^2\,(0.94-1)}{2} = -52{,}907$

See Fig. 13.3.4 (d); ϕ is the lagging pf angle ; $-52907 = -V_t I_a \sin\phi$

or $I_a \sin\phi = \frac{52907}{1328} = 39.84$ or $I_a = \sqrt{227^2 + 39.84^2} = 230.5$ A ←

$\phi = \tan^{-1}(39.84/227) = 9.95°$ or $\cos\phi = 0.985$ lagging ← which is the same as (227/230.5)

CHECK: $\bar{E}_f = 1328 \angle 0° - j\,230.5 \angle -9.95° \,(2)$ or $E_f = 1328$ V.

13.3.10.

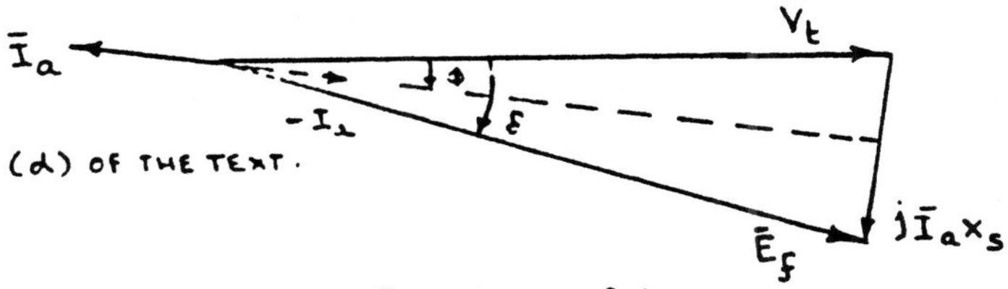

SEE FIG. 13.3.4 (d) OF THE TEXT.

(a) FROM THE PHASOR DIAGRAM, $I_a = \frac{2V}{X_s} \left| \sin\frac{\delta}{2} \right|$

WHERE $E_f = V_t = V = 2300/\sqrt{3} = 1328$ V ; $X_s = 3\,\Omega$; $\delta = -15°$

$\therefore\ I_a = \frac{2\,(1328)}{3} \sin 7.5° = 115.6$ A ←

(b) NEGLECTING R_a, $|P_d| = V_t I_a \cos\phi = \frac{E_f V_t}{X_s} |\sin\delta|$

OR $\cos\phi = \frac{1328}{(115.6)(3)} \sin 15° = 0.991$ LAGGING ←

13.3.11.

Induction motor: real power = 350 kW which is power absorbed.

reactive power = $\frac{350}{0.8}(0.6) = 262.5$ kVAR which is the capacitive vars received or inductive vars delivered.

Syn. Motor: real power = 150 kW which is the power absorbed.

Let the reactive power be Q kVAR representing the capacitive vars delivered or inductive vars recd.

Then for the overall system:

real power: 350 + 150 = 500 kW which is the power received by the system.

reactive power = $\frac{500}{0.95}\left[\sin(\cos^{-1}0.95)\right] = \frac{500}{0.95}(0.3122) = 164.34$ kVAR,

which is the capacitive vars received, which must be equal to

$(262.5 - Q)$; or $Q = 262.5 - 164.34 = 98.16$ kVAR

$\therefore$ kVA rating of the synchronous motor = $\sqrt{150^2 + 98.16^2} = 179.26$ ←

13.3.12.

$\cos^{-1}0.6 = 53.13°$; $\cos^{-1}0.9 = 25.8°$; from the diagram drawn below

kVA rating of a synchronous capacitor = 424.67 kVAR ←

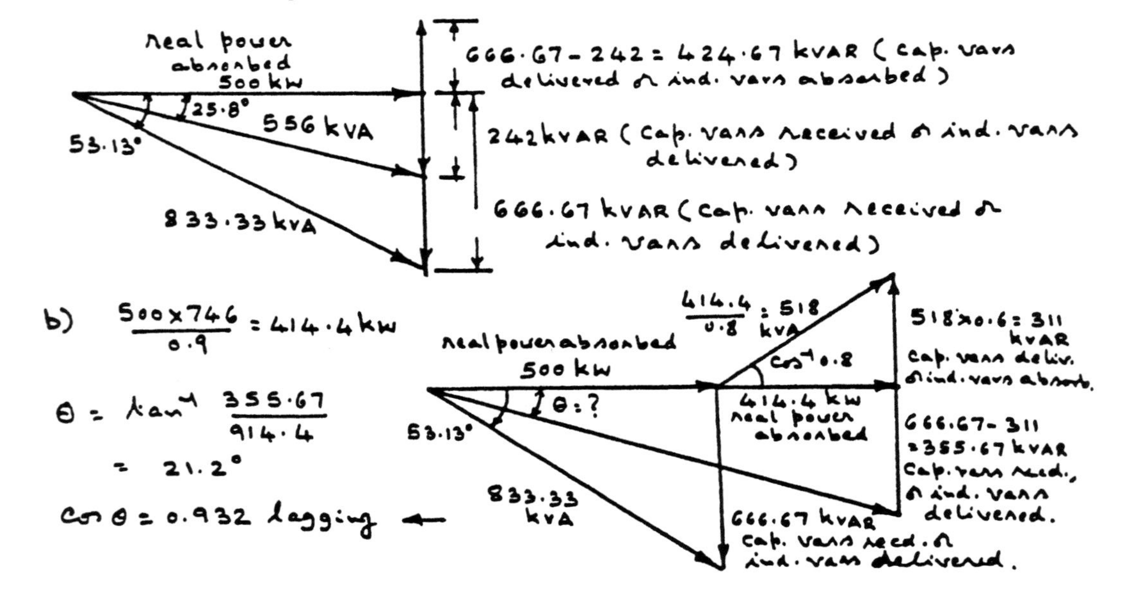

b) $\frac{500 \times 746}{0.9} = 414.4$ kW

$\theta = \tan^{-1}\frac{355.67}{914.4}$

$= 21.2°$

$\cos\theta = 0.932$ lagging ←

13.3.13.

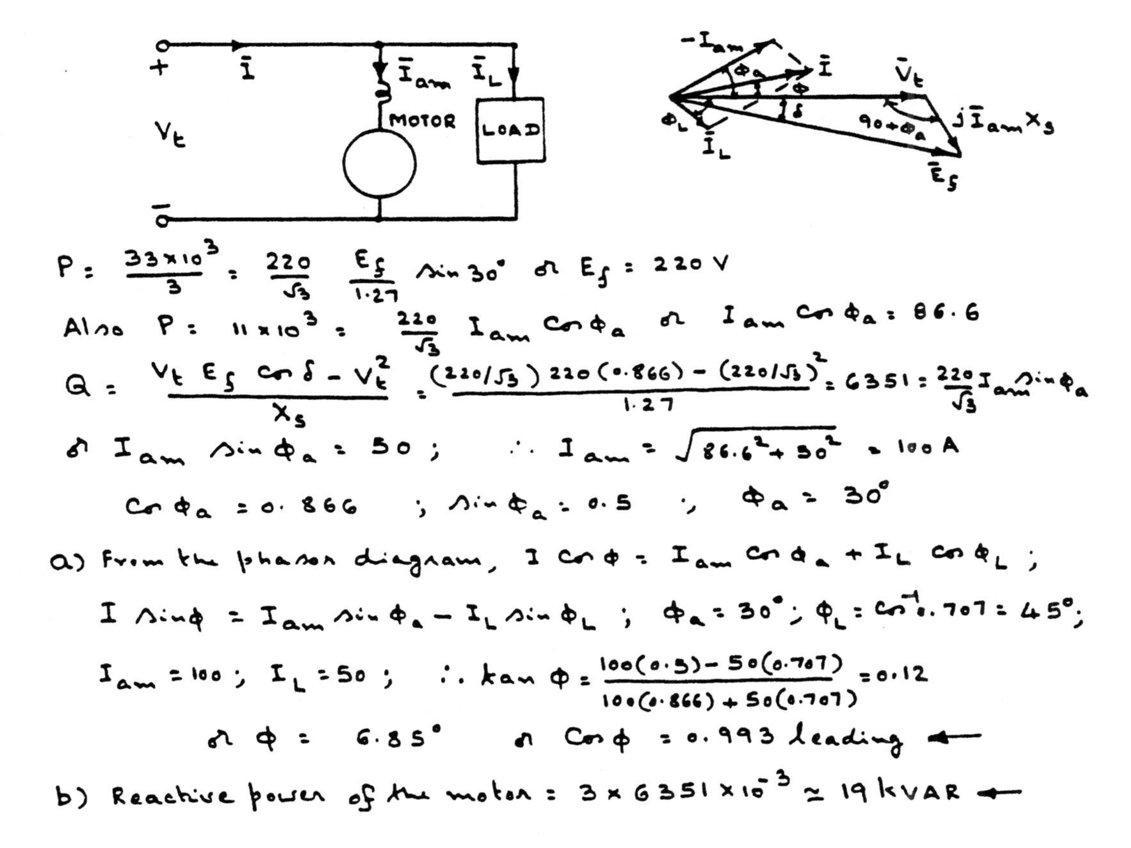

$P = \frac{33\times10^3}{3} = \frac{220}{\sqrt{3}}\frac{E_f}{1.27}\sin 30°$ or $E_f = 220$ V

Also $P = 11\times10^3 = \frac{220}{\sqrt{3}} I_{am}\cos\phi_a$ or $I_{am}\cos\phi_a = 86.6$

$Q = \frac{V_t E_f \cos\delta - V_t^2}{X_s} = \frac{(220/\sqrt{3})\,220\,(0.866) - (220/\sqrt{3})^2}{1.27} = 6351 = \frac{220}{\sqrt{3}} I_{am}\sin\phi_a$

or $I_{am}\sin\phi_a = 50$; $\therefore I_{am} = \sqrt{86.6^2 + 50^2} = 100$ A

$\cos\phi_a = 0.866$; $\sin\phi_a = 0.5$; $\phi_a = 30°$

a) From the phasor diagram, $I\cos\phi = I_{am}\cos\phi_a + I_L\cos\phi_L$;

$I\sin\phi = I_{am}\sin\phi_a - I_L\sin\phi_L$; $\phi_a = 30°$; $\phi_L = \cos^{-1}0.707 = 45°$;

$I_{am} = 100$; $I_L = 50$; $\therefore \tan\phi = \frac{100(0.5) - 50(0.707)}{100(0.866) + 50(0.707)} = 0.12$

or $\phi = 6.85°$ or $\cos\phi = 0.993$ leading ←

b) Reactive power of the motor $= 3\times6351\times10^{-3} \simeq 19$ kVAR ←

13.3.14.

a) $I_1 \cos\phi_1 = I_2 \cos\phi_2 = \frac{1}{2} I \cos\phi$

$\bar{V}_t$, ϕ_2, ϕ, $\bar{I}_2$, ϕ_1, $\bar{I}_1$, $\bar{I}$

$$I = \frac{12 \times 10^6}{\sqrt{3} \times 33 \times 10^3 \times 0.8} = 262.44\,A \;;\; I_1 = 125\,A \text{ (given)}$$

$I_1 \cos\phi_1 = I_2 \cos\phi_2 = \frac{1}{2}(262.44)(0.8) = 105$

$I_1 |\sin\phi_1| = \sqrt{125^2 - 105^2} = 67.8 \;;\; \tan\phi_1 = \frac{67.8}{105} = 0.6457$ or $\phi_1 = 32.85°$

or $pf = \cos\phi_1 = 0.84$ lagging ←

total reactive current $= I|\sin\phi| = 262.44(0.6) = 157.5\,A$

$I_2 |\sin\phi_2| = 157.5 - 67.8 = 89.7 \;;\; \therefore I_2 = \sqrt{105^2 + 89.7^2} = 138.1\,A$ ←

$\tan\phi_2 = \frac{89.7}{105} = 0.854$ or $\phi_2 = 40.5°$; $pf = \cos\phi_2 = 0.76$ lagging ←

$\bar{E}_{f1} = \bar{V}_t + j\bar{I}_{a1} X_s = \frac{33 \times 10^3}{\sqrt{3}} \angle 0° + j125 \angle -32.85° (8) = 19613.4 \angle 2.45°$

Line-to-line excitation voltage $= 33.97\,kV$; $\delta_1 = 2.45°$ ←

$\bar{E}_{f2} = \frac{33 \times 10^3}{\sqrt{3}} \angle 0° + j138.1 \angle -33° (8) = 19677 \angle 2.7°$

Line-to-line excitation voltage $= 34.08\,kV$; $\delta_2 = 2.7°$ ←

b) Load: real power $= 12 \times 10^6\,W$; apparent power $= \frac{12 \times 10^6}{0.8} = 15 \times 10^6\,VA$;
reactive power $= 15 \times 10^6 \times 0.6 = 9 \times 10^6$ VAR.

Machine 1: real power $= 6 \times 10^6\,W$; $\phi_1 = \cos^{-1} 0.9 = 25.8°$ lagging
reactive power $= 6 \times 10^6 \times \tan\phi_1 = 2.901$ MVAR

Machine 2: Real power $= 6 \times 10^6\,W$; Reactive power $= 9 - 2.901 \simeq 6.1$ MVAR
$\tan\phi_2 = 6.1/6 = 1.0166$ or $\phi_2 = 45.5°$; $pf = \cos\phi_2 = 0.701$ lagging

$$I_2 = \frac{6 \times 10^6}{\sqrt{3} \times 33 \times 10^3 \times 0.701} = 149.75\,A$$ ←

13.3.15.

FIGURE
(*a*) Circuit diagram.
(*b*) Phasor diagram.

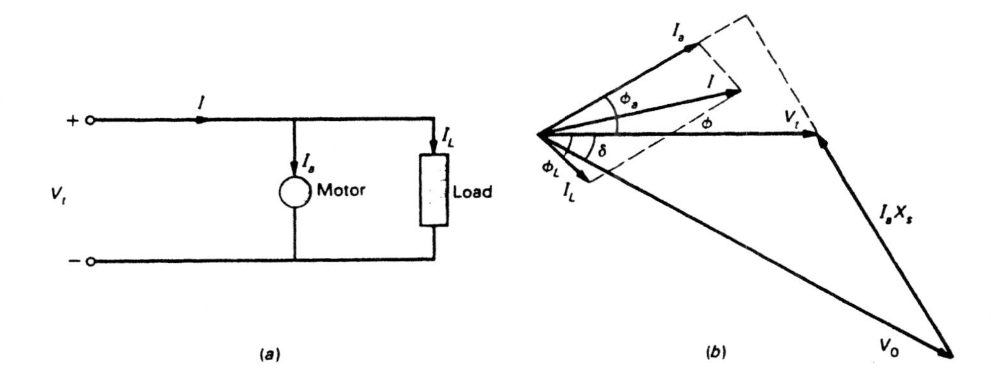

Solution The per-phase circuit and the phasor diagram on a per-phase basis are shown in Fig. above. From Eq. we have $P_d = \frac{V_0 V_t}{X_s} \sin\delta$

$$P_d = \frac{1}{3} \times 33{,}000$$

$$= \frac{220}{\sqrt{3}} \frac{V_0}{1.27} \sin 30^\circ$$

which yields $V_0 = 220$ V. From the phasor diagram, $I_aX_s = 127$ or $I_a = 127/1.27 = 100$ A and $\phi_a = 30^\circ$. The reactive kilovoltamperes (kVAr) of the motor $= \sqrt{3} V_t I_a \sin\phi_a = \sqrt{3} \times (220/1000) \times 100 \times \sin 30^\circ = 19$ kVAr.

The overall power-factor angle is given by

$$\tan\phi = \frac{I_a \sin\phi_a - I_L \sin\phi_L}{I_a \cos\phi_a + I_L \cos\phi_L} = 0.122$$

or $\phi = 7^\circ$ and $\cos\phi = 0.992$ leading.

13.3.16.

Solution Let ϕ be the power-factor angle. Then

$$\mathbf{I}_a\mathbf{Z}_s = 218.7 \times 10.4\underline{/\phi + 89.6}$$
$$= 2274.48\underline{/\phi + 89.6} \quad \text{V}$$

For voltage regulation to be zero, $|V_0| = |V_t|$. Hence

$$|3810| = |3810 + 2274.48[\cos(\phi + 89.6) + j\sin(\phi + 89.6)]$$
$$3810^2 = [3810 + 2274.48\cos(\phi + 89.6)]^2 + [2274.48\sin(\phi + 89.6)]^2$$

from which

$$\phi = 17.76^\circ \quad \text{and} \quad \cos\phi = 0.95 \text{ leading}$$

13.4.1.

$V_t = E_a - I_a R_a$ (9.2.2); $I_f = 250/60 = 4.17$ A

a) $I_{L\,FL} = 100 \times 10^3 / 250 = 400$A; $I_a = I_L + I_f = 404.17$ A

$E_a = V_t + I_a R_a = 250 + (404.17)(0.05) = 270.2$ V ←

b) $I_{L\,\frac{1}{2}FL} = 200$A; $I_a = 204.17$A; $E_a = 250 + (204.17)(0.05) = 260.2$V ←

13.4.2.

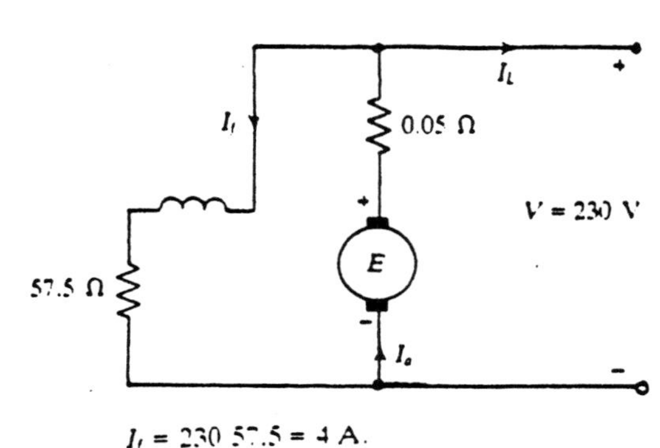

$I_f = 230/57.5 = 4$ A.

(a)
$$I_L = \frac{100 \times 10^3}{230} = 434.8 \text{ A}$$
$$I_a = I_L + I_f = 434.8 + 4 = 438.8 \text{ A}$$
$$I_a R_a = (438.8)(0.05) = 22 \text{ V}$$
$$E = V + I_a R_a = 230 + 22 = 252 \text{ V} \leftarrow$$

(b)
$$I_L = 217.4 \text{ A}$$
$$I_a = 217.4 + 4 = 221.4 \text{ A}$$
$$I_a R_a = 11 \text{ V}$$
$$E = 230 + 11 = 241 \text{ V} \leftarrow$$

13.4.3.

$R_a = 0.5\Omega$; $R_f = 200\Omega$; At no load, $I_a R_a = 3 \times 0.5 = 1.5$ V

$E_{a_{NL}} = V_t - I_a R_a = 250 - 1.5 = 248.5$ V; $E_{a_{NL}}/RPM = \frac{248.5}{1200} = 0.207$

$I_f = 250/200 = 1.25$A; At full load, $I_a = 40 - 1.25 = 38.75$A;

$E_{a_{FL}} = 250 - (38.75)(0.5) = 230.6$V; Flux is 0.95 of the no-load value.

$\therefore \frac{E_a}{RPM} = (0.95)(0.207) = 0.197$ and FL speed $= \frac{230.6}{0.197} = 1170.6$ rpm ←

13.4.4.

$$V_t = 230\,V;\ I_{L_{FL}} = \frac{10 \times 10^3}{230} = 43.5\,A;\ I_f = 0.04 \times 43.5 = 1.74\,A$$

$$I_{a_{FL}} = I_L + I_f = 43.5 + 1.74 = 45.24\,A;\ I_a R_a = 0.06 \times 230 = 13.8\,V;$$

$$R_a = \frac{13.8}{45.24} = 0.305\,\Omega;\ R_f I_f = 230 \text{ or } R_f = \frac{230}{1.74} = 132.2\,\Omega \leftarrow$$

13.4.5.

$$I_f = 250/200 = 1.25\,A;\ I_{a_{NL}} = 4.5 - 1.25 = 3.25\,A$$

$$E_{a_{NL}} = 250 - (3.25)(0.25) = 249.2\,V;\ E_{a_{FL}} = 250 - (65 - 1.25)(0.25) = 234\,V$$

$$\because \omega_m \propto E_a/\Phi;\ RPM_{FL} = 1200\left(\frac{234}{249.2}\right)\left(\frac{1}{0.94}\right) = 1199 \leftarrow$$

13.4.6.

Since voltage drops across R_f and R_a are neglected, using subscripts 1 and 2 for the cases without the diverter and with diverter, respectively,

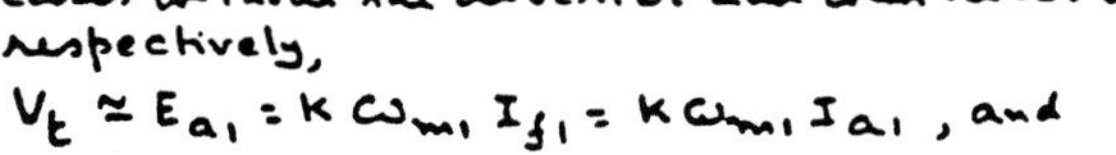

$V_t \simeq E_{a1} = K\omega_{m1} I_{f1} = K\omega_{m1} I_{a1}$, and

$V_t \simeq E_{a2} = K\omega_{m2} I_{f2} = K\omega_{m2} I_{a2}/2$;

Also $T_{e1} = K I_{f1} I_{a1} = K I_{a1}^2 = T_{L1} = A\omega_{m1}^2$; $T_{e2} = K I_{f2} I_{a2} = K \frac{I_{a2}^2}{2} = T_{L2} = A\omega_{m2}^2$ where A is a constant.

$$\therefore \omega_{m2}/\omega_{m1} = \sqrt{\frac{1}{2}\frac{I_{a2}^2}{I_{a1}^2}} = \frac{1}{\sqrt{2}}\frac{I_{a2}}{I_{a1}} = \frac{2 I_{a1}}{I_{a2}} \Rightarrow I_{a2} = 2^{3/4} I_{a1} = 2^{3/4} \times 20 = 33.63\,A$$

$$\omega_{m2} = 2\left(\frac{20}{33.63}\right)\omega_{m1} = 2^{1/4}\omega_{m1} \text{ or } RPM_2 = 2^{1/4}\cdot(500) = 594.6 \leftarrow$$

13.4.7.

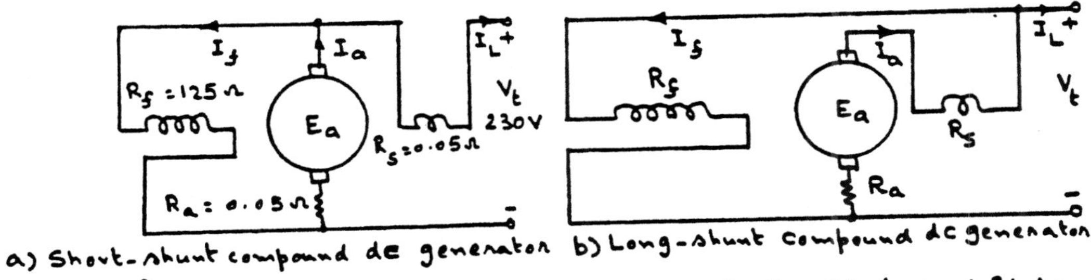

a) Short-shunt compound dc generator b) Long-shunt compound dc generator

a)

$$I_L = \frac{50 \times 10^3}{230} = 217.4\,A;\ I_L R_s = 11.5\,V;$$

$$V_f = 230 + 11.5 = 241.5\,V;\ I_f = \frac{241.5}{125} = 1.93\,A;$$

$$I_a = I_L + I_f = 219.33\,A;\ I_a R_a = 11\,V;$$

$$E_a = 230 + 11.5 + 11 + 2 = 254.5\,V \leftarrow$$

b)

$$I_L = 217.4\,A;\ I_f = 230/125 = 1.84\,A;$$

$$I_a = I_L + I_f = 219.24\,A$$

$$I_a(R_a + R_s) = 21.92\,V$$

$$E_a = 230 + 21.9 + 2 = 253.9\,V \leftarrow$$

13.4.8.

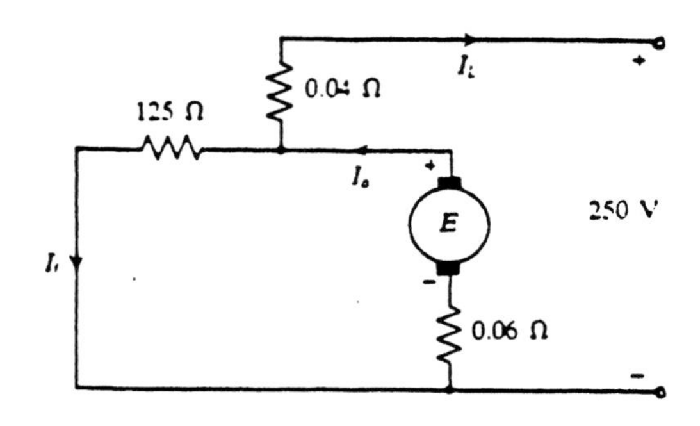

$$I_L = \frac{50 \times 10^3}{250} = 200 \text{ A}$$

$$I_L R_{se} = (200)(0.04) = 8 \text{ V}$$

$$V_f = 250 + 8 = 258 \text{ V}$$

$$I_f = \frac{258}{125} = 2.06 \text{ A}$$

$$I_a = 200 + 2.06 = 202.06 \text{ A}$$

$$I_a R_a = (202.06)(0.06) = 12.12 \text{ V}$$

$$E = 250 + 12.12 + 8 + 2 = 272.12 \text{ V} \leftarrow$$

13.4.9.

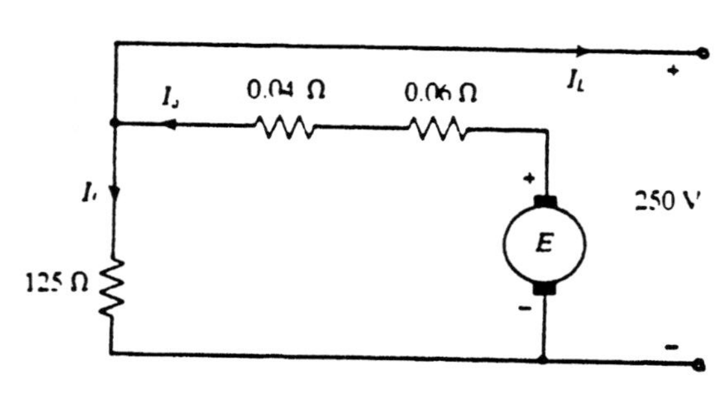

$$I_L = 200 \text{ A}$$

$$I_f = \frac{250}{125} = 2 \text{ A}$$

$$I_a = 200 + 2 = 202 \text{ A}$$

$$I_a(R_a + R_s) = 202(0.06 + 0.04) = 20.2 \text{ V}$$

$$E = 250 + 20.2 + 2 = 272.2 \text{ V} \leftarrow$$

13.4.10.

As a generator: $I_f = 230/230 = 1A$; $I_L = \dfrac{10 \times 10000}{230} = 43.5A$.

$I_a = 43.5 + 1 = 44.5A$; $I_a R_a = (44.5)(0.1) = 4.45V$;

$$E_a = 230 + 4.45 = 234.45V$$

As a motor: $I_L = 43.5A$; $I_f = 1A$; $I_a = 43.5 - 1 = 42.5A$; $I_a R_a = 4.25V$

$$E_a = 230 - 4.25 = 225.75V$$

Now $N_m / N_g = E_{am} / E_{ag}$ or $N_m = \dfrac{225.75}{234.45} \times 1000 = 963$ rpm ←

13.4.11.

$E_a = K_a \Phi \omega_m$ or $E_{a1}/E_{a2} = (\Phi_1 \omega_{m1})/(\Phi_2 \omega_{m2}) = (\Phi_1 N_1)/(\Phi_2 N_2)$

a) When $I_a = 25A$, $E_{a1} = 236V$ and $N_1 = 1000$ rpm

$E_{a2} = V_t - I_a R_a = 200 - (25)(0.25 + 0.25) = 187.5V$

$\therefore 236/187.5 = 1000/N_2$ or speed $N_2 = \frac{187.5}{236} \times 1000 = 794.5$ rpm

b) When $T_e = 36$ N.m, since $T_e = E_a I_a / \omega_m$,

at 1000 rpm, $E_a I_a = 36\left(\frac{2\pi}{60} \times 1000\right) = 3770$

Plot the rectangular hyperbola $E_a I_a = 3770$ on the OCC;

At the point of intersection with the OCC, $I_a = 17.5A$ & $E_{a1} = 215.4V$ at $N_1 = 1000$ rpm

$E_{a2} = 200 - (17.5)(0.5) = 191.25V$

Speed $N_2 = \frac{191.25}{215.4} \times 1000 = 888$ rpm ←

13.4.12.

At 100A and 250V, $E_a = 250 - (100)(0.15 + 0.1) = 225V$

$E_a/RPM = 225/750 = 0.3$ V/rpm

At 25A and 250V, $E_a' = 250 - (25)(0.15 + 0.1) = 243.75V$

Since flux is 40% of that corresponding to 100A,

$E_a'/RPM = (0.4)(0.3) = 0.12$ V/rpm

or $RPM = \frac{E_a'}{0.12} = \frac{243.75}{0.12} = 2{,}031$ rpm ←

13.4.13.

While operating at rated load and rated speed, the starting resistance in series with the armature circuit is cut out; the counter emf is $E_a = V_t - I_a R_a$ and $I_a = I_L - I_f$, where

$I_L = 26.0A$ and $I_f = 250/350 = 0.71A$;

$I_a = 26.0 - 0.71 \cong 25.3A$; Hence $E_a = 250 - (0.48 \times 25.3) = 237.8V$ and the speed is then the rated speed of 1,800 rpm.

13.4.13.

(CONTD.)

FIRST STEP: With the entire resistance of the starting box in series with the armature circuit,

$$R_{T1} = 2.24 + 1.47 + 0.95 + 0.62 + 0.4 + 0.26 + 0.48 = 6.42\,\Omega$$

Since the counter emf is zero at starting, the armature starting current is $I_{st} = V/R_{T1} = 250/6.42 = 38.9\,A$. By the time the armature current drops to its rated value of 25.3 A, the counter emf is $E_a = 250 - (6.42 \times 25.3) = 88\,V$, and the speed is then $N = (88/237.8)\,1800 = 667\,rpm$

SECOND STEP: With the 2.24-ohm step suddenly cut out, the armature-circuit resistance now becomes $6.42 - 2.24 = 4.18\,\Omega$; with the motor speed still at 667 rpm, the counter emf is still 88 V if the effect of armature reaction is neglected. Therefore the resistance drop in the armature circuit is still 162 V; i.e. $I_a R_{T2} = 162\,V$. Neglecting the armature inductance, the initial current is $I_a = 162/R_{T2} = 162/4.18 = 38.7\,A$; with the final current at 25.3 A, the counter emf is

$$E_a = 250 - (25.3)(4.18) = 144.3\,V$$

and the speed is then $N = \dfrac{144.3}{237.8} \times 1800 = 1093\,rpm$

Following the same procedure for the remaining steps, the following values are obtained:

13.4.13. CONTD.

Step No.	Current A Initial	Current A Final	Speed rpm Initial	Speed rpm Final
3	39.0	25.3	1093	1370
4	39.0	25.3	1370	1557
5	39.0	25.3	1557	1675
6	39.0	25.3	1675	1755
7	39.0	25.3	1755	1800

13.4.14

$$V_t = E_a - I_a R_a \; ; \; I_a = I_L + I_f = I_L + \frac{V_t}{R_f} = I_L + \frac{V_t}{100}$$

$$E_{a1} = V_t + \left(I_{L1} + \frac{V_t}{100}\right)(0.05) = 200 = 1.0005\,V_t + 0.05\,I_{L1} \quad \text{— ①}$$

$$E_{a2} = V_t + \left(I_{L2} + \frac{V_t}{100}\right)(0.05) = 210 = 1.0005\,V_t + 0.05\,I_{L2} \quad \text{— ②}$$

$I_{L1} + I_{L2} = 3000$ — ③ ; Solving $I_{L1} = 1400\,A$; $I_{L2} = 1600\,A$; $V_t = 130\,V$ ←

13.4.15.

a) $I_{L1} + I_{L2} + I_{L3} = 4350\,A$; $V_{tG1} = 492.5 - I_{L1}\dfrac{492.5 - 482.5}{2000}$

or $V_t = V_{tG1} = 492.5 - \frac{1}{200} I_{L1}$; $V_t = V_{tG2} = 510 - \frac{40}{2000} I_{L2}$;

$V_t = V_{tG3} = 525 - \frac{50}{2000} I_{L3}$; Solving $I_{L1} = 1500\,A$; $I_{L2} = 1250\,A$; $V_t = 485\,V$ ←
$I_{L3} = 1600\,A$

b) $I_{L1} + I_{L2} + I_{L3} = 0$; The three terminal voltage equations are the same as in case (a) ; Solving $V_t = 500\,V$;
$I_{L1} = -1500\,A$; $I_{L2} = 500\,A$; $I_{L3} = 1000\,A$ ←

13.4.16.

a) Let the terminal voltage be V_t ; $I_{L1} + I_{L2} = V_t / 6$
Plot the external characteristics.
It can be seen that
$V_t = 240\,V$; $I_{L1} = 15\,A$; $I_{L2} = 25\,A$

b) $I_{L1} + I_{L2} = (V_t - 200)/1.5$
From the plot of ext. characteristics it can be seen that
$V_t = 248$; $I_{L1} = 12\,A$; $I_{L2} = 22\,A$

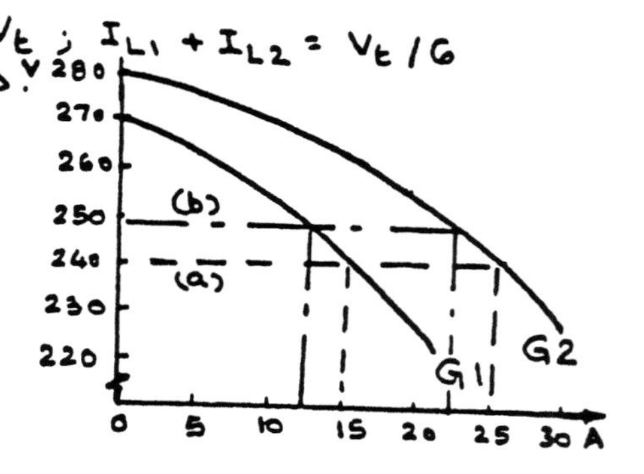

13.4.17.

Output $= V_t I_a$; Input $= V_t I_a + I_a^2 R_a + P_c$

$$\eta = (V_t I_a) / (V_t I_a + I_a^2 R_a + P_c)$$

For η to be a maximum, $d\eta / d I_a = 0$;

i.e., $V_t \, (V_t I_a + I_a^2 R_a + P_c) - V_t I_a \, (V_t + 2 I_a R_a) = 0$

$$\text{or} \quad I_a = \sqrt{P_c / R_a} \quad \leftarrow$$

i.e., η is a maximum when the armature copper loss $I_a^2 R_a$ equals the constant loss P_c.

13.4.18.

$I_f = 4$ A and $I_a = 438.8$ A. so that

$$I_f^2 R_f = (16)(57.5) = 0.92 \text{ kW}$$

$$I_a^2 R_a = (438.8)^2 (0.05) = 9.63 \text{ kW}$$

and total losses $= 0.92 + 9.63 + 1.8 = 12.35$ kW

(a)

output $= 100$ kW

input $= 100 + 12.35 = 112.35$ kW

$$\text{efficiency} = \frac{100}{112.35} = 89\% \quad \leftarrow$$

(b)

$$\text{prime mover output} = \frac{112.35 \times 10^3 \text{ W}}{746 \text{ W/hp}} = 150.6 \text{ hp} \leftarrow$$

13.4.19.

$$\phi = 0.04615 \text{ Wb}$$

(a) $$T_e = k_a \phi I_a = (40)(0.04615)(325) = 600 \text{ N} \cdot \text{m}$$ ←

(b) $$E = 200 - (325)(0.025 - 0.050) = 175.6 \text{ V}$$

$$\omega_m = \frac{E}{k_a \phi} = \frac{175.6}{(40)(0.04615)} = 95.14 \text{ rad/s}$$

or $n = 908$ rpm. ←

(c) output power = (600)(95.14) = 57.084 kW

or 76.52 hp. ←

(d)

ohmic loss = $(325)^2(0.025 + 0.050)$	=	7922 W
core loss	=	220 W
windage and friction loss	=	40 W
total losses	=	8182 W

$$\text{efficiency} = \frac{57\,084}{57\,084 + 8182} = 87.5\%$$ ←

13.4.20.

(a) $$\text{input power} = \frac{\text{output}}{\text{efficiency}} = \frac{(30)(746)}{0.87} = 25.72 \text{ kW}$$ ←

(b) $$\text{input current} = \frac{\text{input power}}{\text{input voltage}} = \frac{25\,720}{230} = 111.8 \text{ A}$$ ←

(c) $$\text{output torque} = \frac{\text{output power}}{\text{angular velocity}} = \frac{(30)(746)}{(2\pi \times 1120)/60} = 190.8 \text{ N} \cdot \text{m}$$

$$\text{developed torque} = (1.07)(190.8) = 204.2 \text{ N} \cdot \text{m}$$ ←

13.4.21.

As a generator:

$$I_f = \frac{250}{250} = 1\ \text{A} \qquad I_L = \frac{10 \times 10^3}{250} = 40\ \text{A}$$

$$I_a = 40 - 1 = 41\ \text{A} \qquad I_aR_a = (41)(0.1) = 4.1\ \text{V}$$

$$E_g = 250 + 4.1 = 254.1\ \text{V}$$

As a motor:

$$I_L = \frac{10 \times 10^3}{250} = 40\ \text{A} \qquad I_f = \frac{250}{250} = 1\ \text{A}$$

$$I_a = 40 - 1 = 39\ \text{A} \qquad I_aR_a = (39)(0.1) = 3.9\ \text{V}$$

$$E_m = 250 - 3.9 = 246.1\ \text{V}$$

Now $\dfrac{n_m}{n_g} = \dfrac{E_m}{E_g}$ or $n_m = \dfrac{E_m}{E_g} n_g = \dfrac{246.1}{254.1}(800) = 774.8$ rpm ←

13.4.22.

input = (40)(230)	= 9200 W
field-resistance loss = $\left(\frac{230}{230}\right)^2(230)$	= 230 W
armature-resistance loss = $(40 - 1)^2(0.25)$	= 380 W
core loss and friction loss	= 380 W
brush-contact loss = (2)(39)	= 78 W
stray-load loss = $\frac{10}{100} \times 746$	= 75 W
total losses	= 1143 W
power output = 9200 − 1143	= 8057 W

$$\text{efficiency} = \frac{8057}{9200} = 87.6\%$$ ←

CHAPTER 14

14.1.1.

(a)

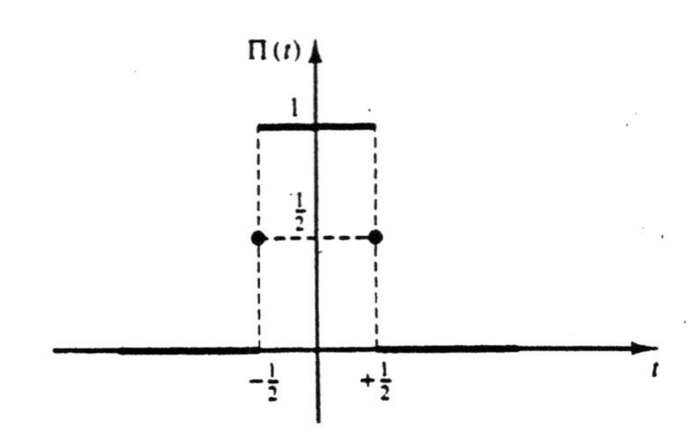

$\Pi(t) = u_{-1}(t + \frac{1}{2}) - u_{-1}(t - \frac{1}{2})$ ←

(b)

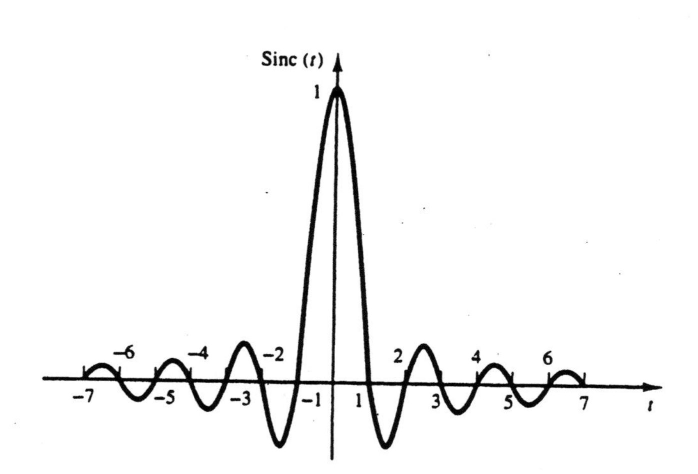

The sinc signal.

SINC SIGNAL REACHES ITS MAXIMUM AT $t = 0$ WITH A VALUE OF 1.

ZEROS OF THE SIGNAL: $t = \pm 1, \pm 2, \pm 3, \ldots$

NOTE: LOCAL EXTREMA ARE NOT AT $t = \pm \frac{3}{2}, \pm \frac{5}{2}, \ldots$

14.1.1.(C)

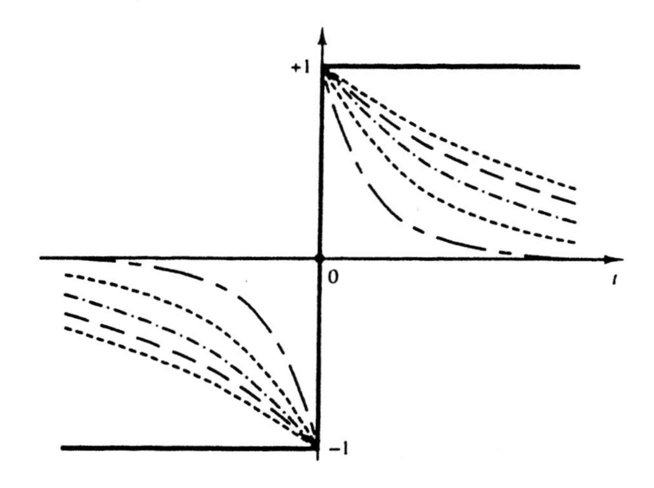

The signum signal as the limit of $x_n(t)$.

14.1.2.

SINCE INTEGRATION IS LINEAR, THIS SYSTEM IS LINEAR.

ALSO, THE RESPONSE TO $x(t-t_0)$ IS

$$y_1(t) = \int_{-\infty}^{t} x(\tau - t_0)\, d\tau = \int_{-\infty}^{t-t_0} x(u)\, du = y(t-t_0)$$

WHERE WE HAVE USED THE CHANGE OF VARIABLE $u = \tau - t_0$.

$\therefore$ THE SYSTEM IS LTI. ←

THE IMPULSE RESPONSE IS OBTAINED BY APPLYING AN IMPULSE AT THE INPUT, i.e.

$$h(t) = \int_{-\infty}^{t} \delta(\tau)\, d\tau = u_{-1}(t)$$

14.1.3.

We have

$$\begin{aligned} x_n &= \frac{1}{T_0} \int_{-\frac{T_0}{2}}^{+\frac{T_0}{2}} x(t) e^{-j2\pi \frac{n}{T_0} t} dt \\ &= \frac{1}{T_0} \int_{-\frac{T_0}{2}}^{+\frac{T_0}{2}} \delta(t) e^{-j2\pi \frac{n}{T_0} t} dt \\ &= \frac{1}{T_0} \end{aligned}$$

With these coefficients we have the following expansion:

$$\sum_{n=-\infty}^{+\infty} \delta(t - nT_0) = \frac{1}{T_0} \sum_{n=-\infty}^{+\infty} e^{j2\pi \frac{n}{T_0} t}$$

14.1.3(a) CONTD.

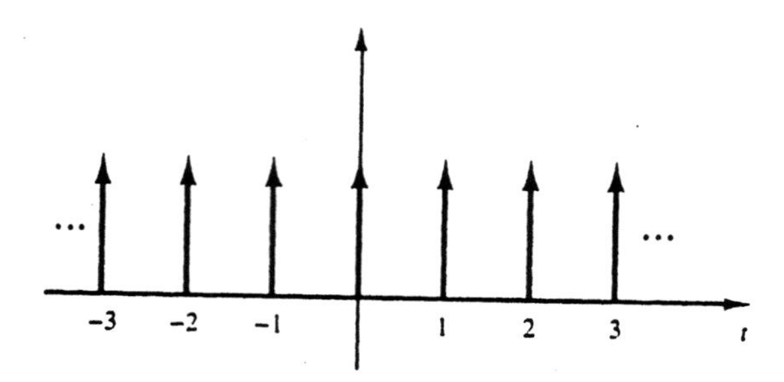

An impulse train.

14.1.3.(b)

Here $T_0 = 2$ and it is convenient to choose $\alpha = -\frac{1}{2}$. Then,

$$
\begin{aligned}
x_n &= \frac{1}{2}\int_{-\frac{1}{2}}^{\frac{3}{2}} x(t)e^{-jn\pi t}dt \\
&= \frac{1}{2}\int_{-\frac{1}{2}}^{\frac{1}{2}} e^{-jn\pi t}dt - \frac{1}{2}\int_{\frac{1}{2}}^{\frac{3}{2}} e^{-jn\pi t}dt \\
&= -\frac{1}{j2\pi n}\left[e^{-j\frac{n\pi}{2}} - e^{j\frac{n\pi}{2}}\right] - \frac{1}{-j2\pi n}\left[e^{-jn\frac{3\pi}{2}} - e^{-jn\frac{\pi}{2}}\right] \\
&= \frac{1}{n\pi}\sin\left(\frac{n\pi}{2}\right) - \frac{1}{n\pi}e^{-jn\pi}\sin\left(\frac{n\pi}{2}\right) \\
&= \frac{1}{n\pi}(1-\cos(n\pi))\sin\left(\frac{n\pi}{2}\right) \\
&= \begin{cases} \frac{2}{n\pi} & n = 4k+1 \\ -\frac{2}{n\pi} & n = 4k+3 \\ 0 & n \text{ even} \end{cases}
\end{aligned}
$$

From these values of x_n we have the following Fourier series expansion for $x(t)$:

$$
\begin{aligned}
x(t) &= \frac{2}{\pi}(e^{j\pi t} + e^{-j\pi t}) - \frac{2}{3\pi}(e^{j3\pi t} + e^{-j3\pi t}) + \frac{2}{5\pi}(e^{j5\pi t} + e^{-j5\pi t}) - \cdots \\
&= \frac{4}{\pi}\cos(\pi t) - \frac{4}{3\pi}\cos(3\pi t) + \frac{4}{5\pi}\cos(5\pi t) - \cdots \\
&= \frac{4}{\pi}\sum_{k=0}^{\infty}\frac{(-1)^k}{2k+1}\cos(2k+1)\pi t
\end{aligned}
$$

14.1.4.

Let $x_1[n]$ and $x_2[n]$ be two discrete periodic signals with periods N_1 and N_2 respectively. Thus,

$$\begin{aligned} x_1[n] &= x_1[n+kN_1] \\ x_2[n] &= x_2[n+kN_2] \end{aligned}$$

Let N be the Least Common Multiplier of N_1, N_2. Then $N = nN_1$, $N = mN_2$ for some integers n, m mutually prime. We will show that the sum $y[n] = x_1[n] + x_2[n]$ is periodic with period N. Clearly

$$\begin{aligned} y[n+kN] &= x_1[n+kN] + x_2[n+kN] \\ &= x_1[n+knN_1] + x_2[n+kmN_2] \\ &= x_1[n] + x_2[n] = y[n] \end{aligned}$$

Thus the sum of two discrete periodic signals is periodic. Assume now two continuous periodic signals with periods T_1 and T_2 respectively. Then,

$$\begin{aligned} x_1(t) &= x_1(t+kT_1) \\ x_2(t) &= x_2(t+kT_2) \end{aligned}$$

For the sum $y(t) = x_1(t) + x_2(t)$ to be periodic, there must exist a real number T such that

$$y(t+kT) = x_1(t+kT) + x_2(t+kT) = x_1(t) + x_2(t) = y(t)$$

Since the previous must hold for every t we must have $x_1(t+kT) = x_1(t)$ and $x_2(t+kT) = x_2(t)$. This implies that

$$\begin{aligned} kT &= nT_1 \\ kT &= mT_2 \end{aligned}$$

for some integers n and m. Thus,

$$nT_1 = mT_2 \quad \text{or} \quad \frac{n}{m} = \frac{T_2}{T_1}$$

Hence, in order for the sum signal to be periodic the ratio of the periods of the two component signals must be rational. This is not true in general for arbitrary real numbers T_1 and T_2.

14.1.5.

(a)

$$x_1(t) = \begin{cases} e^{-t} & t > 0 \\ -e^{t} & t < 0 \\ 0 & t = 0 \end{cases} \Longrightarrow x_1(-t) = \begin{cases} -e^{-t} & t > 0 \\ e^{t} & t < 0 \\ 0 & t = 0 \end{cases} = -x_1(t)$$

Thus, $x_1(t)$ is an odd signal

(b)

$$x_2(t) = e^{-|t|} \Longrightarrow x_2(-t) = e^{-|(-t)|} = e^{-|t|} = x_2(t)$$

Hence, the signal $x_2(t)$ is even.

(c)

$$x_3(t) = \begin{cases} \frac{t}{|t|} & t \neq 0 \\ 0 & t = 0 \end{cases} \Longrightarrow x_3(-t) = \begin{cases} \frac{-t}{|t|} & t \neq 0 \\ 0 & t = 0 \end{cases} = -x_3(t)$$

Thus, the signal $x_3(t)$ is odd.

14.1.6.

(a)

$$x_4(t) = \begin{cases} t & t \geq 0 \\ 0 & t < 0 \end{cases} \Longrightarrow x_4(-t) = \begin{cases} 0 & t \geq 0 \\ -t & t < 0 \end{cases}$$

The signal $x_4(t)$ is neither even nor odd. The even part of the signal is

$$x_{4,e}(t) = \frac{x_4(t) + x_4(-t)}{2} = \begin{cases} \frac{t}{2} & t \geq 0 \\ \frac{-t}{2} & t < 0 \end{cases} = \frac{|t|}{2}$$

The odd part is

$$x_{4,o}(t) = \frac{x_4(t) - x_4(-t)}{2} = \begin{cases} \frac{t}{2} & t \geq 0 \\ \frac{t}{2} & t < 0 \end{cases} = \frac{t}{2}$$

(b)

$$x_5(t) = \sin t + \cos t \Longrightarrow x_5(-t) = -\sin t + \cos t$$

Clearly $x_5(-t) \neq x_5(t)$ for every t since otherwise $2\sin t = 0\ \forall t$. Similarly $x_5(-t) \neq -x_5(t)$ for every t since otherwise $2\cos t = 0\ \forall t$. Thus $x_5(t)$ is neither even or odd. The even and the odd parts of $x_5(t)$ are given by

$$\begin{aligned} x_{5,e}(t) &= \frac{x_5(t) + x_5(-t)}{2} = \cos t \\ x_{5,o}(t) &= \frac{x_5(t) - x_5(-t)}{2} = \sin t \end{aligned}$$

14.1.7.

For (a) AND (b) : we will need the integral $I = \int e^{ax} \cos^2 x dx$.

$$\begin{aligned} I &= \frac{1}{a}\int \cos^2 x \, de^{ax} = \frac{1}{a}e^{ax}\cos^2 x + \frac{1}{a}\int e^{ax}\sin 2x \, dx \\ &= \frac{1}{a}e^{ax}\cos^2 x + \frac{1}{a^2}\int \sin 2x \, de^{ax} \\ &= \frac{1}{a}e^{ax}\cos^2 x + \frac{1}{a^2}e^{ax}\sin 2x - \frac{2}{a^2}\int e^{ax}\cos 2x \, dx \\ &= \frac{1}{a}e^{ax}\cos^2 x + \frac{1}{a^2}e^{ax}\sin 2x - \frac{2}{a^2}\int e^{ax}(2\cos^2 x - 1) \, dx \\ &= \frac{1}{a}e^{ax}\cos^2 x + \frac{1}{a^2}e^{ax}\sin 2x - \frac{2}{a^2}\int e^{ax} \, dx - \frac{4}{a^2}I \end{aligned}$$

Thus,

$$I = \frac{1}{4+a^2}\left[(a\cos^2 x + \sin 2x) + \frac{2}{a}\right]e^{ax}$$

WHICH IS GIVEN IN PROBLEM STATEMENT.

14.1.7. CONTD.

(a)

$$\begin{aligned}E_x &= \lim_{T\to\infty}\int_{-\frac{T}{2}}^{\frac{T}{2}} x_1^2(t)dx = \lim_{T\to\infty}\int_0^{\frac{T}{2}} e^{-2t}\cos^2 t dt \\ &= \lim_{T\to\infty}\frac{1}{8}\left[(-2\cos^2 t+\sin 2t)-1\right]e^{-2t}\Big|_0^{\frac{T}{2}} \\ &= \lim_{T\to\infty}\frac{1}{8}\left[(-2\cos^2\frac{T}{2}+\sin T-1)e^{-T}+3\right]=\frac{3}{8}\end{aligned}$$

Thus $x_1(t)$ is an energy-type signal and the energy content is 3/8

(b)

$$\begin{aligned}E_x &= \lim_{T\to\infty}\int_{-\frac{T}{2}}^{\frac{T}{2}} x_2^2(t)dx = \lim_{T\to\infty}\int_{-\frac{T}{2}}^{\frac{T}{2}} e^{-2t}\cos^2 t dt \\ &= \lim_{T\to\infty}\left[\int_{-\frac{T}{2}}^{0} e^{-2t}\cos^2 t dt+\int_0^{\frac{T}{2}} e^{-2t}\cos^2 t dt\right]\end{aligned}$$

$$\begin{aligned}\lim_{T\to\infty}\int_{-\frac{T}{2}}^{0} e^{-2t}\cos^2 t dt &= \lim_{T\to\infty}\frac{1}{8}\left[(-2\cos^2 t+\sin 2t)-1\right]e^{-2t}\Big|_{-\frac{T}{2}}^{0} \\ &= \lim_{T\to\infty}\frac{1}{8}\left[-3+(2\cos^2\frac{T}{2}+1+\sin T)e^{T}\right]=\infty\end{aligned}$$

since $2+\cos\theta+\sin\theta>0$. Thus, $E_x=\infty$ since as we have seen from the first question the second integral is bounded. Hence, the signal $x_2(t)$ is not an energy-type signal. To test if $x_2(t)$ is a power-type signal we find P_x.

$$P_x = \lim_{T\to\infty}\frac{1}{T}\int_{-\frac{T}{2}}^{0} e^{-2t}\cos^2 dt+\lim_{T\to\infty}\frac{1}{T}\int_0^{\frac{T}{2}} e^{-2t}\cos^2 dt$$

But $\lim_{T\to\infty}\frac{1}{T}\int_0^{\frac{T}{2}} e^{-2t}\cos^2 dt$ is zero and

$$\begin{aligned}\lim_{T\to\infty}\frac{1}{T}\int_{-\frac{T}{2}}^{0} e^{-2t}\cos^2 dt &= \lim_{T\to\infty}\frac{1}{8T}\left[2\cos^2\frac{T}{2}+1+\sin T\right]e^{T} \\ &> \lim_{T\to\infty}\frac{1}{T}e^{T} > \lim_{T\to\infty}\frac{1}{T}(1+T+T^2) > \lim_{T\to\infty} T=\infty\end{aligned}$$

Thus the signal $x_2(t)$ is not a power-type signal.

(c)

$$\begin{aligned}E_x &= \lim_{T\to\infty}\int_{-\frac{T}{2}}^{\frac{T}{2}} x_3^2(t)dx = \lim_{T\to\infty}\int_{-\frac{T}{2}}^{\frac{T}{2}}\text{sgn}^2(t)dt = \lim_{T\to\infty}\int_{-\frac{T}{2}}^{\frac{T}{2}} dt = \lim_{T\to\infty} T=\infty \\ P_x &= \lim_{T\to\infty}\frac{1}{T}\int_{-\frac{T}{2}}^{\frac{T}{2}}\text{sgn}^2(t)dt = \lim_{T\to\infty}\frac{1}{T}\int_{-\frac{T}{2}}^{\frac{T}{2}} dt = \lim_{T\to\infty}\frac{1}{T}T=1\end{aligned}$$

The signal $x_3(t)$ is of the power-type and the power content is 1.

14.1.8. (a)

$$E_x = \lim_{T\to\infty} \int_{-T/2}^{T/2} \left(A^2 \cos^2 2\pi f_1 t + B^2 \cos^2 2\pi f_2 t + 2AB \cos 2\pi f_1 t \cos 2\pi f_2 t\right) dt$$

$$= \lim_{T\to\infty} \int_{-T/2}^{T/2} A^2 \cos^2 2\pi f_1 t\, dt + \lim_{T\to\infty} \int_{-T/2}^{T/2} B^2 \cos^2 2\pi f_2 t\, dt + AB \lim_{T\to\infty} \int_{-T/2}^{T/2} \left[\cos 2\pi (f_1+f_2)t + \cos 2\pi (f_1-f_2)t\right] dt$$

$$= \infty + \infty + 0 = \infty$$

∴ THE SIGNAL IS NOT OF THE ENERGY TYPE ←

TO CHECK IF IT IS OF THE POWER TYPE

(i) $f_1 = f_2$

$$P_x = \lim_{T\to\infty} \frac{1}{T} \int_{-T/2}^{T/2} (A+B)^2 \cos^2 2\pi f_1 t\, dt$$

$$= \lim_{T\to\infty} \frac{1}{2T} (A+B)^2 \int_{-T/2}^{T/2} dt = \frac{1}{2} (A+B)^2$$

(ii) $f_1 \neq f_2$

$$P_x = \lim_{T\to\infty} \int_{-T/2}^{T/2} \left[A^2 \cos^2 2\pi f_1 t + B^2 \cos^2 2\pi f_2 t + 2AB \cos 2\pi f_1 t \cos 2\pi f_2 t\right] dt$$

$$= \lim_{T\to\infty} \frac{1}{T} \left[\frac{A^2 T}{2} + \frac{B^2 T}{2}\right] = \frac{A^2}{2} + \frac{B^2}{2} = \frac{1}{2}(A^2 + B^2)$$

∴ SIGNAL IS OF THE POWER TYPE ←

(b)

For $f_1 = f_2$, $P_x = (A+B)^2/2$ ←

For $f_1 \neq f_2$, $P_x = \frac{1}{2}(A^2 + B^2)$ ←

14.1.9.

(a) $$P_x = \lim_{T\to\infty} \frac{1}{T} \int_{-T/2}^{T/2} A^2 \left|e^{j(2\pi f_0 t + \theta)}\right|^2 = \lim_{T\to\infty} \frac{1}{T} \int_{-T/2}^{T/2} A^2 dt = \lim_{T\to\infty} \frac{1}{T} A^2 T = A^2$$ ←

(b)

$$P_x = \lim_{T\to\infty} \frac{1}{T} \int_{-T/2}^{T/2} u_{-1}^2(t)\, dt = \lim_{T\to\infty} \frac{1}{T} \int_0^{T/2} dt = \lim_{T\to\infty} \frac{1}{T} \frac{T}{2} = \frac{1}{2}$$ ←

14.1.10.

Let $x_e^1(t)$, $x_e^2(t)$ be two even signals and $x_o^1(t)$, $x_o^2(t)$ be two odd signals. Then,

$$
\begin{aligned}
y(t) = x_e^1(t)x_e^2(t) \quad &\Longrightarrow \quad y(-t) = x_e^1(-t)x_e^2(-t) = x_e^1(t)x_e^2(t) = y(t) \\
z(t) = x_o^1(t)x_o^2(t) \quad &\Longrightarrow \quad z(-t) = x_o^1(-t)x_o^2(-t) = (-x_o^1(t))(-x_o^2(t)) = z(t)
\end{aligned}
$$

Thus the product of two even or odd signals is an even signal. For $v(t) = x_e^1(t)x_o^1(t)$ we have

$$
v(-t) = x_e^1(-t)x_o^1(-t) = x_e^1(t)(-x_o^1(t)) = -x_e^1(t)x_o^1(t) = -v(t)
$$

Thus the product of an even and an odd signal is an odd signal.

14.1.11.

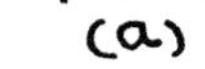

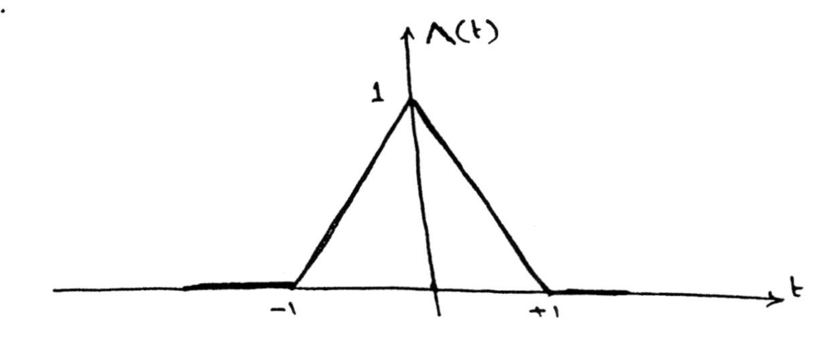

(b)

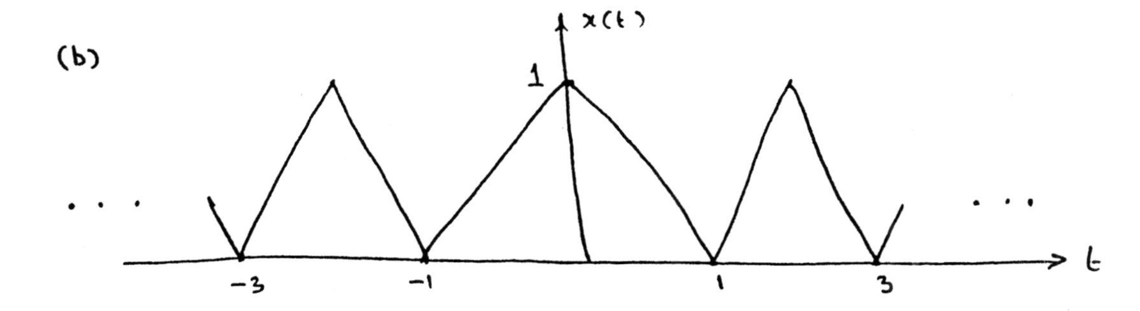

(c)

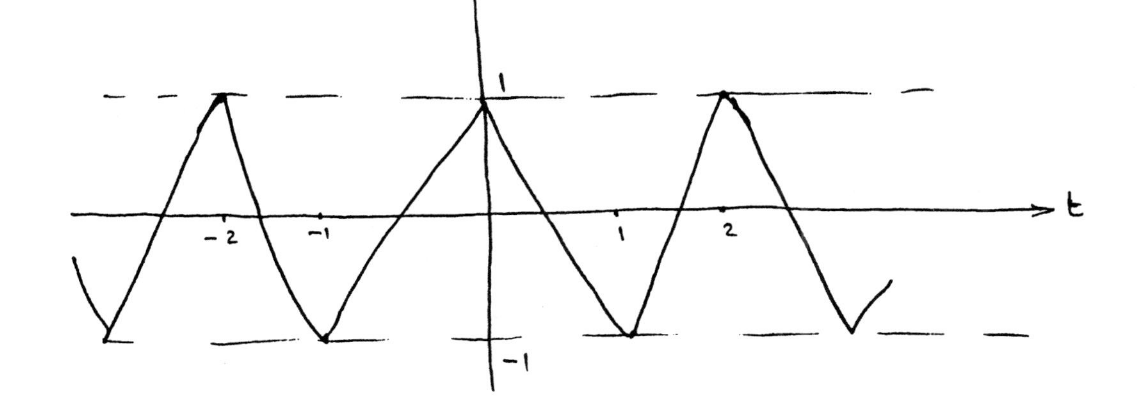

14.1.12.

(a) $T_0 = 1/2f_0$

$$a_n = 2f_0 \int_{-\frac{1}{4f_0}}^{\frac{1}{4f_0}} (\cos 2\pi f_0 t)(\cos 2\pi n\, 2f_0 t)\, dt$$

$$= f_0 \int_{-\frac{1}{4f_0}}^{\frac{1}{4f_0}} \{\cos 2\pi f_0 (1+2n)t\}\, dt + f_0 \int_{-\frac{1}{4f_0}}^{\frac{1}{4f_0}} \{\cos 2\pi f_0 (1-2n)t\}\, dt$$

$$= \frac{1}{2\pi(1+2n)} \sin\{2\pi f_0 (1+2n)t\}\Big]_{-\frac{1}{4f_0}}^{1/4f_0} + \frac{1}{2\pi(1-2n)} \sin\{2\pi f_0 (1-2n)t\}\Big]_{-\frac{1}{4f_0}}^{1/4f_0}$$

$$= \frac{(-1)^n}{\pi}\left[\frac{1}{1+2n} + \frac{1}{1-2n}\right] \leftarrow$$

(b) $T_0 = 1/f_0$; $x(t) = 2\cos(2\pi f_0 t)$ IN THE INTERVAL $\left[-\frac{1}{4f_0}, \frac{1}{4f_0}\right]$

$= 0$ IN THE INTERVAL $\left[\frac{1}{4f_0}, \frac{3}{4f_0}\right]$

$$a_n = 2f_0 \int_{-\frac{1}{4f_0}}^{\frac{1}{4f_0}} (\cos 2\pi f_0 t \cos 2\pi n f_0 t)\, dt$$

$$= f_0 \int_{-\frac{1}{4f_0}}^{1/4f_0} (\cos 2\pi f_0 (1+n)t)\, dt + f_0 \int_{-\frac{1}{4f_0}}^{\frac{1}{4f_0}} (\cos 2\pi f_0 (1-n)t)\, dt$$

$$= \frac{1}{2\pi(1+n)} \sin 2\pi f_0 (1+n)t \Big|_{-\frac{1}{4f_0}}^{\frac{1}{4f_0}} + \frac{1}{2\pi(1-n)} \sin 2\pi f_0 (1-n)t \Big|_{-\frac{1}{4f_0}}^{\frac{1}{4f_0}}$$

$$= \frac{1}{\pi(1+n)} \sin\left(\frac{\pi}{2}(1+n)\right) + \frac{1}{\pi(1-n)} \sin\left[\frac{\pi}{2}(1-n)\right]$$

$a_n = 0$ FOR ODD VALUES OF n UNLESS $n = \pm 1$

IN WHICH CASE $a_{\pm 1} = \frac{1}{2}$

WHEN n IS EVEN, $n = 2l$, THEN $a_{2l} = \frac{(-1)^l}{\pi}\left[\frac{1}{1+2l} + \frac{1}{1-2l}\right]$ $\leftarrow$

14.1.13.

$$x_e(t) = a_0 + \sum_{n=1}^{\infty} a_n \cos n\omega t \leftarrow$$

$$x_o(t) = \sum_{n=1}^{\infty} b_n \sin n\omega t \leftarrow$$

14.1.14.

a) The signal is periodic with period T. Thus

$$\begin{aligned}
x_n &= \frac{1}{T}\int_0^T e^{-t}e^{-j2\pi\frac{n}{T}t}dt = \frac{1}{T}\int_0^T e^{-(j2\pi\frac{n}{T}+1)t}dt \\
&= -\frac{1}{T\left(j2\pi\frac{n}{T}+1\right)}e^{-(j2\pi\frac{n}{T}+1)t}\Big|_0^T = -\frac{1}{j2\pi n+T}\left[e^{-(j2\pi n+T)}-1\right] \\
&= \frac{1}{j2\pi n+T}[1-e^{-T}] = \frac{T-j2\pi n}{T^2+4\pi^2n^2}[1-e^{-T}]
\end{aligned}$$

If we write $x_n = \frac{a_n - jb_n}{2}$ we obtain the trigonometric Fourier series expansion coefficients as

$$a_n = \frac{2T}{T^2+4\pi^2n^2}[1-e^{-T}], \qquad b_n = \frac{4\pi n}{T^2+4\pi^2n^2}[1-e^{-T}]$$

b) The signal is periodic with period $2T$. Since the signal is odd we obtain $x_0 = 0$. For $n \neq 0$

$$x_n = \frac{1}{2T}\int_{-T}^{T} x(t)e^{-j2\pi\frac{n}{2T}t}dt = \frac{1}{2T}\int_{-T}^{T}\frac{t}{T}e^{-j2\pi\frac{n}{2T}t}dt$$

14.1.14. CONTD.

$$\begin{aligned}
&= \frac{1}{2T^2}\int_{-T}^{T} te^{-j\pi\frac{n}{T}t}dt \\
&= \frac{1}{2T^2}\left(\frac{jT}{\pi n}te^{-j\pi\frac{n}{T}t} + \frac{T^2}{\pi^2 n^2}e^{-j\pi\frac{n}{T}t}\right)\Big|_{-T}^{T} \\
&= \frac{1}{2T^2}\left[\frac{jT^2}{\pi n}e^{-j\pi n} + \frac{T^2}{\pi^2 n^2}e^{-j\pi n} + \frac{jT^2}{\pi n}e^{j\pi n} - \frac{T^2}{\pi^2 n^2}e^{j\pi n}\right] \\
&= \frac{j}{\pi n}(-1)^n
\end{aligned}$$

The trigonometric Fourier series expansion coefficients are:

$$a_n = 0, \qquad b_n = (-1)^{n+1}\frac{2}{\pi n}$$

c) The signal is periodic with period T. For $n = 0$

$$x_0 = \frac{1}{T}\int_{-\frac{T}{2}}^{\frac{T}{2}} x(t)dt = \frac{3}{2}$$

If $n \neq 0$ then

$$\begin{aligned}
x_n &= \frac{1}{T}\int_{-\frac{T}{2}}^{\frac{T}{2}} x(t)e^{-j2\pi\frac{n}{T}t}dt \\
&= \frac{1}{T}\int_{-\frac{T}{2}}^{\frac{T}{2}} e^{-j2\pi\frac{n}{T}t}dt + \frac{1}{T}\int_{-\frac{T}{4}}^{\frac{T}{4}} e^{-j2\pi\frac{n}{T}t}dt \\
&= \frac{j}{2\pi n}e^{-j2\pi\frac{n}{T}t}\Big|_{-\frac{T}{2}}^{\frac{T}{2}} + \frac{j}{2\pi n}e^{-j2\pi\frac{n}{T}t}\Big|_{-\frac{T}{4}}^{\frac{T}{4}} \\
&= \frac{j}{2\pi n}\left[e^{-j\pi n} - e^{j\pi n} + e^{-j\pi\frac{n}{2}} - e^{-j\pi\frac{n}{2}}\right] \\
&= \frac{1}{\pi n}\sin(\pi\frac{n}{2}) = \frac{1}{2}\text{sinc}(\frac{n}{2})
\end{aligned}$$

Note that $x_n = 0$ for n even and $x_{2l+1} = \frac{1}{\pi(2l+1)}(-1)^l$. The trigonometric Fourier series expansion coefficients are:

$$a_0 = 3, \quad , a_{2l} = 0, \quad , a_{2l+1} = \frac{2}{\pi(2l+1)}(-1)^l, \quad , b_n = 0, \ \forall n$$

d) The signal is periodic with period T. For $n = 0$

$$x_0 = \frac{1}{T}\int_0^T x(t)dt = \frac{2}{3}$$

If $n \neq 0$ then

$$\begin{aligned}
x_n &= \frac{1}{T}\int_0^T x(t)e^{-j2\pi\frac{n}{T}t}dt = \frac{1}{T}\int_0^{\frac{T}{3}} \frac{3}{T}te^{-j2\pi\frac{n}{T}t}dt \\
&\quad + \frac{1}{T}\int_{\frac{T}{3}}^{\frac{2T}{3}} e^{-j2\pi\frac{n}{T}t}dt + \frac{1}{T}\int_{\frac{2T}{3}}^{T}(-\frac{3}{T}t + 3)e^{-j2\pi\frac{n}{T}t}dt \\
&= \frac{3}{T^2}\left(\frac{jT}{2\pi n}te^{-j2\pi\frac{n}{T}t} + \frac{T^2}{4\pi^2 n^2}e^{-j2\pi\frac{n}{T}t}\right)\Big|_0^{\frac{T}{3}} \\
&\quad - \frac{3}{T^2}\left(\frac{jT}{2\pi n}te^{-j2\pi\frac{n}{T}t} + \frac{T^2}{4\pi^2 n^2}e^{-j2\pi\frac{n}{T}t}\right)\Big|_{\frac{2T}{3}}^{T} \\
&\quad + \frac{j}{2\pi n}e^{-j2\pi\frac{n}{T}t}\Big|_{\frac{T}{3}}^{\frac{2T}{3}} + \frac{3}{T}\frac{jT}{2\pi n}e^{-j2\pi\frac{n}{T}t}\Big|_{\frac{2T}{3}}^{T} \\
&= \frac{3}{2\pi^2 n^2}[\cos(\frac{2\pi n}{3}) - 1]
\end{aligned}$$

14.1.14 CONTD.

The trigonometric Fourier series expansion coefficients are:

$$a_0 = \frac{4}{3}, \quad a_n = \frac{3}{\pi^2 n^2}[\cos(\frac{2\pi n}{3}) - 1], \quad b_n = 0, \ \forall n$$

e) The signal is periodic with period T. Since the signal is odd $x_0 = a_0 = 0$. For $n \neq 0$

$$\begin{aligned}
x_n &= \frac{1}{T}\int_{-\frac{T}{2}}^{\frac{T}{2}} x(t)dt = \frac{1}{T}\int_{-\frac{T}{2}}^{-\frac{T}{4}} -e^{-j2\pi\frac{n}{T}t}dt \\
&+ \frac{1}{T}\int_{-\frac{T}{4}}^{\frac{T}{4}} \frac{4}{T}te^{-j2\pi\frac{n}{T}t}dt + \frac{1}{T}\int_{\frac{T}{4}}^{\frac{T}{2}} e^{-j2\pi\frac{n}{T}t}dt \\
&= \frac{4}{T^2}\left(\frac{jT}{2\pi n}te^{-j2\pi\frac{n}{T}t} + \frac{T^2}{4\pi^2 n^2}e^{-j2\pi\frac{n}{T}t}\right)\Big|_{-\frac{T}{4}}^{\frac{T}{4}} \\
&- \frac{1}{T}\left(\frac{jT}{2\pi n}e^{-j2\pi\frac{n}{T}t}\right)\Big|_{-\frac{T}{2}}^{-\frac{T}{4}} + \frac{1}{T}\left(\frac{jT}{2\pi n}e^{-j2\pi\frac{n}{T}t}\right)\Big|_{\frac{T}{4}}^{\frac{T}{2}} \\
&= \frac{j}{\pi n}\left[(-1)^n - \frac{2\sin(\frac{\pi n}{2})}{\pi n}\right] = \frac{j}{\pi n}\left[(-1)^n - \text{sinc}(\frac{n}{2})\right]
\end{aligned}$$

For n even, $\text{sinc}(\frac{n}{2}) = 0$ and $x_n = \frac{j}{\pi n}$. The trigonometric Fourier series expansion coefficients are:

$$a_n = 0, \ \forall n, \quad b_n = \begin{cases} -\frac{1}{\pi l} & n = 2l \\ \frac{2}{\pi(2l+1)}[1 + \frac{2(-1)^l}{\pi(2l+1)}] & n = 2l+1 \end{cases}$$

f) The signal is periodic with period T. For $n = 0$

$$x_0 = \frac{1}{T}\int_{-\frac{T}{2}}^{\frac{T}{2}} x(t)dt = 1$$

For $n \neq 0$

$$\begin{aligned}
x_n &= \frac{1}{T}\int_{-\frac{T}{2}}^{0}(\frac{3}{T}t + 2)e^{-j2\pi\frac{n}{T}t}dt + \frac{1}{T}\int_{0}^{\frac{T}{2}}(-\frac{3}{T}t + 2)e^{-j2\pi\frac{n}{T}t}dt \\
&= \frac{3}{T^2}\left(\frac{jT}{2\pi n}te^{-j2\pi\frac{n}{T}t} + \frac{T^2}{4\pi^2 n^2}e^{-j2\pi\frac{n}{T}t}\right)\Big|_{-\frac{T}{2}}^{0} \\
&- \frac{3}{T^2}\left(\frac{jT}{2\pi n}te^{-j2\pi\frac{n}{T}t} + \frac{T^2}{4\pi^2 n^2}e^{-j2\pi\frac{n}{T}t}\right)\Big|_{0}^{\frac{T}{2}} \\
&+ \frac{2}{T}\frac{jT}{2\pi n}e^{-j2\pi\frac{n}{T}t}\Big|_{-\frac{T}{2}}^{0} + \frac{2}{T}\frac{jT}{2\pi n}e^{-j2\pi\frac{n}{T}t}\Big|_{0}^{\frac{T}{2}} \\
&= \frac{3}{\pi^2 n^2}\left[\frac{1}{2} - \cos(\frac{2\pi n}{3})\right] + \frac{1}{\pi n}\sin(\frac{2\pi n}{3})
\end{aligned}$$

The trigonometric Fourier series expansion coefficients are:

$$a_0 = 2, \quad a_n = 2\left[\frac{3}{\pi^2 n^2}\left(\frac{1}{2} - \cos(\frac{2\pi n}{3})\right) + \frac{1}{\pi n}\sin(\frac{2\pi n}{3})\right], \quad b_n = 0, \ \forall n$$

14.1.15.

(a) Let $x_d(t)$ be the square wave in Fig. E14.1.4 with A = 2, which has $b_n = 8/\pi n$ for $n = 1, 3, 5, \ldots$ Let $x_c(t)$ be the sawtooth wave in Fig. E14.1.4 with A = 1, which has $b_n = 2/\pi n$ for $n = 1, 2, 3, \ldots$

Then $x(t) = x_d(t) + x_c(t)$ so

$$b_n = 8/\pi n + 2/\pi n = 10/\pi n \quad n = 1, 3, 5, \ldots$$
$$= 0 + 2/\pi n = 2/\pi n \quad n = 2, 4, 6, \ldots$$

(b) Let $x_d(t)$ be the square wave in Fig. E14.1.4 with A = 4, which has $b_n = 16/\pi n$ for $n = 1, 3, 5, \ldots$ Let $x_a(t)$ be the pulse train in Fig. E14.1.4 with A = 4 and D = T/2, which has $a_0 = 2$ and $a_n = (8/\pi n) \sin (\pi n/2)$ for $n = 1, 2, 3, \ldots$
Then $x(t) = x_d(t) + x_a(t)$ so

$$a_0 = 0 + 2 = 2$$
$$a_n = 0 + (8/\pi n) \sin (\pi n/2)$$
$$= 8/\pi n \quad n = 1, 5, 9, \ldots$$
$$= -8/\pi n \quad n = 3, 7, 11, \ldots$$
$$b_n = 16/\pi n + 0 = 16/\pi n \quad n = 1, 3, 5, \ldots$$

(c) Let $x_d(t)$ be the square wave in Fig. E14.1.4 with A = 6, which has $b_n = 24/\pi n$ for $n = 1, 3, 5, \ldots$ Let $x_e(t)$ be the half-rectified sine wave in Fig. E14.1.4 with A = 6, which has $a_0 = 6/\pi$, $b_1 = 3$, and $a_n = -12/\pi(n^2 - 1)$ for $n = 2, 4, 6, \ldots$
Then $x(t) = x_d(t) + x_e(t)$ so

$$a_0 = 0 - 6/\pi = -6/\pi$$
$$b_1 = 24/\pi - 3$$
$$b_n = 24/\pi n - 0 = 24/\pi n \quad n = 3, 5, 7, \ldots$$
$$a_n = 0 - [-12/\pi(n^2 - 1)]$$
$$= 12/\pi(n^2 - 1) \quad n = 2, 4, 6, \ldots$$

14.1.16.

(a)

$f_1 = 1/800\ \mu s = 1.25$ kHz

$A_0 = \pi/2 \qquad n = 0$

$A_n = |(2/n)\sin(\pi n/2)|$

$= 2/n \quad n = 1, 3, 5, \ldots$

$= 0 \quad n = 2, 4, 6, \ldots$

n	0	1	3	5	7	9
A_n	1.57	2	0.67	0.4	0.29	0.22

n	11	13	15	17	19	21
A_n	0.18	0.15	0.13	0.12	0.11	0.095

$(A_n)_{max}/10 = 0.2 \Rightarrow W = 11f_1 = 22$ MHz

$(A_n)_{max}/20 = 0.1 \Rightarrow W = 19f_1 = 38$ MHz

(b)

$f_1 = 1/2.5\ \mu s = 400$ kHz

$A_n = 0 \quad$ n even

$= 8/n^2 \quad$ n odd

n	1	3	5
A_n	8	0.89	0.32

$(A_n)_{max}/10 = 0.8 \Rightarrow W = 3f_1 = 1.2$ MHz

$(A_n)_{max}/20 = 0.4 \Rightarrow W = 3f_1 = 1.2$ MHz

(c)

$f_1 = 1/10$ ms $= 100$ Hz

$A_n = 0 \quad$ n even

$= 4/n \quad$ n odd

n	1	3	5	7	9	11
A_n	4	1.33	1.25	0.57	0.44	0.36

n	13	15	17	19	21
A_n	0.31	0.27	0.24	0.21	0.19

$(A_n)_{max}/10 = 0.4 \Rightarrow W = 9f_1 = 900$ Hz

$(A_n)_{max}/20 = 0.2 \Rightarrow W = 19f_1 = 1900$ Hz

(d)

$f_1 = 1/800\ \mu s = 1.25$ kHz

$A_n = 1 \qquad n = 0$

$= \pi/2 \qquad n = 1$

$= 2/(n^2 - 1) \quad n = 2, 4, 6, \ldots$

n	0	1	2	4	6
A_n	1	1.57	0.67	0.13	0.057

$(A_n)_{max}/10 = 0.157 \Rightarrow W = 2f_1 = 2.5$ kHz

$(A_n)_{max}/20 = 0.079 \Rightarrow W = 4f_1 = 5.0$ kHz

14.1.17.

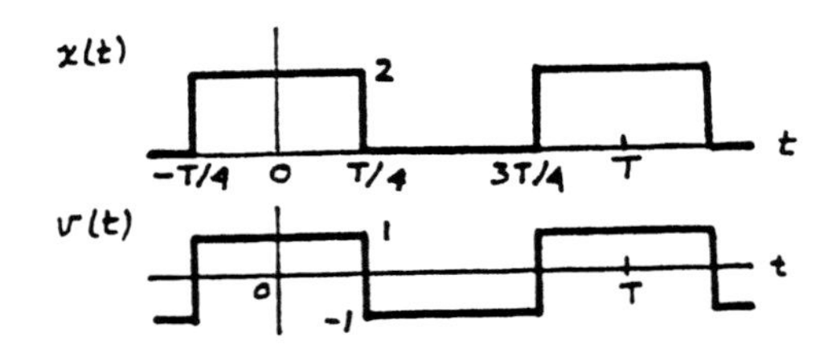

$x(t) = 1 + \Sigma\ (4/\pi n)\ \sin\ (\pi n/2)\ \cos\ n\omega_1 t$

$$\text{where } \sin\ (\pi n/2) = \begin{cases} 0 & n = 2,\ 4,\ 6,\ldots \\ 1 & n = 1,\ 5,\ 8,\ldots \\ -1 & n = 3,\ 7,\ 11,\ldots \end{cases}$$

Thus, $v(t) = \Sigma\ a_n\ \cos\ n\omega_1 t$; EVEN SYMMETRY

$$\text{where } a_n = \begin{cases} 4/\pi n & n = 1,\ 5,\ 9,\ldots \\ -4/\pi n & n = 3,\ 7,\ 11,\ldots \end{cases}$$

ALL OTHER COEFFICIENTS ARE ZERO.

14.1.18.

$f_1 = 5$ kHz $\quad a_o = 0$

Triangular: $A_n = |a_n| = 8/\pi n^2$ for n odd

W = 5 kHz

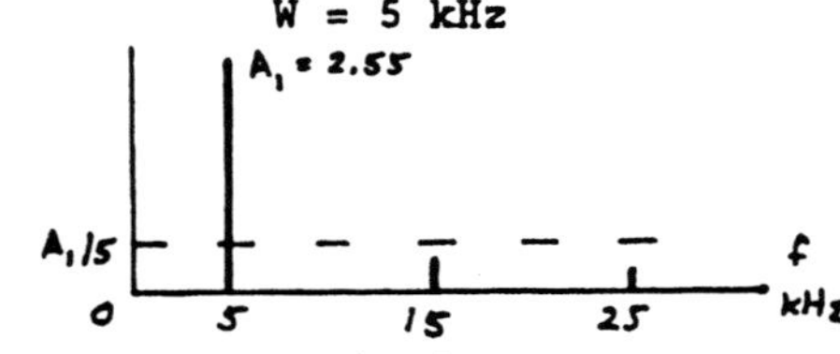

Sawtooth: $A_n = |b_n| = 2/n \quad n = 1,2,3,\ldots$

W = 25 kHz

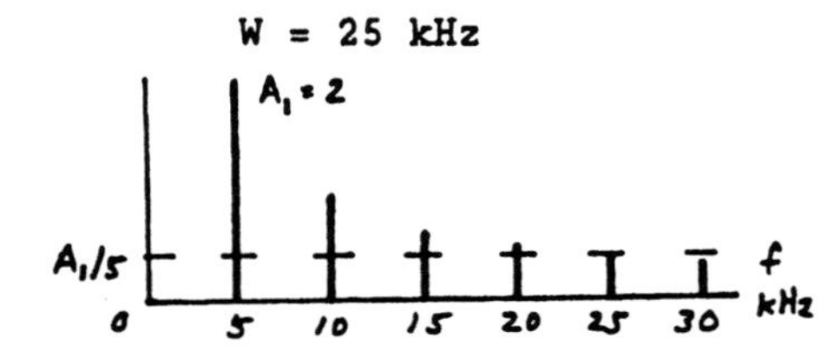

14.1.19.

$x(t) = 3 \cos 2\pi t - \cos 2\pi 3t$

$y(t) = 3 \cos (2\pi t - 90^\circ) - \cos (2\pi 3t - 90^\circ)$

$= 3 \sin 2\pi t - \sin 2\pi 3t$

(continued)

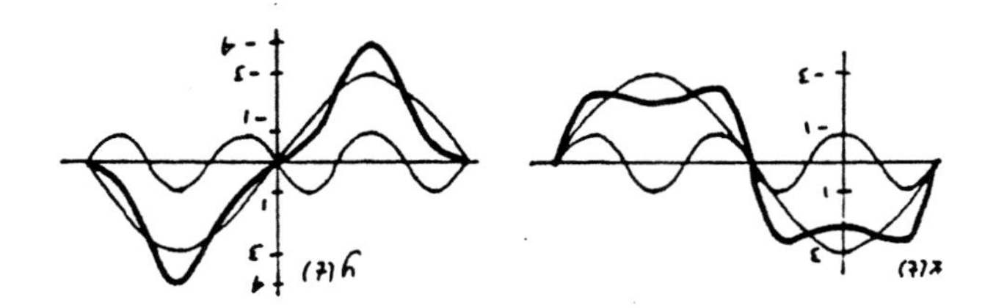

14.1.20.

$|H_{eq}(f)| = \sqrt{1 + (f/f_{co})^2}$

$\theta_{eq}(f) = +\arctan (f/f_{co})$

|Heq|
2.2
1.4
1.0
0
5 kHz
10 kHz

θeq
45°
63°
0
5 kHz
10 kHz

14.1.21.

$A_n = 1/n^2$ for $n = 1, 3, 5, \ldots$

$\phi_n = 0$ since $b_n = 0$

$f_1 = 1/25$ ms $= 40$ Hz, $2\pi n f_1/\omega_{co} = 0.4n$

$|H(f_n)| = 0.4n/\sqrt{1 + (0.4n)^2}$

$\theta(f_n) = 90^\circ - \arctan (0.4n)$

n	A_n	$\|H(f_n)\|$	$A_n\|H(f_n)\|$	$\theta(f_n)$
1	1	0.371	0.371	68.2°
3	1/9	0.768	0.085	39.8°
5	1/25	0.894	0.036	26.6°

$y(t) = 0.371 \cos (2\pi f_1 t + 68.2^\circ)$

$+ 0.085 \cos (2\pi 3 f_1 t + 39.8^\circ)$

14.2.1.

(a) $x_c(t) = (12 \cos 2\pi 100t + 8 \cos 2\pi 150t) \cos 2\pi 600t$

$= 6 [\cos 2\pi 500t + \cos 2\pi 700t] + 4 [\cos 2\pi 450t + \cos 2\pi 750t]$

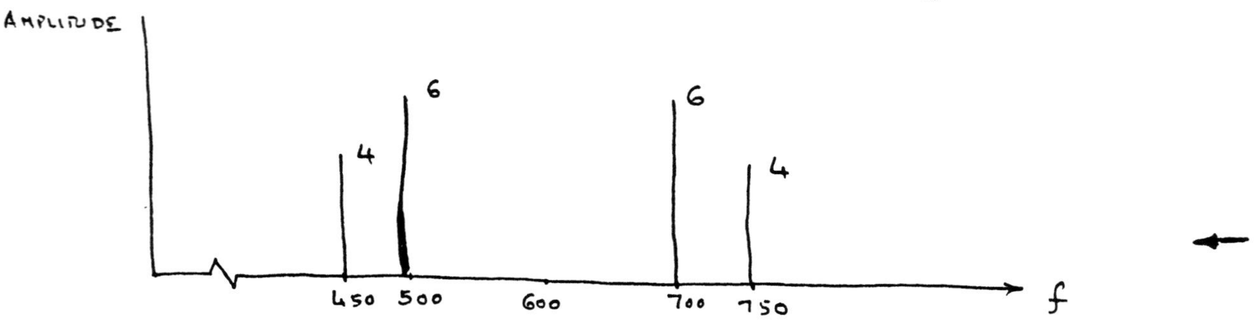

(b) $|450 \pm 500|$, $|500 \pm 500|$, $|700 \pm 500|$, $|750 \pm 500|$

i.e. 0, 50, 200, 250, 950, 1000, 1200, 1250 Hz ←

14.2.2.

$|f_c \pm 3000|$; $|f_c \pm 7000|$

$f_c = 6$ kHz

9 kHz, 3 kHz, 13 kHz, 1 kHz

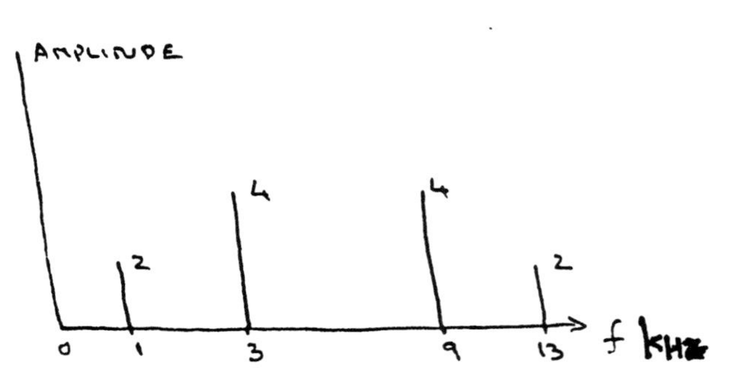

14.2.3.

$x(t) \cos 2\pi f_a t = 6 \cos 2\pi 900t + 6 \cos 2\pi 700t + 2 \cos 2\pi 1100t + 2 \cos 2\pi 500t$

$x_a(t) = 6 \cos 2\pi 900t + 2 \cos 2\pi 1100t$

$x_a(t) \cos 2\pi f_b t = 3 \cos 2\pi 2100t + 3 \cos 2\pi 300t + \cos 2\pi 2300t + \cos 2\pi 100t$

$x_b(t) = \cos 2\pi 100t + 3 \cos 2\pi 300t$ ←

14.2.4.(a)

$x_c(t) \cos [2\pi (f_c + \Delta f)t + \Delta\phi]$

$= 2 \cos 2\pi f_m t \cos [2\pi (2f_c + \Delta f)t + \Delta\phi] + 2 \cos 2\pi f_m t \cos (-2\pi \Delta f t - \Delta\phi)$

$\therefore z(t) = 2 \cos 2\pi f_m t \cos (2\pi \Delta f t + \Delta\phi)$

$= 2 \cos 2\pi 1000t \cos 2\pi 200t$

$= \cos 2\pi 800t + \cos 2\pi 1200t$ ←

14.2.4. (b) $x_c(t)\cos[2\pi(f_c+\Delta f)t+\Delta\phi]$

$= \cos[2\pi(2f_c+f_m+\Delta f)t+\Delta\phi] + \cos[2\pi(f_m-\Delta f)t-\Delta\phi]$

$\therefore z(t) = \cos[2\pi(f_m-\Delta f)t-\Delta\phi]$

$= \cos 2\pi\, 800t$ ←

(c) $x_c(t)\cos[2\pi(f_c+\Delta f)t+\Delta\phi]$

$= \cos[2\pi(2f_c-f_m+\Delta f)t+\Delta\phi] + \cos[2\pi(-f_m-\Delta f)t-\Delta\phi]$

$\therefore z(t) = \cos[2\pi(f_m+\Delta f)t+\Delta\phi]$

$= \cos 2\pi\, 1200t$ ←

14.2.5. (a) $z(t) = 2\cos 2\pi\, 1000t\cos 90^\circ = 0$ ←

(b) $z(t) = \cos(2\pi\, 1000t - 90^\circ)$

$= \sin 2\pi\, 1000t$ ←

(c) $z(t) = \cos(2\pi\, 1000t + 90^\circ)$

$= -\sin 2\pi\, 1000t$ ←

14.2.6.

(a) $a_0 = 1/2$; $a_1 = 2/\pi$; $a_2 = 0$

$x_s(t) = x(t)s(t) = a_0 x(t) + a_1 x(t)\cos 2\pi f_s t$

$= \frac{1}{2}[18\cos 2\pi 20t + 12\cos 2\pi 60t]$

$+ \frac{2}{\pi}[18\cos 2\pi 20t + 12\cos 2\pi 60t]\cos 2\pi f_s t$

$f_s = 100$

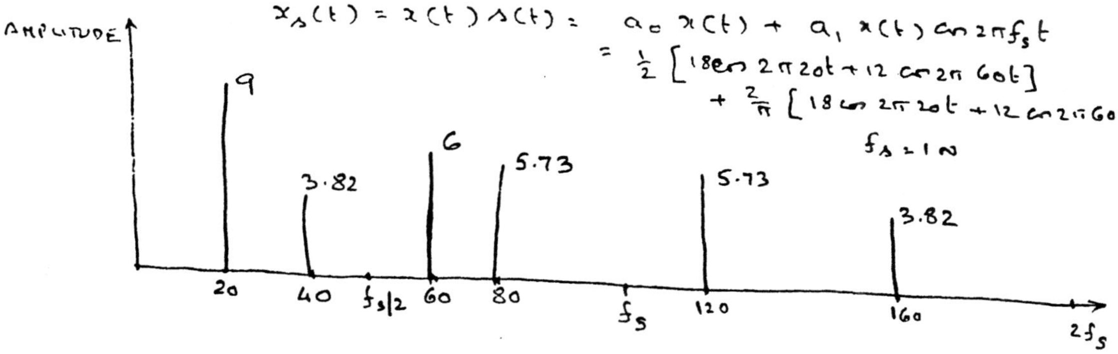

←

(b) $y(t) = 9\cos 2\pi 20t + 3.82\cos 2\pi 40t$ ←

14.2.7. (a)

$a_0 = 1/4$; $a_1 = 1.41/\pi$; $a_2 = 1/\pi$

$f_s - 40 = 30$; $2f_s - 40 = 100$

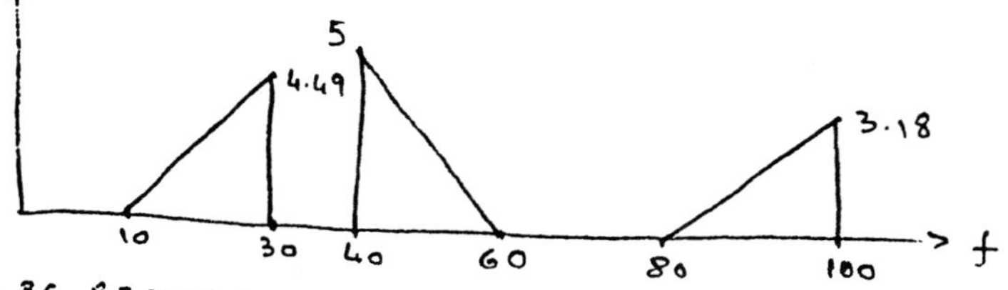

(b)

$x(t)$ CAN BE RECONSTRUCTED USING A BANDPASS FILTER THAT REJECTS $f \leq 30$ AND $f \geq 80$

14.2.8.

$$D + 2 \times 2D \leq T_s$$

$$\therefore \quad D_{max} = \frac{1}{5f_s} = 10\,\mu s \quad \leftarrow$$

$$B \geq 1/D = 5f_s = 100\text{ kHz} \quad \leftarrow$$

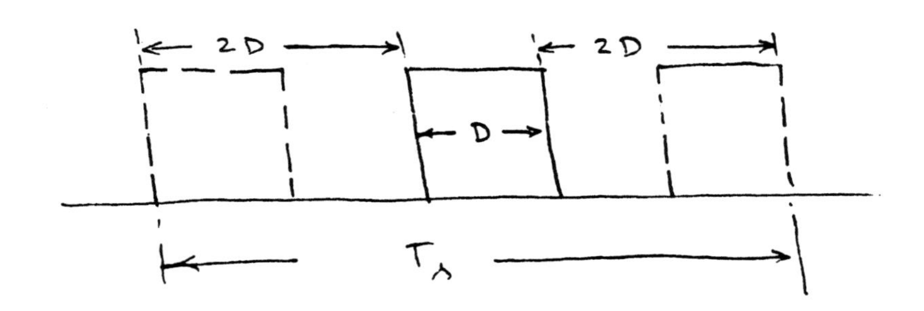

14.2.9.

(a). $B \geq 1/D$ AND n SIGNALS REQUIRE

$$nD \leq T_s \leq 1/2W$$

$$\therefore \quad n/B \leq 1/2W$$

THUS $n \leq 250\text{ kHz}/8\text{ kHz} = 31.25$

$$\Longrightarrow n = 31 \quad \leftarrow$$

(b) $B \geq 1/D$ AND n SIGNALS REQUIRE

$$nD \leq T_s/2 \quad \text{WITH } T_s \leq 1/2W$$

$$\therefore \quad n/B \leq 1/4W$$

THUS $n \leq 250\text{ kHz}/16\text{ kHz} = 15.625$

$$\Longrightarrow n = 15 \quad \leftarrow$$

14.3.1.

$$G = 50\text{ dB} \triangleq 10^5$$

$$N_{out} = 10^5\,(25+1)\,kT_o \times 9 \times 10^6 = 10^5 \times 26 \times 4 \times 10^{-21} \times 9 \times 10^6$$

$$= 936 \times 10^{-10} = 9.36 \times 10^{-8}\text{ W}$$

$[G\,k\,(T+T_a)\,B =]$

$$n_{rms} = \sqrt{R\,N_{out}} = \sqrt{(100)\,9.36 \times 10^{-8}} = \sqrt{9.36 \times 10^{-6}} = 3.06\text{ mV} \leftarrow$$

14.3.2.

WITH $T_a = 0$, $\frac{P_{out}}{N_{out}} = \frac{120 \times 10^{-12}}{4 \times 10^{-21} \times 600 \times 10^{3}} = 5 \times 10^{4}$

WITH $T_a \neq 0$, $P_{out}/N_{out} = \frac{5 \times 10^{4}}{\left(1 + \frac{T_a}{T_0}\right)}$

SO $\frac{P_{out}}{N_{out}} \geq 10^{5} \Rightarrow T_a \leq -0.5\,T_0$ WHICH IS IMPOSSIBLE.

14.3.3.

(i) $\alpha = 0.25$

$$|H(\omega)|^2 = \frac{1}{1 + (\omega/\omega_{co})^2} = 0.25^2 \Rightarrow 1 + (\omega/\omega_{co})^2 = 16$$

$$\therefore \omega_{co} = \omega/\sqrt{15} \Rightarrow f_{co} = 120/\sqrt{15} \simeq 31 \leftarrow$$

$$|H(\omega)|^2 = \frac{1}{1 + (\omega/\omega_{co})^4} = 0.25^2 \Rightarrow 1 + (\omega/\omega_{co})^4 = 16.$$

$$\therefore \omega_{co} = \omega/(15)^{1/4} \Rightarrow f_{co} = 120/(15)^{1/4} \simeq 61 \leftarrow$$

SINCE BOTH FILTERS HAVE $f_{co} > 30$ HZ TO PASS THE INFORMATION, USE THE SIMPLER AND LESS EXPENSIVE RC FILTER. ←

(ii) $\alpha = 0.1$

(a) $$|H(\omega)|^2 = \frac{1}{1 + (\omega/\omega_{co})^2} = 0.1^2 \Rightarrow 1 + (\omega/\omega_{co})^2 = 100$$

$$\therefore \omega_{co} = \omega/\sqrt{99} \Rightarrow f_{co} = 120/\sqrt{99} \simeq 12.1 \leftarrow$$

(b) $$|H(\omega)|^2 = \frac{1}{1 + (\omega/\omega_{co})^4} = 0.1^2 \Rightarrow 1 + (\omega/\omega_{co})^4 = 100$$

$$\therefore \omega_{co} = \omega/(99)^{1/4} \Rightarrow f_{co} = 120/(99)^{1/4} = 38 \leftarrow$$

MUST USE THE BUTTERWORTH FILTER TO HAVE $f_{co} > 30$ Hz TO PASS THE INFORMATION ←

14.3.4.
(a)

ω (LOG SCALE)

$$H(j\omega) = \left[\frac{10}{10+j\omega}\right]^2 \left[1 + j2 \times 0.05(\omega/10) - (\omega/10)^2\right]$$

FOR $\omega = 10$, $H(j10) = \dfrac{100 + j10 - 100}{(10+j10)^2} = \dfrac{j10}{(10+j10)^2}$

$$|H(j10)| = \frac{10}{200} = \frac{1}{20}$$

$$H_{dB} = 20 \log\left(\frac{1}{20}\right) = -26$$

(b) AT $f = 0$ AND $f = \infty$, $|H(f)| = 1$

AT $f = f_\ell$ AND $f = f_u$, $|H(f)| = 0.707$ FOR $Q >> 1$

(c) $H(j\omega) = \dfrac{j\omega L + 1/j\omega C}{R + j\omega L + 1/j\omega C} = \dfrac{j(1/R)(\omega L - 1/\omega C)}{1 + j(1/R)(\omega L - 1/\omega C)}$

$$(1/R)(\omega L - 1/\omega C) = Q(\omega/\omega_0 - \omega_0/\omega)$$

(d) $f_0 - f_\ell = f_u - f_0 = 20\text{ Hz} = f_0/2Q$

$\therefore Q = 1000/40 = 25 >> 1$

$(1/R)\sqrt{L/C} = 25 \Rightarrow L/C = (50 \times 25)^2$

$1/\sqrt{LC} = 2\pi \times 1000 \Rightarrow LC = 1/(2000\pi)^2$

SOLVING, $L = 0.199\text{ H}$, $C = 0.127\ \mu\text{F}$ ←

14.3.4.(e)

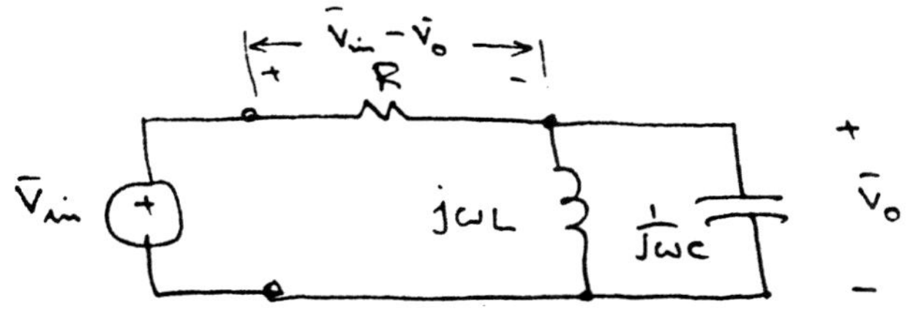

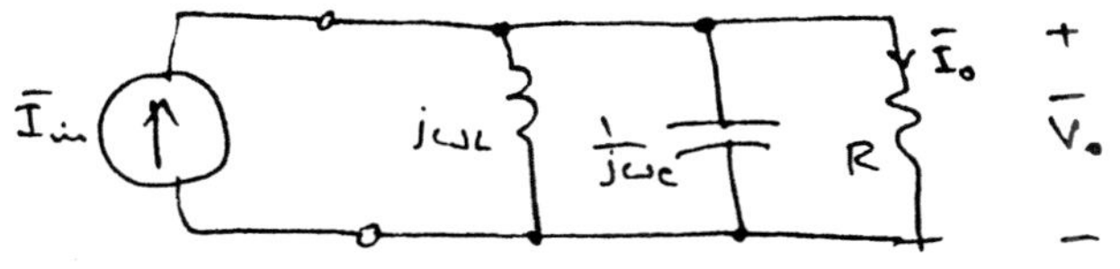

$$\frac{\bar{V}_o}{\bar{V}_{in}} = \frac{\bar{I}_o}{\bar{I}_{in}} = \frac{1}{R\,Y(j\omega)} = \frac{1}{1+j(\omega CR - R/\omega L)}$$

NORTON'S CKT

$$H(j\omega) = \frac{\bar{V}_{in}-\bar{V}_o}{\bar{V}_{in}} = 1 - \frac{\bar{V}_o}{\bar{V}_{in}} = 1 - \frac{1}{1+j(\omega CR - R/\omega L)} = \frac{j(\omega CR - R/\omega L)}{1+j(\omega CR - R/\omega L)}$$

$$= \frac{jR(\omega C - 1/\omega L)}{1+jR(\omega C - 1/\omega L)}$$

$$R(\omega C - 1/\omega L) = Q(\omega/\omega_0 - \omega_0/\omega)$$

(f) $f_o - f_\ell = f_u - f_o = 2\text{ Hz} = f_o/2Q$

$\therefore\ Q = 60/4 = 15 \quad >>1$

$R\sqrt{C/L} = 15 \Rightarrow C/L = (15/1000)^2$

$1/\sqrt{LC} = 2\pi \times 60 \Rightarrow LC = 1/(120\pi)^2$

SOLVING, $C = 39.8\,\mu F$, $L = 0.177\,H$ ←

14.3.5. $N_{out} = Gk(T+T_a)B$

$Gk(290 + T_a)B = 600\,\mu W$

$Gk(80 + T_a)B = 480\,\mu W$

$$\frac{290+T_a}{80+T_a} = \frac{600}{480} = \frac{10}{8} \text{ or } 800 + 10T_a = 2320 + 8T_a$$

or $2T_a = 1520$

or $T_a = 760\text{ K}$ ←

14.3.6.

$N = kTB$ → [G_1, B, T_{a1}] → N_1 → [G_2, B, T_{a2}] → N_{out}

$$N_1 = G_1 k (T + T_{a1}) B$$

$$N_{out} = G_2 N_1 + G_2 k T_{a2} B$$

$$= G_2 G_1 k (T + T_{a1}) B + G_2 k T_{a2} B$$

$$= G_2 G_1 k \left(T + T_{a1} + T_{a2}/G_1 \right) B \quad \leftarrow$$

$$= G k (T + T_a) B$$

$$\therefore \quad T_a = T_{a1} + (T_{a2}/G_1) \quad \leftarrow$$

14.3.7.

For $T = T_0$, $\quad N_{out} = N_1 = Gk(T_0 + T_a) B$

For $T = T_R$, $\quad N_{out} = 2N_1 = Gk(T_R + T_a) B$

$$\therefore \quad \frac{T_0 + T_a}{T_R + T_a} = \frac{1}{2} \implies T_a = T_R - 2T_0 \quad \leftarrow$$

14.3.8.

$$\frac{P_{out}}{N_{out}} = \frac{10^{-6}}{4 \times 10^{-21} (2 + T_a/T_0)\; 250 \times 10^3} \geq 1.2 \times 10^8$$

$$\therefore \quad 2 + \frac{T_a}{T_0} \leq 12 \implies T_a \leq 10\, T_0 \quad \leftarrow$$

14.3.9.

$$F = 1 + T_a/T_0 \implies T_a = (F-1) T_0$$

$$\therefore N_{out} = Gk(T + T_a) B = GkTB + Gk(F-1)T_0 B$$

$$\frac{P_{out}}{N_{out}} = \frac{G P_{in}}{GkTB + Gk(F-1)T_0 B}$$

$$= \frac{P_{in}/kTB}{1 + (F-1) T_0/T} = \frac{P_{in}/N}{1 + (F-1) T_0/T} \quad \leftarrow$$

If $T = T_0$, $\quad \dfrac{P_{out}}{N_{out}} = \dfrac{P_{in}}{FN} \quad \leftarrow$

14.3.10.

$$\epsilon = x \Big/ \sqrt{P_{out}/N_{out}} = x \Big/ \sqrt{10^4} = 0.01\,x$$

$$\epsilon_M = \epsilon / \sqrt{M} \quad ; \quad \epsilon_M / x = 0.01 / \sqrt{M} = 0.2 \times 10^{-2}$$

$$\Rightarrow M = 25$$

TIME REQUIRED TO OBSERVE M SAMPLES SPACED BY $1/B$ IS

$$(M-1) \times 1/B = (25-1)\ 1/8 = 3 \text{ SECONDS} \leftarrow$$

14.3.11.

$$N_{out} = Gk\,(T + T_a)\,B = 10^8\,k\,T_0$$

WITH $B = 25 \times 10^3$, $G = 10^3$, $T = T_0$

$$10^3\,(T_0 + T_a)\,25 \times 10^3 = 10^8\,T_0$$

$$T_0 + T_a = T_0 \cdot \frac{10^8}{25 \times 10^6} = 4T_0$$

$$\text{or } T_a = 3T_0 \leftarrow$$

NOISE FIGURE $F = 1 + \dfrac{T_a}{T_0} = 1 + 3 = 4 \leftarrow$

14.3.12.

$$F = F_1 + \frac{F_2 - 1}{G_1} + \frac{F_3 - 1}{G_1 G_2}$$

$$F_1 = F_2 = F_3 = 6 \quad ; \quad G_1 = G_2 = 5$$

$$\therefore \quad F = 6 + 1 + 0.2 = 7.2$$

$$\text{OR} \quad 8.57\,\text{dB} \leftarrow$$

NOTE THAT THE DOMINANT TERM IS F_1 WHICH IS THE NOISE FIGURE OF THE FIRST AMPLIFIER STAGE

REMAINING TERMS IN THE SUM WILL BE NEGLIGIBLE IF THE STATEMENT IS TRUE.

HENCE JUSTIFICATION OF THE STATEMENT ←

14.3.13.

(a) NOISE FIGURE IS 2 dB

$$10 \log (1 + T_a/T_o) = 2$$

WITH $T_o = 290$

$$\therefore \quad T_a = 169.62\,K$$

(b) $N_{out} = G k B (T + T_a)$

$10 \log G = 30 \quad \therefore G = 3000$

$$N_{out} = 3000 \times 1.38 \times 10^{-23} \times 10 \times 10^6 (169.62 + 60)$$

$$= 9.5 \times 10^{-11} \text{ WATTS}$$

$$= 0.95 \text{ pW}$$

CHAPTER 15

15.1.1.

$$v_g = \frac{c}{\sqrt{\epsilon_r}} = \frac{3\times10^8}{\sqrt{2.26}} \simeq 0.665\times3\times10^8 \simeq 2\times10^8 \text{ m/s} \leftarrow$$

$$\text{DELAY} = \tau = \frac{L}{v_g} \simeq \frac{30}{2\times10^8} = 0.15\,\mu s \leftarrow$$

15.1.2.

$$\ln\frac{b}{a} = \frac{\sqrt{\epsilon_r}\,R_0}{60} = \frac{\sqrt{2.26}\;50}{60} = 1.253$$

$$\therefore \frac{b}{a} = e^{1.253} = 3.5 \leftarrow$$

15.1.3.

$\epsilon_r = 1$ FOR AIR

$$a = b\,e^{-50/60} = 10\,e^{-5/6} \simeq 4.346 \text{ mm}$$

$$f_c = \frac{3\times10^8}{\pi(10+4.346)10^{-3}} \simeq 6.656 \text{ GHz} \leftarrow$$

15.1.4.

$$\text{ATTENUATION}|_c = \frac{1.373\times10^{-3}\sqrt{1.72\times10^{-8}\times3\times10^9}}{50}\left(\frac{1}{0.01}+\frac{1}{0.004346}\right)$$

VALUES USED FROM 15.1.3

$$\simeq 65.113\,(10^{-3}) \text{ dB/m}$$

$$\therefore \text{LENGTH} \le \frac{3}{0.065113} \simeq 46.074 \text{ m} \leftarrow$$

15.1.5.

$$f_c = \frac{c}{2a} = \frac{3\times10^8}{2(0.5842)} = 256.76 \text{ MHz}$$

$$2f_c = 513.52 \text{ MHz}$$

THEORETICAL BAND IS $256.76 < f < 513.52$ MHz $\leftarrow$

PRACTICAL BAND IS $1.25 f_c \le f \le 1.9 f_c$

$$320.9517 \le f \le 487.8466 \text{ MHz} \leftarrow$$

15.1.6. (a)

$$f_c = \frac{c}{2a} = \frac{3\times10^8}{2(0.02286)} = 6.5617 \text{ GHz}$$

RANGE: $1.25 f_c \le f \le 1.9 f_c$

$f_1 = 1.25 f_c = 8.2021$ GHz ; $f_2 = 1.9 f_c = 12.4672$ GHz

$$Z_0 = R_0 = \frac{377}{\sqrt{1-(f_c/f)^2}} = \frac{377}{\sqrt{1-\left(\frac{1}{1.25}\right)^2}} = 628.33\,\Omega \text{ AT } 1.25 f_c \leftarrow$$

15.1.6.(a) CONTD. $Z_0 = \dfrac{377}{\sqrt{1-\left(\frac{1}{1.9}\right)^2}} = 443.38\,\Omega$ at $1.9 f_c$. ←

(b) $$\text{ATTENUATION}\Big|_c = \frac{(0.458\times10^{-4})\sqrt{1.25 f_c \rho}\left[1+\frac{2b}{a}\left(\frac{1}{1.25}\right)^2\right]}{b\sqrt{1-\left(\frac{1}{1.25}\right)^2}}$$

WITH $f_c = 6.5617$ GHz AND $\rho = 3.9\times10^{-8}\ \Omega\cdot m$

AND $b = 0.01016$ m AND $a = 0.02286$ m

$\text{ATT.}|_c = 0.21082$ dB/m $= 21.082$ dB/100m AT $1.25 f_c$ ←

AT $1.9 f_c$, $\text{ATT.}|_c = 14.568$ dB/100m ←

15.1.7. $f_c = 2.5678\times10^8$ Hz FROM PR. 15.1.5

USING THE FORMULA AND SUBSTITUTING VALUES, WE GET

$\text{ATT.}|_c = 0.001292$ dB/m or 0.1292 dB/100m AT $1.25 f_c$ ←

$= 0.000875$ dB/m or 0.0875 dB/100m AT $1.9 f_c$ ←

15.1.8.

$$f_c = \frac{3\times10^8}{2(0.8636)10^{-3}} = 173.6915\text{ GHz}$$

RANGE $f_c < f < 2f_c$ i.e. $173.6915 < f < 347.3830$ GHz ←

PRACTICAL RANGE $1.25 f_c \le f \le 1.9 f_c$ i.e. $217.1144 \le f \le 330.0139$ GHz ←

15.1.9. $f_c = \dfrac{0.293c}{a}$ OR $2a = \text{DIA} = \dfrac{2(0.293)3\times10^8}{10(10^9)} = 17.58$ mm ←

UPPER LIMIT $= 1.307 f_c = 13.07$ GHz ←

15.1.10.

$$v = \frac{c}{\sqrt{\epsilon_r}} = \frac{3\times10^8}{\sqrt{2.3}} = 1.98\times10^8\ \text{m/s}$$

$$\lambda = \frac{v}{f} = \frac{1.98\times10^8}{150\times10^6} = 1.32\text{ m};\quad \frac{\lambda}{4} = \frac{1.32}{4} = 0.33\text{ m}$$ ←

15.1.11.

$$\lambda = \frac{c}{f} = \frac{3\times10^8}{f}$$

AT 1kHz, $\lambda = 3\times10^5$ m ; L IN WAVELENGTHS $= \dfrac{25}{3\times10^5} = 8.33\times10^{-5}$ ←

AT 10MHz, $\lambda = 30$ m ; L " $= \dfrac{25}{30} = 0.833$ ←

AT 100MHz, $\lambda = 3$ m ; L " $= \dfrac{25}{3} = 8.33$ ←

15.1.12. $v = c/\sqrt{\epsilon_r} = 3\times10^8/\sqrt{3.5} = 1.6\times10^8$ m/s
AT 10 GHz, $\lambda = v/f = 1.6\times10^8/10\times10^9 = 0.016$ m
L IN WAVELENGTH = 100/0.016 = 6250 WAVELENGTHS LONG ←

15.1.13.(a) AT LOW FREQUENCIES, WAVEGUIDES BECOME EXCESSIVELY LARGE ←

(b) AT HIGH FREQUENCIES, OPEN-WIRE LINES CAN BECOME VERY LOSSY DUE TO THE RADIATION THAT CAN OCCUR. ←

(c) $$v = \frac{c}{\sqrt{\epsilon_r}} = \frac{3\times10^8}{\sqrt{2.1}} = 2.07\times10^8 \text{ m/s} \leftarrow$$

15.1.14 (a)

TEM : BOTH ELECTRIC AND MAGNETIC FIELDS ARE PERPENDICULAR TO THE DIRECTION OF PROPAGATION

TE : ELECTRIC FIELD IS $\perp$r TO THE DIR. OF PROPAGATION

(b)
CUTOFF WAVELENGTH : THAT WAVELENGTH, IN FREE SPACE OR IN THE UNBOUNDE GUIDE MEDIUM ABOVE WHICH A TRAVELING WAVE IN A GIVEN MODE CANNOT BE MAINTAINED IN A GUIDE

FOR AN AIR-FILLED RECTR. WAVEGUIDE, FOR THE PROPAGATION BY THE DOMINANT MODE, $f_c = \frac{c}{2a}$ AND $\lambda_c = \frac{c}{f_c} = 2a$

DOMINANT MODE: THE MODE OF PROPAGATION WITH THE LOWEST CUTOFF FREQUENCY

15.1.15

$$f_c = \frac{c}{2a} = 3\times10^9$$
$$\therefore\ a = \frac{3\times10^{10}}{6\times10^9} = 5 \text{ cm}$$

WITH $\epsilon_r = 3.24$, $$f_c = \frac{c}{\sqrt{\epsilon_r}\,2a} = \frac{3\times10^{10}}{\sqrt{3.24}\,(10)} = 1.67 \text{ GHz} \leftarrow$$

15.1.16.
THE REFLECTION COEFFICIENT CAUSES A REFLECTED-WAVE AMPLITUDE OF 0.3 TIMES THE INCIDENT WAVE'S AMPLITUDE. THE REFLECTED POWER BECOMES $(0.3)^2(15) = 1.35$ kW. SINCE THE TRANSMITTER IS MATCHED ON THE LINE, ALL THE REFLECTED POWER IS DISSIPATED IN ITS OUTPUT IMPEDANCE. HOWEVER, SINCE IT IS LESS THAN 2 kW, THE TRANSMITTER SURVIVES. POWER RADIATED BY ANTENNA BECOMES $15 - 1.35 = 13.65$ kW ←

15.1.17. FROM (15.1.3) AND (15.1.4)

$$v_1(t) = Re\left[\frac{75}{100+75}\; 12.5\, e^{j\omega_0 t}\right] \simeq Re\left[5.357\, e^{j\omega_0 t}\right] = 5.357 \cos\omega_0 t \leftarrow$$

$$v_L(t) = Re\left[\frac{1}{100+75}\; 12.5\, e^{j\omega_0 t}\right] \simeq Re\left[71.429\,(10^{-3})\, e^{j\omega_0 t}\right]$$

$$= 71.429\,(10^{-3}) \cos\omega_0 t \leftarrow$$

15.1.18.

(a)

$$\bar{Z}_0 = \sqrt{\frac{R+j\omega L}{G+j\omega C}} = \sqrt{\frac{32.1\angle 64.2^\circ}{62.8\times 10^{-6}\angle 89.7^\circ}} = 715\angle -12.8^\circ\ \Omega \leftarrow$$

$$\bar{\gamma} = \sqrt{\bar{Z}\bar{Y}} = \sqrt{(32.1\angle 64.2^\circ)(62.8\times 10^{-6}\angle 89.7^\circ)} = 0.0449\angle 77^\circ$$

$$= (0.01 + j\,0.0438) \text{ PER MILE}$$

$$\alpha = 0.01 \text{ Np/mi};\quad \beta = 0.0438 \text{ rad/mi} \leftarrow$$

(b) $\bar{Z}_0 = 715\angle -12.8^\circ = (697 - j\,158)\ \Omega$

$$\bar{\gamma} = \alpha + j\beta = 0.01 + j\,0.0438 \text{ PER MILE}$$

MATCHED LINE $\bar{Z}_S = \bar{Z}_0$

$$\bar{E}_S = \bar{E}_g \frac{\bar{Z}_0}{\bar{Z}_g + \bar{Z}_0} \quad \text{FROM} \rightarrow$$

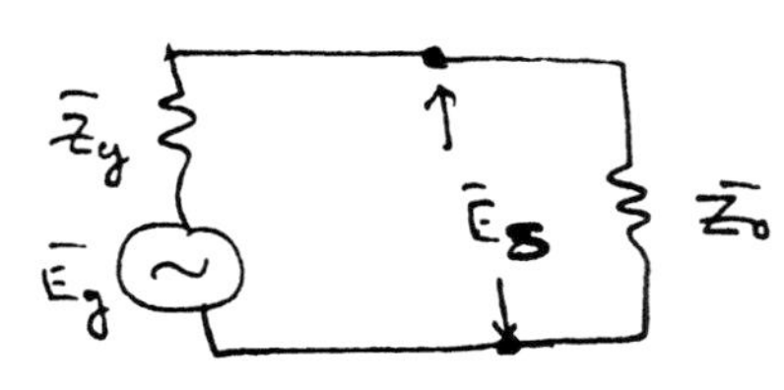

EQUIVALENT SENDING-END CIRCUIT FOR A MATCHED LINE

(i) $$\bar{I}_S = \frac{\bar{E}_g}{\bar{Z}_g + \bar{Z}_0} = \frac{10\angle 0^\circ}{600 + 697 - j158}$$

$$= \frac{10\angle 0^\circ}{1307\angle -6.96^\circ} = 7.65\angle 6.96^\circ \text{ mA} \leftarrow$$

(ii) $\bar{E}_S = \bar{I}_S \bar{Z}_S = 7.65\angle 6.96^\circ \times 10^{-3} \times 715\angle -12.8^\circ = 5.47\angle -5.8^\circ \leftarrow$

(iii) $P_S = E_S I_S (pf) = 5.47 \times 7.65 \times 10^{-3} \cos(6.96^\circ + 5.8^\circ) = 40.8 \text{ mW} \leftarrow$

OR $I_S^2 R_S = (7.65\times 10^{-3})^2\, 697 = 40.8 \text{ mW}$

(iv) $$\bar{E}_R = \bar{E}_S\, e^{-\alpha l}\, e^{-j\beta l} = 5.47\angle -5.8^\circ\, e^{-0.01\times 100}\, e^{-j0.438\times 100}$$

$$= 5.47\, e^{-1.0} \angle -5.8^\circ - 4.38 \text{ rad}$$

$$= 5.47 \times 0.368 \angle -5.8^\circ - 251^\circ$$

$$= 2.02\angle -257^\circ \leftarrow$$

15.1.18. (b) CONTD.

(v) $\bar{I}_R = \frac{\bar{E}_R}{\bar{Z}_R} = \frac{2.02 \angle -257°}{715 \angle -12.8°} = 2.83 \angle -244° \text{ mA}$ ←

(vi) $P_R = |\bar{I}_R|^2 R_R = (2.83 \times 10^{-3})^2 \, 697 = 5.58 \text{ mW}$ ←

(vii) $10 \log (40.8/5.58) = 8.7 \text{ dB}$

(c)

$\lambda = 2\pi/\beta = \frac{2\pi}{0.0438} = 143.5 \text{ mi}$; l in λ's $= \frac{100}{143.5} = 0.697\lambda$ ←

Loss in Nepers $= \alpha l = 0.01 \times 100 = 1 \text{ Np}$ ←

Loss in dB $= 8.686 \times 1 = 8.7 \text{ dB}$ ←

(d) LOSSLESS LINE : $\alpha = 0$; $\beta = \omega\sqrt{LC}$

SO THAT $\bar{\gamma} = j\omega\sqrt{LC}$; $\alpha = 0$; $\beta = 2\pi(1000)\sqrt{4.6\times10^{-3}\times 0.01\times10^{-6}} = 0.0426$ rad/mi

$v_p = \omega/\beta = \frac{2\pi(1000)}{0.0426} \simeq 14.75\times10^4$ mi/s WHICH IS $\frac{14.75\times10^4}{18.6\times10^6} \simeq 0.793$ OF VEL. OF LIGHT ←

15.1.19. (a) A TRAVELING WAVE MOVING IN THE +Z DIRECTION, EXPERIENCES CONTINUED ATTENUATION AS IT PROGRESSES, THE RATE BEING DEPENDENT UPON α. ←

(b) AN EXPONENTIALLY INCREASING TRAVELING WAVE MOVING IN −Z DIRECTION. ←

15.1.20.

$L_1 = 10^{1.5\times20/10} = 1000$; $L_2 = 10^{1.5\times18/10} = 500$

$G = G_1 G_2 G_3 = (L_1 L_2) P_{out}/P_{in} = 1000 \times 500 \times 200/5 = 2\times10^7$ ←

$G_1 = 200/5 = 40$; $G_2 = \frac{200}{200/L_1} = \frac{200\times L_1}{200} = L_1 = 1000$

$G_3 = \frac{200}{200/L_2} = L_2 = 500$

15.1.21.

$\alpha l = 0.6\times50 = 30 \text{ dB} \Rightarrow L = 10^3$

$G_1 G_2 / L = P_{out}/P_{in} = 200$

$\therefore G_1 G_2 = 200\times10^3 = 2\times10^5$

TAKING $G_1 = 1\text{W}/P_{in} = 500$ or 27 dB ←

$G_2 = 2\times10^5/500 = 400$ or 26 dB ←

15.1.22. $G_1 P_{in} = 10^{2.3} \times 0.001 = 0.2$

$L_1 = 10^{2.5 l_1 / 10}$

$G_1 P_{in} / L_1 = 20 \mu W = 2 \times 10^{-5} \Rightarrow 10^{0.25 l_1} = \frac{0.2}{2 \times 10^{-5}} = 10^4$

$\therefore l_1 = 4/0.25 = 16 \text{ km} \leftarrow$

$l_2 = 30 - 16 = 14 \text{ km} \leftarrow \qquad L_2 = 10^{3.5} = 3162$

$G_2 \times 20 \mu W / L_2 = 20 \mu W \Rightarrow G_2 = L_2 = 3162$ or $35 dB \leftarrow$

$G_3 = P_{out} / 20 \mu W = 2500$ or $34 dB \leftarrow$

15.1.23. $f_c = 30 \times 100 \text{ MHZ} = 3 \text{ GHZ}; \ \lambda = 10^{-4} \text{ km} = 10 \text{ cm};$

$G_t = G_r = 4\pi (\pi 50)^2 / 10^2$ USING (15.1.17) $\simeq 1000$

$\therefore L = \frac{1}{10^6} \left(\frac{4\pi \times 40}{10^{-4}} \right)^2 = 2.5 \times 10^7$ OR 74 dB $\leftarrow$ [USING (15.1.25)]

15.1.24. $\lambda = c/f_c = 3 \times 10^8 / (500 \times 10^6) = 6 \times 10^{-4} \text{km} = 0.6 \text{ m}$

$G_t G_r = (10^{0.2})^2 = 2.5; \ P_{out} = G_t G_r \left(\frac{\lambda}{4\pi R} \right)^2 P_{in} = 1.41 \times 10^{-10} W = 141 pW \leftarrow$

15.1.25. $G_t = 10^3; \ G_r = 4\pi (4\pi r^2)/\lambda^2$

$P_{out} = 10^3 \, 4\pi (\pi r/\lambda)^2 (\lambda / 4\pi R)^2 \, 3 = 30 \times 10^{-12}$

$r = 2 \times 10^{-7} R = 2 \times 10^{-7} \times 30 \times 10^6 = 6 \text{ m} \leftarrow$

15.1.26. $\lambda = c/f_c = 5 \times 10^{-5} \text{ km}; \ G_t = G_r = 4\pi A_e / \lambda^2$

$L = (\lambda^2 / 4\pi A_e)^2 (4\pi R/\lambda)^2 = 10^6$

$A_e = \lambda R / 10^3 = 5 \times 10^{-2} \times 40 \times 10^3 / 10^3 = 2 m^2 \leftarrow$

15.1.27.

(a) $G = \frac{4\pi A_e}{\lambda^2} = \frac{4\pi \rho_a A}{\lambda^2} = \frac{4\pi (0.6) 10 (5 \times 10^9)^2}{(3 \times 10^8)^2} = 20.94 \times 10^3 \leftarrow \left[\because \lambda = \frac{c}{f} \right]$

(b) $S = \frac{P_t G_t}{4\pi R^2} = \frac{(2 \times 10^3)(20.94 \times 10^3)}{4\pi \ (20 \times 10^3)^2} = 0.0083$ OR $8.3 \times 10^{-3} \ W/m^2 \leftarrow$

15.1.28.

MAX. RAD. INTENSITY $= \dfrac{PG}{4\pi}$ FROM EQ. (15.1.18)

$$= \frac{(1000)\,(10{,}000)}{4\pi} = 0.796 \times 10^6 \text{ W/steradian} \leftarrow$$

15.1.29.

FROM (15.1.27), $G = \dfrac{41.3 \times 10^3}{3\,(10)} = 1.38 \times 10^3$ or 31.4 dB

(NOTE: $G_{dB} = 10 \log G$)

IF ALL POWER WERE RADIATED, THEN MAXM. INTENSITY BY EQ. (15.1.18) IS

$$\frac{(1.38 \times 10^3)\,(10^3)}{4\pi} = 109.8 \text{ kW/sr.}$$

HOWEVER ONLY 0.6 OF THAT IS EFFECTIVELY TRANSMITTED:

$\therefore$ MAX. INTENSITY $= 0.6\,(109.8) = 65.88$ kW/sr. $\leftarrow$

15.1.30. (a)

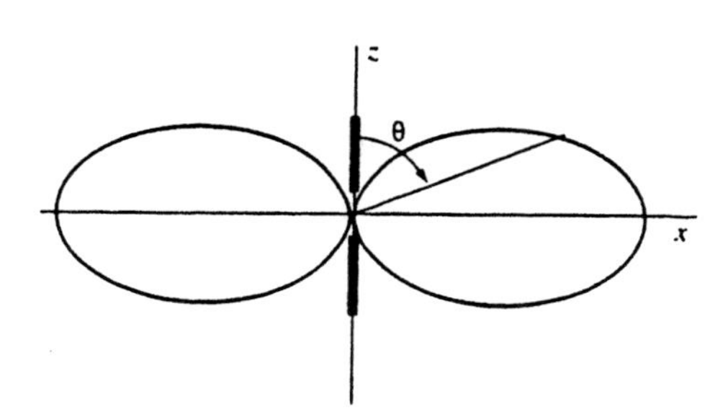

(b) $\cos^2\left[\frac{\pi}{2}\cos\theta\right] / \sin^2\theta = 0.1$; BY TRIAL & ERROR $\theta = 22.82°$; $\therefore \theta_B = 2\,(90 - 22.82°) = 134.36°$ $\leftarrow$

15.1.31.

$\lambda = \dfrac{c}{f} = \dfrac{3 \times 10^8}{3 \times 10^9} = 0.1$ m HALF-POWER BEAMWIDTH $= \theta_B = 78.08°$ $\leftarrow$

$A_e = 0.13\,\lambda^2 = 0.13\,(0.1)^2 = 0.0013\ m^2$ or $13\ cm^2$ $\leftarrow$

15.1.32. $\lambda = \dfrac{c}{f} = \dfrac{3 \times 10^8}{5 \times 10^8} = 0.6$ m ; $\therefore C = \lambda = 0.6$ m $\leftarrow$

$S = C \tan\alpha = 0.6 \tan(14\pi/180) \simeq 0.15$ m $\leftarrow$

$\therefore L = NS = 1.5$ m $\leftarrow$

$$\theta_B \simeq \frac{52\,(0.6)^{3/2}}{0.6\sqrt{1.5}} \simeq 32.9° \leftarrow$$

$$G_D \simeq \frac{12\,(10)\,(0.6)^2\,(0.15)}{(0.6)^3} \simeq 30 \text{ or } 14.8 \text{ dB} \leftarrow$$

$$Z_a \simeq \frac{140\,(0.6)}{0.6} = 140\,\Omega \leftarrow$$

15.1.33. $G_D = \dfrac{2.05 \times \pi \times 6 \times 4.86 \times \lambda^2}{\lambda^2} = 187.8$ OR 22.74 dB ←

$\Theta_B = \dfrac{54\lambda}{4.86\lambda} = 11.11°$ ← ; $\Phi_B = \dfrac{78\lambda}{6\lambda} = 13°$ ← ; YES, OPTIMUM, ∵ $B = 0.81A$. ←

15.1.34. $d = 0.025$ m; $\lambda = \dfrac{3\times10^8}{8\times10^9} = 0.0375$ m; $D = 5.771\lambda = 0.216413$ m;

(a) $L_1 = 10\lambda = 0.375$ m; $L \simeq L_1\left(1 - \dfrac{d}{D}\right) = 0.375\left(1 - \dfrac{0.025}{0.216413}\right) = 0.3317$ m ←

(b) $G_D \simeq 5.13(5.771)^2 = 170.8518$ OR 22.33 dB ←

(c) $\Theta_B \simeq \dfrac{60}{5.771} = 10.397°$ ← ; $\Phi_B = \dfrac{70}{5.771} = 12.13°$ ←

15.1.35.

$$G = \frac{4\pi \rho_a A}{\lambda^2} = 4\pi(0.6)\pi(50\lambda)^2/\lambda^2 = 5.9218\times10^4$$

OR 47.72 dB

EQ. (15.1.27):

$$G = \frac{41.3\times10^3}{\Theta_B \Phi_B} \Rightarrow \Theta_B = \Phi_B = \sqrt{\frac{41.3\times10^3}{5.9218\times10^4}} = 0.835°$$ ←

15.1.36.

$$G \simeq \frac{41.3\times10^3}{1^2} = 4.13\times10^4 = G_t = G_r$$

$$4.13\times10^4 = \frac{4\pi\rho_a A}{\lambda^2} = 4\pi\rho_a\pi\left(\frac{D}{2}\right)^2\frac{1}{\lambda^2} = \pi^2\rho_a\left(\frac{D}{\lambda}\right)^2$$

$$\therefore D = \sqrt{\frac{41.3\times10^3}{\pi^2(0.9)}}\lambda = 72.32\lambda = 72.32\,\frac{3\times10^8}{8(10^9)} = 2.7121 \text{ m}$$ ←

EQ. (15.1.24):

$$S_r = \frac{P_t G_t G_r \lambda^2}{(4\pi)^2 R^2 L} = \frac{60(4.13)^2 10^8 (0.3)^2/64}{16\pi^2(16\times10^8)(2)\left(\frac{1}{0.85}\right)^2(2.5)(1.5)}$$

(NOTE: $L = L_t L_{ch} L_r$)

$$= 54.9\times10^{-6} \text{ W}$$ ←

15.1.37.(a)

FROM (15.1.29) : $T_{sys} = 85 + 285(1.33-1) + 250(1.33) = 511.55\text{K}$

$$G_t = G_r = \frac{4\pi \rho_a A}{\lambda^2} = 4\pi(0.6)\,\frac{\pi(50\lambda)^2}{4}\,\frac{1}{\lambda^2} = 14804.41$$

$$\lambda = \frac{3\times10^8}{35\times10^9} = 0.008571 \text{ m} = 8.571\times10^{-3} \text{ m}$$

RAIN ATTENUATION: $3.9\times6 = 23.4$ dB

CLEAR-AIR ATTENUATION: $0.072\times30 = 2.16$ dB

$\therefore L_{ch}$ (CLEAR) $= 1.64437$; L_{ch} (RAIN) $= 218.7762$

FROM EQ. (15.1.33): [Note: 45 dB $\rightarrow 3.16228\times10^4$]

$$\left(\frac{S}{N}\right)_A = 3.16228\times10^4 = \frac{P_t\times(1.48\times10^4)^2\,(8.571\times10^{-3})^2}{(4\pi)^2(30\times10^3)^2(1.33)(218.7762)(1.64437)\times(1.38\times10^{-23})(511.55)(12\times10^6)}$$

$$= 0.28\times10^4\,P_t$$

$P_t = 11.3$ W $\leftarrow$

(b)

$$N_{ao} = \frac{(1.38\times10^{-23})(511.55)(12\times10^6)\,10^7}{1.33} \quad [\text{EQ } (15.1.30)$$

$$= 6.37\times10^{-7} \text{ W} \leftarrow$$

(c)

WITH RAIN: $S_{ao} = \left(\frac{S}{N}\right)_A N_{ao} = (3.16228\times10^4)(6.37\times10^{-7})$

$= 20.14\times10^{-3}$ W

or 20.14 mW $\leftarrow$

WITHOUT RAIN: S_{ao} IS LARGER BY 218.7762 TIMES

$S_{ao} = 20.14(10^{-3})\,218.7762 = 4.41$ W $\leftarrow$

15.1.38.

$$T_a = \frac{dN_{ao}}{k\,df} = \frac{1.6\times10^{-15}}{(1.38\times10^{-23})10^6} = 1.16\times10^2 = 116 \text{ K} \leftarrow$$

15.1.39.

$$T_e = T_L(L_n - 1) = 280(2.1878-1) = 332.57 \text{ K} \leftarrow$$

(↑ NUMERIC)

(NOTE: $G_{dB} = 10\log G$; $3.4 = 10\log G$; $G = 10^{0.34} = 2.1878$)

15.1.40.

$$dN_{ao} = kT_a\,df\,\frac{1}{L_r} + kT_L(L_r-1)\,df = kT_s\,df$$

$$\therefore\ T_s = \frac{T_a}{L_r} + \frac{T_L(L_r-1)}{L_r}$$

$$= \frac{130}{1.2023} + \frac{280(1.2023-1)}{1.2023} = 108.13 + 47.11$$

$$= 155.24\text{ K} \leftarrow$$

15.1.41.

$$T_{sys} = T_a + T_L(L_r-1) + \bar{T}_R L_r$$

$$= 130 + 280\,(1.2023-1) + 300\,(1.2023)$$

$$= 130 + 56.64 + 360.69 = 547.33\text{K} \leftarrow$$

$$N_{ao} = k\,T_{sys}\,B_N\,G_a(f_0)\,/\,L_r$$

$$= \frac{1.38\times10^{-23}\,(547.33)\,10^{7}\times10^{12}}{1.2023} = 0.0628\text{ W} \leftarrow$$

15.1.42.
(a)

Eq. (15.1.28) : $\bar{T}_R = 290(3-1) = 580\text{ K}$

Eq. (15.1.29) : $T_{sys} = 200 + 250\,(1.45-1) + 580(1.45)$

$$= 1153.5\text{ K}$$

Eq. (15.1.32) : $N_{aA} = (1.38\times10^{-23})(1153.5)(14\times10^{6})$

$$\simeq 2.23\times10^{-13}\text{ W}$$

$$0.223\times10^{-12}\text{ W} \leftarrow$$

(b)

$$S_A = L_r S_r = (1.45)(1.1\times10^{-8})\text{ W}$$

$$N_{aA} = 0.223\times10^{-12}\text{ W}$$

$$\left(\frac{S}{N}\right)_A = \frac{S_A}{N_{aA}} = \frac{1.45\times1.1\times10^{-8}}{0.223\times10^{-12}}$$

$$= 7.152\times10^{4}\text{ OR } 48.54\text{ dB} \leftarrow$$

15.2.1. $(1605-535)/10 = 107$ STATION FREQUENCIES ←

BY ALLOWING 10kHz CHANNELS TO BE CENTERED AT THE END FREQUENCIES OF 1600 AND 540 kHz.

15.2.2.

(a)

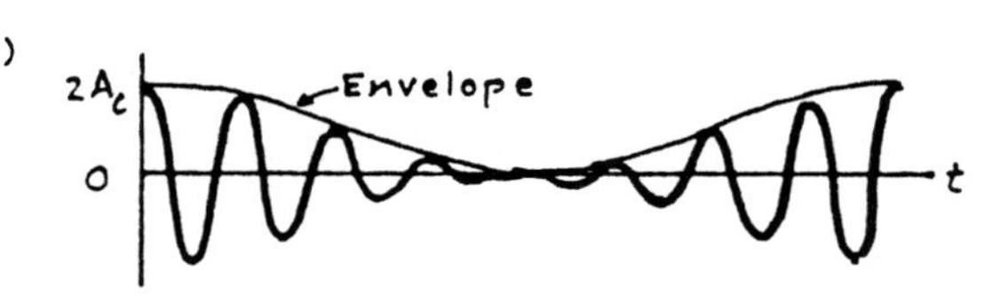

(b)

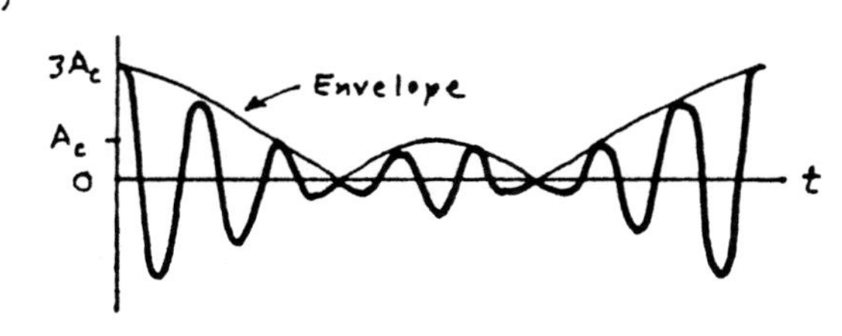

15.2.3. $y = x + A_c \cos 2\pi f_c t$

$$z = y + ay^2 = x + A_c \cos 2\pi f_c t + ax^2 + 2axA_c \cos 2\pi f_c t + aA_c^2 \cos^2 2\pi f_c t$$

$$\cos^2 2\pi f_c t = \frac{1}{2} + \frac{1}{2}\cos 4\pi f_c t$$

$$x^2 = A_m^2 \cos^2 2\pi f_m t = \frac{A_m^2}{2}[1 + \cos 4\pi f_m t]$$

$$\therefore z = A_m \cos 2\pi f_m t + A_c \cos 2\pi f_c t$$
$$+ \frac{A_m^2 a}{2}[1 + \cos 4\pi f_m t]$$
$$+ 2a(A_m \cos 2\pi f_m t)(A_c \cos 2\pi f_c t)$$
$$+ \frac{aA_c^2}{2}[1 + \cos 4\pi f_c t]$$
$$= A_c \cos 2\pi f_c t + A_c\, 2a\, x(t) \cos 2\pi f_c t$$
$$+ \frac{A_m^2 a}{2}\cos 4\pi f_m t + \frac{A_c^2 a}{2}\cos 4\pi f_c t$$
$$+ \frac{a}{2}(A_m^2 + A_c^2) + A_m \cos 2\pi f_m t$$

IF BPF REJECTS $f \le 2f_m$ AND $f \ge 2f_c$, THEN ←

$$x_{out}(t) = A_c[1 + 2a\,x(t)]\cos 2\pi f_c t$$

WHICH IS OF THE FORM OF EQ. (15.2.2) WITH $\phi_c = 0$ AND $m_A = 2a$.

15.2.4.

$y = x + A_c \cos 2\pi f_c t$

$$z = ay^2 = ax^2 + 2axA_c \cos 2\pi f_c t + aA_c^2 \cos^2 2\pi f_c t$$

$$\cos^2 2\pi f_c t = \frac{1}{2}[1 + \cos 4\pi f_c t]$$

$$x^2 = A_m^2 \cos^2 2\pi f_m t = \frac{A_m^2}{2}[1 + \cos 4\pi f_m t]$$

15.2.4. CONTD.

$$z = \frac{a}{2}\left(A_m^2 + A_c^2\right) + \frac{a A_m^2}{2}\cos 4\pi f_m t + 2ax\, A_c \cos 2\pi f_c t + \frac{a A_c^2}{2}\cos 4\pi f_c t$$

IF BPF REJECTS $f \le 2f_m$ AND $f \ge 2f_c$, THEN

$$x_{out}(t) = A_c (2a)\, x(t) \cos 2\pi f_c t$$

ADDING CARRIER TO x_{out} ←

THEN GIVES $A_c \cos 2\pi f_c t + A_c (2a) x(t) \cos 2\pi f_c t = x_c(t)$

WHICH IS OF THE FORM OF EQ. (15.2.2) WITH $\phi = 0$ AND $m_A = 2a$.

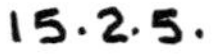
15.2.5.

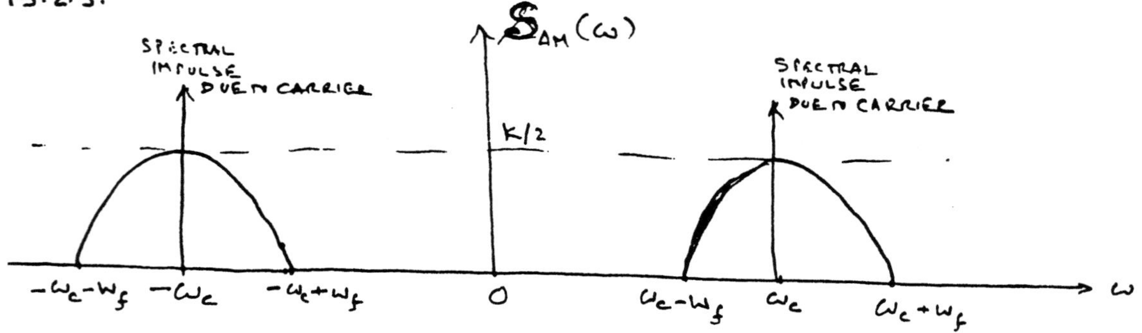

$K \propto$ AMPLITUDE OF $f(t)$; SO RELATES TO TOTAL SIDEBAND POWER ←

W_f IS THE SPECTRAL EXTENT AND DETERMINES BANDWIDTH ←

15.2.6.

$$\eta_{AM} = \frac{\overline{f^2}}{A_c^2 + \overline{f^2}} = \frac{50^2}{200^2 + 50^2} \simeq 0.06 \leftarrow$$

$$P_c = \frac{A_c^2}{2Z_0} = \frac{200^2}{2 \times 50} = 400\,W \leftarrow$$

$$P_f = \frac{\overline{f^2}}{2Z_0} = \frac{50^2}{100} = 25\,W \leftarrow$$

$$P_{AM} = P_c + P_f = 425\,W \leftarrow$$

15.2.7.

$$P_{AM} = 10^3 = \frac{A_c^2 + \overline{f^2(t)}}{2 \times 75}\ ; \quad \eta_{AM} = 0.1 = \frac{1}{1 + A_c^2/\overline{f^2(t)}}$$

SOLVING SIMULTANEOUSLY, $\overline{f^2(t)} = 150(10^3)\,0.1 = 15{,}000\ V^2 \leftarrow$

$$A_c^2 = \left(\frac{1}{0.1} - 1\right)15000 = 135{,}000\,V^2\ ; \ \therefore\ P_c = A_c^2/2Z_0 = 135000/2(75) = 900\,W \leftarrow$$

$$P_f = \overline{f^2(t)}/2Z_0 = 15000/2(75) = 100\,W \leftarrow$$

15.2.8. $\frac{\pi}{\omega_{IF}} \ll RC < \frac{1}{\omega_{m,max}}$

$$\frac{\pi}{2\pi(455)10^3} \ll (5\times10^3)C < \frac{1}{2\pi(5)10^3}$$

$$\frac{10^{-6}}{(910)(5)} \ll C < 6366\times10^{-12}$$

$$220\times10^{-12} \ll C < 6366\times10^{-12}$$

FOR A FACTOR OF 10 TO REPRESENT $\ll$,

$$2200\ pF < C < 6366$$

C CAN BE CHOSEN TO BE IN THE CENTER OF THIS RANGE : 4283 pF ←

15.2.9.

$P_c = 850\ W$; $P_c + P_f = 1000\ W \Rightarrow P_f = 150\ W$

$$\eta_{AM} = \frac{P_f}{P_c + P_f} = 0.15$$

$$\therefore \left(\frac{S_i}{N_i}\right)_{AM} = \frac{10^3}{0.3} \simeq 3333 \text{ or } 35.2\ dB \leftarrow$$

15.2.10.

$$\left(\frac{S_o}{N_o}\right)_{AM} = 250 = 2\eta_{AM}\left(\frac{S_i}{N_i}\right)_{AM} = 2\eta_{AM}(3000)$$

$$\therefore \eta_{AM} = \frac{250}{2\times3000} = 0.0417$$

$$\eta_{AM} = \frac{P_f}{P_c + P_f} = \frac{50}{P_c + 50} = 0.0417$$

$$\therefore P_c = \left(\frac{50}{0.0417} - 50\right) = 1149\ W \leftarrow$$

$$P_{AM} = P_c + P_f = 1149 + 50 = 1199\ W \leftarrow$$

15.2.11.

$$f_{IMAGE} = f_c + 2f_{IF} = 1030 + 2(455) = 1940\ kHz = 1.94\ MHz \leftarrow$$

15.2.12.

$$f_{image} = f_c - 2f_{IF} = f_c - 2(455) = f_c - 910\ kHz$$

$$-370\ kHz \le f_{image} \le 690\ kHz \leftarrow$$

$$f_{LO} = f_c - f_{IF}$$

$$85\ kHz \le f_{LO} \le 1145\ kHz \leftarrow$$

15.2.13.

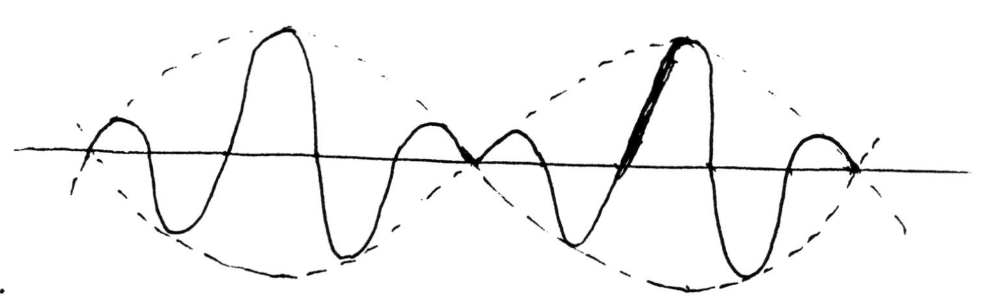

15.2.14.

IF $|f(t)|$ AT THE SECONDARY OF T_1 IS SMALL RELATIVE TO UNITY, THEN FOR $\cos(\omega_c t) > 0$, Q_1 AND Q_2 CONDUCT:

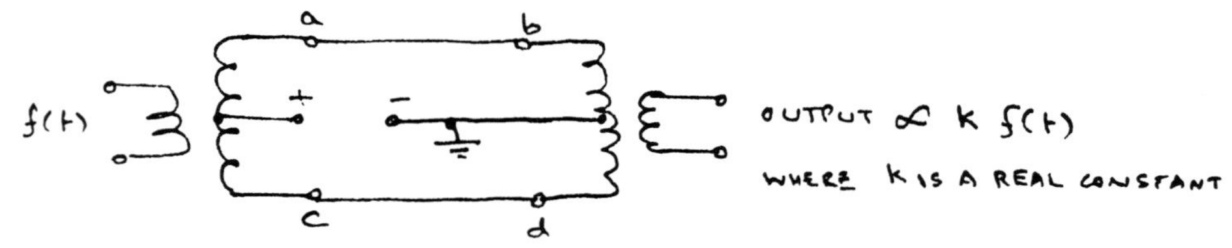

FOR $\cos(\omega_c t) < 0$, Q_3 AND Q_4 CONDUCT:

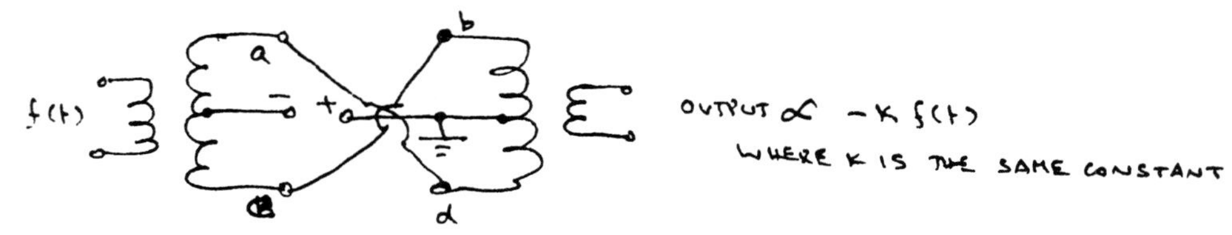

THE COMPOSITE IS $K f(t)$ MULTIPLIED BY ± 1 AS SHOWN BELOW:

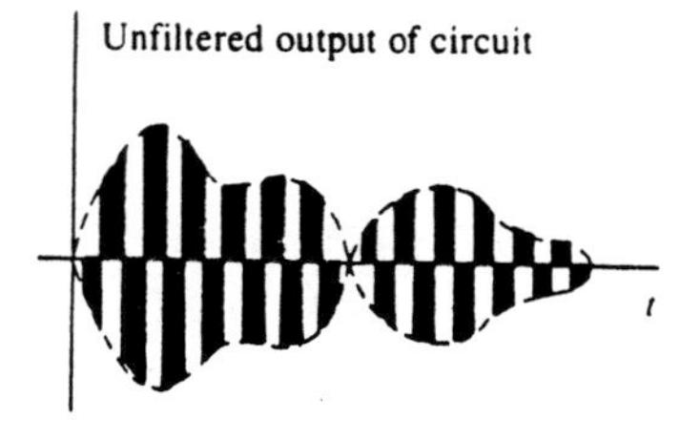

15.2.15.

$$u(t) = 5\cos 1800\pi t + 20\cos 2000\pi t + 5\cos 2200\pi t$$

$$= 20\left[1 + \tfrac{1}{2}\cos 200\pi t\right]\cos 2000\pi t$$

MODULATING SIGNAL IS $m(t) = \cos 2\pi 100t$ ←; CARRIER SIGNAL $= 20\cos 2\pi(1000)t$ ←

MODULATION INDEX $= \frac{1}{2}$ ←

$P_{CARRIER} \propto \frac{400}{2} = 200$; $P_{SIDEBANDS} \propto \frac{400\alpha^2}{2} = 50$

$$\frac{P_{sidebands}}{P_{carrier}} = \frac{50}{200} = \frac{1}{4} \quad \leftarrow$$

15.2.16.(a)

FOR USSB, BANDWIDTH OF MODULATED SIGNAL = BANDWIDTH OF MESSAGE SIGNAL

$$W_{USSB} = W = 10^4 \text{ Hz or } 10 \text{ kHz} \leftarrow$$

(b)

FOR DSB,
$$W_{DSB} = 2W = 2\times 10^4 \text{ Hz or } 20 \text{ kHz} \leftarrow$$

(c) FOR CONVENTIONAL AM:
$$W_{AM} = 2W = 2\times 10^4 \text{ Hz or } 20 \text{ kHz} \leftarrow$$

(d)

USING CARSON'S RULE (EQ. 15.2.12), EFFECTIVE BANDWIDTH OF THE FM MODULATED SIGNAL IS ESTIMATED AS

$$W_{FM} \simeq 2(k_f + W) = 2(60000 + 10000) = 140{,}000 \text{ Hz or } 140 \text{ kHz} \leftarrow$$

OR USING EQ. 15.2.13, FOR WIDEBAND FM,

$$W_{FM} \simeq 2(k_f + 2W) = 2(60000 + 20000)$$

$$= 160{,}000 \text{ Hz or } 160 \text{ kHz} \leftarrow$$

15.2.17.

$A_c = 50\text{V} \leftarrow$; $\omega_c = 185\pi(10^6)\ \frac{\text{rad}}{\text{s}} \leftarrow$; $\phi_c = \frac{\pi}{3} \leftarrow$; $\omega_m = \pi(10^4)\ \frac{\text{rad}}{\text{s}} \leftarrow$;

$$k_{FM}\int 2\cos(\omega_m t)\,dt = (2k_{FM}/\omega_m)\sin\omega_m t = 6\sin\omega_m t$$

$$\therefore k_{FM} = 6\pi(10^4)/2 = 3(10^4)\pi = 9.425\times 10^4\ \frac{\text{rad/s}}{\text{V}} \leftarrow$$

$$\beta_{FM} = \Delta\omega/\omega_m = A_f k_{FM}/\omega_m = 2(9.425)10^4/(\pi\times 10^4) = 6 \leftarrow$$

15.2.18. $W_{FM} = 2\pi(180)10^3 = 2(2\pi)(\Delta f + 30\text{kHz})$ AS PER CARSON'S RULE EQ. (15.2.12)

$\therefore \Delta f = 60$ kHz; EQ.(15.2.11): $k_{FM} = 2\pi(60)10^3/2 = 60\pi(10^3)$ rad/s-V $\leftarrow$

15.2.19. $s_d(t) = k_D k_{FM} f(t) = (10^{-5}/\pi)\, 5\pi(10^4) f(t) = f(t)/2 \leftarrow$

15.2.20.

$$R_{FM} = \frac{(3/2.12)^3}{3\left[\frac{3}{2.12} - \tan^{-1}\left(\frac{3}{2.12}\right)\right]} = \frac{2.833}{3(1.415-0.956)} = 2.06 \text{ OR } 3.13\text{ dB}$$

$$R_{FM \text{ WITH BANDWIDTH LIMITATION}} = \frac{2.06}{1+\left(\frac{1}{3}\right)^2\left(\frac{3}{2.12}\right)^2} = 1.685 \text{ OR } 2.27\text{ dB}$$

BANDWIDTH LIMITATION CAUSES $(1-0.818) = 0.182$

OR 18.2% CHANGE $\leftarrow$

15.2.21.

HIGH-SIDE LO: $f_{image} = f_c + 2f_{IF} = f_c + 2(10.7)$ MHz

$$= \begin{cases} 88.1 + 21.4 = 109.5 \rightarrow \text{1.6 MHz HIGHER THAN UPPER EDGE} \\ 107.9 + 21.4 = 129.3 \end{cases}$$

LOW-SIDE LO: $f_{image} = f_c - 2f_{IF} = f_c - 21.4$ MHz

$$= \begin{matrix} 88.1 - 21.4 = 66.7 \\ 107.9 - 21.4 = 86.5 \rightarrow \text{1.6 MHz BELOW LOWER EDGE.} \end{matrix}$$

Q.E.D. ←

15.2.22.

R: $m_I = 0.6\, m_R$; $m_Q = 0.21\, m_R$ WITH $m_R = 1$

AMPL. $= \left[m_I^2 + m_Q^2\right]^{1/2} = \sqrt{0.6^2 + 0.21^2} = 0.636$

ANGLE $= \tan^{-1}\left[\frac{-0.21}{0.6}\right] = -19.29°$

G: AMPL: $\left[0.28^2 + 0.52^2\right]^{1/2} = 0.591$

ANGLE: $\tan^{-1}\left[\frac{0.52}{-0.28}\right] = 118.3°$

B: AMPL: $\left(0.32^2 + 0.31^2\right)^{1/2} = 0.446$

ANGLE: $\tan^{-1}\left[\frac{-0.31}{-0.32}\right] = 224.09°$

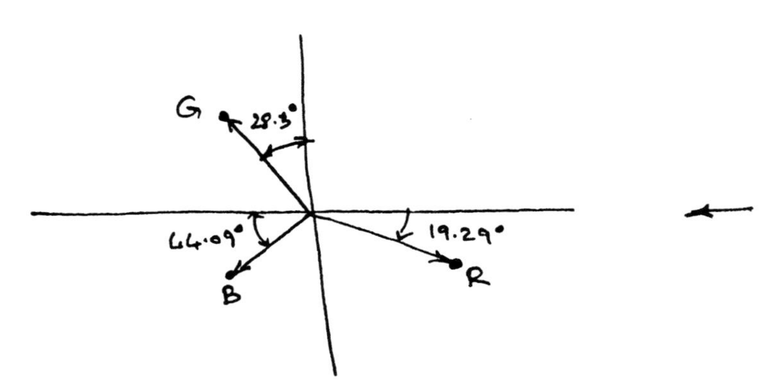

15.2.23. SEE FIG. 15.2.29.

VISUAL CARRIER IS 1.25 MHz ABOVE THE LOWER EDGE OF THE CHANNEL.

∴ STATION'S VISUAL CARRIER FREQUENCY = 506 + 1.25 = 507.25 MHz. ←

15.2.24.

DSB MODULATED SIGNAL:

$$s_{DSB}(t) = f(t)\cdot \cos(2\pi f_c t + \phi_c) \quad \text{EQ.}(15.2.7)$$

$$= f(t)\cos\phi_c \cos 2\pi f_c t - f(t)\sin\phi_c \sin 2\pi f_c t$$

$$\therefore \quad x_d(t) = f(t)\cos\phi_c \; ; \quad x_q(t) = f(t)\sin\phi_c \leftarrow$$

$$V(t) = \sqrt{f^2(t)\cos^2\phi_c + f^2(t)\sin^2\phi_c} = |f(t)| \leftarrow$$

$$\Phi(t) = \tan^{-1}\left(\frac{f(t)\sin\phi_c}{f(t)\cos\phi_c}\right) = \tan^{-1}(\tan\phi_c) = \phi_c \leftarrow$$

CONVENTIONAL AM: $s_{AM}(t) = A_c(1 + f(t))\cos(2\pi f_c t + \phi_c)$ Eq (15.2.2)

$$= A_c(1+f(t))\cos\phi_c \cos 2\pi f_c t - A_c(1+f(t))\sin\phi_c \sin 2\pi f_c t$$

$$\therefore \quad x_d(t) = A_c(1+f(t))\cos\phi_c \leftarrow$$

$$x_q(t) = A_c(1+f(t))\sin\phi_c \leftarrow$$

$$V(t) = A_c|1+f(t)| \leftarrow$$

$$\Phi(t) = \tan^{-1}(\tan\phi_c) = \phi_c \leftarrow$$

15.2.25.

(a) $s_{AM}(t) = A_c[1 + m_A x(t)]\cos 2\pi f_c t$

$$m_A = 0.6 \; ; \quad \frac{A_c^2}{2} = P_c = 200\,\text{W} \; ; \quad P_x = 0.5\,\text{W}$$

$$P_{AM} = \frac{A_c^2}{2} + \frac{A_c^2 m_A^2 P_x}{2} = 200 + (200)\,0.6^2\,(0.5) = 236\,\text{W} \leftarrow$$

$$W_{AM} = 2W = 20\,\text{kHz} \leftarrow$$

(b) $s_{DSB-SC}(t) = A_c x(t)\cos 2\pi f_c t$

$$P_{DSB} = \frac{A_c^2 P_x}{2} = 200(0.5) = 100\,\text{W} \leftarrow$$

$$W_{SDB} = 2W = 20\,\text{kHz} \leftarrow$$

(c) $s_{SSB} = A_c x(t)\cos 2\pi f_c t \mp A_c \hat{x}(t)\sin 2\pi f_c t$

$$P_{SSB} = \frac{A_c^2 P_x}{2} + \frac{A_c^2 P_x}{2} = 200(0.5) + 200(0.5) = 200\,\text{W}.$$

$$W_{SSB} = W = 10\,\text{kHz} \leftarrow$$

15.2.25.(d) $P_{FM} = A_c^2/2 = 200\,W$ ←

$B_e \simeq 2(k_f + W) = 2(50{,}000 + 10{,}000) = 120\,kHz$ (USING CARSON'S RULE) ←

15.2.26. (a)

$$s_{AM} = A_c[1 + \alpha \cos 2\pi f_m t]\cos(2\pi f_c t + \phi_c)$$

$$= A_c \cos(2\pi f_c t + \phi_c) + \frac{A_c\alpha}{2}\cos[2\pi(f_c - f_m)t + \phi_c] + \frac{A_c\alpha}{2}\cos[2\pi(f_c + f_m)t + \phi_c]$$

LOWER-SIDEBAND COMPONENT: $\frac{A_c\alpha}{2}\cos[2\pi(f_c - f_m)t + \phi_c]$

UPPER - " " : $\frac{A_c\alpha}{2}\cos[2\pi(f_c + f_m)t + \phi_c]$

SPECTRUM OF THE DSB AM (CONVENTIONAL) SIGNAL IS

$$S(f) = \frac{A_c}{2}\left[e^{j\phi_c}\delta(f - f_c) + e^{-j\phi_c}\delta(f + f_c)\right] + \frac{A_c\alpha}{4}\left[e^{j\phi_c}\delta(f - f_c + f_m) + e^{-j\phi_c}\delta(f + f_c - f_m)\right] + \frac{A_c\alpha}{4}\left[e^{j\phi_c}\delta(f - f_c - f_m) + e^{-j\phi_c}\delta(f + f_c + f_m)\right]$$

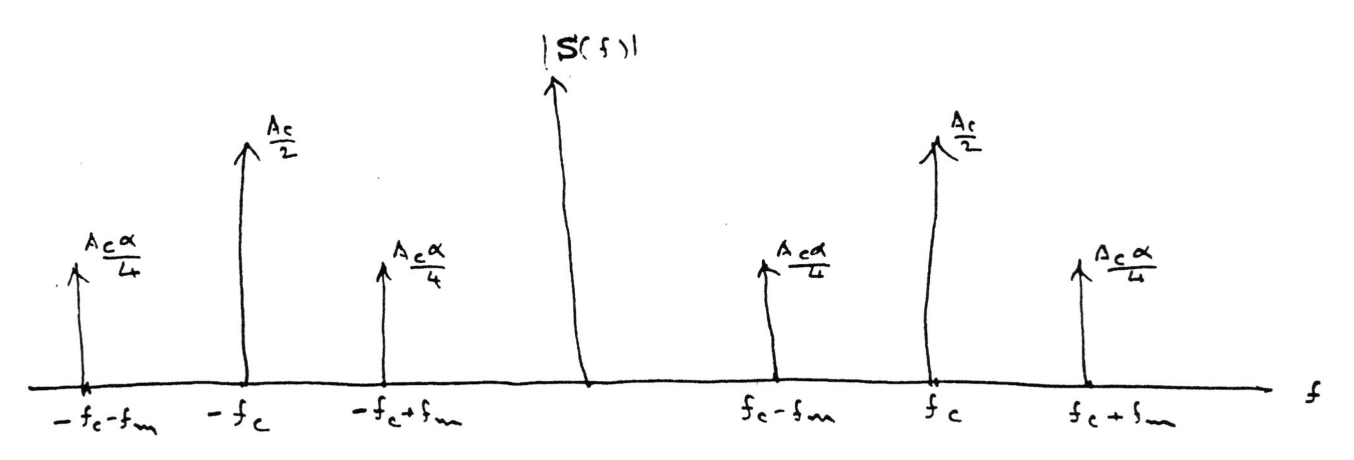

(b)

$$s_{DSB-SC}(t) = A_c \alpha \cos 2\pi f_m t \cos(2\pi f_c t + \phi_c)$$

$$= \frac{A_c\alpha}{2}\cos[2\pi(f_c - f_m)t + \phi_c] + \frac{A_c\alpha}{2}\cos[2\pi(f_c + f_m)t + \phi_c]$$

$$S(f) = \frac{A_c\alpha}{4}\left[e^{j\phi_c}\delta(f - f_c + f_m) + e^{-j\phi_c}\delta(f + f_c - f_m)\right] + \frac{A_c\alpha}{4}\left[e^{j\phi_c}\delta(f - f_c - f_m) + e^{-j\phi_c}\delta(f + f_c + f_m)\right]$$

LOWER SIDEBAND: $\frac{A_c\alpha}{2}\cos[2\pi(f_c - f_m)t + \phi_c]$

UPPER " : $\frac{A_c\alpha}{2}\cos[2\pi(f_c + f_m)t + \phi_c]$

15.2.26.(b) CONTD.

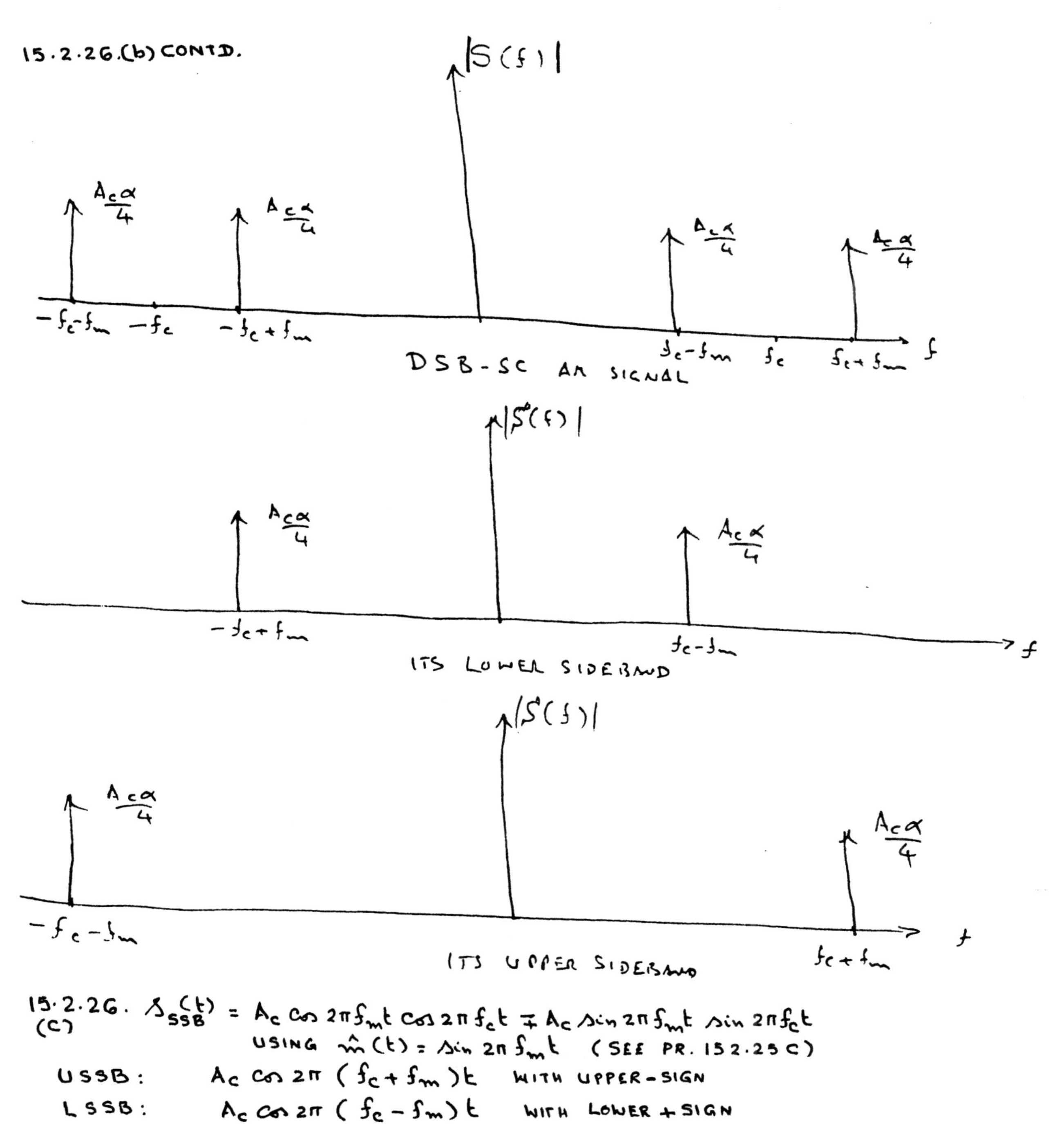

15.2.26. (C) $s_{SSB}(t) = A_c \cos 2\pi f_m t \cos 2\pi f_c t \mp A_c \sin 2\pi f_m t \sin 2\pi f_c t$

USING $\hat{m}(t) = \sin 2\pi f_m t$ (SEE PR. 15.2.25 C)

USSB: $A_c \cos 2\pi (f_c + f_m) t$ WITH UPPER − SIGN

LSSB: $A_c \cos 2\pi (f_c - f_m) t$ WITH LOWER + SIGN

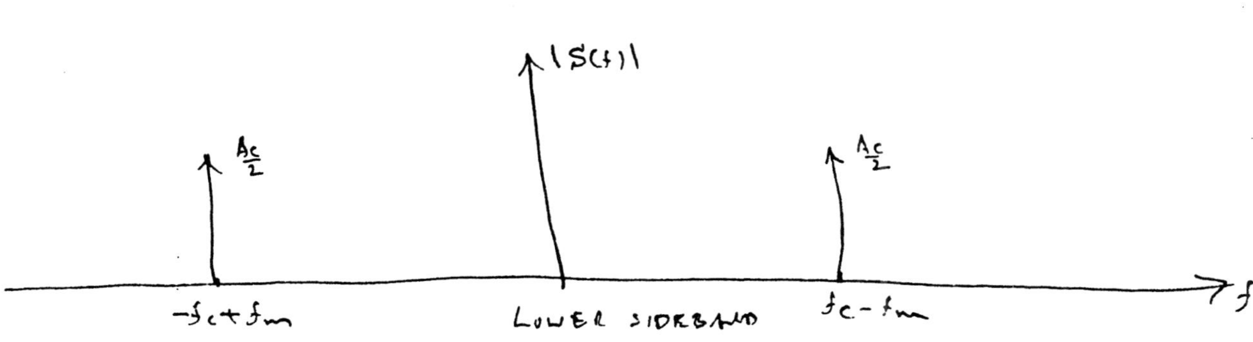

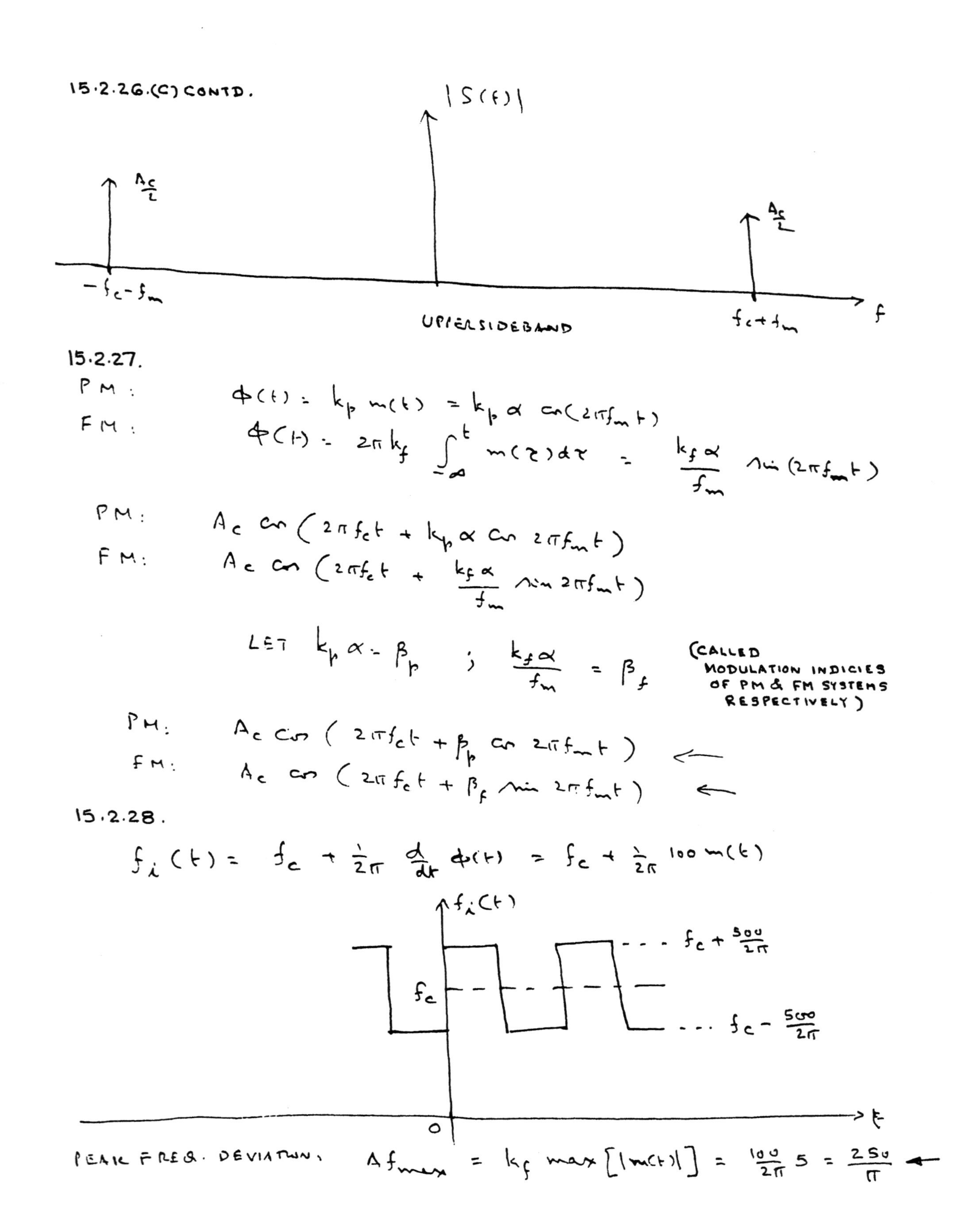

15.2.26.(C) CONTD.

15.2.27.

PM: $\phi(t) = k_p\, m(t) = k_p \alpha \cos(2\pi f_m t)$

FM: $\phi(t) = 2\pi k_f \int_{-\infty}^{t} m(\tau)\, d\tau = \frac{k_f \alpha}{f_m} \sin(2\pi f_m t)$

PM: $A_c \cos(2\pi f_c t + k_p \alpha \cos 2\pi f_m t)$

FM: $A_c \cos\left(2\pi f_c t + \frac{k_f \alpha}{f_m} \sin 2\pi f_m t\right)$

LET $k_p \alpha = \beta_p$; $\frac{k_f \alpha}{f_m} = \beta_f$ (CALLED MODULATION INDICIES OF PM & FM SYSTEMS RESPECTIVELY)

PM: $A_c \cos(2\pi f_c t + \beta_p \cos 2\pi f_m t)$ ←

FM: $A_c \cos(2\pi f_c t + \beta_f \sin 2\pi f_m t)$ ←

15.2.28.

$$f_i(t) = f_c + \frac{1}{2\pi} \frac{d}{dt} \phi(t) = f_c + \frac{1}{2\pi} 100\, m(t)$$

PEAK FREQ. DEVIATION, $\Delta f_{max} = k_f \max[|m(t)|] = \frac{100}{2\pi} 5 = \frac{250}{\pi}$ ←

15.2.29. FREQ. DEVIATION $f_d(t) = f_i(t) - f_c = k_f\, m(t)$

PHASE DEVIATION $\phi_d(t) = 2\pi k_f \int_{-\infty}^{t} m(\tau)\, d\tau$

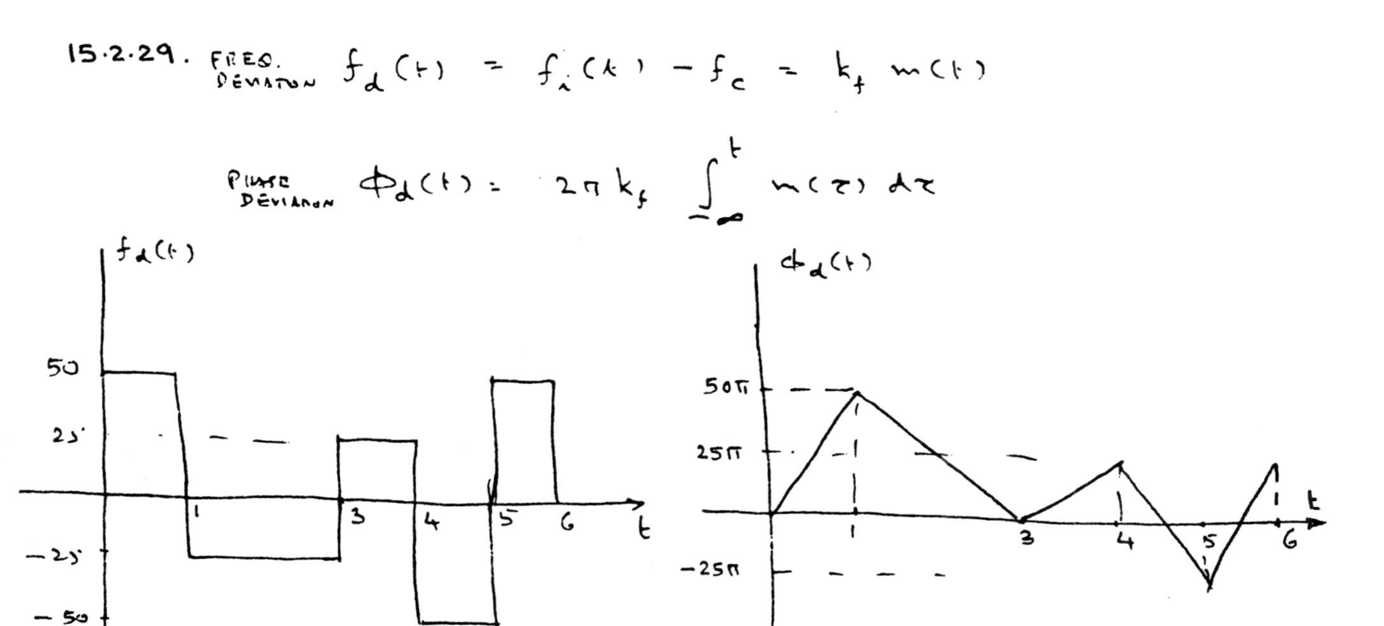

15.2.30.(a)

FM: $u(t) = 100 \cos\left(2\pi f_c t + 2\pi k_f \int_{-\infty}^{\infty} \alpha \cos(2\pi f_m \tau)\, d\tau\right)$

$= 100 \cos\left(2\pi f_c t + \frac{k_f \alpha}{f_m} \sin 2\pi f_m t\right)$

$\beta_f = \frac{k_f \alpha}{f_m}$ IS MODULATION INDEX

$\beta_f = 4$; ←

$B_{FM} = 2(\beta_f + 1) f_m = 2(4+1)\, 1\,\text{kHz} = 10\,\text{kHz}$ ←

(b)

PM: $\beta_p = \Delta\phi_{max} = \max\left[4 \sin(2\pi f_m t)\right] = 4$ ←

$B_{PM} = 2(\beta_p + 1) f_m = 10\,\text{kHz}$ ←

15.2.31.(a)

Let $m(t) = m_1(t) + m_2(t)$

DSB MODULATE THE CARRIER $A_c \cos 2\pi f_c t$

RESULTING SIGNAL $u(t) = A_c\, m(t) \cos 2\pi f_c t$

$= A_c (m_1(t) + m_2(t)) \cos 2\pi f_c t$

$= A_c\, m_1(t) \cos 2\pi f_c t + A_c\, m_2(t) \cos 2\pi f_c t$

$= u_1(t) + u_2(t)$ ←

15.2.31. (b)

WHEN $m(t) = m_1(t) + m_2(t)$ FREQUENCY MODULATES A CARRIER $A_c \, 2\pi f_c t$, THE RESULT IS

$$u(t) = A_c \cos\left(2\pi f_c t + 2\pi k_f \int_{-\infty}^{\infty} (m_1(\tau) + m_2(\tau))\, d\tau\right)$$

$$\neq A_c \cos\left(2\pi f_c t + 2\pi k_f \int_{-\infty}^{\infty} m_1(\tau)\, d\tau\right)$$

$$+ A_c \cos\left(2\pi f_c t + 2\pi k_f \int_{-\infty}^{\infty} m_2(\tau)\, d\tau\right)$$

$$= u_1(t) + u_2(t)$$

WHERE THE INEQUALITY FOLLOWS FROM THE NONLINEARITY OF THE COSINE FUNCTION.

15.2.32. TRANSFER FUNCTION OF THE FM DISCRIMINATOR IS

$$H(s) = \frac{R}{R + Ls + \frac{1}{Cs}} = \frac{\frac{R}{L} s}{s^2 + \frac{R}{L} s + \frac{1}{LC}}$$

THUS

$$|H(f)|^2 = \frac{4\pi^2 \left(\frac{R}{L}\right)^2 f^2}{\left(\frac{1}{LC} - 4\pi^2 f^2\right)^2 + 4\pi^2 \left(\frac{R}{L}\right)^2 f^2}$$

$$|H(f)|^2 \leq 1 \quad \text{IF} \quad f = \frac{1}{2\pi\sqrt{LC}}$$

SINCE THIS FILTER IS TO BE USED AS A SLOPE DETECTOR, THE FREQUENCY-CONTENTS OF THE SIGNAL, WHICH IS $[80-6, 80+6]$ MHz, MUST FALL INSIDE THE REGION OVER WHICH $|H(f)|$ IS ALMOST LINEAR. SUCH A REGION CAN BE CONSIDERED THE INTERVAL $[f_{10}, f_{90}]$ WHERE f_{10} IS THE FREQUENCY SUCH THAT $|H(f_{10})| = 10\% \max[|H(f)|]$ AND f_{90} IS THE FREQUENCY SUCH THAT $|H(f_{90})| = 90\% \max[|H(f)|]$

WITH $\max|H(f)| = 1$, $f_{10} = 74 \times 10^6$ AND $f_{90} = 86 \times 10^6$, WE OBTAIN SYSTEM OF EQUATIONS

$$4\pi^2 f_{10}^2 + \frac{50\times10^3}{L} 2\pi f_{10} \left[1 - 0.1^2\right]^{1/2} - \frac{1}{LC} = 0$$

$$4\pi^2 f_{90}^2 + \frac{50\times10^3}{L} 2\pi f_{90} \left[1 - 0.9^2\right]^{1/2} - \frac{1}{LC} = 0$$

SOLVING, ONE GETS $L = 14.98$ mH; $C = 0.018013$ pF. ←

15.2.33. USING CARSON'S RULE

$$B_c = 2(\beta+1)W$$

$$= 2\left(\frac{k_f \max[|m(t)|]}{W} + 1\right)W$$

or $W_{FM} \simeq 2(\Delta\omega + W_f)$ Eq. (15.2.12)

WITH $k_f = 10$, $B_c = 20020$

$k_f = 100$, $B_c = 20200$

$k_f = 1000$, $B_c = 22000$ } ←

15.2.34.

SINCE $88 \text{ MHz} < f_c < 108 \text{ MHz}$, AND $|f_c - f_c'| = 2f_{IF}$ if $f_{IF} < f_{LO}$, WE CONCLUDE THAT, IN ORDER FOR THE IMAGE FREQUENCY f_c' TO FALL OUTSIDE THE INTERVAL $[88, 108]$ MHz, THE MINIMUM f_{IF} IS SUCH THAT

$$2f_{IF} = 108 - 88 \Rightarrow f_{IF} = 10 \text{ MHz} \leftarrow$$

IF $f_{IF} = 10$ MHz, THEN THE RANGE OF f_{LO} IS

$$[88+10, 108+10] = [98, 118] \text{ MHz}. \leftarrow$$

15.2.35. $W_{FM} \simeq 2(25 + 2\times10) = 90$ kHz; 90 kHz / 6 MHz = 1.5% ←

15.2.36.

$$B = 2\Delta f + 4W = 60 \text{ MHz}$$

FOR $1/100 \leq B/f_c \leq 1/10$

WE WANT $600 \text{ MHz} \leq f_c \leq 6 \text{ GHz}$. ←

15.2.37. $D \leq T_s/5$ SO $W = 1/D \geq 5f_s = 40$ kHz

IF $\Delta f \geq 2W$, THEN $B = 2\Delta f + 4W$

$\therefore \Delta f \leq \frac{1}{2}(400 - 160) = 120$ kHz ←

WHICH AGREES WITH $\Delta f \geq 2W$.

15.2.38.

$D = d_{min}/c = 0.03\text{ km}/(3\times10^5)\text{ km/s} = 0.1\,\mu s$

$W = 1/D$; $B \geq 2W = 20$ MHz

SO $f_c \geq 10B = 200$ MHz ←

15.2.39.

$D \leq T_s/M$ so $W = 1/D \geq M f_s$; $B = 2W \geq 2M f_s$

$B \leq f_c/10 \Rightarrow M \leq f_c/20 f_s = 25$ ←

15.3.1.

(a) $f_s = 2 \times 3 = 6$ kHz ←

(b) $f_s = 2 \times 4.5 = 9$ MHz ←

15.3.2. $f_s = 3 \times 2 \times 15 = 90$ kHz ←

15.3.3.(a)

$-3.3, -3.1, -2.9, -2.7, \ldots -0.1, 0.1, 0.3, \ldots 3.3$ V ←

(b) MIDRISER ←

(c)

$-2.1, -0.5, 0.9, 0.1, -0.7$ V ←

(d) FROM EQ.(15.3.2), $|f| \leq |f|_{max} = \frac{L}{2} \delta v = \frac{34}{2}(0.2) = 3.4$V ←

15.3.4.

A GAIN OF k_q MEANS THE QUANTUM LEVEL STEP SIZE IS $k_q \cdot \delta v$,

WHERE δv IS THE STEP SIZE OF $f(t)$;

ALTERNATIVELY, FOR QUANTUM STEP SIZE δv, THE STEP SIZE OF $f(t)$ IS $\delta v / k_q$

15.3.5. (a)

$\overline{f^2(t)} = 2.25\,V^2$; $k_{cr} = 3$; $(S_o/N_q) = 2700$

FROM EQ (15.3.8), $|f(t)|_{max} = k_{cr}\sqrt{\overline{f^2(t)}}$

$= 3\sqrt{2.25} = 4.5\ V$

FROM EQ (15.3.5), $\delta v = \sqrt{\dfrac{12\,\overline{f^2(t)}}{(S_o/N_q)}} = \sqrt{\dfrac{12(2.25)}{2700}}$

(b) FROM EQ (15.3.8), $L = \dfrac{2\,|f(t)|_{max}}{\delta v} = \dfrac{2 \times 4.5\sqrt{2700}}{\sqrt{12(2.25)}}$

$= 90$ LEVELS ←

NEAREST BINARY NUMBER IS $L_b = 128 = 2^7$ ←

(c) EXTREME QUANTUM LEVELS ARE GIVEN BY

$\pm\left(\dfrac{L_b - 1}{2}\right)\delta v = \pm\dfrac{127}{2}(0.1) = \pm 6.35\ V$ ←

15.3.6.

FROM EQ.(15.3.2),

(a) $|f(t)|_{max} = \dfrac{L}{2}\delta v = \dfrac{130}{2}(0.04) = 2.6V$ ←

(b) FROM EQ (15.3.9), $k_{cr} = \sqrt{\dfrac{3L^2}{(S_o/N_q)}} \le \sqrt{\dfrac{3 \times (130)^2}{5500}}$

$= 3.036$ ←

15.3.7.

FROM EQ (15.3.9) IT FOLLOWS: dB IMPROVEMENT $= 10\log_{10}\left[\dfrac{(S_o/N_q)_2}{(S_o/N_q)_1}\right]$

$= 10\log_{10}\left[\dfrac{k_{cr1}^2}{k_{cr2}^2}\right] = 20\log_{10}\left[\dfrac{3.2}{2}\right] \simeq 4.08\ dB$ ←

15.3.8.

(a) $L = 1 + \dfrac{2(3.1)}{0.2} = 32 \Rightarrow N_b = 5$ BITS ←

(b) $L = 128 \Rightarrow N_b = 7$ BITS ←

15.3.9.

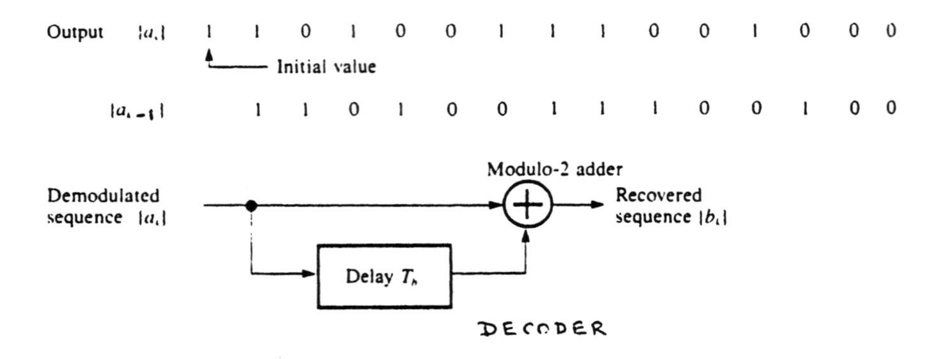

15.3.10.

(a)

$N_b = 4$ so $L_b = 2^4 = 16$ levels

level	Folded Binary Codewords	
15	1 1 1 1	Positive Amplitudes
14	1 1 1 0	
13	1 1 0 1	
12	1 1 0 0	
11	1 0 1 1	
10	1 0 1 0	
9	1 0 0 1	
8	1 0 0 0	
7	0 1 1 1	Negative Amplitudes
6	0 1 1 0	
5	0 1 0 1	
4	0 1 0 0	
3	0 0 1 1	
2	0 0 1 0	
1	0 0 0 1	
0	0 0 0 0	

First digit: sign digit; remaining digits: Magnitude digits

(b)

$b_4\ b_3\ b_2\ b_1$	level	Gray Codewords $g_4\ g_3\ g_2\ g_1$
1 1 1 1	15	1 0 0 0
1 1 1 0	14	1 0 0 1
1 1 0 1	13	1 0 1 1
1 1 0 0	12	1 0 1 0
1 0 1 1	11	1 1 1 0
1 0 1 0	10	1 1 1 1
1 0 0 1	9	1 1 0 1
1 0 0 0	8	1 1 0 0
0 1 1 1	7	0 1 0 0
0 1 1 0	6	0 1 0 1
0 1 0 1	5	0 1 1 1
0 1 0 0	4	0 1 1 0
0 0 1 1	3	0 0 1 0
0 0 1 0	2	0 0 1 1
0 0 0 1	1	0 0 0 1
0 0 0 0	0	0 0 0 0

NOTE: Codewords change in only 1 digit between adjacent levels.

15.3.11. 1 0 0 1 1 1 0 0 1 0 1 1 0 1 1 0 0 0

t Polar

t unipolar

There are 9 1s and 9 0s so about half the time we have 1s and 0s. Prob. of binary 1 in next interval ≈ 1/2.

	1 0 0 1 1 1 0 0 1 0 1 1 0 1 1 0 0 0	$\{b_k\}$
(i)	1 0 0 0 1 0 1 1 1 0 0 1 0 0 1 0 0 0 0	$\{a_k\}$ initial value = 1
(ii)	0 1 1 1 0 1 0 0 0 1 1 0 1 1 0 1 1 1 1	$\{a_k\}$ initial value = 0
(iii)	1 0 0 1 1 1 0 0 1 0 1 1 0 1 1 0 0 0	Decoded, initial value=1
	1 0 0 1 1 1 0 0 1 0 1 1 0 1 1 0 0 0	Decoded, initial value=0

Original sequences are recovered.

1 0 0 1 1 1 0 0 1 0

t Manchester format

15.3.12.

SAMPLING AT TWICE THE NYQUIST RATE; SO 4 TIMES THE SPECTRAL EXTENT.

$$4 \times (15 \times 10^3) = 60 \times 10^3 \text{ samples/sec.}$$

PERIOD BETWEEN SAMPLES $= \dfrac{1}{60 \times 10^3} = \dfrac{100}{6}\ \mu s$

THIS PERIOD MUST CONTAIN 12 PULSES, ONE FOR EACH BIT IN THE 12-BIT CODE

$$\therefore T_b = \frac{100}{6(12)}\ \mu s$$

FIRST-NULL BANDWIDTH $= \omega_b/2\pi = 1/T_b = 0.72$ MHz ←

FROM EQ. (15.3.9), $(S_0/N_q) = \dfrac{3(4096)^2}{(3.8)^2} \simeq 3.486 \times 10^6$ or 65.4 dB ←

15.3.13.

FROM EQ. (15.3.24)

$$10^{-5} = \frac{1}{2}\,\text{erfc}\sqrt{\frac{6^2 \times 0.5 \times 10^{-6}}{N_0}}$$

BY TRIAL & ERROR, $\text{erfc}(x) = 2 \times 10^{-5} \Rightarrow x = 3.0234$

$$\simeq \frac{e^{-x^2}}{x\sqrt{\pi}} \quad \text{for } x > 2 \quad (15.3.26)$$

$$\therefore N_0 = \frac{36 \times 0.5 \times 10^{-6}}{(3.0234)^2} = 1.9692 \times 10^{-6}$$

OR $N_0/2 = 0.9846 \times 10^{-6}\ V^2/\text{Hz}$ ←

15.3.14.

$$\sqrt{\frac{1}{8}\left(\frac{\hat{S}_i}{N_i}\right)} = \sqrt{\frac{32}{8}} = 2 \quad \text{UNIPOLAR PCM}$$

$$\sqrt{\frac{1}{2}\left(\frac{\hat{S}_i}{N_i}\right)} = \sqrt{\frac{32}{2}} = 4 \quad \text{POLAR PCM}$$

BY USING EQ. (15.3.26) $(\because x > 2)$

$$P_e \simeq \frac{1}{2}\,\text{erfc}\,(2.0) \simeq \frac{e^{-2^2}}{2(2)\sqrt{\pi}} = 2.58 \times 10^{-3} \quad \text{UNIPOLAR PCM}$$

$$P_e \simeq \frac{1}{2}\,\text{erfc}\,(4.0) \simeq \frac{e^{-4^2}}{2(4)\sqrt{\pi}} = 7.94 \times 10^{-9} \quad \text{POLAR PCM}$$

BIT ERRORS IN A UNIPOLAR SYSTEM OCCUR MORE FREQUENTLY THAN IN A POLAR SYSTEM.

15.3.15.

(a) $2 \times 10^{-4} = \text{erfc}\,(x)$ ~~WHERE~~ WHEN $x = \sqrt{\frac{1}{2}\left(\frac{\hat{S}_i}{N_i}\right)_{PCM}} = 2.64079$

So $(\hat{S}_i/N_i)_{PCM} \geq 13.9475$ ←

(b) $\sqrt{\frac{A^2 T_b}{N_0}} = 2.64079$ FOR $P_e = 10^{-4}$

So $$A \geq \sqrt{\frac{(2.64079)^2 N_0}{T_b}} = \sqrt{\frac{(2.64079)^2\, 8 \times 10^{-7}}{0.4 \times 10^{-6}}}$$

$\simeq 3.735$ V ←

15.3.16.

$N_b = 8$; $L_b = 2^8 = 256$

FROM EQ (15.3.9), $$\left(\frac{S_0}{N_q}\right) = \frac{3 \times 256^2}{3^2} = 21845.33 \text{ or } 43.4 \text{ dB}$$

$$\left(\frac{S_0}{N_0}\right)_{PCM} = \frac{21845.33}{1 + 2^{17}\,\text{erfc}\,(\sqrt{10})} = \frac{21845.33}{1 + 1.0611672}$$

$$= \frac{21845.33}{2.0611672} = 10598.5 \text{ or } 40.25 \text{ dB} \leftarrow$$

THE SYSTEM IS BELOW THRESHOLD BECAUSE THE DENOMINATOR IS SIGNIFICANTLY LARGER THAN 1.0 ←

15.3.17.

(a)

THRESHOLD EFFECT OCCURS WHERE $(\hat{S}_i / N_i)_{PCM}$ IS IN THE RANGE OF 8 TO 15 dB FOR N_b FROM 2 TO 10. ←

(b)

WITH $(\hat{S}_i / N_i)_{PCM} = 32 \wedge 15.05$ dB, WITH $N_b = 8$, OPERATION IS IN THE FLAT PART OF THE CURVE.

∴ RECEIVER-NOISE EFFECTS ARE NEGLIGIBLE AND PERFORMANCE IS LIMITED BY QUANTIZATION ERROR. ←

15.3.18.

FOR NYQUIST SAMPLING OF A VOICE MESSAGE $T_{s\,\text{VOICE}} = \frac{1}{2(3)10^3}$ $= \frac{1}{6}$ ms. DURING $\frac{1}{6}$ ms ALL 4 VOICE SIGNALS MUST BE SAMPLED ONCE. FOR 2× NYQUIST SAMPLING OF A MONITORING SIGNAL $T_{sM} = \frac{1}{2(2)(\frac{1}{2})10^3} = \frac{1}{2}$ ms.

DURING $\frac{1}{2}$ ms ALL 6 MONITORING SIGNALS ARE TO BE SAMPLED ONCE.

HENCE

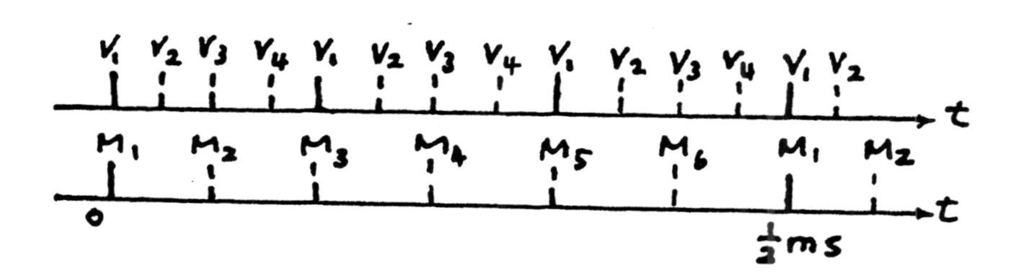

WE NEED TO GET 3 SETS OF SAMPLES OF VOICE SIGNALS FOR EVERY 1 SET FROM MONITOR. HENCE A FRAME EQUAL TO $\frac{1}{2}$ mS ←

MAY STAGGER SAMPLES AS SHOWN BELOW:

15.3.18. CONTD.

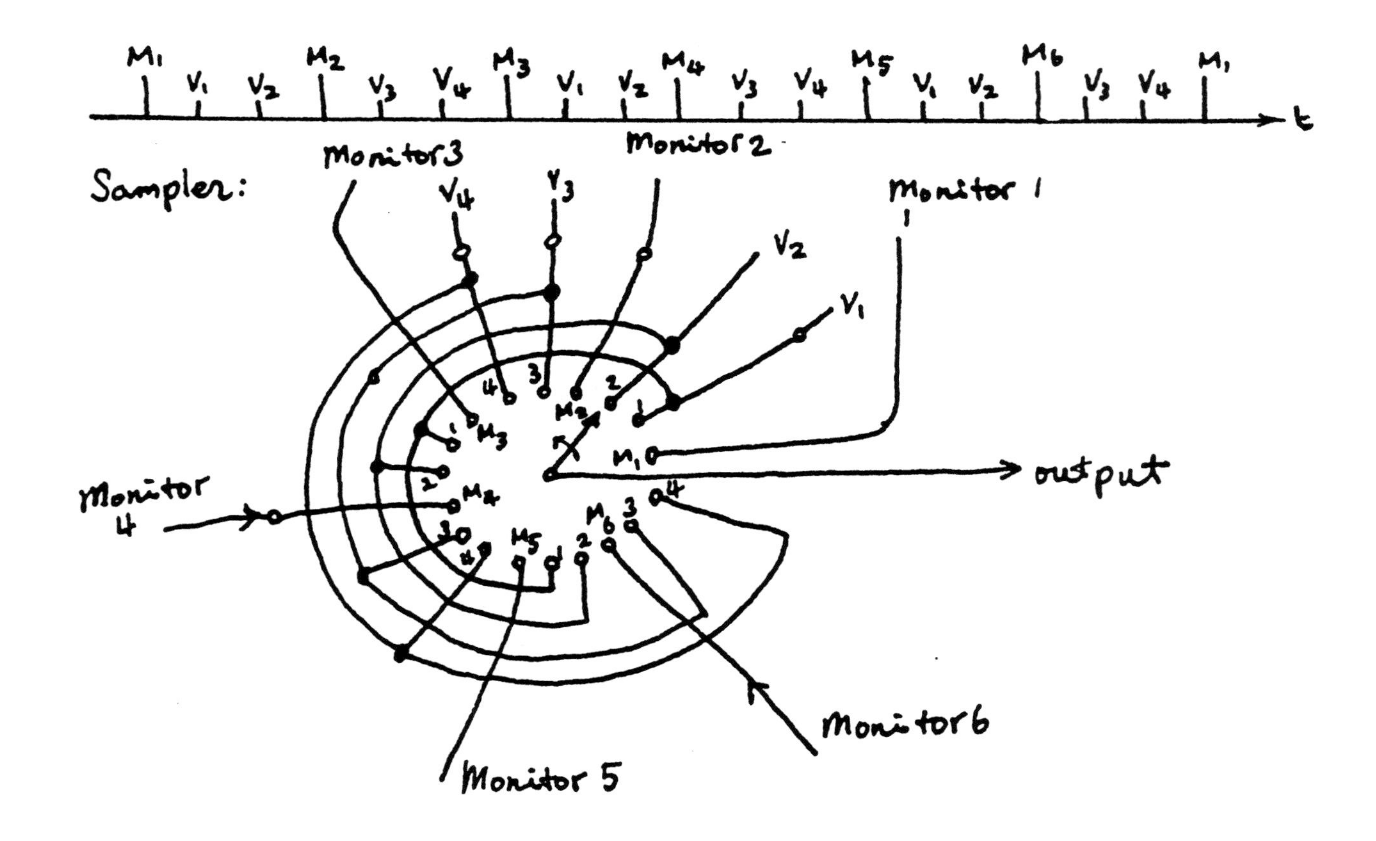

M1
V1
V2
M2
V3
V4
M3
V1
V2
M4
V3
V4
M5
V1
V2
M6
V3
V4
M1
t
Sampler:
Monitor 3
Monitor 2
Monitor 1
V4
V3
V2
V1
M1
M2
M3
M4
M5
M6
output
Monitor 4
Monitor 5
Monitor 6

15.3.19. $\tau_g = 2\tau$, $T_s = \dfrac{1}{2(2)200} = 1.25$ mS = FRAME TIME TO SAMPLE EACH MESSAGE ONCE; 86 TIME SLOTS REQD.: TIME SLOT $= \dfrac{1.25\text{ ms}}{86} = 3\tau$;

BECAUSE TIME SLOT $= \tau + \tau_g$, $\tau = 1.25\times10^{-3}/(86\times3) \simeq 4.845\,\mu S$ ←

15.3.20. $\omega_0 T_b = 6\pi = 3(2\pi)$ SO 3 CYCLES AT ω_0 OCCUR EACH T_b.

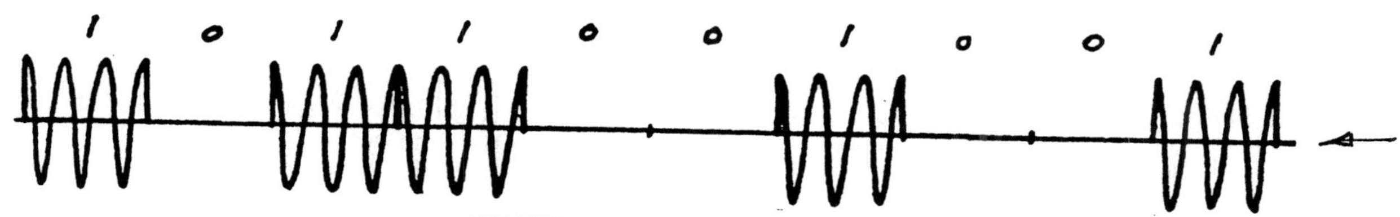

15.3.21. $P_e = \frac{1}{2}\,\text{erfc}\left[\sqrt{\dfrac{A^2T_b}{8N_0}}\right] = \frac{1}{2}\,\text{erfc}\left[\sqrt{\dfrac{2.2^2(2.6)10^{-6}}{16\times10^{-7}}}\right] = 3.862\times10^{-5}$ ←

$E_b = A^2 T_b/4 = 2.2^2 \times 2.6\times10^{-6}/4 = 3.146\,\mu J$ ←

15.3.22. FIG. 15.3.14: $D = -\dfrac{A^2T_b}{4} + \int_0^{T_b} A^2\cos^2(\omega_c t + \phi_c)\,dt$

$$= -\frac{A^2T_b}{4} + \frac{A^2}{2}\int_0^{T_b}[1+\cos(2\omega_c t + 2\phi_c)\,dt$$

$$= -\frac{A^2T_b}{4} + \frac{A^2T_b}{2} + \frac{A^2}{2}\int_{2\phi_c}^{2\omega_cT_b+2\phi_c}\cos x\,\frac{dx}{2\omega_c}$$

$$= \frac{A^2T_b}{4} + \frac{A^2}{4\omega_c}\{\sin(2\omega_cT_b+2\phi_c) - \sin 2\phi_c\}$$

$$= \frac{A^2T_b}{4}\left[1 + \frac{1}{\omega_cT_b}\{\sin(2\omega_cT_b+2\phi_c) - \sin 2\phi_c\}\right]$$

NOTING THAT $\dfrac{1}{\omega_cT_b} \ll 1$, $\simeq A^2T_b/4$ ←

15.3.23. $E_b/N_0 = 20$ OR 13.01 dB

$P_e = \frac{1}{2}\,\text{erfc}\sqrt{\frac{1}{2}\left(\frac{E_b}{N_0}\right)}$; COHERENT ASK; USE EQ. (15.3.26);

$P_e = 4.04996\times10^{-6}$

$P_e = \frac{1}{2}\left[1 + \sqrt{\dfrac{1}{2\pi(E_b/N_0)}}\right]e^{-(E_b/N_0)/2}$; NONCOHERENT ASK;

BY TRIAL AND ERROR $P_e = 4.04996\times10^{-6}$ WHEN $\dfrac{E_b}{N_0} = 23.60514$ OR 13.73 dB ←

IN THE NONCOHERENT SYSTEM WHICH IS 0.72 dB LARGER THAN THE COHERENT SYSTEM.

15.3.24. $P_e = \frac{1}{2}\,\text{erfc}\sqrt{E_b/N_0}$; COHERENT PSK

$P_e = \frac{1}{2}\,\text{erfc}\sqrt{\frac{1}{2}(E_b/N_0)}$; COHERENT ASK

TO BE EQUAL, ARGUMENTS MUST BE EQUAL, SO

$$\frac{1}{2}\left(\frac{E_b}{N_0}\right)_{ASK} = \left(\frac{E_b}{N_0}\right)_{PSK} \quad \text{OR} \quad \left(\frac{E_b}{N_0}\right)_{ASK} = 2\left(\frac{E_b}{N_0}\right)_{PSK}$$ ←

15.3.25.

$$P_e = \frac{1}{2} e^{-E_b/N_0} = \frac{1}{2} e^{-A^2 T_b/2N_0} \; ; \quad A^2 T_b/2N_0 = \ln(1/2P_e) \; ;$$

$$A^2 = 2N_0 \ln(1/2P_e)/T_b$$

THUS, $P_{e1} = 3\times10^{8}$ WHEN $A = A_1 = 2V$; $P_{e2} = 2\times10^{-6}$ WHEN $A = A_2$

$$\therefore \left(\frac{A_2}{A_1}\right)^2 = \frac{A_2^2}{(2)^2} = \ln\left(\frac{1}{2P_{e2}}\right) / \ln\left(\frac{1}{2P_{e1}}\right) = \frac{\ln 250{,}000}{\ln(10{,}000/6)} = 1.675417$$

$$A_2 = \sqrt{2^2 \times 1.675417} = 2.59\,V \longleftarrow$$

15.3.26. FOR A 1 : UPPER INTEGRATOR PRODUCES $\frac{A^2 T_b}{2}$; LOWER INTEGRATOR PRODUCES $\simeq 0$ } $D = \frac{A^2 T_b}{2} > 0$ $\therefore$ 1 DECLARED

FOR A 0 : UPPER INTEGRATOR PRODUCES $\simeq 0$; LOWER INTEGRATOR PRODUCES $\frac{A^2 T_b}{2}$ } D WOULD BE > 0 $\therefore$ 1 DECLARED

UNLESS A SIGN INVERSION IS USED.

WITH THE SIGN INVERSION $D = -\frac{A^2 T_b}{2}$ AND A 0 IS DECLARED.

15.3.27. (a) UPPER INTEGRATOR : OUTPUT $= \int_0^{T_b} A^2 \cos^2(\omega_2 t + \phi_2)\,dt$

$$= \frac{A^2}{2}\int_0^{T_b}\{1 + \cos(2\omega_2 t + 2\phi_2)\}\,dt \simeq A^2 T_b/2 \longleftarrow$$

SINCE $\cos(2\omega_2 t + 2\phi_2) \simeq 0$ FOR $\omega_c >> \frac{1}{T_b}$.

LOWER INTEGRATOR: OUTPUT $= \int_0^{T_b} A^2 \cos(\omega_1 t + \phi_1)\cos(\omega_2 t + \phi_2)\,dt$

$$= \frac{A^2}{2}\int_0^{T_b}\{\cos(\omega_1 t - \omega_2 t + \phi_1 - \phi_2)\,dt + \cos(\omega_1 t + \omega_2 t + \phi_1 + \phi_2)\,dt\}$$

$$\simeq \frac{A^2}{2}\int_0^{T_b}\cos(2\Delta\omega t + \phi_2 - \phi_1)\,dt \quad \text{SINCE } \cos(\omega_1 t + \omega_2 t + \phi_1 + \phi_2) \simeq 0$$

NOTING THAT $\Delta\omega T_b = \pi$, $= \frac{A^2}{2}\left\{\frac{\sin(2\Delta\omega T_b + \phi_2 - \phi_1)}{2\Delta\omega} - \frac{\sin(\phi_2 - \phi_1)}{2\Delta\omega}\right\} = 0 \longleftarrow$

(b)

UPPER INTEGRATOR : OUTPUT $= \int_0^{T_b} A^2 \cos(\omega_1 t + \phi_1)\cos(\omega_2 t + \phi_2)\,dt = 0 \longleftarrow$

SAME AS IN LOWER INTEGRATOR OF (a)

LOWER INTEGRATOR: OUTPUT $= \int_0^{T_b} A^2 \cos^2(\omega_1 t + \phi_1)\,dt = \frac{A^2 T_b}{2} \longleftarrow$

(SAME FORM AS ABOVE EXCEPT IN ω_1 AND ϕ_1.)

15.3.28.

$E_b/N_o = 12$; NONCOHERENT ASK ; $P_e = \frac{1}{2}\left[1+\sqrt{\frac{1}{2\pi(12)}}\right]e^{-6} = 1.382\times10^{-3}$

NONCOHERENT FSK: $P_e = \frac{1}{2}e^{-6} = 1.239\times10^{-3}$

COHERENT ASK AND FSK, UNIPOLAR PCM:

$$P_e = \frac{1}{2}\,\text{erfc}\,\sqrt{6} \simeq e^{-6}/2\sqrt{6\pi} = 2.855\times10^{-4}$$

DPSK: $P_e = \frac{1}{2}e^{-12} = 3.072\times10^{-6}$

PSK, POLAR PCM, MANCHESTER PCM: $P_e = \frac{1}{2}\,\text{erfc}\,\sqrt{12} \simeq \frac{e^{-12}}{2\sqrt{12\pi}} = 5.003\times10^{-7}$

USING PSK AS THE REFERENCE THESE ARE LARGER BY FACTORS

$$\frac{P_e}{P_{e\,PSK}} = 6.14, \quad \text{DPSK}$$

$= 570.7$, COHERENT ASK AND FSK, UNIPOLAR PCM

$= 2{,}476.5$ NONCOHERENT FSK

$= 2{,}762.3$ NONCOHERENT ASK ←

15.3.29.

$$\text{SNR AT DETECTOR} = \frac{E_b}{N_0} = \frac{P_b T}{N_o} = \frac{P_b}{N_o W} = 30\ \text{dB}$$

IT IS DESIRED TO EXPAND THE BANDWIDTH BY A FACTOR OF 10/3, WHILE MAINTAINING THE SAME SNR.

∴ THE RECEIVED POWER P_b MUST INCREASE BY THE SAME FACTOR

THUS, ADDL. POWER NEEDED $P_a = 10\log_{10}\left(\frac{10}{3}\right) = 5.2288\ \text{dB}$

IF THE LOSS IN TRANSMISSION IS L_s dB, THEN THE RELATION BETN THE TRANSMITTED AND RECEIVED POWER IS

$$10\log_{10} P_s = L_s + 10\log_{10} P_b$$

∴ REQD. TRANSMITTED POWER IS

$$P_s = -3 + 5.2288 = 2.2288\ \text{dBW}.$$ ←

CHAPTER 16

16.1.1.

(a) WHEN THE SWITCH IS CLOSED AT $t=0$,

$$V_S = v_R + v_C = v_R + \frac{1}{C}\int i\,dt + v_C(t=0)\;;\; v_R = Ri$$

WITH INITIAL CONDITION $v_C(t=0)=0$, THE CHARGING CURRENT IS GIVEN BY

$$i(t) = \frac{V_S}{R}\, e^{-t/RC} \quad \longleftarrow$$

THE CAPACITOR VOLTAGE IS $v_C(t) = \frac{1}{C}\int_0^t i\,dt = V_S\left(1 - e^{-t/RC}\right)$ ⟵

WHERE $RC = \tau$ IS THE TIME CONSTANT OF THE RC LOAD.

$$\frac{dv_C}{dt} = \frac{V_S}{RC}\, e^{-t/RC}\;;\quad \left.\frac{dv_C}{dt}\right|_{t=0} = \frac{V_S}{RC}$$

WAVEFORMS OF VOLTAGE AND CURRENT ARE SHOWN IN FIGURES BELOW:

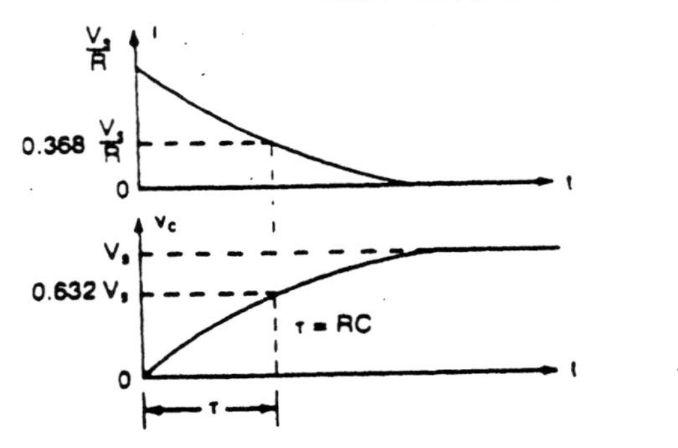

⟵

(b) WHEN THE SWITCH IS CLOSED AT $t=0$, THE CURRENT THROUGH THE INDUCTOR INCREASES.

$$V_S = v_L + v_R = L\frac{di}{dt} + iR$$

WITH INITIAL CONDITION $i(t=0)=0$, $\quad i(t) = \frac{V_S}{R}\left(1 - e^{-tR/L}\right)$ ⟵

$$\frac{di}{dt} = \frac{V_S}{L}\, e^{-tR/L}\;;\quad \left.\frac{di}{dt}\right|_{t=0} = \frac{V_S}{L}$$

VOLTAGE ACROSS THE INDUCTOR $v_L(t) = L\frac{di}{dt} = V_S e^{-tR/L}$ ⟵

WHERE $L/R = \tau$ IS THE TIME CONSTANT OF AN RL LOAD.

WAVEFORMS OF VOLTAGE AND CURRENT ARE SHOWN BELOW:

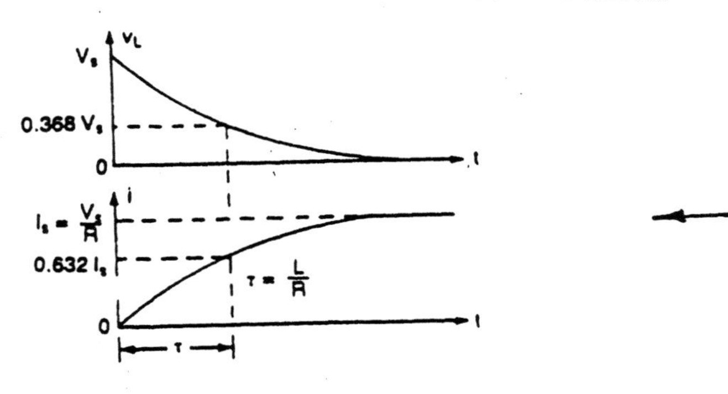

⟵

16.1.1. (CONTD.)

(c) IF $t >> L/R$, THE VOLTAGE ACROSS THE INDUCTOR TENDS TO BE ZERO AND ITS CURRENT REACHES A STEADY-STATE VALUE OF $I_S = V_S/R$.
IF AN ATTEMPT IS THEN MADE TO OPEN SWITCH S, THE ENERGY STORED IN THE INDUCTOR $(= \frac{1}{2}Li^2)$ WILL BE TRANSFORMED INTO A HIGH REVERSE VOLTAGE ACROSS THE SWITCH AND DIODE. THIS ENERGY WILL BE DISSIPATED IN THE FORM OF SPARKS ACROSS THE SWITCH. DIODE D IS LIKELY TO BE DAMAGED IN THE PROCESS. [NOTE THAT, TO OVERCOME SUCH A SITUATION, A FREE-WHEELING DIODE IS CONNECTED ACROSS AN INDUCTIVE LOAD. SEE PROB. 16.1.2. (a)]

(d) WHEN THE SWITCH S IS CLOSED AT $t=0$,

$$V_S = L\frac{di}{dt} + \frac{1}{C}\int i\,dt + v_c(t=0)$$

WITH INITIAL CONDITION $i(t=0)=0$, AND $v_c(t=0)=0$,

$$i(t) = V_S\sqrt{\frac{C}{L}}\sin\omega t = I_p \sin\omega t, \text{ WHERE } \omega = 1/\sqrt{LC}$$

AND PEAK CURRENT $I_p = V_S\sqrt{\frac{C}{L}}$.

$$\frac{di}{dt} = \frac{V_S}{L}\cos\omega t\,; \quad \left.\frac{di}{dt}\right|_{t=0} = \frac{V_S}{L}$$

$$v_c(t) = \frac{1}{C}\int_0^t i\,dt = V_s(1-\cos\omega t)$$

AT A TIME $t = t_1 = \pi\sqrt{LC}$, THE DIODE CURRENT i FALLS TO ZERO, AND THE CAPACITOR IS CHARGED TO $2V_S$.
WAVEFORMS OF VOLTAGE AND CURRENT ARE SHOWN BELOW:

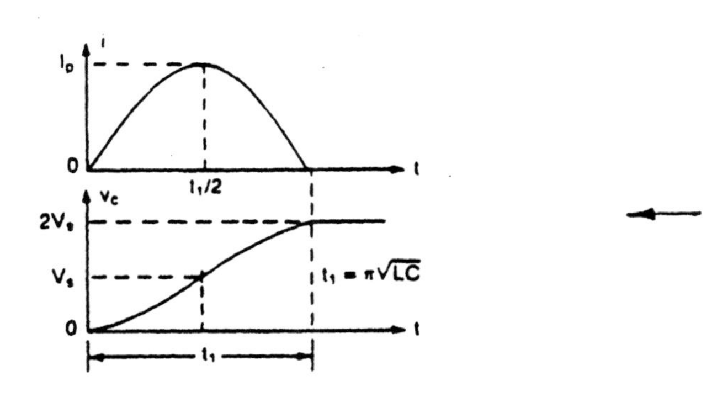

16.1.2.

EQUIVALENT CIRCUITS FOR THE MODES ARE SHOWN BELOW:

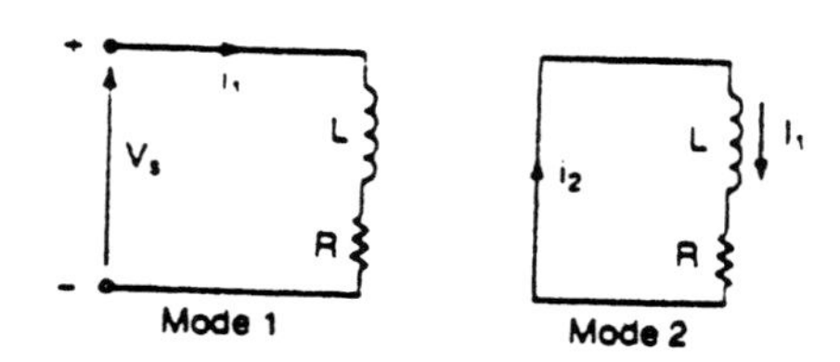

i_1 AND i_2 ARE THE INSTANTANEOUS CURRENTS FOR MODES 1 AND 2, RESPECTIVELY. t_1 AND t_2 ARE THE CORRESPONDING DURATIONS OF THOSE MODES.

(a) MODE 1: $i_1(t) = \frac{V_s}{R}(1 - e^{-tR/L})$

WHEN THE SWITCH IS OPENED AT $t = t_1$ (AT THE END OF THIS MODE), THE CURRENT AT THAT TIME $I_1 = i_1(t = t_1) = \frac{V_s}{R}(1 - e^{-t_1R/L})$

IF t_1 IS SUFFICIENTLY LONG, THE CURRENT REACHES A STEADY-STATE VALUE AND A STEADY-STATE CURRENT OF $I_s = V_s/R$ FLOWS THROUGH THE LOAD.

MODE 2: THIS MODE BEGINS WHEN THE SWITCH IS OPENED AND THE LOAD CURRENT STARTS TO FLOW THROUGH THE FREEWHEELING DIODE D_m. REDEFINING THE TIME ORIGIN AT THE BEGINNING OF THIS MODE, THE CURRENT THROUGH THE FREEWHEELING DIODE IS FOUND FROM $L\frac{di_2}{dt} + Ri_2 = 0$

WITH INITIAL CONDITION $i_2(t = 0) = I_1$, THE FREEWHEELING CURRENT IS THEN GIVEN BY $i_2(t) = I_1 e^{-tR/L}$

AND THIS DECAYS EXPONENTIALLY TO ZERO AT $t = t_2$, PROVIDED THAT $t_2 \gg L/R$.

WAVEFORMS FOR THE CURRENTS ARE GIVEN BELOW:

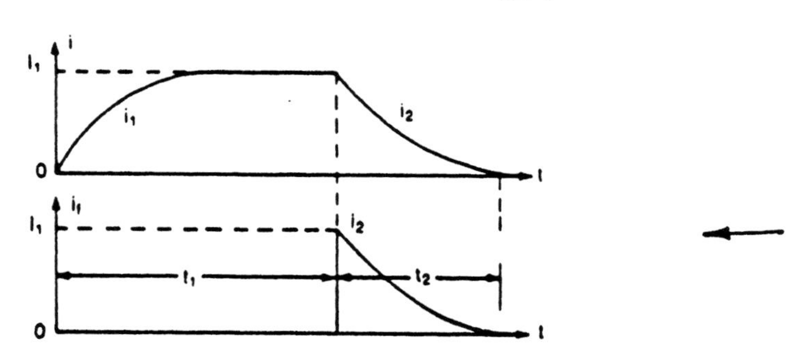

16.1.2. CONTD.

(b) WITH A MAGNETIZING INDUCTANCE OF L_m FOR THE TRANSFORMER, THE EQUIVALENT CIRCUIT IS GIVEN BELOW:

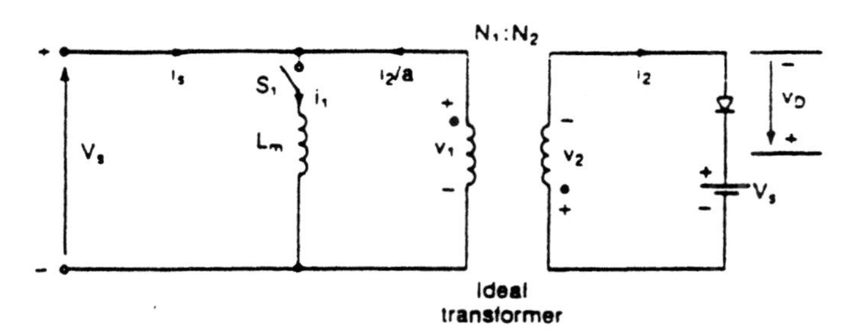

REFERRING THE DIODE AND SECONDARY VOLTAGE (SOURCE VOLTAGE) TO THE PRIMARY SIDE, THE EQUIVALENT CIRCUIT IS SHOWN BELOW:

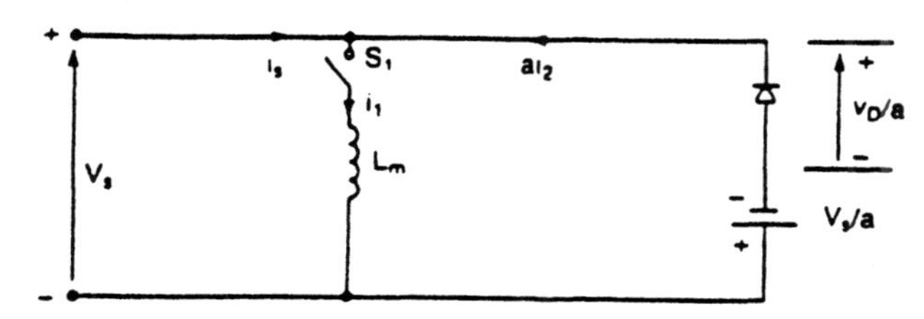

EQUIVALENT CIRCUITS FOR THE MODES ARE SHOWN BELOW:

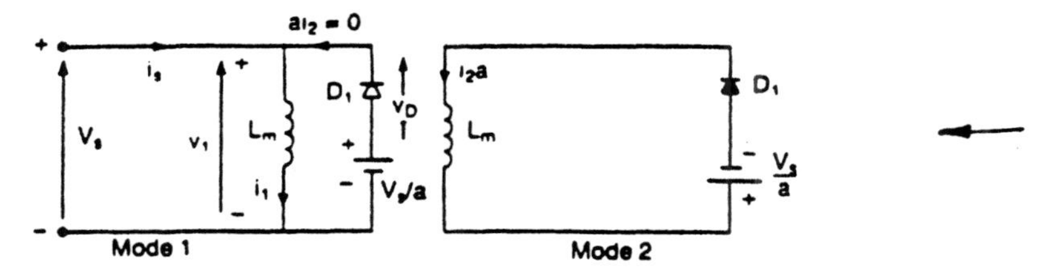

LET t_1 AND t_2 BE THE DURATIONS OF MODES 1 AND 2, RESPECTIVELY.

MODE 1: DURING THIS MODE, THE SWITCH IS CLOSED AT $t=0$. THE DIODE D_1 IS REVERSE BIASED AND THE CURRENT THROUGH THE DIODE (SECONDARY CURRENT) IS $ai_2 = 0$ OR $i_2 = 0$.

FROM MODE 1 - EQUIVALENT CIRCUIT, $V_S = (v_D - V_S)/a$

AND THE REVERSE DIODE VOLTAGE $v_D = V_S(1+a)$.

WITH NO INITIAL CURRENT IN THE CIRCUIT, THE PRIMARY CURRENT IS THE SAME AS THE SWITCH CURRENT i_S:

$$V_S = L_m \frac{di_1}{dt} \quad ; \quad \text{AND } i_1(t) = i_S(t) = \frac{V_S}{L_m} t$$

THIS MODE IS VALID FOR $0 \le t \le t_1$ AND ENDS WHEN THE SWITCH IS OPENED AT $t = t_1$.

AT THE END OF THIS MODE 1, THE PRIMARY CURRENT IS

$$I_0 = \frac{V_S}{L_m} t_1$$

16.1.2.(b) CONTD.

MODE 2: DURING THIS MODE THE SWITCH IS OPENED; THE VOLTAGE ACROSS THE INDUCTOR IS REVERSED AND THE DIODE D_1 IS FORWARD BIASED. A CURRENT FLOWS THROUGH THE TRANSFORMER SECONDARY AND THE ENERGY STORED IN THE INDUCTOR IS RETURNED TO THE SOURCE. REDEFINING THE TIME ORIGIN AT THE START OF THIS MODE,

$$L_m \frac{di_1}{dt} + \frac{V_s}{a} = 0$$

WITH INITIAL CONDITION $i_1(t=0) = I_o$

$$i_1(t) = -\frac{V_s}{a L_m} t + I_o$$

THE CONDUCTION TIME OF DIODE D_1 IS OBTAINED FROM THE CONDITION

$$i_1(t=t_2) = 0$$

THUS $$t_2 = \frac{a L_m I_o}{V_s} = a t_1$$

MODE 2 IS VALID FOR $0 \le t \le t_2$. AT THE END OF THIS MODE AT $t = t_2$, ALL THE ENERGY STORED IN THE INDUCTOR L_m IS RETURNED TO THE SOURCE.

WAVEFORMS FOR CURRENTS AND VOLTAGES ARE SHOWN BELOW:

16.1.2. (b) CONTD.

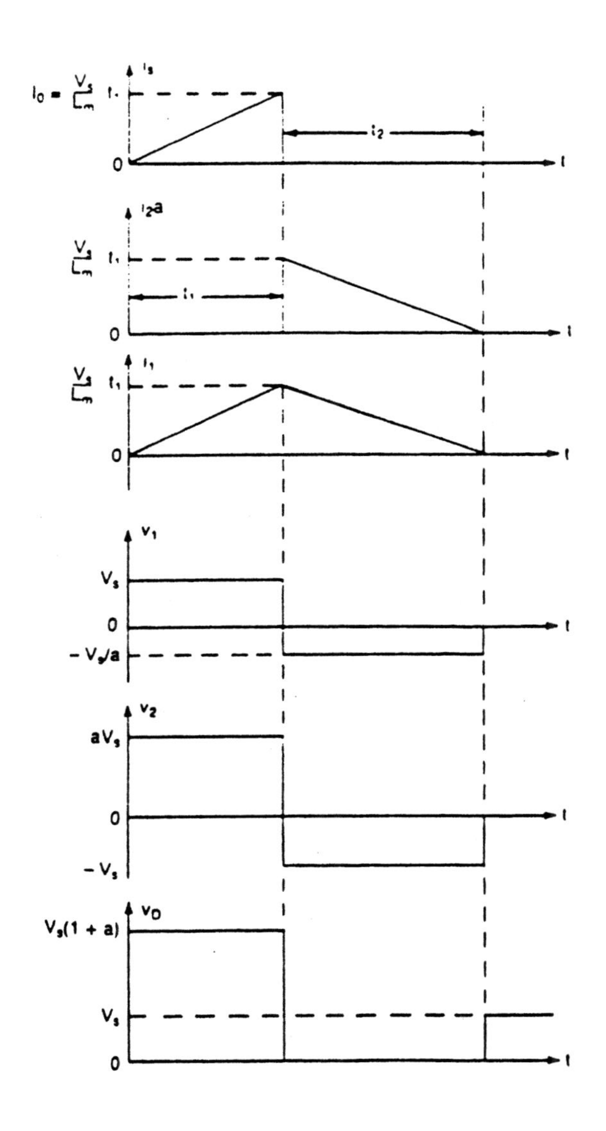

i_s
$I_0 = \frac{V_s}{L_m} t_1$
t_2
0
t
$i_2 a$
$\frac{V_s}{L_m} t_1$
t_1
0
t
i_1
$\frac{V_s}{L_m} t_1$
0
t
v_1
V_s
0
t
$-V_s/a$
v_2
aV_s
0
t
$-V_s$
v_D
$V_s(1 + a)$
V_s
0
t

16.1.3.

THE OUTPUT OF A FULL-WAVE RECTIFIER IS SHOWN BELOW:

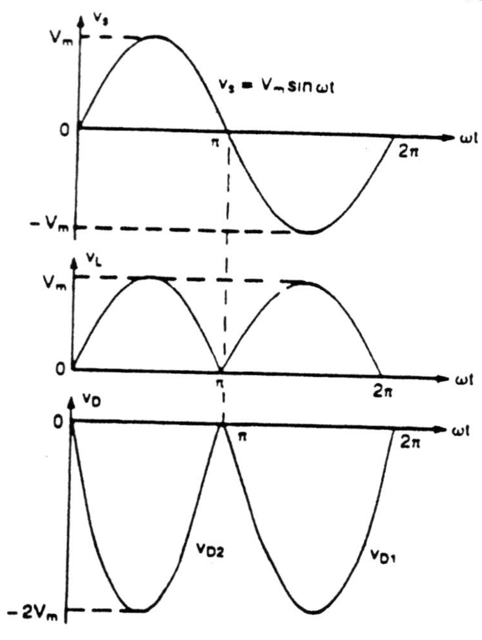

(a) AVERAGE OUTPUT VOLTAGE $V_{dc} = \frac{2}{T}\int_0^{T/2} V_m \sin\omega t \, dt = \frac{2V_m}{\pi} = 0.6366\, V_m$

AVERAGE LOAD CURRENT $I_{dc} = V_{dc}/R = 0.6366\, V_m / R$

$V_{RMS} = \left[\frac{2}{T}\int_0^{T/2} (V_m \sin\omega t)^2 dt\right]^{1/2} = V_m/\sqrt{2} = 0.707\, V_m$

$I_{RMS} = V_{RMS}/R = 0.707\, V_m/R$

$P_{dc} = V_{dc} I_{dc} = (0.6366\, V_m)^2/R$

$P_{ac} = V_{RMS} I_{RMS} = (0.707\, V_m)^2/R$

(i) $\eta = P_{dc}/P_{ac} = 0.6366^2/0.707^2 = 81\%$ ←

(ii) $FF = V_{RMS}/V_{dc} = 0.707/0.6366 = 1.11$ ←

(iii) $RF = \sqrt{FF^2 - 1} = \sqrt{1.11^2 - 1} = 0.482$ OR 48.2% ←

(iv) RMS VOLTAGE OF TRANSFORMER SECONDARY $V_S = V_m/\sqrt{2} = 0.707\, V_m$

RMS TRANSFORMER SECONDARY CURRENT $I_S = 0.5\, V_m/R$

VOLTAMPERE RATING OF TRANSFORMER $VA = 2 V_S I_S = 2(0.707 V_m)(0.5 V_m/R)$

$TUF = \frac{P_{dc}}{VA} = \frac{0.6366^2}{2 \times 0.707 \times 0.5} = 0.5732$ OR 57.32% ←

(v) $PIV = 2V_m$ ←

THE PERFORMANCE OF A FULL-WAVE RECTIFIER IS SIGNIFICANTLY IMPROVED COMPARED TO THAT OF A HALF-WAVE RECTIFIER. ←

16.1.3.(b) CONTD.

THE RECTIFIER OUTPUT VOLTAGE CAN BE DESCRIBED BY A FOURIER SERIES AS

$$v_L(t) = V_{dc} + \sum_{n=1,2,\ldots}^{\infty} (a_n \cos n\omega t + b_n \sin n\omega t)$$

WHERE $V_{dc} = \frac{1}{2\pi}\int_0^{2\pi} v_L(t)\, d(\omega t) = \frac{2}{2\pi}\int_0^{\pi} V_m \sin\omega t\, d(\omega t) = \frac{2V_m}{\pi}$

$$a_n = \frac{1}{\pi}\int_0^{2\pi} v_L \cos n\omega t\, d(\omega t) = \frac{2}{\pi}\int_0^{\pi} V_m \sin\omega t \cos n\omega t\, d(\omega t)$$

$$b_n = \frac{1}{\pi}\int_0^{2\pi} v_L \sin n\omega t\, d(\omega t) = \frac{2}{\pi}\int_0^{\pi} V_m \sin\omega t \sin n\omega t\, d(\omega t) = 0$$

$$a_n = \frac{4V_m}{\pi} \sum_{n=2,4,\ldots}^{\infty} \frac{1}{(n-1)(n+1)}$$

$$\therefore v_L(t) = \frac{2V_m}{\pi} - \frac{4V_m}{3\pi}\cos 2\omega t - \frac{4V_m}{15\pi}\cos 4\omega t - \frac{4V_m}{35\pi}\cos 6\omega t - \ldots$$

LOAD IMPEDANCE $Z = R + j(n\omega L) = \sqrt{R^2 + (n\omega L)^2}\ \angle\theta_n$

WHERE $\theta_n = \tan^{-1}\frac{n\omega L}{R}$

INSTANTANEOUS CURRENT $i_L(t) = I_{dc} - \frac{4V_m}{\pi\sqrt{R^2 + (n\omega L)^2}}\Big[$

$$\left[\frac{1}{3}\cos(2\omega t - \theta_2) - \frac{1}{15}\cos(4\omega t - \theta_4) \cdots\right]$$

WHERE $I_{dc} = V_{dc}/R = 2V_m/(\pi R)$.

16.1.4.

(i) INPUT CURRENT MAY BE EXPRESSED IN A FOURIER SERIES AS

$$i_1(t) = I_{dc} + \sum_{n=1}^{\infty} (a_n \cos n\omega t + b_n \sin n\omega t)$$

WHERE $I_{dc} = \frac{1}{2\pi}\int_0^{2\pi} i_1(t)\, d(\omega t) = \frac{1}{2\pi}\int_0^{2\pi} I_a\, d(\omega t) = 0$

$$a_n = \frac{1}{\pi}\int_0^{2\pi} i_1(t)\cos n\omega t\, d(\omega t) = \frac{2}{\pi}\int_0^{\pi} I_a \cos n\omega t\, d(\omega t) = 0$$

$$b_n = \frac{1}{\pi}\int_0^{2\pi} i_1(t)\sin n\omega t\, d(\omega t) = \frac{2}{\pi}\int_0^{\pi} I_a \sin n\omega t\, d(\omega t) = \frac{4 I_a}{n\pi}$$

$$\therefore i_1(t) = \frac{4 I_a}{\pi}\left(\sin \omega t + \frac{1}{3}\sin 3\omega t + \frac{1}{5}\sin 5\omega t + \dots\right)$$

RMS VALUE OF FUNDAMENTAL COMPONENT OF INPUT CURRENT IS

$$I_1 = \frac{4 I_a}{\pi\sqrt{2}} = 0.9\, I_a$$

RMS VALUE OF INPUT CURRENT IS $I_s = \frac{4}{\pi\sqrt{2}} I_a \left[1 + \left(\frac{1}{3}\right)^2 + \left(\frac{1}{5}\right)^2 + \left(\frac{1}{7}\right)^2 + \left(\frac{1}{9}\right)^2 + \dots\right]^{1/2} = I_a$

$$\therefore HF = \left(\frac{I_s^2 - I_1^2}{I_1^2}\right)^{1/2} = \left[\left(\frac{I_s}{I_1}\right)^2 - 1\right]^{1/2} = \left[\left(\frac{1}{0.9}\right)^2 - 1\right]^{1/2} = 0.4843$$

or 48.43% ←

(ii) ϕ, THE DISPLACEMENT ANGLE = 0

DISPLACEMENT FACTOR, $DF = \cos\phi = 1$

$$PF = \frac{V_s I_1}{V_s I_s}\cos\phi = \frac{I_1}{I_s}\cos\phi = 0.9 \text{ LAGGING}$$ ←

16.1.5.

(a)

(i) THE AVERAGE OUTPUT VOLTAGE FOR A 3-PHASE RECTIFIER IS GIVEN BY

$$V_{dc} = \frac{2}{2\pi/3}\int_0^{\pi/3} V_m \cos\omega t\, d(\omega t) = V_m \frac{3}{\pi}\sin\frac{\pi}{3} = 0.827\, V_m$$

$$I_{dc} = 0.827\, V_m / R$$

$$V_{RMS} = \left[\frac{2}{2\pi/3}\int_0^{\pi/3} V_m^2 \cos^2\omega t\, d(\omega t)\right]^{1/2}$$

$$= V_m\left[\frac{3}{2\pi}\left(\frac{\pi}{3} + \frac{1}{2}\sin\frac{2\pi}{3}\right)\right]^{1/2} = 0.84068\, V_m$$

$$I_{RMS} = 0.84068\, V_m / R$$

$$P_{dc} = V_{dc} I_{dc} = (0.827\, V_m)^2 / R$$

$$P_{ac} = V_{RMS} I_{RMS} = (0.84068\, V_m)^2 / R$$

$$\eta = P_{dc} / P_{ac} = 0.827^2 / 0.84068^2 = 96.77\%$$ ←

16·1·5·(a) CONTD.

(ii) FORM FACTOR, $FF = V_{RMS} / V_{dc} = 0.84068 / 0.827 = 1.0165$ OR 101.65% ←

(iii) RIPPLE FACTOR, $RF = \sqrt{FF^2 - 1} = \sqrt{1.0165^2 - 1} = 0.1824$ OR 18.24% ←

(iv) RMS VOLTAGE OF THE TRANSFORMER SECONDARY $V_S = V_m / \sqrt{2} = 0.707 V_m$

RMS TRANSFORMER SECONDARY CURRENT $I_S = \left[\frac{2}{2\pi}\int_0^{\pi/3} I_m^2 \cos^2\omega t \, d(\omega t)\right]^{1/2}$

OR $I_S = I_m\left[\frac{1}{2\pi}\left(\frac{\pi}{3} + \frac{1}{2}\sin\frac{2\pi}{3}\right)\right]^{1/2} = 0.4854 I_m = 0.4854 V_m / R$

VOLTAMPERE RATING OF THE TRANSFORMER $VA = 3 V_S I_S$

$= 3 \times 0.707 V_m \times 0.4854 V_m / R$

$\therefore$ $TUF = P_{dc} / VA = \frac{0.827^2}{3 \times 0.707 \times 0.4854} = 0.6643$ ←

(v) $PIV = \sqrt{3} V_m$, WHICH IS THE PEAK VALUE OF THE SECONDARY LINE-TO-LINE VOLTAGE. ←

(b) FIGURE BELOW SHOWS A MULTIPHASE (q-PHASE) STAR RECTIFIER AND THE ASSOCIATED WAVEFORMS FOR q PULSES :

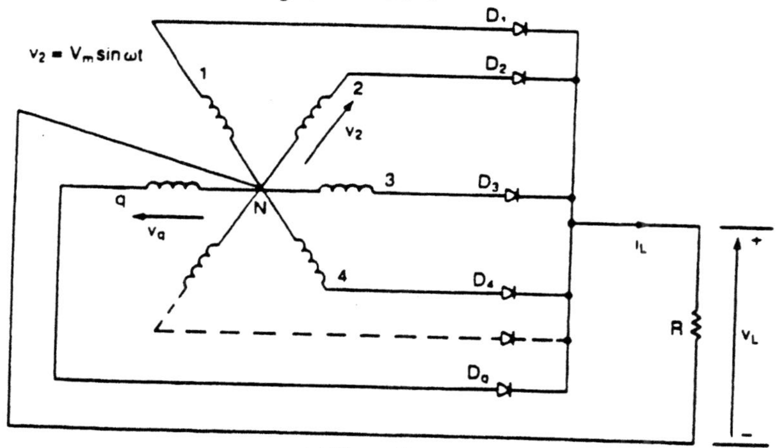

(a) Circuit diagram

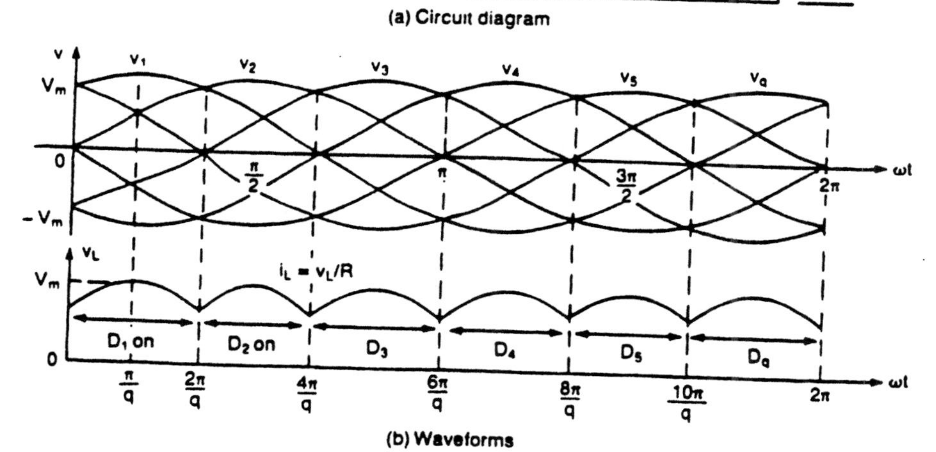

(b) Waveforms

16.1.5.(b) CONTD.

FREQUENCY OF THE OUTPUT IS q TIMES THE FUNDAMENTAL COMPONENT (qf)

FOURIER SERIES COEFFICIENTS $b_n = 0$

$$a_n = \frac{1}{\pi/q} \int_{-\pi/q}^{\pi/q} V_m \cos\omega t \cos n\omega t \; d(\omega t)$$

$$= \frac{q V_m}{\pi} \left[\frac{\sin\{(n-1)\pi/q\}}{n-1} + \frac{\sin\{(n+1)\pi/q\}}{n+1} \right]$$

$$= \frac{q V_m}{\pi} \; \frac{(n+1)\sin[(n-1)\pi/q] + (n-1)\sin[(n+1)\pi/q]}{n^2-1}$$

USING TRIG. IDENTITIES, AFTER SIMPLIFICATION, ONE GETS

$$a_n = \frac{2 q V_m}{\pi (n^2-1)} \left(n \sin\frac{n\pi}{q} \cos\frac{\pi}{q} - \cos\frac{n\pi}{q} \sin\frac{\pi}{q} \right) \quad \ldots \quad (A)$$

FOR A RECTIFIER WITH q PULSES PER CYCLE, THE HARMONICS OF THE OUTPUT VOLTAGE ARE q TH, $2q$ TH, $3q$ TH, $4q$ TH, AND EQ. (A) IS VALID FOR $n = 0, 1q, 2q, 3q$.

THE TERM $\sin(n\pi/q) = \sin\pi = 0$

AND

$$\therefore \; a_n = \frac{-2 q V_m}{\pi (n^2-1)} \left(\cos\frac{n\pi}{q} \sin\frac{\pi}{q} \right)$$

THE DC COMPONENT IS FOUND BY LETTING $n = 0$;

$$V_{dc} = \frac{a_0}{2} = V_m \frac{q}{\pi} \sin\frac{\pi}{q}$$

THUS, THE FOURIER SERIES OF THE OUTPUT VOLTAGE IS GIVEN BY

$$v_L(t) = \frac{a_0}{2} + \sum_{n=q,2q,\ldots}^{\infty} a_n \cos n\omega t$$

$$= V_m \frac{q}{\pi} \sin\frac{\pi}{q} \left(1 - \sum_{n=q,2q,\ldots}^{\infty} \frac{2}{n^2-1} \cos\frac{n\pi}{q} \cos n\omega t \right)$$

FOR $q = 3$,

$$v_L(t) = V_m \frac{3}{\pi} \sin\frac{\pi}{3} \left(1 - \sum_{n=3,6,\ldots}^{\infty} \frac{2}{n^2-1} \cos\frac{n\pi}{3} \cos n\omega t \right)$$

$$= 0.827 \, V_m \left(1 + \frac{2}{8} \cos 3\omega t - \frac{2}{35} \cos 6\omega t + \ldots \right) \longleftarrow$$

NOTE THAT THE THIRD HARMONIC IS THE DOMINANT ONE.

16.1.6.

THE WAVEFORMS AND CONDUCTION TIMES OF DIODES ARE SHOWN BELOW:

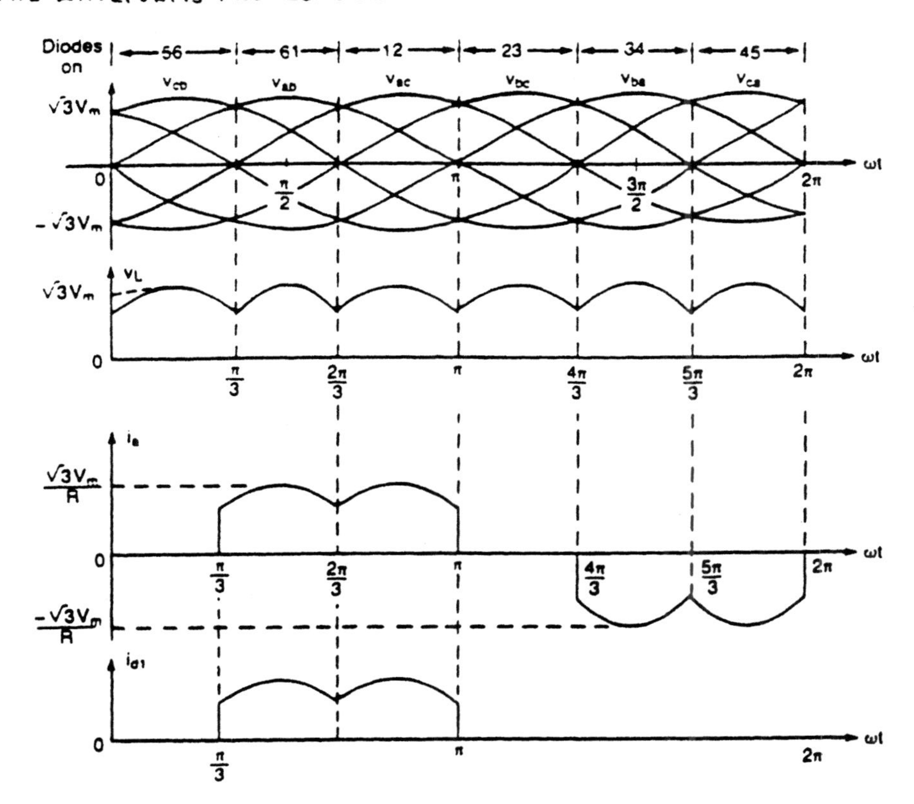

(i) THE AVERAGE OUTPUT VOLTAGE $V_{dc} = \frac{2}{2\pi/6}\int_0^{\pi/6} \sqrt{3}\, V_m \cos\omega t \, d(\omega t)$

$= \frac{3\sqrt{3}}{\pi} V_m = 1.6542\, V_m$

WHERE V_m IS THE PEAK PHASE VOLTAGE.

$I_{dc} = 1.6542\, V_m / R$

$V_{RMS} = \left[\frac{2}{2\pi/6} \int_0^{\pi/6} 3\, V_m^2 \cos^2\omega t \, d(\omega t) \right]^{1/2}$

$= \left(\frac{3}{2} + \frac{9\sqrt{3}}{4\pi} \right)^{1/2} V_m = 1.6554\, V_m$

$I_{RMS} = 1.6554\, V_m / R$

OUTPUT DC POWER $P_{dc} = V_{dc} I_{dc} = (1.6542\, V_m)^2 / R$

OUTPUT AC POWER $P_{ac} = V_{RMS} I_{RMS} = (1.6554\, V_m)^2 / R$

$\eta = P_{dc} / P_{ac} = 1.6542^2 / 1.6554^2 = 99.86\%$ ←

16.1.6. CONTD.

(ii) FORM FACTOR, $FF = V_{RMS}/V_{dc} = 1.6554/1.6542 = 1.0007$ OR 100.07% ←

(iii) RIPPLE FACTOR, $RF = \sqrt{\left(\frac{V_{RMS}}{V_{dc}}\right)^2 - 1} = \sqrt{FF^2 - 1} = \sqrt{1.0007^2 - 1} = 0.0374$ OR 3.74% ←

(iv) RMS VOLTAGE OF THE TRANSFORMER SECONDARY $V_S = V_m/\sqrt{2} = 0.707\, V_m$

RMS TRANSFORMER SECONDARY CURRENT $I_S = \left[\frac{8}{2\pi}\int_0^{\pi/6} I_m^2 \cos^2\omega t\, d(\omega t)\right]^{1/2}$

$= 0.7804\, I_m = 0.7804\sqrt{3}\, V_m/R$

VOLTAMPERE RATING OF TRANSFORMER $VA = 3 V_S I_S = 3(0.707 V_m)(0.7804\sqrt{3} V_m/R)$

$TUF = P_{dc}/VA = 1.6542^2/(3\sqrt{3} \times 0.707 \times 0.7804) = 0.9545$ ←

(v) PEAK LINE-TO-NEUTRAL VOLTAGE $= V_m$

$PIV = \sqrt{3}\, V_m$, WHICH IS THE PEAK VALUE OF THE SECONDARY LINE-TO-LINE VOLTAGE. ←

16.1.7.

(a) FROM EQ. (16.1.19), $\omega_m = \frac{2V_m}{\pi K_m}\cos\alpha - \frac{R_a}{K_m^2}T_a$

$$-\frac{1350\times\pi}{30} = \frac{2\times 230\sqrt{2}}{\pi\times 1.075}\cos\alpha - \frac{1.5}{1.075^2}(35)$$

$\cos\alpha = -0.4981$ or $\alpha = 119.9°$ ←

(ω–T axes sketch)

FOURTH-QUADRANT OPERATION ←

(b) FROM EQ. (16.1.18), FOR CONITINUOUS CONDUCTION, $\frac{2V_m}{\pi}\cos\alpha = I_aR_a + E$

FOR $\alpha = 160°$, $I_a = 11.56\,A$, $\frac{2\times 230\sqrt{2}}{\pi}\cos 160° = (11.56\times 1.5) + E$

OR $E = -211.9\ V$

FORWARD REGENERATION IS OBTAINED EITHER BY THE FIELD REVERSAL OR THE ARMATURE REVERSAL, FOR WHICH $K_m = -1.075$

$\therefore\ \omega_m = E/K_m = -211.9/(-1.075) = 197.1$ rad/s.

OR 1882 rpm ←

16.1.8.

THE IDEAL NO-LOAD OPERATION IS OBTAINED WHEN $E = V_m$ FOR $0 \le \alpha \le \pi/6$ AND $E = V_m \sin(\alpha + \frac{\pi}{3})$ FOR $\pi/6 \le \alpha \le \pi$.

$\therefore$ NO-LOAD SPEEDS ARE GIVEN BY

$$\omega_{mo} = V_m / K_m, \quad 0 \le \alpha \le \pi/6$$

AND $$\frac{V_m \sin(\alpha + \frac{\pi}{3})}{K_m}, \quad \frac{\pi}{6} \le \alpha \le \pi$$

UNLIKE IN SINGLE-PHASE CASE, NO-LOAD SPEEDS CAN BE NEGATIVE FOR $\frac{2\pi}{3} < \alpha < \pi$. ←

YES, A CONSIDERABLE REDUCTION IN THE ZONE OF DISCONTINUOUS CONDUCTION WILL RESULT. ←

16.1.9.

(a) $\alpha = 120°$; $T_a = 25$ N·m

FROM EQ. (12.2.23) $\omega_m = \frac{3V_m}{\pi K_m} \cos\alpha - \frac{R_a}{K_m^2} T_a$, ONE HAS

$$\omega_m = \frac{3 \times 230.5}{\pi \times 1.075} \cos 120° - \frac{1.5}{1.075^2} \times 25 = -102.4 - 32.45$$
$$= -134.85 \text{ rad/s OR } -1288 \text{ rpm} \leftarrow$$

(b) $\alpha = 60°$; $T_a = 5$ N·m

FROM EQ. (12.2.23), $$\omega_m = \frac{3 \times 230.5}{\pi \times 1.075} \cos 60° - \frac{1.5}{1.075^2} (5)$$
$$= 102.4 - 6.5 = 95.9 \text{ rad/s. OR } 916 \text{ rpm} \leftarrow$$

16.1.10.

AT 29.2 Hz, $V = a \times 440/\sqrt{3} = a \times 254 = 0.4864 \times 254 = 123.5\,V$

$$I_2' = \frac{V}{\left[\left(R_1 + \frac{R_2'}{S}\right)^2 + a^2 (X_{\ell 1} + X_{\ell 2}')^2\right]^{1/2}} \angle -\tan^{-1} \frac{a(X_{\ell 1} + X_{\ell 2}')}{\left(R_1 + \frac{R_2'}{S}\right)}$$

$$= \frac{123.5}{\left[\{0.2 + (0.1/0.0235)\}^2 + (0.4864 \times 1.45)^2\right]^{1/2}} \angle -\tan^{-1} \frac{0.4864 \times 1.45}{\left(0.2 + \frac{0.1}{0.0235}\right)}$$

$$= 27.4 \angle -\tan^{-1} 0.1583 = 27.4 \angle -9^\circ$$

$$\bar{I}_m = \frac{V}{a X_m} \angle -90^\circ = \frac{123.5}{0.4864 \times 20} \angle -90^\circ = 12.7 \angle -90^\circ$$

$$I_S = \bar{I}_2' + \bar{I}_m = 27.4 \angle -9^\circ + 12.7 \angle -90^\circ \quad \text{OR} \quad I_S = 32\,A \longleftarrow$$

16.1.11.

$a = 15/60 = 0.25$

(a) $$T_{max} = \frac{3}{2\omega_s}\left[\frac{V_{rated}^2}{\frac{R_1}{a} + \sqrt{(R_1/a)^2 + (X_{\ell 1} + X_{\ell 2}')^2}}\right], \quad a < 1,$$ FOR MOTORING

RATIO OF MAXIMUM TORQUES FOR $a = 0.25$ AND $a = 1$ IS THEN GIVEN BY

$$\frac{T_{max}(a = 0.25)}{T_{max}(a = 1)} = \frac{0.025 + \sqrt{0.025^2 + 0.25^2}}{\frac{0.025}{0.25} + \sqrt{\left(\frac{0.025}{0.25}\right)^2 + (0.25)^2}} = 0.75 \longleftarrow$$

(b) $$T = \frac{3}{\omega_s}\left[\frac{V_{rated}^2 \; R_2'/(aS)}{\left(\frac{R_1}{a} + \frac{R_2'}{aS}\right)^2 + (X_{\ell 1} + X_{\ell 2}')^2}\right], \quad a < 1$$

WITH $S = 1$ FOR STARTING

$$T_{st} = \frac{3}{\omega_s}\left[\frac{V_{rated}^2 \; R_2'/a}{\left(\frac{R_1 + R_2'}{a}\right)^2 + (X_{\ell 1} + X_{\ell 2}')^2}\right]$$

RATIO OF STARTING TORQUES FOR $a = 0.25$ AND $a = 1$ IS THEN

$$\frac{T_{st}(a = 0.25)}{T_{st}(a = 1)} = \frac{(0.025/0.25)/\left[(0.05/0.25)^2 + (0.25)^2\right]}{0.025/\{(0.05)^2 + (0.25)^2\}} = 2.54 \longleftarrow$$

16.1.12.

$$a = 30/60 = 0.5$$

$$T_{rated} = \frac{3}{\omega_s}\left[\frac{V_{rated}^2 \,(R_2'/s)}{\left(R_1 + \frac{R_2'}{s}\right)^2 + (X_{l1} + X_{l2}')^2}\right]$$

FOR 60 HZ, $T_{rated} = \frac{3}{\omega_s}\left[\frac{V_{rated}^2\,(0.025/0.05)}{\left(0.025 + \frac{0.025}{0.05}\right)^2 + (0.125+0.125)^2}\right]$

$$= \frac{3\,V_{rated}^2}{\omega_s}\,(1.48)$$

FOR 30 HZ, $T_{rated} = \frac{3}{\omega_s}\left[\frac{V_{rated}^2\; R_2'/(as)}{\left(\frac{R_1}{a} + \frac{R_2'}{as}\right)^2 + (X_{l1} + X_{l2}')^2}\right]$, FOR $a<1$ (HERE $a = 0.5$)

$$= \frac{3}{\omega_s}\left[\frac{V_{rated}^2\,(0.025/0.5s)}{\left(\frac{0.025}{0.5} + \frac{0.025}{0.5s}\right)^2 + (0.125+0.125)^2}\right]$$

THUS $\dfrac{0.025/0.5s}{\left(\frac{0.025}{0.5} + \frac{0.025}{0.5s}\right)^2 + (0.25)^2} = 1.48$

OR $\dfrac{0.05\,\frac{1}{s}}{\left(0.05 + 0.05\,\frac{1}{s}\right)^2 + 0.0625} = 1.48$

OR $s = 0.119$ or 0.324

THE SLIP ON THE STABLE PART OF THE SPEED-TORQUE CURVE WILL BE 0.119

$\therefore$ MOTOR SPEED $= 600\,(1 - 0.119) = 529$ rpm ←

16.1.13.

SYNCHRONOUS SPEED $= 120 \times 60/6 = 1200$ rpm or 125.66 rad/s.

SLIP $= (1200 - 1170)/1200 = 30/1200 = 0.025$

ROTOR IMPEDANCE $Z_r' = \frac{R_2'}{s} + jX_{l2}' = \frac{0.1}{0.025} + j0.7 = 4 + j0.7$
$= 4.06\angle 9.93^\circ\ \Omega$

STATOR IMPEDANCE $Z_s = 0.1 + j0.75 = 0.757\angle 82.4^\circ$

MACHINE IMPEDANCE $Z_{in} = Z_s + \dfrac{Z_r' Z_M}{Z_r' + Z_M} = 0.1 + j0.75 + \dfrac{(4.06\angle 9.93^\circ)(j20)}{4 + j20.7} = 3.7 + j2.12 = 4.26\angle 29.8^\circ\ \Omega$

$I_s = \dfrac{440/\sqrt{3}}{4.26} = 59.6\text{ A}$

$I_2' = \left|\dfrac{Z_M}{Z_M + Z_r'}\right| I_s = \dfrac{20}{21.08} \times 59.6 = 56.55\text{ A}$

$E_1 = I_2'\,|Z_r'| = 56.55 \times 4.06 = 229.6\text{ V}$

RATED TORQUE $= \dfrac{3}{\omega_s}(I_2')^2 R_2'/s = \dfrac{3}{125.66} \times 56.55^2 \times \dfrac{0.1}{0.025} = 305.4$ N·m

16.1.13. CONTD.

(a) AT 30 HZ, $a = 30/60 = 0.5$

$$T = \frac{3}{\omega_s}\left[\frac{E^2_{1\,rated}\; R_2'/(as)}{\{R_2'/(as)\}^2 + (x_{\ell 2}')^2}\right]$$

or $\frac{305.4}{2} = \frac{3}{125.66}\left[\frac{229.6^2 \times 0.1/(0.5S)}{\{0.1^2/(0.5S)^2\} + 0.7^2}\right]$ or $S = 0.0239$

$\omega_m = a\omega_s(1-S)$ or $N = 0.5(1200)(1-0.0239) = 585.7$ rpm ←

AT 30 HZ, $E_1 = 0.5 \times 229.6 = 114.8$ V

$Z_r' = \frac{R_2'}{S} + jax_{\ell 2}' = \frac{0.1}{0.0239} + j0.35 = 4.184 + j0.35 = 4.2\angle 4.8°\ \Omega$

WITH $\bar{E}_1$ AS REFERENCE, $\bar{I}_2' = \bar{E}_1/Z_r' = 114.8/(4.2\angle 4.8°) = 27.3\angle -4.8°$ A

$\bar{I}_m = \bar{E}_1/(jax_M) = 114.8/(j0.5 \times 20) = 11.48\angle -90°$

$\bar{I}_S = \bar{I}_2' + \bar{I}_m = 27.3\angle -4.8° + 11.48\angle -90°$ OR $I_S = 30.5$A ←

(b) SLIP SPEED IN RPM AT RATED TORQUE AND FREQUENCY = $N_{S\ell} = 0.025 \times 1200 = 30$ rpm

SINCE THE SPEED-TORQUE CHARACTERISTIC IS A STRAIGHT LINE, SLIP-SPEED AT ONE-HALF THE RATED TORQUE IS $0.5 \times 30 = 15$ rpm.

AT 30 HZ, SYNCHRONOUS SPEED = 600 rpm.

BECAUSE THE SLIP-SPEED REMAINS CONSTANT FOR A GIVEN TORQUE,

MOTOR SPEED = $600 - 15 = 585$ rpm ←

FOR A CONSTANT FLUX, THE (E_1/f) RATIO WILL BE A CONSTANT.

∴ AT 30 HZ, $E_1 = 0.5 \times 229.6 = 114.8$ V; SLIP = $15/600 = 0.025$

$Z_r' = \frac{R_2'}{S} + jax_{\ell 2}' = \frac{0.1}{0.025} + (j0.5 \times 0.7) = 4 + j0.35 = 4.015\angle 5°\ \Omega$

WITH $\bar{E}_1$ AS REFERENCE, $\bar{I}_2' = \bar{E}_1/Z_r' = 114.8/(4.015\angle 5°) = 28.6\angle -5°$ A

$\bar{I}_m = \bar{E}_1/(jax_M) = 114.8/j10 = 11.48\angle -90°$ A

∴ $\bar{I}_S = 28.6\angle -5° + 11.48\angle -90°$ OR $I_S = 31.7$A ←

16.1.14.

REFERRING TO FIG. 13.2.6 OF THE TEXT AND ITS NOTATION, IN A FUNDAMENTAL EQUIVALENT CIRCUIT, POWER TRANSFERRED ACROSS THE AIRGAP $P_g = 3E_1 I_2' \cos\phi_2$, WHERE ϕ_2 IS THE PHASE ANGLE BETWEEN PHASORS $\bar{E}_1$ AND $\bar{I}_2'$.

FOR THE DRIVE UNDER CONSIDERATION, TOTAL POWER CONSUMED IN THE ROTOR CIRCUIT

$$P_g' = 3 I_{rms}^2 (R_2 + R_{eff}) + P_m$$

WITH $I_{rms} = \frac{\pi}{3} I_2$, $P_g' = \frac{\pi^2}{3} I_2^2 (R_2 + R_{eff}) + P_m$

THE CONDITION $P_g = P_g'$ MUST BE SATISFIED BY THE FUNDAMENTAL EQUIVALENT CIRCUIT.

$$\therefore \quad E_1 I_2' \cos\phi_2 = \frac{\pi^2}{9} I_2^2 (R_2 + R_{eff}) + \frac{P_m}{3}$$

THE SLIP-POWER DUE TO FUNDAMENTAL ROTOR CURRENT IN THE DRIVE IS

$$S P_{g1} = 3 I_2^2 (R_2 + R_{eff})$$

WHERE P_{g1} IS THE FUNDAMENTAL AIR-GAP POWER IN THE DRIVE.

MECHANICAL POWER DEVELOPED BY THE FUNDAMENTAL ROTOR CURRENT IS

$$P_m = (1-S) P_{g1} = 3 I_2^2 (R_2 + R_{eff}) \frac{1-S}{S}$$

USING THE ABOVE, $E_1 I_2' \cos\phi_2 = \left[\left(\frac{\pi^2}{9} - 1\right)(R_2 + R_{eff}) + \frac{(R_2 + R_{eff})}{S} \right] I_2^2$

$$= \left(R_b + \frac{R_a}{S}\right) I_2^2 \quad \ldots \text{ EQ. (1)}$$

WHERE $R_b = \left(\frac{\pi^2}{9} - 1\right)(R_2 + R_{eff})$, AND $R_a = R_2 + R_{eff}$.

REFERRING ALL PARAMETERS ON THE RIGHT SIDE OF EQ. (1) ALSO TO THE STATOR SIDE,

$$E_1 I_2' \cos\phi_2 = \left(R_b' + \frac{R_a'}{S}\right) (I_2')^2$$

WHERE R_b' AND R_a' ARE THE REFERRED VALUES OF R_b AND R_a, RESPECTIVELY.

THE PER-PHASE FUNDAMENTAL EQUIVALENT CIRCUIT OF THE DRIVE REFERRED TO THE STATOR IS THEN GIVEN BY FIG. 16.1.25. ⟵

RESISTANCE (R_a'/S) ACCOUNTS FOR THE MECHANICAL POWER DEVELOPED AND THE FUNDAMENTAL ROTOR COPPER LOSS. RESISTANCE R_b' ACCOUNTS FOR THE ROTOR HARMONIC COPPER LOSS.

16.1.15.

FROM EQS. (16.1.31) AND (16.1.32)

$$R_{eff} = 0.5 R_t = 0.5 \left[R_d + (1-\delta) R \right]$$

$$= 0.5 \left[0.02 + (1 - 0.65)\, 1 \right] = 0.185\ \Omega$$

$$R_a' = R_2' + R_{eff}\ (\text{TURNS RATIO})^2 = 0.5 + (0.185)(2.5)^2 = 1.656\ \Omega$$

$$R_b' = \left(\frac{\pi^2}{9} - 1\right) R_a' = 0.0966 \times 1.656 = 0.16\ \Omega$$

$$T = \frac{3}{\omega_s} \left[\frac{V^2\, R_a'/s}{\{R_1 + R_b' + (R_a'/s)\}^2 + (X_{\ell 1} + X_{\ell 2}')^2} \right]$$

OR

$$1.5 \times 70.6 = \frac{3}{125.7} \left[\frac{(254)^2\, (1.656/s)}{\left(0.5 + 0.16 + \frac{1.656}{s}\right)^2 + (4)^2} \right]$$

WHICH GIVES $\left(\frac{1}{s}\right)^2 - 7.98\left(\frac{1}{s}\right) + 5.99 = 0$

OR $\frac{1}{s} = 7.16$ OR 0.84 ; OR $S = 0.14$ OR 1.19

DISCARDING THE LATTER UNFEASIBLE VALUE,

SPEED $= 1200\,(1 - 0.14) = 1032$ rpm. ←

16.1.16.

(a)

$\bar{I}_1$ jX_1 $V\angle 0°$ $V_1 \angle -\delta$

(b)

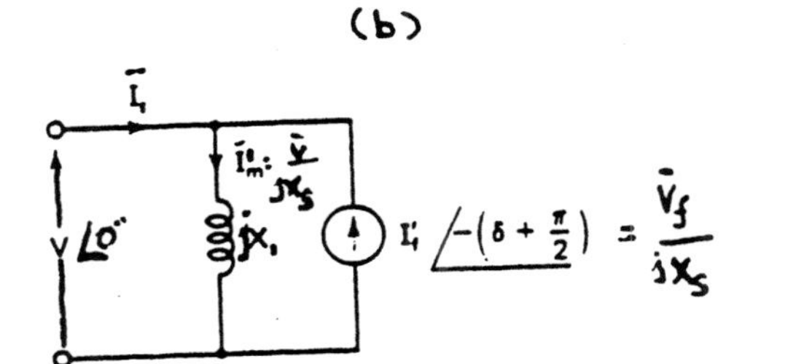

←

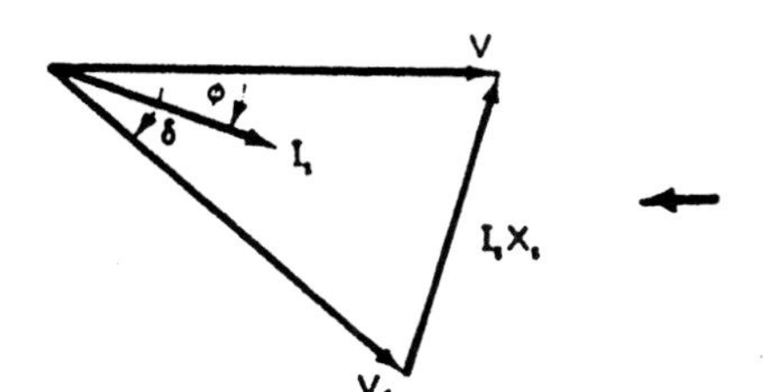

←

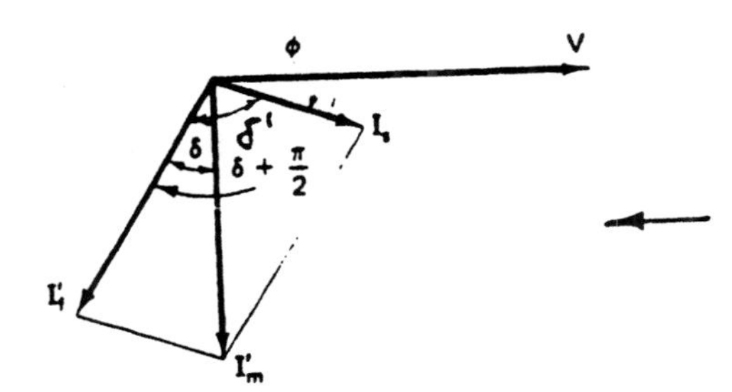

←

16.1.16. CONTD.

(c) $P_{in} = P_m = 3 V I_s \cos\phi$

$$\bar{I}_s = \frac{V\angle 0° - V_f \angle -\delta}{j X_s} = \frac{V}{X_s}\angle -\pi/2 - \frac{V_f}{X_s}\angle\left(-\delta - \frac{\pi}{2}\right)$$

$$I_s \cos\phi = \frac{V}{X_s}\cos(-\pi/2) - \frac{V_f}{X_s}\cos\left(-\delta - \frac{\pi}{2}\right) = \frac{V_f}{X_s}\sin\delta$$

$$\therefore \quad P_m = \frac{3 V V_f}{X_s}\sin\delta \; ; \quad T = \frac{P_m}{\omega_s} = \frac{3}{\omega_s}\frac{V V_f}{X_s}\sin\delta \quad \longleftarrow$$

FOR PART (b): $P_m = 3 V I_f' \sin\delta = 3 X_s I_m' I_f' \sin\delta \quad \longleftarrow$

$$T = \frac{3}{\omega_s} V I_f' \sin\delta = \frac{3 X_s}{\omega_s} I_m' I_f' \sin\delta \quad \longleftarrow$$

[NOTE: HERE $\bar{V}_f$ LAGS BEHIND $\bar{V}$; δ SHOULD BE POSITIVE WHEN $\bar{V}_f$ LAGS $\bar{V}$.]

(d)

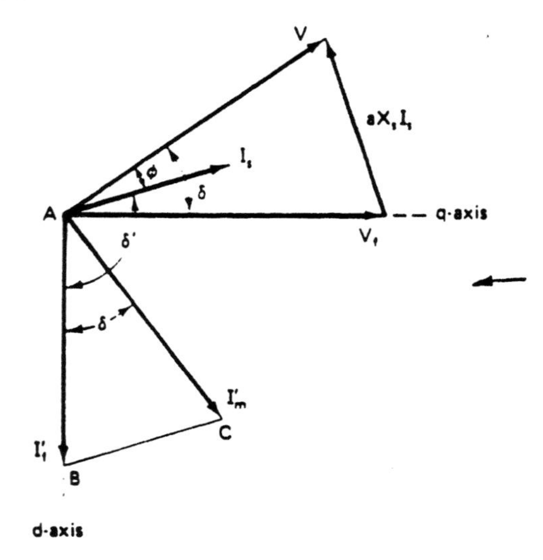

CURRENT PHASORS ARE INDEPENDENT OF FREQUENCY. INDUCED VOLTAGES ARE GIVEN BY

$$V = a X_s I_m' \; ; \quad V_f = a X_s I_f'$$

WHERE X_s IS THE SYNCHRONOUS REACTANCE AT RATED FREQUENCY.

$\longleftarrow$ WHEN OPERATING AT GIVEN VALUES OF I_s, I_f', AND δ OR δ', THE PARAMETERS I_m', FLUX, AND TORQUE HAVE FIXED VALUES.

INDUCED VOLTAGES V AND V_f, AND THE REACTANCE DROP INCREASE LINEARLY WITH FREQUENCY; BUT THEIR PHASE RELATIONSHIP REMAINS INDEPENDENT OF FREQUENCY.

WITH NEGLIGIBLE ARMATURE RESISTANCE, V REPRESENTS BOTH TERMINAL VOLTAGE AND INDUCED VOLTAGE.

16.1.17.

(a) HALF THE RATED TORQUE $= \frac{1}{2} \times \frac{1500 \times 746}{125.7} = 4451$ N.m

$\because T = \frac{3X_s}{\omega_s} I_m' \, I_f' \sin\delta$, $4451 = \frac{3 \times 0.5 \times 36}{0.5 \times 125.7} \times 105.9 \times 144.2 \sin\delta$

or $\sin\delta = 0.3392$ or $\delta = 19.83°$; $\cos\delta = 0.9407$

NOTING THAT $\bar{I}_S = \bar{I}_m' - \bar{I}_f'$; $I_S \sin\phi = -I_m' + I_f' \cos\delta$; $I_S \cos\phi = I_f' \sin\delta$

$I_S \sin\phi = -105.9 + (144.2 \times 0.9407) = 29.75$

$I_S \cos\phi = 144.2 \times 0.3392 = 48.91$

WHICH ARE INDEPENDENT OF FREQUENCY.

$\therefore I_S = \sqrt{29.75^2 + 48.91^2} = 57.25$ A ←

$\cos\phi = 0.854$ LEADING ←

(b) AT CONSTANT RATED TERMINAL VOLTAGE, RATED CURRENT, AND UNITY POWER FACTOR,

POWER DEVELOPED $= 1500 \times 746$ W ; $\omega_m = 1.25 \times 125.7 = 157.1$ rad/s.

TORQUE $= \frac{1500 \times 746}{157.1} = 7123$ N.m ←

AT 1.25 TIMES THE RATED SPEED, $X_S = 1.25 \times 36 = 45\,\Omega$;

$I_m' = \frac{6600/\sqrt{3}}{45} = 84.7$ A

WITH $\phi = 0$, $I_f' \cos\delta = I_m' = 84.7$ A ; $I_f' \sin\delta = I_S = 97.9$ A

$\therefore I_f' = 129.5$ A ←

16.2.1.

$$v_f = R_f i_f + L_f p i_f = R_f(1 + \tau_f p) i_f \quad \ldots (1)$$

WHERE $\tau_f = L_f / R_f$.

WITH ZERO INITIAL CONDITIONS,

$$V_f(s) = I_f(s) R_f + L_f s I_f(s) = R_f(1 + \tau_f s) I_f(s) \quad \ldots (2)$$

or $$s I_f(s) = \frac{1}{\tau_f}\left[\frac{V_f(s)}{R_f} - I_f(s)\right] \quad -- (3) \leftarrow$$

$$\frac{E_a(s)}{V_f(s)} = \frac{K_g I_f(s)}{V_f(s)} = \frac{K_g}{R_f(1 + \tau_f s)} \quad -- (4) \leftarrow$$

$(\because e_a = k_g i_f \omega_m$ or $K_g i_f$ for constant ω_m.)

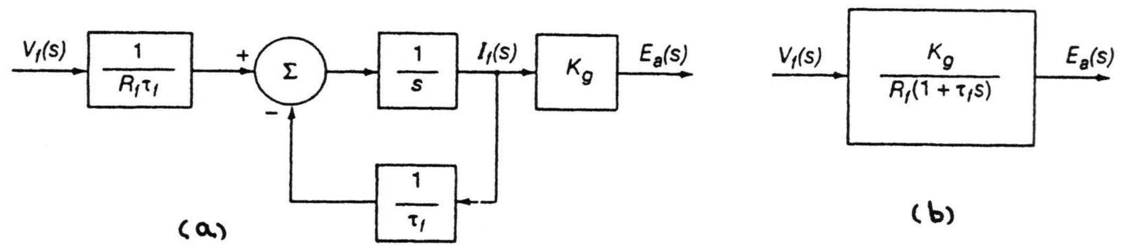

FIG. S16.2.1 BLOCK DIAGRAM REPRESENTING EQUATIONS (3) AND (4)

$$e_a = i_a R_a + L_a p i_a + i_a R_L + L_L p i_a$$

$$= i_a(R_a + R_L) + (L_a + L_L) p i_a \quad \ldots (5)$$

$$E_a(s) = I_a(s)(R_a + R_L)(1 + s\tau_{at}) \quad (6)$$

WHERE $\tau_{at} = \dfrac{L_{at}}{R_{at}} = \dfrac{L_a + L_L}{R_a + R_L}$

$$\therefore \frac{I_a(s)}{E_a(s)} = \frac{1}{R_{at}(1 + s\tau_{at})} \quad \ldots (7)$$

$$\frac{I_a(s)}{V_f(s)} = \frac{E_a(s)}{V_f(s)} \frac{I_a(s)}{E_a(s)} = \frac{K_g}{R_f(1+\tau_f s)} \frac{1}{R_{at}(1 + s\tau_{at})} \quad \ldots (8) \leftarrow$$

ELECTROMAGNETIC TORQUE: $T_e = K_g i_f i_a = \dfrac{e_a}{\omega_m} i_a$

16.2.1. CONTD.

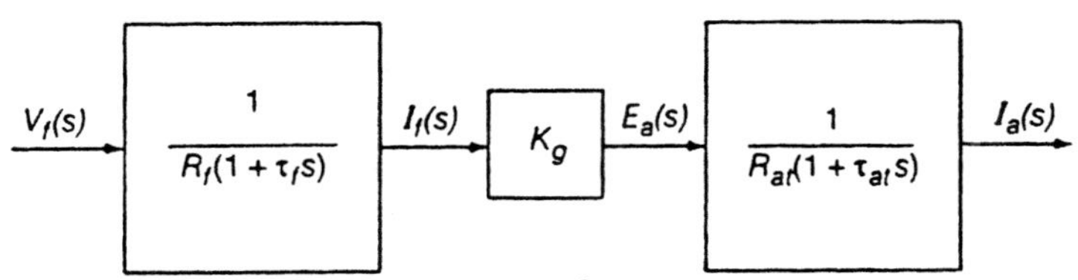

Figure S16.2.1 Block diagram of Equation (8) representing the generator dynamics including the effect of the load.

16.2.2.

The transfer function relating the armature current to the field voltage is given by Equation (8) in the solution of Pr. (16.2.1)

$$\frac{I_a(s)}{V_f(s)} = \frac{K_g}{R_f(1 + \tau_f s)} \frac{1}{R_{at}(1 + \tau_{at} s)}$$

With $R_{at} = R_a + R_L = 0.1 + 4.5 = 4.6\ \Omega$; $\tau_f = L_f/R_f = 20/100 = 0.2$; and $\tau_{at} = L_{at}/R_{at} = (L_a + L_L)/(R_a + R_L) = (0.1 + 2.2)/(0.1 + 4.5) = 0.5$,

$$\frac{I_a(s)}{V_f(s)} = \frac{100}{100(1 + 0.2s)} \frac{1}{4.6(1 + 0.5s)}$$

The Laplace transform of the field voltage for a step change of 230 V is

$$V_f(s) = 230/s$$

Hence

$$I_a(s) = \frac{230}{s} \frac{100}{100(1 + 0.2s)} \frac{1}{4.6(1 + 0.5s)}$$

or

$$I_a(s) = \frac{500}{s(s + 5)(s + 2)} = \frac{K_1}{s} + \frac{K_2}{s + 5} + \frac{K_3}{s + 2}$$

where

$$K_1 = \left.\frac{500}{(s + 5)(s + 2)}\right|_{s=0} = 50$$

$$K_2 = \left.\frac{500}{s(s + 2)}\right|_{s=-5} = 33.3$$

$$K_3 = \left.\frac{500}{s(s + 5)}\right|_{s=-2} = -83.3$$

Taking the inverse Laplace transform, we get the current buildup of

$$i_a(t) = 50 + 33.3e^{-5t} - 83.3e^{-2t}$$

The effect of the smaller of the two time constraints, τ_f, on the current buildup can be ignored when its value is less than about one-quarter of the longer one.

16.2.3.

a. Modifying Figure E16.2.1b for the negligible armature winding inductance, one has the diagram shown in Figure S16.2.3

$$\frac{\Omega_m(s)}{V_t(s)} = \frac{K_m/[R_aB(1 + \tau_m s)]}{1 + K_m^2/[R_aB(1 + \tau_m s)]} = \frac{K_m}{K_m^2 + R_aB}\frac{1}{1 + \tau'_m s}$$

where

$$\tau'_m = \tau_m \frac{R_aB}{R_aB + K_m^2} = \frac{JR_a}{R_aB + K_m^2}$$

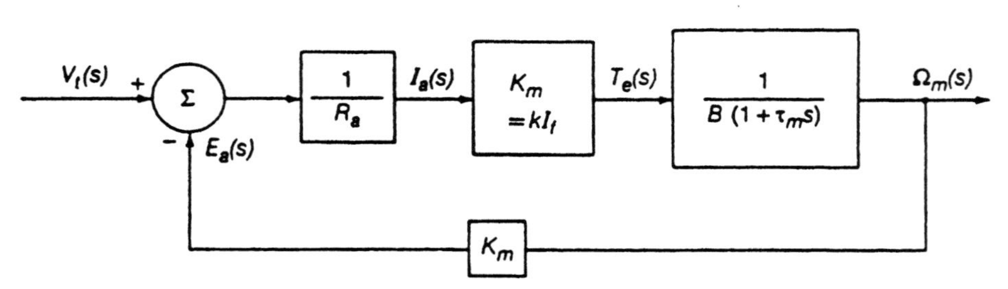

Figure S16.2.3

b. $\omega_{ss} = \lim_{s \to 0} s\Omega_m(s)$
Noting that $V_t(s) = \frac{220}{s}$, we obtain

$$\omega_{ss} = 220\frac{K_m}{R_aB + K_m^2} = (220)(2)/(0.5 \times 0.3 + 4) = 440/4.15 = 106 \text{ rad/s}$$

c. $\tau'_m = JR_a/(R_aB + K_m^2) = 3 \times 0.5/4.15 = 0.36$
For the motor to reach 0.95 of ω_{ss},

$$e^{-t/0.36} = 0.05 \text{ or } t = 0.36 \times 3 = 1.08 \text{ seconds.}$$

d. Total effective viscous damping $= B + (K_m^2/R_a) = 0.3 + (4/0.5) = 8.3 \text{ kg} \cdot \text{m}^2/\text{s}$

16.2.4.

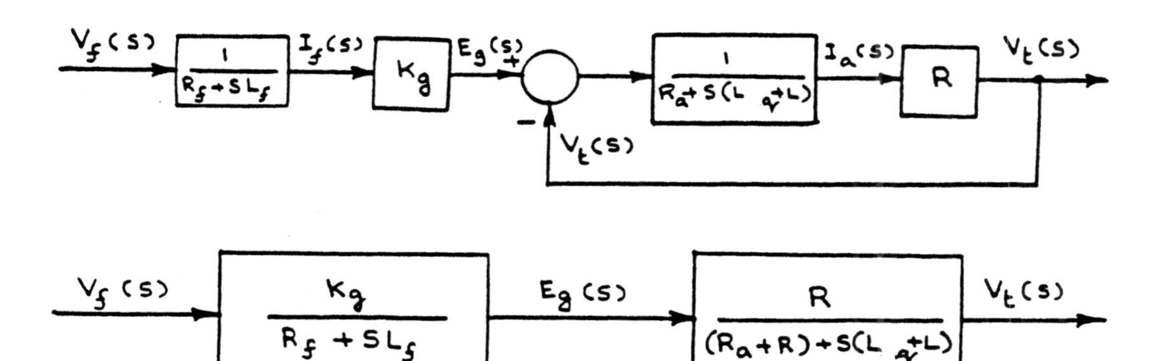

Substituting the given values, one gets

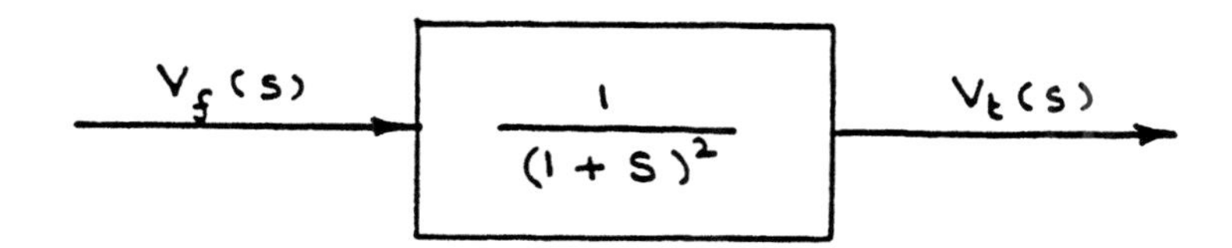

16.2.5.

Following Example 11.2.1, $\tau_f = 5/50 = 0.1$; $1/\tau_f = 10$;

$R_{at} = 1.10$; $L_{at} = 1$; $\tau_{at} = 1/1.1 = 0.91$ or $1/\tau_{at} = 1.1$

$$e_a(t) = \frac{100 \times 50}{50}\left(1 - e^{-10t}\right) = 100\left(1 - e^{-10t}\right) \leftarrow$$

and
$$i_a(t) = \frac{100}{1.1} + \frac{100}{1.1} \times \frac{1}{8.9 \times 0.91} e^{-10t} - \frac{100}{1.1}\,\frac{1}{0.1 \times 8.9} e^{-1.1t}$$

$$= 90.9\left(1 + 0.1235\, e^{-10t} - 1.1236\, e^{-1.1t}\right) \text{ A} \leftarrow$$

16.2.6.

a) $L_f = R_f \tau_f = 150 \times 0.5 = 75 \text{ h} \leftarrow$

b) $L_q = R_a \tau_a = 0.15 \times 0.05 = 0.0075 \text{ h} \leftarrow$

c) $k = e_a/(i_f \omega_m) = 250/[(2)(\frac{600}{60} \times 2\pi)] = 1.99 \text{ h} \leftarrow$

d) $P_L = T_L \omega_m = B_L \omega_m^2$ or $B_L = P_L/\omega_m^2 = \frac{20 \times 746}{(20\pi)^2} = 3.78 \leftarrow$

16.2.7.

$V_f(s) = (R_f + L_f\, s)\, I_f(s)$

$E_a(s) = k\omega_m\, I_f(s)$

Also $E_a(s) = (R_a + R_L + L_q s)\, I_a(s)$

and $V_L(s) = R_L\, I_a(s)$

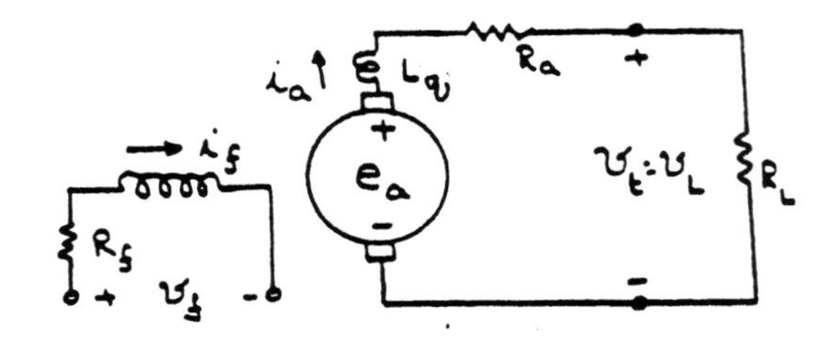

$$\therefore \quad \frac{V_L(s)}{V_f(s)} = \frac{R_L k\omega_m}{R_f(R_a + R_L)} \cdot \frac{1}{(1+\tau_f s)(1+\tau_a s)}$$, where $\tau_f = L_f / R_f$ and $\tau_a = L_q / (R_a + R_L)$

Under steady state, d/dt terms go to zero; i.e. $s \to 0$

$$\text{Voltage gain} = \frac{R_L k\omega_m}{R_f(R_a + R_L)} = \frac{R_L K_g}{R_f(R_a + R_L)} = \frac{1 \times 100}{10(0.1+1)} = 9.1 \leftarrow$$

Input power to the field $= 25^2/10 = 62.5$ W

$E_a = k\omega_m I_f = K_g I_f = (100)(25/10) = 250$ V

$I_a = 250/(0.1+1) = 227.3$ A; So, output power $= (227.3)^2\, 1 = 51{,}665.3$ W

$\therefore$ power gain $= 51665.3 / 62.5 = 826.6 \leftarrow$

16.2.8.

a) $v_t = k\omega_m i_f + L_q\, p\, i_a + R_a i_a$ (11.1.4)

Under no load and no losses, $i_a = 0$; $k = 1.99$ from Pr. 11-3 (p.108) Soln.

$\therefore \quad \omega_m = 250/[(1.99)(2)] = 62.8$ rad/s $\leftarrow$

b) $$\Delta I_a(s) = \frac{s\, \Delta V_t(s)}{R_a\left(s + \frac{1}{\tau_m}\right)}$$, where $\tau_m = \dfrac{J R_a}{K_m^2}$ and $K_m = k i_f$

$$\Delta V_t(s) = \frac{260 - 250}{s} = \frac{10}{s}$$

$$\Delta I_a(s) = \frac{10}{0.15\left[s + \frac{(1.99 \times 2)^2}{3 \times 0.15}\right]} = \frac{66.7}{s + 35.2}$$

or $\Delta I_a(t) = 66.7\, e^{-35.2t}$; Since the initial current

$i_a = 0$, $i_a(t) = 66.7\, e^{-35.2t}$ $\leftarrow$

16.2.8.

CONTD.

$$\Delta \Omega_m(s) = \frac{(k I_f / J R_a)\, V_t(s)}{s + (1/\tau_m)} = \frac{1.99 \times 2 \times 10/(3 \times 0.15)}{s(s+35.2)} = \frac{88.44}{s(s+35.2)}$$

$$\therefore \omega_m(t) = 62.8 + \frac{88.44}{35.2}(1 - e^{-35.2t})$$

$$= 65.3 - 2.5\, e^{-35.2t} \leftarrow$$

c) From Eq. (11.1.4) and $J p \omega_m = k I_f i_a$,

$$\Delta i_a(s) = \frac{s\, \Delta V_t(s)/L_{aq}}{s^2 + (R_a/L_{aq})s + (k I_f)^2/J L_{aq}}$$

$$= \frac{10/0.0075}{s^2 + \frac{0.15}{0.0075}s + \frac{(1.99 \times 2)^2}{3 \times 0.0075}} = \frac{1333.3}{s^2 + 20s + 704}$$

Since $s_{1,2} = -10 \pm j\,24.58$, in terms of partial fractions,

$$\Delta I_a(s) = \frac{C_1}{s + 10 - j\,24.6} + \frac{C_2}{s + 10 + j\,24.6}$$; on evaluation,

$C_1 = -j\,27.1$ and $C_2 = C_1^* = +j\,27.1$

$$\therefore \Delta i_a = -j\,27.1\, e^{-10t}(e^{j24.6t} - e^{-j24.6t})$$

or $$i_a(t) = 54.2\, e^{-10t} \sin 24.6t \leftarrow$$

For $B = 0$,

$$\Delta \Omega_m(s) = \frac{k I_f\, \Delta V_t(s)/J L_{aq}}{s^2 + (R_a/L_{aq})s + \{(k I_f)^2/J L_{aq}\}}$$

$$= \frac{1.99 \times 2 \times 10/(3 \times 0.0075)}{s(s^2 + 20s + 704)}$$

$$= \frac{1768.9}{s(s + 10 - j\,24.6)(s + 10 + j\,24.6)}$$

$$= \frac{C_1}{s} + \frac{C_2}{s + 10 - j\,24.6} + \frac{C_3}{s + 10 + j\,24.6}$$

where $C_1 = 2.51$; $C_2 = -1.35\angle{-22.2°}$; $C_3 = C_2^* = -1.35\angle{22.2°}$;

$$\Delta \omega_m(t) = 2.51 - 2.7\, e^{-10t}\cos(24.6t - 22.2°)$$

$$\omega_m(t) = 62.8 + \Delta\omega_m(t) = 65.31 - 2.7 e^{-10t}\cos(24.6t - 22.2°) \leftarrow$$

16.2.9.

$T_e = K_m i_a = 300\,\dot{\omega}_m + \omega_m$; $e_a = K_m \omega_m$;

$v_t = K_m \omega_m + R_a i_a$ (Note that $L_{aq} p i_a = 0$)

$$\therefore V_t(s) = K_m \Omega_m(s) + R_a \left[\frac{300\, s\Omega_m + \Omega_m}{K_m} \right]$$

$$\text{or } \Omega_m(s) = \frac{K_m V_t(s)}{(300 R_a) s + (K_m^2 + R_a)} = \frac{10\,(100/s)}{300 s + 101} = \frac{1000}{300 s\,(s + \frac{101}{300})} = \frac{10/3}{s\,(s + \frac{101}{300})}$$

$$= \frac{C_1}{s} + \frac{C_2}{s + \frac{101}{300}}, \text{ where } C_1 = 9.9 \text{ and } C_2 = -9.9$$

$$\therefore \omega_m(t) = 9.9\,(1 - e^{-0.337t}) \text{ rad/s} \leftarrow$$

16.2.10

For a generator, $v_t = e_a - L_{aq} p i_a - R_a i_a$ (11.1.7)

$$= K_m \omega_m - (L_{aq} p + R_a)\, i_a$$

$$i_a = i_L = \frac{K_m \omega_m - v_t}{(L_{aq} p + R_a)} \; ; \quad v_t = Z_L i_L = (L_L p + R_L)\, i_L$$

Substituting the above in the Eqn. for i_L, one gets

$$i_L = \frac{K_m \omega_m - (L_L p + R_L)\, i_L}{(L_{aq} p + R_a)} \text{ or } i_L = \frac{K_m \omega_m}{(L_{aq} + L_L)\left(p + \frac{R_a + R_L}{L_{aq} + L_L}\right)}$$

Laplace transforming the eqn. above with zero initial arm. curr. and with $i_f = I_f$ = constant, one has

$$I_L(s) = \frac{k I_f \omega_m}{(L_{aq} + L_L)} \; \frac{1}{s\left[s + \frac{R_a + R_L}{L_{aq} + L_L}\right]} = \frac{(2)(1.5)(30\pi)}{(0.01 + 0.1)} \; \frac{1}{s\left[s + \frac{0.5 + 11.5}{0.01 + 0.1}\right]}$$

Note that $\omega_m = \frac{900}{60} \times 2\pi = 30\pi$ rad/s

$$\text{or } I_L(s) = \frac{2571.4}{s(s + 109.1)} \text{ or } i_L(t) = 23.57\,(1 - e^{-109.1t}) \leftarrow$$

$$v_t = Z_L i_L = (L_L p + R_L)\, i_L \text{ or } V_t(s) = (0.1s + 11.5)\, I_L(s)$$

$$= \frac{257.14}{s + 109.1} + \frac{257.14 \times 115}{s(s + 109.1)}$$

$$v_t(t) = 257.14\, e^{-109.1t} + \frac{257.14 \times 115}{109.1}(1 - e^{-109.1t})$$

$$= 271.04 - 13.9\, e^{-109.1t} \leftarrow$$

16.2.11.

$$e_g = e_m + Ri + L\frac{di}{dt}$$

or $$\frac{e_g - e_m}{R} = (1 + \tau_a p)i$$

where $\tau_a = L/R = 0.008/0.04 = 0.2$

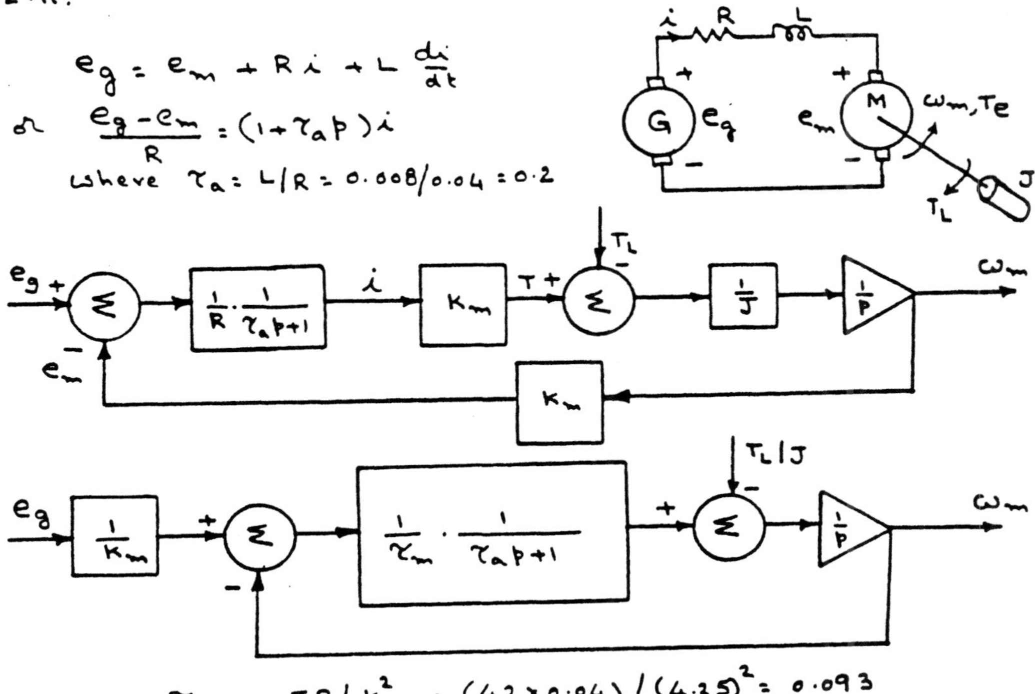

$$\tau_m = JR/K_m^2 = (42 \times 0.04)/(4.25)^2 = 0.093$$

The frequencies can be obtained from: $\tau_m p(\tau_a p + 1) + 1 = 0$ or

$$\tau_m \tau_a p^2 + \tau_m p + 1 = 0 \text{ or } p^2 + \frac{1}{\tau_a}p + \frac{1}{\tau_m \tau_a} = 0$$

i) The undamped natural frequency ω_n is

$$\omega_n = \frac{1}{\sqrt{\tau_m \tau_a}} = \frac{1}{\sqrt{0.093 \times 0.2}} = 7.33 \text{ rad/s} \leftarrow$$

The damping factor $\alpha = 1/(2\tau_a) = 1/(2 \times 0.2) = 2.5 \leftarrow$

The damping ratio $\xi = \alpha/\omega_n = 2.5/7.33 = 0.341 \leftarrow$

ii) initial $e_m = 400 - (35)(0.04) = 398.6$ V

initial $\omega_m = 398.6/4.25 = 93.8$ rad/s or 14.92 rps or 895 rpm

final current $= 35 + \frac{2000}{4.25} = 505.6$ A ; final $e_m = 400 - (505.6)(0.04) = 379.8$ V

final $\omega_m = 379.8/4.25 = 89.36$ rad/s or 14.2 rps or 853 rpm $\leftarrow$

Speed drop in rpm $= 895 - 853 = 42 \leftarrow$

16.2.12.

With $B = 0$, $G_{eq} = 0$ or $R_{eq} = \infty$.

$i_a(\infty) = 0$; $q(\infty) = C_{eq} V$

energy stored $= Vq/2 = \frac{1}{2} C_{eq} V^2$

energy input $= V \int_0^\infty i\,dt = Vq = C_{eq} V^2$

energy lost in $R_a = W_{in} - W_{stored} = \frac{1}{2} C_{eq} V^2$

$V = k I_f \omega_m$; $C_{eq} = J/k_m^2 = J/(k I_f)^2$

Energy lost in $R_a = \frac{1}{2} \cdot \frac{J}{(k I_f)^2} \cdot k^2 I_f^2 \omega_m^2 = \frac{1}{2} J \omega_m^2$ ←

i_a, R_a, L, q; $v_t = V$ (+, −); $C_{eq} = \frac{J}{k_m^2}$

16.2.13.

$L_{aq} = 0.01$; $R_a = 0.5$; $C_{eq} = \frac{J}{k_m^2} = \frac{3.0}{(k\, I_f)^2} = \frac{3.0}{(2\times 1)^2} = 0.75$

$G_{eq} = \frac{B}{k_m^2} = \frac{0.3}{4} = 0.075$ or $R_{eq} = 1/0.075 = 13.33$

$\left[4 + 0.01S + \frac{13.33/0.75S}{13.33 + \frac{1}{0.75S}}\right] i_a = 220$

+ i_a 3.5Ω 0.5Ω 0.01h 13.33 Ω 0.75 f 220V −

or

$I_a(S) = \frac{22000\,(S+0.1)}{S(S+0.45)(S+399.6)}$

$= \frac{12.23}{S} + \frac{42.87}{S+0.45} - \frac{55.1}{S+399.6}$ or $i_a = 12.23 + 42.87\, e^{-0.45t} + - 55.1 e^{-399.6t}$ ←

16.2.14.

$\tau = R_{par.} C_{eq} = \left(\frac{R_a R_{eq}}{R_a + R_{eq}}\right) C_{eq}$

R_a C_{eq} $R_{eq} = \frac{1}{G_{eq}}$

16.2.15.

a) See solution of Prob. 11-12.

$i_a(\infty) = \frac{220}{(0.5 + 13.33)} = 15.91$ A ←

+ V: 220V 0.5Ω 0.75 f 13.33 Ω −

b) Since $C_{eq} \propto (1/i_f)^2$, $C'_{eq} = 0.75 \times (1/0.8)^2 = 1.17 f$

Since $R_{eq} \propto (i_f)^2$, $R'_{eq} = (0.8)^2\, 13.33 = 8.53\Omega$

Initial voltage across C_{eq} : $E = 220 - (15.91)(0.5) = 212$ V

Energy stored in C_{eq} is then $W_c = \frac{C_{eq} \cdot E^2}{2} = \frac{1}{2}(0.75)(212)^2$

At $i'_f = 0.8$ A , $C'_{eq} = 1.17 f$

Since the stored energy is unchanged,

$1.17 (E')^2 = (0.75)(212)^2$ or $E' \simeq 170$ V

$i'_a(0) = (220 - 170)/0.5 = 100$ A ←

c) $i'_a(\infty) = 220/(0.5 + 8.53) = 24.4$ A ←

d) $\tau'_{am} = R_a R'_{eq} C'_{eq} / (R_a + R_{eq}) = \frac{0.5(8.53)(1.17)}{(0.5 + 8.53)} = 0.553S$

$i'_a = 24.4 + (100 - 24.4)e^{-1.81t} = 24.4 + 75.6\, e^{-1.81t}$ ←

16.2.16.

The equations of motion are:

$$v = R_a i_a + L_a \frac{di_a}{dt} + e$$

$$e = kI_f \omega_m$$

$$T_e = kI_f i_a = J\dot{\omega}_m$$

These equations yield

$$v = R_a i_a - L_a \frac{di_a}{dt} + \frac{(kI_f)^2}{J} \int i_a \, dt$$

which is similar to

$$v = Ri + L\frac{di}{dt} + \frac{1}{C}\int i \, dt$$

corresponding to the circuit of Fig. (b). For equivalence: $R \leftrightarrow R_a$, $L \leftrightarrow L_a$, and $C \leftrightarrow J/(kI_f)^2$. ⟵

16.2.17.

EQUATIONS OF MOTION ARE:

$$v_{fg} = R_{fg} i_{fg} - L_{fg} \frac{di_{fg}}{dt} \quad \text{or} \quad V_{fg} = (R_{fg} - L_{fg}s)I_{fg}$$

$$e_g = k_g \omega_g i_{fg} = Ri - k_m I_{fm} \omega_m \quad \text{or} \quad k_g \omega_g I_{fg} = RI + k_m I_{fm} \Omega_m$$

$$T_m = k_m I_{fm} i = J\dot{\omega}_m + b\omega_m \quad \text{or} \quad k_m I_{fm} I = (b - Js)\Omega_m$$

Hence.

$$G(s) \equiv \frac{\Omega_m(s)}{V_{fg}(s)} = \frac{k_g \omega_g k_m I_{fm}}{(R_{fg} - L_{fg}s)(k_m^2 I_{fm}^2 - bR + JRs)}$$ ⟵

16.2.18.

(a)

Field circuit time constant $\tau_f = 25/100 = 0.25$ sec.

(i)

$$e_a(t) = \frac{100 \times 200}{100}(1 - e^{-t/0.25})$$

$$= 200(1 - e^{-4t})$$ ⟵

(ii)

$$e_a(\infty) = 200 \text{ V}$$ ⟵

(iii)

$$0.9 \times 200 = 200(1 - e^{-4t})$$

$$t = 0.575 \text{ sec}$$ ⟵

16.2.18.

(CONTD.)

(b)

$$\tau_f = 0.25 \text{ sec}$$

$$\tau_{at} = \frac{0.15 + 0.02}{1 + 0.25} = 0.136 \text{ sec}$$

$$I_a(s) = \frac{100 \times 200}{100 \times 1.25 \times 0.25 \times 0.136 s(s+4)(s+7.35)}$$

$$= \frac{4705.88}{s(s+4)(s+7.35)}$$

$$= \frac{A_1}{s} + \frac{A_2}{s+4} + \frac{A_3}{s+7.35}$$

where $A_1 = \left.\dfrac{4705.88}{(s+4)(s+7.35)}\right|_{s=0} = 160$

$$A_2 = \left.\frac{4705.88}{s(s+7.35)}\right|_{s=-4} = -351$$

$$A_3 = \left.\frac{4705.88}{s(s-4)}\right|_{s=-7.35} = 191$$

$$\therefore \quad i_a(t) = 160 - 351e^{-4t} + 191e^{-7.35t} \leftarrow$$

16.2.19.

(a) THE BLOCK DIAGRAM IS SHOWN BELOW IN FIG. (a), WHICH CAN BE SIMPLIFIED TO THAT OF FIG. (b).

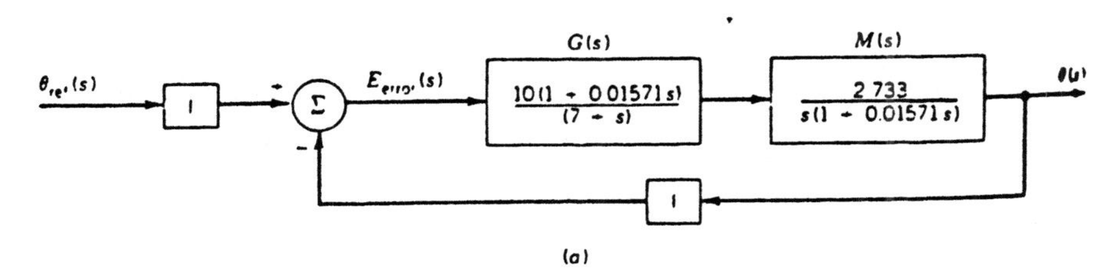

(a)

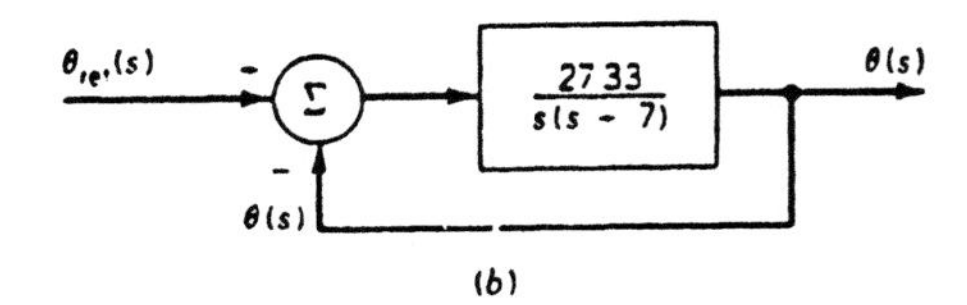

(b)

16.2.19.

(a) (CONTD.)

FROM FIG. (b), IT FOLLOWS:

$$\frac{\theta(s)}{\theta_{ref}(s)} = \frac{27.33/s(s+7)}{1 + 27.33/s(s+7)} = \frac{27.33}{s^2 + 7s + 27.33}$$

WHICH REPRESENTS A SECOND-ORDER SYSTEM. THE CORRESPONDING BLOCK DIAGRAM IS SHOWN BELOW IN FIG. (C).

$\theta_{ref}(s)$ → $\boxed{\dfrac{27.33}{s^2 + 7s + 27.33}}$ → $G(s)$

(c)

(b)

$$\theta_{ref}(s) = \frac{\pi}{s}$$

$$\theta(s) = \frac{27.33}{s^2 + 7s + 27.33}\,\frac{\pi}{s}$$

$$= \pi\,\frac{27.33}{s(s^2 + 7s + 27.33)}$$

$$= \pi\,\frac{\omega_n^2}{s(s^2 + 2\xi\omega_n s + \omega_n^2)}$$

where $\omega_n = \sqrt{27.33} = 5.228$ rad/sec

$$\xi = \frac{7}{2\omega_n} = \frac{7}{2 \times 5.288} = 0.67$$

The time response is

$$\theta(t) = \pi\left[1 - \frac{e^{-\xi\omega_n t}}{\sqrt{1-\xi^2}}\sin(\omega_n\sqrt{1-\xi^2}\,t + \cos^{-1}\xi)\right]$$

$$= \pi[1 - 1.347e^{-3.5t}\sin(3.88t + 48°)] \text{ radian}$$

THE POSITION RESPONSE IS SHOWN BELOW IN FIG. (d).

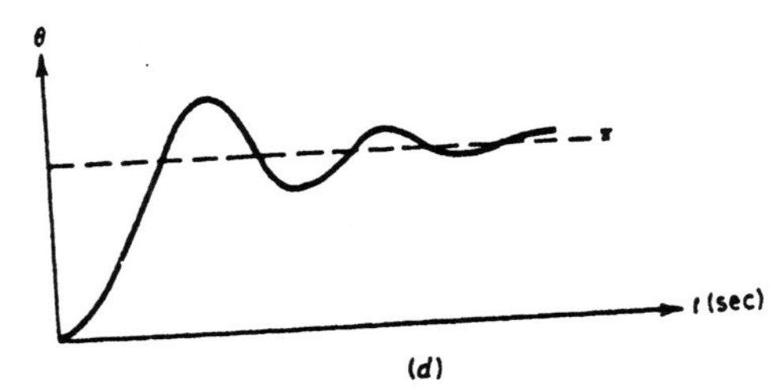

(d)

16.2.20.

(a)

$$E_a = K_m \omega_m$$

$$K_m = \frac{220}{(2000/60) \times 2\pi} = 1.05 \text{ V/rad/sec}$$

$$V_t = e_a + i_a R_a = K_m \omega_m + i_a R_a$$

$$T = K_m i_a = J \frac{d\omega_m}{dt} + T_L$$

From the last two equations,

$$V_t = K_m \omega_m + R_a \left(\frac{J}{K_m} \frac{d\omega_m}{dt} + \frac{T_L}{K_m} \right)$$

$$= K_m \omega_m + \frac{R_a J}{K_m} \frac{d\omega_m}{dt} + \frac{R_a T_L}{K_m}$$

$$= 1.05\omega_m + \frac{0.5 \times 2.5}{1.05} \frac{d\omega_m}{dt} + \frac{0.5 \times 25}{1.05}$$

$$= 1.05\omega_m + 1.19 \frac{d\omega_m}{dt} + 11.9$$

$$V_t(s) = \frac{220}{s} = 1.05\omega_m(s) + 1.19 s\omega_m(s) + \frac{11.9}{s}$$

$$\omega_m(s) = \frac{220 - 11.9}{s(1.05 + 1.19s)}$$

$$= \frac{174.874}{s(s + 0.8824)}$$

$$= \frac{198.2}{s} - \frac{198.2}{s + 0.8824}$$

$$\omega_m(t) = 198.2(1 - e^{-0.8824t}) \longleftarrow$$

$$i_a = \frac{V_t - K_m \omega_m}{R_a}$$

$$= \frac{220 - 1.05\omega_m}{0.5}$$

$$= 440 - 2.1 \times 198.2(1 - e^{-0.8824t})$$

$$= 23.8 + 416.2e^{-0.8824t} \longleftarrow$$

(b) Steady-state speed is $\omega_m(\infty) = 198.2$ rad/sec. ←
Steady-state current is $I_a = i_a(\infty) = 23.8$ A. ←

16.2.21.

On no-load, $E_a = V_t = 200 = K_m\omega_o = 2\omega_o$ or $\omega_o = 100\ \frac{rad}{s}$ (No-load speed)

Plugging corresponds to applying a step voltage $-200\, u(t)$ to the armature with the initial speed of $\omega_o = 100\ \frac{rad}{s}$.

Then $v_t = e_a + i_a R_a = K_m\omega_m + i_a R_a$; $T_e = K_m i_a = J p\omega$

so that $v_t = K_m\omega_m + \frac{R_a J}{K_m} p\omega_m$. Taking the Laplace transform

$$-\frac{200}{s} = K_m\Omega_m(s) + \frac{R_a J}{K_m}\left[s\Omega_m(s) - \omega_o\right]$$

$$= \left(2 + \frac{0.5\times 4}{2} s\right)\Omega_m(s) - \frac{0.5\times 4}{2}\times 100 = (2+s)\Omega_m(s) - 100$$

or $\Omega_m(s) = \frac{100}{s+2} - \frac{100}{s} + \frac{100}{s+2} = \frac{200}{s+2} - \frac{100}{s}$

$$\omega_m(t) = 200\, e^{-2t} - 100 \leftarrow$$

CHECK: at $t=0$, $\omega_m(0) = \omega_o = 100$; at $t=\infty$, $\omega_m(\infty) = -100$

When $\omega_m = 0$, $0 = 200e^{-2t} - 100$ so that $2t = \log_e 2$

$\rightarrow$ or $t = 0.278\,s$ which is the time taken for the machine to stop.

16.2.22.

Let $K = k_G\omega_{mG}$; $\tau = L/R$; $\tau_{fG} = L_{fG}/R_{fG}$; $\tau_M = J/B$

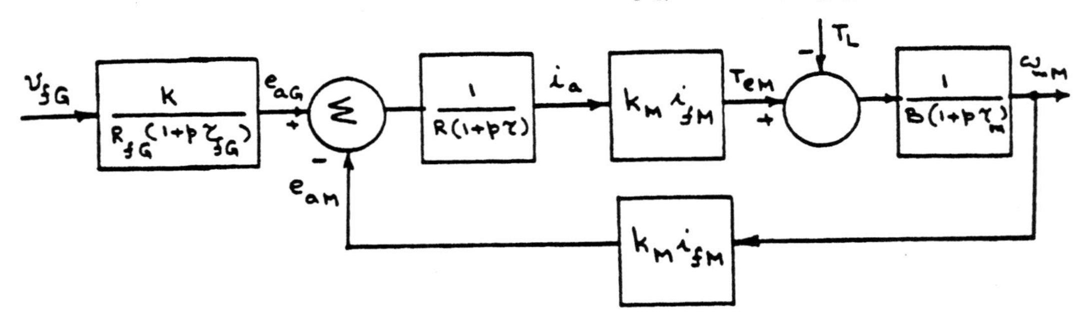

$$G(s) = \frac{\Omega_{mM}(s)}{V_{fG}(s)} = \frac{k_G\,\omega_{mG}\,k_M\,I_{fM}}{(R_{fG} + L_{fG}s)\left[k_M^2 I_{fM}^2 + (R+Ls)(B+Js)\right]}$$

16.2.23.

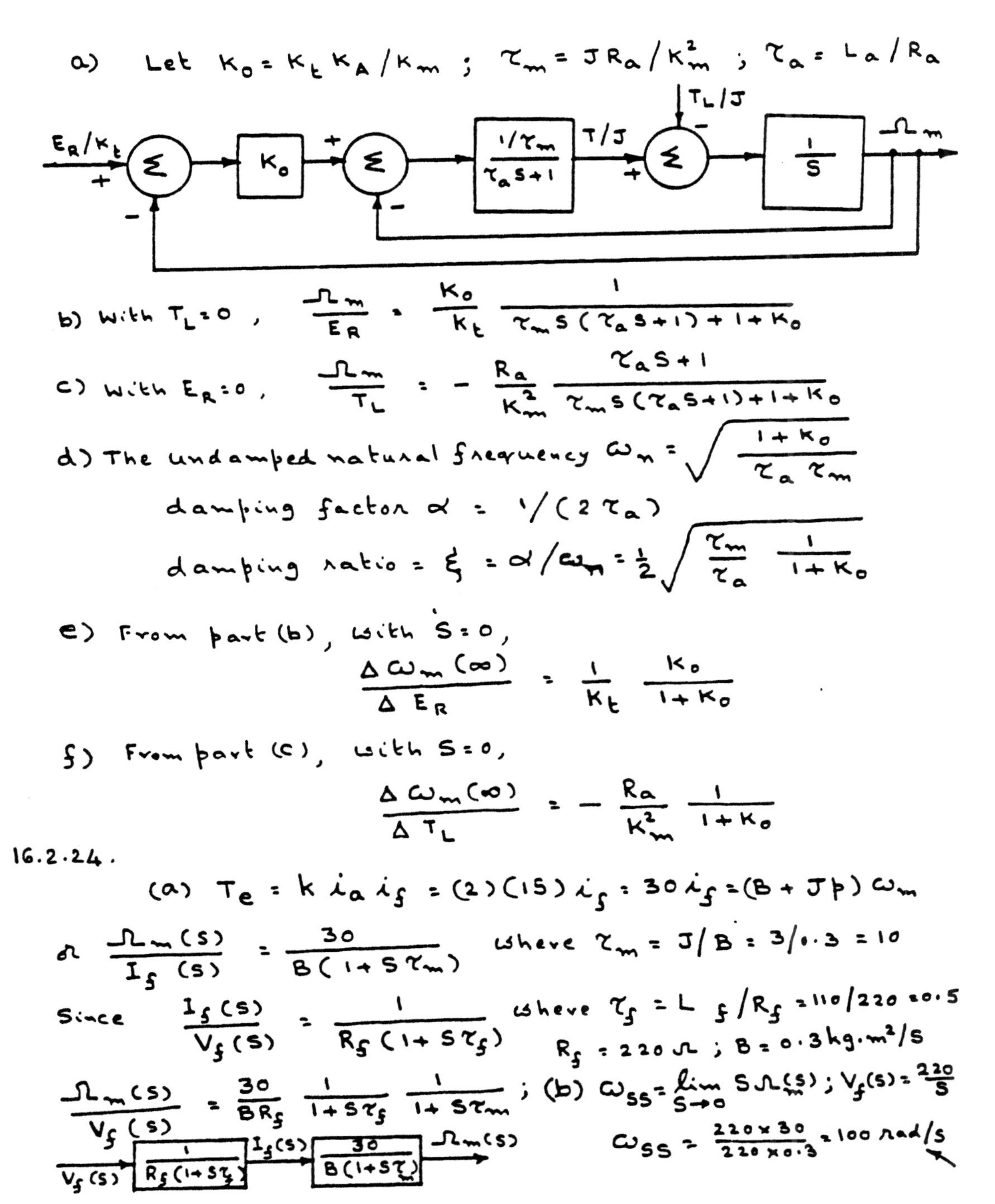

a) Let $K_0 = K_t K_A / K_m$; $\tau_m = J R_a / K_m^2$; $\tau_a = L_a / R_a$

b) With $T_L = 0$, $\quad \dfrac{\Omega_m}{E_R} = \dfrac{K_0}{K_t} \dfrac{1}{\tau_m s(\tau_a s+1)+1+K_0}$

c) With $E_R = 0$, $\quad \dfrac{\Omega_m}{T_L} = -\dfrac{R_a}{K_m^2} \dfrac{\tau_a s+1}{\tau_m s(\tau_a s+1)+1+K_0}$

d) The undamped natural frequency $\omega_n = \sqrt{\dfrac{1+K_0}{\tau_a \tau_m}}$

damping factor $\alpha = 1/(2\tau_a)$

damping ratio $= \xi = \alpha/\omega_n = \dfrac{1}{2}\sqrt{\dfrac{\tau_m}{\tau_a} \dfrac{1}{1+K_0}}$

e) From part (b), with $s = 0$,

$$\frac{\Delta \omega_m(\infty)}{\Delta E_R} = \frac{1}{K_t} \frac{K_0}{1+K_0}$$

f) From part (c), with $s = 0$,

$$\frac{\Delta \omega_m(\infty)}{\Delta T_L} = -\frac{R_a}{K_m^2} \frac{1}{1+K_0}$$

16.2.24.

(a) $T_e = k\, i_a i_f = (2)(15)\, i_f = 30\, i_f = (B + Jp)\,\omega_m$

or $\dfrac{\Omega_m(s)}{I_f(s)} = \dfrac{30}{B(1+s\tau_m)}$ where $\tau_m = J/B = 3/0.3 = 10$

Since $\dfrac{I_f(s)}{V_f(s)} = \dfrac{1}{R_f(1+s\tau_f)}$ where $\tau_f = L_f/R_f = 110/220 = 0.5$

$R_f = 220\ \Omega$; $B = 0.3\ \text{kg}\cdot\text{m}^2/\text{s}$

$\dfrac{\Omega_m(s)}{V_f(s)} = \dfrac{30}{BR_f} \dfrac{1}{1+s\tau_f} \dfrac{1}{1+s\tau_m}$; (b) $\omega_{ss} = \lim_{s\to 0} s\Omega_m(s)$; $V_f(s) = \dfrac{220}{s}$

$\omega_{ss} = \dfrac{220 \times 30}{220 \times 0.3} = 100$ rad/s

16.2.24.

(CONTD.)

(c) $\Omega_m(s) = \dfrac{220}{s}\dfrac{30}{66(1+0.5s)(1+10s)} = 20\left[\dfrac{5}{s} + \dfrac{0.263}{s+2} - \dfrac{5.263}{s+0.1}\right]$

$$\omega_m(t) = 20\left[5 + 0.263\,e^{-2t} - 5.263\,e^{-0.1t}\right]u(t) \simeq 100\left[1 - \frac{5.263}{5}e^{-0.1t}\right]u(t)$$

For the motor to reach $0.95\,\omega_{ss}$,

$$t \simeq -\frac{1}{0.1}\ln\left(0.05 \times \frac{5}{5.263}\right) = 10\,(3.047) = 30.47\ \text{s}$$

16.2.25.

$$v_f = A\,v_e = Aa\,(v_r - v_a)$$

$$V_a(s) = \frac{K\left[V_r(s) - V_a(s)\right]}{1 + s\tau_f}, \quad \text{where } K = \frac{aAK_E R_L}{R_f(R_a + R_L)}$$

$$\therefore\ V_a(s) = \frac{K\,V_r(s)}{1 + K + s\tau_f} = \frac{K}{1+K}\,\frac{V_r(s)}{(1 + s\tau_f')}, \text{where } \tau_f' = \frac{\tau_f}{K+1}$$

a) At no load, $K = \dfrac{aAK_E}{R_f} = \dfrac{1(10)(150)}{150} = 10$

$$V_a(s) = \frac{K}{K+1}\,\frac{V_r(s)}{1+s\tau_f'}\ ;\ \text{With } s=0,\ V_a = \frac{K}{K+1}V_r = \frac{10}{11}\times 250 = 227.3\ \text{V}$$

With $I_a = 42$ A, $n = 1140$ rpm, $K_E = 150 \times 1140/1200 = 142.5$

With $R_L = \dfrac{227.3}{42} = 5.4\ \Omega$ and $R_a = 0.5\ \Omega$, $K = \dfrac{1(10)\,142.5}{150}\,\dfrac{R_L}{R_a + R_L} = 8.7$

$$V_a = \frac{8.7}{9.7} \times 250 = 224.2\ \text{V} \leftarrow$$

(b) $\tau_f = \dfrac{L_f}{R_f} = \dfrac{75}{150} = 0.5\ ;\ \tau_f' = \dfrac{\tau_f}{K+1} = \dfrac{0.5}{11} = 0.045\ \text{s}$ at no load

and $\tau_f' = \dfrac{0.5}{9.7} = 0.05\ \text{s}$ at $I_a = 42$ A and $n = 1140$ rpm.

(c) $V_f = A\,(V_r - V_a) = 10\,(250 - 227.3) = 227\ \text{V}$

$$I_f = \frac{227}{150} = 1.51\ \text{A}$$

$$E_a = 1.51\,K_E = 1.51 \times 150 \times 1140/1200 = 215.2\ \text{V}$$

$$V_a = 215.2 - (0.5)(42) = 194.2 \leftarrow$$

16.2.26.

(a)

$$e_a = k_a \omega_m = i_a (R_a + R_L) + L_a \frac{di_a}{dt}$$

$$v_o = i_a R_L$$

$$\therefore \quad \frac{V_o(s)}{\Omega_m(s)} = \frac{R_L k_a}{R_a + R_L + sL_a} \quad \leftarrow$$

NOTE: IF $R_L \to \infty$, THE ABOVE TRANSFER FUNCTION WILL APPROACH K_a. ←

(b)

$$e_i = i_a R_a + L_a \frac{di_a}{dt} + e_a$$

$$e_a = k_a \omega_m$$

$$T_m = k_a i_a = B_L \omega_m + J_L \frac{d\omega_m}{dt}$$

WHERE T_m IS THE MOTOR ELECTROMAGNETIC TORQUE

$$\frac{\Omega_m(s)}{E_i(s)} = \frac{k_a}{(B_L + sJ_L)(R_a + sL_a) + k_a^2} \quad \leftarrow$$

(c)

$$e_i = i_a R_a + L_a \frac{di_a}{dt} + e_a$$

$$e_a = k_a \omega_m$$

$$T_m = k_a i_a = B_L \omega_m + J_L \frac{d\omega_m}{dt} + T_L$$

ADDING THE REFERENCE SIGNAL $E_R(s)$, WE HAVE THE FOLLOWING:

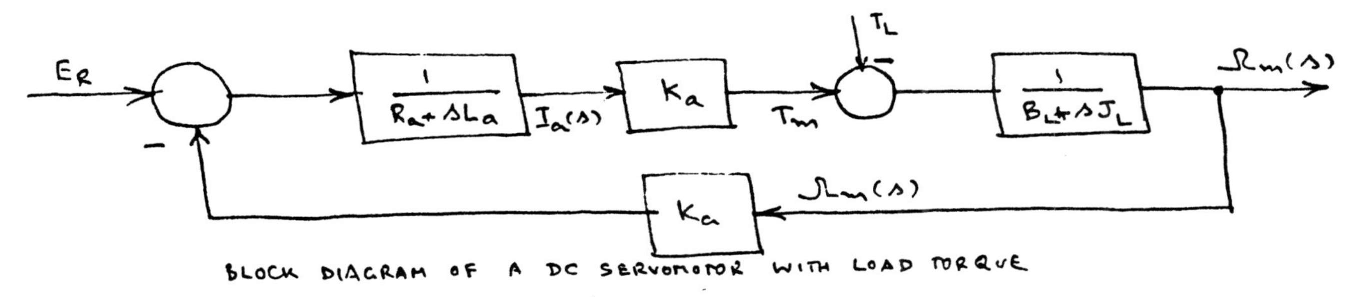

BLOCK DIAGRAM OF A DC SERVOMOTOR WITH LOAD TORQUE

16.2.27.

(a)

$$\frac{C(s)}{R(s)} = \frac{G(s)}{1 - G(s)H(s)}$$

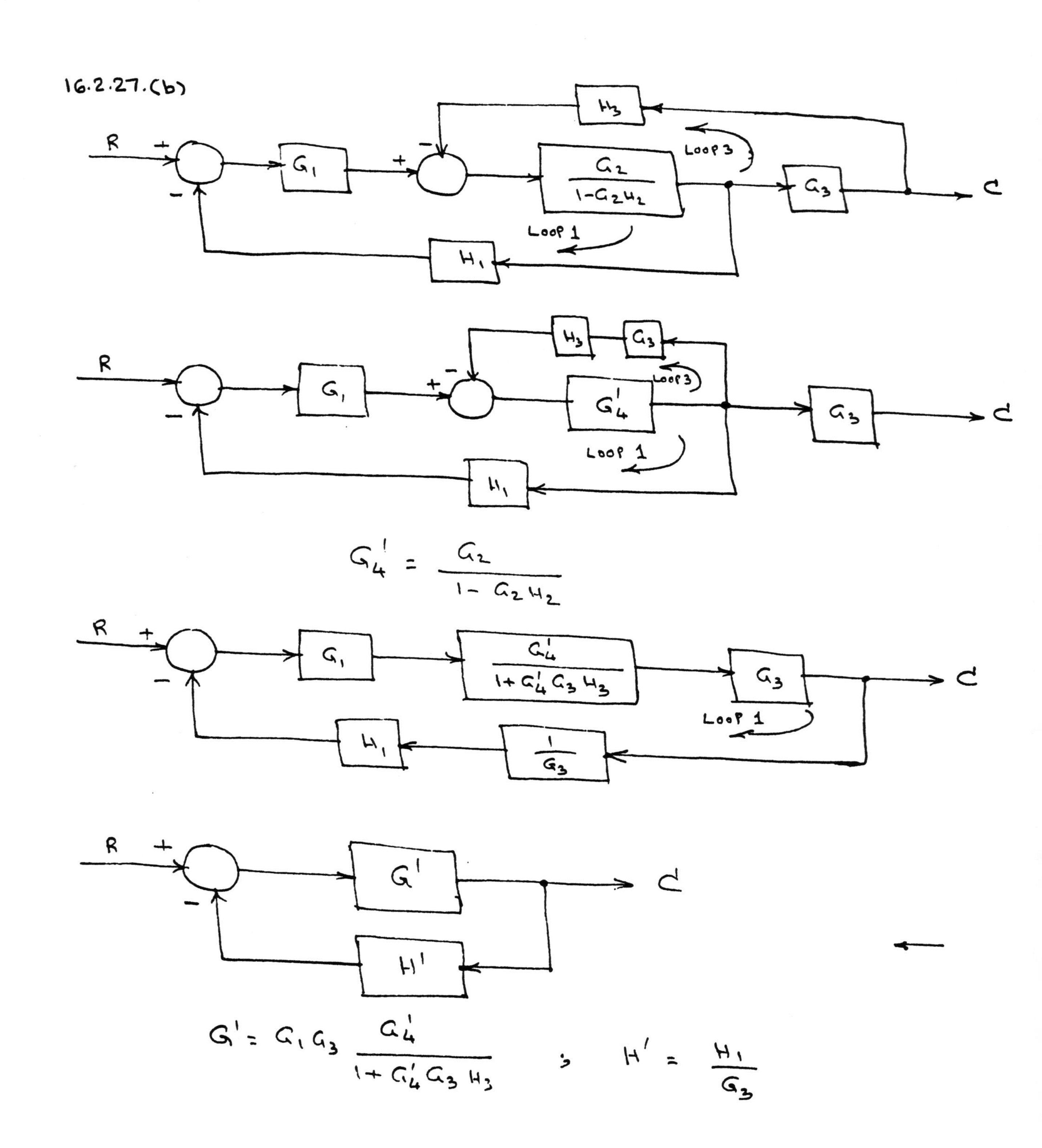

16.2.27.(b)
R
+
-
G1
G2
1-G2H2
H3
Loop 3
G3
C
Loop 1
H1
G'4
G'4 = G2 / (1 - G2H2)
G'4
1+G'4G3H3
1/G3
G'
H'
G' = G1G3 G'4 / (1+G'4G3H3) ; H' = H1/G3

16.2.28.

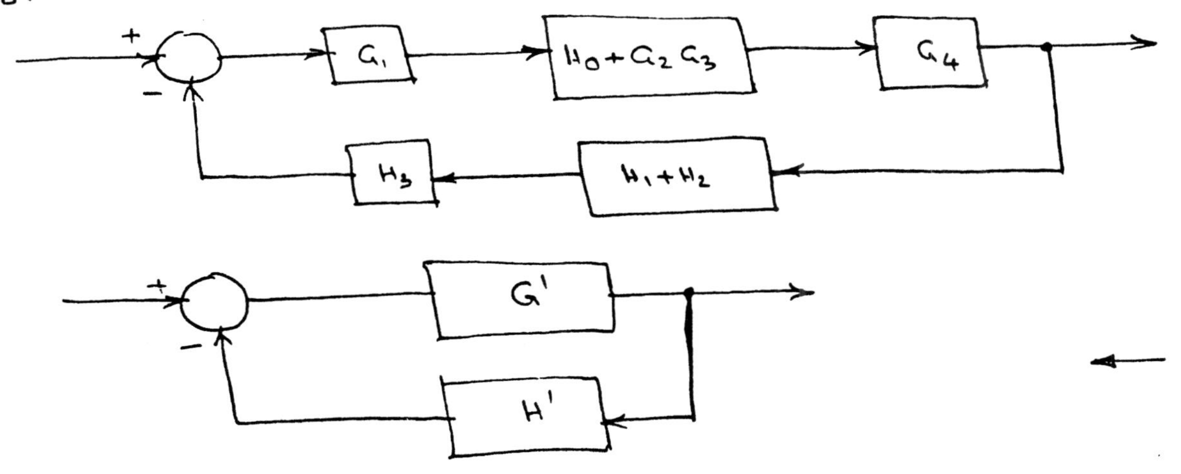

$$G' = G_1 G_4 (H_0 + G_2 G_3) \; ; \quad H' = H_3 (H_1 + H_2)$$

16.2.29.

REPLACE INNER LOOP WITH $\quad G_2' = \dfrac{G_2}{1 + G_2 H_2}$

RESPONSE TO REF. FUNCTION R_1 : $\quad C_1 = \dfrac{G_1 G_2'}{1 + G_1 G_2' H_1} R_1$

RESPONSE TO R_2 : $\quad C_2 = \dfrac{G_2'}{1 + G_1 G_2' H_1} R_2$

SYSTEM RESPONSE $\quad C = C_1 + C_2 = \dfrac{G_1 G_2' R_1 + G_2' R_2}{1 + G_1 G_2' H_1}$

16.2.30.

(i) MOVE 3rd SUMMING JUNCTION TO RIGHT

+ − G_1 I_1 + − I_4 $G_2 G_3 G_4$ I_2 + I_3 G_6 H_1 G_5 H_2

16.2.30. CONTD.

(ii) LET $G_x = G_2 G_3 G_4$

(iii) SET UP EQS. AT JUNCTIONS:

$$I_4 = I_1 - H_1 I_3$$
$$I_2 = G_x I_4$$
$$I_3 = I_2 + G_5 I_1$$

(iv) ELIMINATE I_4 AND I_2

$$\frac{I_3}{I_1} = \frac{G_x + G_5}{1 + G_x H_1} = G_y$$

(v) SIMPLIFIED BLOCK DIAGRAM

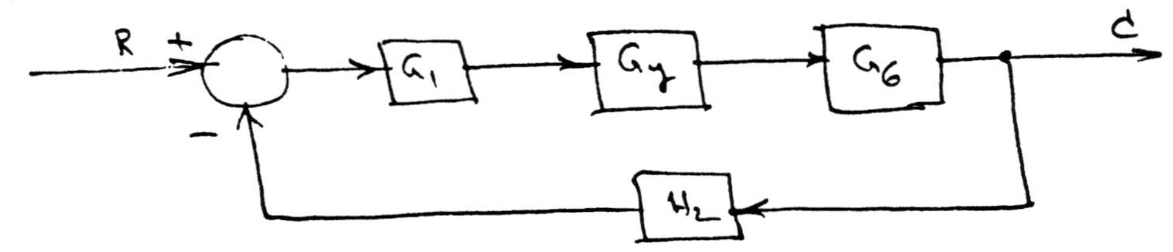

(vi)

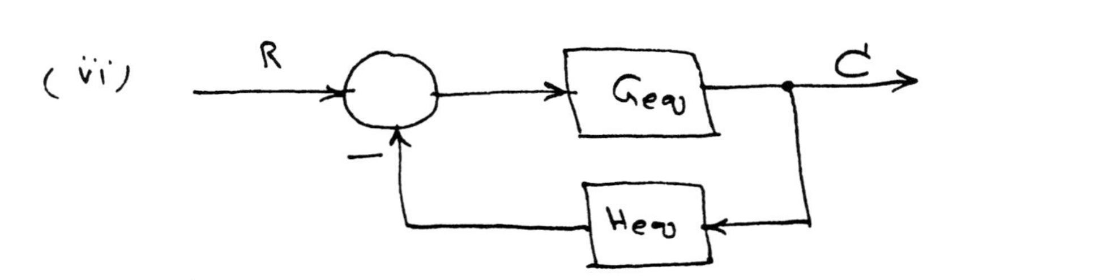

$$G_{eq} = G_1 G_6 G_y \quad ; \quad H_{eq} = H_2$$

16.2.31.

$$\text{LOOP GAIN} = G(s)H(s) = \frac{K_a K_g K_m K_\omega}{(R_f + sL_f)\left[(R_a + sL_a)(B_m + sJ_m) + K_m^2\right]}$$

WHICH IS A TYPE 0 SYSTEM.

$$R(s) = \frac{1}{s}$$

$$E(s) = \frac{R(s)}{1 + G(s)H(s)} = \frac{(R_f + sL_f)\left[(R_a + sL_a)(B_m + sJ_m)\right] + K_m^2}{s\left\{K_a K_g K_m K_\omega + (R_f + sL_f)\left[(R_a + sL_a)(B_m + sJ_m)\right] + K_m^2\right\}}$$

APPLYING FINAL-VALUE THEOREM,

$$\text{STEADY-STATE ERROR} = \lim_{t \to \infty} e(t) = \lim_{s \to 0} sE(s) = \frac{R_f(R_a B_m + K_m^2)}{K_a K_g K_m K_\omega + R_f(R_a B_m + K_m^2)}$$

16.2.32.

TYPE 0: $\eta = 0$ in Eq. (16.2.21): $G(s)H(s) = \dfrac{K N(s)}{s^0 D(s)} = \dfrac{K N(s)}{D(s)}$

$$E(s) = \frac{R(s)}{1 + G(s)H(s)}$$

LET $$E(s) = R(s) \frac{D(s)}{D(s) + K N(s)}$$

WHERE $D(s) = s^n \pm d_1 s^{n-1} \pm d_2 s^{n-2} \pm \cdots \pm d_i s^{n-i} \pm \cdots \pm d_{n-1} s \pm d_n$

$N(s) = s^n \pm c_1 s^{n-1} \pm c_2 s^{n-2} \pm \cdots \pm c_i s^{n-i} \pm \cdots \pm c_{n-1} s \pm c_n$

(WITH ALL COEFFICIENTS POSITIVE, AND K AS A NUMERICAL CONSTANT)

FOR UNIT STEP: $R(s) = \frac{1}{s}$

$$\lim_{s \to 0} sE(s) = \frac{d_n}{d_n + K c_n}$$ WHICH IS FINITE ERROR

FOR UNIT RAMP, $R(s) = \frac{1}{s^2}$; $\lim_{s \to 0} s(E(s)) = \infty$

FOR UNIT ACCELERATION, $R(s) = \frac{1}{s^3}$; $\lim_{s \to 0} sE(s) = \infty$

TYPE 1: $\eta = 1$

$$E(s) = \frac{R(s)\, s D(s)}{s D(s) + K N(s)}$$

FOR UNIT STEP, $\lim_{s \to 0} sE(s) = 0$

FOR UNIT RAMP, $\lim_{s \to 0} sE(s) = \frac{d_n}{K c_n}$ WHICH IS FINITE

FOR UNIT ACCLN, $\lim_{s \to 0} sE(s) = \infty$

TYPE 2: $\eta = 2$

$$E(s) = \frac{R(s)\, s^2 D(s)}{s^2 D(s) + K N(s)}$$

FOR UNIT STEP, $\lim_{s \to 0} sE(s) = 0$

FOR UNIT RAMP, $\lim_{s \to 0} sE(s) = 0$

FOR UNIT ACCLN, $\lim_{s \to 0} sE(s) = \frac{d_n}{K c_n}$ WHICH IS FINITE

16.2.33.
(a)

$$G(s) = \frac{\Omega_o(s)}{E(s)} = \frac{K_p K_t}{B + sJ}$$

(FORWARD-LOOP TRANSFER FUNCTION)

IT IS A TYPE 0 SYSTEM ←

$$\text{ERROR RESPONSE} = \frac{E(s)}{\Omega_R(s)} = \frac{B + sJ}{(K_p K_t + B + sJ)}$$

FOR A UNIT-STEP REFERENCE FUNCTION, $\Omega_R(s) = \frac{1}{s}$

BY FINAL-VALUE THEOREM,

$$\text{STEADY-STATE ERROR} = \frac{B}{B + K_p K_t} \quad \leftarrow$$

$$\text{OUTPUT RESPONSE} = \frac{\Omega_o(s)}{\Omega_R(s)} = \frac{K_p K_t}{K_p K_t + B + sJ}$$

THE TIME CONSTANT OF THE CHARACTERISTIC EQ (i.e. DENOMINATOR = 0) IS

$J/(K_p K_t + B)$ ←

(b) FORWARD LOOP TRANSFER FUNCTION = $G(s) = \dfrac{K_i K_t}{s(B + sJ)} = \dfrac{\Omega_o(s)}{E(s)}$

TYPE 1 SYSTEM ←

$$\text{ERROR RESPONSE} = \frac{E(s)}{\Omega_R(s)} = \frac{s(B + sJ)}{K_i K_t + sB + s^2 J}$$

FOR A UNIT-STEP REFERENCE FUNCTION, $\Omega_R(s) = \frac{1}{s}$

BY FINAL-VALUE THEOREM,

STEADY-STATE ERROR = 0 (AS EXPECTED FROM TABLE 16.2.2) ←

$$\text{OUTPUT RESPONSE} = \frac{\Omega_o(s)}{\Omega_R(s)} = \frac{K_i K_t}{K_i K_t + sB + s^2 J} = \frac{K_i K_t / J}{s^2 + \frac{sB}{J} + \frac{K_i K_t}{J}}$$

$$\text{CHARACTERISTIC EQ:} \quad s^2 + s\frac{B}{J} + \frac{K_i K_t}{J} = 0$$

SECOND ORDER EQ OF THE TYPE $s^2 + 2\xi\omega_n s + \omega_n^2 =$

$$\text{DAMPING RATIO } \xi = \frac{B}{2\sqrt{K_i K_t J}} \quad \leftarrow$$

16.2.33.(c)

$\Omega_R(s)$ (+, −) → $E(s)$ → $K_p + \frac{K_i}{s}$ → K_t → $\frac{1}{sJ + B}$ → $\Omega_o(s)$ (unity feedback)

NEW BLOCK DIAGRAM

OUTPUT RESPONSE : $\dfrac{G(s)}{1 + G(s)}$

CHARACTERISTIC EQ : $Js^2 + (B + K_t K_p)s + K_t K_i = 0$

$$s^2 + \frac{B + K_t K_p}{J} s + \frac{K_t K_i}{J} = 0$$

SECOND ORDER EQ OF THE FORM: $s^2 + 2\xi \omega_n s + \omega_n^2 = 0$

DAMPING RATIO $= \xi = \dfrac{B + K_t K_p}{2\sqrt{K_t K_i J}}$

16.2.34.

$G(s) = \dfrac{K_p K_t}{Js}$; $H(s) = 1$

LET $k = \dfrac{K_p K_t}{J}$

$$\frac{C(s)}{R(s)} = \frac{\Omega_c(s)}{\Omega_r(s)} = \frac{G(s)}{1 + G(s)H(s)} = \frac{k}{s + k}$$

$$C(s) = R(s)\frac{k}{s+k} = \frac{1}{s}\,\frac{k}{s+k} = \frac{1}{s} + \frac{-1}{(s+k)}$$

$$c(t) = 1 - e^{-kt} = 1 - e^{-(K_p K_t / J)t} \leftarrow$$

16.2.35.

$G(s) = \dfrac{1}{Js}$; $H(s) = K_t\left(K_p + \dfrac{K_i}{s}\right)$; $T_d(s) = \dfrac{T_d}{s}$

$$\frac{\Omega_c(s)}{T_d(s)} = \frac{G(s)}{1 + G(s)H(s)} = \frac{s}{J\left[s^2 + \frac{K_t K_p}{J}s + \frac{K_i K_t}{J}\right]}$$

$$\Omega_c(s) = T_d(s)\frac{G(s)}{1 + G(s)H(s)} = \frac{T_d}{J\left[s^2 + \frac{K_t K_p}{J}s + \frac{K_i K_t}{J}\right]}$$

$$\lim_{s \to 0} s\,\Omega_c(s) = 0 \leftarrow$$

16.2.36.

(a)

$$\frac{\epsilon(s)}{U(s)} = \frac{G(s)}{1+G(s)H(s)} \quad ; \quad G(s) = \frac{K_t}{1+\tau s} \; ; \; H(s) = \frac{K_i}{s}$$

$$\epsilon(s) = U(s) \frac{s K_t}{s(1+\tau s) + K_t K_i} \quad ; \quad U(s) = \frac{1}{s}$$

$$\lim_{s \to 0} s\epsilon(s) = 0 \quad \leftarrow$$

ZERO STEADY-STATE ERROR DUE TO A STEP-CHANGE IN $U(s)$ ←

(b)

$$G(s) = \frac{K_t}{s(1+\tau s)} \quad ; \quad H(s) = K_p$$

$$\epsilon(s) = U(s) \frac{K_t}{s(1+\tau s) + K_p K_t} \quad ; \quad U(s) = \frac{1}{s}$$

$$\lim_{s \to 0} s\epsilon(s) = \frac{1}{K_p} \quad \leftarrow$$

THUS $\epsilon(s)$ HAS A STEADY-STATE ERROR DUE TO A STEP CHANGE IN $U(s)$. ←

16.2.37. LET

M_o = ORIGINAL CLOSED LOOP TRANSMISSION = 980

M_f = FINAL CLOSED LOOP TRANSMISSION = 1020

$\therefore \delta M = M_f - M_o = 40$

G_o = ORIGINAL OPEN LOOP TRANSMISSION = 10^n

WHERE n = no. of stages

G_f = FINAL Open Loop Transmission = $2^n 10^n$

$\therefore \delta G = (2^n - 1) 10^n$

$$\frac{\delta M}{M_f} = \frac{\delta G}{G_f}\left(\frac{1}{1+HG_o}\right) = \frac{\delta G}{G_f}\left(\frac{M_o}{G_o}\right)$$

$$\therefore \frac{40}{1020} \geq \frac{(2^n-1)10^n}{2^n 10^n}\left(\frac{980}{10^n}\right)$$

THIS IS SATISFIED WHEN $n = 5$ ←

WITH $n = 5$, $H = \dfrac{10^5 - 980}{980 \times 10^5} = 0.001$ ←

16.2.38.

(a)

FROM EQ. (16.2.7)

$$\frac{\partial M}{M} = \frac{\partial G}{G}\left(\frac{1}{1+HG}\right)$$

$$0.001 = 0.25\left(\frac{1}{1+HG}\right)$$

$\therefore 1 + HG = 250$

ALSO $M = \dfrac{G}{1+GH}$ WHERE G IS THE DIRECT GAIN OF THE CLOSED-LOOP SYSTEM

$$\therefore 80 = \frac{G}{1+GH}$$

$\therefore G = 80(1+GH) = 80 \times 250 = 20{,}000$

$\therefore 80^n \geq 20{,}000$ WHERE n IS THE NUMBER OF STAGES

$\therefore n = 3$ ←

(b) G IS THEN $80^3 = 512{,}000$

$$\therefore H = \frac{250-1}{512{,}000} = \frac{249}{512} \times 10^{-3} = 0.487 \times 10^{-3}.$$

16.2.39.

(a) (i) $s^2 + s - 2 = (s+2)(s-1) = 0$

ONE ROOT OF THE CHARACTERISTIC EQ. OF THIS SYSTEM IS POSITIVE. HENCE ANY DISTURBANCE IN THE SYSTEM IS ACCOMPANIED BY A TRANSIENT TERM ke^{t} THAT INCREASES WITH TIME.

∴ UNSTABLE ←

(ii)

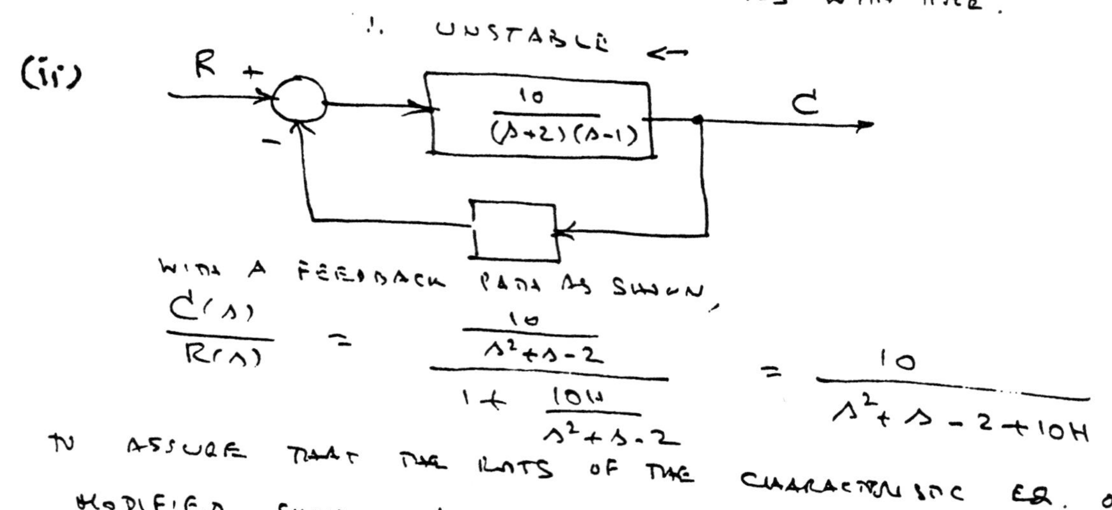

WITH A FEEDBACK PATH AS SHOWN,

$$\frac{C(s)}{R(s)} = \frac{\dfrac{10}{s^2+s-2}}{1 + \dfrac{10H}{s^2+s-2}} = \frac{10}{s^2 + s - 2 + 10H}$$

TO ASSURE THAT THE ROOTS OF THE CHARACTERISTIC EQ. OF THE MODIFIED SYSTEM LIE INSIDE THE LEFT-HALF s-plane, IT IS NECESSARY THAT

$$10H > 2$$

$$\text{or } H > 1/5 \quad \leftarrow$$

(b)

(i) SAME ANSWER AS a(i)

UNSTABLE

(ii)

R, +, −, $\dfrac{10}{s(s+2)(s-1)}$, H

$$\frac{C(s)}{R(s)} = \frac{10}{s^3 + s^2 - 2s + 10H}$$

∴ CHOOSE $H = as$ AND IMPOSE THE CONDITION THAT

$$as > 2s$$

$$\text{or } a > 2$$

16.2.40. THE GENERAL FORM OF THE SECOND-ORDER SYSTEM IS GIVEN BY

$$c(t) = r_0 \left[1 - \frac{e^{-\xi \omega_n t}}{\sqrt{1-\xi^2}} \sin(\omega_d t + \theta) \right]$$

(a) $\therefore \omega_d = 6$ rad/s ←

(b) $\xi \omega_n = 8$

$$\frac{1}{\sqrt{1-\xi^2}} = 1.66 = \frac{5}{3} \Rightarrow \xi^2 = 1 - \frac{9}{25} = \frac{16}{25} \text{ or } \xi = 0.8 \leftarrow$$

(c) $\omega_n = 8/0.8 = 10$ rad/s ←

(d) SINCE $\xi = \frac{F}{2\sqrt{KJ}}$ or $\sqrt{K} = \frac{F}{2\xi\sqrt{J}} = \frac{0.2}{2(0.8)\sqrt{0.01}}$

$$= \frac{0.2}{2 \times 0.8 \times 0.1} = \frac{5}{4} = 1.25$$

or $K = (1.25)^2 = 1.5625$ ←

(e) $\xi \propto \frac{1}{\sqrt{K}}$

$$\therefore \frac{K_2}{K_1} = \left(\frac{\xi_1}{\xi_2}\right)^2 = \left(\frac{0.8}{0.4}\right)^2 = 4$$

$\therefore K_2 = 1.5625 \times 4 = 6.25$ ←

(f)

$$M(s) = \frac{C(s)}{R(s)} = \frac{\omega_n^2}{s^2 + 2\xi\omega_n s + \omega_n^2}$$

WITH $\xi = 0.8$ AND $\omega_n = 10$

$$M(s) = \frac{C(s)}{R(s)} = \frac{100}{s^2 + 16s + 100} = \frac{G(s)}{1 + G(s)}$$

(NOTE: WITH $H=1$)

(g)

$$\frac{1}{G(s)} = \frac{s^2 + 16s + 100}{100} - 1 = \frac{s^2 + 16s}{100}$$

$$\therefore G(s) = \frac{100}{s(s+16)} \leftarrow$$

16.2.40.(h)

$$C(s) = R(s)\,G(s) = \frac{100}{s^2(s+16)} = \frac{k_{01}}{s^2} + \frac{k_{02}}{s} + \frac{k_1}{s+16}$$

$$k_{01} = \frac{100}{s+16}\bigg|_{s=0} = \frac{100}{16}$$

$$k_{02} = \frac{d}{ds}\left[\frac{100}{s+16}\right]\bigg\}_{s=0} = -\frac{100}{256}$$

$$k_1 = \frac{100}{s^2}\bigg|_{s=-16} = \frac{100}{256}$$

$$\therefore C(s) = \frac{100}{16}\frac{1}{s^2} - \frac{100}{256}\frac{1}{s} + \frac{100}{256}\left(\frac{1}{s+16}\right)$$

THEN $$c(t) = \frac{100}{16}t - \frac{100}{256}\left(1 - e^{-16t}\right) \leftarrow$$

16.2.41. (a)

$$\frac{C(s)}{R(s)} = \frac{\frac{10}{s(s+2)}}{1 + \frac{10}{s(s+2)}} = \frac{10}{s^2+2s+10} = \frac{\omega_n^2}{s^2 + 2\xi\omega_n s + \omega_n^2}$$

(b)

$$s_{1,2} = -1 \pm \sqrt{1-10} = -1 \pm j3$$

GENERAL $s_{1,2} = -\xi\omega_n \pm j\omega_d$ FOR A SECOND-ORDER SYSTEM

$$\therefore \omega_d = 3 \text{ rad/s} \leftarrow$$

(c) $2\xi\omega_n = 2$ or $\xi = \frac{1}{\omega_n} = \frac{1}{\sqrt{10}} = 0.31 \leftarrow$

PERCENT MAXM OVERSHOOT = 35% FROM FIG. 16.2.10 $\leftarrow$

(d) $t_s = \frac{5}{\xi\omega_n} = 5$ sec $\leftarrow$

16.2.42. $$\frac{d^2c}{dt^2} + 6.4\frac{dc}{dt} + 0.4(160)\,c = 160r$$

CHAR. EQ : $s^2 + 6.4s + 64 = 0$

COMPARING WITH GENERAL FORM $s^2 + 2\xi\omega_n s + \omega_n^2 = 0$

$$\xi = \frac{6.4}{2\sqrt{64}} = 0.4 \leftarrow$$

$\omega_n = 8$ rad/sec $\leftarrow$ FROM FIG 16.2.10, PERCENT MAX. OVERSHOOT = 25% $\leftarrow$

16.2.43.

(a) $$\zeta = \frac{F}{2\sqrt{kJ}} = \frac{220\times10^{-6}}{2\sqrt{24\times10^{-4}\times1.4\times10^{-5}}} = 0.6 \leftarrow$$

(b) $$k' = 250\times10^{-4}$$

$$0.6 = \frac{F+Q_e}{2\sqrt{k'J}} \quad \text{or} \quad F+Q_e = 1.2\sqrt{250\times10^{-4}\times1.4\times10^{-5}}$$

$$= 7.1\times10^{-4} = 710\times10^{-6}$$

$$\therefore Q_e = (710-220)\,10^{-6} = 490\times10^{-6}\ \text{kg}\cdot\text{m}^2/\text{s} \leftarrow$$

16.2.44.

THE CHARACTERISTIC EQ. IS GIVEN BY $1+G(s)=0$

$$1+\frac{10(1+sK_e)}{s(s+2)}=0$$

$$s^2+(2+10K_e)s+10=0$$

$$\zeta = 0.6 = \frac{2+10K_e}{2\sqrt{10}} = \frac{2+10K_e}{6.32}$$

$$\therefore K_e = 0.179 \leftarrow$$

16.2.45.

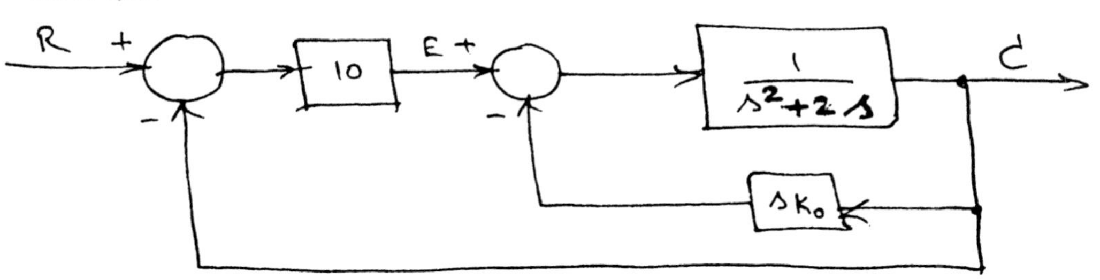

COMPARING WITH FIG. 16.2.12 OF THE TEXT, WE GET THE FOLLOWING (SIMILAR TO FIG. 16.2.13)

R → ( ) → [10] → E → $\left[\frac{1}{s^2+(2+K_o)s}\right]$ → C (unity negative feedback, −)

$$M(s) = \frac{C(s)}{R(s)} = \frac{10}{s^2+(2+K_o)s+10}$$

CHARACTERISTIC EQ: $s^2+(2+K_o)s+10=0$

$\omega_n=\sqrt{10}$; $\zeta = 0.6$ FROM FIG. 16.2.10 FOR PERCENT MAX OVERSHOOT OF 10

$$2\zeta\omega_n = 2+K_o \quad \text{or} \quad 2(0.6)\sqrt{10} = 2+K_o$$

$$\therefore K_o = 3.79-2 = 1.79 \leftarrow$$

16.2.46.

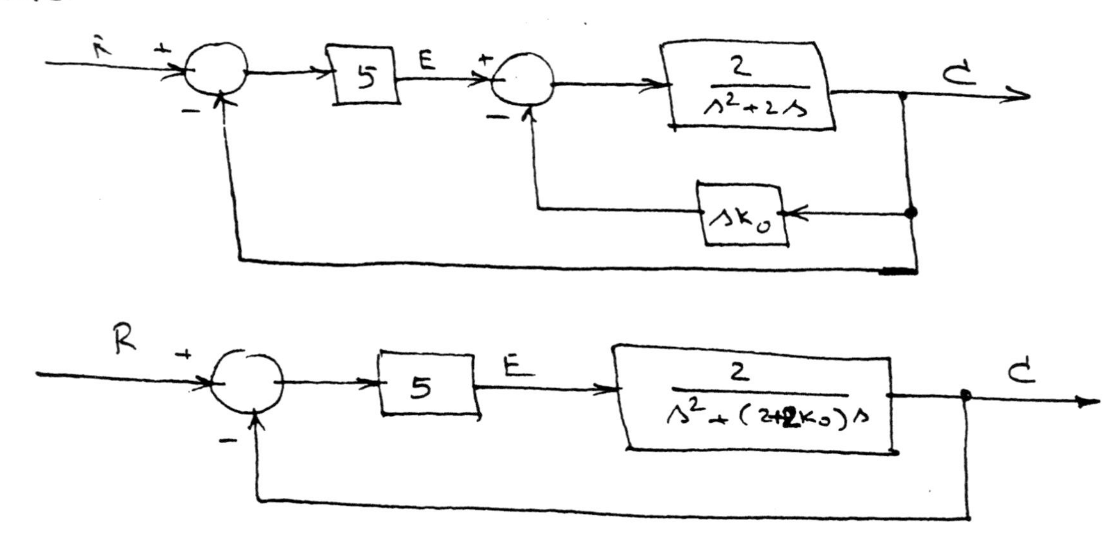

$$M(s) = \frac{C(s)}{R(s)} = \frac{10}{s^2 + (2+2K_o)s + 10}$$

CHAR. EQ : $s^2 + (2+2K_o)s + 10 = 0$

$\omega_n = \sqrt{10}$; $\xi = 0.6$ FROM FIG 16.2.10 FOR PERCENT MAX. OVERSHOOT OF 10%.

$2\xi\omega_n = 2 + 2K_o = 2(1+K_o)$

or $K_o = 1.9 - 1 = 0.9$ ←